AF760307

Modern Topics in the Phototrophic Prokaryotes

Patrick C. Hallenbeck
Editor

Modern Topics in the Phototrophic Prokaryotes

Environmental and Applied Aspects

Springer

Editor
Patrick C. Hallenbeck
Department of Biology
Life Sciences Research Center
United States Air Force Academy
Colorado Springs, CO, USA

Département de microbiologie, infectiologie et immunologie
Université de Montréal
Montréal, Québec, Canada

ISBN 978-3-319-46259-2 ISBN 978-3-319-46261-5 (eBook)
DOI 10.1007/978-3-319-46261-5

Library of Congress Control Number: 2017935069

© Springer International Publishing Switzerland 2017
This work is subject to copyright. All rights are reserved by the Publisher, whether the whole or part of the material is concerned, specifically the rights of translation, reprinting, reuse of illustrations, recitation, broadcasting, reproduction on microfilms or in any other physical way, and transmission or information storage and retrieval, electronic adaptation, computer software, or by similar or dissimilar methodology now known or hereafter developed.
The use of general descriptive names, registered names, trademarks, service marks, etc. in this publication does not imply, even in the absence of a specific statement, that such names are exempt from the relevant protective laws and regulations and therefore free for general use.
The publisher, the authors and the editors are safe to assume that the advice and information in this book are believed to be true and accurate at the date of publication. Neither the publisher nor the authors or the editors give a warranty, express or implied, with respect to the material contained herein or for any errors or omissions that may have been made. The publisher remains neutral with regard to jurisdictional claims in published maps and institutional affiliations.

Printed on acid-free paper

This Springer imprint is published by Springer Nature
The registered company is Springer International Publishing AG
The registered company address is: Gewerbestrasse 11, 6330 Cham, Switzerland

Contents

Part I
Phylogeny, Taxonomy, and Diversity

Diversity of the Cyanobacteria

Petr Dvořák, Dale A. Casamatta, Petr Hašler, Eva Jahodářová, Alyson R. Norwich, and Aloisie Poulíčková

Abstract The cyanobacteria are an ancient lineage of photo-oxygenic bacteria. Globally responsible for much of the primary productivity and nitrogen fixation, they are also evolutionarily significant as the photosynthetic members of serial endosymbiotic events leading to the establishment of chloroplasts. Traditionally classified based on morphological characters, recent research revealed an abundance of cryptic diversity evidenced by molecular analyses, most notably the 16S rDNA gene sequence. Explorations of seldom sampled habitats, such as tropics environments, aerophytic habitats, soil crusts, etc., have also revealed a tremendous new diversity of taxa. This increase in the alpha-level diversity, coupled with new molecular techniques, has greatly altered our perceptions of the evolutionary relationships within this clade. Many of the traditional genera have proven to be polyphyletic, but revisions are underway.

Keywords Cyanobacteria • Phylogeny • Taxonomy • Biodiversity • Morphology

Introduction

Cyanobacteria (Cyanophytes, Cyanoprokaryotes, blue-green algae) are an ancient group of prokaryotic microorganisms with the ability to undertake oxygenic photosynthesis. The earliest estimates of their origin lie at the beginning of Archean 3.5–3.8 BYA (reviewed in Sleep 2010 and Schopf 2001), while more conservative estimates suggest a later appearance at the end of Archean 2.7 BYA (reviewed in Blank and Sanchez-Baracaldo 2010). Cyanobacteria have greatly impacted global ecosystems, as their photosynthetic activity provided much of the oxygen necessary for the proliferation of aerobic life forms (Bekker et al. 2004). Moreover, cyanobacteria are among

P. Dvořák (✉) • P. Hašler • E. Jahodářová • A. Poulíčková
Department of Botany, Faculty of Science, Palacký University Olomouc,
Šlechtitelů 27, CZ-783 71 Olomouc, Czech Republic
e-mail: p.dvorak@upol.cz

D.A. Casamatta • A.R. Norwich
Department of Biology, University of North Florida,
1 UNF Drive, 32224 Jacksonville, FL, USA

© Springer International Publishing Switzerland 2017

P.C. Hallenbeck (ed.), *Modern Topics in the Phototrophic Prokaryotes*,
DOI 10.1007/978-3-319-46261-5_1

the most abundant and potent primary producers on Earth. They occur in freshwater, marine, cold (e.g., polar), hot (e.g., thermal springs), and terrestrial habitats. Cyanobacteria are also commonly encountered in symbiotic relationships with plants, fungi (lichens), and eukaryotic algae (Whitton and Potts 2000). Through the process of endosymbiosis, ancient organisms (likely protistans) engulfed a cyanobacterium, resulting in the origin of the chloroplast of algae and plants (McFadden 1999). The current number of cyanobacterial species is subject to debate, with estimates ranging from 2783 (Nabout et al. 2013) to 4484 (Guiry and Guiry 2015). However, the expected total number of species may reach 8000 (Guiry 2012).

Besides primary metabolism, cyanobacteria produce myriad secondary metabolites, or bioactive compounds (chemically, mostly alkaloids and oligopeptides), which are toxic to the environment and humans (collectively referred to as cyanotoxins; e.g., Carmichael 1992; Dittmann et al. 2013). However, a number of these compounds have been isolated and show promise as drugs for the treatment of cancer and other diseases due to their biological activity, e.g., anti-viral, anti-protistan, and anti-bacterial properties (comprehensively reviewed in Singh et al. 2011). A variety of cyanobacterial strains also promise further advances in biotechnology. Cyanobacteria have been postulated as an alternative source of energy, they may be used in wastewater plants for utilization of macronutrients, for degradation of oil, for fertilization in agriculture, and as food for humans and animals (reviewed in Abed et al. 2008).

On the other hand, cyanobacteria are often noted for their ability to form blooms, which results from an overabundance of planktic forms. This typically occurs in eutrophic habitats (those with elevated nutrient levels, typically nitrogen and/or phosphorus), such as freshwater lakes. Freshwater harmful algal blooms are often accompanied by unfavorable phenomena such as extreme pH fluctuations, anaerobic conditions, and aforementioned toxins (reviewed in Oliver and Ganf 2000). Thus, cyanobacteria may have a large negative impact on both the environment and human endeavors such as fisheries, potable water production, recreational usage of aquatic habitats, etc.

The purpose of this chapter is to provide an overview of recent advances in the taxonomy, phylogeny, and diversity of cyanobacteria. We will focus on taxonomic revisions and new taxa in light of current problems of cyanobacterial species concepts and definitions. Furthermore, we will show that significant gaps persist in our knowledge of cyanobacterial biodiversity. We will discuss geographical regions and habitats where our knowledge of the diversity and ecology of cyanobacteria is most limited.

How to Distinguish Cyanobacteria?

Morphology

Cyanobacteria exhibit a relatively high degree of morphological features compared to other prokaryotes. Until recently, cyanobacteria were identified and categorized using morphological traits such as cell dimensions, shape, color, type of branching, sheath characteristics, and cell contents (summarized in Komárek and Anagnostidis 1998, 2005; Komárek 2013). Cyanobacteria may be unicellular, colonial, or filamentous.

Colonies of unicellular cyanobacteria may have regular (e.g., *Merismopedia*) or irregular distribution of cells (e.g., *Microcystis*). The Pleurocapsales possess relatively complex colony formation, which may resemble filaments, branching, and cells may be heteropolar (see Komárek and Anagnostidis 1998 for review and Fig. 1). The number of cells may vary from two to several thousand per colony.

Filamentous cyanobacteria may exhibit both false and true branching. False branching is present in all orders of filamentous cyanobacteria, while true branching has been observed only in members of Nostocales (see details of cyanobacterial systematics for further details). Similarly, multiseriate growth of trichomes (a parallel succession of multiple trichomes) evolved only in some members of the Nostocales, specifically in the order formerly referred to as the Stigonematales (Komárek 2013).

While cyanobacteria reproduce via binary fission, they may not be considered as fully clonal organisms due to horizontal gene flow and recombination, as seen in other prokaryotes (for detailed discussion see, e.g., Cohan 2001, 2002; Cohan and Perry 2007). However, cyanobacteria have evolved some interesting reproductive strategies. For instance, some unicellular cyanobacteria may produce baeocytes and exocytes, which are differentiated from the mother cell by size, shape, and successive multiple fission with subsequent release to the environment (see details in Komárek and Anagnostidis 1998). Filamentous cyanobacteria may produce short, often motile filaments called hormogonia. Furthermore, Nostocalean cyanobacteria may produce long-term or overwintering reproductive cells called akinetes (see further).

Besides vegetative cells (those cells dedicated to photosynthetic processes), filamentous cyanobacteria of the order Nostocales may produce two types of differentiated cells: heterocytes and akinetes. Heterocytes do not possess functioning photosynthetic apparatus because their primary function is anaerobic fixation of atmospheric nitrogen using the enzyme nitrogenase (Meeks et al. 2002), also developed in some soil bacteria. Heterocytes may be distinguished from vegetative cells by the former's homogenous content and the presence of polar pores, and heterocytes may be situated intercalary or terminally in filaments (Kumar et al. 2010; Komárek 2013). Akinetes are usually larger than vegetative cells, and with large amounts of stored nutrients visible as granules. Akinetes are reproductive cells, which may lie quiescent in the environment (e.g., lake sediments, soils, etc.) during unfavorable conditions (drought, low temperatures, fall turnover in dimictic lakes, etc.; Kaplan-Levy et al. 2010). Olsson-Francis et al. (2009) suggested that akinetes of *Anabaena cylindrica* may survive the environment on Mars.

Cell ultrastructures are typically visualized by transmission electron microscopy (TEM) and may exhibit significant variability among cyanobacteria. The cell walls of cyanobacteria have similar composition to other gram-negative bacteria. Furthermore, some cyanobacteria possess S-layers, a crystalline, proteinaceous layer covering the entire surface of the cell. These S-layers appear to be an important structure for filament motility (Hoiczyk and Hansel 2000). Cells often produce mucilaginous sheaths composed of exopolysaccharides, which range from <1 μm to several times the filament thickness. Sheaths are often environmentally inducible and provide several putative functions, such as protection from UV radiation, desiccation,

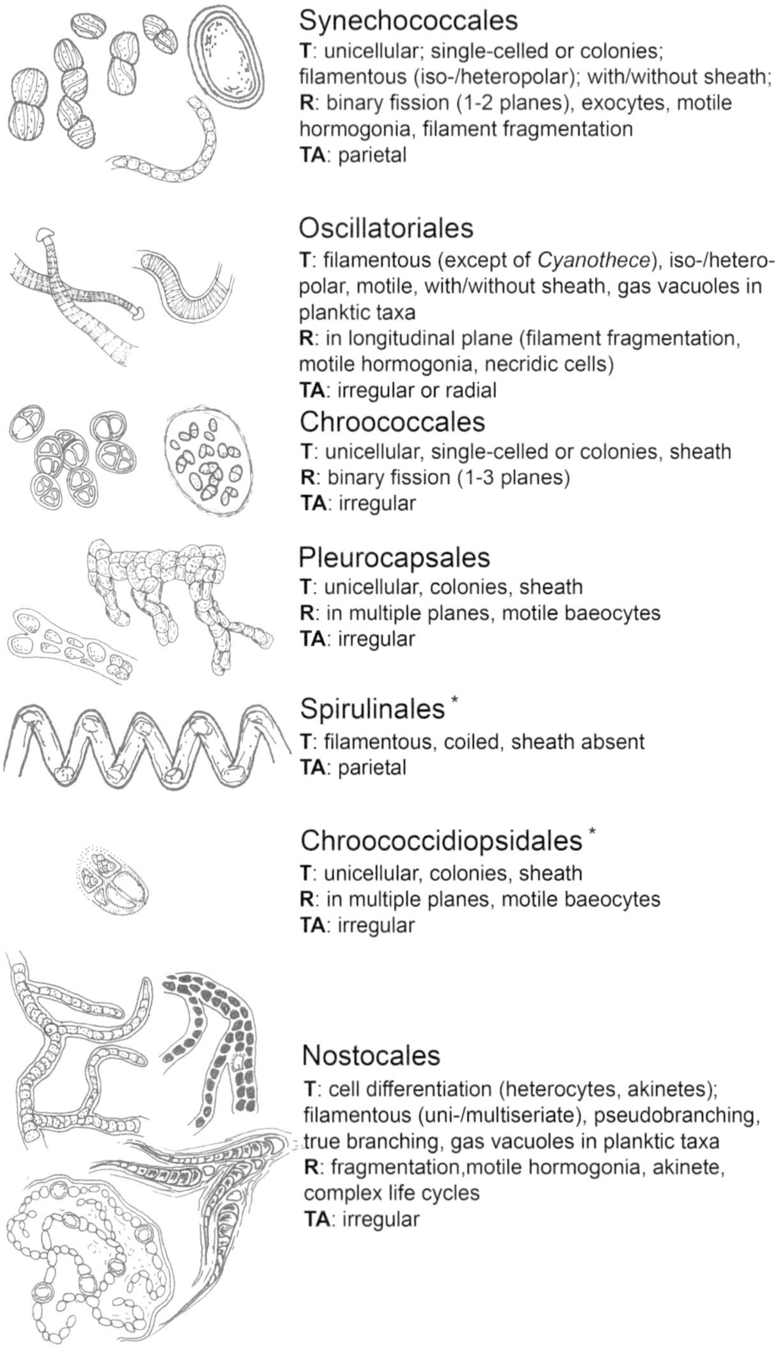

Fig. 1 Ordinal level taxonomic scheme of cyanobacterial taxonomy as proposed by Komárek et al. (2014). The *left side* contains examples of morphotypes for each order, the *right typical features*: T—thallus, R—reproduction type, TA—thylakoid arrangement. *Asterisks* represent the most recently established orders. Note that we did not include the Gloeobacterales, as it is extensively discussed elsewhere and it possesses very simple morphology

and anti-herbivory (Ehling-Schulz and Scherer 1999). Cyanobacteria also possess thylakoids, membrane invaginations used in photosynthetic activity. The placement and arrangement of thylakoids are key taxonomic features. Thylakoids may be located in the cell parietally, radially, or irregularly, where their position is mostly relevant at the family or order level (Fig. 1; Komárek et al. 2014).

Photosynthesis

Cyanobacteria are capable of oxygenic photosynthesis. The main photosynthetic pigment is chlorophyll-*a*, which may be accompanied by carotenoids, xanthophylls, and phycobilisomes, which contain mostly phycoerythrin and phycocyanin. Some cyanobacteria possess divinyl-chlorophyll-*b* (*Prochlorococcus*; Partensky et al. 1999) or chlorophyll-*d* (*Acaryochloris*; Miyashita et al. 2003).

Phylogeny

The advent of molecular biology brought a plethora of tools that have been used to investigate species diversity and evolution. It has allowed researchers to move from a solely phenetic approach to the reconstruction of evolutionary relationships. The most reliable and reproducible approach developed for phylogenetic inference is considered DNA sequencing. Since Woese et al. (1990) have proposed the 16S rDNA gene as the universal marker for all Bacteria, it is the most frequently used gene in phylogenetic reconstructions (Rajendhran and Gunasekaran 2011). However, an expansion of sequenced loci is increasingly inevitable since the 16S itself is too conserved to reflect the true variation of this lineage (Johansen and Casamatta 2005). A combination of various housekeeping and protein-coding loci (usually up to 10), or multilocus sequence typing (MLST; Maiden et al. 1998), has been proposed, as it offers better resolution for recognition of taxa since some of the genes are less conservative than the 16S. Moreover, MLST may provide more robust phylogenetic reconstructions (Maiden et al. 1998; Kämpfer and Glaeser 2012).

Recently, prokaryotic genome sequencing has become relatively inexpensive and widely available to researchers. However, the selection of cyanobacterial taxa to sequence is strongly biased. Most of the sequenced cyanobacteria are marine picoplankton, typically *Synechococcus* and *Prochlorococcus* (Larsson et al. 2011), but efforts to increase genome data coverage of cyanobacterial diversity have been undertaken (Shih et al. 2013). Nevertheless, it should be noted that a number of strains whose genomes have been analyzed are poorly characterized and/or have been in culture collections for extended periods, leading to taxonomic confusion. Fortunately, the taxonomic coverage of genomic sequence data is ever increasing, and thus we anticipate a soon-to-be expansion of whole genome phylogenies for taxonomical purposes outside marine picoplanktic or biotechnologically important taxa.

Challenges of Modern Cyanobacterial Taxonomy

As noted above, the taxonomy of cyanobacteria has traditionally been heavily dependent on morphological data. With a growing number of taxa sequenced, it has become increasingly obvious that the evolutionary relationships of cyanobacteria are more entangled than previously thought (for extensive reviews see Komárek et al. 2014; Dvořák et al. 2015a). First of all, some morphological features (including ultrastructural) appear to be polyphyletic in nature. One of a few exceptions is in the production of specialized, differentiated cells of the Nostocales, which appeared only once, and it is consistently monophyletic in phylogenetic reconstructions based on one or multiple loci. Thus, many (and possibly a majority) of the genera defined using morphology are polyphyletic and therefore need significant taxonomical intervention (Komárek 2010; Komárek et al. 2014). Furthermore, unrelated lineages may be so similar morphologically that it is impossible to distinguish between them. Some authors use the term "cryptogenera" to describe this phenomenon (Komárek et al. 2014). This is also an eminent problem in newly described taxa (Dvořák et al. 2015a). For example, the most enigmatic cyanobacterial genus (measured by number of polyphyletic lineages) is probably *Synechococcus*. Dvořák et al. (2014a) identified 12 polyphyletic lineages within this genus, which all conform to the generic, morphologically derived description of *Synechococcus*. Similar patterns within genera of "cryptic species," or taxa not distinguishable by morphology, may also lead to taxonomic confusion. For example, *Phormidium retzii*, considered the most commonly encountered macro-cyanobacterial taxon in North America (Sheath and Cole 1992), is actually a species complex (Casamatta et al. 2003). This is particularly problematic in the case of morphologically simple cyanobacteria such as *Leptolyngbya* or *Synechococcus* (Osorio-Santos et al. 2014; Dvořák et al. 2015a).

What factors may be responsible for the enigmatic evolutionary relationships among cyanobacteria? Cyanobacterial evolution, as in other prokaryotes, is driven by similar factors (e.g., ecological parameters) to eukaryotes but with different selection pressures. For example, cyanobacteria exhibit vast population sizes, relatively fast generation times, and immense dispersal abilities (reviewed in Achtman and Wagner 2008). Moreover, albeit largely asexual, cyanobacteria horizontally exchange DNA via homologous recombination and horizontal (lateral) gene transfer (e.g., Zhaxybayeva et al. 2006; Fraser et al. 2007; Polz et al. 2013; Dvořák et al. 2014a, and many others). Thus, these complicated evolutionary trajectories lead to an equally complicated species concept with no consensus among scientists (reviewed cf. Achtman and Wagner 2008).

A majority of recent cyanobacterial taxonomical studies use species definitions based on a monophyletic species concept coupled with a "polyphasic" or total evidence approach, which is defined as a combination of morphological, molecular, ecological, and other data (Vandamme et al. 1996; Johansen and Casamatta 2005). Cyanobacterial systematics is further muddled since species descriptions differ between bacteriological and botanical nomenclature, both of which are appropriate for cyanobacteria. The International Code for Nomenclature of Prokaryotes (ICNP) has very strict requirements compared to the botanical code (International Code for Algae, Fungi and Plants; ICN). For example, the ICNP requires an axenic culture for a valid species description, something exceedingly difficult to achieve with cyanobacteria. This problem is discussed in detail elsewhere (Oren 2011; Pinevich 2015).

Higher taxonomical ranks (e.g., Orders and Families) have undergone drastic changes over the past 20 years due to advances in reconstructions of molecular evolution. A few alternative systems for cyanobacteria have been proposed. Although there is no strict consensus among researchers as to which system is the most reliable, for the sake of clarity and congruence throughout the chapter we will employ the recent scheme advocated by Komárek and collaborators in 2014. Komárek's system relies largely on total evidence reconstructions, and thus should be a good approximation of evolutionary relationships. For example, Fig. 1 is a schematic representation of Komárek's system on a rank of order with the most important morphological features. Figure 2 shows a phylogenetic reconstruction with a designation to orders based on Komárek et al. (2014). However, it should be noted that the system has only recently been proposed and it is far from being generally accepted.

Unfortunately, we unable to capture the entirety of the diversity of all cyanobacteria in a single chapter and thus we will focus on the most important genera with an exception of Nostocales, where we use family level, because we possess a significant amount of data from each family designed by Komárek et al. (2014). Familial designations for the other cyanobacterial lineages (e.g., the Chroococcales) are not as well defined at this time.

Synechococcales

This large group (Fig. 3) contains the most abundant, ecologically significant, and the oldest cyanobacteria, which are characterized by parietal thylakoid arrangement (Komárek et al. 2014). Annual global abundances of marine picoplanktic *Prochlorococcus* sp. and *Synechococcus* sp. in oceans may reach up $2.9 \pm 0.1 \times 10^{27}$ and $7.0 \pm 0.3 \times 10^{26}$ cells, respectively (Flombaum et al. 2013). Although there is no direct evidence in paleontological data, molecular dating methods allowed reconstruction of the earliest events in cyanobacterial evolution. It has been suggested that the first *Synechococcus* cells may have appeared in hot springs before 3 BYA (Dvořák et al. 2014a). While the order contains 70+ genera, the most abundant are *Synechococcus*, *Prochlorococcus*, *Leptolyngbya*, and *Pseudanabaena* (Whitton and Potts 2000). The order Synechococcales is not a morphologically coherent group, containing both unicellular and filamentous cyanobacteria. Since it is the largest order of the non-heterocystous cyanobacteria in terms of number of genera, we cannot cover the whole diversity of this order. Thus, we will focus on the most commonly encountered and abundant taxa.

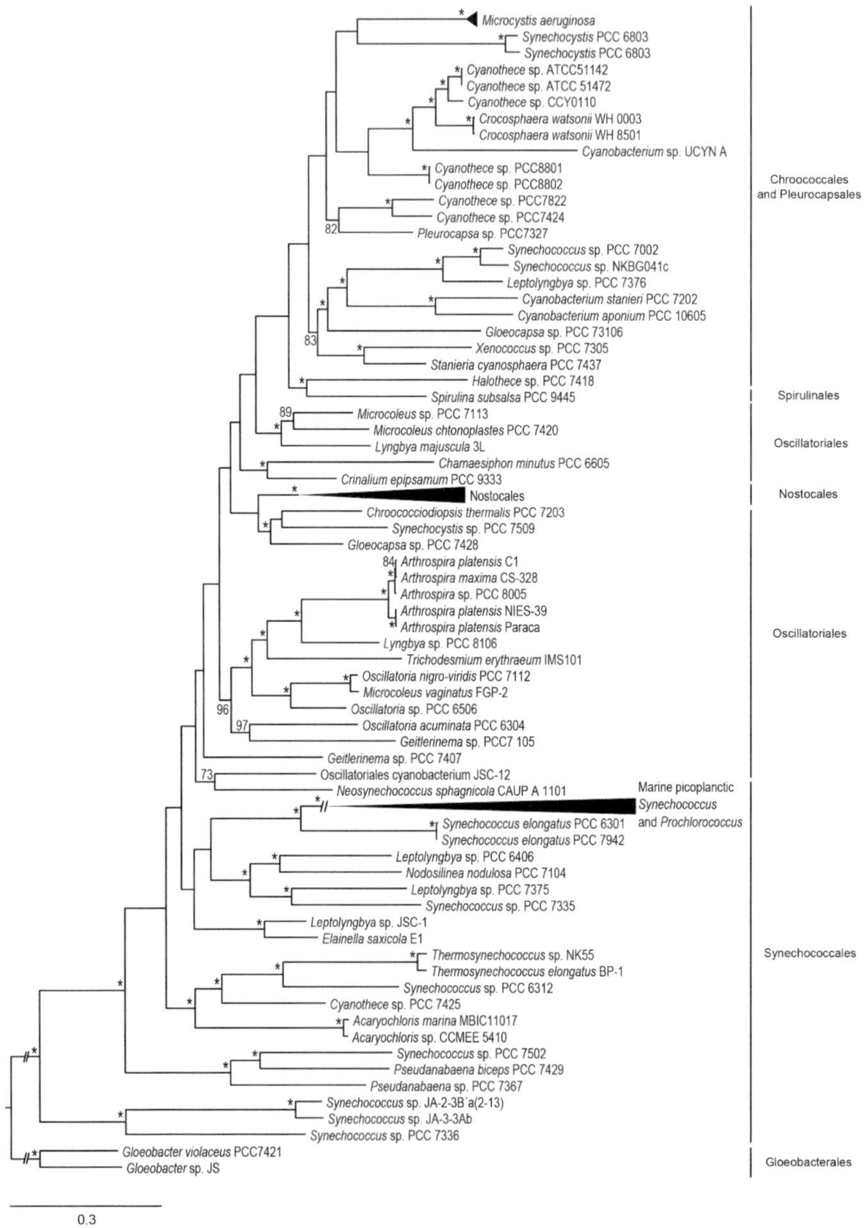

Fig. 2 Phylogenomic reconstruction of cyanobacterial evolution based on 69 concatenated, orthologous protein coding loci. These loci were selected, aligned, and concatenated using Hal (Robbertse et al. 2011). The tree was reconstructed using maximum likelihood optimality criterion and the CAT + LG model in RaxML 8.0.2 (Stamatakis 2006). The topology was tested by 500 rapid bootstrap replicates. Bootstrap support is located at the nodes with an *asterisk* representing 100 % bootstrap support. Orders of cyanobacteria *sensu* Komárek et al. (2014) are labeled

Synechococcus *and* Prochlorococcus

Under the name *Synechococcus* is hidden an example of morphologically similar yet genetically distant lineages. *Synechococcus* is a morphologically simple, yet polyphyletic genus. Described by Nägeli (1849), members are unicellular, small (<4 μm), sometimes forming pseudofilaments, and may possess involuted cells (unusual elongation of cells without cross-wall formation). On the other hand, this lineage exhibits immense genetic variability and extreme polyphyletic origins. For example, Dvořák et al. (2014a) found 12 unrelated lineages with congruent morphology. In subsequent paragraphs, we will focus on some of them according to Fig. 1 in Dvořák et al. (2014a). It should be noted that all of these clades have apparently very distinct evolutionary histories. However, only minor efforts have been made to taxonomically revise this genus, even though the polyphyletic nature of this lineage was amongst the first elucidated with molecular methods (Honda et al. 1999; Robertson et al. 2001). Komárek and Anagnostidis (1998) recognized at least 13 species of *Synechococcus*, which are not linked to type cultures. Moreover, we cannot assign them plausibly to 12 clades from Dvořák et al. (2014a) and to Bergey's Manual (Boone et al. 2001), which recognizes only five, unnamed clusters (see supplementary materials from Dvořák et al. 2014a).

The oldest clade of *Synechococcus* (clade 1) is composed of cyanobacteria inhabiting hot springs, specifically in Yellowstone National Park (Ferris and Ward 1997). This also supports an origin of cyanobacteria in hot springs, currently a prevalent opinion (Butterfield 2015) contrary to a freshwater origin proposed by Blank and Sanchez-Baracaldo (2010). Thermal strains later diverged and are mixed with freshwater strains in clade 6. They are often called the *Thermosynechococcus* (Katoh et al. 2001), but this has not been formally accepted under any code of nomenclature. *Synechococcus* strains dwelling in peat-bogs (*Sphagnum* bogs) are located in clades 8 and 9. *Neosynechococcus sphagnicola* (clade 9) comprises an exception among unnamed or incorrectly named clades, being only recently described (Dvořák et al. 2014b; Fig. 3).

Clade 10 contains the freshwater taxon *Synechococcus elongatus* (PCC 7942, PCC 7943, and PCC 6301), marine and freshwater picoplanktic *Synechococcus*, *candidatus Synechococcus spongiarum*, and *Prochlorococcus* (see details on *Prochlorococcus* below). All these taxa comprise a coherent monophyletic group (Dvořák et al. 2014a), which is, however, highly ecologically and genetically diverse. This group is composed of two clades, which contain either freshwater or marine strains. However, among freshwater strains, there are mixed marine (e.g., WH5701) and thermal (CCAP 1479/1B) strains. It was previously thought that picoplanktic strains of *Synechococcus* were strictly marine, but recent studies (Callieri et al. 2013) showed also their significance and high abundance in a plankton of freshwater lakes. Thus, we may conclude that some strains of *Synechococcus* have high salinity tolerance or were repeatedly introduced to the marine environment. Clade 10 contains *candidatus Synechococcus spongiarum*, which is the most common cyanobacterial symbiont of marine sponges (Erwin and Thacker 2008). It exhibits a very similar

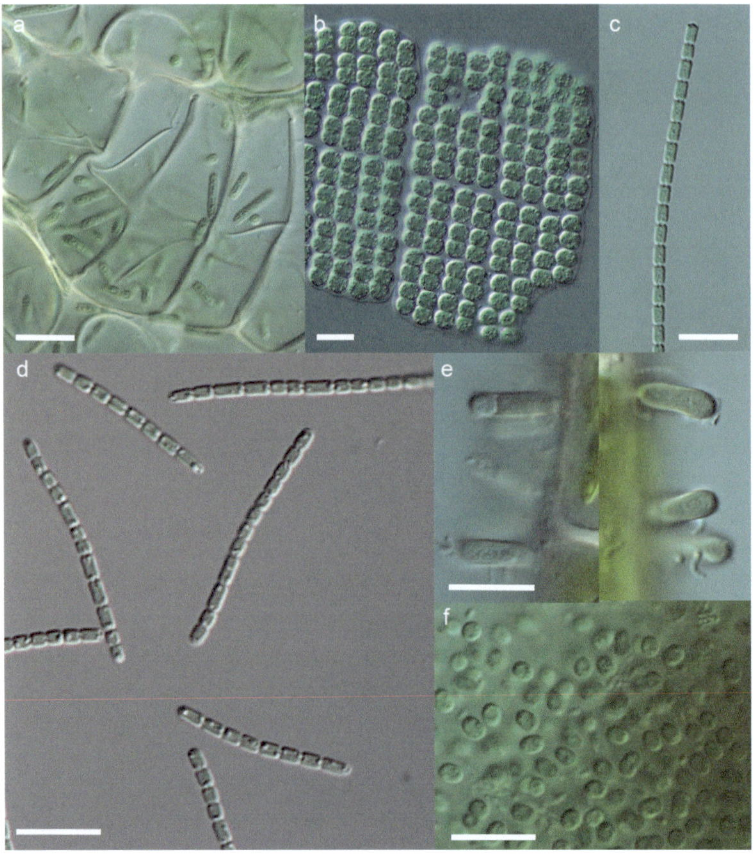

Fig. 3 Microphotographs of Synechococcales. (**a**) *Neosynechococcus sphagnicola*, (**b**) *Merismopedia glauca*, (**c**) *Pseudanabaena galeata*, (**d**) *Pinocchia polymorpha*, (**e**) *Chamaesiphon* sp., (**f**) *Aphanocapsa* sp. Scale bar 10 μm

genome to other *Synechococcus* strains from this clade; however, the genome has been undergoing a streamlining, which is typical for symbiotic bacteria due to more stable environment in the host (Gao et al. 2014).

Prochlorococcus is probably the most abundant oligotrophic cyanobacterium, and exhibits unique physiological features. First of all, it is very small (<2 μm), with simple rod shaped cells and an unusual photosynthetic pigment composition: divinyl-chlorophyll *a* and *b*. Further, it lacks phycobilisomes (specialized cyanobacterial and rhodophyte light harvesting antennae). The genome of *Prochlorococcus* is one of the smallest among cyanobacteria as it has undergone intensive genome reduction since it diverged from *Synechococcus* (Partensky and Garczarek 2010). *Prochlorococcus* is one of the most intensively studied cyanobacteria, and thus its ecological importance and evolutionary impact is extensively reviewed elsewhere: Partensky et al. (1999), Garcıa-Fernandez et al. (2004), and Biller et al. (2015).

Although *Prochlorococcus* seems to be monophyletic, there are four distinct ecotypes. Two of them are more abundant in high light conditions (abbreviated HL). HLII is more abundant in lower latitudes and HLI in higher latitudes. Two other ecotypes are adapted to low light conditions (LL). Ecotype LLI is more abundant in higher latitudes in upper layers of oceans and in lower latitudes in deeper layers around the thermocline. The last ecotype, LLIV, lives only in low latitudes and in the deepest areas of occurrence under the thermocline (Moore et al. 1998; West and Scanlan 1999).

Leptolyngbya

Leptolyngbya is possibly the largest genus of filamentous cyanobacteria in terms of number of species, and is characterized by overall simple morphology: thin filaments (<2 μm) with narrow cells and occasional false branching (reviewed in Komárek and Anagnostidis 2005). *Leptolyngbya* may be found in benthic, aerophytic, subaerophytic, periphytic, and even artificial habitats such as unsanitized urinals (Rulík et al. 2003). Due to their simple morphology, species have been classified based on their ecological preferences. Phylogenetic analyses repeatedly revealed extensive polyphyly within this genus (e.g., Casamatta et al. 2005). Interestingly, some lineages of "*Leptolyngbya*" and "*Synechococcus*" *sensu lato* exhibit similar phylogenetic relationships (e.g., Dvořák et al. 2014a, b). Thus, although they are not coherent in morphology, they often form monophyletic clusters.

The majority of the described species of *Leptolyngbya* have not been sequenced, so molecular revisions are nearly impossible at this stage. Moreover, there are new species with morphology similar to *Leptolyngbya* (based only on morphology), but they are unrelated to the type species. We will discuss a few examples of recently discovered and revised taxa.

The genus *Oculatella* has been proposed by Zammit et al. (2012). Initially isolated from hypogean environments in Malta and Italy, it has a conspicuous apomorphy: a rhodopsin-like reddish inclusion in a terminal cell. Subsequently, Osorio-Santos et al. (2014) proposed another seven species within this monophyletic cluster mostly from desert environments. Some of them lack any unique morphological feature, thus they can be recognized only using sequence data and therefore they are cryptic species.

Ecologically even more diverse are species within the genus *Nodosilinea* (formerly cluster 3 based on Bergey's manual; Boone et al. 2001) which has been established by Perkerson et al. (2011). Initially, this genus contained only four species, isolated from marine and freshwater lakes, but further isolates have been obtained from rocks and desert soils. All strains, though, have a distinctive apomorphy: the production of nodules when grown under low light levels. Interestingly, this appears to have nothing to do with nitrogen fixation (Li and Brand 2007).

Oculatella and *Nodosilinea* are both excellent examples of ecological, biogeographical, and evolutionary differentiation with available molecular data besides the marine picoplanktic *Synechococcus*.

The most comprehensive phylogeny of *Leptolyngbya* has been presented in the paper of Osorio-Santos et al. (2014). It suggests that the monophyly of the genus *Leptolyngbya* will be significantly more disturbed in the future, because there are at least four other species of *Leptolyngbya* outside the *Leptolyngbya sensu stricto* cluster, which contains *L. boryana*, *L. faveolarum*, *L. tenerrima*, and *L. angustata*.

Pseudanabaena

The genus *Pseudanabaena* consists of 36 species, which predominately occur in the plankton and benthos of freshwater or brackish water bodies all around the world. It is filamentous with small cells (mostly <2 μm), distinctive constrictions at cross-walls, rare sheath production, and often with distinctive chromatoplasma (Fig. 3; reviewed in Komárek and Anagnostidis 2005). This chromatoplasm refers to the parietal part of the cell, where thylakoid are concentrated, thus appearing to be darker than the centroplasm, or transparent, central portion of the cell.

Recent investigations have revealed that *Pseudanabaena* is also polyphyletic (Acinas et al. 2009; Dvořák et al. 2015a; Yu et al. 2015). Yu et al. (2015) have attempted to revise the genus by sequencing *P. mucicola*, *P. galeata*, *P. limnetica*, and *P. minima,* showing that they form a monophyletic clade together with *P. catenata*, the type species. Further revisions are forthcoming. For example, Dvořák et al. (2015a) investigated two strains similar to *P. galeata* from a freshwater lake in Vietnam. Although the morphological similarity of these strains to *P. galeata* was remarkable, the strains formed a separate cluster far from *P. galeata.* This represents another evidence of polyphyletic origin of *Pseudanabaena* and new monospecific genus *Pinocchia polymorpha* (Fig. 3) has been described.

The genus *Pseudanabaeana* is also evolutionarily related to the common planktic cyanobacterium *Limnothrix redekei*. While *L. redekei* (a non-toxic, planktic, filamentous cyanobacterium) largely differ in morphology from *P. catenata* and *P. galeata*, they are very closely related in phylogeny of 16S rRNA as shown by, e.g., Suda et al. 2002).

Synechocystis

One of the most popular model organisms for molecular biology is *Synechocystis* sp. strain PCC 6803. Commonly employed by researchers investigating photosynthesis, it was the first sequenced cyanobacterial genome (Kaneko et al. 1995). *Synechocystis* sp. PCC 6803 is also considered as a reference strain for the genus in Bergey's manual (Boone et al. 2001).

Recently, it has been shown that *Synechocystis*, like many other genera, is polyphyletic. Based on phylogenetic position and thylakoid arrangement, the genus *Geminocystis*, containing two species *G. herdmanii* and *G. papuanica*, was split from *Synechocystis* (Korelusová et al. 2009).

Merismopedia

There are 40 species within this genus, but molecular data are available only for three of them. *Merismopedia* is a group of unicellular cyanobacteria, which has a unique colony formation where cells divide in a single plane and create flat colonies (Komárek and Anagnostidis 1998; Fig. 3). *Merismopedia* is most commonly found in periphytic habitats. Palinska et al. (1996) analyzed strains of *M. punctata*, *M. glauca*, *M. elegans* and found that although these strains exhibited a degree of high morphological variability, their near complete 16S rRNA sequence was identical. Thus, it contradicts the vastly more common pattern of the hidden genetic "cryptic diversity" without observable phenotype variability (see an example of *Oculatella* above).

Acaryochloris marina

Acaryochloris marina is a unicellular marine cyanobacterium with simple morphology and has a unique metabolic feature and an unusual photosynthetic pigment: chlorophyll-*d*. First described in 2003 (Miyashita et al. 2003), it has an unexpectedly large genome of 8.3 Mb, which is predominantly typical for morphologically more complex cyanobacteria (Shih et al. 2013). *A. marina* is a monospecific genus, which deserves attention due to its importance as a model organism for understanding photosystem modification and genome expansion in cyanobacteria (Swingley et al. 2008).

Chamaesiphon

This asymmetrically dividing, chroococcalean genus commonly inhabits both submerged and periodically wetted substrates (Fig. 3), in tropical and temperate regions such as on stones, plants, and filamentous algae, especially in running waters. The genus can be divided into three subgenera: *Chamaesiphon sensu stricto*, *Chamaesiphonopsis*, and *Godlewskia* (Sant'Anna et al. 2011), but this is not supported via molecular data. Interestingly, Honda et al. (1999) showed phylogenetic affinity of *Chamaesiphon subglobosus* PCC 7430 to *Leptolyngbya boryanum* PCC 73110 and *Leptolyngbya foveolarum* Komárek 1964/112. Loza et al. (2013) suggested a molecular similarity of *Chamaesiphon subglobosus* PCC 7340 and *Ch. investiens* UAM 386 with genera *Synechococcus* and *Cyanobium*. Komárek et al.

(2014) showed that *Ch. minutus* PCC 6605 forms a separate clade with *Crinalium epipsammum* PCC 9333 within the order Oscillatoriales. Members of subgenus *Godlewskia* probably belong to the order Chroococcales, family Stichosiphonaceae, but it has not been confirmed by molecular methods.

Chroococcales

Members of the order Chroococcales represent unicellular cyanobacteria commonly encountered in aquatic (planktic and periphytic) and terrestrial environments (aerophytic, subaerophytic, soil, epilithic, epiphytic, etc.). Most taxa possess relatively simple morphology, leading to often times indistinct boundaries between species and genera. The order Chroococcales includes eight families, whose members live in single or colonial mode of live, dividing in one or more planes and usually forming irregular type of thylakoid arrangement (Fig. 4; Komárek et al. 2014).

Unicellular cyanobacteria represent an important group in aquatic habitats with respect to their abundance, diversity, and ecological role.

Microcystis

Microcystis is a well-defined genus and possibly the most commonly cited and studied cosmopolitan, bloom-forming, and toxin producing genus (Fig. 4). Planktic *Microcystis* is usually dominated by a few species such as *M. aeruginosa*, *M. wesenbergii*, *M. ichtyoblabe*, often concurrently present with other taxa such as *Woronichinia*, *Snowella*, and *Merismopedia*. The most frequently occurring species (e.g., *M. aeruginosa*, *M. wesenbergii*, or *M. ichtyoblabe*) may be polyphyletic based on 16S rRNA sequence data (Neilan et al. 1997; Komárek and Komárková 2002). Interestingly, *M. aeruginosa*, possibly the most commonly encountered eutrophic cyanobacterium, is quite consistent morphologically across freshwater lakes worldwide. However, it has high 16S-23S ITS diversity and it exhibits geographical barriers on gene flow among populations, with a concurrent lack of phylogeographical pattern (van Gremberghe et al. 2011). Current research reveals specialized clones (chemotypes) of *Microcystis* producing various types of toxic microcystins (Walker et al. 2004). Thus, if we take into account secondary metabolite production as an additional feature, the real biodiversity within *Microcystis* is likely higher than currently suggested.

Aphanothece sensu lato

Members of the genus *Aphanothece* occur worldwide from tropical to polar areas (Gardner 1927; Copeland 1936; Komárek 2003, 2014; Hašler and Poulíčková 2005; Whitton 2005). The genus includes morphospecies with elongated to rod-like cells,

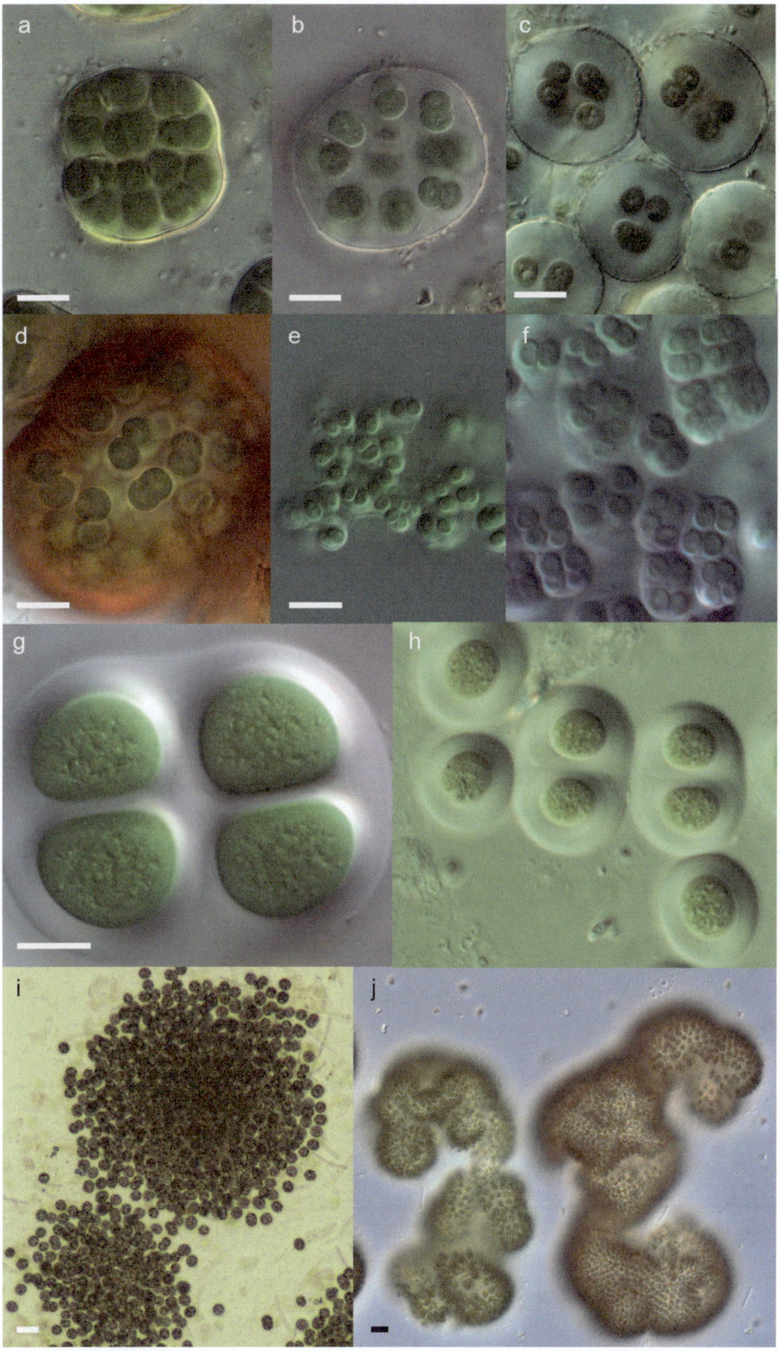

Fig. 4 Microphotographs of Chroococcales. (**a**) *Asterocapsa divina*, (**b**) *Asterocapsa* sp., (**c**) *Asterocapsa* sp., (**d**) *Gloeocapsa novacekii*, (**e**, **f**) *Gloeocapsa* sp., (**g**) *Chroococcus subnudus*, (**h**) *Chroococcus* sp., (**i**) *Microcystis novacekii*, (**j**) *Woronichinia naegeliana*. Scale bar 10 μm

dividing in one plane (perpendicular to main axis) and forming small gelatinous colonies. Although only few records of 16S rRNA exist in GenBank, the genus seems to have a polyphyletic origin. For instance, Komárek et al. (2011) pointed out wide genetic and morphological variability within the genus *Aphanothece*.

Several subgenera (or separate genera if subsequently elevated) of the genus *Aphanothece* can be distinguished: *Aphanothece sensu stricto*, *Anathece*, *Cyanogastrum*, *Halothece-Euhalothece* complex. The new genus *Anathece* represents morphospecies forming oval or rod-like forms, based on the type species *Anathece clathrata* (basionym: *Aphanothece clathrata*). Molecular and morphological features indicate an *Anathece* affiliation with members of the order Synechococcales (Komárek et al. 2011). The remaining species of *Aphanothece* group belong to the order Chroococcales.

Gloeothece

Coccoid cyanobacteria from the genus *Gloeothece* form oval, elongated, or rod-like cells, enveloped with distinct mucilaginous sheaths. *Gloeothece* plays a substantial role within aerophytic or subaerophytic communities or on submerged substrates all over the world (Komárek and Anagnostidis 1998). Their morphology, type of cell division, and molecular features are similar to *Aphanothece*. Sequence similarity of the 16S rRNA from GenBank ranges from 88 to 100 %, clearly falling outside of the traditionally accepted levels for within generic variation. *Gloeothece fuscolutea* was recently designated as a new type of the genus (Mareš et al. 2013) because of similarity between the former type (*Gloeothece linearis*) and *Gloeobacter violaceus*. However, the proposed neotype is still in conflict with Rippka et al. (2001b) who propose *G. membranacea* PCC 6501 as a reference strain and neotype for a botanical nomenclature.

Chroococcus

Chroococcus is commonly encountered on aerophytic, subaerophytic or submerged substrates, where it typically forms small microscopic colonies consisting of hemispherical cells, covered with mucilaginous envelopes, often layered (Fig. 4). Members of this unmistakable genus occur all over the world where they frequently occur in tropical or temperate zones, and less towards the poles (Komárek and Anagnostidis 1998). Molecular analyses show that this genus consists of ca. 60 species in a polyphyletic lineage (e.g., Komárková et al. 2010; Kováčik et al. 2011). Recently, *Chroococcus* was split into two new genera: *Limnococcus* and *Chroococcus sensu stricto*. Species of *Limnococcus* are planktic and differ morphologically and molecularly from the members of *Chroococcus sensu stricto*. Despite a molecular characterization of *Chroococcus sensu stricto,* it is still not clear what the "true"

Chroococcus is, because sequences of the type species, *Ch. rufescens,* are lacking. Further, the phylogenetic placement of some small species is unclear due to molecular similarity with other genera such as *Eucapsis* or *Synechocystis*. For example, Rippka et al. (2001a) note that there is morphological and molecular similarity among several PCC strains of small *Chroococcus* and *Gloeocapsa*. Thus, the true diversity of *Chroococcus* in nature across its range remains unclear and is still based on morphological investigation only.

Gloeocapsa

Several members of the genus *Gloeocapsa* possess morphological similarity with *Chroococcus*. *Gloeocapsa* represents a relatively heterogeneous group of colonial cyanobacteria forming small or medium sized spherical or oval cells, covered with distinct or indistinct envelopes of various colors (Fig. 4). The majority of species prefer aerophytic or subaerophytic habitats such as moist rocks or bark of trees (Komárek and Anagnostidis 1998). *Gloeocapsa* exhibits a specific life cycle including several distinct morphological stages, the observance of which is essential in the process of species identification. Available sequences of "*Gloeocapsa*" correspond to different genera such as *Chroococcidiopsis*, *Gloeocapsopsis*, *Cyanothece*, *Gloeothece*, but whether or not this is due to convergent evolution of simple morphologies or as a result of misidentifications of strains remains to be seen. For instance, *Gloeocapsa* sp. PCC 7428 (isolated from a moderate hot spring in Sri Lanka) corresponds to the genus *Gloeocapsopsis* (Azua-Bustos et al. 2014). *Gloeocapsa* "*alpicola*" FACHB-400 (=*G. atrata*?) shares 99 % identity with *Gloeothece* sp. PCC 6909 or *Synechocystis* sp. LEGE 06083, and probably does not represent the *Gloeocapsa* at all. Several morphologically distinct lineages within *Gloeocapsa* exist, which can be recognized based on cell size and color of mucilaginous envelopes. Small species (cells <6 μm in diameter) such as *G. atrata*, *G. aeruginosa*, *G. punctata*, *G. compacta* or *G. fusco-lutea* occur frequently among aerophytic habitats, where they form various eco- and morphospecies. Species delimitations without molecular confirmation can be problematic because of the overlapping morphologies. Our knowledge about the diversity of species producing large cells (>6 μm) is incomplete. Current research is based only on floristic data and ecologies of aerophytic/subaerophytic populations.

Peculiar Chroococcales with No Sequence Data

The genera *Cyanophanon*, *Clastidium*, *Stichosiphon* and probably *Chamaecalyx* belong to the currently established family Stichosiphonaceae (Komárek et al. 2014). They represent typical, globally distributed epilithic or epiphytic taxa in streams, rivers, stagnant water bodies or artificial water bodies such as aquaria, channels or basins (Komárek and Anagnostidis 1998). There is currently no molecular data for the family as a whole.

The family Gomphospaericeae has also been established only on a basis of morphological features. Members of this lineage typically occur in plankton from tropical to temperate zones, and rarely cold arctic waters. Only one sequence is available: *Gomphosphaeria aponina* SAG 52.96 (freshwater, Austria). The family Entophysalidaceae represents the most morphologically distinctive group of chroococcalean cyanobacteria, but the molecular diversity within the family and species is not known. However, the few deposited sequences of *Chlorogloea* exhibit a high level of heterogeneity. The majority of these species inhabit stony substrates or plants in both aquatic and terrestrial environments, especially in tropical or temperate zones (Komárek and Anagnostidis 1998).

Chroococcidiopsidales

Coccoid cyanobacteria from the genus *Chroococcidiopsis*, formerly placed within the order Chroococcales, now form a separate cluster (Fewer et al. 2002; Azua-Bustos et al. 2014). Interestingly, the genus *Chroococcidiopsis sensu stricto* is more similar in 16S sequence to heterocystous cyanobacteria such as *Fischerella*, *Nostoc*, *Scytonema* than morphologically similar genera from the order Chroococcales or Pleurocapsales. Molecular diversity indicates heterogeneity within the genus *Chroococcidiopsis* (Rippka et al. 2001c; Donner 2013). The reference strain of *Chroococcidiopsis thermalis* PCC 7203 (Rippka et al. 2001b) represents a cluster of freshwater or soil species with cells <5 μm in diameter producing baeocytes <4 μm. Strain PCC 6712 represents freshwater species forming larger cells than members of first cluster. Donner (2013) showed that the genus consists of more lineages than previously found by Rippka et al. (2001c). *Chroococcidiopsis* occurs worldwide, often in extreme environments such as hot and cold deserts, aerophytic, epi/endolithic, on soil or symbiotic in lichens (e.g., Boison et al. 2004; Sompong et al. 2005; Büdel et al. 2009).

Pleurocapsales

Members of the order Pleurocapsales are closely related to the order Chroococcales. They exhibit irregular cell division, specific formations of pseudofilamentous or pseudoparenchymatous thalli, and various types of polarized cells (Fig. 5). Members of the order form a monophyletic group closely related to the Chroococcales (e.g., Komárek et al. 2014). On the other hand, molecular analysis based on genome sequencing placed pleurocapsalean cyanobacteria in the same clade together with members of the order Chroococcales (Shih et al. 2013).

Rippka et al. (2001d) distinguished three clusters based on phenotypic features of PCC strains. *Pleurocapsa minor* strain PCC 7327 seems to be separated from the pleurocapsalean clade (Shih et al. 2013, Fig. 2). It was isolated from a hot spring

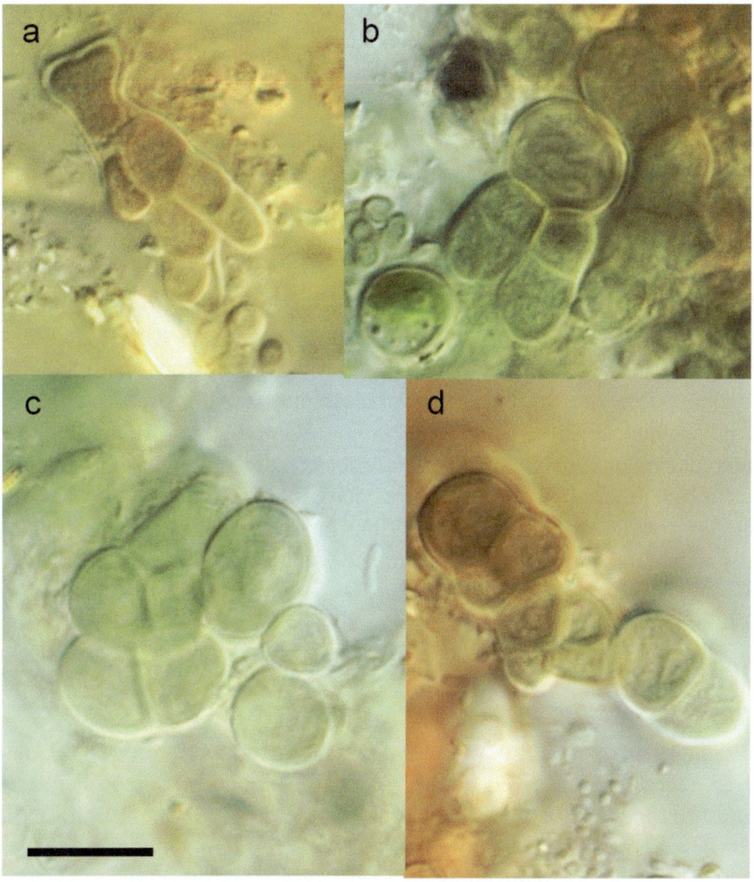

Fig. 5 Microphotographs of Pleurocapsales. (**a–d**) *Pleurocapsa minor*. Scale bar 10 μm

and probably represents a different species than designated. The most frequently used strain, PCC 7319, was isolated from a snail shell at intertidal zone in the Northern Mexico. However, the description of *P. minor* originates from streams near Prague (Hansgirg 1890, 1892). Records of *P. minor* from biotopes other than streams in temperate zones should be confirmed by molecular analysis. The genus *Pleurocapsa* includes almost 40 species occurring worldwide (Fig. 5), but only 16S rRNA sequences of *P. minor* and *Pleurocapsa* sp. are available.

The genus *Myxosarcina* is characterized by packet-like colonies and baeocyte production. The type species has not been sequenced and thus we are unable to thoroughly evaluate phylogenetic relationships. However, results from analyses of two complete genomes place *Myxosarcina* in the same clade as *Staineria, Pleurocapsa,* and *Dermatocarpella* (Yu et al. 2015).

Members of the genus *Stanieria* (family Dermocarpellaceae) represent coccoid, baeocyte forming taxa, which inhabit submerged substrates both in fresh and salt waters. Most of the sequences in GenBank belong to *S. cyanosphaera* and *Stanieria* spp. The phylogeny of Yu et al. (2015) based on 16S rRNA sequence data divides *Stanieria* into two groups including freshwater (e.g., strain PCC 7437) and salt-water species (e.g., strains PCC 7301 and 7302).

Dermocarpella contains six periphytic species with only scarce molecular data available. The most frequent records belong to species originated from stromatolites in Shark Bay (Goh et al. 2009), which may belong to *D. incrassata* (currently designated as *Chamaecalyx incrassatus*) inhabiting snail shells at intertidal zone in Mexico.

Likewise, all available sequences of *Xenococcus* originate from stromatolites at Shark Bay (Goh et al. 2009), geothermal springs in Costa Rica, symbiosis with marine sponges, epiphytic species of seagrasses in East Africa, periphytic in salt-water aquaria, or species inhabiting rock chips (unpublished, only stored in GenBank). Sequences of the type, *X. schousboei*, are lacking. Both molecular and morphological/ecological features of *Xenococcus* indicate a genus with polyphyletic character. Few genera of Pleurocapsales have been sequenced because of their inability to grow under laboratory conditions or rare occurrence. Thus, the real biodiversity within the order must be reexamined using field studies and molecular data.

Oscillatoriales

The Oscillatoriales were introduced in the monograph of Gomont (1892), and contained 15 genera characterized by the type of sheath and trichomes characteristics (Anagnostidis and Komárek 1988; Fig. 6). The order has been significantly expanded since its description and now contains 47 genera (Komárek et al. 2014). The order Oscillatoriales contains filamentous taxa with mostly fasciculated, radial, or irregular thylakoid arrangement (Fig. 6). One coccoid genus (*Cyanothece*; Fig. 6) has been transferred to Oscillatoriales based on thylakoid arrangement (Komárek et al. 2014). As in the other cyanobacteria, the majority of Oscillatoriales is polyphyletic.

Geitlerinema *sensu lato*

Geitlerinema (Fig. 6) consists of the filamentous cyanobacteria with terminal cell prolonged or hooked, parietal thylakoid arrangement, tricheomes <4 μm wide, prominent granules (usually), and the formation of mats on submersed substrates (wood, plants, stones, etc.) or in soils (Komárek and Anagnostidis 2005; Hašler et al. 2012). Originally this taxon was described as a subgenus of genus *Phormidium*, the type species is *Geitlerinema splendidum* (Anagnostidis and Komárek 1988). Hašler et al (2012) suggest *Geitlerinema* is polyphyletic with a subsequent revision by Strunecký et al. (2017) erecting a new genus *Anagnostidinema*, named in honor of the original author of the genus.

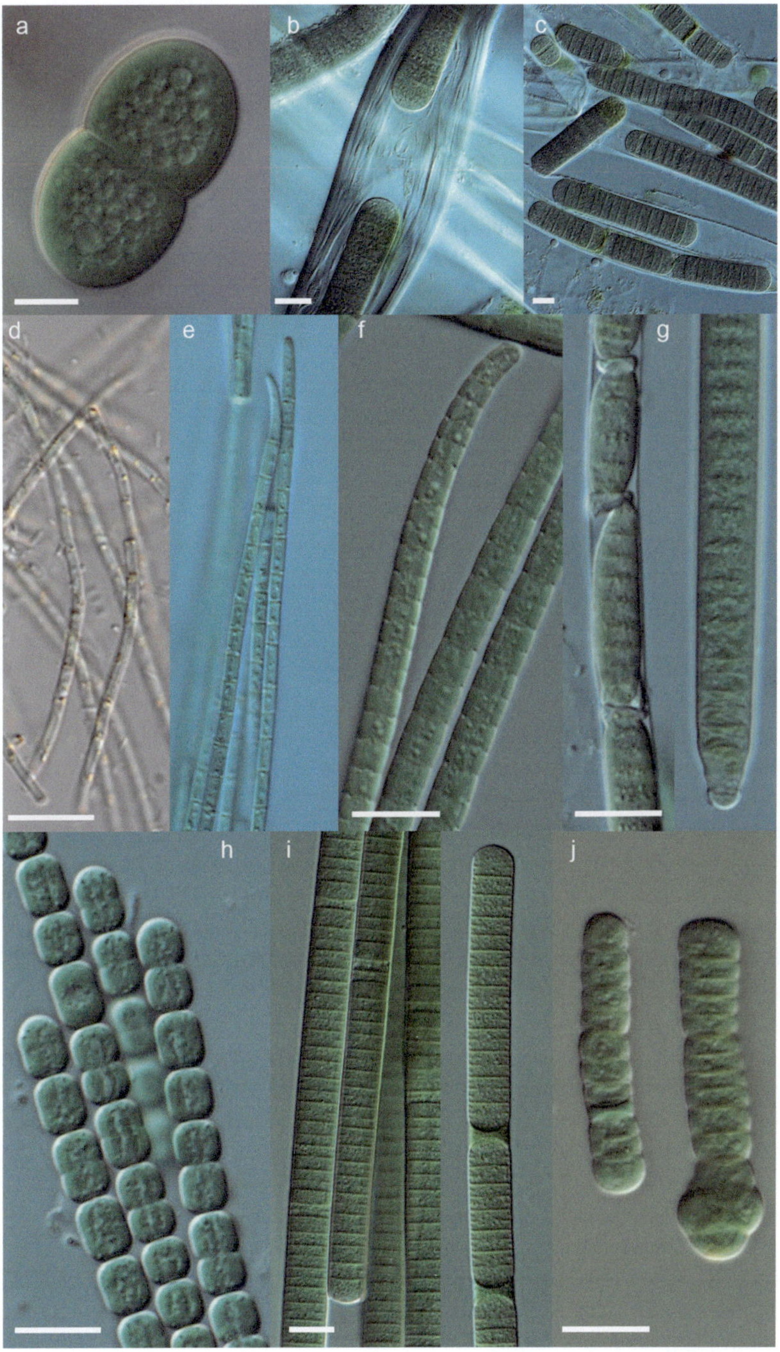

Fig. 6 Microphotographs of Oscillatoriales. (**a**) *Cyanothece aeruginosa*, (**b**, **c**) *Blennothrix* sp., (**d**) *Geitlerinema carotinosum*, (**e**) *Geitlerinema pseudacutissimum*, (**f**) *Kamptonema animale*, (**g**) *Phormidium* sp., (**h**) *Johanseninema constrictum*, (**i**) *Oscillatoria* sp., (**j**) *Crinalium* sp. Scale bar 10 μm

Microcoleus and *Phormidium*

Microcoleus was described by Gomont in 1892, and recent studies suggest that this taxon is polyphyletic. Type species, *M. vaginatus*, is one of the most abundant cyanobacterial species in soil crusts (e.g., Garcia-Pichel et al. 2001). *Microcoleus* forms multiple filaments in one sheath. Trichomes are cylindrical with usually narrowed, straight ends and may be terminated by calyptra. *Microcoleus* is morphologically quite similar to *Phormidium* and their morphological characters can overlap. Traditionally they have been separated on the basis of sheath formation (Komárek and Anagnostidis 2005), though this is not a stable feature (Dvořák et al. 2012; Hašler et al. 2012).

Phormidium sensu lato includes more than 200 species but it is polyphyletic, likely containing several unrecognized genera (Casamatta et al. 2003). Still, this genus is related to *Microcoleus* (Hašler et al. 2012; Strunecký et al. 2013).

Strunecký et al. (2013) revised and transferred *Phormidium autumnale* to *Microcoleus sensu stricto* based on phylogenetic analysis. It should be noted that two species, *M. vaginatus* and *M. atumnalis,* may be recognized only using a stable molecular feature: an 11 base pair insert within the 16S rRNA, which occurs only in *M. vaginatus* (Boyer et al. 2002).

Coleofasciculus originated by a revision of the *Microcoleus chtonoplastes*. It has been erected according to molecular, ecological, and secondary structure data. *Coleofasciculus* is a filamentous cyanobacterium with multiple trichomes covered by unlamellated and colorless sheath. Trichomes are non-tapering, terminal cells are conically rounded, and without calyptra. A typical habitat for *Coleofascicculus* is a littoral of brackish or marine habitats, and never occupies freshwater or terrestrial environments (Siegesmund et al. 2008).

Another genus separated from the genus *Phormidium* is *Wilmottia.* It has been derived from *Phormidium murrayi* (Strunecký et al. 2011), which was originally described from the Antarctica in 1911 as *Leplolyngbya murrayi* (West and West 1911). It is a simple filamentous cyanobacterium with rounded terminal cells, conspicuous granulation, cells <5 μm, and parietal thylakoids (Strunecký et al. 2011).

Larger and more complex species of the genus *Phormidium* have been recently revised as well. For example, Strunecký et al. (2014) erected *Kamptonema* (Fig. 6) by revising *P. animale*, a common freshwater littoral and epipelic cyanobacterium with tapering terminal part of the filament (Hašler et al. 2012). *Oxynema,* derived from the *Phormidium* of the group I *sensu* Komárek and Anagnostidis (2005) and characterized by a pointed terminal cell, was erected by Chatchawan et al. (2012). The genus *Phormidium* as a whole still has unexplored species diversity. For instance, a morphologically indistinguishable, yet genetically distinct, new cyanobacterium *Ammassolinea* (Hašler et al. 2014a) was isolated from the epipelon of the subtropical lakes in Florida.

Planktothrix

This genus was described by Anagnostidis and Komárek (1988) with type species *Planktothrix agardhii*. It consists of filamentous cyanobacteria characterized by the presence of aerotopes, straight filaments with rounded terminal cells, and without sheath (Komárek and Anagnostidis 2005). *Planktothrix* belongs to a major group of water-bloom forming cyanobacteria, which have a cosmopolitan distribution in the freshwater, eutrophic habitats (Komárek and Komárková 2004). Some species produce a variety of toxic, bioactive secondary metabolites, such as *P. agardhii* and *P. rubescens* (Walker et al. 2004). Suda et al. (2002) recognized four species based on a combination of molecular and morphological data: *P. agardhii*, *P. rubescens*, *P. mougeotii*, and *P. pseudoaghardii*. The genus as a whole was revised early, after the introduction of the molecular methods and it seems to be monophyletic.

Komvophoron

Anagnostidis and Komárek (1988) separated the genus *Komvophoron* from *Pseudanabaena* due to spherical or barrel-shaped cells and with different organization. On the basis of the shape of the vegetative and apical cells, this genus has been separated into two subgenera: *Alyssophoron* (type = *A. minutum*) and *Komvophoron* (type = *K. schmidlei*) (Anagnostidis and Komárek 1988; Hašler and Poulíčková 2010). *Komvophoron* mainly inhabits benthic areas, growing on the sand and muddy sediments in freshwater reservoirs (Komárek and Anagnostidis 2005). *Komvophoron* is likely largely overlooked or ignored and has only scant molecular data because benthic cyanobacteria in general are poorly explored (Hašler and Poulíčková 2010; Pulíčková et al. 2014) and *Komvophoron* is resistant to conventional cultivation techniques. However, Hašler and Poulíčková (2010) were able to employ single-filament PCR techniques, obtaining 16S rRNA and 16S-23S ITS sequences of *K. hindakii* and *K. constrictum*. This work showed that *Komvophoron* is polyphyletic, leading to the description of a new genus *Johanseninema* (Fig. 6; Hašler et al 2014b).

Oscillatoria

Oscillatoria is filamentous cyanobacterium with discoid cells (Fig. 6). Trichomes are slightly waved or straight, never branched. Trichomes are >8 μm, usually without sheath. *Oscillatoria, Phormidium,* and *Lyngbya* were distinguished from each other (Geitler 1932) based on sheath properties, but sheath production is not necessarily phylogenetically informative as this character heavily depends on local conditions (Whitton 1992). *Oscillatoria* often creates macroscopic layered, smooth mats. The type species is *Oscillatoria princeps* (Anagnostidis and Komárek 1988).

Although *Oscillatoria* is a very common cyanobacterium, we lack enough molecular data for wholesale revisions at this time. However, we may only infer from our phylogeny (Fig. 2) that *Oscillatoria* is polyphyletic as previously suggested by Ishida et al. (2001).

Lyngbya

This genus was described by Gomont (1892), with the type *L. confervoides*. Marine strains of *Lyngbya* are very important diazotrophic (fix atmospheric nitrogen) organisms and primary producers, but they are also rich in bioactive secondary metabolites, which are mostly toxic (Hoffmann 1994). Water blooms of *Lyngbya* can have adverse effects on coral reefs, especially on coral larvae recruitment (Kuffner and Paul 2004), because *Lyngbya* filaments are not consumed by herbivores and very quickly exploit available surfaces (Paul et al. 2005). On the basis of morphological similarities *Lyngbya, Phormidium*, and *Plectonema* were classified as the "LPP group" (Rippka et al. 1979), subsequently shown to be polyphyletic (Komárek et al. 2014). Although the genus *Lyngbya* is largely coherent in morphological features, it is polyphyletic based on 16S rRNA phylogeny (Engene et al. 2010, 2013). *Lyngbya* may be separated into three distant clades according to ecology: a mixed halophilic/brackish/freshwater lineage, a lineage more closely related to the genus *Oscillatoria* (freshwater), and a marine lineage (Engene et al. 2010). *Moorea* (Engene et al. 2012) and *Okeania* (Engene et al. 2013) were recently erected to include the marine members, which are potent producers of bioactive secondary metabolites. *Moorea* contains two species and *Okeania* five. Finally, *Limnoraphis* has been proposed by Komárek et al. (2013) for some of the freshwater strains, which are responsible for heavy water-blooms in some tropical reservoirs in South America.

Symplocastrum

Gomont (1892) was originally described *Symplocastrum* as a subgenus of the genus *Schizotrix*. Recently elevated to genus level (Anagnostidis 2001; Komárek and Anagnostidis 2005), the type species, *Sy. friesii. Symplocastrum*, is a relatively poorly researched taxon of Oscillatoriaceae (Pietrasiak et al. 2014). Morphologically, this taxon is similar to both *Hydrocoleum* and *Microcoleus*. Only recently sequenced (Pietrasiak et al. 2014), *Symplocastrum* is phylogenetically related to the newly erected genus *Kastovskya* (Mühlsteinová et al. 2014).

Spirulinales

Members of the Spirulinales are solitary or colonial (mats), have trichomes without sheathes, are regularly screw-like coiled, and usually possess an intense motility. Trichomes are not constricted at the cross walls, without branching, and without necridic cells. They reproduce by a disintegration of trichomes or by motile hormogonia (Komárek and Anagnostidis 2005). *Spirulina* is an important organism in biotechnology, as animal fodder, and as a human dietary supplement (Khan et al. 2005). However, this generic name is actually merely conserved; *Spirulina* was revised and the species *S. platensis*, the taxon of most human interest, now belongs to the genus *Arthrospira* (Vonshak 1997), which clusters with the order Oscillatoriales (Komárek et al. 2014). The genus *Halospirulina* has been split from *Spirulina* on the basis of morphology and high halotolerance (salinity between 3 and 13 %+; Nübel et al. 2000).

Nostocales

The Nostocales represent a species rich, diverse lineage of cyanobacteria typified by the ability (obligatory or not) to produce specialized cells (Fig. 7), mainly heterocytes (dedicated to nitrogen fixation) and akinetes (overwintering cells). This class was historically broken into two major groups: those taxa which undergo cell division in multiple planes (the Stigonematales *sensu stricto*, e.g., *Hapalosiphon*, *Mastigocladus*, *Stigonema*) and those taxa that never show reproduction in multiple planes (the Nostocales *sensu stricto*, e.g., *Nostoc*, *Anabaena*, *Calothrix*). These distinctions based solely on morphological features arose from the earliest works of cyanobacterial researchers (e.g., Bornet and Flahault 1886; Geitler 1932; Desikachary 1959) and while useful at a gross level, there were some inherent limitations to these classification schemes. First, some of the features, such as the production of heterocytes, have been shown to be environmentally inducible (in the classic case of the Nostocales) and thus might not be present in all examined specimens, potentially leading to taxonomic confusion. Second, cell division in >1 plane has been shown to be present in several families of the heterocytous clades, making its use in phylogenetic reconstructions questionable.

The relatively recent advent and propagation of modern molecular markers (mainly of the 16S rDNA gene sequence, secondary folding structures of the 16S-23S Internally Transcribed Spacer {ITS} region and comparison of the genes used in nitrogen fixation {e.g., *hetR*, *nifH*}) have allowed more finely nuanced, robust phylogenetic assessments. Rather than relying upon potentially environmentally plastic (e.g., sheath production) or inducible characters (e.g., specialized cells), a wealth of potentially phylogenetically informative characters may be obtained. While many members of this clade are difficult to fully assess due to difficulties in culturing or sequencing, modern approaches are yielding new

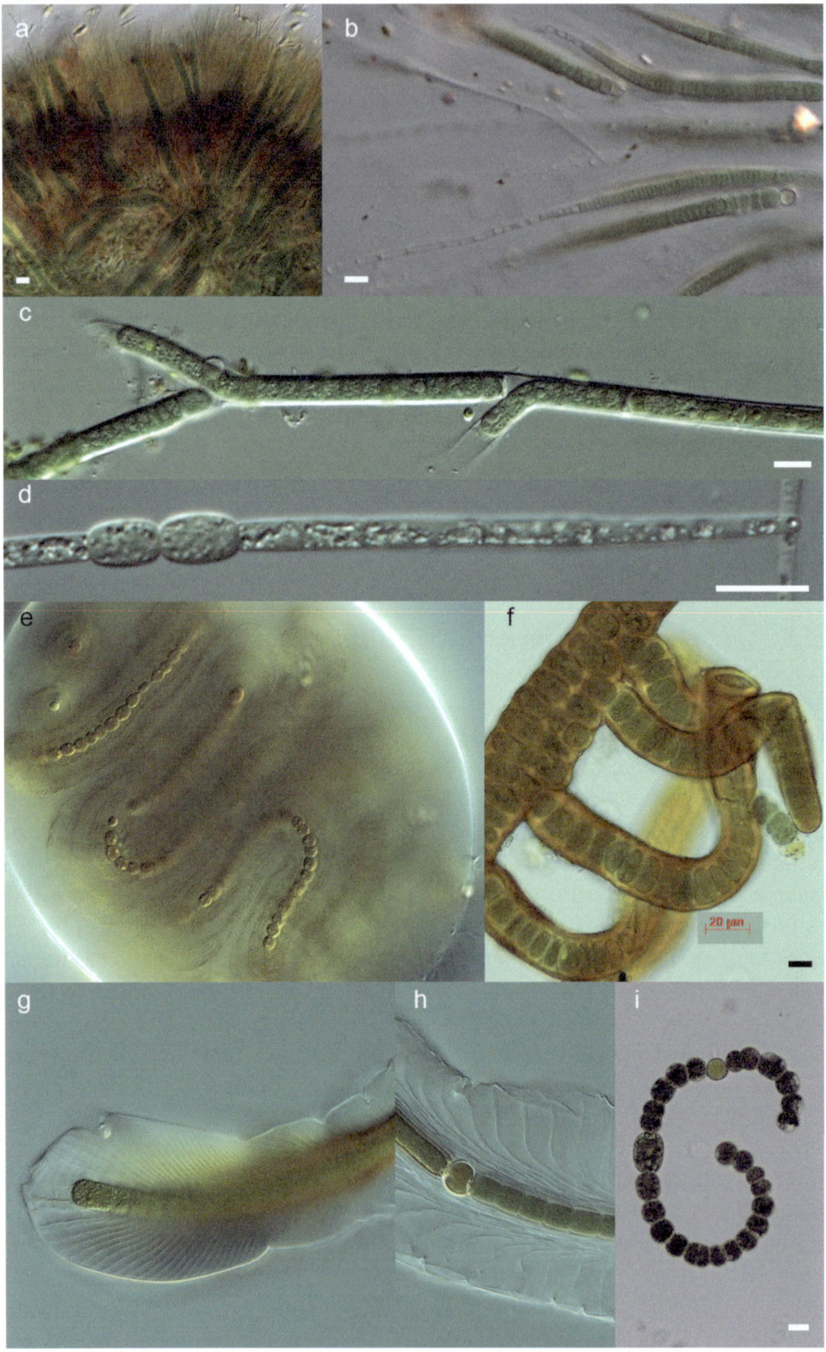

Fig. 7 Microphotographs of Nostocales. (**a**, **b**) *Rivularia* sp., (**c**) *Tolypothrix* sp., (**d**) *Cuspidothrix issatschenkoi*, (**e**) *Nostoc microscopicum*, (**f**) *Stigonema* sp., (**g**, **h**) *Petalonema alatum*, (**i**) *Dolichospermum* sp. Scale bar 10 µm

methods for elucidating phylogenetic relationships (e.g., Mareš et al. 2015). Even though this lineage represents one of the most morphologically character-rich lineages of cyanobacteria, there are most likely not enough unique morphologies to truly differentiate this diverse lineage. Thus, characters such as the structures of the ITS region (e.g., *Roholtiella, sensu* Bohunická et al. 2015), physiology (e.g., *Halotia, sensu* Genuário et al. 2015), or polyphasic approaches also utilizing ecology (e.g., *Dapisostemon, sensu* Hentschke et al. 2016) are increasingly being employed.

While the Nostocolaes are widely distributed and found in terrestrial, aquatic, subaerial, and symbiotic environments, the exploration of the molecular phylogenies has been rather unevenly applied. As Komárek et al. (2014) point out, 50 new cyanobacterial genera have been erected since 2000, with an additional 16 proposed just at the 2013 IAC meeting. However, many of these new taxa are from the other cyanobacterial lineages. Many Nostocalean taxa are difficult to culture and thus difficult to describe. Also, many lineages possess thick, copious mucilaginous sheaths, necessitating additional steps before successful sequencing (e.g., Mareš et al. 2015). Numerous Nostocalean taxa may be endemic or have restricted ranges (e.g., *Rexia sensu* Casamatta et al. 2006). Thus, the total diversity of this lineage has only been cursorily examined, especially in tropical or seldom sampled habitats (Dvořák et al. 2015b; Hentschke et al. 2016). Lastly, certain lineages, especially the more commonly encountered lentic taxa (e.g., *Anabaena*, *Dolicospermum*, *Cylindrospermopsis*), have received much greater attention than others, and thus have greater phylogenetic resolution. Recent expansion of geographic ranges, such as seen in *Cylindrospermopsis* (Padisák 1997), or the discovery of novel, bio-active secondary metabolites (e.g., *Aetokthonos*, *sensu* Wilde et al. 2014) point to the need for much more comprehensive assessments of the phylogeny of this lineage.

In a recent paper, Komárek et al. (2014) proposed a complete revision of the cyanobacteria as a whole. After examining molecular data sets from all available lineages, the researchers note that the Nostocales is only monophyletic when all taxa capable of producing specialized cells are included, and propose the order Nostocales to encompass these taxa. While all of these taxa have irregular thylakoids, familial level designations are posited based on a number of characters.

The Nostocales represents a diverse lineage in terms of morphology, genetics, and ecological preferences. Many members are associated with aquatic habitats, where they may be planktic (e.g., *Cylindrospermopsis* and *Dolichospermum*), benthic (e.g., *Anabaena*) or associated with the margins of the aquatic and terrestrial landscapes (e.g., *Mastigocladus*). The Nostocales are also commonly encountered as terrestrial microbes, where they have been described from tropical (e.g., *Dapistostemon*) to arctic habitats (e.g., *Nostoc*). Further, they have been described from polar (Vincent 2007), hot, arid (Řeháková et al. 2007), and temperate soils (Lukešová et al. 2009). They are also commonly involved in symbiotic associations (e.g., *Trichormus* and *Azolla*). Many Nostocales are also known to produce a wide array of bio-active compounds, which may be potent neuro-, hepato-, and dermatotoxins (for a review, see Codd et al. 1997).

Type of Cell Division

The majority of cyanobacteria, filamentous, coccoid, or unicellular, undergo binary fission as the main form of reproduction. Some Nostocales exhibit false-branching, where a new filament is not formed as a result of cell division and does not result in the thallus exhibiting division in multiple planes, leading to the appearance of filaments which appear to pass each other (e.g., *Tolypothrix*). However, unique among the Nostocales is a second type of cell division known as true-branching. Generally defined as cell division in which one or more of the cells change the polarity of growth, true-branching allows some cyanobacteria the capability of erect or creeping growth (although this is not a necessary condition for such growth). True branching is differentiated into three main forms: T-, V-, and Y- (for a review, see Golubic et al. 1996). Traditionally considered an important phylogenetic character, and serving to differentiate the Stigonematales, recent works have indicated that this might be a useful feature for familial assignments (Gugger and Hoffman 2004), but not higher levels.

While some Nostocalean taxa exhibit cell division in multiple planes, many taxa reproduce by the liberation of solitary cells generated by division perpendicular to the trichome axis. Other taxa employ hormogonia, which are distinct segments of the trichome, often arising from the formation of adjacent necridial cells or by the result of fragmentation. Those taxa capable of also generating gas vesicles often employ this method to propagate themselves in lentic habitats (e.g., *Dolichospermum*, formerly *Anabaena sensu lato*).

Ultrastructural Features

Recent work has shown this to be a polyphyletic character, but thus far all known Nostocalean taxa have irregularly arranged thylakoids (for a review of thylakoids see Komárek 2013).

Another important cell ultrastructural feature is the presence of gas vesicles, a form of inclusion body which may be filled with atmospheric gases. While they may be induced in some lineages, the presence or absence of gas vesicles is being employed in a phylogenetic sense. For example, gas vesicles were recently employed when the polyphyletic genus *Anabaena* was split into *Anabaena sensu stricto* (containing mainly periphytic species without gas vesicles), *Dolichospermum* (mainly planktic species with gas vesicles), and *Sphaerospermum* (planktic with gas vesicles).

Heterocytes and Akinetes

Perhaps the most distinguishing aspect of the Nostoclaes is their specialized cells, the most common of which are heterocytes and akinetes. The size, shape, and placement of heterocytes are frequently employed in identifications (e.g., *Anabaena*).

Heterocytes may be apoheterocytic (developing from vegetative cells between heterocytes) or paraheterocytic (developing from vegetative cells outside of heterocytes). Heterocytes may form at the ends of trichomes (terminal or basal), within a trichome (intercalary), at a right angle from a trichome (lateral), or within a multiseriate trichome (lateral). The formation of heterocytes along the trichome may also be solitary, in pairs, or as several in a row. Further, the actual shape and size of heterocytes differs by taxa. Although the formation and frequency of heterocytes may be environmentally inducible, the size and position in trichomes appears to be genetically controlled.

Akinetes are thick-walled resting cells, often environmentally inducible, typically used to survive adverse environments. These may also be used in phylogenetic reconstructions. For example, *Gloeotrichia* is separated from other members of the Rivulariaceae by the obligatory presence of akinetes (it should be noted that Komárek et al. 2014 separated this into a new family, the Gloeotrichiaceae). Akinetes, like heterocytes, have been used in phylogenetic assessments as the size, shape, and placement of akinetes seem genetically fixed (but the number produced seems environmentally influenced).

Modern Systematic Scheme

The clade containing the Nostocales/Stigonematales has long been recognized as taxonomically challenging. While monophyletic on the grossest level (e.g., the clade containing taxa with the capability of forming specialized cells), relationships within this clade have proven to be confusing, leading to calls from numerous researchers to resolve some of the phylogenetic uncertainties (Fig. 8; e.g. Kaštovský and Johansen 2008; Lukešová et al. 2008; Hauer et al. 2014; Komárek 2015, etc.). Numerous changes have been proposed and as newly discovered taxa are isolated, sequenced, and examined, our understanding of the relationships within this lineage is ever changing. A recent paper from Komárek et al. (2014) has proposed a taxonomic scheme integrating molecular, cellular, ecological, and morphological data in a total evidence (polyphasic) approach. This scheme involved the examination of the largest data set of cyanobacteria published thus far and represents an excellent, testable hypothesis of evolutionary relationships going forward.

Familial Designations

Aphanizomenonaceae

Characterized by unbranched, isopolar or subsymmetric filaments with akinetes and typically aerotopes (e.g., *Aphanizomenon*), this is a widely distributed, commonly encountered lineage, especially in eutrophic systems. A rather confusing clade in

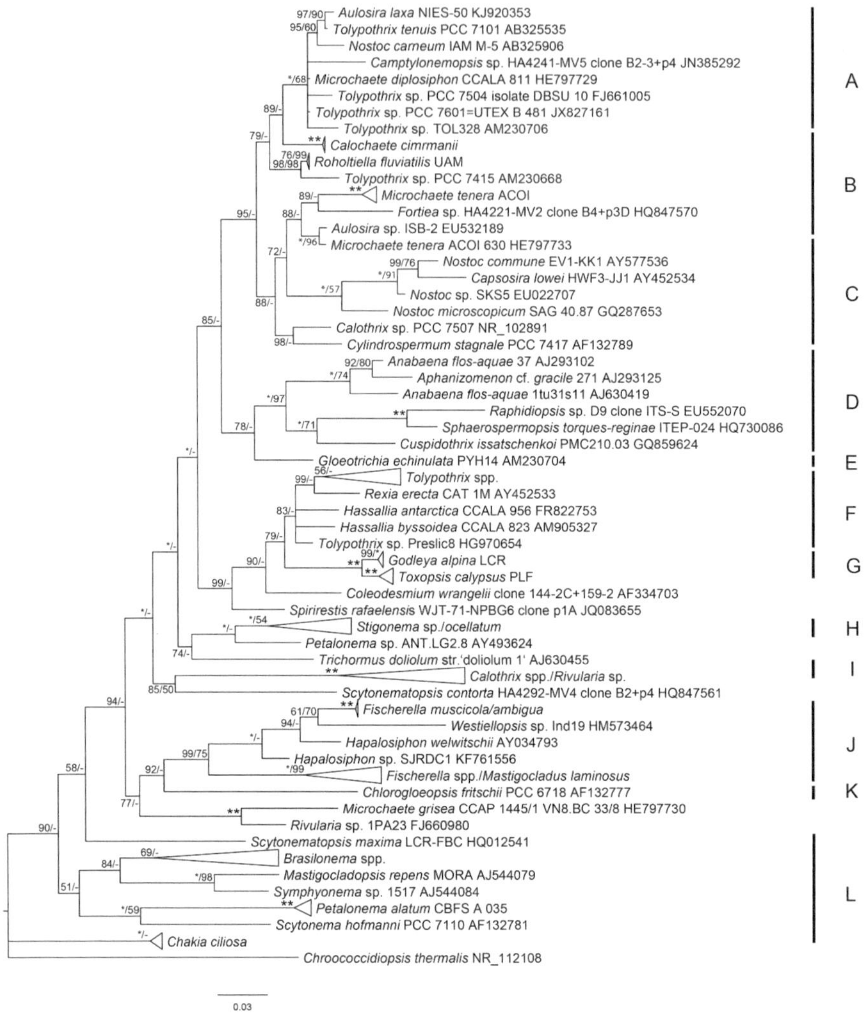

Fig. 8 Phylogenetic reconstruction of the main Nostocalean lineages based on 16S rRNA gene sequence data. The tree was inferred in MrBayes 3.2.1 (Ronquist and Huelsenbeck 2003) using the F81+G+I model. Two separate runs, each with 4 chains (1 cold and 3 heated) were run for 4,115,000 generations and sampled every 1000th generation. The bootstrap analysis was performed in RaxML 8.0.2 (Stamatakis 2006) with 1000 bootstrap replicates and GTR+GAMMA model. Posterior probabilities and bootstrap supports are at the nodes and *asterisk* represents 100 % posterior probability and bootstrap supports. A—Tolypotrichaceae, B—Fortieaceae, C—Nostocaceae, D—Aphanizomenonaceae, E—Gloeotrichiaceae, F—Tolypotrichaceae, G—Godleyaceae, H—Stigonemataceae, I—Rivulariaceae, J—Hapalosiphonaceae, K—Chlorogloeopsidaceae, L—Scytonemataceae and Symphyonemataceae

terms of morphological assessments, this includes the common planktic genera *Aphanizomenon*, *Cylindrospermopsis*, *Dolichospermum*, and *Raphidiopsis*. Many of these genera have long been considered polyphyletic, but recent investigations of morphology, coupled with 16S gene sequence data have allowed researchers to create monophyletic lineages for several genera. For example, Wacklin et al. (2009) were able to find a stable morphological character, the presence of aerotopes, which separated the planktic forms of "*Anabaena*" into the monophyletic genus *Dolichospermum*. However, it should be noted that other relationships within the Aphanizomenonaceae itself remain less clear, but monophyletic groupings are being elucidated and described, especially for clades of human interest (e.g., Werner et al. 2012).

Dapisostemonaceae

This family was recently described from examination of aerophytic taxa from Brazil. Not included in the Komárek et al. (2014) paper, Hentschke et al. (2016) separated this family from the Tolypothricaceae based on 16S rDNA sequence data. Characterized by heteropolar filaments, intercalary, bipolar heterocytes, lack of akinetes and single false branching (e.g., *Dapisostemon*). This mono-generic family probably represents a larger lineage of taxa that have not yet been identified or sequenced. The original materials were collected from seldom sampled tropical, aerophytic habitats and thus sister taxa may still need to be described, leading to uncertain familial assignments. For example, Hentschke et al. (2016) note that a new, closely related genus *Streptostemon*, putatively placed in the Scytonemataceae or Tolypothrichaceae, may actually be elevated to the first member of a heretofore undescribed family.

Scytonemataceae

A species-rich lineage characterized by isopolar filaments and false branching (e.g., *Brasilonema*, Fiore et al. 2007). There exists a wide range of morphological and ecological variability within this lineage, and numerous genera have been poorly or not at all characterized molecularly (e.g., *Chakia* and *Petalonema*). Some generic designations appear to be well supported (e.g., *Brasilonema*), while others are poorly resolved. Note: Komárek et al. (2014) caution that the relationship of this family to the Symphyonemataceae needs to be further elucidated.

Symphyonemataceae

A small family contains at least two genera (*Mastigocladopsis* and *Symphyonema*), characterized by both isopolar filaments and Y-type true branching (but see note above). Lamprinou et al. (2011) erected two new genera of cave dwelling cyanobacteria (*Iphinoe* and *Loriellopsis*) that may or may not belong to this family; current phylogenies lack resolution to clearly place them.

Rivulariaceae

A well-characterized, widely distributed, commonly encountered lineage whose taxa possess tapering, heteropolar filaments with facultative false branching, and the presence of a long, thin hair-like projection (e.g., *Calothrix*). The current disposition of some of the most known genera (e.g., *Calothrix*) from this lineage is in a state of flux. The original type of *Calothrix* was isolated from marine habitats, but has not been sequenced. Berrendero et al. (2016) have set about to resolve these issues by erecting new genera and attempting to create smaller, monophyletic units within this lineage, but do not propose any wholesale alterations to the family.

Chlorogloeopsidaceae

A poorly understood lineage, characterized by isopolar filaments or cell aggregates, the capacity for true-branching remains unclear (e.g., *Chlorogloeopsis*). Rarely described from nature, very little is known of the ecology of this potentially monotypic family (*Chlorogloeopsis fritschii* is the only taxon thus described). The majority of the work on this family has come from isolated strains in culture (and mainly of strains from PCC), so phenotypic plasticity remains a question.

Hapalosiphonaceae

A morphologically character-rich, yet difficult to properly identify lineage, characterized by isopolar filaments with T-type true branching, filaments and branches which may be uni- or multiseriate. A difficult clade to sequence due to recalcitrant mucilaginous sheaths, certain genera appear polyphyletic (e.g., *Fischerella*, a separate family in some phylogenies) while others appear monophyletic (e.g., *Westiellopsis*). However, assessments of this lineage are hampered by misidentifications of deposited cultures and sequences in GenBank. Komárek et al. (2014) suggest that this family is in need of revision with more sister taxa and strains in culture collections sequenced.

Stigonemataceae

A rarely reported (except for *Stigonema*), morphologically complex clade possessing filaments and branches that may be uni- or multiseriate with T-type true branching (e.g., *Stigonema*). Difficult to work with since strains do not survive well in culture, this lineage is challenging to investigate. Strains exhibit a wide array of morphological variation, further adding to taxonomic confusion.

Godleyaceae

A recently erected family containing two genera (*Toxopsis* and *Godleya*), characterized by both iso- and heteropolar tapering filaments and false branching (e.g., *Toxopsis*). This lineage may be sister to the Tolypothrichaceae (Lamprinou et al. 2012). Thus far, all known isolates have been from subaerial or terrestrial habitats.

Tolypothrichaceae

Recently revised in order to create monophyletic lineages from former Microchaetacean taxa (Hauer et al. 2014), this family is now monophyletic and characterized by non-tapering trichomes with heteropolar filaments and frequent false branching (e.g., *Tolypothrix*). Members of the Tolypothrichaceae are typically found in terrestrial or freshwater habitats (never marine) and are commonly encountered (Hauer et al. 2014).

Capsosiraceae

A poorly understood lineage, characterized by polar growth of colonies, which may be filamentous or aggregates, and possessing the ability for true-branching. Caution is warranted with this lineage, as the only sequenced member thus far, *Capsosira lowei*, may actually represent a taxon from a phylogenetically related lineage (Casamatta et al. 2006). Further collection and sequencing is warranted.

Gloeotrichiaceae

Previously assigned to the Rivulariaceae, this morphologically distinct lineage possesses tapering trichomes with heteropolar filaments, and spherical colonies that form akinetes (e.g., *Gloeotrichia*). This is a phylogenetically unsettled lineage, and future revisions might employ akinete characteristics (certain taxa exhibit akinetes in pairs, others as solitary cells) and the presence or absence of gas vesicles. Thus far only a single taxon has been sequenced (*G. echinulata*), so additional data will help resolve final phylogenetic placements.

Nostocaceae

Perhaps the most well known of the noctocalean lineages, typified by iso- or heteropolar filaments with facultative false-branching and akinete production (e.g., *Anabaena*). Species rich and ecologically permissive, these taxa can be found worldwide. Well studied, it is also clear that many genera are both morphologically character-poor and polyphyletic. While some members are difficult to identify, total evidence approaches employing characters such as ecology, ITS structures or slight morphological differences are increasingly being employed to separate monophyletic clusters (e.g., *Mojavia* isolated from desert soils or *Halotia* from Antarctica soils). This is also an interesting clade as many members have definable, diagnosable development sequences (e.g., *Nostoc*).

The Dark Matter of Cyanobacterial Diversity

Cyanobacteria represent 23.4 % of known prokaryotes and are among the most morphologically distinct prokaryotes. Nabout et al. (2013) note 2698 described species, with ca. 15 new species per year. Many cyanobacterial species remain to be described, with some models predicting at least 6280 species (Nabout et al. 2013). Some cyanobacterial groups are more studied than others (number of papers in ISI Web of Knowledge database; Thomson Reuters, New York; accessed 19.1.2016; search based on order name in title: Synechococcales 3 papers, Chroococcales 99, Pleurocapsales 7, Oscillatoriales 99, Nostocales 126 and Stigonematales 36 papers published), so additional work is needed. In addition, there exist gaps in our knowledge of cyanobacterial diversity from certain geographic regions (e.g., tropics) and specific habitats (e.g., benthic or aerophytic habitats).

There is a long-standing joke that the distribution of microalgae depends on the distribution of phycologists. For example, the vast majority of cyanobacteria have been described from Europe, the center of floristic research over the last three centuries. For example, Hauer et al. (2015) have gathered floristic records of terrestrial cyanobacteria dwelling on rocks and found 401 taxa recorded from Europe, 155 from North America, 175 from South America, 72 from Africa, 280 taxa from Asia, 86 from Australia and Oceania, and 27 taxa from Antarctica and Arctic regions. These habitats host approximately 30 % of the entire described cyanobacterial diversity (Nabout et al. 2013). Likewise, all studies of hypogeal cyanobacteria are from the Mediterranean area (Hauer et al. 2015). Reliable comparisons of diversity among geographical regions are not available, due to the lack of studies from different areas of the world. One of the questions that can now be addressed is the geographic distribution of the cyanobacteria and existence of endemic taxa (Taton et al. 2003; Finlay 2002; Dvořák et al. 2012). For example, a recent molecular study (Taton et al. 2003) provided evidence that cyanobacterial diversity and endemism in Antarctica is greater than assumptions of diversity/endemism based on microscopic analysis. Nadeau et al. (2001) suggest a bipolar distribution for several oscillatorians taxa, in congruence with results of some other authors (Jungblut et al. 2010; Comte et al. 2007). Globally dispersed microorganisms have been reported from geothermal environments (Papke et al. 2003; Ward et al. 2008). On the other hand, some genera (e.g., *Rexia*) have been postulated as being endemic to very specific, limited geographic distributions (Casamatta et al. 2005).

Tropical regions represent large geographic areas with a variety of habitats. Coupled with high humidity and low seasonality, this may enable many species to coexistence (Mittelbach et al. 2007). The estimated proportion of possibly undescribed microalgae was about 60 % (Neustupa and Škaloud 2008). Indeed, the exploration of little-known habitats in tropical regions has led to discoveries of new taxa (Sant'Anna et al. 2011; Dadheech et al. 2014; Hašler et al. 2014a; Dvořák et al. 2015b). Neustupa and Škaloud (2010) concluded that tropical, corticolous habitats harbor higher diversity than corresponding temperate habitats. Their results indicate that the microhabitat conditions, in the case of terrestrial phototrophs, typically humidity and light, may play a crucial role in determining algal and cyanobacterial diversity.

Although freshwater habitats in general have received more attention than terrestrial ones, the studies focused on microalgal assemblages are not equally distributed. The ratio of papers on planktic, epiphytic, and epipelic microalgal assemblages was 62:32:4 as of January 2013 (Pulíčková et al. 2014). Since 2007, when the epipelon was proposed as a major unexplored freshwater cyanobacterial habitat (Poulíčková et al. 2008), several new cyanobacterial taxa were distinguished using a polyphasic approach (Hašler et al. 2012, 2014a, b).

Additional Information

For further information and copious illustrations, the authors suggest the works of Komárek and Anagnostidis (1998, 2005) for Synechococcales, Chrooccales, Oscillatoriales, Spirulinales and Chroococcidiopsidales, Komárek (2013) and Komárek and Johansen 2015 for Nostocales. The most comprehensive web sources are CyanoDB (http://www.cyanodb.cz/) and AlgaeBase (http://www.algaebase.org/).

Conclusions and Future Prospects

The diversity of cyanobacteria is immense. While we have already discovered thousands of species, it seems to be but a glimpse of the real biodiversity, illustrated by the growing number of the new taxa erected every year, especially from tropical habitats. Further, only a minority of described species has been sequenced. However, phylogenetic reconstructions based on obtained sequences mostly exhibited entangled relationships with polyphyletic genera. This problem is amplified by the fact that there is debate about species concepts and the inability to cultivate a majority of cyanobacteria.

Nevertheless, we remain optimistic, because recent advances in genomics, metagenomics, single cell genomics, and related fields promise development of tools which may allow us to tackle the problems outlined above. We expect that the number of newly described taxa will only increase in the future. With a growing body of whole genome data, we will be able to recognize more nuanced differences among lineages, thus precisely resolving species relationships.

Acknowledgements We would like to express our gratitude to Zdenka Giacintová, who made drawings for Fig. 1. This work was supported by Internal Grant Agency of Palacký University Prf-2016-001.

References

Abed RMM, Dobretsov S, Sudesh K (2008) Application of cyanobacteria in biotechnology. J Appl Microbiol 106:1–12. doi:10.1007/s00253-011-3394-0

Achtman M, Wagner M (2008) Microbial diversity and the genetic nature of microbial species. Nat Rev Microbiol 6:431–440. doi:10.1038/nrmicro1872

Acinas SG, Haverkamp THA, Huisman J et al (2009) Phenotypic and genetic diversification of *Pseudanabaena* spp. (cyanobacteria). ISME J 3:31–46. doi:10.1038/ismej.2008.78

Anagnostidis K (2001) Nomenclatural changes in cyanoprokaryotic order Oscillatoriales. Preslia 73:359–375

Anagnostidis K, Komárek J (1988) Modern approach to the classification system of cyanophytes, 3-Oscillatoriales. Arch Hydrobiol 80/Algol Stud 50:327–472

Azua-Bustos A, Zúñiga J, Arenas-Fajardo C et al (2014) *Gloeocapsopsis* AAB1, an extremely desiccation-tolerant cyanobacterium isolated from the Atacama Desert. Extremophiles 18:61–74. doi:10.1007/s00792-013-0592-y

Bekker A, Holland HD, Wang P-L et al (2004) Dating the rise of atmospheric oxygen. Nature 427:117–120. doi:10.1038/nature02260

Berrendero E, Arenas C, Mateo P et al (2016) Cyanobacterial diversity and related sedimentary facies as a function of water flow conditions: example from the monasterio de piedra natural park (Spain). Sediment Geol 337:12–28. doi:10.1016/j.sedgeo.2016.03.003

Biller SJ, Berube PM, Lindell D et al (2015) *Prochlorococcus*: the structure and function of collective diversity. Nat Rev Microbiol 13:13–27. doi:10.1038/nrmicro3378

Blank CE, Sanchez-Baracaldo P (2010) Timing of morphological and ecological innovations in the cyanobacteria – a key to understanding the rise in atmospheric oxygen. Geobiology 8:1–23. doi:10.1111/j.1472-4669.2009.00220.x

Bohunická M, Pietrasiak N, Johansen JR et al (2015) *Roholtiella*, gen. nov. (Nostocales, Cyanobacteria) – a tapering and branching cyanobacteria of the family Nostocaceae. Phytotaxa 197:84–103. doi:10.11646/phytotaxa.197.2.2

Boison G, Mergel A, Jolkver H et al (2004) Bacterial life and dinitrogen fixation at a gypsum rock. Appl Environ Microbiol 70:7070–7077. doi:10.1128/AEM.70.12.7070-7077.2004

Boone DR, Castenholz RW, Garrity GM (2001) Bergey's manual of systematic bacteriology, vol 1, The archaea and the deeply branching and phototrophic bacteria. Springer, New York

Bornet E, Flahault C (1886–1888) Révision des Nostocacées heterocystées continues dans les principaux herbiers de France. Ann Sci Nat Bot 3:323–381, 4:343–373, 5:51–129, 7:177–262

Boyer SL, Johansen JR, Flechiner VR et al (2002) Phylogeny and genetic variance in terrestrial *Microcoleus* (Cyanophyceae) species based on sequence analysis of the 16S rRNA gene and associated 16S-23S ITS region. J Phycol 38:1222–1235. doi:10.1046/j.1529-8817.2002.01168.x

Büdel B, Darienko T, Deutschewitz K et al (2009) Southern African biological soil crusts are ubiquitous and highly diverse in drylands, being restricted by rainfall frequency. Microb Ecol 57:229–247. doi:10.1007/s00248-008-9449-9

Butterfield NJ (2015) Proterozoic photosynthesis – a critical review. Palaeontology 58:95–972. doi:10.1111/pala.12211

Callieri C, Coci M, Corno G et al (2013) Phylogenetic diversity of nonmarine picocyanobacteria. FEMS Microbiol Ecol 85:293–301. doi:10.1111/1574-6941.12118

Carmichael WW (1992) Cyanobacteria secondary metabolites – the cyanotoxins. J Appl Bacteriol 72:445–459. doi:10.1111/j.1365-2672.1992.tb01858.x

Casamatta DA, Vis ML, Sheath RG (2003) Cryptic species in cyanobacterial systematics: a case study of *Phormidium retzii* (*Oscillatoriales*) using RAPD molecular markers and 16S rDNA sequence data. Aquat Bot 74:295–309. doi:10.1016/j.aquabot.2003.08.005

Casamatta DA, Johansen JR, Vis ML et al (2005) Molecular and morphological characterization of ten polar and near-polar strains within the oscillatoriales (cyanobacteria). J Phycol 41:421–433. doi:10.1111/j.1529-8817.2005.04062.x

Casamatta DA, Gomez SR, Johansen JR (2006) *Rexia erecta* gen. et sp. nov. and *Capsosira lowei* sp. nov., two newly described cyanobacterial taxa from the Great Smoky Mountains National Park (USA). Hydrobiologia 561:13–26. doi:10.1007/s10750-005-1602-6

Chatchawan T, Komárek J, Strunecký O et al (2012) *Oxynema*, a new genus separated from the genus *Phormidium* (Cyanophyta). Cryptog Algol 33:41–59. doi:10.7872/crya.v33.iss1.2011.041

Codd GA, Ward CJ, Bell SG (1997) Cyanobacterial toxins: occurrence, modes of action, health effects and exposure routes. Arch Toxicol Suppl 19:399–410. doi:10.1016/j.chemosphere.2013.07.056

Cohan FM (2001) Bacterial species and speciation. Syst Biol 50:513–524. doi:10.1016/j.cub.2015.10.022

Cohan FM (2002) What are bacterial species? Annu Rev Microbiol 56:457–487. doi:10.1146/annurev.micro.56.012302.160634

Cohan FM, Perry EB (2007) A systematics for discovering the fundamental units of bacterial diversity. Curr Biol 17:R373–R386. doi:10.1016/j.cub.2007.03.032

Comte K, Šabacká M, Carré-Mlouka A et al (2007) Relationships between the Arctic and the Antarctic cyanobacteria; three *Phormidium*-like strains evaluated by a polyphasic approach. FEMS Microbiol Ecol 59:366–376. doi:10.1111/j.1574-6941.2006.00257.x

Copeland JJ (1936) Yellowstone thermal Myxophyceae. Ann N Y Acad Sci 36:1–232. doi:10.1111/j.1749-6632.1936.tb56976.x

Dadheech PK, Mahmoud H, Kotut K et al (2014) *Desertifilum fontinale* sp. nov. (Oscillatoriales, Cyanobacteria) from a warm spring in East Africa, based on conventional and molecular studies. Fottea 14:129–140. doi:10.5507/fot.2014.010

Desikachary TV (1959) Cyanophyta. ICAR monographs on algae. Indian Council of Agricultural research, New Delhi, p 686

Dittmann E, Fewer DP, Neilan BA (2013) Cyanobacterial toxins: biosynthetic routes and evolutionary roots. FEMS Microbiol Rev 37:23–43. doi:10.1111/j.1574-6976.2012.12000.x

Donner A (2013) The case of *Chroococcidiopsis*: new phylogenetic and morphological insights into ecologically important Cyanobacteria. PhD Thesis, Technischen Universität Kaiserslautern, Kaiserslautern, p 147

Dvořák P, Hašler P, Poulíčková A (2012) Phylogeography of the *Microcoleus vaginatus* (cyanobacteria) from three continents – a spatial and temporal characterization. PLoS One 7, e40153. doi:10.1371/journal.pone.0040153

Dvořák P, Casamatta DA, Hašler P et al (2014a) *Synechococcus*: 3 billion years of global dominance. Mol Ecol 23:5538–5551. doi:10.1111/mec.12948

Dvořák P, Hindák F, Hašler P et al (2014b) Morphological and molecular studies of *Neosynechococcus sphagnicola*, gen. et sp. nov. (Cyanobacteria, Synechococcales). Phytotaxa 170:24–34. doi:10.11646/phytotaxa.170.1.3

Dvořák P, Casamatta DA, Hašler P et al (2015a) Putative extremely long evolutionary stasis in bacteria might be explained by serial convergence. Proc Natl Acad Sci U S A 112, E2229. doi:10.1073/pnas.1502924112

Dvořák P, Jahodářová E, Hašler P et al (2015b) A new tropical cyanobacterium *Pinocchia polymorpha* gen. et sp. nov. derived from genus *Pseudanabaena*. Fottea 15:113–120. doi:10.5507/fot.2015.010

Ehling-Schulz M, Scherer S (1999) UV protection in cyanobacteria. Eur J Phycol 34:329–338. doi:10.1080/09670269910001736392

Engene N, Coates C, Gerwick WH (2010) 16S rRNA gene heterogeneity in the filamentous marine cyanobacterial genus *Lyngbya*. J Phycol 46:591–601. doi:10.1111/j.1529-8817.2010.00840.x

Engene N, Rottacker EC, Kaštovský J et al (2012) *Moorea producens* gen. nov. sp. nov. and *Moorea bouillonii* comb. nov. tropical marine cyanobacteria rich in bioactive secondary metabolites. Int J Syst Evol Microbiol 62:1171–1178. doi:10.1099/ijs.0.033761-0

Engene N, Paul VJ, Byrum T et al (2013) Five chemically rich species of tropical marine cyanobacteria of the genus *Okeania* gen. nov. (Oscillatoriales, Cyanoprokaryota). J Phycol 49:1096–1106. doi:10.1111/jpy.12115
Erwin PM, Thacker W (2008) Cryptic diversity of the symbiotic cyanobacterium *Synechococcus spongiarum* among sponge hosts. Mol Ecol 17:2937–2947. doi:10.1111/j.1365-294X.2008.03808.x
Ferris MJ, Ward DM (1997) Seasonal distributions of dominant 16S rRNA-defined populations in a hot spring microbial mat examined by denaturing gradient gel electrophoresis. Appl Environ Microbiol 63:1375–1381
Fewer D, Friedl T, Büdel B (2002) *Chroococcidiopsis* and heterocyst-differentiating cyanobacteria are each other's closest living relatives. Mol Phylogenet Evol 23:82–90. doi:10.1006/mpev.2001.1075
Finlay BJ (2002) Global dispersal of free-living microbial eukaryote species. Science 296:1061–1063. doi:10.1126/science.1070710
Fiore MF, Sant'Anna CL, Azevedo MTP (2007) The cyanobacterial genus *Brasilonema*, gen. nov., a molecular and phenotype evaluation. J Phycol 43:789–798. doi:10.1111/j.1529-8817.2007.00376.x
Flombaum P, Gallegos JL, Gordillo RA et al (2013) Present and future global distributions of the marine cyanobacteria *Prochlorococcus* and *Synechococcus*. Proc Natl Acad Sci U S A 110:9824–9829. doi:10.1073/pnas.1307701110
Fraser C, Hanage WP, Spratt BG (2007) Recombination and the nature of bacterial speciation. Science 315:476–480. doi:10.1126/science.1127573
Gao Z-M, Wang Y, Tian R-M et al (2014) Symbiotic adaptation drives genome streamlining of the cyanobacterial sponge symbiont "Candidatus *Synechococcus spongiarum*." mBio 5:e00079-14. doi:10.1128/mBio.00079-14
Garcıa-Fernandez JM, de Marsac NT, Diez J (2004) Streamlined regulation and gene loss as adaptive mechanisms in *Prochlorococcus* for optimized nitrogen utilization in oligotrophic environments. Microbiol Mol Biol Rev 68:630–638. doi:10.1128/MMBR.68.4.630-638.2004
Garcia-Pichel F, Lopez-Cortez A, Nubel U (2001) Phylogenetic and morphological diversity of cyanobacteria in soil desert crusts from the Colorado Plateau. Appl Environ Microbiol 67:1902–1910. doi:10.1128/AEM.67.4.1902-1910.2001
Gardner NL (1927) New Myxophyceae from Porto Rico. Mem N Y Bot Gard 7:1–144
Geitler L (1932) Cyanophyceae. In: Rabenhorst L (ed) Kryptogamen von Deutschland, Österreich und der Schweiz, vol 14. Akademische Verlagsgesellschaft, Leipzig, pp 673–1196
Genuário DB, Vaz MGMV, Hentschke GS et al (2015) *Halotia* gen. nov., a phylogenetically and physiologically coherent cyanobacterial genus isolated from marine coastal environments. Int J Syst Evol Microbiol 65:663–675. doi:10.1099/ijs.0.070078-0
Goh F, Allen MA, Leuko S et al (2009) Determining the specific microbial populations and their spatial distribution within the stromatolite ecosystem of Shark Bay. ISME J 3:383–396. doi:10.1038/ismej.2008.114
Golubic S, Hernandez-Marine M, Hoffman L (1996) Developmental aspects of branching in filamentous cyanophyta/cyanobacteria. Algol Stud 83:303–329
Gomont M (1892 '1893') Monographie des Oscillariées (Nostocacées Homocystées). Deuxième partie. Lyngbyées. Botanique et biologie végétale, Série 7 16:91–264, p 1–7
Gugger MF, Hoffman L (2004) Polyphyly of true branching cyanobacteria (Stigonematales). Int J Syst Evol Microbiol 54:349–357. doi:10.1099/ijs.0.02744-0
Guiry MD (2012) How many species of algae are there? J Phycol 48:1057–1063. doi:10.1111/j.1529-8817.2012.01222.x
Guiry MD, Guiry GM (2015) AlgaeBase. World-wide electronic publication, National University of Ireland, Galway. http://www.algaebase.org; searched on 02 September 2015
Hansgirg A (1890) Über neue Süsswasser- und Meeres-Algen und Bacterien, mit Bemerkungen zur Systematik dieser Phycophyten und über den Einfluss des Lichtes auf die Ortsbewegungen des Bacillus pfefferi nob. Sitzungsberichte der Königlichen Böhmischen Gesellschaft der Wissenschaften, Mathematisch-naturwissenschaftliche Classe 1890:3–34
Hansgirg A (1892) Prodromus der Algenflora von Böhmen. Zweiter Theil welcher die blaugrünen Algae (Myxophyceen, Cyanophyceen), nebst Nachträgen zum ersten Theile und einer systmatischen Bearbeitung der in Böhmen verbreiten saprophytischen Bacterien und Euglenen enhält.

Mit den opiz-preise Gekrönte Arbeit auf Kosten des Opizfondes. Archiv für die naturwissenschaftliche Landesdurchforschung von Böhmen 8:1–268
Hašler P, Poulíčková A (2005) Cyanobacteria of the West Carpathian spring fens: single sampling. Czech Phycol 5:43–55
Hašler P, Poulíčková A (2010) Diversity, taxonomy and autecology of autochtonous epipelic cyanobacteria of the genera *Komvophoron* and *Isocystis* (Borziaceae, Oscillatoriales). Biologia 65:7–16. doi:10.2478/s11756-009-0214-4
Hašler P, Dvořák P, Johansen JR et al (2012) Morphological and molecular study of epipelic filamenous genera *Phormidium, Microcoleus* and *Geitlerinema* (Oscillatoriales, Cyanophyta/Cyanobacteria). Fottea 12:341–356. doi:10.5507/fot.2012.024
Hašler P, Dvořák P, Poulíčková A et al (2014a) A novel genus *Ammasolinea* gen. nov. (cyanobacteria) isolated from sub–tropical epipelic habitats. Fottea 14:241–248. doi:10.5507/fot.2014.018
Hašler P, Dvořák P, Poulíčková A (2014b) A new genus of filamentous epipelic cyanobacterium, *Johansenia*. Preslia 86:1–15
Hauer T, Bohunická M, Johansen JR et al (2014) Reassessment of the cyanobacterial family Microchaetaceae and establishment of the new families Tolypothrichaceae and Godleyaceae. J Phycol 50:1089–1100. doi:10.1111/jpy.12241
Hauer T, Mühlsteinová R, Bohunická M et al (2015) Diversity of cyanobacteria on rock surfaces. Biodivers Conserv 24:759–779. doi:10.1007/s10531-015-0890-z
Hentschke GS, Johansen JR, Pietrasiak N et al (2016) Phylogenetic placement of *Dapisostemon* gen. nov. and *Streptostemon*, two tropical heterocytous genera (Cyanobacteria). Phytotaxa 245:129–143. doi:10.11646/phytotaxa.245.2.4
Hoffmann L (1994) Marine cyanophyceae of Papua New Guinea. VI. The genus *Lyngbya* S.L. Belg J Bot 127:79–86
Hoiczyk E, Hansel A (2000) Cyanobacterial cell walls: news from an unusual prokaryotic envelope. J Bacteriol 182:1191–1199. doi:10.1128/JB.182.5.1191-1199.2000
Honda D, Yokota A, Sugiyama J (1999) Detection of seven major evolutionary lineages in cyanobacteria based on the 16S rRNA gene sequence analysis with new sequences of five marine *Synechococcus* strains. J Mol Evol 48:723–739. doi:10.1007/PL00006517
Ishida T, Watanabe MM, Sugiyama J et al (2001) Evidence for polyphyletic origin of the members of the orders of Oscillatoriales and Pleurocapsales se determined by 16S rDNA analysis. FEMS Microbiol Lett 201:79–82. doi:10.1111/j.1574-6968.2001
Johansen JR, Casamatta DA (2005) Recognizing cyanobacterial diversity through adoption of a new species paradigm. Algol Stud 117:71–93
Jungblut AD, Lovejoy C, Vincent WF (2010) Global distribution of cyanobacterial ecotypes in the cold biosphere. ISME J 4:191–202. doi:10.1038/ismej.2009.113
Kämpfer P, Glaeser SP (2012) Prokaryotic taxonomy in the sequencing era—the polyphasic approach revisited. Environ Microbiol 14:291–317. doi:10.1111/j.1462-2920.2011.02615.x
Kaneko T, Tanaka A, Sato S et al (1995) Sequence analysis of the genome of the unicellular cyanobacterium *Synechocystis* sp. strain PCC 6803. I. Sequence features in the 1 Mb region from map position 64% to 92% of the genome. DNA Res 2(153–166):191–198. doi:10.1093/dnares/2.4.153
Kaplan-Levy RN, Hadas O, Summers ML et al (2010) Akinetes - dormant cells of cyanobacteria. In: Lubzens E, Cerda J, Clark MS (eds) Topics in current genetics. Springer, Berlin/Heidelberg
Kaštovský J, Johansen JR (2008) *Mastigocladus laminosus* (Stigonematales, Cyanobacteria): phylogenetic relationship to the soil-inhabiting genera of the order, and taxonomic implications for the genus. Phycologia 47:307–320. doi:10.2216/PH07-69.1
Katoh H, Itoh S, Shen J-R (2001) Functional analysis of psbV and a novel c-type cytochrome gene psbV2 of the thermophilic cyanobacterium *Thermosynechococcus elongatus* strain BP-1. Plant Cell Physiol 42:599–607. doi:10.1093/pcp/pce074
Khan Z, Bhadouria P, Bisen PS (2005) Nutritional and therapeutic potential of *Spirulina*. Curr Pharm Biotechnol 6:373–379. doi:10.3923/ijp.2016.36.51
Komárek J (2003) Coccoid and colonial cyanobacteria. In: Wehr JD, Sheath RG (eds) Freshwater algae of North America. Academic, San Diego, pp 59–116

Komárek J (2010) Recent changes (2008) in cyanobacteria taxonomy based on a combination of molecular background with phenotype and ecological consequences (genus and species concept). Hydrobiologia 1:245–259. doi:10.1007/s10750-009-0031-3
Komárek J (2013) Cyanoprokaryota. 3. Heterocytous genera. In: Büdel B, Gärtner G, Krienitz L, Schagerl M (eds) Süswasserflora von Mitteleuropa/Freshwater flora of Central Europe. Springer Spektrum, Berlin/Heidelberg/Germany, pp 1–1130
Komárek J (2014) Phenotypic and ecological diversity of freshwater coccoid cyanobacteria from maritime Antarctica and Islands of NW Weddell Sea. II. Czech Polar Rep 4:17–39. doi:10.5817/CPR2013-2-14
Komárek J (2015) About endemism of cyanobacteria in freshwater habitats of maritime Antarctica. Algol Stud 148:15–32. doi:10.1127/algol_stud/2015/0219
Komárek J, Anagnostidis K (1998) Cyanoprokaryota 1. Teil: Chroococcales. In: Ettl H, Gärtner G, Heynig H et al (eds) Süsswasserflora von Mitteleuropa 19/1. Gustav Fischer, Jena-Stuttgart-Lübeck-Ulm, Germany, pp 1–548
Komárek J, Anagnostidis K (2005) Cyanoprokaryota 2. Teil Oscillatoriales. In: Büdel B, Gärtner G, Krienitz et al (eds) Süßwasserflora von Mitteleuropa, vol 19/2. Elsevier, München, pp 1–759
Komárek J, Johansen JR (2015) Filamentous cyanobacteria. In: Wehr JD, Sheath RG, Kociolek JP (eds) Freshwater algae of North America: ecology and classification. Elsevier, London, pp 135–235
Komárek J, Komárková J (2002) Review of the European *Microcystis* morphospecies (Cyanoprokaryotes) from nature. Czech Phycol 2:1–24
Komárek J, Komárková J (2004) Taxonomic review of the cyanoprokaryotic genera *Planktothrix* and *Planktothricoides*. Czech Phycol 4:1–18
Komárek J, Kaštovský J, Jezberová J (2011) Phylogenetic and taxonomic delimitation of the cyanobacterial genera *Aphanothece* Nägeli and *Anathece* (Komárek et Anagnostidis) comb. nova. Eur J Phycol 46:315–326. doi:10.1080/09670262.2011.606373
Komárek J, Zapomělová E, Šmarda J et al (2013) Polyphasic evaluation of *Limnoraphis robusta*, a water–bloom forming cyanobacterium from Lake Arirlán, Guatemala, with a descrition of *Limnoraphis* gen. nov. Fottea 13:39–52. doi:10.5507/fot.2013.004
Komárek J, Kaštovský J, Mareš J et al (2014) Taxonomic classification of cyanoprokaryotes (cyanobacterial genera) 2014 using a polyphasic approach. Preslia 86:295–335
Komárková J, Jezberová J, Komárek O et al (2010) Variability of *Chroococcus* (Cyanobacteria) morphospecies with regard to phylogenetic relationships. Hydrobiologia 639:69–83. doi:10.1007/s10750-009-0015-3
Korelusová J, Kaštovský J, Komárek J (2009) Heterogeneity of the cyanobacterial genus *Synechocystis* and description of a new genus *Geminocystis*. J Phycol 45:928–937. doi:10.1111/j.1529-8817.2009.00701.x
Kováčik L, Jezberová J, Komárková J et al (2011) Ecological characteristics and polyphasic taxonomic classification of stable pigment-types of the genus *Chroococcus* (Cyanobacteria). Preslia 83:145–166
Kuffner IB, Paul VJ (2004) Effects of the benthic cyanobacterium *Lyngbya majuscula* on larval recruitment of the reef corals *Acropora surculosa* and *Pocillopora diamicornis*. Coral Reefs 24:455–458. doi:10.1007/s00338-004-0416-8
Kumar K, Mella-Herrera RA, Golden JW (2010) Cyanobacterial heterocysts. Cold Spring Harb Perspect Biol 2:a000315. doi:10.1101/cshperspect.a000315
Lamprinou V, Hernández-Mariné M, Canals T et al (2011) Morphology and molecular evaluation of *Iphinoe spelaeobios* gen. nov., sp. nov. and *Loriellopsis cavernicola* gen. nov., sp. nov., two stigonematalean cyanobacteria from Greek and Spanish caves. Int J Syst Evol Microbiol 61:2907–2915. doi:10.1099/ijs.0.029223-0
Lamprinou V, Skaraki K, Kotoulas G et al (2012) *Toxopsis calypsus* gen. nov., sp. nov. (*Cyanobacteria*, Nostocales) from cave 'Francthi', Peloponnese, Greece: a morphological and molecular evaluation. Int J Syst Evol Microbiol 62:2870–2877. doi:10.1099/ijs.0.038679-0
Larsson J, Nylander JA, Bergman B (2011) Genome fluctuations in cyanobacteria reflect evolutionary, developmental and adaptive traits. BMC Evol Biol 11:187. doi:10.1186/1471-2148-11-187

Li Z, Brand J (2007) *Leptolyngbya nodulosa* sp. nov. (Oscillatoriaceae), a subtropical marine cyanobacterium that produces a unique multicellular structure. Phycologia 46:396–401. doi:10.2216/06-89.1

Loza V, Berrendero E, Perona E et al (2013) Polyphasic characterization of benthic cyanobacterial diversity from biofilms of the Guadarrama river (Spain): morphological, molecular, and ecological approaches. J Phycol 49:282–297. doi:10.1111/jpy.12036

Lukešová A, Hrouzek P, Harding K et al (2008) Deployment of the encapsulation/dehydration protocol to cryopreserve diverse microalgae held at the Institute of Soil Biology, Academy of Sciences of the Czech Republic. Cryo-Letters 29:21–26

Lukešová A, Johansen JR, Martin MP et al (2009) *Aulosira bohemensis* sp. nov.: further phylogenetic uncertainty at the base of the Nostocales (Cyanobacteria). Phycologia 48:118–129. doi:10.2216/08-56.1

Maiden MC, Bygraves JA, Feil EJ et al (1998) Multilocus sequence typing: a portable approach to the identification of clones within populations of pathogenic microorganisms. Proc Natl Acad Sci U S A 95:3140–3145

Mareš J, Hauer T, Komárek J et al (2013) Proposal to conserve the name *Gloeothece* (Cyanophyceae) with a conserved type. Taxon 62:1056. doi:10.12705/625.24

Mareš J, Lara Y, Dadáková I et al (2015) Phylogenetic analysis of cultivation-resistant terrestrial cyanobacteria with massive sheaths (*Stigonema* spp. and *Petalonema alatum*, Nostocales, Cyanobacteria) using single-cell and filament sequencing of environmental samples. J Phycol 51:288–297. doi:10.1111/jpy.12273

McFadden GI (1999) Endosymbiosis and evolution of the plant cell. Curr Opin Plant Biol 2:513–519

Meeks JC, Campbell EL, Summers ML (2002) Cellular differentiation in the cyanobacterium *Nostoc punctiforme*. Arch Microbiol 178:395–403. doi:10.1007/s00203-002-0476-5

Mittelbach GG, Schemske DW, Cornell HV et al (2007) Evolution and the latitudinal diversity gradient: speciation, extinction and biogeography. Ecol Let 10:325–331. doi:10.1111/j.1461-0248.2007.01020.x

Miyashita H, Ikemoto H, Kurano N et al (2003) *Acaryochloris marina* gen. et sp. nov. (Cyanobacteria), an oxygenic photosynthetic prokaryote containing chl d as a major pigment. J Phycol 39:1247–1253. doi:10.1111/j.0022-3646.2003.03-158.x

Moore LR, Rocap G, Chisholm SW (1998) Physiology and molecular phylogeny of coexisting *Prochlorococcus* ecotypes. Nature 393:464–467. doi:10.1038/30965

Mühlsteinová R, Johansen JR, Pietrasiak N et al (2014) Polyphasic characterization of *Kastovskya adunca* gen. nov. et comb. nov. (Cyanobacteria: Oscillatoriales), from desert soils of the Atacama Desert, Chile. Phytotaxa 163:216–228. doi:10.11646/phytotaxa.163.4.2

Nabout JC, da Silva RB, Carneiro FM et al (2013) How many species of cyanobacteria are there? Using a discovery curve to predict the species number. Biodivers Conserv 22:2907–2918. doi:10.1007/s10531-013-0561-x

Nadeau TL, Milbrandt EC, Castenholz RW (2001) Evolutionary relationships of cultivated Antarctic oscillatoriaceans (cyanobacteria). J Phycol 37:650–654. doi:10.1046/j.1529-8817.2001.037004650.x

Nägeli C (1849) Gattungen einzellinger Algen. Neue Denkschriften der Allgemeinen Schweizerischen Gesellschaft für die Gesammten Naturwissenschaften 10:1–139

Neilan BA, Jacobs D, Del Dot T et al (1997) rRNA sequences and evolutionary relationships among toxic and nontoxic cyanobacteria of the genus *Microcystis*. Int J Syst Bacteriol 47:693–697. doi:10.1099/00207713-47-3-693

Neustupa J, Škaloud P (2008) Diversity of subaerial algae and cyanobacteria on tree bark in tropical mountain habitats. Biologia 63:806–812. doi:10.2478/s11756-008-0102-3

Neustupa J, Škaloud P (2010) Diversity of subaerial algae and cyanobacteria growing on bark and wood in the lowland tropical forests of Singapore. Plant Ecol Evol 143:51–62. doi:10.5091/plecevo.2010.417

Nübel U, Garcia-Pichel F, Muyzer G (2000) The halotolerance and phylogeny of cyanobacteria with tightly coiled trichomes (*Spirulina* Turpin) and the description of *Halospirulina tapeticola* gen. nov., sp. nov. Int J Syst Evol Microbiol 50:1265–1277. doi:10.1099/00207713-50-3-1265

Oliver RL, Ganf GG (2000) Freshwater blooms. In: Whitton BA, Potts M (eds) The ecology of cyanobacteria, their diversity in time and space. Kluwer, Dordrecht, pp 149–194

Olsson-Francis K, de la Torre R, Towner MC et al (2009) Survival of akinetes (resting-state cells of cyanobacteria) in low Earth orbit and simulated extraterrestrial conditions. Orig Life Evol Biosph 39:565–579. doi:10.1007/s11084-009-9167-4

Oren A (2011) Cyanobacterial systematics and nomenclature as featured in the International Bulletin of Bacteriological Nomenclature and Taxonomy/International Journal of Systematic Bacteriology/International Journal of Systematic and Evolutionary Microbiology. Int J Syst Evol Microbiol 61:10–15. doi:10.1099/ijs.0.018838-0

Osorio-Santos K, Pietrasiak N, Bohunická M et al (2014) Seven new species of *Oculatella* (Pseudanabaenales, Cyanobacteria). Eur J Phycol 49:450–470. doi:10.1080/09670262.2014.976843

Padisák J (1997) *Cylindrospermopsis raciborskii* (Woloszynska) Seenayya et Subba Raju, an expanding, highly adaptive cyanobacterium: worldwide distribution and review of its ecology. Arch Hydrobiol Suppl 4:563–593

Palinska KA, Liesack W, Rhiel E et al (1996) Phenotype variability of identical genotypes: the need for a combined approach in cyanobacterial taxonomy demonstrated on *Merismopedia*-like isolates. Arch Microbiol 166:226–233. doi:10.1007/s002030050378

Papke RT, Ramsing NB, Bateson MM et al (2003) Geographical isolation in hot spring cyanobacteria. Environ Microbiol 5:650–659. doi:10.1046/j.1462-2920.2003.00460.x

Partensky F, Garczarek L (2010) *Prochlorococcus*: advantages and limits of minimalism. Ann Rev Mar Sci 2:305–331. doi:10.1146/annurev-marine-120308-081034

Partensky F, Hess WR, Vaulot D (1999) *Prochlorococcus*, a marine photosynthetic prokaryote of global significance. Microbiol Mol Biol Rev 63:106–127

Paul VJ, Thacker RW, Banks K et al (2005) Benthic cyanobacterial bloom impacts the reefs of South Florida (Broward County, USA). Coral Reefs 24:693–697. doi:10.1007/s00338-005-0061-x

Perkerson RB, Johansen JR, Kováčik L et al (2011) A unique pseudanabaenalean (cyanobacteria) genus *Nodosilinea* gen. nov. based on morphological and molecular data. J Phycol 47:1397–1412. doi:10.1111/j.1529-8817.2011.01077.x

Pietrasiak N, Mühlsteinová R, Siegesmund MA et al (2014) Phylogenetic placement of *Symplocastrum* (Phormidiaceae, Cyanophyceae) with a new combination *S. californicum* and two new species: *S. flechtnerae* and *S. torsivum*. Phycologia 53:529–541. doi:10.2216/14-029.1

Pinevich AV (2015) Proposal to consistently apply the International Code of Nomenclature of Prokaryotes (ICNP) to names of the oxygenic photosynthetic bacteria (cyanobacteria), including those validly published under the International Code of Botanical Nomenclature (ICBN)/International Code of Nomenclature for algae, fungi and plants (ICN), and proposal to change Principle 2 of the ICNP. Int J Syst Evol Microbiol 65:1070–1074. doi:10.1099/ijs.0.000034

Polz MF, Alm EJ, Hanage WP (2013) Horizontal gene transfer and the evolution of bacterial and archeal population structure. Trends Genet 29:170–175. doi:10.1016/j.tig.2012.12.006

Poulíčková A, Hašler P, Lysáková M (2008) The ecology of freshwater epipelic algae an update. Phycologia 47:437–450. doi:10.2216/07-59.1

Pulíčková A, Dvořák P, Mazalová P et al (2014) Epipelic microphytotrophs: an overlooked assemblage in lake ecosystems. Freshwat Sci 33:513–523. doi:10.1086/676313

Rajendhran J, Gunasekaran P (2011) Microbial phylogeny and diversity: small subunit ribosomal RNA sequence analysis and beyond. Microbiol Res 166:99–110. doi:10.1016/j.micres.2010.02.003

Řeháková K, Johansen JR, Casamatta DA et al (2007) Morphological and molecular characterization of selected desert soil cyanobacteria: Three species new to science including *Mojavia pulchra* gen. et sp. nov. Phycologia 46:481–502. doi:10.2216/06-92.1

Rippka R, Deruelles J, Waterbury J et al (1979) Genetic assignments, strain histories and properties of pure cultures of cyanobacteria. J Gen Microbiol 111:1–61. doi:10.1099/00221287-111-1-1

Rippka R, Castenholz RW, Iteman I (2001a) Form-genus II. *Chroococcus* Nägeli 1849. In: Boone DR, Castenholz RW, Garrity GM (eds) Bergey's manual of systematic bacteriology, vol 1. Springer, New York, pp 496–497

Rippka R, Castenholz RW, Waterbury JB et al (2001b) Form-genus IX. Gloeothece Nägeli 1849. In: Boone DR, Castenholz RW, Garrity GM (eds) Bergey's manual of systematic bacteriology, vol 1. Springer, New York, pp 504–505

Rippka R, Waterbury JB, Herdman M et al (2001c) Form-genus I. *Chroococcidiopsis* Geitler 1933, emend. Waterbury and Stanier 1978. In: Boone DR, Castenholz RW, Garrity GM (eds) Bergey's manual of systematic bacteriology, vol 1. Springer, New York, pp 528–531

Rippka R, Waterbury JB, Herdman M et al (2001d) *Pleurocapsa*-group sensu Waterbury und Stanier 1978, Waterbury 1989. In: Boone DR, Castenholz RW, Garrity GM (eds) Bergey's manual of systematic bacteriology, vol 1. Springer, New York, pp 533–539

Robbertse B, Yoder RJ, Boyd A et al (2011) Hal: and automated pipeline for phylogenetic analyses of genomic data. PLoS Currents 3, RRN1213. doi:10.1371/currents.RRN1213

Robertson BR, Tezuka N, Watanabe M (2001) Phylogenetic analyses of *Synechococcus* strains (cyanobacteria) using sequences of 16S rDNA and part of the phycocyanin operon reveal multiple evolutionary lines and reflect phycobilin content. Int J Syst Evol Microbiol 51:861–871. doi:10.1099/00207713-51-3-861

Ronquist F, Huelsenbeck JP (2003) Mrbayes 3: Bayesian phylogenetic inference under mixed models. Bioinformatics 19:1572–1574. doi:10.1093/bioinformatics/btg180

Rulík M, Tóthová L, Koutný J et al (2003) Mikrobiální společenstva pisoárů. pp. 68–71. In: Mikrobiológia vody. Zbornik prednasok a posterov. Poprad 8–10.10.2003

Sant'Anna CL, Gama Junior WA, Azevedo MTP et al (2011) New morphospecies of *Chamaesiphon* (Cyanobacteria) from the Atlantic Rainforest, Brazil. Fottea 11:25–30. doi:10.5507/fot.2011.004

Schopf JW (2001) The fossil record: tracing the roots of the cyanobacterial lineage. In: Whitton BA, Potts M (eds) The ecology of cyanobacteria: their diversity in time and space. Springer, Berlin, pp 13–35

Sheath RG, Cole KM (1992) Biogeography of stream macroalgae in North America. J Phycol 28:448–460. doi:10.1111/j.0022-3646.1992.00448.x

Shih PM, Wu D, Latifi A et al (2013) Improving the coverage of the cyanobacterial phylum using diversity-driven genome sequencing. Proc Natl Acad Sci U S A 110:1053–1058. doi:10.1073/pnas.1217107110

Siegesmund MA, Johansen JR, Karsten U et al (2008) *Coleofasciculus* gen. nov. (*Cyanobacteria*): morphological and molecular criteria for revision of the genus *Microcoleus* Gomont. J Phycol 44:1572–1585. doi:10.1111/j.1529-8817.2008.00604.x

Singh RK, Tiwari SP, Rai AK et al (2011) Cyanobacteria: an emerging source for drug discovery. J Antibiot 64:401–412. doi:10.1038/ja.2011.21

Sleep NH (2010) The Hadean-Archean environment. Cold Spring Harb Perspect Biol 2:a002527. doi:10.1101/cshperspect.a002527

Sompong U, Hawkins PR, Besley C (2005) The distribution of cyanobacteria across physical and chemical gradients in hot springs in northern Thailand. FEMS Microbiol Ecol 52:365–376. doi:10.1016/j.femsec.2004.12.007

Stamatakis A (2006) RAxML-VI-HPC: maximum likelihood-based phylogenetic analyses with thousands of taxa and mixed models. Bioinformatics 22:2688–2690. doi:10.1093/bioinformatics/btl446

Strunecký O, Bohunická M, Johansen JR et al (2017) A revision of the genus *Geitlerinema* and a description of the genus *Anagnostidinema* gen. nov. Fottea in press

Strunecký O, Elster J, Komárek J (2011) Taxonomic revision of the fresh–water cyanobacterium "*Phormidium*" *murrayi* = *Wilmottia murrayi*. Fottea 11:57–71. doi:10.5507/fot.2011.007

Strunecký O, Komárek J, Johansen JR et al (2013) Molecular and morphological criteria for revision of the genus *Microcoleus* (Oscillatoriales, Cyanobacteria). J Phycol 49:1167–1180. doi:10.1111/jpy.12128

Strunecký O, Komárek J, Šmarda J (2014) *Kamptonema* (Microcoleaceae, Cyanobacteria), a new genus derivated from the polyphyletic *Phormidium* on the basis of combined molecular and cytomorphological markers. Preslia 86:193–207

Suda S, Watanabe MM, Otsuka S et al (2002) Taxonomic revision of water-bloom-forming species of oscillatorioid cyanobacteria. Int J Syst Evol Microbiol 52:1577–1595. doi:10.1099/00207713-52-5-1577

Swingley WD, Chen M, Cheung PC et al (2008) Niche adaptation and genome expansion in the chlorophyll d-producing cyanobacterium *Acaryochloris marina*. Proc Natl Acad Sci U S A 105:2005–2010. doi:10.1073/pnas.0709772105

Taton A, Grubisic S, Brambilla E et al (2003) Cyanobacterial diversity in natural and artificial microbial mats of Lake Fryxell (McMurdo Dry Valleys, Antarctica): a morphological and molecular approach. Appl Environ Microbiol 69:5157–5169. doi:10.1128/AEM.69.9.5157-5169.2003

van Gremberghe I, Leliaert F, Mergeay J et al (2011) Lack of phylogeographic structure in the freshwater cyanobacterium *Microcystis aeruginosa* suggests global dispersal. PLoS One 6, e19561. doi:10.1371/journal.pone.0019561

Vandamme P, Pot B, Gillis M et al (1996) Polyphasic taxonomy, a consensus approach to bacterial systematics. Microbiol Rev 60:407–438

Vincent WF (2007) Cold tolerance in cyanobacteria and life in the cryo-sphere. In: Seckbach J (ed) Algae and cyanobacteria in extreme environments. Springer, Heidelberg, pp 289–304

Vonshak A (1997) *Spirulina platensis* (*Arthrospira*): physiology, cell-biology and biotechnology. Taylor & Francis Ltd, London

Wacklin P, Hoffmann L, Komárek J (2009) Nomenclatural validation of the genetically revised cyanobacterial genus *Dolichospermum* (Ralfs ex Bornet et Flahault) comb. nova. Fottea 9:59–64. doi:10.5507/fot.2009.005

Walker M, Christiansen G, Von Döhrem H (2004) Diversity of coexisting *Planktothrix* (Cyanobacteria) chemotypes deduced by mass spectral analysis of microcystins and other oligopeptides. Arch Microbiol 182:288–298. doi:10.1007/s00203-004-0711-3

Ward DM, Cohan FM, Bhaya D et al (2008) Genomics, environmental genomics and the issue of microbial species. Heredity 100:207–219. doi:10.1038/sj.hdy.6801011

Werner VR, Laughinghouse HD IV, Fiore MF et al (2012) Morphological and molecular studies of *Sphaerospermopsis torques-reginae* (Cyanobacteria, Nostocales) from South American water blooms. Phycologia 51:228–238. doi:10.2216/11-32.1

West NJ, Scanlan DJ (1999) Niche-partitioning of *Prochlorococcus* populations in a stratified water column in the eastern North Atlantic Ocean. Appl Environ Microbiol 65:2585–2591

West M, West GS (1911) Freshwater algae in British Antarctic Expedition 1907-09. Biol., part 7, 1:263–298

Whitton BA (1992) Diversity, ecology and taxonomy of the cyanobacteria. In: Mann NH, Carr NG (eds) Photosynthetic prokaryotes. Plenum Press, New York, pp 1–51

Whitton BA (2005) Phylum Caynophyta (cyanobacteria). In: John DM, Whitton BA, Brook AJ (eds) The freshwater algal flora of the British Isles: an identification guide to freshwater and terrestrial algae. Natural History Museum, London/Cambridge University Press, Cambridge, pp 25–122

Whitton BA, Potts M (2000) The ecology of cyanobacteria. Their diversity in time and space. Springer, Berlin

Wilde SB, Johansen JR, Wilde HD et al (2014) *Aetokthonos hydrillicola* gen. et sp. nov.: epiphytic cyanobacteria associated with invasive aquatic plants and implicated in bird deaths from avian vacuolar myelinopathy. Phytotaxa 181:243–260. doi:10.11646/phytotaxa.181.5.1

Woese CR, Kandler O, Wheelis ML (1990) Towards a natural system of organisms: proposal for the domains Archaea, Bacteria, and Eucarya. Proc Natl Acad Sci U S A 87:4576–4579

Yu G, Zhu M, Chen Y (2015) Polyphasic characterization of four species of *Pseudanabaena* (Oscillatoriales, cyanobacteria) from China and insights into polyphyletic divergence within the *Pseudanabaena* genus. Phytotaxa 192:1–12. doi:10.11646/phytotaxa.192.1.1

Zammit G, Billi D, Albertano P (2012) The subaerophytic cyanobacterium *Oculatella subterranea* (Oscillatoriales, Cyanophyceae) gen. et sp. nov.: a cytomorphological and molecular description. Eur J Phycol 47:341–354. doi:10.1080/09670262.2012.717106

Zhaxybayeva O, Gogarten JP, Charlebois RL et al (2006) Phylogenetic analysis of cyanobacterial genomes: quantification of horizontal gene transfer events. Genome Res 16:1099–1108. doi:10.1101/gr.5322306

Diversity of Anaerobic Anoxygenic Phototrophic Purple Bacteria

Johannes F. Imhoff

Abstract The anaerobic phototrophic purple bacteria are a phylogenetically highly diverse group of bacteria with the common physiological property of performing anoxygenic photosynthesis under anaerobic conditions. They are representatives of the *Alpha-*, *Beta-*, and *Gammaproteobacteria*. More than 160 known species are classified into 57 genera, 12 families, and 7 orders of the *Proteobacteria*. A short historical review on the systematic treatment of the phototrophic purple bacteria and the actual state is presented. The phylogenetic relationship of the cultured phototrophic purple bacteria based upon 16S rRNA gene sequences of the type strains is shown in a comprehensive phylogenetic tree. Correlation and differences between taxonomic treatment and phylogenetic relatedness are discussed. As the 16S rRNA gene is only of limited value for biodiversity studies of functional groups of bacteria, the application of functional genes for these purposes gains importance and genes coding for bacteriochlorophyll synthesis (*bchY*), and reaction center proteins (*pufL* and *pufM*) have been applied for biodiversity studies of phototrophic purple bacteria. The correlation of phylogenetic relationships based on *pufLM* gene sequences with that of 16S rRNA gene sequences, the appropriate information content in the *pufLM* sequences (>1400 nt), and a database of *pufLM* sequences from many of the cultured purple bacteria make these genes a preferred target to study environmental communities of phototrophic purple bacteria.

Studies on the diversity of phototrophic purple bacteria in three representative habitats are discussed: an exceptionally well-developed microbial mat in a salt marsh from which a number of new phototrophic purple bacteria have been isolated (Sippewissett Salt Marsh, MA), the chemocline of a meromictic freshwater lake from which new species have been isolated and in which genetic diversity studies have been performed (Lake Cadagno, Switzerland), and a coastal lagoon (Baltic Sea lagoon at Stein, Germany) in which the diversity of phototrophic purple bacteria and the impact of changes in temperature and salinity on the community composition was studied using almost complete sequences of the *pufLM* genes.

J.F. Imhoff (✉)
GEOMAR Helmholtz Center for Ocean Research Kiel, Marine Microbiology, 24105 Kiel, Germany
e-mail: jimhoff@geomar.de

© Springer International Publishing Switzerland 2017

P.C. Hallenbeck (ed.), *Modern Topics in the Phototrophic Prokaryotes*,
DOI 10.1007/978-3-319-46261-5_2

Keywords Phototrophic purple bacteria • Biodiversity • Species list • Phylogeny • Functional genes • *pufLM* genes • Biodiversity analysis • Environmental selection • Purple sulfur bacteria • Purple nonsulfur bacteria • Chromatiaceae • Ectothiorhodospiraceae

Introduction

Among the bacterial phyla with phototrophic bacteria, the *Proteobacteria* harbor the phototrophic purple bacteria with representatives of phototrophic purple sulfur bacteria in the *Gammaproteobacteria* and members of the purple nonsulfur bacteria in the *Alpha*- and *Betaproteobacteria* (Imhoff 2001d; Imhoff et al. 2005; Imhoff 2006a, b, c, d). The so-called aerobic phototrophic purple bacteria are close relatives of the anaerobic anoxygenic phototrophic purple bacteria. They have a primarily chemoheterotrophic metabolism and are aerobic respiring bacteria forming bacteriochlorophyll and a photosynthetic apparatus in the presence of oxygen.

Anaerobic anoxygenic phototrophic bacteria are major players in ecological niches, which primarily are strictly anoxic but extend to microoxic and even oxic conditions in the light (Pfennig 1977, 1989; Imhoff 1995; Imhoff et al. 2005; Madigan 1988; van Gemerden and Mas 1995). Their primary ecological niche is in stratified environments with countercurrent gradients of sulfide and oxygen, in freshwater, marine, and hypersaline habitats; in salt and soda lakes; in hot springs, cold waters, and sea ice; in coastal sediments; and in the chemocline of many lakes, fjords, and stratified water bodies. The habitats of purple sulfur bacteria generally are the lower part of the chemocline in sediments and waters, in which sulfide is present and light is available. More oxygen-sensitive representatives develop in the lower part where oxygen is absent, while more oxygen-tolerant species may occur in the upper part and even may make use out of the oxygen for respiratory purposes.

In this chapter, we will focus on the anaerobic phototrophic purple sulfur and purple nonsulfur bacteria and the advances in studies of their biodiversity over the past decades. We will use the term purple nonsulfur bacteria for all anaerobic anoxygenic purple *Alpha*- and *Betaproteobacteria* and the term purple sulfur bacteria for the *Gammaproteobacteria*.

A Short Historical Review on Systematic Studies of Phototrophic Purple Bacteria

Since the pioneering work of Winogradsky (1888) and the systematic treatments of phototrophic purple bacteria by Molisch (1907), who for the first time distinguished between the purple sulfur (*Thiorhodaceae*) and purple nonsulfur bacteria (*Athiorhodaceae*), more than a century ago, several important key steps in advancing our knowledge on the diversity of phototrophic purple bacteria have to be mentioned. After the important contributions to the physiology of phototrophic bacteria by van Niel (van Niel 1931, 1944), the formulation of a suitable culture medium for phototrophic sulfur bacteria by Norbert Pfennig (Pfennig 1965; Pfennig and Lippert 1966; Pfennig and Trüper 1992; Imhoff 2006c) significantly stimulated the work on the physiology, ecology, and biodiversity of these bacteria (Pfennig 1967, 1977, 1989). The studies of Pfennig over the following decades resulted in the description of a number of new purple sulfur and purple nonsulfur bacteria (see Tables 1, 2, and 3) and also green sulfur bacteria. Together with Hans-Georg Trüper, a consequent systematic treatment of these bacteria based on phenotypic properties was implemented. Reevaluation of historical data on the taxonomy of phototrophic bacteria, formal description of species and definition of type and neotype strains of the phototrophic bacteria formed the fundament for all future taxonomic work on these bacteria (Pfennig and Trüper 1969, 1971, 1974). A key for the recognition of genera and species was established which basically used easily recognized phenotypical properties that could in part also be determined by microscopic examination of environmental samples or enrichment cultures. Oxidation of sulfide and deposition of elemental sulfur globules (inside or outside the cells) were major criteria to distinguish not only purple nonsulfur from purple sulfur bacteria but also the genus *Ectothiorhodospira* from other purple sulfur bacteria. Motility by flagella, formation of gas vesicles, shape and size of the cells, and structure of internal photosynthetic membrane systems were other criteria used for differentiation of the species and genera. A culture collection of type strains and reference cultures of anoxygenic phototrophic bacteria established by Pfennig and Trüper still is a reference for studies today (most of these strains are maintained in the author's lab and also available through DSMZ, Braunschweig, Germany). It is interesting to see that at that time only the genera *Rhodopseudomonas*, *Rhodospirillum*, and *Rhodomicrobium* were recognized in the *Rhodospirillaceae* family (purple nonsulfur bacteria) and ten genera (*Ectothiorhodospira*, *Thiospirillum*, *Chromatium*, *Thiocystis*, *Thiosarcina*, *Thiocapsa*, *Lamprocystis*, *Thiodictyon*, *Thiopedia*, *and Amoebobacter*) in the *Chromatiaceae* family (Pfennig and Trüper 1971, 1974). Altogether less than 40 species of phototrophic purple bacteria were known at that time (Pfennig and Trüper 1971). Today, more than 160 species of anoxygenic phototrophic purple bacteria are recognized in 28 genera (95 species) of purple nonsulfur bacteria and 29 genera (74 species) of purple sulfur bacteria. They are members of seven orders and more than 11 families of the *Proteobacteria* (Tables 1, 2, and 3).

The introduction of ribosomal RNA sequences (at the beginning oligonucleotide catalogues, later complete sequences) by Carl Woese into microbial phylogeny opened up completely new possibilities in diversity studies and systematics also of the phototrophic bacteria. Woese actually considered the *Proteobacteria* as "phototrophic purple bacteria and their relatives," presuming a crucial role of phototrophic purple bacteria in the evolution of *Proteobacteria* and the origin of *Proteobacteria* from phototrophic ancestors (Woese et al. 1984a, b, 1985; Woese 1987). The purple sulfur bacteria were *Gammaproteobacteria* and the purple nonsulfur bacteria representatives of *Alpha-* and *Betaproteobacteria* (Woese 1987).

The implementation of 16S rRNA gene sequences and phylogenetic considerations into the systematic treatment of phototrophic bacteria led to major changes in the

Table 1 *Alphaproteobacteria*: Purple nonsulfur bacteria

Order	Family	Genus abbr[a]	Species name	Reference	Previous names	Reference	Habitat
Rhodospirillales							
Rhodospirillaceae							
		Rsp.	***Rhodospirillum rubrum***	Molisch (1907)			Stagnant freshwater habitats
		Phs.	***Phaeospirillum fulvum***	Imhoff et al. (1998a)	*Rhodospirillum fulvum*	van Niel (1944)	Stagnant freshwater ponds
			Phaeospirillum chandramohanii	Kumar et al. (2009a)			Freshwater reservoir, India
			Phaeospirillum molischianum	Imhoff et al. (1998a)	*Rhodospirillum molischianum*	Giesberger (1947)	Stagnant freshwater ponds
			Phaeospirillum oryzae	Lakshmi et al. (2011c)			Paddy rhizosphere soil, India
			Phaeospirillum tilakii	Raj et al. (2012)			Freshwater habitat, bird sanctuary, India
		Ros.	***Roseospira mediosalina***	Imhoff et al. (1998a)	*Rhodospirillum mediosalinum*	Kompantseva and Gorlenko (1984)	Saline hot sulfur spring, Azerbaijan
			Roseospira visakhapatnamensis	Chakravarthy et al. (2007)			Marine saltern, India
			Roseospira goensis	Chakravarthy et al. (2007)			Marine saltern, India
			Roseospira marina	Guyoneaud et al. (2002)			Brackish-water lagoon of Certes, France
			Roseospira navarrensis	Guyoneaud et al. (2002)			Saline springs, Salinas de Oro, Spain
			Roseospira thiosulfatophila	Guyoneaud et al. (2002)			Coastal marine microbial mats
			Pararhodospirillum photometricum	Lakshmi et al. (2014)	*Rhodospirillum photometricum*	Molisch (1907)	Stagnant freshwater ponds
		Pas.	***Pararhodospirillum oryzae***	Lakshmi et al. (2014)	*Rhodospirillum oryzae*	Lakshmi et al. (2013)	Paddy rhizospere soil, India
			Pararhodospirillum sulfurexigens	Lakshmi et al. (2014)	*Rhodospirillum sulfurexigens*	Kumar et al. (2008b)	Freshwater reservoir, India
		Phv.	***Phaeovibrio sulfidiphilus***	Lakshmi et al. (2011b)			Brackish-water, Nagapattinam, India
		Rsa.	***Rhodospira trueperi***	Pfennig et al. (1997)			Sippewissett Salt Marsh
		Rcs.	***Rhodocista centenaria***	Kawasaki et al. (1992)	*Rhodospirillum centenum*	Favinger et al. (1989)	Thermopolis Hot Springs, Wyoming
			Rhodocista pekingensis	Zhang et al. (2003)			Wastewater treatment plant

Acetobacteraceae						
	Rpi.	***Rhodopila globiformis***	Imhoff et al. (1984)	*Rhodopseudomonas globiformis*	Pfennig (1974)	Yellowstone acidic sulfur spring, Thermopolis Hot Springs microbial mat
	Rva.	***Rhodovastum atsumiense***	Okamura et al. (2009a)			Paddy soil, Japan
Rhizobiales						
Bradyrhizobiaceae						
	Rps.	***Rhodopseudomonas palustris***	van Niel (1944)	*Rhodobacillus palustris*	Molisch (1907)	Freshwater and terrestrial habitats
		Rhodopseudomonas rutila	Akiba et al. (1983)			Rice field soil
		Rhodopseudomonas faecalis	Zhang et al. (2002)			Chicken feces
		Rhodopseudomonas harwoodiae	Ramana et al. (2012)			Sediment of soil pan, India
		Rhodopseudomonas julia	Kompantseva (1989)	*Rps. palustris*	Hiraishi et al. (1992)	Acidic sulfur spring
		Rhodopseudomonas parapalustris	Ramana et al. (2012)			Soil sample, India
		Rhodopseudomonas pentothenatexigens	Kumar et al. (2013)			Paddy soil, India
		Rhodopseudomonas pseudopalustris	Ramana et al. (2012)			Freshwater habitats
		Rhodopseudomonas rhenobacensis	Hougardy et al. (2000)			Eutrophic freshwater pond
		Rhodopseudomonas thermotolerans	Kumar et al. (2013)			Paddy soil, India
	Rbl.	***Rhodoblastus acidophilus***	Imhoff (2001b)	*Rhodopseudomonas acidophila*	Pfennig (1969a)	Acidic freshwaters, peat bog pools
		Rhodoblastus sphagnicola	Kulichevskaya et al. (2006)			Acidic sphagnum peat bog, Russia

(continued)

Table 1 (continued)

Order	Family	Genus abbr[a]	Species name	Reference	Previous names	Reference	Habitat
Hyphomicrobiaceae							
		Rpl.	***Rhodoplanes roseus***	Hiraishi and Ueda (1994b)	*Rhodopseudomonas rosea*	Janssen and Harfoot (1991)	Freshwater pond
			Rhodoplanes elegans	Hiraishi and Ueda (1994b)			Wastewater
			Rhodoplanes oryzae	Srinivas et al. (2014b)			Paddy soil
			Rhodoplanes piscinae	Chakravarthy et al. (2012)			Freshwater pond
			Rhodoplanes pokkaliisoli	Lakshmi et al. (2009)			Rice field soil, India
			Rhodoplanes serenus	Okamura et al. (2009b)			Freshwater pond
			Rhodoplanes cryptolactis	Okamura et al. (2007)	*Rhodopseudomonas cryptolactis*	Stadtwald-Demchick et al. (1990)	Thermopolis Hot Spring
		Blc.	***Blastochloris viridis***	Hiraishi (1997)	*Rhodopseudomonas viridis*	Drews and Giesbrecht (1966)	Stagnant freshwater bodies
			Blastochloris sulfoviridis	Hiraishi (1997)	*Rhodopseudomonas sulfoviridis*	Keppen and Gorlenko (1975)	Acidic sulfur spring
			Blastochloris gulmargensis	Ramana et al. (2011)			Cold sulfur spring at Gulmarg, India
		Rmi.	***Rhodomicrobium vannielii***	Duchow and Douglas (1949)			Freshwater, brackish and marine habitats
			Rhodomicrobium udaipurense	Ramana et al. (2013)			River water, India
Rhodobiaceae							
		Rbi.	***Rhodobium orientis***	Hiraishi et al. (1995)			Seawater pools
			Rhodobium gokarnense	Srinivas et al. (2007d)			Saltern, India
		Afi.	***Afifella marina***	Urdiain et al. (2008)	*Rhodopseudomonas marina*	Imhoff (1983)	Marine coastal habitats
					Rhodobium marinum	Hiraishi et al. (1995)	
			Afifella pfennigii	Urdiain et al. (2008)	*Rhodobium pfennigii*	Caumette et al. (2007)	Brackish-water pond, Rangiroa Atoll, French Polynesia

Rhodobacterales						
Rhodobacteraceae						
	Ceb.	***Cereibacter changlensis***	Suresh et al. (2015)	*Rhodobacter changlensis*	Kumar et al. (2007d)	Snow from Changla Pass, Himalaya, India
	Rba.	***Rhodobacter capsulatus***	Imhoff et al. (1984)	*Rhodopseudomonas capsulata*	Molisch (1907)	Stagnant freshwater
		Rhodobacter aestuarii	Ramana et al. (2009)			Brackish-water microbial mat, mangrove forest, India
		Rhodobacter azotoformans	Hiraishi et al. (1996)			Wastewater sewage plant
		Rhodobacter blasticus	Kawasaki et al. (1993)	*Rhodopseudomonas blastica*	Eckersley and Dow (1980)	Eutrophic freshwater pond
		Rhodobacter johrii	Girija et al. (2010)			Jowar (sorghum) rhizosphere, India
		Rhodobacter maris	Ramana et al. (2008)			Seashore sediment, India
		Rhodobacter megalophilus	Arunasri et al. (2008)			Soil of Indian Himalaya
		Rhodobacter ovatus	Srinivas et al. (2008)			"Pink Pond" polluted freshwater pond, India
		Rhodobacter sphaeroides	Imhoff et al. (1984)	*Rhodopseudomonas sphaeroides*	van Niel (1944)	Stagnant freshwater bodies
		Rhodobacter veldkampii	Hansen and Imhoff (1985)			Freshwater habitats
		Rhodobacter vinaykumarii	Srinivas et al. (2007b)			Marine tidal waters, India
		Rhodobacter viridis	Raj et al. (2013)			Stagnant freshwater, India
	Rca.	***Rhodobaca bogoriensis***	Milford et al. (2000)			Saline Soda Lake Bogoria, Kenya
		Rhodobaca barguzinensis	Boldareva et al. (2008)			Saline Soda Lake Algin Russia
	Rhb.	***Rhodobaculum claviforme***	Bryantseva et al. (2015)			Saline Soda Lake Doroninskoe, Siberia
	Rdv.	***Rhodovulum sulfidophilum***	Hiraishi and Ueda (1994a)	*Rhodopseudomonas sulfidophila*	Hansen and Veldkamp (1973)	Marine coastal waters and sediments
				Rhodobacter sulfidophilus	Imhoff et al. (1984)	
		Rhodovulum adriaticum	Hiraishi and Ueda (1994a)	*Rhodopseudomonas adriatica*	Neutzling et al. (1984)	Marine waters, Malo Jezero, Adria
		Rhodovulum aestuarii	Divyasree et al. (2016)			Estuarine brackish-water, India
		Rhodovulum bhavnagarense	Srinivas et al. (2012)			Colored pond, India

(continued)

Table 1 (continued)

Order	Family	Genus abbr[a]	Species name	Reference	Previous names	Reference	Habitat
			Rhodovulum euryhalinum	Hiraishi and Ueda (1994a)	*Rhodobacter euryhalinus*	Kompantseva (1985)	Marine coastal waters
			Rhodovulum imhoffii	Srinivas et al. (2007a)			Aquaculture pond, India
			Rhodovulum iodosum	Straub et al. (1999)			Marine coastal habitat
			Rhodovulum kholense	Kumar et al. (2008a)			Mangrove forest in Khola, India
			Rhodovulum lacipunicei	Chakravarthy et al. (2009)			Saline pond near Satpada, India
			Rhodovulum mangrovi	Nupur et al. (2014)			Mangrove forest
			Rhodovulum marinum	Srinivas et al. (2006)			Tidal waters, India
			Rhodovulum phaeolacus	Lakshmi et al. (2011a)			Saltern at Kanyakumari, India
			Rhodovulum robiginosum	Straub et al. (1999)			Marine coastal sediment
			Rhodovulum salis	Srinivas et al. (2014a)			Salt pan, India
			Rhodovulum steppense	Kompantseva et al. (2010)			Soda Lakes, Siberia, Russia
			Rhodovulum strictum	Hiraishi and Ueda (1995)			Tidal seawater pools
			Rhodovulum tesquicola	Kompantseva et al. (2012)			Brackish-water Soda Lake, Siberia, Russia
			Rhodovulum viride	Srinivas et al. (2014a)			Colored pond at Chirala, India
			Rhodovulum visakhapatnamense	Srinivas et al. (2007c)			Tidal waters, India
		Rub.	***Rubribacterium polymorphum***	Boldareva et al. (2009)			Soda Lake, Siberia
Rhodothalassiales							
Rhodothalassiaceae							
		Rts.	***Rhodothalassium salexigens***	Imhoff et al. (1998a)	*Rhodospirillum salexigens*	Drews (1981)	Evaporated seawater pools, Oregon
Genera of uncertain affiliation							
		Rhv.	***Rhodovibrio salinarum***	Imhoff et al. (1998a)	*Rhodospirillum salinarum*	Nissen and Dundas (1984)	Salty pond of solar saltern, Portugal
			Rhodovibrio sodomensis	Imhoff et al. (1998a)	*Rhodospirillum sodomense*	Mack et al. (1993)	Sediment from the Dead Sea, Israel
		Rss.	***Roseospirillum parvum***	Glaeser and Overmann (1999)			Sippewissett Salt Marsh

Note: All type species are shown in bold

[a]Genus abbreviations are used in conformance with recommendations of the subcommittee on the taxonomy of phototrophic bacteria of the International Committee on Systematics of Prokaryotes (Imhoff and Madigan 2004; Madigan and Imhoff 2007)

Table 2 *Betaproteobacteria*: Purple nonsulfur bacteria

Order	Family	Genus abbr[a]	Species name	Reference	Previous names	Reference	Habitat
Rhodocyclales							
Rhodocyclaceae							
		Rcy.	***Rhodocyclus purpureus***	Pfennig (1978)			Wastewater lagoon
			Rhodocyclus tenuis	Imhoff et al. (1984)	*Rhodospirillum tenue*	Pfennig (1969b)	Freshwater ponds
Burkholderiales							
Comamonadaceae							
		Rfx.	***Rhodoferax fermentans***	Hiraishi et al. (1991)			Ditchwater, activated sludge
			Rhodoferax antarcticus	Madigan et al. (2000)			Antarctic freshwater Lake Fryxell
		Rvi.	***Rubrivivax gelatinosus***	Willems et al. (1991)	*Rhodocystis gelatinosa*	Molisch (1907)	Stagnant freshwater
					Rhodopseudomonas gelatinosa	van Niel (1944)	
					Rhodocyclus gelatinosus	Imhoff et al. (1984)	
			Rubrivivax benzoatilyticus	Ramana et al. (2006)			Paddy soil, India

Note: All type species are shown in bold

[a]Genus abbreviations are used in conformance with recommendations of the subcommittee on the taxonomy of phototrophic bacteria of the International Committee on Systematics of Prokaryotes (Imhoff and Madigan 2004; Madigan and Imhoff 2007)

Table 3 *Gammaproteobacteria*: Purple sulfur bacteria - *Chromatiaceae* and *Ectothiorhodospiraceae*

	Genus abbr[a]	Species name	Reference	Previous name	Reference	Habitat[b]
Chromatiales						
Chromatiaceae						
	Tca.	***Thiocapsa roseopersicina***	Winogradsky (1888)			Common in estuarine habitats, sewage lagoons
		Thiocapsa imhoffii	Asao et al. (2007)			Soda Lake, Soap Lake, USA
		Thiocapsa litoralis	Puchkova et al. (2000)			White Sea coastal sediment
		Thiocapsa marina	Caumette et al. (2004)			Marine coastal sediments
		Thiocapsa pendens	Guyoneaud et al. (1998)	*Amoebobacter pendens*	Pfennig and Trüper (1971)	Mud and stagnant water, sulfur springs
		Thiocapsa rosea	Guyoneaud et al. (1998)	*Amoebobacter roseus*	Winogradsky (1888)	Mud and stagnant water, sulfur springs
	Tlp.	***Thiolamprovum pedioforme***	Guyoneaud et al. (1998)	*Amoebobacter pedioformis*	Eichler and Pfennig (1986)	Wastewater pond, Taiwan
	Tba.	***Thiobaca trueperi***	Rees et al. (2002)			Eutrophic freshwater lake, Australia
	Lpc.	***Lamprocystis roseopersicina***	Schroeter (1886)			Mud and stagnant water
		Lamprocystis purpurea	Imhoff (2001a)	*Amoebobacter purpureus*	Eichler and Pfennig (1988)	Chemocline of freshwater lakes
	Tdc.	***Thiodictyon elegans***	Winogradsky (1888)			Mud and stagnant freshwater, sulfur springs
		Thiodictyon bacillosum	Pfennig and Trüper (1971)	*Amoebobacter bacillosus*	Winogradsky (1888)	Mud and stagnant freshwater, sulfur springs
	Isc.	***Isochromatium buderi***	Imhoff et al. (1998b)	*Chromatium buderi*	Trüper and Jannasch (1968)	Estuarine sediments, salt flats
	Pha.	***Phaeobacterium nitratireducens***	Nupur et al. (2015)			Coringa mangrove forest sediment, India

	Chr.	***Chromatium okenii***	Perty (1852)	*Monas okenii*	Ehrenberg (1838)	Stagnant freshwater
		Chromatium weissei	Perty (1852)			Stagnant freshwater
	Tcs.	***Thiocystis violacea***	Winogradsky (1888)			Stagnant freshwater and marine habitats
		Thiocystis cadagnonensis	Peduzzi et al. (2011)			Chemocline Lake Cadagno
		Thiocystis chemoclinalis	Peduzzi et al. (2011)			Chemocline Lake Cadagno
		Thiocystis gelatinosa	Pfennig and Trüper (1971)	*Thiothece gelatinosa*	Winogradsky (1888)	Stagnant freshwater and marine habitats
		Thiocystis minor	Imhoff et al. (1998b)	*Chromatium minus*	Winogradsky (1888)	Stagnant freshwaters
		Thiocystis violascens	Imhoff et al. (1998b)	*Chromatium violascens*	Perty (1852)	Stagnant freshwaters
	Trc.	***Thiorhodococcus minor***	Guyoneaud et al. (1997)			Sediment of coastal lagoon, Arcachon Bay
		Thiorhodococcus bheemlicus	Kumar et al. (2007b)			Marine aquaculture pond, India
		Thiorhodococcus drewsii	Zaar et al. (2003)			Microbial mat, Sippewissett Salt Marsh
		Thiorhodococcus fuscus	Lakshmi et al. (2015)			Sediment of Chilika Lagoon, India
		Thiorhodococcus kakinadensis	Kumar et al. (2007b)			Marine tidal waters, India
		Thiorhodococcus mannitoliphagus	Rabold et al. (2006)			White Sea estuary
		Thiorhodococcus modestalkaliphilus	Sucharita et al. (2010a)			Sediment of Chilika Lagoon, India
	Tpc.	***Thiophaeococcus mangrovi***	Kumar et al. (2008d)			Brackish-water sediment, mangrove forest, India
		Thiophaeococcus fuscus	Divyasree et al. (2014)			Sediment from Chilika Lagoon, India
	Alc.	***Allochromatium vinosum***	Imhoff et al. (1998b)	*Chromatium vinosum*	Winogradsky (1888)	Stagnant freshwater and marine habitat
		Allochromatium minutissimum	Imhoff et al. (1998b)	*Chromatium minutissimum*	Winogradsky (1888)	Stagnant waters and sediments
		Allochromatium phaeobacterium	Srinivas et al. (2009)			Brackish-water near Bheemili, India

(continued)

Table 3 (continued)

Genus abbr[a]	Species name	Reference	Previous name	Reference	Habitat[b]
	Allochromatium renukae	Kumar et al. (2008c)			Brackish-water pool, India
	Allochromatium warmingii	Imhoff et al. (1998b)	*Chromatium warmingii*	Migula (1900)	Stagnant freshwater with sulfide
	Allochromatium humboldtianum	Serrano et al. (2015)			Coastal sediment, Callao Bay Peru
Tch.	***Thermochromatium tepidum***	Imhoff et al. (1998b)	*Chromatium tepidum*	Madigan (1986)	Hot springs
Pch.	***Phaeochromatium fluminis***	Shivali et al. (2012)	*Marichromatium fluminis*	Sucharita et al. (2010b)	Baitarani River sediment, India
Mch.	***Marichromatium gracile***	Imhoff et al. (1998b)	*Chromatium gracile*	Strzeszewski (1913)	Stagnant marine and estuarine habitats
	Marichromatium bheemlicum	Kumar et al. (2007a)			Marine aquaculture pond, India
	Marichromatium chrysaorae	Shivali et al. (2011)			Jellyfish from seashore, India
	Marichromatium indicum	Arunasri et al. (2005)			Mangrove forest soil, India
	Marichromatium litoris	Shivali et al. (2011)			Sandy beach, India
	Marichromatium purpuratum	Imhoff et al. (1998b)	*Chromatium purpuratum*	Imhoff and Trüper (1980)	Marine habitats
Trv.	***Thiorhodovibrio winogradskyi***	Overmann et al. (1992)			Saline Lake Mahoney, Canada
Rbc.	***Rhabdochromatium marinum***	Dilling et al. (1995)			Sippewissett Salt Marsh
Thc.	***Thiohalocapsa halophila***	Imhoff et al. (1998b)	*Thiocapsa halophila*	Caumette et al. (1991)	Solar saltern microbial mat, Salin-de-Giraud, France
	Thiohalocapsa marina	Kumar et al. (2009b)			Marine aquaculture pond near Bheemili, India
Lpb.	***Lamprobacter modestohalophilus***	Gorlenko et al. (1979)			Saline lakes, Turkmenia, Ukraine
Hch.	***Halochromatium salexigens***	Imhoff et al. (1998b)	*Chromatium salexigens*	Caumette et al. (1988)	Solar saltern microbial mat Salin-de-Giraud, France
	Halochromatium glycolicum	Imhoff et al. (1998b)	*Chromatium glycolicum*	Caumette et al. (1997)	Microbial mats of Solar Lake, Sinai

		Halochromatium roseum	Kumar et al. (2007c)			Marine solar saltern at Kakinada, India
	Tco.	***Thiococcus pfennigii***	Imhoff et al. (1998b)	*Thiocapsa pfennigii*	Eimhjellen (1970)	Mud and water with sulfide
	Tac.	***Thioalkalicoccus limnaeus***	Bryantseva et al. (2000)			Sediments of saline Soda Lakes, Siberia, Russia
	Tfc.	***Thioflavicoccus mobilis***	Imhoff and Pfennig (2001)			Sippewissett Salt Marsh
	Tpd.	***Thiopedia rosea***	Winogradsky (1888)			Freshwater, brackish and marine habitats
	Tsp.	***Thiospirillum jenense***	Migula (1900)	*Ophidomonas jenensis*	Ehrenberg (1838)	Mud and freshwater containing sulfide
Ectothiorhodospiraceae						
	Ect.	***Ectothiorhodospira mobilis***	Trüper (1968)			Salt Lakes, estuaries, salt flats containing sulfide
		Ectothiorhodospira haloalkaliphila	Imhoff and Süling (1996)			Saline salt and Soda Lakes
		Ectothiorhodospira marina	Imhoff and Süling (1996)			Marine coastal sediments
		Ectothiorhodospira marismortui	Oren et al. (1989)			Dead Sea
		Ectothiorhodospira shaposhnikovii	Cherni et al. (1969)			Ponds and lakes containing sulfide
		Ectothiorhodospira vacuolata	Imhoff et al. (1981)			Saline salt and Soda Lakes
		Ectothiorhodospira salini	Ramana et al. (2010)			Solar saltern sediment, India
		Ectothiorhodospira variabilis	Gorlenko et al. (2009)			Saline Soda Lake Um-Risha, Wadi Natrun, Egypt
		Ectothiorhodospira magna	Bryantseva et al. (2010)			Saline Soda Lake Doroninskoe, Russia
	Ets.	***Ectothiorhodosinus mongolicus***	Gorlenko et al. (2004)			Saline Soda Lake Dzun Uldziin, Mongolia
	Trs.	***Thiorhodospira sibirica***	Bryantseva et al. (1999)			Saline Soda Lakes, Siberia, Russia

(continued)

Table 3 (continued)

Genus abbr[a]	Species name	Reference	Previous name	Reference	Habitat[b]
Hlr.	***Halorhodospira halophila***	Imhoff and Süling (1996)	*Ectothiorhodospira halophila*	Raymond and Sistrom (1969)	Highly saline Soda Lakes, Wadi Natrun
	Halorhodospira abdelmalekii	Imhoff and Süling (1996)	*Ectothiorhodospira abdelmalekii*	Imhoff and Trüper (1981)	Highly saline Soda Lakes, Wadi Natrun
	Halorhodospira halochloris	Imhoff and Süling (1996)	*Ectothiorhodospira halochloris*	Imhoff and Trüper (1977)	Highly saline Soda Lakes, Wadi Natrun
	Halorhodospira neutriphila	Hirschler-Réa et al. (2003)			Microbial mat, saltern Salin-de-Giraud, France

Note: All type species are shown in bold

[a]Genus abbreviations are used in conformance with recommendations of the subcommittee on the taxonomy of phototrophic bacteria of the International Committee on Systematics of Prokaryotes (Imhoff and Madigan 2004; Madigan and Imhoff 2007)

[b]Common property of all habitats of purple sulfur bacteria is the presence of sulfide and light

taxonomy of these bacteria. First of all, in a reclassification of the purple nonsulfur bacteria, the *Betaproteobacteria* were separated from the *Alphaproteobacteria*. Several new genera were formed, with the recognition of some bacteria previously classified as *Rhodopseudomonas* and *Rhodospirillum* species included into the new genera *Rhodocyclus*, *Rhodopila*, and *Rhodobacter* (Imhoff et al. 1984). Also the genus *Ectothiorhodospira* was recognized as a separate family, the *Ectothiorhodospiraceae* besides the *Chromatiaceae* (Imhoff 1984), and later the two genera *Ectothiorhodospira* and *Halorhodospira* were distinguished in this family (Imhoff and Süling 1996). After these first fundamental changes in systematics of phototrophic purple bacteria and based on more extensive analyses of the phylogeny of 16S rRNA genes in these bacteria, a number of additional modifications have been proposed in the years that followed. The *Chromatiaceae* were rearranged according to their phylogenetic relations, and several new genera were proposed to distinguish the known species according to both phenotypic and genetic similarities (Imhoff et al. 1998b; Guyoneaud et al. 1998). Also a number of purple nonsulfur bacteria were reclassified in order to achieve better congruence between systematic treatment and phylogeny. The great heterogeneity of species treated in the genus *Rhodopseudomonas* was recognized, and two species with bacteriochlorophyll b were transferred to a new genus as *Blastochloris viridis* and *Blastochloris sulfoviridis* (Hiraishi 1997). *Rhodopseudomonas acidophila* was transferred to *Rhodoblastus acidophilus* (Imhoff 2001b), and *Rhodopseudomonas blastica* was transferred to *Rhodobacter blasticus* (Kawasaki et al. 1993). Also the marine species of *Rhodobacter* were removed to the genus *Rhodovulum* (Hiraishi and Ueda 1994a); *Rhodopseudomonas rosea* was recognized as a member of the new genus *Rhodoplanes* (Hiraishi and Ueda 1994b), and *Rhodospirillum centenum* was transferred to *Rhodocista centenaria* (Kawasaki et al. 1992). The heterogeneity of species recognized as members of the genus *Rhodospirillum* was further resolved by removing a number of these species into new genera (Imhoff et al. 1998a). More recently, *Rhodospirillum photometricum* and related species were transferred to the new genus *Pararhodospirillum* as *Pararhodospirillum photometricum* (Lakshmi et al. 2014). These reclassifications and the general conformity of the new systematic treatment with phylogenetic relations paved the way for biodiversity studies and species recognition in the environment based on genetic sequence information. The current systematic treatment of the species of phototrophic purple bacteria is shown in Tables 1, 2, and 3.

The Diversity of Phototrophic Purple Bacteria

Much of the motivation to study bacterial systematics comes from the desire to understand the phylogeny and evolution of the bacteria; their species-specific diversity in the environment, including aspects of adaptation of species to changing environmental conditions; their competition in the environment; and the biogeographic distribution on a species-specific level.

Several approaches are used to obtain information on the biodiversity of microbial communities. The classical approach involves the separation and isolation of microbial strains, the characterization of pure cultures, and the identification and description of species. This approach is appropriate to identify and characterize components of the community and to determine their physiological properties but is less suited to depict the species diversity within a sample.

The molecular, metagenetic approach uses genetic tools to separate DNA molecules or amplified PCR products and uses the sequence information to determine the genetic diversity within a sample. With this approach, the diversity of a community can be approached and known species can be recognized on the basis of their sequence, but identification and characterization of unknown species are not possible.

Ideally diversity studies of phototrophic bacteria combine aspects of both approaches, and culture-dependent studies that include the characterization of the isolates are accompanied by genetic analyses determining the phylogenetic relationship of the bacteria within the sample. If untreated environmental samples are diluted in agar deeps or on agar plates with a proper medium for phototrophic bacteria, the cultured biodiversity can be estimated by selection of representative colonies and their characterization by microscopic examination and sequence analysis of the 16S rRNA gene or other suitable genes. If the number of colonies are counted and associated to phylotypes according to sequence information and phenotypic properties, a rough estimation of the diversity of phototrophic bacteria is possible with this method (Imhoff 2006c). Although not commonly used, this combined approach is a valuable alternative to pure metagenetic approaches with the advantage of better resolution and higher specificity and the possibility to approach ecological questions.

The Cultured Diversity of Phototrophic Purple Bacteria

Over the decades, the presence, abundance, and activities of phototrophic purple bacteria were studied in freshwater and marine habitats on the basis of microscopic observations and culture studies. For these analyses, a systematic treatment based on morphological and easily recognizable phenotypic characteristics (motility by flagella, formation of gas vesicles, shape and size of the cells, and structure of internal photosynthetic membrane systems, gas vesicles, absorption spectra and photosynthetic pigments, substrate utilization in particular oxidation of sulfide and thiosulfate) was a solid basis (Pfennig and Trüper 1974; Imhoff 2001d). These criteria maintain to be of major relevance for the taxonomic characterization of species and are included in the guidelines for the description of new species of phototrophic bacteria (Imhoff and Caumette 2004). Most of the species descriptions since the 1970s used these criteria, later with additional support by molecular analysis of cellular components and sequence information of the 16S ribosomal RNA molecule.

Much of the work on the systematics of phototrophic purple bacteria over the past decades has been made, and the overwhelming majority of species have been described by a few experts in this field, with a clear habitat focus related to the location of their research laboratory. In the lab of Pierre Caumette (Arcachon and Pau, France), the focus was on the work on coastal lagoons and marine salterns in southern France and on the isolation of new moderately halophilic purple sulfur bacteria including, among others, *Halochromatium salexigens*, *Thiohalocapsa halophila*, and *Halorhodospira neutriphila* from marine salterns and *Thiorhodococcus minor* and *Roseospira* species from coastal lagoons (Tables 1 and 3). The work of Mike Madigan (Carbondale, IL) had a clear focus on phototrophic bacteria from hot springs in the USA but included other extreme habitats such as permanently cold Antarctic lakes, soda lakes, and the Dead Sea with new species descriptions of *Thermochromatium tepidum*, *Thiocapsa imhoffii*, *Rhodoferax fermentans*, *Rhodobaca bogoriensis*, and *Rhodovibrio sodomensis* (Tables 1, 2, and 3). Japanese freshwater and marine sources were studied by Akira Hiraishi (Toyohashi, Japan), and several new genera and species were described such as *Rhodoferax fermentans*, *Rhodovastum atsumiense*, *Rhodobium orientis*, *Rhodoplanes elegans*, and *Rpl. serenus* (Tables 1 and 2). Many of the saline and hypersaline waters, alkaline soda lakes, and other extreme environments of Russia (Siberia) and also of Mongolia and Kasachstan were studied by the group of Vladimir Gorlenko (Moscow, Russia) (to a great part together with Elena Kompantseva and Irina Bryantseva). Most of his isolates of new genera and species were derived from alkaline soda lakes and from salt lakes, e.g., *Thiorhodospira sibirica*, *Ectothiorhodosinus mongolicus*, *Thioalkalicoccus limnaeus*, and *Rubribacterium polymorphum* (Tables 1, 2, and 3). The focus of the author's group (Bonn and Kiel, Germany) was on marine habitats and African soda lakes, and a number of species were described, many in cooperation with P. Caumette, V. Gorlenko, and Ch. Sasikala (Tables 1, 2, and 3). In more recent years since 2005, more than 40 new species have been isolated from Indian sources by the groups of Ch. Sasikala and Ch.V. Ramana (Hyderabad, India) (in part together with A.P. Kumar and T.N.R. Srinivas) (Tables 1, 2, and 3).

Based on 16S rRNA gene sequences of the type strains, the phylogenetic relationship of the species is depicted in Fig. 1. Data on the species including references, previous names, and habitats are compiled in Tables 1, 2, and 3 for the *Alpha-*, *Beta-*, and *Gammaproteobacteria*. Also the higher taxonomic ranks, families, and orders in which the genera are grouped are included. The type species of the genera are shown in bold face and recommended genus abbreviations are depicted. This three-letter code is in conformance with the recommendations of the subcommittee on the taxonomy of phototrophic bacteria of the International Committee on Systematics of Prokaryotes (Imhoff and Madigan 2004; Madigan and Imhoff 2007). The following abbreviations were added as suggestions to the three-letter code for the recently described genera: *Ceb. Cereibacter*, *Rhb. Rhodobaculum*, *Rub. Rubribacterium*, *Afi. Afifella*, *Pas. Pararhodospirillum*, *Phv. Phaeovibrio*, *Rva. Rhodovastum*, *Pha. Phaeobacterium*, *Tpc. Thiophaeococcus*, and *Pch. Phaeochromatium*.

The high diversity of phototrophic *Alphaproteobacteria* is demonstrated by their presence in three orders and a number of families. Their phylogenetic relations are well represented by the distribution in the *Rhodobacterales* (*Rhodobacteriaceae*), *Rhizobiales* (*Bradyrhizobiaceae*, *Hyphomicrobiaceae*, and *Rhodobiaceae*), and *Rhodospirillales* (*Rhodospirillaceae* and *Acetobacteraceae*) as shown in the phylogenetic tree in Fig. 1 and in Table 1. While most of the species and genera are well aligned phylogenetically, an apparent lack of clear association to any of the etsablished families within the *Alphaproteobacteria* is given for *Rts. salexigens*, *Rss.*

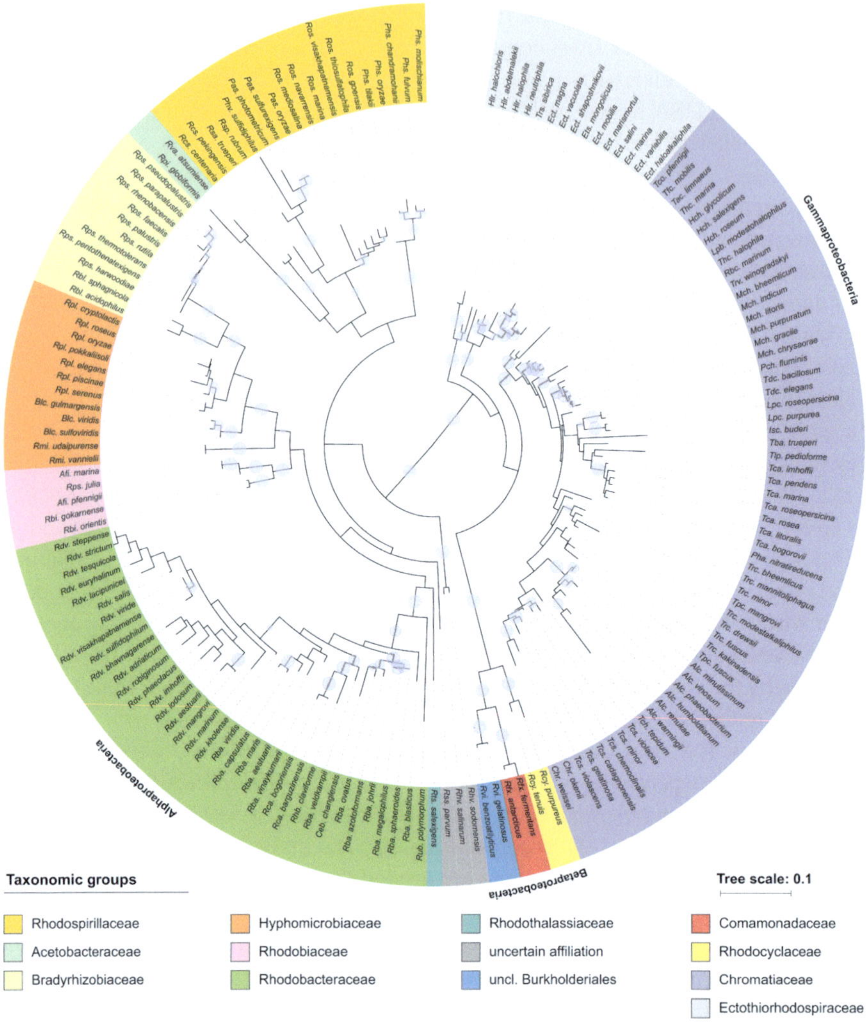

Fig. 1 Phylogenetic relationship of the species of anaerobic anoxygenic phototrophic purple bacteria based on 16S rRNA gene sequences of their type or neotype strains. Sequences used were 1459 nt in length and were aligned using SILVA and the alignment was corrected manually. A maximum likelihood (ML) phylogenetic tree was calculated from the edited alignment with the program IQ-TREE6 v1.4.2. The optimal substitution model given the data under consideration was determined by IQ-TREE to be TN+F+R5. "Ultrafast bootstrap approximation" 8 (UFBoot) was used to provide branch support values with 1000 bootstrap replicates. Branch support values were assigned onto the original ML tree. The calculated phylogenetic tree was midpoint-rooted using the R package phangorn 9 v2.0.3 and bootstrap values >95 % are indicated in the tree by *light blue circles*

parvum, and the two *Rhodovibrio* species, all of which are well adapted to marine/saline or even highly saline habitats. Their exact affiliation in the phylogenetic tree remains unclear. The genera of the *Betaproteobacteria* are assigned to the families *Rhodocyclaceae* (*Rhodocyclus*) and the *Comamonadaceae* (*Rhodoferax*), while

Rubrivivax is not clearly affiliated to a family so far (Imhoff 2005). Also the separation of the families *Chromatiaceae* and *Ectothiorhodospiraceae* in the *Chromatiales* is well depicted in the phylogenetic relationship of the 16S rRNA gene (Fig. 1).

However, the taxonomic treatment is not always well represented in the phylogenetic relations: On the basis of the depicted phylogenetic relationship, *Ectothiorhodosinus mongolicus* clusters between species of *Ectothiorhodospira*, *Lamprobacter modestohalophilus* is highly similar to *Halochromatium* species, the genus *Thiorhodococcus* forms two or three clusters, and the two species of *Thiophaeococcus* are clustering within two of these groups. Also, species of *Rhodobaca*, *Rhodobaculum*, and *Cereibacter* cluster within the *Rhodobacter* clade (Fig. 1). These examples need careful further studies for clarification of their systematic treatment.

There are also some ambiguities on the level of the species identification. Several pairs of species have almost identical sequences. *Rps. rutila* has a sequence identical to that of *Rps. palustris* (0.03 % difference) and is regarded as a later subjective synonym of *Rps. palustris* (Hiraishi et al. 1992). Quite similar the sequence of *Rps. julia* is identical to that of *Afi. marina*, and this species should as well be regarded as a subjective synonym of *Afi. marina*. Also 16S rRNA gene sequences of *Alc. vinosum* and *Alc. minutissimum* are identical (0.00 % difference) as has been noted earlier (Serrano et al. 2011). Further couples of species with highly similar 16S rRNA gene sequences (less than 0.05 % dissimilarity) are *Rdv. viride*/*Rdv. visakhapatnamense*, *Rba. sphaeroides*/*Rba. megalophilus*, *Rba. megalophilus*/*Rba. johrii*, *Rps. thermotolerans*/*Rps. pentothenatexigens*, *Rmi. vannielii*/*Rmi. udaipurense*, and *Pas. oryzae*/*Pas. sulfurexigens*. As far as the used sequence information is concerned, the species within the indicated couples may be regarded as identical at the species level, i.e., represent subjective synonyms. A careful reevaluation of these species is required in order to confirm the existence of two separate species in the given examples.

In parts, the phylogenetic relationship correlates with some phenotypic properties that are common to close neighbors. For example, three species of bacteriochlorophyll b-containing *Chromatiaceae* are close relatives, and the internal photosynthetic membrane structure is a distinctive property of these three bacteria (tubules) to other *Chromatiaceae* (vesicles) and to the *Ectothiorhodospiraceae* (lamellae). Though cell morphology is not a core property for identification of species, even morphological properties maybe common features at least of some phylogenetically related groups, e.g., the budding mode of cell division (*Rhodopseudomonas* and *Rhodoblastus* species) and the spiral shape in some purple nonsulfur bacteria (*Rhodospirillum*, *Phaeospirillum*, *Roseospira*, *Pararhodospirillum*, *Rhodospira*, and *Rhodocista* species).

In addition, habitat preferences are depicted in the phylogenetic relatedness. For many of the genera, a specific preference for a special type of habitat is visible (Tables 1, 2, and 3). A few examples of the *Alphaproteobacteria* can demonstrate this observation: *Phaeospirillum*, *Rhodoplanes*, and *Rhodobacter* species prefer freshwater habitats; *Roseospira*, *Rhodovibrio*, *Rhodovulum*, and *Rhodobium* species preference marine and saline habitats; *Rhodoblastus* species live in acidic peat bogs; and *Rhodobaca* species live in alkaline soda lakes. The preference of all *Ectothiorhodospiraceae* for saline and alkaline conditions and of *Halorhodospira* species for highly saline and alkaline habitats is well known. Many *Chromatiaceae*, e.g., *Marichromatium* and *Thiorhodococcus* species, have a clear preference for marine habitats, while species of other genera such as *Thiocystis*, *Chromatium*, and *Lamprocystis* were found in freshwater habitats and a group around *Halochromatium* and *Thiohalocapsa* species includes species with elevated salt tolerance (see Table 3). Obviously, the preference for a specific set of physicochemical conditions in the environment preselects species of phototrophic purple bacteria with a potential to thrive in a particular habitat. Most obvious in this context is the clear difference between purple bacteria found in the chemocline of freshwater lakes and in marine coastal sediments or lagoons.

The Biodiversity of Environmental Communities of Purple Bacteria

Sequence information is predestined to link bacterial systematics and environmental biodiversity studies of phototrophic bacteria because sequence information is now well established as a property in systematics of phototrophic bacteria. A first important step for the possible identification of species of phototrophic purple bacteria in environmental DNA sequences was the establishment of a phylogenetic-based taxonomy supported by 16S rRNA gene sequences (Imhoff 1984; Imhoff et al. 1984; Imhoff and Süling 1996; Imhoff et al. 1998a, b; Guyoneaud et al. 1998; Kawasaki et al. 1992, 1993; Hiraishi 1997; Hiraishi and Ueda 1994a). Because sequence information becomes easily available from environmental communities, from individual clones, as well as from complete metagenomes, it can supply information about the diversity of a particular genetically targeted function in a species-specific resolution.

In order to establish a measure for the species diversity in environmental samples based on genetic sequence information, phylotypes can be defined on the basis of environmental 16S rRNA gene sequences. If a distinction of phylotypes is made at a sequence level that compares to the level of distinction between species with pure cultures (Stackebrandt and Ebers 2006), phylotypes can be used to approach the species diversity of environmental communities. If environmental clone sequences are sufficiently similar to known species, represented by their type strains, it is quite likely that they are belonging to this species or are close relatives thereof. If considerations concerning sequence similarities as a rough guide for species differentiation of pure cultures are transferred to sequences from the environment, species recognition and an estimate of the species diversity in environmental samples can be achieved with phylotypes as an equivalent to the taxonomic defined species.

However, pitfalls of applying 16S rRNA-based approaches to the analysis of communities of anoxygenic phototrophic bacteria, in particular the phototrophic *Proteobacteria*, were realized and are due to the close phylogenetic relationship between phototrophic and non-phototrophic *Proteobacteria*. Even close phylogenetic neighbors may perform different physiological functions. Specific sequence stretches of the 16S rRNA gene that would clearly allow identification of phototrophic representatives in complex mixtures of environmental samples and distinguish these from non-phototrophic relatives could not be identified. Therefore, many metagenetic studies that focus on sequences of the 16S rRNA gene do not supply clear information on the presence and diversity of phototrophic bacteria in the studied samples. In consequence, increasing research activities are concerned with the application of functional genes to characterize the diversity of functional microbial groups, including anoxygenic phototrophic bacteria (Imhoff 2016).

Functional Gene Studies

In order to specifically study the biodiversity of phototrophic bacteria and their responses to environmental factors, genetic tools were established targeting *pufL* and *pufM* genes of the reaction center proteins (Nagashima et al. 1997; Achenbach et al. 2001; Karr et al. 2003; Tank et al. 2009) and the *bchY* gene of the biosynthesis of bacteriochlorophyll (Yutin et al. 2009).

The *bchY* gene, encoding the Y subunit of chlorophyllide reductase, is at a branch point in the biosynthesis of chlorophyll and bacteriochlorophyll (Chew and Bryant 2007). This gene is present in all known anoxygenic phototrophic bacteria, but absent in oxygenic phototrophs, and therefore, it is suited for targeting the bacteriochlorophyll-containing anoxygenic phototrophic bacteria (Yutin et al. 2009). The comparable low information in the amplified sequences (approx. 500 nt) is certainly a limitation of the used primers. The phylogenetic diversity of phototrophic bacterial communities based on these bchY gene sequences was studied in Lake Kinneret and in the Mediterranean Sea (Yutin et al. 2009), but extended studies on environmental samples and in particular a comprehensive database with reference sequences from type strains of cultured anoxygenic phototrophic bacteria are so far lacking.

The *pufLM* genes encode for the light (L) and medium (M) subunit of the photosynthetic reaction center type II structural proteins of all phototrophic *Proteobacteria* (purple sulfur bacteria, purple nonsulfur bacteria, as well as aerobic phototrophic purple bacteria producing bacteriochlorophyll and forming a photosynthetic apparatus) and in addition of the phototrophic members of the Chloroflexi (Nagashima et al. 1997; Tank et al. 2009). A primer system which amplifies the combined sequence information of *pufL* and *pufM* genes (products of >1450 nt length) was used to build a comprehensive database of *pufLM* gene sequences of most of the recognized type strains of the purple sulfur bacteria (Tank et al. 2009) and to study the biodiversity of phototrophic purple bacteria in the environment (Tank et al. 2011; Thiel et al. 2010). The phylogenetic relationship demonstrated by *pufLM* gene sequences of the purple sulfur bacteria (*Gammaproteobacteria*) was in good correlation to that of 16S rRNA gene sequences (Tank et al. 2009). In context with the phylogenetically based taxonomy of the purple sulfur bacteria (Imhoff et al. 1998a, b; Guyoneaud et al. 1998) and based on the established *pufLM* sequence data, this correlation very much supports the recognition of species in environmental samples using *pufLM* gene sequences.

Molecular genetic studies to characterize the communities of phototrophic purple bacteria based on sequences of the *pufM* gene revealed a remarkable high diversity in different habitats (Achenbach et al. 2001; Karr et al. 2003; Asao et al. 2011; Hirose et al. 2012), although the sequence information obtained was quite low (less than 400 nt). Two case studies of salt lakes in the Chilean highland (Thiel et al. 2010, see Chap. 13) and of a coastal lagoon of the Baltic Sea (Tank et al. 2011) highlight the possibilities of this approach to study the diversity of communities of phototrophic purple bacteria. In order to compare the sequence data from environmental samples with those from type and reference strains on a systematic level, phylotypes were defined for the *pufLM* sequences. Considering different evolutionary rates of the *pufLM* genes compared to the 16S rRNA gene, borderlines of 86 and 95 % sequence similarity of the *pufLM* genes were proposed for the distinction of genera and species of the purple sulfur bacteria (Tank et al. 2009, 2011).

Selected Habitats of Phototrophic Purple Bacteria

Much of the work on the ecology and the communities of anoxygenic phototrophic bacteria has been made in freshwater lakes; in stratified water bodies of fjords and even the Black Sea; also in coastal habitats including coastal lagoons, microbial mats, and sediments; and of course in extreme habitats such as hot springs, salt and soda lakes, and cold habitats such as Antarctic waters and sea ice. From these habitats but also from peat bogs and from waste water treatment plants, paddy soils, and others, purple bacteria have been isolated (Tables 1, 2, and 3). The work on the general ecology of phototrophic bacteria, their ecological relevance, their occurrence in various types of habitats, and their physiology have been discussed in a number of reviews before (Madigan 1988; Imhoff 1988, 1992, 2001c; Pfennig 1989; van Gemerden and Mas 1995). In the following, representative examples are presented

1. of microbial mats from coastal marine habitats, which are well characterized, and from which a diverse number of new species have been isolated (Sippewissett Salt Marsh, MA),
2. of a freshwater chemocline in which cultivation-dependent approaches and molecular diversity studies have been successfully combined (Lake Cadagno, Switzerland), and
3. of a coastal lagoon, in which a most comprehensive analysis of the diversity of phototrophic purple sulfur bacteria has been made using *pufLM* gene sequences specifically targeting the phototrophic purple bacteria (Baltic Sea lagoon at Stein, Germany).

Sippewissett Salt Marsh, MA, USA

Phototrophic purple bacteria are common to marine coastal habitats, and in tidal sediments of estuaries and the Wadden Sea, diverse communities of these bacteria develop worldwide. One of the most prominent examples of such a marine habitat where phototrophic bacteria occur in visible masses and from which several new species have been isolated is represented by the sandy sediments of the Great Sippewissett Salt Marsh and its intertidal flats (MA, USA) (Trüper 1970; Nicholson et al. 1987; Pfennig et al. 1997; Glaeser and Overmann 1999; Imhoff and Pfennig 2001). A detailed analysis of the different layers of these well-developed microbial mats based on microscopic studies and pigment analysis revealed three distinct layers of phototrophic bacteria between a top layer of algae and cyanobacteria and the black bottom sediment with actively sulfate-reducing bacteria (Nicholson et al. 1987).

The uppermost of these three layers was pink and dominated by spherical purple sulfur bacteria most likely representatives of *Thiocapsa* (*Thiocapsa roseopersicina* and *Thiocapsa rosea*, including former *Amoebobacter* species) and also forms resembling *Marichromatium* and *Thiocystis* species (Nicholson et al. 1987). The layer below was distinctive peach-colored and largely contained bacteria with bacteriochlorophyll b, which were supposed to represent *Thiococcus pfennigii* (the only purple sulfur bacterium with bacteriochlorophyll b known at this time). The bacteria of the lowermost green and thinnest layer, which was not always present, were identified as green sulfur bacteria of the genus *Prosthecochloris* (Nicholson et al. 1987). This described pattern of layers very well reflects the properties of the different phototrophic bacteria, in particular concerning their pigmentation and physiological properties.

Thiocapsa roseopersicina and other purple bacteria of the top layer contain bacteriochlorophyll a, while the peach-colored layer contains predominantly bacteria with bacteriochlorophyll b (which have a special advantage in sandy sediments where long wavelength radiation penetrates especially deep) and the green sulfur bacteria in the lower layer contain bacteriochlorophyll c. All of these pigments have different absorption windows (700–750 nm for bchl c, 805–930 nm for bchl a, >1000 nm for bchl b), and the bacteria can therefore easily develop below the chlorophyll a-containing top layer (absorption maximum at 680 nm) and independent from each other. As not only the quality but also the quantity of light matters, it is interesting to see that the bacteriochlorophyll b-containing bacteria with their long-range absorption maxima beyond 1000 nm have a special niche in the sandy sediments below those purple bacteria having bchl a. In addition, the lowermost position of the green sulfur bacteria matches with known experiences in sediments and waters and relates to the special antenna organelles, the chlorosomes, which enable these bacteria to grow at the lowermost amounts of light available for photosynthesis. Also, the relations to sulfide and oxygen perfectly match with the distribution of the phototrophic bacteria in different layers. While the green sulfur bacteria are the most sulfide-tolerant and most sensitive to oxygen, many purple sulfur bacteria not only tolerate oxygen but also can perform oxic respiration (Kämpf and Pfennig 1980, 1986). In particular *Thiocapsa roseopersicina* is metabolically highly flexible and known to be well adapted to diurnal changes of oxic and anoxic conditions, performing photosynthesis in the presence of sulfide during (the onset of) daytime and performing aerobic respiration after depletion of sulfide, also growing as a chemolithotroph or chemoorganotroph in the dark (Kämpf and Pfennig 1980, 1986; De Wit and Van Gemerden 1987, 1990; Schaub and Van Gemerden 1993). These properties predestine this bacterium as a major player in the topmost layer of the phototrophic bacterial mats. The possibility to consume diffusing oxygen, including the oxygen produced in the top layer of the microbial mat, by the upper layer of purple sulfur bacteria protects the layers below with more oxygen-sensitive bacteriochlorophyll b-containing purple sulfur bacteria from the oxygen (Pfennig 1989).

In a preliminary characterization of the diversity of the green sulfur bacteria in samples from Sippewissett Salt Marsh using *fmoA* and 16S rRNA gene sequences, clone sequences related to *Prosthecochloris aestuarii* and to *Chlorobaculum* species were identified (Alexander and Imhoff 2006), supporting and extending the previous findings. A corresponding analysis of the phototrophic purple bacteria in these mats has not been made so far. However, a number of new phototrophic purple bacteria were isolated from these mats suggesting a high, so far unrecognized diversity of phototrophic purple bacteria in the mats. The isolated purple sulfur bacteria from this habitat include strains of *Thiocystis violascens* (formerly *Chromatium violascens*) and *Thiocystis violacea* (Trüper 1970) and a number of new species. Bacteriochlorophyll b-containing bacteria first isolated from these microbial mats include the purple sulfur bacterium *Thioflavicoccus mobilis* (Imhoff and Pfennig 2001), which is a close relative of *Thiococcus pfennigii*, and the purple nonsulfur bacterium *Rhodospira trueperi* (Pfennig et al. 1997). Also *Rhabdochromatium marinum*, which is an obligate phototrophic and strictly anaerobic bacterium, was isolated from a microbial mat of Great Sippewissett Salt Marsh (Dilling et al. 1995). In addition, *Thiorhodococcus drewsii* was isolated from Sippewissett Salt Marsh (Zaar et al. 2003), and finally the purple nonsulfur bacterium *Roseospirillum parvum* with antenna bacteriochlorophyll complexes absorbing at approx. 930 nm originated from this salt marsh (Glaeser and Overmann 1999). It is amazing to see the high degree of novelty in the isolated phototrophic bacteria from this habitat, and a comprehensive study on the metagenomic diversity of the communities of phototrophic bacteria in this salt marsh would be a highly demanding task.

Lake Cadagno, Switzerland

Meromictic lakes represent the most important freshwater habitat of phototrophic sulfur bacteria. In these lakes a more or less stable chemocline is formed, and if light penetrates down to this zone, massive developments of phototrophic sulfur bacteria (generally in company with purple nonsulfur bacteria) develop. In a number of recent studies, phototrophic purple sulfur bacteria from Lake Cadagno were isolated, and in addition the community was characterized by sequence analysis of the 16S rRNA gene.

Lake Cadagno is a small meromictic lake located at 1923 m elevation in the Swiss Alps. It is 21 m deep, and during spring and summer time a permanent chemocline exists at approx. 10 m depth in which phototrophic purple bacteria develop (Fischer et al. 1996; Schanz et al. 1998). Based on microscopic studies and photopigment analysis, the predominance of the okenone-containing *Chromatium okenii* and *Lamprocystis purpurea* forming a 2 m deep zone at the sulfide/oxygen interface was found (Schanz et al. 1998). In agreement with these studies, the molecular analysis of the bacterial community composition at the chemocline revealed clone sequences closely related to *Chromatium okenii*, *Lamprocystis purpurea*, and *Lamprocystis roseopersicina* (Bosshard et al. 2000a; Tonolla et al. 1999) and a seasonal dominance of *Chromatium okenii* during the summer months and of *Lamprocystis purpurea* during autumn (Bosshard et al. 2000b). Later, also clone sequences related to *Thiocystis* species were found (Tonolla et al. 2005), and two new species were described from isolates of this lake, *Thiocystis chemoclinalis* and *Thiocystis cadagnonensis* (Peduzzi et al. 2011). Despite seasonal and long-term changes, the diversity of the phototrophic bacterial community of Lake Cadagno appears typical of low-light chemocline habitats with the dominance of okenone-containing purple bacteria and green sulfur bacteria. More detailed molecular studies targeting specifically the phototrophic purple and green bacteria are expected to demonstrate a more diverse community in this lake as is known till now.

Baltic Sea Coastal Lagoon, Germany

Coastal lagoons are habitats related to typical coastal sediment habitats not permanently covered by water (coastal sediments in estuaries and the Wadden Sea) and also to marine salterns. Whenever sea water and organic matter is locked into coastal lagoons, the onset of active sulfate reduction rapidly creates conditions suitable for the development of anoxygenic phototrophic sulfur bacteria not only in the sediment but also in the water body. The reservoirs of phototrophic bacteria within the coastal sediments provide the initial "feeding" of the populations that develop in the lagoons. Often phototrophic sulfur bacteria are forming massive colored blooms in coastal sediments and lagoons accompanied by purple nonsulfur bacteria (Imhoff 1988, 2001c). Depending on the strength of the sun light and the water regime (evaporation versus dilution with fresh seawater or rain), the concentration of salts in the lagoons may increase and species with increased salt tolerance or salt requirement may be favored. Such changes can create conditions comparable to the situation in the intermediate concentration range of marine salterns. Therefore, it is expected that the three different types of habitats share a number of common species of phototrophic bacteria. Indeed, a number of moderately halophilic *Chromatiaceae* have been isolated from and are considered common inhabitants of both marine coastal lagoons and salterns (Table 3).

In a coastal lagoon of the Baltic Sea in the Kiel Bight, the biodiversity of phototrophic purple bacteria was studied on the basis of *pufLM* gene sequences (Tank et al. 2009, 2011). In order to establish the species diversity, the sequences were arranged in phylotypes. The great majority of *pufLM* phylotypes of the community from this lagoon belonged to the purple sulfur bacteria. A few single clones from aerobic phototrophic *Proteobacteria* were also present in the habitat. The purple sulfur bacteria were affiliated to genera and species typically found in such or similar habitats, including the genera *Marichromatium*, *Thiorhodovibrio*, *Thiorhodococcus*, *Allochromatium*, *Thiocapsa*, and *Thiocystis* (Fig. 2). Some sequences were also related to moderately halophilic *Halochromatium* and *Thiohalocapsa* species first isolated from solar salterns (Caumette et al. 1988, 1991, 1997). From a total of 20

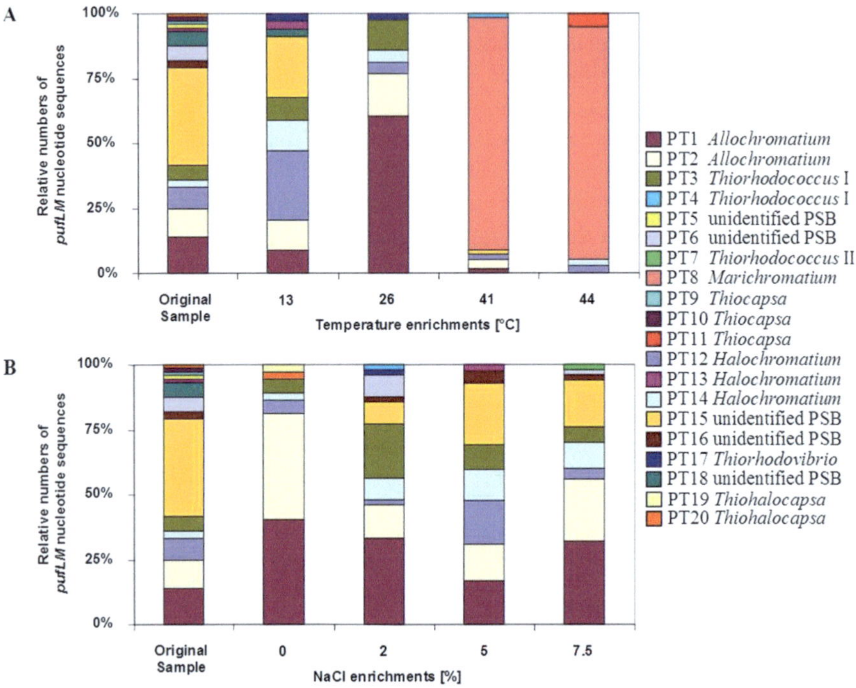

Fig. 2 The contribution of different phylotypes to the community of purple sulfur bacteria in a Baltic Sea lagoon on the basis of *pufLM* gene sequences is shown. The figure depicts the relative composition in the sample and under experimental conditions with variation of temperature and salt concentration as indicated (From Tank et al. 2011)

identified phylotypes of purple sulfur bacteria, five could be clearly assigned to type strains of known species, ten additional phylotypes to a genus, and only five phylotypes had sequence similarities (83.4–85.6 %) slightly below the proposed limit of 86 % *pufLM* sequence similarity to the closest known type strain, which was proposed as a borderline value for the inclusion into genera (Tank et al. 2011). Thus, most of the purple sulfur bacteria of this lagoon more or less were known at the genus level, but novelty of these bacteria was high at the species level. In contrast, *pufLM* sequences of aerobic phototrophic purple bacteria of *Alpha*- and *Gammaproteobacteria* were generally below 84 % similar to the next known relative type strain, in most cases even far below 80 % similarity (Tank et al. 2011).

As this type of habitat is subjected to considerable changes in temperature and salt content during daily and seasonal cycles, the impact of these parameters on the community composition was measured under controlled conditions in the laboratory (Tank et al. 2011). In these experiments, a considerable impact on the community structure was found within studied ranges of temperature (13–44 °C, at 2 % salts) and salinity (0–7.5 % NaCl, at 23.5 °C). The highest diversity of identified phylotypes was observed in the natural sample (at 23.5 °C and 2 % salinity), the lowest diversity at temperatures of 26 °C and higher and in the absence of salt (Fig. 2). With the exception of three phylotypes found as single clones in the environmental sample, all others were retrieved at least from one of the experimental conditions. In addition, six phylotypes that were not detected in the environmental sample showed up alongside the applied salt and temperature gradients (Tank et al. 2011). Among these were phylotypes most similar to the type strains of *Trc. mannitoliphagus* (99.8 % similar), *Trc. kakinadensis* (98.2 % similar), and *Mch. gracile* (100 % similar). This result is quite remarkable and indicates an even higher diversity in the environmental sample than resolved by its direct molecular analysis. It also points out the general limitation of molecular, metagenetic/metagenomic biodiversity studies in environmental samples.

Significant changes in the relative composition of the phylotypes were seen along the range of salt concentrations but more drastic changes in response to the temperature (Fig. 2). Five phylotypes related to *Allochromatium vinosum* (PT1, PT2), *Halochromatium roseum* (PT12, PT14), and *Trc. mannitoliphagus* (PT3) represented approx. 45 % of the clone sequences in the habitat and the majority of sequences under all conditions, except at temperatures above 40 °C (Fig. 2). They were the exclusive representatives found at 26 °C, with the exception of a single clone related to *Thiorhodovibrio winogradskyi* (PT17). The major phylotype in the environmental sample (40 % abundance of PT15) was distantly related to *Thiorhodovibrio winogradskyi* and apparently was dependent on the presence of salt (not found at 0 % NaCl) and low temperatures (only present at 13 °C).

Most dramatic changes of the community were seen with the increase of temperature. A single phylotype of *Mch. gracile* (100 % similar), which was not detected at lower temperatures and not in the habitat sample, absolutely dominated the communities at temperatures above 40 °C (Fig. 2, Tank et al. 2011). *Mch. gracile* has been repeatedly isolated from marine coastal habitats before, but the preference for elevated temperatures was not noted as a general property of this species (Imhoff 1988, 2001a). However, the dominance at 40 °C supports findings of Serrano et al. (2009) who characterized a slightly thermophilic strain of this species as a biotype. The clear preference of *Mch. gracile* for elevated temperatures points to an obvious competitive advantage in shallow-water habitats which are heated during daytime.

A remarkable aspect of this work is the finding that media and cultivation conditions used were appropriate for all purple sulfur bacteria which were found in the environmental sample by the genetic approach. It also demonstrates the great advantage of functional genes in biodiversity studies but also the high flexibility and diversity of purple sulfur bacteria communities in the coastal habitats and their potential to adapt to changing environmental conditions.

Conclusions

A common property of all phototrophic purple bacteria is the presence of a photosynthetic apparatus and the performance of phototrophic growth under anaerobic conditions, though phylogenetically they are members of *Alpha-*, *Beta-*, and *Gammaproteobacteria* and found in several orders and families of these classes (Fig. 1, Tables 1, 2, and 3).

They are important ecological players at the light-receiving chemocline in all types of habitats. Different phototrophic purple bacteria are living in freshwater and in marine and hypersaline habitats; special species are adapted to alkaline or acidic conditions, to cold or hot temperatures. In addition, different light conditions, concentrations of sulfide and oxygen, and different physicochemical conditions define their ecological niches. In reflection of all of these factors that determine distribution and occurrence of the species and in consideration of the results of first detailed metagenetic studies specifically targeting the phototrophic purple bacteria, a much higher diversity of species very likely exists than known to date.

The possibility to approach the diversity of phototrophic purple bacteria with genetic methods specifically targeting the photosynthetic reaction center proteins opens up possibilities to specifically study the environmental diversity of this group of bacteria with high resolution. Probably one of the most detailed diversity studies of a community of phototrophic purple sulfur bacteria and also one with the highest resolution toward detection of members of this group of bacteria (longest used sequence stretch and highest specificity for the group) has been performed with *pufLM* genes as a target (Tank et al. 2011). The results on the analysis of the community from a coastal lagoon are remarkable for different reasons.

1. Despite the highly selective specificity of the *pufLM* target, almost a third of the total phylotypes recognized was detected only after experimental modification of the environmental conditions but not in the environmental sample itself. This incomplete coverage of the environmental community even by highly specific functional gene probes quite likely demonstrates a general limitation of metagenetic and metagenomic biodiversity studies.
2. Changes in physicochemical conditions, especially of the temperature, can cause dramatic shifts in the community composition. This indicates a quite specific adaptation of individual species to selected environmental conditions. It also reflects a high flexibility of the environmental community to adapt to changing conditions and thereby may explain the incomplete coverage of the metagenetic approach, which comes to its limits if certain bacteria are present at very low abundance.

A few relevant case studies using *pufLM* sequences demonstrate that the broad application of this approach is suited to deliver diversity profiles of many different habitats and to study the dynamic changes therein. In a long-term perspective, the comparison of communities using *pufLM* genes allows to address questions of biogeographic distribution, habitat specificity, and ecological niche identification of species on a global scale. We are currently at the beginning of this era.

Acknowledgment Bioinformatic and statistical consulting was provided by Dr. S. Neulinger (omics2view.consulting GbR, Kiel, Germany).

References

Achenbach LA, Carey JR, Madigan MT (2001) Photosynthesis and phylogenetic primers for the detection of anoxygenic phototrophs in natural environments. Appl Environ Microbiol 67:2922–2926

Akiba T, Usami R, Horikoshi K (1983) *Rhodopseudomonas rutila*, a new species of nonsulfur purple photosynthetic bacteria. Int J Syst Bacteriol 33:551–556

Alexander B, Imhoff JF (2006) Communities of green sulfur bacteria in different marine and saline habitats analyzed by gene sequences of 16S rRNA and of the Fenna-Matthews-Olson protein. Int Microbiol 9:259–266

Arunasri K, Sasikala C, Ramana CV, Süling J, Imhoff JF (2005) *Marichromatium indicum* sp. Nov., a novel purple sulfur gammaproteobacterium from mangrove soil of India. Int J Syst Evol Microbiol 55:673–679

Arunasri K, Venkata Ramana V, Spröer C, Sasikala C, Ramana CV (2008) *Rhodobacter megalophilus* sp. nov., a phototroph from the Indian Himalayas possessing a wide temperature range for growth. Int J Syst Evol Microbiol 58:1792–1796

Asao M, Shinichi T, Madigan MT (2007) *Thiocapsa imhoffii*, sp. nov., an alkaliphilic purple sulfur bacterium of the family Chromatiaceae from Soap Lake, Washington (USA). Arch Mikrobiol 188:665–675

Asao M, Pinkart HC, Madigan MT (2011) Diversity of extremophilic purple phototrophic bacteria in Soap Lake, a central Washington (USA) soda lake. Environ Microbiol 13:2146–2157

Boldareva EN, Akimov VN, Boychenko VA, Stadnichuk IN, Moskalenko AA, Makhneva ZK, Gorlenko VM (2008) *Rhodobaca barguzinensis* sp. nov., a new alkaliphilic purple nonsulfur bacterium isolated from a soda lake of the Barguzin Valley (Buryat Republic, Eastern Siberia). Mikrobiologiia 77:206–218

Boldareva EN, Moskalenko AA, Makhneva ZK, Tourova TP, Kolganova TV, Gorlenko VM (2009) *Rubribacterium polymorphum* gen. nov., sp. nov., a novel alkaliphilic nonsulfur purple bacterium from an eastern Siberian soda lake. Microbiology 78:732–740

Bosshard PP, Santini Y, Grüter D, Stettler R, Bachofen R (2000a) Bacterial diversity and community composition in the chemocline of the meromictic alpine Lake Cadagno as revealed by 16S rDNA analysis. FEMS Microbiol Ecol 31:173–182

Bosshard PP, Stettler R, Bachofen R (2000b) Seasonal and spatial community dynamics in the meromictic Lake Cadagno. Arch Microbiol 174:168–174

Bryantseva IA, Gorlenko VM, Kompantseva EI, Imhoff JF, Süling J, Mityushina L (1999) *Thiorhodospira sibirica*, gen. nov. and sp. nov., a new alkaliphilic purple sulfur bacterium from a Siberian soda lake. Int J Syst Bacteriol 49:697–703

Bryantseva IA, Gorlenko VM, Kompantseva EI, Imhoff JF (2000) *Thioalkalicoccus limnaeus* gen. nov., sp. nov., a new alkaliphilic purple sulfur bacterium with bacteriochlorophyll b. Int J Syst Evol Microbiol 50:2157–2163

Bryantseva IA, Tourova TP, Kovaleva OL, Kostrikina NA, Gorlenko VM (2010) *Ectothiorhodospira magna* sp. nov., a new large alkaliphilic purple sulfur bacterium. Mikrobiologiia 79:780–790

Bryantseva IA, Gaisin VA, Gorlenko VM (2015) *Rhodobaculum claviforme* gen. nov., sp. nov., a new alkaliphilic nonsulfur purple bacterium. Mikrobiologiia 84:247–255

Caumette P, Baulaigue R, Matheron R (1988) Characterization of *Chromatium salexigens* sp. nov., a halophilic Chromatiaceae isolated from Mediterranean salinas. Syst Appl Microbiol 10:284–292

Caumette P, Baulaigue R, Matheron R (1991) *Thiocapsa halophila* sp. nov., a new halophilic phototrophic purple sulfur bacterium. Arch Microbiol 155:170–176

Caumette P, Imhoff JF, Süling J, Matheron R (1997) *Chromatium glycolicum* sp. nov., a moderately halophilic purple sulfur bacterium that uses glycolate as substrate. Arch Microbiol 167:11–18

Caumette P, Guyoneaud R, Imhoff JF, Süling J, Gorlenko V (2004) *Thiocapsa marina* sp. nov., a novel, okenone-containing, purple sulfur bacterium isolated from brackish coastal and marine environments. Int J Syst Evol Microbiol 54:1031–1036

Caumette P, Guyoneaud R, Duran R, Cravo-Laureau C, Matheron R (2007) *Rhodobium pfennigii* sp. nov., a phototrophic purple non-sulfur bacterium with unusual bacteriochlorophyll a antennae, isolated from a brackish microbial mat on Rangiroa atoll, French Polynesia. Int J Syst Evol Microbiol 57:1250–1255

Chakravarthy SK, Srinivas TNR, Anil Kumar P, Sasikala C, Ramana CV (2007) *Roseospira visakhapatnamensis* sp. nov. and *Roseospira goensis* sp. nov. Int J Syst Evol Microbiol 57:2453–2457

Chakravarthy SK, Sucharitha K, Sasikala C, Ramana CV (2009) *Rhodovulum lacipunicei* sp. nov., an obligate sulfide-demanding phototrophic alphaproteobacterium isolated from a purple pond in India. Int J Syst Evol Microbiol 59:1615–1619

Chakravarthy SK, Ramprasad EVV, Shoba E, Sasikala C, Ramana CV (2012) *Rhodoplanes piscinae* sp. nov. isolated from pond water. Int J Syst Evol Microbiol 62:2828–2834

Cherni NE, Solovieva JV, Fedorov VD, Kondratieva EN (1969) Ultrastructure of two species of purple sulfur bacteria. Mikrobiologiia 38:479–484

Chew AGM, Bryant DA (2007) Chlorophyll biosynthesis in bacteria: the origins of structural and functional diversity. Annu Rev Microbiol 61:113–129

De Wit R, Van Gemerden H (1987) Chemolithotrophic growth of the phototrophic sulfur bacterium *Thiocapsa roseopersicina*. FEMS Microbiol Ecol 45:117–126

De Wit R, Van Gemerden H (1990) Growth and metabolism of the purple sulfur bacterium *Thiocapsa roseopersicina* under combined light/dark and oxic/anoxic regimens. Arch Microbiol 154:459–464

Dilling W, Liesack W, Pfennig N (1995) *Rhabdochromatium marinum* gen. nom. rev., sp. nov., a purple sulfur bacterium from a salt marsh microbial mat. Arch Microbiol 164:125–131

Divyasree B, Lakshmi KV, Sasikala C, Ramana CV (2014) *Thiophaeococcus fuscus* sp. nov., isolated from a lagoon. Int J Syst Evol Microbiol 64:2528–2533

Divyasree B, Lakshmi KVNS, Dave B, Sasikala C, Ramana CV (2016) *Rhodovulum aestuarii* sp. nov., isolated from a brackish water body. Int J Syst Evol Microbiol 66:165–171

Drews G (1981) *Rhodospirillum salexigens,* spec. nov., an obligatory halophilic phototrophic bacterium. Arch Microbiol 130:325–327

Drews G, Giesbrecht P (1966) *Rhodopseudomonas viridis*, nov. spec., ein neu isoliertes, obligat phototrophes Bakterium. Arch Mikrobiol 53:255–262

Duchow E, Douglas HC (1949) *Rhodomicrobium vannielii*, a new photoheterotrophic bacterium. J Bacteriol 58:407–416

Eckersley K, Dow CS (1980) *Rhodopseudomonas blastica* sp. nov.: A member of the Rhodospirillaceae. J Gen Microbiol 119:465–473

Ehrenberg CG (1838) Die Infusionstierchen als vollkommene Organismen. Voss L, Leipzig, I–XVIII:1–547

Eichler B, Pfennig N (1986) Characterization of a new platelet-forming purple sulfur bacterium, *Amoebobacter pedioformis* sp. nov. Arch Microbiol 146:295–300

Eichler B, Pfennig N (1988) A new purple sulfur bacterium from stratified fresh-water lakes, *Amoebobacter purpureus* sp. nov. Arch Microbiol 149:395–400

Eimhjellen KE (1970) *Thiocapsa pfennigii* sp. nov., a new species of the phototrophic sulfur bacteria. Arch Mikrobiol 73:193–194

Favinger J, Stadtwald R, Gest H (1989) *Rhodospirillum centenum* sp. nov., a thermotolerant cyst-forming photosynthetic bacterium. Antonie Van Leeuwenhoek 55:291–296

Fischer C, Wiggli M, Schanz F, Hanselmann KW, Bachofen R (1996) Light environment and synthesis of bacteriochlorophyll by populations of *Chromatium okenii* under natural environmental conditions. FEMS Microbiol Ecol 21:1–9

Giesberger G (1947) Some observations on the culture, physiology and morphology of some brown-red Rhodospirillum species. Antonie van Leeuwenhoek J Microbiol Serol 13:135–148

Girija KR, Sasikala C, Ramana CV, Spröer C, Takaichi S, Thiel V, Imhoff JF (2010) *Rhodobacter johrii* sp. nov., an endospore-producing cryptic species isolated from semi-arid tropical soils. Int J Syst Evol Microbiol 60:2099–2107

Glaeser J, Overmann J (1999) Selective enrichment and characterization of *Roseospirillum parvum*, gen. nov. and sp. nov., a new purple nonsulfur bacterium with unusual light absorption properties. Arch Microbiol 171:405–416

Gorlenko VM, Krasilnikova EN, Kikina OG, Tatarinova NJ (1979) The new motile purple sulphur bacteria *Lamprobacter modestohalophilus* nov. gen., nov. spec. with gas vacuoles. Biol Bull Acad Sci USSR 6:631–642 (in Russian)

Gorlenko VM, Bryantseva IA, Panteleeva EE, Tourova TP, Kolganova TV, Makhneva ZK, Moskalenko AA (2004) *Ectothiorhodosinus mongolicum* gen. nov., sp. nov., a new purple bacterium from a soda lake in Mongolia. Microbiology 73:80–88

Gorlenko VM, Bryantseva IA, Rabold S, Tourova TP, Rubtsova D, Smirnova E, Thiel V, Imhoff JF (2009) *Ectothiorhodospira variabilis* sp. nov., an alkaliphilic and halophilic purple sulfur bacterium from soda lakes. Int J Syst Evol Microbiol 59:658–664

Guyoneaud R, Matheron R, Liesack W, Imhoff JF, Caumette P (1997) *Thiorhodococcus minus*, gen. nov., sp. nov., A new purple sulfur bacterium isolated from coastal lagoon sediments. Arch Microbiol 168:16–23

Guyoneaud R, Süling J, Petri R, Matheron R, Caumette P, Pfennig N, Imhoff JF (1998) Taxonomic rearrangements of the genera *Thiocapsa* and *Amoebobacter* on the basis of 16S rDNA sequence analyses and description of *Thiolamprovum* gen. nov. Int J Syst Bacteriol 48:957–964

Guyoneaud R, Mouné S, Eatock C, Bothorel V, Hirschler-Réa A, Willison J, Duran R, Liesack W, Herbert R, Matheron R, Caumette P (2002) Characterization of three spiral-shaped purple nonsulfur bacteria isolated from coastal lagoon sediments, saline sulfur springs, and microbial mats: emended description of the genus Roseospira and description of Roseospira marina sp. nov., Roseospira navarrensis sp. nov., and Roseospira thiosulfatophila sp. nov. Arch Microbiol 178:315–324

Hansen TA, Imhoff JF (1985) *Rhodobacter veldkampii*, a new species of phototrophic purple nonsulfur bacteria. Int J Syst Bacteriol 35:115–116

Hansen TA, Veldkamp H (1973) *Rhodopseudomonas sulfidophila* nov. spec., a new species of the purple nonsulfur bacteria. Arch Mikrobiol 92:45–58

Hiraishi A (1997) Transfer of bacteriochlorophyll b-containing phototrophic bacteria *Rhodopseudomonas viridis* and *Rhodopseudomonas sulfoviridis* to the genus *Blastochloris* gen. nov. Int J Syst Bacteriol 47:217–219

Hiraishi A, Ueda Y (1994a) Intrageneric structure of the genus *Rhodobacter*: transfer of *Rhodobacter sulfidophilus* and related marine species to the genus *Rhodovulum* gen. nov. Int J Syst Bacteriol 44:15–23

Hiraishi A, Ueda Y (1994b) *Rhodoplanes* gen. nov., a new genus of phototrophic bacteria including *Rhodopseudomonas rosea* as *Rhodoplanes roseus* comb. nov. and *Rhodoplanes elegans* sp. nov. Int J Syst Bacteriol 44:665–673

Hiraishi A, Ueda Y (1995) Isolation and characterization of *Rhodovulum strictum* sp. nov. and some other purple nonsulfur bacteria from colored blooms in tidal and seawater pools. Int J Syst Bacteriol 45:319–326

Hiraishi A, Hoshino Y, Satoh T (1991) *Rhodoferax fermentans* gen. nov., sp. nov., a phototrophic purple nonsulfur bacterium previously referred to as the "Rhodocyclus gelatinosus- like" group. Arch Microbiol 155:330–336

Hiraishi A, Santos TS, Sugiyama J, Komagata K (1992) *Rhodopseudomonas rutila* is a later subjective synonym of *Rhodopseudomonas palustris*. Int J Syst Bacteriol 42:186–188

Hiraishi A, Urata K, Satoh T (1995) A new genus of marine budding phototrophic bacteria, *Rhodobium* gen. nov., which includes *Rhodobium orientis* sp. nov. and *Rhodobium marinum* comb. nov. Int J Syst Bacteriol 45:226–234

Hiraishi A, Muramatsu K, Ueda Y (1996) Molecular genetic analyses of *Rhodobacter azotoformans* sp. nov. and related species of phototrophic bacteria. Syst Appl Microbiol 19:168–177

Hirose S, Nagashima KVP, Matsuura K, Haruta S (2012) Diversity of purple phototrophic bacteria, inferred from *pufM* gene, within epilithic biofilm in Tama river, Japan. Microbes Environ 27:327–329

Hirschler-Réa A, Matheron R, Riffaud C, Mouné S, Eatock C, Herbert RA, Willison JC, Caumette P (2003) Isolation and characterization of spirilloid purple phototrophic bacteria forming red layers in microbial mats of Mediterranean salterns; description of *Halorhodospira neutriphila* sp. nov. and emendation of the genus *Halorhodospira*. Int J Syst Evol Microbiol 53:153–163

Hougardy A, Tindall BJ, Klemme J-H (2000) *Rhodopseudomonas rhenobacensis* sp. nov., a new nitrate-reducing purple non-sulfur bacterium. Int J Syst Evol Microbiol 50:985–992

Imhoff JF (1983) *Rhodopseudomonas marina* sp. nov., a new marine phototrophic purple bacterium. Syst Appl Microbiol 4:512–521

Imhoff JF (1984) Reassignment of the genus *Ectothiorhodospira* Pelsh 1936 to a new family Ectothiorhodospiraceae fam. nov., and emended description of the Chromatiaceae Bavendamm 1924. Int J Syst Bacteriol 34:338–339

Imhoff JF (1988) Anoxygenic phototrophic bacteria. In: Methods in Aquatic Bacteriology, B. Austin B (ed.), John Wiley and Sons Ltd. chapter 9, pp. 207–240

Imhoff JF (1992) Taxonomy, phylogeny and general ecology of anoxygenic phototrophic bacteria. In: Biotechnology Handbook Photosynthetic Prokaryotes. Carr NG, Mann NH (eds.), Plenum Press, London, New York, pp. 53–92

Imhoff JF (1995) Taxonomy and physiology of phototrophic purple bacteria and green sulfur bacteria. In: Blankenship RE, Madigan MT, Bauer CE (eds) Anoxygenic photosynthetic bacteria. Kluwer Academic, Dordrecht, pp 1–15

Imhoff JF (2001a) Transfer of *Pfennigia purpurea* Tindall 1999 *(Amoebobacter purpureus* Eichler and Pfennig 1988) to the genus *Lamprocystis* as *Lamprocystis purpurea*. Int J Syst Evol Microbiol 51:1699–1701

Imhoff JF (2001b) Transfer of *Rhodopseudomonas acidophila* to the new genus *Rhodoblastus* as *Rhodoblastus acidophilus* comb. nov. Int J Syst Evol Microbiol 51:1863–1866

Imhoff JF (2001c) Minireview - true marine and halophilic anoxygenic phototrophic bacteria. Arch Microbiol 176:243–254

Imhoff JF (2001) The anoxygenic phototrophic purple bacteria. In: Bergey's manual of systematic bacteriology, 2nd edn, vol 1, pp 621–627

Imhoff JF (2005) Genus *Rubrivivax*. In: Brenner DJ, Krieg NR, Staley JT (eds) Bergey's manual of systematic bacteriology, 2nd edn, vol 2, Part C, pp 749–750

Imhoff JF (2006a) The phototrophic Alphaproteobacteria. In: Dworkin M, Falkow S, Rosenberg E, Schleifer KH, Stackebrandt E (eds) The Prokaryotes. A handbook on the biology of bacteria, vol 5, 3rd edn. Springer, New York, pp 41–64

Imhoff JF (2006b) The phototrophic Betaproteobacteria. In: Dworkin M, Falkow S, Rosenberg E, Schleifer KH, Stackebrandt E (eds) The Prokaryotes. A handbook on the biology of bacteria, vol 5, 3rd edn. Springer, New York, pp 593–601

Imhoff JF (2006c) The Chromatiaceae. In: Dworkin M, Falkow S, Rosenberg E, Schleifer KH, Stackebrandt E (eds) The Prokaryotes. A handbook on the biology of bacteria, vol 6, 3rd edn. Springer, New York, pp 846–873

Imhoff JF (2006d) The family Ectothiorhodospiraceae. In: Dworkin M, Falkow S, Rosenberg E, Schleifer KH, Stackebrandt E (eds) The Prokaryotes. A handbook on the biology of bacteria, vol 6, 3rd edn. Springer, New York, pp 874–886

Imhoff JF (2016) New dimensions in microbial ecology - functional genes in studies to unravel the biodiversity and role of functional microbial groups in the environment. Microorganisms 4, article 19

Imhoff JF, Caumette P (2004) Recommended standards for the description of new species of anoxygenic phototrophic bacteria. Int J Syst Evol Microbiol 54:1415–1421

Imhoff JF, Madigan M (2004) International Committee on Systematics of Prokaryotes. Subcommittee on the taxonomy of phototrophic bacteria. Minutes of the meetings 2003, Tokyo, Japan. Int J Syst Evol Microbiol 54:1000–1001

Imhoff JF, Pfennig N (2001) *Thioflavicoccus mobilis* gen. nov., sp. nov., a novel purple sulfur bacterium with bacteriochlorophyll *b*. Int J Syst Evol Microbiol 51:105–110

Imhoff JF, Süling J (1996) The phylogenetic relationship among Ectothiorhodospiraceae. A reevaluation of their taxonomy on the basis of 16S rDNA analysis. Arch Microbiol 165:106–113

Imhoff JF, Trüper HG (1977) *Ectothiorhodospira halochloris* sp. nov., a new extremely halophilic phototrophic bacterium containing bacteriochlorophyll b. Arch Microbiol 114:115–121

Imhoff JF, Trüper HG (1980) Chromatium purpuratum sp. nov., a new species of the Chromatiaceae. Zbl Bakt I. Abt Orig C1:61–69

Imhoff JF, Trüper HG (1981) Ectothiorhodospira abdelmalekii sp. nov., a new halophilic and alkaliphilic phototrophic bacterium. Zbl. Bakt., I. Abt. Orig. C2:228–234

Imhoff JF, Tindall B, Grant WD, Trüper HG (1981) *Ectothiorhodospira vacuolata* sp. nov., a new phototrophic bacterium from soda lakes. Arch Microbiol 130:238–242

Imhoff JF, Trüper HG, Pfennig N (1984) Rearrangement of the species and genera of the phototrophic "Purple Nonsulfur Bacteria". Int J Syst Bacteriol 34:340–343

Imhoff JF, Petri R, Süling J (1998a) Reclassification of species of the spiral-shaped phototrophic purple nonsulfur bacteria of the alpha-proteobacteria: description of the new genera *Phaeospirillum* gen. nov., *Rhodovibrio* gen. nov., *Rhodothalassium* gen. nov. and *Roseospira* gen. nov. as well as transfer of *Rhodospirillum fulvum* to *Phaeospirillum fulvum* comb. nov., of *Rhodospirillum molischianum* to *Phaeospirillum molischianum* comb. nov., of *Rhodospirillum salinarum* to *Rhodovibrio salinarum* comb. nov., of *Rhodospirillum sodomense* to *Rhodovibrio sodomensis* comb. nov., of *Rhodospirillum salexigens* to *Rhodothalassium salexigens* comb. nov., and of *Rhodospirillum mediosalinum* to *Roseospira mediosalina* comb. nov. Int J Syst Bacteriol 48:793–798

Imhoff JF, Süling J, Petri R (1998b) Phylogenetic relationships among the *Chromatiaceae*, their taxonomic reclassification and description of the new genera *Allochromatium, Halochromatium, Isochromatium, Marichromatium, Thiococcus, Thiohalocapsa* and *Thermochromatium*. Int J Syst Bacteriol 48:1129–1143

Imhoff JF, Hiraishi A, Süling J (2005) Anoxygenic phototrophic purple bacteria. In: Bergey's manual of systematic bacteriology, 2nd edn, vol 2, Part A, pp 119–132

Janssen PH, Harfoot CG (1991) *Rhodopseudomonas rosea* sp. nov., a new purple nonsulfur bacterium. Int J Syst Bacteriol 41:26–30

Kämpf C, Pfennig N (1980) Capacity of Chromatiaceae for chemotrophic growth. Specific respiration rates of *Thiocystis violacea* and *Chromatium vinosum*. Arch Microbiol 127:125–135

Kämpf C, Pfennig N (1986) Isolation and characterization of some chemoautotrophic Chromatiaceae. J Basic Microbiol 9:507–515

Karr EA, Sattley WM, Jung DO, Madigan MT, Achenbach LA (2003) Remarkable diversity of phototrophic purple bacteria in a permanently frozen Antarctic lake. Appl Environ Microbiol 69:4910–4914

Kawasaki H, Hoshino Y, Kuraishi H, Yamasato K (1992) *Rhodocista centenaria* gen. nov., sp. nov., a cyst-forming anoxygenic photosynthetic bacterium and its phylogenetic position in the Proteobacteria alpha group. J Gen Appl Microbiol 38:541–551

Kawasaki H, Hoshino Y, Hirata A, Yamasato K (1993) Is intracytoplasmic membrane structure a generic criterion? It does not coincide with phylogenetic interrelationships among phototrophic purple nonsulfur bacteria. Arch Microbiol 160:358–362

Keppen OI, Gorlenko VM (1975) A new species of purple budding bacteria containing bacteriochlorophyll b. Microbiol (Moscow) 44:258–264

Kompantseva EJ (1985) *Rhodobacter euryhalinus* sp. nov., a new halophilic purple bacterial species. Mikrobiologiia 54:974–982 (in Russian)

Kompantseva EJ (1989) A new species of budding purple bacterium: *Rhodopseudomonas julia* sp. nov. Mikrobiologiia 58:254–259

Kompantseva EJ, Gorlenko VM (1984) A new species of moderately halophilic purple bacterium *Rhodospirillum mediosalinum* sp. nov. Mikrobiologiia 53:775–781

Kompantseva EI, Komova AV, Kostrikina NA (2010) *Rhodovulum steppense* sp. nov., an obligatory haloalkaliphilic purple nonsulfur bacterium widespread in saline soda lakes of Central Asia. Int J Syst Evol Microbiol 60:1210–1214

Kompantseva EI, Komova AV, Novikov AA, Kostrikina NA (2012) *Rhodovulum tesquicola* sp. nov., a haloalkaliphilic purple non-sulfur bacterium from brackish steppe soda lakes. Int J Syst Evol Microbiol 62:2962–2966

Kulichevskaya IS, Guzev VS, Gorlenko VM, Liesack W, Dedysh SN (2006) *Rhodoblastus sphagnicola* sp. nov., a novel acidophilic purple non-sulfur bacterium from Sphagnum peat bog. Int J Syst Evol Microbiol 56:1397–1402

Kumar PA, Jyothsna TS, Srinivas TNR, Sasikala C, Ramana CV, Imhoff JF (2007a) *Marichromatium bheemlicum* sp. nov., a non-diazotrophic, photosynthetic gammaproteobacterium from a marine aquaculture pond. Int J Syst Evol Microbiol 57:1261–1265

Kumar PA, Sasi Jyothsna TS, Srinivas TNR, Sasikala C, Ramana CV, Imhoff JF (2007b) Two novel species of marine phototrophic Gammaproteobacteria: *Thiorhodococcus bheemlicus* sp. nov. and *Thiorhodococcus kakinadensis* sp. nov. Int J Syst Evol Microbiol 57:2458–2461

Kumar PA, Srinivas TNR, Sasikala C, Ramana CV (2007c) *Halochromatium roseum* sp. nov., a non-motile phototrophic gammaproteobacterium with gas vesicles, and emended description of the genus *Halochromatium*. Int J Syst Evol Microbiol 57:2110–2113

Kumar PA, Srinivas TNR, Sasikala C, Ramana CV (2007d) *Rhodobacter changlensis* sp. nov., a psychrotolerant, phototrophic alphaproteobacterium from the Himalaya of India. Int J Syst Evol Microbiol 57:2568–2571

Kumar PA, Aparna P, Srinivas TNR, Sasikala C, Ramana CV (2008a) *Rhodovulum kholense* sp. nov. Int J Syst Evol Microbiol 58:1723–1726

Kumar PA, Aparna P, Srinivas TNR, Sasikala C, Ramana CV (2008b) *Rhodospirillum sulfurexigens* sp. nov., a phototrophic alphaproteobacterium requiring a reduced sulfur source for growth. Int J Syst Evol Microbiol 58:2917–2920

Kumar PA, Srinivas TNR, Sasikala C, Ramana CV (2008c) *Allochromatium renukae* sp. nov. Int J Syst Evol Microbiol 58:404–407

Kumar PA, Srinivas TNR, Sasikala C, Ramana CV, Imhoff JF (2008d) *Thiophaeococcus mangrovi* gen. nov., sp. nov., a photosynthetic, marine gammaproteobacterium isolated from the Bhitarkanika mangrove forest of India. Int J Syst Evol Microbiol 58:2660–2664

Kumar PA, Srinivas TNR, Takaichi S, Maoka T, Sasikala C, Ramana CV (2009a) *Phaeospirillum chandramohanii* sp. nov., a phototrophic alphaproteobacterium with carotenoid glycosides. Int J Syst Evol Microbiol 59:2089–2093

Kumar PA, Srinivas TNR, Thiel V, Tank V, Sasikala C, Ramana CV, Imhoff JF (2009b) *Thiohalocapsa marina* sp. nov., from an Indian marine aquaculture pond. Int J Syst Evol Microbiol 59:2333–2338

Kumar BV, Ramprasad EVV, Sasikala C, Ramana CV (2013) *Rhodopseudomonas pentothenatexigens* sp. nov. and *Rhodopseudomonas thermotolerans* sp. nov., isolated from paddy soils. Int J Syst Evol Microbiol 63:200–207

Lakshmi KV, Sasikala C, Ramana CV (2009) *Rhodoplanes pokkaliisoli* sp. nov., a phototrophic alphaproteobacterium isolated from a waterlogged brackish paddy soil. Int J Syst Evol Microbiol 59:2153–2157

Lakshmi KV, Sasikala C, Ramana VV, Ramaprasad EV, Ramana CV (2011a) *Rhodovulum phaeolacus* sp. nov. a phototrophic alphaproteobacterium isolated from a brown pond. J Gen Appl Microbiol 57:145–151

Lakshmi KV, Sasikala C, Ashok Kumar GV, Chandrasekaran R, Ramana CV (2011b) *Phaeovibrio sulfidiphilus* gen. nov., sp. nov., phototrophic alphaproteobacteria isolated from brackish water. Int J Syst Evol Microbiol 61:828–833

Lakshmi KV, Sasikala C, Takaichi S, Ramana CV (2011c) *Phaeospirillum oryzae* sp. nov., a spheroplast-forming, phototrophic alphaproteobacterium from a paddy soil. Int J Syst Evol Microbiol 61:1656–1661

Lakshmi KV, Sasikala C, Ramaprasad EV, Ramana CV (2013) *Rhodospirillum oryzae* sp. nov., a phototrophic bacterium isolated from rhizosphere soil of paddy. Int J Syst Evol Microbiol 63:3050–3055

Lakshmi KV, Divyasree B, Ramprasad EVV, Sasikala C, Ramana CV (2014) Reclassification of *Rhodospirillum photometricum* Molisch 1907, *Rhodospirillum sulfurexigens* Anil Kumar et al. 2008 and *Rhodospirillum oryzae* Lakshmi et al. 2013 in a new genus, *Pararhodospirillum photometricum* comb. nov., *Pararhodospirillum sulfurexigens* comb. nov. and *Pararhodospirillum oryzae* comb. nov., respectively, and emended description of the genus *Rhodospirillum*. Int J Syst Evol Microbiol 64:1154–1159

Lakshmi KV, Divyasree B, Sucharita K, Sasikala C, Ramana CV (2015) *Thiorhodococcus fuscus* sp. nov., isolated from a lagoon. Int J Syst Evol Microbiol 65:3938–3943

Mack EE, Mandelco L, Woese CR, Madigan MT (1993) *Rhodospirillum sodomense,* sp. nov., a Dead Sea *Rhodospirillum* species. Arch Microbiol 160:363–371

Madigan MT (1986) *Chromatium tepidum* sp. nov. a thermophilic photosynthetic bacterium of the family Chromatiaceae. IJSEM 36:222–227

Madigan MT (1988) Microbiology, physiology, and ecology of phototrophic bacteria. In: Zehnder AJB (ed) Biology of anaerobic microorganisms. Wiley-Liss, New York, pp 39–111

Madigan M, Imhoff JF (2007) International Committee on Systematics of Prokaryotes. Subcommittee on the taxonomy of phototrophic bacteria. Minutes of the meetings 2006, Pau, France. Int J Syst Evol Microbiol 57:1169–1171

Madigan MT, Jung DO, Woese CR, Achenbach LA (2000) *Rhodoferax antarcticus* sp. nov., a moderately psychrophilic purple nonsulfur bacterium isolated from an Antarctic microbial mat. Arch Microbiol 173:269–277

Migula W (1900) Das System der Bakterien, vol 2. Gustav Fischer, Jena

Milford AD, Achenbach LA, Jung DO, Madigan MT (2000) *Rhodobaca bogoriensis* gen. nov. and sp. nov., an alkaliphilic purple nonsulfur bacterium from African Rift Valley soda lakes. Arch Microbiol 174:18–27

Molisch H (1907) Die Purpurbakterien nach neuen Untersuchungen. G. Fischer, Jena

Nagashima KV, Hiraishi A, Shimada K, Matsuura K (1997) Horizontal transfer of genes coding for the photosynthetic reaction centers of purple bacteria. J Mol Evol 45:131–136

Neutzling O, Imhoff JF, Trüper HG (1984) *Rhodopseudomonas adriatica* sp. nov., a new species of the Rhodospirillaceae, dependent on reduced sulfur compounds. Arch Microbiol 137:256–261

Nicholson JAM, Stolz JF, Pierson BK (1987) Structure of a microbial mat at Great Sippewissett Marsh, Cape Cod, Massachusetts. FEMS Microbiol Ecol 45:343–364

Nissen H, Dundas ID (1984) *Rhodospirillum salinarum* sp. nov., a halophilic photosynthetic bacterium isolated from a Portuguese saltern. Arch Microbiol 138:251–256

Nupur P, Srinivas TNR, Takaichi S, Kumar PA (2014) *Rhodovulum mangrovi* sp. nov., a phototrophic alphaproteobacterium isolated from a mangrove forest sediment sample. Int J Syst Evol Microbiol 64:3168–3173

Nupur P, Srinivas NGT, Takaishi S, Kumar PA (2015) *Phaeobacterium nitratireducens* gen. nov., sp. nov., a phototrophic gammaproteobacterium isolated from a sediment sample of Coringa mangrove forest, India. Int J Syst Evol Microbiol 65:2357–2364

Okamura K, Hisada T, Hiraishi A (2007) Characterization of thermotolerant purple nonsulfur bacteria isolated from hot-spring *Chloroflexus* mats and the reclassification of "*Rhodopseudomonas cryptolactis*" Stadtwald-Demchick et al. 1990 as *Rhodoplanes cryptolactis* nom.rev., comb. nov. J Gen Appl Microbiol 53:357–361

Okamura K, Hisada T, Kanbe T, Hiraishi A (2009a) *Rhodovastum atsumiense* gen. nov., sp. nov., a phototrophic alphaproteobacterium isolated from paddy soil. J Gen Appl Microbiol 55:43–50

Okamura K, Kanbe T, Hiraishi A (2009b) *Rhodoplanes serenus* sp. nov., a purple non-sulfur bacterium isolated from pond water. Int J Syst Evol Microbiol 59:531–535

Oren A, Kessel M, Stackebrandt E (1989) *Ectothiorhodospira marismortui* sp. nov., an obligately anaerobic, moderately halophilic purple sulfur bacterium from a hypersaline sulfur spring at the shore of the Dead Sea. Arch Microbiol 151:524–529

Overmann J, Fischer U, Pfennig N (1992) A new purple sulfur bacterium from saline littoral sediments, *Thiorhodovibrio winogradskyi* gen. nov. and sp. nov. Arch Microbiol 157:329–335

Peduzzi S, Welsh A, Demarta A, Decristophoris P, Peduzzi R, Hahn D, Tonolla M (2011) *Thiocystis chemoclinalis* sp. nov. and *Thiocystis cadagnonensis* sp. nov., motile purple sulfur bacteria isolated from the chemocline of a meromictic lake. Int J Syst Evol Microbiol 61:1682–1687

Perty M (1852) Zur Kenntnis kleinster Lebensformen. Jent and Reinert, Bernpp 1–228

Pfennig N (1965) Anreicherungskulturen für rote und grüne Schwefelbakterien. Zentralbl Bakteriol Parasitenkd Infektionskrankh. Hyg Abt 1(Orig Suppl. 1): 179–189, 503–505
Pfennig N (1967) Photosynthetic bacteria. Ann Rev Microbiol 21:285–324
Pfennig N (1969a) *Rhodopseudomonas acidophila*, sp. n., a new species of the budding purple nonsulfur bacteria. J Bacteriol 99:597–602
Pfennig N (1969b) *Rhodospirillum tenue* sp. n., a new species of the purple nonsulfur bacteria. J Bacteriol 99:619–620
Pfennig N (1974) *Rhodopseudomonas globiformis*, sp. n., a new species of the Rhodospirillaceae. Arch Microbiol 100:197–206
Pfennig N (1977) Phototrophic green and purple bacteria: a comparative systematic survey. Ann Rev Microbiol 31:275–290
Pfennig N (1978) *Rhodocyclus purpureus* gen. nov. and sp. nov., a ring-shaped, vitamin B_{12}-requiring member of the family Rhodospirillaceae. Int J Syst Bacteriol 28:283–288
Pfennig N (1989) Ecology of purple and green sulfur bacteria. In: Schlegel HG, Bowien B (eds) Autotrophic bacteria. Springer, New York, pp 97–116
Pfennig N, Lippert KD (1966) Über das Vitamin B_{12}-Bedürfnis phototropher Schwefelbakterien. Arch Mikrobiol 55:245–256
Pfennig N, Trüper HG (1969) Phototrophic bacteria. GSF-Bericht M32, Gesellschaft für Strahlenforschung mbH, München
Pfennig N, Trüper HG (1971) Type and neotype strains of the species of phototrophic bacteria maintained in pure culture. Int J Syst Bacteriol 21:19–24
Pfennig N, Trüper HG (1974) The phototrophic bacteria. In: Buchanan RE, Gibbons NE (eds) Bergey's manual of determinative bacteriology, 8th edn. The Williams & Wilkins Co., Baltimore, pp 24–75
Pfennig N, Trüper HG (1992) The family Chromatiaceae. In: Balows A, Trüper HG, Dworkin M, Harder W, Schleifer KH (eds) The Prokaryotes. A handbook on the biology of bacteria. Ecophysiology, isolation, identification, applications, 2nd edn. Springer, New York, pp 3200–3221
Pfennig N, Lünsdorf H, Süling J, Imhoff JF (1997) *Rhodospira trueperi*, gen. nov. and spec. nov., a new phototrophic Proteobacterium of the alpha group. Arch Microbiol 168:39–45
Puchkova NN, Imhoff JF, Gorlenko VM (2000) *Thiocapsa litoralis* sp. nov., a new purple sulfur bacterium from microbial mats from the White Sea. Int J Syst Evol Microbiol 50:1441–1447
Rabold S, Gorlenko VM, Imhoff JF (2006) *Thiorhodococcus mannitoliphagus* sp. nov., a new purple sulfur bacterium from the White Sea. Int J Syst Evol Microbiol 56:1945–1951
Raj PS, Chakravarthy SK, Ramaprasad EV, Sasikala C, Ramana CV (2012) *Phaeospirillum tilakii* sp. nov., a phototrophic alphaproteobacterium isolated from aquatic sediments. Int J Syst Evol Microbiol 62:1069–1074
Raj PS, Ramaprasad EV, Vaseef S, Sasikala C, Ramana CV (2013) *Rhodobacter viridis* sp. nov., a phototrophic bacterium isolated from mud of a stream. Int J Syst Evol Microbiol 63:181–186
Ramana CV, Sasikala C, Arunrasi K, Kumar PA, Srinivas TNR, Shivaji S, Süling J, Imhoff JF (2006) *Rubrivivax benzoatilyticus* sp. nov., an aromatic hydrocarbon-degrading purple betaproteobacterium. Int J Syst Evol Microbiol 56:2157–2164
Ramana VV, Sasikala C, Ramana CV (2008) *Rhodobacter maris* sp. nov., a phototrophic alphaproteobacterium isolated from a marine habitat of India. Int J Syst Evol Microbiol 58:1719–1722
Ramana VV, Kumar PA, Srinivas TNR, Sasikala C, Ramana CV (2009) *Rhodobacter aestuarii* sp. nov., a phototrophic alphaproteobacterium isolated from an estuarine environment. Int J Syst Evol Microbiol 59:1133–1136
Ramana VV, Sasikala C, Ramaprasad EVV, Ramana CV (2010) Description of *Ectothiorhodospira salini* sp. nov. J Gen Appl Microbiol 56:313–319
Ramana VV, Kapoor S, Shobha E, Ramprasad EV, Sasikala C, Ramana CV (2011) *Blastochloris gulmargensis* sp. nov., isolated from an epilithic phototrophic biofilm. Int J Syst Evol Microbiol 61:1811–1816

Ramana VV, Chakravarthy SK, Raj PS, Kumar BV, Shobha E, Ramaprasad EVV, Sasikala C, Ramana CV (2012) Descriptions of *Rhodopseudomonas parapalustris* sp. nov., *Rhodopseudomonas harwoodiae* sp. nov. and *Rhodopseudomonas pseudopalustris* sp. nov., and emended description of *Rhodopseudomonas palustris*. Int J Syst Evol Microbiol 62:1790–1798

Ramana VV, Raj PS, Tushar L, Sasikala C, Ramana CV (2013) *Rhodomicrobium udaipurense* sp. nov., a psychrotolerant, phototrophic alphaproteobacterium isolated from a freshwater stream. Int J Syst Evol Microbiol 63:2684–2689

Raymond JC, Sistrom WR (1969) *Ectothiorhodospira halophila*: a new species of the genus Ectothiorhodospira. Arch Mikrobiol 69:121–126

Rees GN, Harfoot CG, Janssen PH, Schoenborn L, Kuever J, Lünsdorf H (2002) *Thiobaca trueperi* gen. nov., sp. nov., a phototrophic purple sulfur bacterium isolated from freshwater lake sediment. Int J Syst Evol Microbiol 52:671–678

Schanz F, Fischer-Romero C, Bachofen R (1998) Photosynthetic production and photoadaptation of phototrophic sulfur bacteria in Lake Cadagno (Switzerland). Limnol Oceanogr 43:1262–1269

Schaub BEM, Van Gemerden H (1993) Simultaneous phototrophic and chemotrophic growth in the purple sulfur bacterium *Thiocapsa roseopersicina*. FEMS Microbiol Ecol 13:185–196

Schroeter J (1886) In Cohn (1889) Kryptogamen-Flora von Schlesien 3:1–814

Serrano W, Amann R, Fischer U (2009) A new moderately thermophilic and high sulfide tolerant biotype of *Marichromatium gracile*, isolated from tidal sediments of the German Wadden Sea: *Marichromatium gracile* biotype thermosulfidiphilum. Syst Appl Microbiol 32:1–7

Serrano W, Amann R, Rossello-Mora R, Herbert RA, Fischer U (2011) The genus *Allochromatium* (Chromatiales Chromatiaceae revisited: A study on its intragenic structure based on multilocus sequence analysis (MLSA) and DNA-DNA hybridisation (DDH). Syst Appl Microbiol 34:590–594

Serrano W, Schrübbers J, Amann R, Fischer U (2015) *Allochromatium humboldtianum* sp. nov., isolated from soft coastal sediments. Int J Syst Evol Microbiol 65:2980–2985

Shivali K, Ramana VV, Ramaprasad EV, Sasikala C, Ramana CV (2011) *Marichromatium litoris* sp. nov. and *Marichromatium chrysaorae* sp. nov. isolated from beach sand and from a jelly fish (Chrysaora colorata). Syst Appl Microbiol 34:600–605

Shivali K, Sasikala C, Ramana CV (2012) MLSA barcoding of *Marichromatium* spp. and reclassification of *Marichromatium fluminis* (Sucharita et al., 2010) as *Phaeochromatium fluminis* gen. nov. comb. nov. Syst Appl Microbiol 35:221–225

Srinivas TNR, Kumar PA, Sasikala C, Ramana CV, Süling J, Imhoff JF (2006) *Rhodovulum marinum* sp. nov., a novel phototrophic purple non-sulfur alphaproteobacterium from marine tides of Visakhapatnam, India. Int J Syst Evol Microbiol 56:1651–1656

Srinivas TNR, Kumar PA, Sasikala C, Ramana CV (2007a) *Rhodovulum imhoffii* sp. nov. Int J Syst Evol Microbiol 57:228–232

Srinivas TNR, Kumar PA, Sasikala C, Ramana CV, Imhoff JF (2007b) *Rhodobacter vinaykumarii* sp. nov., a marine phototrophic alphaproteobacterium from tidal waters, and emended description of the genus Rhodobacter. Int J Syst Evol Microbiol 57:1984–1987

Srinivas TNR, Kumar PA, Sasikala C, Ramana CV, Imhoff JF (2007c) *Rhodovulum visakhapatnamense* sp. nov. Int J Syst Evol Microbiol 57:1762–1764

Srinivas TNR, Kumar PA, Sasikala C, Ramana CV, Imhoff JF (2007d) *Rhodobium gokarnense* sp. nov., a novel phototrophic alphaproteobacterium from a saltern. Int J Syst Evol Microbiol 57:923–935

Srinivas TNR, Kumar PA, Sasikala C, Spröer C, Ramana CV (2008) *Rhodobacter ovatus* sp. nov., a phototrophic alphaproteobacterium isolated from a polluted pond. Int J Syst Evol Microbiol 58:1379–1383

Srinivas TNR, Kumar PA, Sucharita K, Sasikala C, Ramana CV (2009) *Allochromatium phaeobacterium* sp. nov. Int J Syst Evol Microbiol 59:750–753

Srinivas A, Rahul K, Ramaprasad EV, Sasikala C, Ramana CV (2012) *Rhodovulum bhavnagarense* sp. nov., a phototrophic alphaproteobacterium isolated from a pink pond. Int J Syst Evol Microbiol 62:2528–2532

Srinivas A, Kumar BV, Divyasree B, Tushar L, Sasikala C, Ramana CV (2014a) *Rhodovulum salis* sp. nov. and *Rhodovulum viride* sp. nov., phototrophic Alphaproteobacteria isolated from marine habitats. Int J Syst Evol Microbiol 64:957–962

Srinivas A, Sasikala C, Ramana CV (2014b) *Rhodoplanes oryzae* sp. nov., a phototrophic alphaproteobacterium isolated from the rhizosphere soil of paddy. Int J Syst Evol Microbiol 64:2198–2203

Stackebrandt E, Ebers J (2006) Taxonomic parameters revisited: tarnished gold standards. Microbiol Today 8:6–9

Stadtwald-Demchick R, Turner FR, Gest H (1990) *Rhodopseudomonas cryptolactis* sp. nov., a new thermotolerant species of budding phototrophic purple bacteria. FEMS Microbiol Lett 71:117–122

Straub KL, Rainey FA, Widdel F (1999) *Rhodovulum iodosum* sp. nov. and *Rhodovulum robiginosum* sp. nov., two new marine phototrophic ferrous-iron-oxidising purple bacteria. Int J Syst Bacteriol 49:729–735

Strzeszewski B (1913) Beiträge zur Kenntnis der Schwefelflora in der Umgebung von Krakau. Bull Acad Sci Cracovia Ser B 1913:309–334

Sucharita K, Sasikala C, Ramana CV (2010a) *Thiorhodococcus modestalkaliphilus* sp. nov. a phototrophic gammaproteobacterium from Chilika salt water lagoon, India. J Gen Appl Microbiol 56:93–99

Sucharita K, Shiva Kumar E, Sasikala C, Panda BB, Takaichi S, Ramana CV (2010b) *Marichromatium fluminis* sp. nov., a slightly alkaliphilic, phototrophic gammaproteobacterium isolated from river sediment. Int J Syst Evol Microbiol 60:1103–1107

Suresh G, Sasikala C, Ramana CV (2015) Reclassification of *Gemmobacter changlensis* to a new genus as *Cereibacter changlensis* gen. nov., comb. nov. Int J Syst Evol Microbiol 65:794–798

Tank M, Thiel V, Imhoff JF (2009) Phylogenetic relationship of phototrophic purple sulfur bacteria according to *pufL* and *pufM* genes. Intern Microbiol 12:175–185

Tank M, Blümel M, Imhoff JF (2011) Communities of purple sulfur bacteria in a Baltic Sea coastal lagoon analyzed by *pufLM* gene libraries and the impact of temperature and NaCl concentration in experimental enrichment cultures. FEMS Microbiol Ecol 78:428–438

Thiel V, Tank M, Neulinger SC, Gehrmann L, Dorador C, Imhoff JF (2010) Unique communities of anoxygenic phototrophic bacteria in saline lakes of Salar de Atacama (Chile). Evidence for a new phylogenetic lineage of phototrophic Gammaproteobacteria from *pufLM* gene analyses. FEMS Microbiol Ecol 74:510–522

Tonolla M, Demarta A, Peduzzi R, Hahn D (1999) In situ analysis of phototrophic sulfur bacteria in the chemocline of meromictic Lake Cadagno (Switzerland). Appl Environ Microbiol 65:1325–1330

Tonolla M, Peduzzi R, Hahn D (2005) Long-term population dynamics of phototrophic sulfur bacteria in the chemocline of Lake Cadagno, Switzerland. Appl Environ Microbiol 71:3544–3550

Trüper HG (1968) *Ectothiorhodospira mobilis* Pelsh, a photosynthetic sulfur bacterium depositing sulfur outside the cells. J Bacteriol 95:1910–1920

Trüper HG (1970) Culture and isolation of phototrophic sulfur bacteria from the marine environment. Helgol wiss Meeresunters 20:6–16

Trüper HG, Jannasch HW (1968) *Chromatium buderi* nov. spec., eine neue Art der grossen Thiorhodaceae. Arch Mikrobiol 61:363–372

Urdiain M, López-López A, Gonzalo C, Busse HJ, Langer S, Kämpfer P, Rosselló-Móra R (2008) Reclassification of *Rhodobium marinum* and *Rhodobium pfennigii* as *Afifella marina* gen. nov. comb. nov. and *Afifella pfennigii* comb. nov., a new genus of photoheterotrophic Alphaproteobacteria and emended descriptions of *Rhodobium*, *Rhodobium orientis* and *Rhodobium gokarnense*. Syst Appl Microbiol 31:339–351

Van Gemerden H, Mas J (1995) Ecology of phototrophic sulfur bacteria. In: Blankenship RE, Madigan MT, Bauer CE (eds) Anoxygenic photosynthetic bacteria. Kluwer Academic, Dordrecht, pp 49–85

Van Niel CB (1931) On the morphology and physiology of the purple and green sulfur bacteria. Arch Mikrobiol 3:1–112

Van Niel CB (1944) The culture, general physiology, morphology and classification of the nonsulfur purple and brown bacteria. Bacteriol Rev 8:1–118

Willems A, Gillis M, de Ley J (1991) Transfer of *Rhodocyclus gelatinosus* to *Rubrivivax gelatinosus* gen. nov., comb nov., and phylogenetic relationships with *Leptothrix, Sphaerotilus natans, Pseudomonas saccharophila*, and *Alcaligenes latus*. Int J Syst Bacteriol 41:65–73

Winogradsky S (1888) Beiträge zur Morphology und Physiologie der Bacterien. Heft I – Zur Morphologie und Physiologie der Schwefelbacterien. Arthur Felix, Leipzig, pp 1–120

Woese CR (1987) Bacterial evolution. Microbiol Rev 51:221–271

Woese CR, Stackebrandt E, Weisburg WG, Paster BJ, Madigan MT, Fowler VJ, Hahn CM, Blanz P, Gupta R, Nealson KH, Fox GE (1984a) The phylogeny of purple bacteria: the alpha subdivision. Syst Appl Microbiol 5:315–326

Woese CR, Weisburg WG, Paster BJ, Hahn CM, Tanner RS, Krieg NR, Koops H-P, Harms H, Stackebrandt E (1984b) The phylogeny of purple bacteria: the beta subdivision. Syst Appl Microbiol 5:327–336

Woese CR, Weisburg WG, Hahn CM, Paster BJ, Zablen LB, Lewis BJ, Macke TJ, Ludwig W, Stackebrandt E (1985) The phylogeny of purple bacteria: the gamma subdivision. Syst Appl Microbiol 6:25–33

Yutin N, Suzuki MT, Rosenberg M, Rotem D, Madigan MT, Süling J, Imhoff JF, Beja O (2009) BchY-based degenerate primers target all types of anoxygenic photosynthetic bacteria in a single PCR. Appl Environ Microbiol 75:7556–7559

Zaar A, Fuchs G, Golecki JR, Overmann J (2003) A new purple sulfur bacterium isolated from a littoral microbial mat, *Thiorhodococcus drewsii* sp. nov. Arch Microbiol 179:174–183

Zhang D, Yang H, Huang Z, Zhang W, Liu S-J (2002) *Rhodopseudomonas faecalis* sp. nov., a phototrophic bacterium isolated from an anaerobic reactor that digests chicken faeces. Int J Syst Evol Microbiol 52:2055–2060

Zhang D, Yang H, Zhang W, Huang Z, Liu S-J (2003) *Rhodocista pekingensis* sp. nov., a cyst-forming phototrophic bacterium from a municipal wastewater treatment plant. Int J Syst Evol Microbiol 53:1111–1114

A Panoply of Phototrophs: An Overview of the Thermophilic Chlorophototrophs of the Microbial Mats of Alkaline Siliceous Hot Springs in Yellowstone National Park, WY, USA

Marcus Tank, Vera Thiel, David M. Ward, and Donald A. Bryant

Abstract Chlorophototrophs are organisms that can synthesize chlorophylls or bacteriochlorophylls, and they use these molecules to harvest and convert light energy into stored chemical potential energy. Some of these organisms also perform photosynthesis, in which light provides the energy (ATP) and reducing power (NAD(P)H or reduced ferredoxin) required for inorganic carbon (C_i) fixation. Over the past decade, we have studied the chlorophototrophs found in two alkaline siliceous hot springs in Yellowstone National Park, WY, USA. The microbial mats that occur at temperatures of 40–73 °C in Mushroom and Octopus Springs have proven to contain a surprisingly diverse array of chlorophototrophs. These include members of six of the seven bacterial phyla known to have members capable of synthesizing (bacterio)-chlorophylls: *Acidobacteria*, *Cyanobacteria*, *Chlorobi*, *Chloroflexi*, *Firmicutes*, and *Proteobacteria*. More than 16 chlorophototrophs have now been associated with these microbial mats, and this does not include the many ecotypes of these organisms that occur within these communities. In this chapter we

M. Tank • V. Thiel
Department of Biological Sciences, Tokyo Metropolitan University,
Hachioji, Tokyo 192-0397, Japan

Department of Biochemistry and Molecular Biology, The Pennsylvania State University,
University Park, PA 16802, USA

D.M. Ward
Department of Land Resources and Environmental Sciences, Montana State University,
Bozeman, MT 59717, USA

D.A. Bryant (✉)
Department of Biochemistry and Molecular Biology, The Pennsylvania State University,
University Park, PA 16802, USA

Department of Chemistry and Biochemistry, Montana State University,
Bozeman, MT 59717, USA
e-mail: dab14@psu.edu

© Springer International Publishing Switzerland 2017
P.C. Hallenbeck (ed.), *Modern Topics in the Phototrophic Prokaryotes*,
DOI 10.1007/978-3-319-46261-5_3

will briefly describe the panoply of phototrophic organisms that occur in these mat communities and will provide an introduction to their morphological appearance and other basic properties. Metagenomic analyses have revealed several novel organisms, e.g., *Chloracidobacterium thermophilum*, "*Candidatus* Thermochlorobacter aerophilum," "*Candidatus* Chloranaerofilum corporosum," "*Candidatus* Roseovibrio tepidum," and "*Candidatus* Roseilinea gracile," which were hitherto unknown to microbiologists because they escaped isolation by classical, culture-based methods. However, by combining molecular methods, in situ physiological observations, metabolic reconstruction, and enrichment techniques, we are now making remarkable progress toward the isolation of these chlorophototrophic organisms.

Introduction

Phototrophs are organisms that use light as their energy source. Two types of phototrophs are known: retinalophototrophs and chlorophototrophs (Bryant and Frigaard 2006). Retinalophototrophs employ retinal-binding proteins, including bacteriorhodopsin, proteorhodopsin, halorhodopsin, and/or xanthorhodopsin, to convert light energy into stored chemical potential energy in the form of proton, sodium, or chloride gradients (Kandori 2015). Retinalophototrophs are widespread in nature, especially in marine environments, where blue-green light is readily available and penetrates to significant depths (>200 m). Retinal-binding proteins characteristically have seven transmembrane α-helices and a specific lysine residue that covalently binds the retinal moiety in a Schiff base, and they employ a simple, isomerization-based photocycle mechanism for energy storage (Lanyi 2006). If an organism can synthesize γ- or β-carotene, then energy acquisition by retinalophototrophy only requires the acquisition of two genes: *bop*, encoding the retinal-binding, bacterial opsin; and *blh*, encoding the 15, 15′-β-carotene dioxygenase that produces the retinal chromophore by oxidative cleavage of β-carotene. Light-induced *trans*-to-*cis* isomerization of the retinal chromophore, and its subsequent dark reversion to the resting all-*trans* configuration, establish the photocycle that transfers one ion per photon across the cytoplasmic membrane (Lanyi 2006). Although proton-pumping retinal proteins are the most common type, recent studies have shown that sodium-pumping retinal proteins often occur in marine bacteria (Inoue et al. 2013).

Although retinalophototrophy is a very simple and efficient system for light energy capture and conversion, it presents two major disadvantages. Firstly, the single retinal chromophore has a rather small molar extinction coefficient, and thus these proteins function most efficiently at high irradiance values (xanthorhodopsins use a second carotenoid chromophore to increase the absorbance cross section for light harvesting; Balashov et al. 2005; Balashov and Lanyi 2007). Secondly, retinalophototrophs are only able to produce ion gradients and thus are unable to produce reductants for C_i fixation. No example of an autotrophic organism that couples retinalophototrophy with lithotrophic oxidation of a compound such as sulfide has yet been described. In spite of these limitations, this mechanism of phototrophy is very widespread as a result of horizontal gene transfer (Frigaard et al. 2006; Ugalde et al. 2011; Martinez-Garcia et al. 2011). Physiological studies have established that proteorhodopsins in marine flavobacteria provide an energetic advantage to cells under carbon limitation conditions (Gómez-Consarnau et al. 2007; Steindler et al. 2011). These data as well as others demonstrate that even relatively low levels of retinal proteins can provide an energetic benefit to cells, which in turn would select for organisms that obtain these genes by horizontal gene transfer.

Chlorophototrophs differ fundamentally from retinalophototrophs in essentially all respects. Chlorophototrophs synthesize chlorophylls (Chls), which they use for light energy capture, and they use light-driven redox reactions, i.e., photochemistry, to produce protonmotive force for ATP production and strong, stable reductants for C_i fixation. The chemical diversity of Chls and bacteriochlorophylls (BChls), which are usually embedded in special protein structures, often together with accessory pigments, allows chlorophototrophs collectively to use a wide range of light wavelengths, spanning the range from about 350 to 1100 nm. However, because Chls are complex tetrapyrroles, the number of genes required for Chl biosynthesis, Chl binding, and related processes is large. Thus, the ability to perform chlorophototrophy is rarely transferred horizontally but probably has occurred (see Zeng et al. 2014). At present, 14 major classes of Chls and BChls can be produced by bacteria: Chl *a*, Chl *b*, Chl *d*, Chl *f*, divinyl-Chl *a*, divinyl-Chl *b*, 8-OH-Chl *a* and BChls *a*, BChl *b*, BChl *c*, BChl *d*, BChl *e*, BChl *f*, and BChl *g* (Gomez Maqueo Chew and Bryant 2007; Chen 2014). Additional possibilities exist if one considers that different esterifying alcohols can occur at position C-17 (e.g., phytol, 2,6-phytadienol, geranylgeraniol, 1-octadecanol, 1-hexadecanol, etc.); that Mg can sometimes be replaced by Zn (Jaschke et al. 2011; Tsukatani et al. 2012; Wakao et al. 1996); and that stereoisomers of some Chls and BChls exist (e.g., Chl *a* and Chl *a′*; BChl *a* and BChl *a′*; BChl *g* and BChl *g′*) (Ohashi et al. 2010).

The invention of chlorophototrophy was arguably one of the most important events in biological evolution, because it allowed an inexhaustible source of energy, the sun, to be directly coupled to metabolic energy and reducing power production in bacteria. Chl- and BChl-binding proteins can bind many Chls and can form complex structures with many light-absorbing chromophores, so chlorophototrophy can proceed efficiently even when irradiance is extremely low. Although nature has evolved many different Chls and BChls, and has evolved many different chromophores and light-harvesting antenna systems for light capture, the core photochemical proteins, known as "reaction centers," evolved once, but gene duplication and divergence led to the formation of a few distinctive types of reaction centers (Golbeck 1993; Sadekar et al. 2006; Hohmann-Marriott and Blankenship 2011). Homodimeric type-1 reaction centers, found in members of the *Chlorobi*, *Acidobacteria*, and *Firmicutes*, produce strong reductants and weak oxidants, while heterodimeric type-2 reaction centers, found in *Proteobacteria*, *Chloroflexi*, and *Gemmatimonadetes*, produce weak reductants and strong oxidants. The much more complex, heterodimeric Photosystem (PS) I and PS II reaction centers found in members of the *Cyanobacteria* are divergent variants of the type-1 and type-2 bacterial reaction centers. PS I and PS II can act in series to oxidize water and produce oxygen as a waste product while producing strong reductants and ATP for C_i fixation. The major advantages of chlorophototrophy are that this process provides enormous flexibility and functionality under highly diverse light and redox conditions. In contrast, retinalophototrophy provides exquisite simplicity for the capture and conversion of visible (typically blue or green) light primarily into protonmotive force.

Although current estimates suggest that there are about 100 bacterial phyla, over the past 200 years microbiologists have only identified seven phyla that contain members capable of chlorophototrophy: in order of discovery, *Cyanobacteria*, *Proteobacteria*, *Chlorobi*, *Chloroflexi*, *Firmicutes*, *Acidobacteria*, and *Gemmatimonadetes* (Bryant and Frigaard 2006; Bryant et al. 2007; Zeng et al. 2014). The discovery and characterization of these various types of chlorophototrophs has contributed greatly to our knowledge of the evolution of photosynthesis, the process that combines chlorophototrophy with autotrophic CO_2 fixation. Note that not all chlorophototrophs perform photosynthesis, because some chlorophototrophs (e.g., *Chloracidobacterium* (*Cab.*) *thermophilum*, *Heliobacterium modesticaldum*) can only grow photoheterotrophically.

This chapter will describe the chlorophototrophs that occur in the microbial mats in Mushroom Spring (MS) and Octopus Spring (OS) (also called Pool A in some early studies). These thermal features are chemically similar, alkaline siliceous hot springs found in the Lower Geyser Basin of Yellowstone National Park (YNP). A surprising diversity of chlorophototrophs, representing six of the seven phyla known to contain chlorophototrophs (the exception is *Gemmatimonadetes*) is found in these microbial mats. Two organisms isolated from these mats, *Chloroflexus aurantiacus* (Pierson and Castenholz 1971, 1974) and *Cab. thermophilum* (Bryant et al. 2007; Tank and Bryant 2015a, b), were the first cultivated examples of chlorophototrophs from their respective phyla, *Chloroflexi* and *Acidobacteria*. Several other highly unusual chlorophototrophs, including "*Candidatus* Thermochlorobacter aerophilum," an *Anaerolineae*-like phototroph, ("*Candidatus* Roseilinea gracile"), and an *Oscillochloris*-like phototroph ("*Candidatus* Chloranaerofilum corporosum"), have also been described for the first time from these mat communities. This chapter will describe the different chlorophototrophs found in these mats and will briefly summarize some properties of these organisms.

Introduction to Mushroom Spring and Octopus Spring in Yellowstone National Park

Although microbiological studies of YNP began in the 1930s (e.g., van Niel and Thayer 1930; Brock 1972), extensive microbiological research in Yellowstone National Park (YNP), Wyoming (WY, USA) started in the early 1960s with pioneering work by Thomas D. Brock (Brock 1967a, b, 1998; Brock and Freeze 1969). Hot springs, specifically MS and OS in the Lower Geyser Basin in YNP, were particularly interesting to microbial ecologists because they were viewed as simple, steady-state ecosystems containing large and readily accessible microbial populations, in which microorganisms could be studied in their natural environments (Brock 1967a, b, 1998; Ward et al. 1998). The names of the two springs possibly are derived from their shapes, which resemble a stylized mushroom and an octopus, respectively (Fig. 1). Hot spring water leaves the source pools forming outflow channels, along which relatively stable thermal gradients are created. With temperature as the principal variable (Brock and Brock 1968) they have provided excellent, naturally occurring experimental conditions to address questions about the ecology of organisms at different temperatures.

MS and OS are alkaline, siliceous hot springs located at about ~2000 m above mean sea level in YNP. The springs are approx. 0.3 km NE and SSE of Great Fountain Geyser in the White Creek Valley, respectively, and are separated by about 0.5 km (Fig. 1; Madigan and Brock 1977; Bott and Brock 1969). Over the course of our studies these springs have had source temperatures of 69.0 and 90.5 °C and have slightly alkaline pH values of ~8.0 and ~8.4, respectively. As has been described several times over the course of the last 50 years, their water chemistries are very similar: they are low in total salt concentrations, and have sulfate (<200 to 270 μM), nitrate (<2 to 6 μM), and ammonium (<3 to 35 μM) values approaching detection limits (Table 1; Doemel and Brock 1976; Ball et al. 2004; Papke et al. 2003).

Although most of the initial research had been conducted at OS, MS became famous as one of the source pools for the isolation of the thermophilic heterotroph, *Thermus aquaticus*, the source of Taq polymerase. The thermostability of Taq DNA polymerase revolutionized molecular-based, nucleic acid research, and has made the polymerase chain reaction an indispensable tool in virtually all aspects of modern life sciences (Brock 1997; Brock and Freeze 1969; Gelfand et al. 1989; Saiki et al. 1988).

As soon as the spring water at the rim of the source pool and in the effluent channels of OS cools down to temperatures <75 °C, the development of phototrophic microbial mats can be observed. The prokaryotic community above 75 °C consists of, non-phototrophic, (hyper-) thermophilic *Aquificales* and other chemoheterotrophic and chemolithoautotrophic bacteria (Reysenbach et al. 1994; Huber et al. 1998). Chlorophototrophs occur in the source pool for MS because the water temperature there is only 69 °C. In the effluent channels of both hot springs, chlorophototrophic microbial mats dominated by unicellular, oxygenic *Cyanobacteria* (*Synechococcus*), and filamentous anoxygenic chlorophototrophic members of the *Chloroflexi* develop at temperatures of 40–73 °C (Fig. 1). Photosynthesis extends up to the upper limit for growth of oxygenic chlorophototrophic cyanobacteria, which is 73–75 °C (Brock 1978; Castenholz 1969; Meeks and Castenholz 1971). Because of the dense growth of *Synechococcus* spp. at the upper layer of the mats, the mats have an intense, bluish-green color (Fig. 1b, c). The underlying region, referred to here as the undermat, is bright reddish-orange in color and is dominated by BChl *a*-containing filamentous anoxygenic chlorophototrophic *Chloroflexi* of the genus *Roseiflexus*; this part of the mat is largely free of *Synechococcus* spp. cells (Doemel and Brock 1977; Nübel et al. 2002; Thiel et al. 2016). Thomas Brock not only studied the upper temperature limit of phototrophic life in his early studies, but

Fig. 1 Map with location (Panel **a**) and pictures of OS (Panel **b**) and MS (Panel **c**), which are near Great Fountain Geyser in the Lower Geyser Basin of YNP. The map was generated using QLandkarte GT version 1.2.3, ©2007 Oliver Eichler, qlandkarte.sourceforge.net/using QT Library 4.8.1, GDAL Library 1.7.3 release 2010/11/10, and Proj4 Library 470. The *insets* in Panels **b** and **c** show the appearance of the phototrophic microbial mat. Note the distinct layers of these mats, with an *upper green layer* and a *reddish-orange undermat*

Table 1 Physicochemical parameters of MS and OS source water

Name and/or site description	Octopus Spring	Mushroom Spring
GPS position	44.539/−110.794	44.539/−110.798
Sample code number	01WA145	01WA146
Date collected	9/16/2001	9/16/2001
Time collected	12:20	15:08
Temperature, °C	90.5	69
pH (field/laboratory)	8.04/8.78	7.96/8.70
pH, selected	8.04	7.96
Specific conductance (field/laboratory), μS/cm	1338/1446	1310/1376
Eh, V −0.078	−0.007	0.127
Density, g/mL at 20 °C	0.99911	0.99904
Dissolved oxygen (DO), mg/L	–	–
Calcium (Ca^{++})	0.765	0.627
Magnesium (Mg^{++})	0.011	0.006
Strontium (Sr^{+})	0.001	0.001
Barium (Ba^{++})	<0.001	<0.001
Sodium (Na^{+})	297	278
Potassium (K^{+})	15.4	19.5
Lithium (Li^{+})	3.25	1.94
Sulfate (SO_4^{2-})	17.6	19
Thiosulfate ($S_2O_3^{2-}$)	<0.1	<0.1
Polythionate (S_nO_6), mM/n	<0.002	<0.002
Hydrogen sulfide (H_2S)	0.024	0.003
Alkalinity (HCO_3^-)	343	292
Acidity (free/total), mM	–/–	–/–
Fluoride (F^-)	21.2	14.9
Chloride (Cl^-)	246	257
Bromide (Br^-)	0.8	0.8
Nitrate (NO_3^-)	<0.1	<0.1
Nitrite (NO_2^-)	0.063	0.0151
Ammonium (NH_4^+)	<0.04	<0.04
Silica (SiO_2)	288	296
Boron (B)	2.74	2.74
Aluminum (Al)	0.24	0.14
Iron total (Fe(T))	<0.002	<0.002
Ferrous iron (Fe(II))	<0.002	<0.002
Manganese (Mn)	0.003	0.005
Copper (Cu)	0.001	<0.0005
Zinc (Zn)	0.002	0.003
Cadmium (Cd)	0.0001	0.0001
Chromium (Cr)	<0.0005	<0.0005
Cobalt (Co)	0.0027	<0.0007

(continued)

Table 1 (continued)

Name and/or site description	Octopus Spring	Mushroom Spring
Mercury (Hg), ng/L	–	–
Nickel (Ni)	<0.002	0.002
Lead (Pb)	0.001	0.0012
Beryllium (Be)	0.002	0.002
Vanadium (V)	<0.002	<0.002
Molybdenum (Mo)	–	–
Antimony (Sb)	0.085	0.072
Selenium (Se)	<0.001	<0.001
Arsenic total (As(T))	1.37	1.86
Arsenite (As(III))	0.148	0.0301
Dissolved organic carbon (DOC)	1.4	1.4
δD, per mil	−144	−142
δ18O, per mil	−16.9	−16.3
Sum cations, meq/L	13.7	12.9
Sum anions, meq/L	14	13.2
Charge imbalance, percent	−2	−2.8

If not stated otherwise all concentrations are in mg/L (Ball et al. 2004)
Data in this table are adapted from various sources

he was the first to show that bacterial activities and photosynthesis of natural communities obtained from hot spring environments were most efficient when samples were incubated at the temperature of the environment from which they were isolated (Brock 1967b, 1998).

Although the microbial chlorophototrophic mats that develop at MS and OS at temperatures >55 °C have a maximal depth of a few cm (e.g., 2 cm in Doemel and Brock 1977; 1 cm in Fig. 1c), most of the microbial activity seems to be restricted to the upper few mm. Most studies have been conducted on the upper green layer that extends to a depth of 1–2 mm. Studies including or focusing on the orange-colored undermat have generally focused on the region extending from a depth of about 3–5 mm (Kim et al. 2015; Nübel et al. 2002; Thiel et al. 2016). DNA obtained from the layers beneath ~5 mm was too degraded to allow any analysis (Thiel et al. 2016), and many cells from below 5 mm appeared to be dead (Doemel and Brock 1977).

Studies have shown that these chlorophototrophic mat communities are controlled by horizontal and vertical gradients (spatial) of environmental parameters, which can vary strongly during the diel cycle (temporal) but which change less significantly in response to seasonal or annual changes (Fig. 2). In addition to the temperature gradient due to the cooling of the water in the effluent channels, the diel cycle and vertical gradients of light and oxygen are important controlling factors in the microbial mats (Figs. 2 and 3). Additionally, microsensor studies disclosed high photosynthetic activity in the upper green layer leading to diel patterns and variations in pH, oxygen, and hydrogen concentration (Jensen et al. 2011;

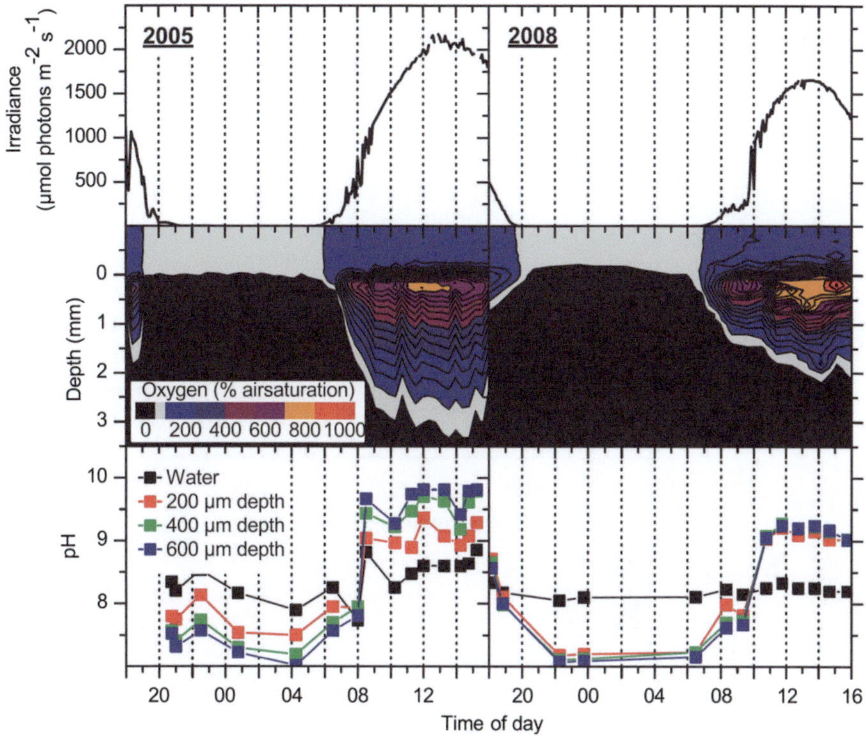

Fig. 2 In situ measurements of incident downwelling irradiance (μmol photons m^{-2} s^{-1}; *top panels*), O_2 levels (% air saturation; *central panels*), and pH (*bottom panels*) at different mat depths, as measured over two diel cycles at a 60 °C sampling site in MS in July 2005 (*left panels*) and September 2008 (*right panels*) (figure from Jensen et al. 2011)

Nübel et al. 2002; Revsbech and Ward 1984; Revsbech et al. 2016). Due to light-dependent autotrophic C_i fixation, the pH rises from ~7.0 under anoxic conditions before sunrise to >9.0 at highlight conditions around noon. Oxygen concentrations in the upper 1 mm region of the mat fluctuate from anoxia during the night, suboxic conditions during a period of dim light in the early morning, O_2 supersaturation during most of the day, and are turn to microoxic conditions during late afternoon (Becraft et al. 2015; Jensen et al. 2011; Revsbech and Ward 1984; Revsbech et al. 2016).

The Chlorophototrophs in Mats of MS and OS

What seemed to be a low-diversity microbial mat consisting of two phototrophic bacteria from the phyla *Cyanobacteria* and *Chloroflexi* over time has been demonstrated to be a much more complex phototrophic microbial community harboring

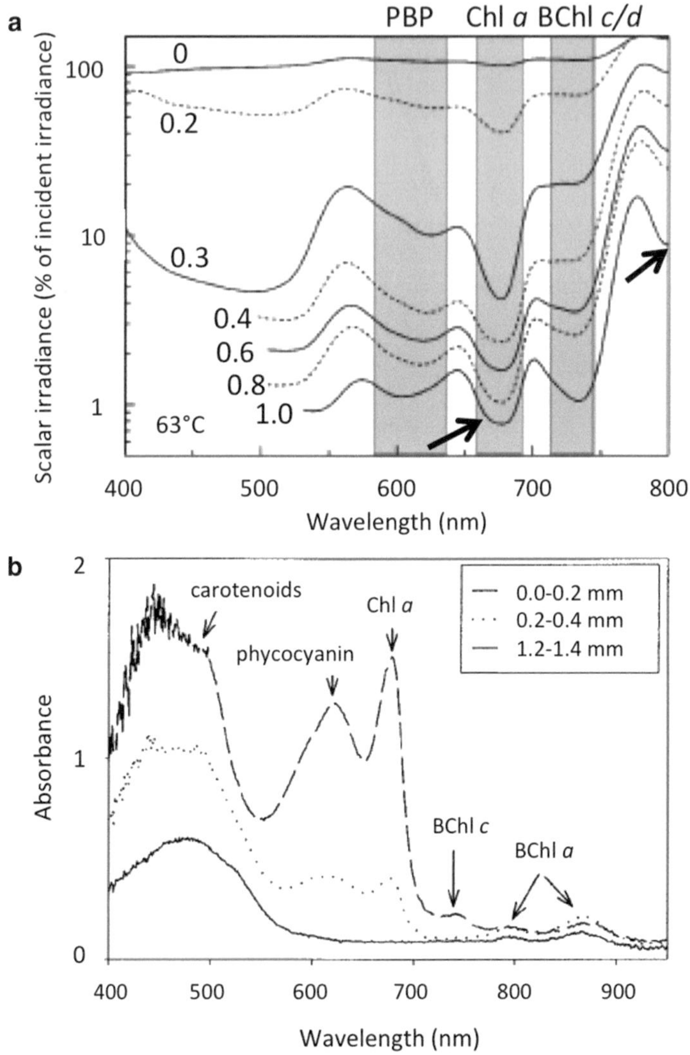

Fig. 3 Spectral scalar irradiance (Panel **a**) and absorbance spectra (Panel **b**) at different depths within the Mushroom Spring mat. Spectral scalar irradiance was measured at 63 °C in September 2008. Numbers on curves denote the depth (mm) below the mat surface. Note log scale showing transmitted light in % of the downwelling irradiance measured at the mat surface. *Shaded areas* indicate absorption ranges of phycobiliproteins (PBP), chlorophyll (Chl) *a*, and bacteriochlorophylls (BChl) *c* and *d*. Also note the relatively higher amount of near-infrared light at 800 nm compared to 680 nm at the depth of 1 mm (*arrows*; modified from Becraft et al. 2015). Wavelengths between 800 and ~950 nm can be absorbed by BChl *a* in anoxygenic members of the mat community. Absorbance spectra (Panel **b**) were obtained by microscopic spectrometry of thin sections of a mat from 60 °C cut perpendicular to the mat surface. Absorbance was calculated for 0.2-mm depth intervals at three depths as indicated. Absorbance peaks: phycocyanin, 622 nm; Chl *a*, 678 nm; BChl *c*, 744 nm; BChl *a*, 794, and 869 nm. Note that cyanobacterial pigments phycocyanin and Chl *a* are not detected at a depth of 1.2–1.4 mm (figure from Nübel et al. 2002)

numerous chlorophototrophic bacteria and members from six out of seven phyla with members that can synthesize (B)Chls (Fig. 4). After decades of study using isolation and/or molecular methods, at least 16 chlorophototrophic bacterial taxa have been detected in the microbial mats of MS and OS. The dominant oxygenic cyanobacteria are members of the *Synechococcus* spp. A/B lineage, many ecotypes of which occur in the mats (see below; Becraft et al. 2015). For instance, 16S rRNA genotypes A″, A′, A, B′, and B are distributed along the thermal gradient from ~72 °C to ~50 °C, respectively. Additional cyanobacteria detected in the mat include uncultured and unidentified cyanobacteria OS type I and type J (*Leptolyngbya* and *Synechococcus* spp., respectively), and the lower abundance isolates *Thermosynechococcus* sp. C1 and *Synechococcus* sp. C9 (most closely related to *Thermosynechococcus elongatus* and "*Candidatus* Gloeomargarita lithophora," respectively; Ferris et al. 1996b). Highly diverse filamentous anoxygenic phototrophs (FAPs) of the phylum *Chloroflexi* inhabit the mat. They include members of the genera *Roseiflexus* and *Chloroflexus* as well as two previously unknown chlorophototrophic *Chloroflexi* (see Sections " "*Candidatus* Roseilinea gracile," a Chlorophototrophic Member of the Class *Anaerolineae* Within the Phylum *Chloroflexi*" and " "*Candidatus* Chloranaerofilum corporosum" MS-FAB1: A New *Oscillochloris*-Like Member of the *Chloroflexineae*" below). The discovery of additional novel and rather unusual anoxygenic but oxygen-requiring chlorophototrophic bacteria came as a surprise, and thanks to metagenomic analyses; the first phototrophic member of the phylum *Acidobacteria*, *Cab. thermophilum* (Bryant et al. 2007; Tank and Bryant 2015a, b) and the first aerobic

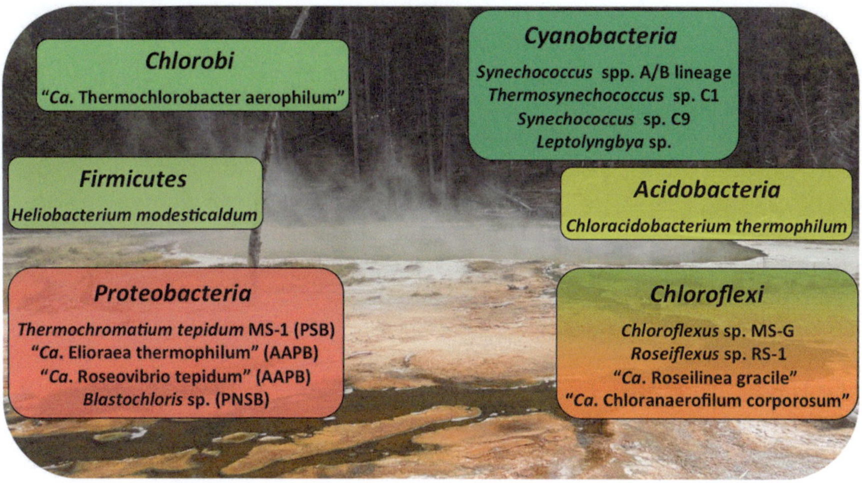

Fig. 4 Summary of the 16 chlorophototrophic taxa found in mats of MS (40–73 °C) by metagenomic and isolation studies. Six out of seven phyla with known chlorophototrophic members have been detected in the mats. Presence and abundance of the phototrophs vary with temperature (see text for details). These mats have been found to be a rich source of chlorophototrophic bacteria with unusual properties. PSB, purple sulfur bacteria; AAPB, aerobic anoxygenic purple bacteria; and PNSB, purple non-sulfur bacteria

photoheterotrophic member of the phylum *Chlorobi*, "*Ca*. T. aerophilum" (Liu et al. 2012) were identified. Isolated chlorophototrophs representing presumably rather small populations in the mats include anoxygenic chlorophototrophic proteobacteria closely related to the purple sulfur bacterium *Thermochromatium tepidum* (PSB; Madigan 1984, 1986) and α-proteobacterial aerobic anoxygenic phototrophic bacteria (AAPB). One isolate was identified as an *Elioraea* sp. ("*Ca*. Elioraea thermophila") and the other is distantly related to *Roseomonas* spp. ("*Ca*. Roseovibrio tepidum"). Additionally, a BChl *b*-containing, purple non-sulfur bacterium (PNSB), *Blastochloris* sp., was obtained from the mats at temperatures between 40 and 50 °C. The mats are also known to harbor *Heliobacterium modesticaldum*, a BChl *g*-containing, chlorophototrophic member of the *Firmicutes* (Kimble et al. 1995; Stevenson et al. 1997).

Oxygenic Chlorophototrophs: Cyanobacteria

Synechococcus spp.

The upper green layers of the MS and OS mats are dominated by cyanobacteria described as the A/B lineage of unicellular *Synechococcus* spp. (see Figs. 5 and 6; Ward et al. 2012a, b). Due to morphological and physiological similarities—and before the age of molecular taxonomy—these cyanobacteria were originally believed to represent strains of the cultivated species *Synechococcus lividus* (Sitz and Schmidt 1973; Doemel and Brock 1977). However, they were later shown to be quite distinct from *S. lividus* on the basis of phylogeny based on 16S rRNA sequences (~90 % nucleotide identity; e.g., Ward et al. 1990; Weller et al. 1992).

The *Synechococcus* spp. found in MS and OS are strictly thermophilic, oxygenic photoautotrophic cyanobacteria. The ability to fix dinitrogen via a molybdenum-iron nitrogenase was predicted and verified for members of *Synechococcus* sp. Type A and B′ (Bhaya et al. 2007; Steunou et al. 2006, 2008; M. Tank, unpublished data). However, one putative A14 ecotype lacks the entire *nif* gene cluster and thus is unable to reduce dinitrogen to ammonia (M. Olsen and D.M. Ward, unpublished results). These cyanobacteria contain PS I and PS II, with chlorophyll (Chl) *a* as their only Chl, and these organisms have phycobilisomes composed of phycocyanin and allophycocyanin; to date none have been found to produce phycoerythrin or phycoerythrocyanin (Bhaya et al. 2007; Nowack et al. 2015; Olsen et al. 2015). Members of the *Synechococcus* spp. Type A and Type B′ lineages dominate the upper ~2-mm thick green layer together with filamentous anoxygenic phototrophic members of the *Chloroflexi*. Unicellular cyanobacteria are found all along the temperature gradient (40–73 °C) but the different lineages (e.g., *Synechococcus* sp. A and B′) and variants (ecotypes, see below) show distinct distribution patterns (Becraft et al. 2011, 2015; Ferris et al. 1996a, 2003; Ferris and Ward 1997; Ramsing et al. 2000). Early studies were able to distinguish two different green layers visually and microscopically: a 0.2- to 1-mm thick first layer and a 0.1- to 0.6-mm thick

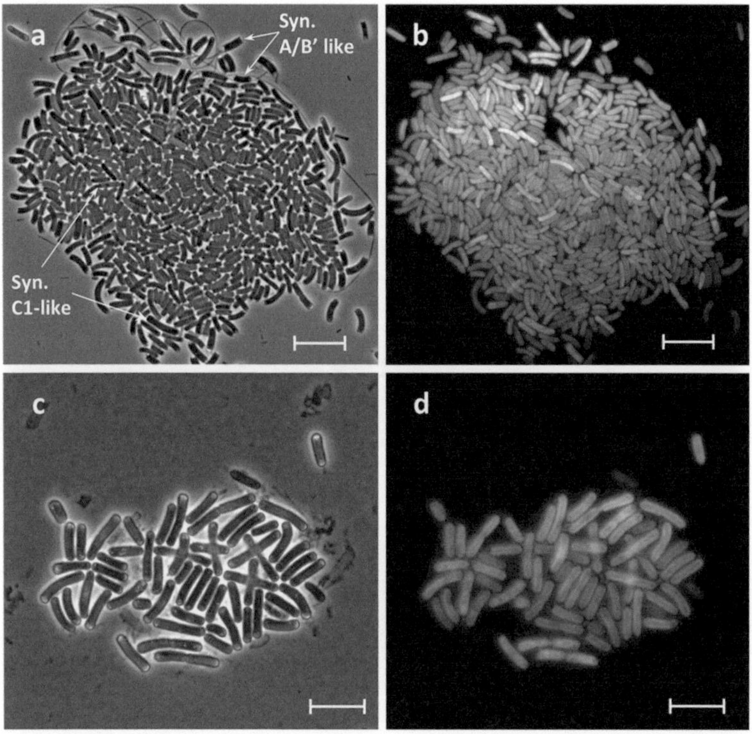

Fig. 5 Phase-contrast (Panels **a** and **c**) and Chl *a* autofluorescence (Panels **b** and **d**) micrographic images of cyanobacterial enrichment cultures from 55 °C MS mat samples. Panels **a** and **b** show a microcolony mainly consisting of A/B lineage *Synechococcus* (Syn.) spp., while panels **c** and **d** show *Thermosynechococcus* sp. C1-like morphotypes. Note the characteristic polar granules in both cell types probably consisting of carbonate deposits (Panels **a** and **c**). Scale bars equal 10 μm

second layer with darker blue-green color and more intense autofluorescence (Doemel and Brock 1977). More recent studies showed vertical distributions of distinct cyanobacterial variants with different light adaptations, which could possibly account for the observed color differences within the upper green layer (Becraft et al. 2015; Nowack et al. 2015). Beneath this region, in the orange-colored undermat, *Synechococcus* is rarely observed, even as lysed cells (Doemel and Brock 1977). The oxygenic cyanobacteria from the genus *Synechococcus* have been shown to be the predominant primary producers in these communities by in situ studies of bicarbonate fixation using stable and radioactive isotopes (Bateson and Ward 1988; Nübel et al. 2002; van der Meer et al. 2007), and they are also a primary source of fixed nitrogen (Steunou et al. 2006, 2008). Their very high biological activity in the upper layer results in extreme values for oxygen concentration and pH in the upper 2 mm of the mat as detected by microsensor studies (Revsbech and Ward 1984).

Synechococcus spp. exist in the mat as several different ecological species, which have nearly identical 16S rRNA gene sequences but different *psaA* sequences—so-called ecotypes (Becraft et al. 2011, 2015; Nowack et al. 2015; Olsen et al. 2015).

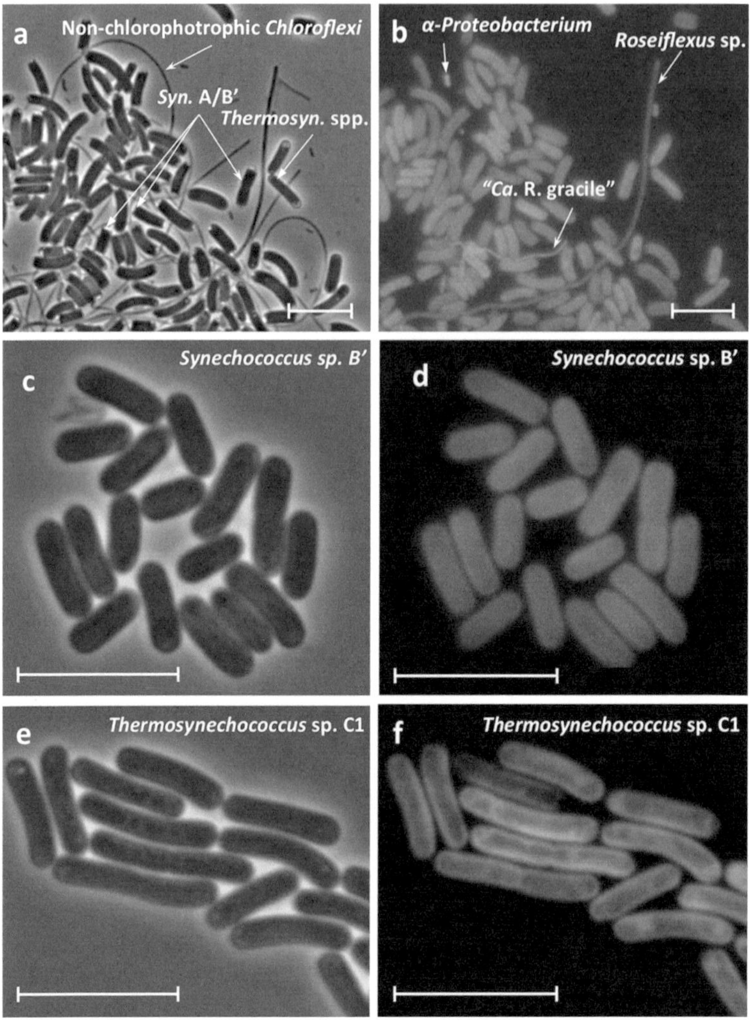

Fig. 6 Phase-contrast (Panels **a**, **c**, and **e**) and fluorescence micrographs (Panels **b**, **d**, and **f**) of *Synechococcus* spp. detected in MS mats. Colony of different unicellular *Synechococcus* spp. under enrichment conditions with characteristic carbonate deposits at the poles (Panels **a** and **b**). *Synechococcus* sp. B′ (Panels **c** and **d**) and *Thermosynechococcus* (*Thermosyn.*) sp. C1-like (Panels **e** and **f**) in agar plates. Note that the carbonate inclusions at the poles that are present in enrichment cultures (panel **a**) are lost when cells are grown on plates in a diel cycle with an N_2 atmosphere during the dark periods (Panel **c**), and that they differ in C1-like *Synechococcus* spp. in which they appear as granules (Panel **e**). Different phototrophs in mat and enrichment samples are visible in autofluorescence microscopy using BChl *a* specific filters (Panel **b**). Scale bars equal 10 μm

Previous studies have detected a number of cyanobacterial (*Synechococcus*) variants not only in 16S rRNA sequence (Ferris and Ward 1997), but also in 16S–23S rRNA internal transcribed spacer region (Ferris et al. 2003) and protein-encoding genes (Becraft et al. 2011, 2015; Melendrez et al. 2011). Different distribution patterns of the putative ecotypes along the effluent channel flow path (Ferris and Ward 1997), or with respect to the vertical dimension within the photic zone of the mat (Ramsing et al. 2000), first suggested that these might be ecological species populations with specific ecological adaptations. Temperature-adapted and light-adapted *Synechococcus* ecotypes have been demonstrated by obtaining representative strains and studying their temperature and light preferences, as well as their genome sequences (Allewalt et al. 2006; Becraft et al. 2015; Bhaya et al. 2007; Nowack et al. 2015; Olsen et al. 2015). Genome sequences of several ecotypes isolated from the dominant cyanobacteria from MS are now available, and these provide comprehensive insights into the physiological and metabolic capacities of the oxygenic chlorophototrophs in the mat (Bhaya et al. 2007; Olsen et al. 2015). Low-light-adapted strains characteristically contain a gene cluster—possibly horizontally acquired—that enhances absorption of far-red light, wavelengths (700–750 nm) that reach deeper into the mat (Gan and Bryant 2015; Olsen et al. 2015). This gene cluster includes allophycocyanin gene variants, *apcB3* and *apcD4*, which are only found in a few cyanobacteria and which produce a variant type of allophycocyanin with enhanced far-red light absorption (λ_{max} = 711 nm; Gan and Bryant 2015; Gan et al. 2014, 2015; Olsen et al. 2015; Sidler 1994). A special member of the PsbC/IsiA superfamily of Chl *a*-binding proteins, IsiX, is also specifically present in these low-light-adapted strains (Gan and Bryant 2015; Olsen et al. 2015; Shen et al. 2016). A metatranscriptomic analysis showed enhanced transcriptional activity for *isiX* specifically in the morning and evening, when low-light conditions predominate, a pattern of expression that is similar to that for the *apcD4* and *apcB3* genes (Olsen et al. 2015). Consistent with the in situ analyses of gene expression, laboratory cultures of low-light-adapted ecotypes only express the *isiX-apcD4-apcB3* genes in cells grown at very low irradiance, but these genes are not expressed when cells are grown at high irradiance (N. Soulier, G. Shen, D.A. Bryant, unpublished results).

Members of the Type A and B′ lineages of *Synechococcus* are the dominant oxygenic chlorophotrophs in these mat systems, and although they have been intensively studied for decades, it has not until now been possible to obtain axenic cultures of these cyanobacteria, which therefore prevented a valid description of the species. Despite this difficulty, many studies conducted in situ and with enrichments or "uni-cyanobacterial" mixed cultures produced important information about their ecology and physiology. Recently, using the information about the life style of these bacteria, in combination with a slight modification of the defined medium developed primarily for *Cab. thermophilum* (CTM-medium; Tank and Bryant 2015a, b), significant progress has been achieved toward obtaining axenic cultures of Type A and B′ *Synechococcus* spp. strains.

Cultivation experiments showed that the unicellular cyanobacteria in the MS and OS mats are very sensitive to reduced sulfur compounds, specifically thioglycolate, and these organisms apparently primarily obtain ammonia from nitrogen fixation, rather than by using ammonium or nitrate added to the CTM-medium as their source of fixed nitrogen. Mimicking a diel cycle of light and oxygen as occurs in situ at a temperature of 52 °C and using the CTM-medium without thioglycolate enabled the growth of axenic colonies of strains of the A/B-lineage of *Synechococcus* sp. and *Thermosynechococcus* sp. C1 in agar plates (Fig. 6c–f). Sulfate present in CTM-medium presumably functions as sulfur source for the cyanobacteria in the enrichments; this observation is supported by the prediction for assimilatory sulfate reduction ability in *Synechococcus* spp. Type A and B′ from analysis of the genomes (Bhaya et al. 2007). Nitrogen fixation is strongly supported by the in situ experiments performed by Steunou et al. (2006, 2008) in which maximal nitrogenase activity was demonstrated in the morning hours. This apparent preference of nitrogen fixation over other sources of inorganic nitrogen is a somewhat surprising finding, because ammonia and nitrate are usually the preferred sources of nitrogen for most cyanobacteria (Rippka et al. 1979; Stanier and Cohen-Bazire 1977). The *Synechococcus* spp. genomes contain a putative ammonium transporter (*amtB*) as well as genes for an ABC transporter for nitrate, nitrate reductase (*narB*), and nitrite reductase (*nirA*). However, active nitrogen fixation in situ is in accordance with the very low concentrations of ammonium and nitrate found in the spring waters (Table 1).

Less Abundant and Unidentified Cyanobacteria

In addition to the dominant A/B-lineages of *Synechococcus* spp., other cyanobacteria have been detected in these mats over the years. Cultivation-independent denaturing gradient gel electrophoresis (DGGE) and 16S rRNA cloning experiments detected cyanobacteria-like sequences, denoted OS Type I and Type J (Ward et al. 1992; Weller et al. 1992). OS Type I is most closely related to *Leptolyngbya* sp. O-77 (AP017367; 97 % nucleotide identity for 16S rRNA) and phylogenetically identified as member of the Subsection III Family I, subgroup "*Leptolyngbya*_3" based on the SILVA Ref database (SSU Ref release NR 123, www.arb-silva.de). OS Type J represents an uncultivated *Synechococcus* sp. (Subsection I Family I, subgroup "*Synechococcus*" within the *Cyanobacteria* based on phylogenetic identification using the SILVA Ref database release NR 123, www.arb-silva.de). It has no closely related relatives in the databases (≤89 % 16S rRNA nucleotide identity to all other sequences in the NCBI Genbank database), and thus it may represent an artifact sequence. Neither isolates nor enrichment cultures representing these genotypes are available, and no representative sequence has been detected in a 16S rRNA amplicon study of 60 °C mat communities (Thiel et al. 2016). However, filamentous *Leptolyngbya laminosa*/*Phormidium laminosum*-like cyanobacteria, which could be representatives of OS Type I, have been detected in several YNP hot springs (Brock 1978; Reysenbach et al. 2001), including MS and OS at temperatures (45–57 °C), lower than the ones studied here (Castenholz 1969; M. Tank, unpublished data; D.M. Ward, personal observations).

In previous cultivation-based studies, the lower abundance organisms *Thermosynechococcus* sp. C1 (formerly "*Synechococcus lividus* C1") (see Fig. 6e, f) and *Synechococcus* sp. C9 were isolated (Ferris et al. 1996a, b). *Synechococcus lividus* was first described by Copeland (1936) and was isolated from Mammoth Hot Springs in YNP. Copeland points out a striking similarity to *Synechococcus elongatus* Naegeli, which is described as the type species for the genus *Synechococcus* in AlgaeBase (www.algaebase.org), but which has been renamed *Thermosynechococcus elongatus* (Katoh et al. 2001). Isolate *Thermosynechococcus* sp. C1 from OS shares highest 16S rRNA sequence identity to species of the genus *Thermosynechococcus* (99 % nucleotide identity, *T. elongatus* strains BP-1 and WFW), and on this basis, we propose to rename "*Synechococcus lividus* C1" as "*Thermosynechococcus* sp. C1". The second isolate obtained, strain C9, represents a member of the *Synechococcus* group (*Cyanobacteria* Section I Family I in SILVA SSU Ref NR 123 database) and is most closely related to uncultivated cyanobacteria detected in Alla hot spring in Russia (Gaisin et al. 2015) as well as to "*Ca.* Gloeomargarita lithophora" (Ragon et al. 2014). The latter cyanobacterium was detected in an alkaline lake in Mexico and forms numerous intracellular Mg-Ca-Sr-Ba carbonate spherules. *Synechococcus* sp. C9 was detected in an amplicon study of MS mats at 60 °C in very low abundance (Thiel et al. 2016) but might be more abundant at other temperatures. Due to its low abundance and predicted close morphological similarity to other *Synechococcus* spp. (A/B lineages, as well as *Thermosynechococcus* sp. C1; see Figs. 5 and 6), the detection of C9-like cyanobacteria in isolation studies might require targeted screening.

Filamentous Anoxygenic Chlorophototrophic Members of the Phylum Chloroflexi

The phylum *Chloroflexi* is a deep-branching lineage of the *Bacteria* that currently comprises eight class-level subgroups: *Chloroflexia*, *Anaerolineae*, *Ardenticatenia*, *Caldilineae*, *Dehalococcoidia*, *Ktedenobacteria*, *Thermoflexia*, and *Thermomicrobia* (Dodsworth et al. 2014; Gupta et al. 2013), containing both chlorophototrophic and non-phototrophic (mostly) filamentous bacteria. Except for the recently discovered filamentous chlorophototroph, "*Ca.* Roseilinea gracile," (see below) that is phylogenetically classified as member of the class *Anaerolineae* (Klatt et al. 2011; Thiel et al. 2016; see below), the well-described chlorophototrophic organisms in the taxon *Chloroflexi* comprise a monophyletic lineage within the order *Chloroflexales*. Although they have previously been called green gliding bacteria or green non-sulfur bacteria, the chlorophototrophic organisms of the *Chloroflexales* are currently known as Filamentous Anoxygenic Phototrophs (FAPs). They are characterized by their multicellular, filamentous morphology with lengths of up to several hundred μm, and their anoxygenic photoheterotrophic/photomixotrophic lifestyle. The "green" FAPs (*Chloroflexineae*) produce bacteriochlorophylls (BChls) *a* and *c*, and type-2 reaction centers and chlorosomes. The "red" FAPs (*Roseiflexineae*) only produce BChl *a* and lack chlorosomes. All FAPs produce a variety of carotenoids (e.g., β- and γ-carotenes in green FAPs and keto-carotenoids in *Roseiflexus* spp.), the exact composition of which varies from organism to organism and with oxygen concentration (Gupta et al. 2013; Hanada et al. 1995, 2002; Hanada and Pierson 2006; Maresca et al. 2008; Takaichi et al. 2001).

In the microbial mat communities of MS and OS, members of the phylum *Chloroflexi* exhibit the greatest diversity of any phylum present in these mats. Four different chlorophototrophic and several non-phototrophic members of this phylum have been detected so far by metagenomic analyses (Klatt et al. 2011; Thiel et al. 2016). The first member of the *Chloroflexaceae* to be isolated and described was *Chloroflexus* (*Cfx.*) *aurantiacus* (Pierson and Castenholz 1971, 1974), one strain of which, Y-400, was obtained from OS (Pool A, Pierson and Castenholz 1974). For some time thereafter, it was assumed that most of the filamentous organisms in the upper green layer of MS and OS mats were members of this species. However, subsequent molecular analyses in the early 1990s showed that there were in fact two very different types of filamentous *Chloroflexi* present: *Chloroflexus* sp. and *Roseiflexus* sp., respectively (Nübel et al. 2002; Ward et al. 1990). The occurrence of the other two chlorophototrophic *Chloroflexi* was revealed much later by metagenomic analyses (Klatt et al. 2011, 2013; Thiel et al. 2016), which have been supported by enrichment, isolation, and microscopy studies (M. Tank, unpublished; see below). The relative abundances of the two most common chlorophototrophic *Chloroflexi*, i.e., *Roseiflexus* spp. and *Chloroflexus* spp., vary with temperature. At lower temperatures, *Roseiflexus* spp. are more abundant, but the more thermophilic *Chloroflexus* spp. predominate at higher temperatures and are the predominant *Chloroflexi* at ~70 °C (Nübel et al. 2002). Both types of organisms show a total gene expression activity that is highest under low-light conditions in the morning and late afternoon and lower at night and under high light conditions (Klatt et al. 2013). However, transcript abundances for some genes, e.g., photosynthetic reaction center genes and BChl biosynthesis genes, are highest at night and showed lowest expression levels during the day.

Roseiflexus sp. RS-1

Roseiflexus spp. are the most abundant members of the undermat community and are also the dominant members of the *Chloroflexi* in the upper green layer. Early in the research history of MS and OS, and soon after the discovery, isolation, and description of the first FAP, *Cfx. aurantiacus* (Pierson and Castenholz 1971, 1974), the filamentous members of the microbial mat resembled this organism (Bauld and Brock 1973; Doemel and Brock 1976, 1977). A pioneering molecular study targeting 16S rRNA gene sequences directly demonstrated not only a greater diversity of uncultivated bacteria in the mat than previously realized, but also showed that none of the sequences obtained from the mat shared more than 95 % 16S rRNA nucleotide identity with *Cfx. aurantiacus* (Ward et al. 1990). Soon afterwards, Weller et al. (1992) showed that two members of the *Chloroflexi* occurred in the mat, neither of which had a 16S rRNA sequence that was identical to *Cfx. aurantiacus* strain Y-400 that had previously been isolated from OS mats (Pierson and Castenholz 1974). 16S rRNA oligonucleotide probes showed that the dominant *Chloroflexi* "Type C" differed from *Cfx. aurantiacus* (Ruff-Roberts et al. 1994). However, the close similarity to 16S rRNA sequences from the then newly isolated and described *Roseiflexus castenholzii* from a Japanese hot spring mat indicated "Type C" to be a phototrophic member of this new genus (Hanada et al. 2002; Nübel et al. 2001). *Roseiflexus*-like "Type C" organisms were shown to dominate the filamentous members of the mat in lower temperatures (e.g., 60 °C) by fluorescence-labeled oligonucleotide probes and 16S rRNA amplicon studies (Nübel et al. 2002; Thiel et al. 2016). While *Chloroflexus* spp. are uncommon at 60 °C, they are the dominant members of the *Chloroflexales* at 70 °C (Nübel et al. 2002).

A representative strain, strain RS-1, of the dominant *Roseiflexus* spp., was isolated, and its genome was sequenced (van der Meer et al. 2010). *Roseiflexus* sp. RS-1 is a filamentous anoxygenic phototroph that synthesizes bacteriochlorophyll (BChl) *a* but not BChl *c* (Fig. 7a–c). This strain grew photoheterotrophically in the light or chemoheterotrophically under dark aerobic conditions (van der Meer et al. 2010). Genomic and metagenomic data suggested the capacity for carbon fixation via the 3-hydroxypropionate bi-cycle (Klatt et al. 2007; van der Meer et al. 2010), and autotrophic and mixotrophic growth has been suggested to occur under in situ conditions in the mat. This has been inferred from stable carbon isotope studies (van der Meer et al. 2005), as well as from metatranscriptomic sequence analyses from the green upper mat layer (Klatt et al. 2013). In the mat photomixotrophy is hypothesized to combine the uptake and incorporation of inorganic carbon from CO_2/bicarbonate and organic carbon, e.g., in the form of acetate or glyoxylate, which is derived from the cyanobacterial photorespiratory by-product, glycolate, by oxidation, and can readily be assimilated by the 3-hydroxypropionate bi-cycle (Klatt et al. 2007). Furthermore, the cyanobacterial fermentation product lactate may be taken up by *Roseiflexus* sp. as suggested by a sharp increase in the transcripts for the lactate permease gene for this organism in the late afternoon (Kim et al. 2015).

Metatranscriptomic analyses support the hypothesis that *Roseiflexus* spp. are the most abundant and active chlorophototrophic *Chloroflexi* in the mats. In the upper green mat layer, the total number of transcripts that uniquely mapped to open reading frames (ORFs) on scaffolds assigned to *Roseiflexus* spp. was 30-fold higher than the total number of transcripts assigned to *Chloroflexus* spp. (Klatt et al. 2013). Preliminary metatranscriptomic data from the undermat also indicate the dominance of *Roseiflexus* spp. in the undermat community, with up to 52 % of all reads mapping to the *Roseiflexus* sp. RS-1 reference genome (V. Thiel and D.A. Bryant, unpublished results). Although overall transcript levels were lowest in the night/early morning hours, the majority of *Roseiflexus*-like genes showed constitutive expression patterns, and a higher proportion of genes with nocturnal expression in comparison to *Chloroflexus* sp. (Klatt et al. 2013). Genes affiliated with photosynthetic metabolic pathways are expressed during the night; e.g., transcripts for *pufLM* encoding for the photosynthetic reaction center proteins and transcripts of genes involved in the biosynthesis of BChls were also most abundant at night. Transcript

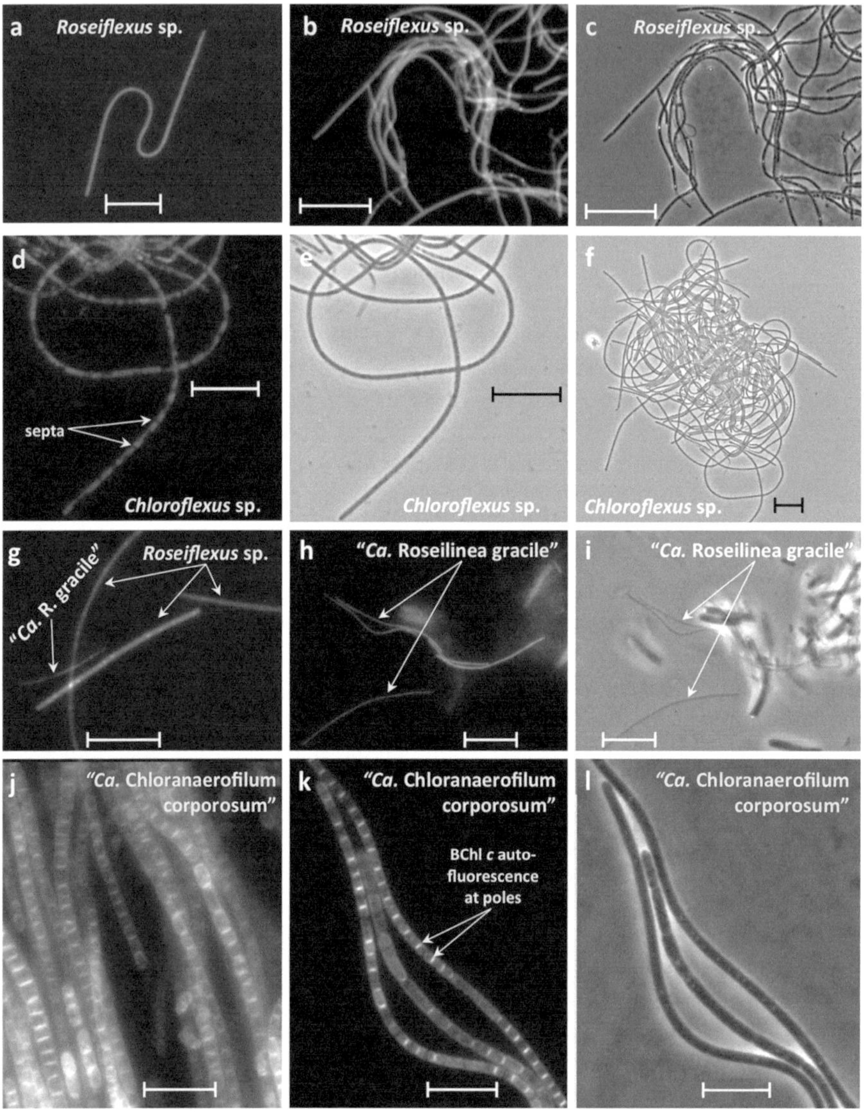

Fig. 7 Phase-contrast (Panels **c**, **e**, **f**, **i**, and **l**) and BChl *a* (Panels **a**, **b**, **g**, **h**) and BChl *c* (Panels **d**, **j**, and **k**) autofluorescence micrographs of phototrophic *Chloroflexi* found in Mushroom Spring mats. Panels **a**–**c**, *Roseiflexus* sp. filaments; Panels **d**–**f**, *Chloroflexus* sp. MS-G filaments; Panels **g**–**i**, "*Ca*. Roseilinea gracile" as very fine filaments exhibiting BChl *a* autofluorescence in fresh mat samples from MS at 55 °C. Panels **j**–**l**, "*Ca*. Chloranaerofilum corporosum" enrichment filaments from an enrichment culture. Note the clearly visible septa in *Chloroflexus* sp. MS-G (Panel **d**) and "*Ca*. Chloranaerofilum corporosum" (Panels **j**–**l**) that is notably larger than the other phototrophic *Chloroflexi* in these mats. Scale bars equal 10 μm

abundances for genes encoding enzymes of the tricarboxylic acid (TCA) cycle, the glyoxylate bypass, and the 3-hydroxypropionate bi-cycle were higher during the day (Klatt et al. 2013).

In addition to BChl *a*-containing, type-2 photosynthetic reaction centers, the RS-1 genome encodes xanthorhodopsin (RoseRS_2966) as well as all genes necessary for β-carotene biosynthesis (RoseRS_2117 (*crtB*), RoseRS_0943 (*crtI*), RoseRS_1811 (*crtI*), RoseRS_2155 (*crtI*), RoseRS_2643 (*crtL*)), which indicates that a possible additional mechanism for the utilization of light energy occurs in this organism (Choi et al. 2014). No carotene oxygenase gene has been identified so far, and further studies are needed to identify the nature and function of this putative bacteriorhodopsin homolog. It is presently uncertain whether retinal is obtained from some other organism in the mat or whether *Roseiflexus* spp. have a carotene oxygenase gene that differs from those found in other bacteria and animals. A recent study showed that a bacteriorhodopsin homolog derived from a heterotrophic member of the *Chloroflexi* functions as a light-driven sodium pump (Y. Nakajima, S. Yoshizawa, V. Thiel and D.A. Bryant, unpublished results).

A high microdiversity and the existence of putative ecological species, as proposed for the *Synechococcus* strains with high 16S rRNA and genome sequence identity, is suggested for *Roseiflexus* spp. based on 16S rRNA amplicon distribution (Ferris and Ward 1997) and metagenome studies of the undermat (Thiel et al. 2016). In addition to a high diversity of operational taxonomic units (OTUs) based on an arbitrary 97 % nucleotide identity of 16S rRNA genes within the phylum *Chloroflexi*, a high microdiversity was found for *Roseiflexus* spp. with 24 abundant (>100 reads) and a total of 6193 dereplicated *Roseiflexus* sp. amplicon sequences, which is supported by a preliminary analysis of *pufLM* amplicon sequence data (Thiel et al. 2016; J. Wood and D.M. Ward, unpublished data). Further molecular analyses of samples taken across temperature, light, and chemical profiles in the mats would provide further evidence for ecotypes of *Roseiflexus* spp. in these mats, similar to the findings of Miller et al. (2009) in White Creek, YNP.

Chloroflexus sp. MS-G

The second most abundant FAP member of the *Chloroflexi* is represented by *Chloroflexus* sp. strain MS-G, which is only distantly (94 % 16S rRNA nucleotide identity) related to described species of this genus, namely *Cfx. aurantiacus* type strain J-10-fl as well as strain Y-400 previously isolated from OS (Pierson and Castenholz 1971, 1974) and *Cfx. aggregans* DSM 9485^T (Hanada et al. 1995). A strain representing 16S rRNA sequences obtained from the mats in molecular studies, *Chloroflexus* sp. MS-G, has been isolated from MS (Thiel et al. 2014). The 16S rRNA sequence of the isolate shares >99 % nucleotide identity to a 16S rRNA sequence obtained from the metagenome (IMG taxon OID 3300002510) of the upper green layer (upper ~1 mm), and both share 100 % identity with the most abundant *Chloroflexus* sp. iTag sequence (253 basepair v4 region of 16S rRNA genes) of the microbial mat at 60 °C (Thiel et al. 2014, 2016). Strain MS-G is an oxygen-tolerant FAP that produces type-2 reaction centers. The whole-cell absorbance maxima at (396 nm), 466 nm, (510 nm), 744 nm, and 866 nm indicate the presence of carotenoids as well as BChl *c*-containing chlorosomes. Genes encoding type-2 (quinone-type) photosynthetic reaction centers (*pufLM*) and light-harvesting complex 1 (*pufAB*) as well as chlorosome proteins are present in the draft genome. The draft genome further includes genes encoding all enzymes required for the synthesis of BChls *a* and *c*, both of which were also detected in cells by high-performance liquid chromatography. A complete set of genes for the enzymes of the 3-hydroxypropionate bi-cycle indicates the potential for autotrophic and mixotrophic growth, which was observed in anaerobic agar-deep cultures of *Chloroflexus* sp. strain MS-G using sulfide-containing growth medium with bicarbonate and CO_2 as the sole carbon sources in repeated transfers. The carotenoid biosynthesis pathway in strain MS-G is probably similar to that in other *Chloroflexus* spp., which produce γ-carotene, β-carotene, echinenone, 1′-OH-γ-carotene, and the glucoside and glucoside esters of 1′-OH-γ-carotene (Takaichi et al. 1995). Consistent with the production of these carotenoids, the MS-G genome contains phytoene synthase (*crtB*), homologs of phytoene saturases and desaturases (*crtI*, *crtD*), lycopene cyclase (*crtY*), a 1′, 2′-hydratase (*cruF*), 1′-OH-glycosyltransferase (*cruC*), carotenoid glycoside acyltransferase (*cruD*), and beta-carotene 4-ketolase (*crtO*) (Bryant et al. 2012; Maresca et al. 2008; Thiel et al. 2014).

In the microbial mats of MS and OS, except for the very high temperatures of ~70 °C, *Chloroflexus* spp. are much less abundant than *Roseiflexus* spp. At 60 °C they represent approximately 1 % of the total microbial mat community based on a 16S rRNA iTag amplicon study, and up to 20 % of all filamentous *Chloroflexi* based on fluorescence in situ hybridization (FISH) (Nübel et al. 2002; Thiel et al. 2016). Although absorption spectral analyses indicate the presence of BChl *c*-containing chlorophototrophic bacteria (i.e., members of the *Chloroflexi* and *Cab. thermophilum*) to be restricted to the upper green layer of the mat, gene expression activity was detected for *Chloroflexus* sp. in both upper green layer and the undermat metatranscriptome (Klatt et al. 2013; V. Thiel and D.A. Bryant unpublished data). The relative number of reads mapping to the *Chloroflexus* sp. reference genome was higher in the undermat than the upper green layer, which does not necessarily indicate a higher activity at greater depths, but could be due to lower overall gene expression activity in the undermat due to the lack of highly active cyanobacteria and less available energy.

In cultivation studies *Chloroflexus* spp. are clearly distinguishable from *Roseiflexus* spp. and other phototrophic *Chloroflexi* based on colony color as well as cellular morphology. Fluorescence microscopy reveals BChl *c* autofluorescence in addition to a weaker BChl *a* autofluorescence. Clearly visible septa distinguish these organisms from *Roseiflexus* spp., which have similar filament sizes (Fig. 7d–f). Several *Chloroflexus* spp. isolates have been obtained which share closely related but not identical 16S rRNA gene sequences. Future studies of their physiology, pigment composition as well as light, sulfide, and temperature optima will reveal if ecological species with distinct niche adaptations also occur for this organism.

"*Candidatus* Roseilinea gracile," a Chlorophototrophic Member of the Class *Anaerolineae* Within the Phylum *Chloroflexi*

Until recently, isolated members of the class *Anaerolineae* were only non-phototrophic, chemoheterotrophic organisms (Grégoire et al. 2011; Imachi et al. 2014; Nunoura et al. 2013; Podosokorskaya et al. 2013; Yamada et al. 2006, 2007). Thus, it came as a surprise that an uncultured, chlorophototrophic *Anaerolineae*-like organism was identified in the MS and OS mats in metagenomic analyses (Klatt et al. 2011; Thiel et al. 2016). We are provisionally naming this organism "*Ca*. Roseilinea gracile." When first reported by Klatt et al. (2011), this uncultured organism was simply identified as "*Anaerolineae*-like," with *Anaerolineae thermophila* strain UNI-1 being its closest cultivated and described relative (85 % 16S rRNA nucleotide identity; Sekiguchi et al. 2003). At the time of this writing [March 2016], a BLAST search identified *Thermanaerothrix daxensis* strain GNS-1^T (Grégoire et al. 2011) and *Thermomarinilinea lacunofontalis* strain SW7(T) (Nunoura et al. 2013) as the closest isolated relatives with 16S rRNA sequence identities of 87 %. Phylogenetic analyses based on the full-length 16S rRNA sequences support a phylogenetic affiliation to the *Anaerolineales* as well as a more distant relationship to known chlorophototrophic *Chloroflexi*. Genes annotated within a metagenomic bin corresponding to this organism suggest that, like *Roseiflexus* spp., this anoxygenic chlorophototroph has the potential to synthesize BChl *a* but not BChl *c*, and it also possesses two putative bacteriorhodopsin-like genes (Klatt et al. 2011). The transcriptional activity of this uncultivated, phototrophic member of the *Chloroflexi* showed that the organism is most active when light is available (Klatt et al. 2011, 2013; Liu et al. 2011). In contrast to other phototrophic *Chloroflexi* in the mats, BChl biosynthesis genes were not down-regulated during the day, but photosynthetic reaction centers genes are similarly up-regulated at night (Klatt et al. 2013).

No isolate of "*Ca*. Roseilinea gracile" is yet available. However, microscopic studies of fresh mat samples and enrichment cultures of chlorophototrophic bacteria from the upper green layer contain very thin filaments (~0.2 μm in width) with a length of approx. 15–50 μm (Fig. 7g–i). These filaments exhibit fluorescence specific for BChl *a* but not BChl *c*. This is in agreement with the predictions based on the gene content of the metagenomic bin assigned to this organism, which contains genes for the synthesis of BChl *a* but no other BChl. These findings, in combination with an observed morphology typical of members of the class *Anaerolineae*, suggest that the observed thin filaments may be the new chlorophototrophic organism, "*Ca*. Roseilinea gracile."

"*Candidatus* Chloranaerofilum corporosum" MS-FAB1: A New *Oscillochloris*-Like Member of the *Chloroflexineae*

Oscillochloris trichoides DG-6^T is the only isolated and described species of the family *Oscillochloridaceae*, which together with the members of the genera *Chloroflexus* and *Chloronema* form the proposed suborder *Chloroflexineae* of the *Chloroflexi* (Gupta et al. 2013). A specific relationship between *Oschillochloris* and

Chloroflexus spp. is supported by seven conserved signature indels (CSIs) that are uniquely shared by the species from these two genera (Gupta et al. 2013). Species of both genera (and also *Chloronema*) also differ from species of the genera *Roseiflexus* and *Heliothrix* (proposed suborder *Roseiflexineae*) by their green color, the presence of BChl *c* and chlorosomes, as well as by their fatty acid profiles, the presence of β- and γ-carotenes, and synthesis of menaquinone-10 (MK-10) (Gupta et al. 2013). *O. trichoides* additionally differs from other chlorophototrophic *Chloroflexi* by having Form-1 ribulose-1,5-bisphosphate carboxylase/oxygenase (Rubisco) and phosphoribulokinase, and thus using the Calvin–Benson–Bassham cycle, for CO_2 fixation (Berg et al. 2005; Kuznetsov et al. 2011).

In addition to three previously described, chlorophototrophic taxa belonging to the phylum *Chloroflexi*, the presence of a fourth chlorophototrophic member of the *Chloroflexi* was indicated by metagenomic analysis in the orange-colored undermat (Thiel et al. 2016). Phylogenetic analyses of the 16S rRNA sequence identified this novel FAP as member of the *Chloroflexineae* (Gupta et al. 2013; Thiel et al. 2016), but showed that this organism is only distantly related (<91 % 16S rRNA nucleotide identity) to any of the known species of the phylum *Chloroflexi*, e.g., *Oscillochloris trichoides* DG-6^T, *Chloroflexus aurantiacus* J-10-fl(T), *Chloronema giganteum*, "*Candidatus* Chlorothrix halophila," and "*Candidatus* Chloroploca asiatica." A partial genome for this organism was obtained by binning of the metagenomic data, and the functional gene content from the corresponding binned sequences indicated that this organism has the capacity to synthesize both BChls *a* and *c*, which corresponds to the observation of filamentous morphotypes similar to *O. trichoides* displaying autofluorescence from BChl *a* and BChl *c* in microbial mat samples and enrichment cultures by fluorescence microscopy (Fig. 7j–l; M. Tank, unpublished results). Interestingly, the highest autofluorescence occurs along the septal ends of cells and not from the longitudinal walls. The filaments are longer than 100 μm and consist of many single cells that are separated by septa (Fig. 7j–l). The cells vary in length but show a constant width of ~2 μm. The cells of this new bacterium are obviously larger than those of all other *Chloroflexi* in these mats. Presently the growth temperature range has not been determined, but the organism was isolated from a 52.5 °C sample from MS and grows in fluffy colonies in the most oxygen-depleted areas in agar deeps made with CTM-medium containing twice the standard concentration of thioglycolate at 52 °C. The colonies are emerald-green in color, and the corpulent filaments seem to grow best at light intensities <50 μmol photons m^{-2} s^{-1}. The whole-cell absorbance spectrum is similar to those of *O. trichoides* (Keppen et al. 1994) and "*Ca.* Chloroploca asiatica" (Gorlenko et al. 2014), but nevertheless is slightly different (Table 2). Based on the partial genomes, this organism lacks Rubisco and uses the 3-hydroxypropionate bi-cycle for CO_2 fixation, unlike *O. trichoides*, which fixes carbon dioxide by the Calvin–Benson–Bassham cycle (Berg et al. 2005; Kuznetsov et al. 2011). Unfortunately, 16S rRNA and *pufLM* gene sequences from the isolate have not yet been obtained. Based on the conserved signature indels specific for different groups within the *Chloroflexi* described by Gupta

Table 2 Overview of physiological and metabolic characteristics of the different chlorophototrophic mat members at MS and OS, YNP

Taxa	Oxygenic/ Anoxygenic	C-source	CO_2 fixation	N_2 fixation	(B)Chl	Photosystems	LH	Oxygen	Whole-cell absorption [nm]
Synechococcus sp. A	Oxygenic	Autotroph	CBB	Yes	Chl *a*	PSI, PSII	PBS	Aerobe	440, 618, 680 440, 620, 680, 712[a]
Synechococcus sp. B′	Oxygenic	Autotroph	CBB	Yes	Chl *a*	PSI, PSII	PBS	Aerobe	440, 614, 680
Synechococcus sp. C9	Oxygenic	Autotroph	CBB	N. D.	Chl *a*	PSI, PSII	PBS	Aerobe	N. D.
Thermosynechococcus sp. C1	Oxygenic	Autotroph	CBB	N. D.	Chl *a*	PSI, PSII	PBS	Aerobe	440, 600, 680
OS type I (*Leptolyngbya* sp.)	Oxygenic	Autotroph	CBB	N. D.	Chl *a*	PSI, PSII	PBS	Aerobe	N. D.
Roseiflexus sp. RS-1	Anoxygenic	Mixotroph	3-OH-PB	No	BChl *a*	Type 2	LH1	Anaerobe, oxygen-tolerant	795, 900
Chloroflexus sp. MS-G	Anoxygenic	Autotroph, mixotroph	3-OH-PB	No	BChl *a*, *c*	Type 2	Csm, LH1	Anaerobe, oxygen-tolerant	465, 742, 866
"*Ca*. Roseilinea gracile"	Anoxygenic	Unknown	3-OH-PB	No	BChl *a*[b]	Type 2[c]	N. D.	Unknown	N. D.
"*Ca*. Chloranaerofilum corporosum"	Anoxygenic	Unknown	3-OH-PB	N. D.	BChl *a*, *c*[b]	Type 2[c]	LH1	Anaerobe	467, 747
Chloracidobacterium thermophilum	Anoxygenic	Heterotroph	None	No	BChl *a*, *c*, Zn-BChl *a*′ Chl *a*,	Type 1	Csm	Microaerophile	461, 745
"*Ca*. Thermochlorobacter aerophilum"	Anoxygenic	Heterotroph	None	No	BChl *a*, *d*[b]	Type 1[b]	Csm	Aerobe	N. D.
Thermochromatium tepidum	Anoxygenic	Autotroph	CBB	Yes[c]	BChl *a*[c]	Type 2[c]	LH1[c] LH2[c]	Anaerobe[c], oxygen-tolerant	505, 599, 808, 858[c]
"*Ca*. Elioraea thermophila"	Anoxygenic	Heterotroph	None	N. D.	BChl *a*[b,d]	Type 2[b]	LH1[b]	Aerobe	N. D.
"*Ca*. Roseovibrio tepidum"	Anoxygenic	Heterotroph	None	N. D.	BChl *a*[d]	Type 2[c]	N. D.	Aerobe	N. D.
Blastochloris sp.	Anoxygenic	Heterotroph	CBB	Yes	BChl *b*	Type 2[c]	LH 1[c]	Anaerobe	1019
Heliobacterium modesticaldum	Anoxygenic	Heterotroph	None	Yes	BChl *g* 8-OH Chl *a*	Type 1	None	Anaerobe	575, 670, 788

PBS phycobilisomes, *Csm* Chlorosomes, *LH* light harvesting complex, *CBB* Calvin–Bassham–Benson cycle, *3-OH-PB* 3-hydroxypropionate bi-cycle, *N. D.* not determined
[a]Only present when grown under low-light irradiance
[b]Predicted from (partial) genome
[c]Based on closely related relative
[d]Predicted from fluorescence microscopy

et al. (2013), and the ribosomal RNA and functional gene sequences derived from the partial genome extracted from the metagenome, this FAP is putatively assigned as a new genus–species of the proposed order of "green nonsulfur bacteria," *Chloroflexales*, suborder *Chloroflexineae*. Our analyses also distinguish this organism from all known members of the genus *Chloroflexus* (proposed family *Chloroflexaceae*) as well as *Oscillochloris* (family *Oscillochloridaceae*) (Thiel et al. 2016). Therefore, we suggest a new genus and species name for this organism, "*Ca.* Chloranaerofilum corporosum."

In summary, the mats of MS contain four distinct chlorophototrophic members of the phylum *Chloroflexi* (Figs. 4 and 7). Two of these organisms are novel phototrophs that were previously undetected by traditional enrichment and cultivation-based methods, but that only first became evident on the basis of metagenomic analyses of the upper and lower mat communities. However, consistent with the properties predicted from the metagenome, both organisms could be observed microscopically, and a stable mixed culture of one of the new *Chloroflexi*, "*Ca.* Chloranaerofilum corporosum" has been obtained (M. Tank, unpublished).

Novel and Unusual Aerobic Chlorophototrophic Bacteria

Bioinformatic analyses of the first metagenomic studies of the upper green layer of the microbial mat communities in MS and OS identified new chlorophototrophic bacteria with predicted characteristics and physiologies that differed from known chlorophototrophs (Bryant et al. 2007; Klatt et al. 2011). Importantly, information about their in situ metabolic activities over a diel cycle was gained from metatranscriptomic analyses of the community (Klatt et al 2013; Liu et al 2011). One of the new chlorophototrophs found in these mats by metagenomic analysis was *Cab. thermophilum*, the first, and to date the only, chlorophototrophic member of the phylum *Acidobacteria* (see Section "*Chloracidobacterium thermophilum*"). Including the discovery of *Cab. thermophilum*, the number of chlorophototrophic bacterial phyla was extended from five to six at that time. Because of the recent discovery of the *Gemmatimonas phototrophica* AP64^{T} (Zeng et al. 2014, 2016), there are now seven phyla that contain chlorophototrophic bacteria: *Acidobacteria*, *Chlorobi*, *Chloroflexi*, *Cyanobacteria*, *Firmicutes*, *Gemmatimonadetes*, and *Proteobacteria*. Interestingly, organisms related to *G. phototrophica* AP64^{T} have been detected in metagenomes from different locations in Yellowstone Lake, YNP (Zeng et al. 2014, 2016). Additional novel and unusual chlorophototrophs detected by metagenomic and metatranscriptomic analyses are "*Ca.* T. aerophilum," the first aerobic, photoheterotrophic member of the phylum *Chlorobi* (Section "*Candidatus* Thermochlorobacter aerophilum"), as well as "*Ca.* Roseilinea gracile," the first phototrophic member of the class *Anaerolineae* (see Section ""*Candidatus* Roseilinea gracile," a Chlorophototrophic Member of the Class *Anaerolineae* Within the Phylum *Chloroflexi*").

Chloracidobacterium thermophilum

One extraordinary outcome of studying the MS and OS microbial mats was the discovery of *Cab. thermophilum*. Detected in early 16S rRNA studies and classified as "OS type D" bacterium with uncertain affiliation (Ward 1998; 99 % nucleotide identity to the latter type strain), *Cab. thermophilum* was identified by metagenomic analyses as a chlorophototroph belonging to the poorly characterized phylum, *Acidobacteria* (Bryant et al. 2007). *Cab. thermophilum* was unusual not only because it was the first chlorophototrophic member of the *Acidobacteria*, but also in its metabolism and other properties. *Cab. thermophilum* is the first characterized aerobic chlorophototrophic bacterium that has a homodimeric type-1 reaction center (RC), the Fenna–Matthews–Olson BChl *a*-binding protein, and chlorosomes similar to those of the *Chlorobi*, nearly all of which are obligately anaerobic (for an exception, see Section "*Ca.* T. aerophilum"). In addition to the PscA and PscB polypeptides of the RC, *Cab. thermophilum* contains a 22-kDa carotenoid-binding protein, CbpA, which might be involved in the photoprotection and/or light harvesting for the RC (Garcia Costas et al. 2011a, b; Tsukatani et al. 2012). BChl *c* and keto-carotenoids are the major light-harvesting pigments in the chlorosomes, whereas BChl *a*, Zn-BChl *a′*, and Chl *a* are found in the RC (Garcia Costas et al. 2012; Tsukatani et al. 2012). The first attempts to isolate *Cab. thermophilum* from an OS cyanobacterial enrichment culture failed; however, it was possible to sequence the genome of the enriched bacterium, which together with the metatranscriptome provided valuable information about the physiology and metabolism of *Cab. thermophilum* (Bryant et al. 2007; Garcia Costas et al. 2012). It was predicted that *Cab. thermophilum* is a chlorophotoheterotroph, as none of the key enzymes of any known CO_2 fixation pathway were present in the genome. *Cab. thermophilum* also lacked sulfate reduction genes, and therefore requires reduced sulfur sources for growth. No nitrogen fixation genes or nitrate reduction genes were found. The genomic data clearly showed that *Cab. thermophilum* lacks the genes for the biosynthesis of branched-chain amino acids, lysine, and vitamin B_{12}. Furthermore, the genomic data clearly indicated a need for oxygen for some enzymatic reactions: e.g., enzymes in bacteriochlorophyll and carotenoid biosynthesis as well as the enzyme for tyrosine biosynthesis require oxygen as a substrate. A metatranscriptomic study from the mats suggested that transcriptional activity of *Cab. thermophilum* is highest at times of the day when the mats have only low oxygen concentrations, namely early in the morning and late in the afternoon (Liu et al. 2011). Considering all these findings, a defined medium for *Cab. thermophilum* was developed, and an axenic culture was achieved. All of the predictions made from the genomic data were tested and verified, including the requirement of reduced oxygen concentration for growth (Tank and Bryant 2015a, b). A species description of the proposed type strain *Cab. thermophilum* strain B^T was published in 2015. *Cab. thermophilum* is indeed a chlorophotoheterotrophic acidobacterium, which needs microoxic conditions and all three branched-chain amino acids, lysine, and vitamin B_{12} for growth. Interestingly, bicarbonate is also essential for growth, and is perhaps used in anaplerotic reactions. 2-Oxoglutarate enhances growth but is not essential. The cells are greenish-brown in color and have widths of 0.8–1.0 μm and lengths of ~2.5 μm (Fig. 8a, b). Light intensities >50 μmol photons m^{-2} s^{-1} are inhibitory for the type strain (Tank and Bryant 2015a, b). *Cab. thermophilum* is found within a temperature range of 40–68 °C (Miller et al. 2009). *Cab. thermophilum* is fairly abundant in the mats and represents about 5–10 % of the community members in the deeper parts of the upper green layer. Interestingly, *Cab. thermophilum* has been shown to have ecotypes adapted to specific temperatures, similar to the proposed *Synechococcus* ecotypes (see Section "*Synechococcus* spp."). Miller et al. (2009) showed in a metagenomic study along a temperature gradient at White Creek, YNP that different *Cab. thermophilum* sequences were found at different temperatures. This finding was supported by cultivation experiments with axenic isolates obtained from MS. Strain E (99 % 16S rRNA nucleotide identity to the type strain) was able to grow at temperatures from 39 to 57 °C whereas the type strain is unable to grow

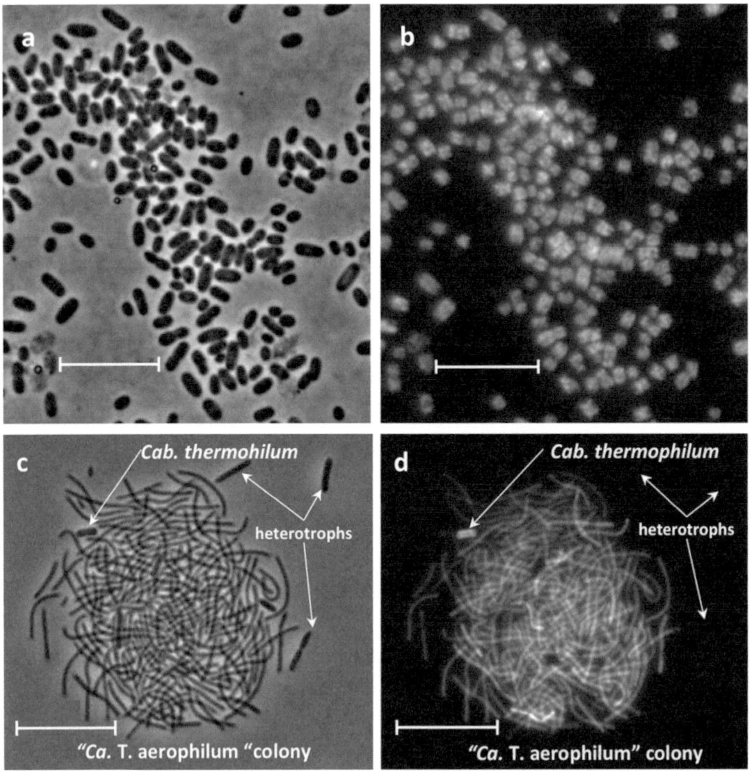

Fig. 8 Phase-contrast (Panels **a** and **c**) and fluorescence (Panels **b** and **d**) micrographs of the novel phototrophs, *Chloracidobacterium (Cab.) thermophilum* (Panels **a** and **b** and *white arrow* in Panels **c** and **d**) and "*Ca.*T. aerophilum" (Panels **c** and **d**). "*Ca.* Thermochlorobacter (T.) aerophilum" are detectable as elongated rods exhibiting autofluorescence under conditions for BChl *c*/*d*. *Cab. thermophilum* also exhibits autofluorescence under conditions for BChl *c*/*d*, notably along the long sides of the cell wall but less on the poles. Heterotrophs do not autofluoresce. Scale bars equal 10 μm

below 44 °C but grows up to 60 °C (Tank and Bryant 2015a). With the CTM-medium specifically developed for *Cab. thermophilum*, it is has been possible to isolate several different strains of *Cab. thermophilum* from MS into axenic culture. Their draft genomes are 95–99 % identical to each other or to the two chromosomes of the type strain (Garcia Costas et al. 2012). Metagenomic studies also detected *Cab. thermophilum* sequences in other hot springs in YNP (e.g., Green Finger Pool), and hot springs in Thailand, Tibet, and Malaysia (Chan et al. 2015; Kanokratana et al. 2004; Lau et al. 2009; Ross et al. 2012; Ward et al. 2006). This broad distribution suggests that this bacterium is a common member of microbial mats in alkaline hot springs around the world. *Cab. thermophilum* enrichment cultures obtained from Ojo Caliente hot spring in New Mexico and Rupite hot spring in Bulgaria are available (Hallenbeck et al. 2016; M. Tank, unpublished results). A draft genome from the New Mexico strain OC-1 was recently published, and it showed a high similarity (≥95 % nucleotide identity for coding sequences) to the type strain genome. The 16S rRNA gene shares 99 % nucleotide identity to the type strain gene (Hallenbeck et al. 2016). The 16S rDNA sequence obtained from the Bulgarian *Cab. thermophilum* isolate differs considerably from that of the type strain (97 % nucleotide identity), which suggests that this bacterium is probably a new species within the genus *Chloracidobacterium* (M. Tank, unpublished).

The discovery and isolation of *Cab. thermophilum* into axenic culture were only possible because of combination of modern molecular techniques and methods of classic microbiology. Interestingly, the CTM-medium developed specifically for *Cab. thermophilum* allows one to grow all of the chlorophototrophic bacteria described in this chapter. The only parameters that must be adjusted for each bacterium are the concentration of oxygen, which implies its high importance in the mats as well, the thioglycolate concentration and/or the reduced sulfur source.

"*Candidatus* Thermochlorobacter aerophilum"

A second special chlorophototrophic member of the mat community is "*Ca.* T. aerophilum," a highly unusual aerobic, photoheterotrophic member of the phylum *Chlorobi* (Liu et al. 2011). Early 16S rRNA gene sequence studies (Ferris et al. 1996a, b; Ferris and Ward 1997) detected sequences (denoted as OS Types E and E′) similar to those of green sulfur bacteria (GSB) in the mats of OS. Furthermore, *pscA* genes in the metagenome of the upper green layer of MS and OS, which were clearly related to those of members of the *Chlorobiales*, indicated the presence of a chlorophototrophic member of the *Chlorobi* (Bryant et al. 2007). In spite of these observations and the isolation of organisms similar to *Chlorobaculum tepidum* from other thermal features in the park (D. Bedard, personal communication), GSB were never isolated from these mats. The inability of "*Ca.* T. aerophilum" to oxidize sulfide, as well as its heterotrophic lifestyle, was revealed through analyses of the metagenome of this organism (Bryant et al. 2007; Klatt et al 2011; Liu et al. 2011). Key metabolic inferences, including an absolute requirement for all three branched-chain amino acids, lysine, and vitamin B_{12}, were essential in defining enrichment conditions that supported the growth of this organism.

"*Ca.* T. aerophilum" is the first aerobic, chlorophototrophic member of the phylum *Chlorobi*, and belongs to the proposed family *Thermochlorobacteriaceae*. On the basis of metagenomic and metatranscriptomic inferences, "*Ca.* T. aerophilum" is a chlorophototroph that synthesizes type-1 reaction centers, the BChl *a*-binding Fenna–Matthews–Olson protein, and chlorosomes containing BChl *d*; the synthesis of BChl *d* is observed in some cultivated relatives among the GSB (e.g., Borrego et al. 1998; Holt and Hughes 1961; Maresca et al. 2004; Steensgaard et al. 1999), but "*Ca.* T. aerophilum" is otherwise very different physiologically. "*Ca.* T. aerophilum" is proposed to be an aerobic photoheterotroph that cannot oxidize sulfur compounds, cannot fix N_2, and does not fix CO_2 (Liu et al. 2012), all of which properties are almost invariant among GSB. Because of these unusual traits, this chlorophototrophic member of the *Chlorobi* differs considerably from all cultivated members of the GSB, which additionally lack swimming motility, are strict anaerobes, and obligate chlorophotoautotrophs (Buchanan and Arnon 1990; Frigaard and Dahl 2009; Gregersen et al. 2011; Overmann 2008; Tang et al. 2011; Wahlund and Madigan 1993; Wahlund and Tabita 1997). The distribution of 16S rRNA sequences with depth revealed that this organism is most abundant at a depth of 300–600 μm in the upper green layer, where the mats become highly oxygenated during daytime due to oxygenic photosynthesis activity of the *Synechococcus* spp. that dominate this region of the mat (Fig. 2; Liu et al. 2012; Ramsing et al. 2000).

Metagenomic binning led to a nearly complete genome for "*Ca.* T. aerophilum." In addition to the 16S rRNA, phylogenetic marker genes and photosynthesis genes clearly established a relationship between "*Ca.* T. aerophilum" and cultivated GSB, especially *Chloroherpeton thalassium* (Gibson et al. 1984), with nucleotide sequence identity values of ~63 % (Klatt et al. 2011; Liu et al. 2011, 2012). Metatranscriptomic analyses further allowed assessment of the in situ activity of this uncultivated, novel phototroph. These analyses showed that in general, transcriptional activity was highest during the day, when not only solar irradiance but also oxygen concentrations in the mat are very high (Liu et al. 2012). However, transcript abundances for genes involved in the biosynthesis of photosynthetic reaction centers and chlorosomes were much higher at night, when oxygen levels are low or absent, whereas BChls biosynthesis genes were found to be transcribed mainly periods that occur in the early morning and late afternoon, when light and oxygen levels are in transition (Liu et al. 2012).

Using the same growth medium that was used to isolate *Cab. thermophilum*, enrichment cultures for "*Ca.* T. aerophilum" have been obtained (M. Tank and D.A. Bryant, unpublished results). "*Ca.* T. aerophilum" forms short filaments of less than ten cells and resembles "over-cooked spaghetti" (i.e., limp filaments; Fig. 8c, d), which are somewhat reminiscent of *Chloroherpeton thalassium* (Gibson et al. 1984). Actively growing cells of "*Ca.* T. aerophilum" have transparent poles and are 0.5 μm wide and ~10–20 μm long (Fig. 8c, d). This organism forms microcolonies in the mat that are often closely associated with members of *Synechococcus* spp. Physiological studies have established that "*Ca.* T. aerophilum" prefers a much higher oxygen concentration than the microaerophilic *Cab. thermophilum*. Interestingly, "*Ca.* T. aerophilum" exhibits gliding motility like *Chloroherpeton thalassium*. "*Ca.* T. aerophilum"-like morphotypes were microscopically detected in mat samples taken at temperatures from 40 to 63 °C and showed highest abundances at temperatures from 50 to 55 °C. The vertical distribution of "*Ca.* T. aerophilum" showed the highest abundances (~20 to 30 % of the amplicon sequences) at the surface of the upper green layer of the mats which also indicates an oxygen-requiring lifestyle of the organism at 60 °C (Liu et al. 2012).

Anoxygenic Phototrophic Proteobacteria

The phylum *Proteobacteria* harbor the greatest variety of anoxygenic phototrophic bacteria. Members exhibiting retinalophototrophy (e.g., the well-known SAR11 cluster strains of *Pelagibacter ubique*) (Béjà et al. 2000; Giovannoni et al. 2005; Rappé et al. 2002; Steindler et al. 2011) and chlorophototrophy (purple bacteria) are found in this phylum. More than 150 years ago, the proteobacterial chlorophototrophs were the first anoxygenic chlorophototrophs to be discovered, whereas the retinalophototophs were only more recently discovered (Béjà et al. 2000, 2001; Finkel et al. 2013). According to their respective phylogenies, the chlorophototrophs are assigned to the α-, β-, and γ-*Proteobacteria*. Because of the production of a variety of carotenoids, these bacteria are mainly reddish-purple in color and therefore are commonly called "purple bacteria." The metabolic versatility of purple bacteria provides useful traits for their functional classification. Three major groups of purple bacteria can be distinguished: the purple sulfur bacteria (PSB), the purple non-sulfur bacteria (PNSB), and the aerobic anoxygenic phototrophic bacteria (AAPB). For a detailed description of these groups, please refer to Chaps. 2 and 6 in this book and to recent reviews (Hunter et al. 2009; Imhoff 2014; Jiao et al. 2007; Koblížek 2015; Oren 2014; Overmann and Garcia-Pichel 2013; Yurkov and Beatty 1998). All species have type-2 reaction centers for the conversion of light into chemical energy. The most common Chl pigment in the RC and the light-harvesting antenna is BChl *a*. However, a few species exclusively use BChl *b* (e.g., members of the PNSB genera *Blastochloris*, and *Rhodospira* as well members of the PSB genera, *Halorhodospira*, *Thiococcus*, *Thioflavicoccus*, and *Thioalkalicoccus*). Still more rarely, Zn-BChl *a* can be synthesized and used as the principal Chl (e.g., by *Acidiphilium rubrum*; Wakao et al. 1996).

PSB belong to the order *Chromatiales* in the γ-*Proteobacteria* (Imhoff 2014), and within the *Chromatiales* the PSB are separated into the families, *Chromatiaceae* and the *Ectothiorhodospiraceae*. All PSB species in both families are able to perform anoxygenic photosynthesis under anoxic conditions and fix CO_2 by Rubisco and the Calvin–Benson–Bassham cycle. Their preferred electron donors are reduced sulfur compounds, which are converted to sulfur (polysulfide) globules that are stored inside the cell of the *Chromatiaceae* and are attached outside the cells of the *Ectothiorhodospiraceae*. With the exceptions of *Halorhodospira halochloris* (Imhoff and Trüper 1977; Imhoff and Süling 1996) and *Thermochromatium tepidum*, which grow at somewhat higher temperatures (>45 °C, see below), the PSB are mesophilic organisms. Many of the PSB can also photoassimilate small organic molecules or grow heterotrophically in the dark. Under favorable conditions and similar to cyanobacteria, PSB can form dense blooms visible to the unaided eye. PSB can serve as primary producers in nature, and many of them have the ability to reduce dinitrogen to ammonia. They are seen as detoxifiers in their habitats, as they convert sulfide, which is highly toxic to many other bacteria, into less toxic compounds (e.g., sulfate).

The purple non-sulfur bacteria (PNSB) are members of the α- and β-proteobacteria. They are generally characterized by their preference for anaerobic, anoxygenic photoheterotrophic, or photomixotrophic growth. PNSB use the Calvin–Benson–Bassham cycle for CO_2 fixation if they are able to grow photoautotrophically or photomixotrophically (Hunter et al. 2009). Nitrogen fixation is widely distributed among the PNSB. Many of them can use sulfide as electron donor but they can typically only tolerate low-sulfide concentrations (< 0.5 mM sulfide). Like PSB, PNSB are ubiquitously found in mesophilic, circum-neutral aquatic, or terrestrial environments. Exceptions occur for some that prefer acidic, alkaline, or hypersaline conditions. Madigan has recently described cold-adapted species (Madigan 2003). Truly thermophilic isolates with growth >50 °C are unknown; only mildly thermophilic PNSB species, e.g., *Blastochloris* sp. and *Rhodocista* sp., with growth up to 47 °C were identified in various slightly alkaline, hot springs (Hoogewerf et al. 2003; Madigan 2003; Resnick and Madigan 1989).

The third physiological group of chlorophototrophic proteobacteria is the aerobic anoxygenic purple bacteria (AAPB), with species belonging to the α-, β-, and γ-*Proteobacteria*. In contrast to the other two groups, the AAPB require oxygen and organic molecules for their growth. They lack the ability to use CO_2 as their primary carbon source because key genes of the classic CO_2 fixation pathways are missing (Koblížek 2015; Tang et al. 2011). However, up to ~15 % of their cell carbon can be obtained by anaplerotic CO_2 fixation reactions (Hauruseu and Koblížek 2012; Tang et al. 2009, 2011). AAPB are found in marine and freshwater aquatic environments, as well as in soil crusts and hot spring microbial mats (Yurkov and Beatty 1998; Yurkov and Csotonyi 2009). Similar to PNSB, AAPB have been isolated from alkaline, acidic, cold, and warm habitats. Two strains (RB3 and RB7) have been obtained from temperatures >50 °C but showed only mesophilic growth properties in the laboratory (Yurkov and Beatty 1998). Compared to other purple bacteria, they are characterized by very low BChl *a* concentrations and high carotenoid contents (Koblížek 2015; Yurkov and Csotonyi 2009). Interestingly, synthesis of BChl *a* is light-sensitive and thereof occurs in the dark (Yurkov and Beatty 1998; Yurkov and Csotonyi 2009), which suggests a cyclic diel mode of metabolism and which might be similar to the diel cycle found in other chlorophototrophs in the MS and OS microbial mats.

Recently, the microbial mats of MS and OS were shown by metagenomic analyses and isolation methods to contain chlorophototrophic proteobacteria belonging to all three common physiological groups: PSB, PNSB, and AAPB (Thiel et al. 2016). It is the first time that members of these bacteria were detected and isolated from MS. A brief description and characterization of each of these new bacteria is given below.

Thermochromatium tepidum MS-P1

Thermochromatium (Tch.) tepidum is the only known thermophilic member of the *Chromatiaceae*. The type strain, MC (ATCC 43061^T), was isolated from Stygian Spring, which is located in the Upper Terraces of Mammoth Hot Springs, YNP. Mammoth Hot Springs are characterized as calcareous, sulfidic springs (Madigan 1986). In the species description it was suggested that *Tch. tepidum* only lives in high-sulfide areas. MS and OS, however, are characterized as non-sulfidic or low-sulfide springs. Therefore, it was a surprise to see dark red lenticular colonies appearing in agar plates inoculated with material from a fresh mat sample taken from MS at 40 °C. In the microscope cells taken from a red colony showed very active motility by swimming, some sulfur globules inside the cells, and the typical morphology of *Tch. tepidum* (Fig. 9a, b). Subsequent 16S rRNA sequence analysis confirmed the microscopic observations, and the new bacterium was assigned to *Tch. tepdium* on the basis of 99 % 16S rRNA nucleotide sequence identity to the type strain. Fresh mat samples taken in February 2016 (Temp. = 40, 45, 50, 55, 60, and 63 °C) were microscopically screened for phototrophs. Cells resembling *Thermochromatium* sp. were only observed in the 40 and 45 °C samples. Metagenomic analyses of the upper green layers at higher temperatures (55 or 60 °C) did not detect any sequences from *Tch. tepidum* (Klatt et al. 2011; Thiel et al. 2016); however, ribosomal sequences representing this organism were detected in very low abundance in 16S rRNA amplicon studies at 60 °C (Thiel et al. 2016). Interestingly, the cells in the fresh mat samples were completely filled with sulfur globules, which indicate that significant sulfide production occurs in the phototrophic mats at lower temperatures. This finding agrees with previous studies, which demonstrated an active sulfur cycle with sulfate reduction activity leading to measurable sulfide accumulation in the mat (approx. 250 μM) despite the low sulfate concentrations (<200 μM) in the source water (Dillon et al. 2007). Initial isolation experiments also showed that *Tch. tepidum* cells only grew from samples collected at temperatures from 40 to 45 °C. This differs from the situation for the type strain, which was isolated from a sample taken at 44 °C, but was enriched at 52 °C and showed growth at temperatures up to 57 °C (Madigan 1986). The MS isolate seems to be more oxygen-tolerant than the type strain, which lost viability shortly after exposure to oxygen; the MS isolate was still viable under microoxic conditions used in combination with the CTM-medium after several days.

"*Candidatus* Elioraea thermophila" MS-P2

An α-proteobacterium, identified as an *Elioraea* sp. with 97 % 16S rRNA nucleotide identity to the type strain, *Elioraea tepidiphila*, from the order *Rhodospirillales*, was first detected in the mats as unidentified clone sequences denoted "OS Type O" (Ward et al. 1992, 96 % 16S rRNA nucleotide identity to the type strain). Representative 16S rRNA sequences with 98 % nucleotide identity to "OS-Type O" and >99 % nucleotide identity to the isolate were also detected in a 16S rRNA

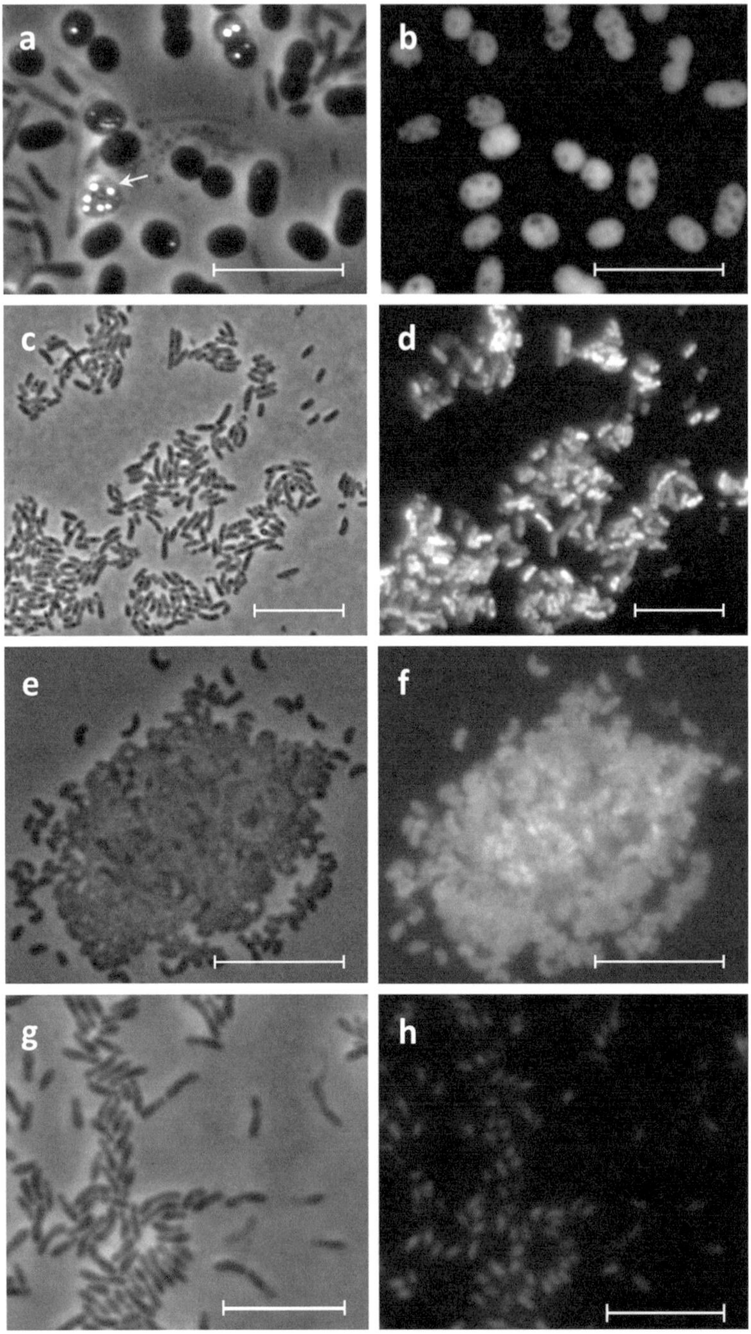

Fig. 9 Phase-contrast (Panels **a**, **c**, **e**, and **g**) and BChl *a* (Panels **b**, **d**, and **f**) and BChl *b* (Panel **h**) autofluorescence micrographs of phototrophic *Proteobacteria* in Mushroom Spring mats: *Thermochromatium tepidum* strain MS-P1 with intracellular sulfur globules (*arrow*, Panels **a** and **b**), "*Ca*. Elioraea thermophila" (Panels **c** and **d**), "*Ca*. Roseovibrio tepidum" (Panels **e** and **f**), and *Blastochloris* sp. (Panels **g** and **h**). Scale bars equal 10 μm

amplicon and metagenomic analysis of a sample from MS at 60 °C (Thiel et al. 2016). A partial genome was recovered from binning of the metagenomic data (Thiel et al. 2016). Suggesting a chlorophototrophic lifestyle, the partial genome as well as the genome for the closest cultured relative, *Elioraea tepidiphila* DSM 17972^T (NCBI acc. no. NZ_KB899965.1), contain genes for anoxygenic chlorophototrophy. Although chlorophototrophy has not yet been described for *E. tepidiphila* (Albuquerque et al. 2008), the ability to synthesize BChl *a* is predicted for "*Ca.* Elioraea thermophila" in the MS mat based on the partial genome. Furthermore, pink-colored cells that have a morphology similar to that of the *E. tepidiphila* type strain and that exhibited BChl *a* autofluorescence have been isolated from the mat. The 16S rRNA sequence of the pink-colored, lenticular colony-forming isolate, which was obtained from an MS mat sample at 52.5 °C, was 99.8 and 99.2 % identical to the 16S rRNA sequences from the metagenome and amplicon studies, respectively (Fig. 9c, d; Thiel et al. 2016). These data establish that the MS mats contain an *Elioraea* sp. that synthesizes BChl *a*, which we provisionally have named "*Ca.* Elioraea thermophila." Further physiological studies with an axenic isolate should establish whether this organism is truly a chlorophototroph. This bacterium showed growth only close below the surface of agar deeps made of CTM-medium with doubled concentration of thioglycolate, which indicates oxygen-dependent aerobic or microaerophilic physiology. The cells are rod-shaped, sometimes show swimming motility, and grow well at 52 °C. The cell dimensions are very similar to those of the type strain, with a width of ~0.5 μm and a length of ~1.0–1.5 μm. According to an iTag analysis, this organism is a low-abundance member (<0.1 % of the reads) of the mat community at 60 °C (Thiel et al. 2016).

"*Candidatus* Roseovibrio tepidum" MS-P3: A *Roseomonas/Rhodovarius*-Like α-*Proteobacterium*

Another novel BChl *a*-containing α-proteobacterial isolate, which shows similarities to *Roseomonas/Rhodovarius*, was obtained from a 52.5 °C mat sample taken at MS in 2015. The taxonomic affiliation of this organism was established from its 16S rRNA sequence, which showed that it represents a new genus within the *Rhodospirillales*. We have provisionally named this organism "*Ca.* Roseovibrio tepidum." Closely related 16S rRNA sequences but with very low read numbers (<0.02 % of all reads) were detected in the mat by an amplicon study of a 60 °C sample (Thiel et al. 2016). The vibrio-shaped isolate exhibits BChl *a* autofluorescence, suggesting a chlorophototrophic lifestyle. This suggestion is further strengthened by the presence of low coverage, unassigned α-proteobacterial *pufLM* sequences in the mat metagenome (Thiel et al. 2016). So far, only *Roseomonas aestuarii* has been reported to produce BChl *a*, but no *pufLM* sequences are available for that isolate (Venkata Ramana et al. 2010). Further physiological studies with this new isolate will establish whether it is truly a chlorophototroph. "*Ca.* Roseovibrio tepidum" showed a very similar growth behavior as the "*Ca.* Elioraea thermophila" isolate. It also forms pink, lenticular colonies close to the surface of CTM-medium agar deeps, which also suggests this organism requires relatively high levels of oxygen. The cells are short vibrioid rods that are 2 μm in length and 1 μm in width (see Fig. 9e, f).

Blastochloris sp. MS-P4

Only species of two genera, *Blastochloris and Rhodospira*, within the α-proteobacterial purple non-sulfur bacteria synthesize BChl *b*. Moderately thermophilic representatives of *B. sulfoviridis* have been isolated from hot spring microbial mats at Ojo Caliente in New Mexico (strain GI; Resnick and Madigan 1989) and Japan (TUT3225; Hisada et al. 2007). Recently, we have detected a *Blastochloris*-like bacterium in MS for the first time. *Blastochloris* sp. MS-P4 colonies were obtained from samples taken at 40 and 45 °C in February 2016. Lenticular, pale green-to-brownish colonies appeared in the anoxic region of CTM-medium agar deeps after 10 days of incubation at 45 °C at a light intensity of ~50 μmol photons m^{-2} s^{-1}. The cells have nearly identical morphology to other *Blastochloris* spp. with a width of ~0.5 μm and length of ~2.0 μm, and the isolates exhibit swimming motility (Fig. 9g, h). Whole-cell absorbance spectra had obvious peaks characteristic for BChl *b* at 400, 600, and 1020 nm (Table 2). *Blastochloris* sp. MS-P4 showed no growth at temperatures ≥50 °C, which is consistent with growth temperature experiments with strain GI1 (Resnick and Madigan 1989), which only grew at temperatures up to 47 °C. As of the writing of this chapter, the 16S rRNA and *pufLM* sequences were not yet available for this new isolate. Nevertheless, we tentatively assign this new bacterium to the genus *Blastochloris* because of the above-mentioned properties.

Chlorophototrophic Firmicutes*:* Heliobacterium modesticaldum

Heliobacterium modesticaldum, a BChl *g*-containing, gram-positive, spore-forming, thermophilic photoheterotrophic anaerobe of the phylum *Firmicutes* was isolated from MS in a previous study (Fig. 10a, b; Kimble et al. 1995). The organism grows at temperatures up to about 56 °C, is a strict anaerobe with the capacity for photoheterotrophic and chemotrophic growth on pyruvate, and fixes dinitrogen (Kimble et al. 1995). Consistent with the previous determination of the maximal growth temperature for this organism, *H. modesticaldum* sequences were not found in the metagenome assembled from samples taken at 60 °C mat in MS (Thiel et al. 2016; Klatt et al. 2011). However, heliobacteria could be enriched and one pure culture (OS-63) was obtained from samples taken in OS at temperatures from 54 to 70 °C (Stevenson et al 1997). This could be due to the presence of spores, which are known to be more heat resistant than vegetative cells, or rare abundance of viable cells below the detection limit in the molecular studies.

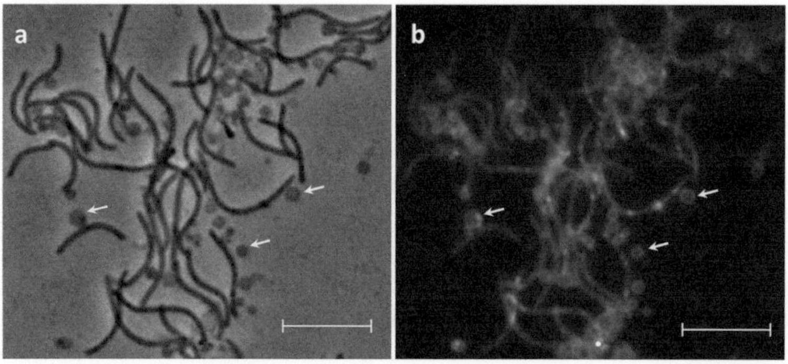

Fig. 10 Phase-contrast (Panel **a**) and fluorescence (Panel **b**) photomicrographs of cells of *Heliobacterium modesticaldum* strain Ice1. Note that the spheroblasts (*white arrows*) exhibit BChl *g* autofluorescence in panel **b**. Scale bars equal 10 μm

Including the isolation of *H. modesticaldum* by Kimble et al. (1995), the microbial mats of MS have now been shown to harbor representatives of six of the seven phyla (*Cyanobacteria*, *Chloroflexi*, *Chlorobi*, *Acidobacteria*, *Proteobacteria*, and *Firmicutes*) that are known to contain chlorophototrophic members (Fig. 4, Table 2). In total, 16 chlorophototrophs have been identified, and efforts are underway to obtain axenic cultures of all of them. No thermophilic member of the seventh phylum, *Gemmatimonadetes* (Zeng et al. 2014, 2016), has yet been reported, but of course this could be because no metagenomic analyses of mat samples taken at <50 °C have been analyzed. *Gemmatimonadetes*, however, have already been found in the Yellowstone Lake (Zeng et al. 2014, 2016), which might suggest their presence also in the mats at lower temperatures.

Ecology of Mushroom and Octopus Spring Mat Systems

Since the initial studies nearly 50 years ago, the principal motivation for research on these microbial mats has been to understand their microbial ecology. Hot spring microbial mats initially appeared to be relatively simple communities, and the thermal gradient that develops in the effluent channels was thought to be the major environmental variable in these systems. While it is still true that—in comparison to other microbial communities, e.g., not only those in soil but also freshwater and marine habitats—the microbial communities in these hot spring mats are relatively "simple," the unexpected complexity of these "simple communities" has become more and more apparent over the past 50 years. As illustrated above, these mats contain a panoply of chlorophototrophic bacteria, each of which ultimately uses the same energy source, namely sunlight. They live together and compete for or provide nutrients and energy sources, inhabiting their own specific niches and interacting with each other and the non-phototrophic members of the mat. At 60 °C the mat has been shown to consist of a highly diverse but very uneven community, which is dominated by only a few highly abundant organisms, and additionally contains a large number of rare members (Thiel et al. 2016). In comparison, less detailed information is available about the community structure at lower temperatures. However, differences in composition and abundances can be assumed. Ultimately, the organisms of this community are dependent upon the primary productivity of cyanobacteria (mainly *Synechococcus* spp.), and to a lesser extent to members of the *Chloroflexi*, but it is likely that all of the organisms maintain metabolic co-dependencies (e.g., branched amino acids and/or vitamins) that help maintain community stability and robustness.

In the discussion below, we will focus on the ecology of the chlorophototrophic mats at temperatures around 50–65 °C. At temperatures below 50 °C not only filamentous cyanobacteria (e.g., *Phormidium* or *Leptolyngbya* spp.) but also phototrophic eukaryotic algae start to become important primary producers (Castenholz 1969). Unicellular cyanobacteria of the genus *Synechococcus* are the main primary producers of these mats between 50 and 74 °C. Their identical morphological appearance long concealed not only the presence of multiple distinct types, e.g., Type A and B′ *Synechococcus* spp., but also a plethora of putative ecotypes, i.e., ecological species that coexist in the mat and inhabit different niches defined by light, temperature, oxygen, and possibly other parameters (e.g., nutrients).

In addition to temperature, other dynamic factors including light, nutrients, and oxygen affect microbial communities and shape the microenvironments in which the microbes live. The temperature of a given location in the hot spring microbial mat is affected by (a relatively constant) source pool temperature, flow velocity, and ambient temperature (summer vs. winter, more than day and night differences). The temperature variation over a diel cycle is only about ±1 °C, but can vary considerably over the course of a year (Nowack 2014). The natural cycling of day and night, and thus light and dark conditions, influences the oxygenic photosynthetic activity of *Synechococcus* spp., which in turn affects the oxygen concentration and pH in the mats. Cellular pigments, like (B)Chls and carotenoids, not only protect the cells or enable phototrophy, but also absorb light, qualitatively and quantitatively shaping the conditions in the layers below (Fig. 3). Blue and red light is mostly absorbed within the first 1.0 to 2.0 mm of the mat, as seen in Fig. 3a; only ~1 % of 660–690 nm light is present at a depth of 1 mm, whereas near-infrared light is still available in deeper layers (750–800 nm; >10 % in 1 mm depth; Fig. 3a). Both the quantity and quality of light reaching particular depths influence which phototrophic organisms thrive there. For example, Chl *a*-containing *Synechococcus* spp. are restricted to the upper layers of the mat, while BChl *a*- and *c*-containing bacteria are also present at greater depths (Fig. 3b; note that phycocyanin and Chl *a* were not detected at a depth of 1.2–1.4 mm). *Synechococcus* spp. perform oxygenic photosynthesis during daylight hours, leading to O_2 supersaturation and C_i depletion at peak daylight hours, as indicated by elevated pH values (Fig. 2).

Oxygen is also an important influencing factor for the community structure in the mats. O_2 concentrations not only vary over the day but also decline with depth (Fig. 2). Phototrophic bacteria with different oxygen needs and tolerances have been detected in the mats, which in addition to the light available probably define their vertical distribution within the mat (M. Tank, unpublished results). When photosynthesis declines in the evening, O_2 uptake by the aerobically respiring community members exceeds O_2 production and the mat becomes anoxic almost up to the mat surface (Revsbech and Ward 1984; Steunou et al. 2008). In the dark, anoxic conditions at night, fermentation genes as well as genes involved in N_2 fixation are expressed. Energy is gained by fermentation of the carbohydrate storage product, glycogen, which is built up during the day, both in *Synechococcus* spp. and the anoxygenic phototrophic *Chloroflexi* (Kim et al. 2015; Klatt et al. 2013; Liu et al. 2011; Steunou et al. 2006). The latter store polyhydroxyalkanoates and wax esters at night, which in daytime are degraded and used as carbon and electron reserves to support photomixotrophy via the 3-hydroxypropionate bi-cycle (Kim et al. 2015; Klatt et al. 2013). Night is also the time of highest expression activity of genes involved in the biosynthesis of BChls and photosynthetic reaction centers for the chlorophototrophic *Chloroflexi*, whereas the aerobic anoxygenic phototrophs, *Cab. thermophilum* and "*Ca.* T. aerophilum," express reaction center genes at night, but BChl biosynthesis related genes under low-light conditions during the day (Klatt et al. 2013; Liu et al. 2011, 2012).

Fixed nitrogen levels in the source water are near the detection limit for NH_3/NH_4^+ and are also very low for NO_2^- and NO_3^- (Table 1; Ball et al. 2004). Putative *nifH*-like genes for the ATPase subunit of nitrogenase have been detected in several mat members, e.g., *Synechococcus* spp., *Roseiflexus* spp., *Thermodesulfovibrio* sp., however, active nitrogen fixation has only been demonstrated for *Synechococcus* spp., which presumably provides the complete mat community with bio-available nitrogen (Table 2). Nitrogen fixation is energetically expensive, consuming large amounts of ATP and electrons. Genes encoding nitrogenase show increased expression starting in the evening, with a subsequent decline over the course of the night (Steunou et al. 2008). In contrast, nitrogenase activity is low during the night, when light energy is not available. Measured enzymatic activity exhibited a small peak in the evening, but the main nitrogen fixation activity was measured in the early morning, when light began to stimulate cyanobacterial photosynthetic activity, but O_2 consumption by respiration still exceeded the rate of O_2 evolution (Steunou et al. 2008). Once the irradiance increases to the point at which the mat becomes oxic, nitrogenase activity is strongly inhibited (Steunou et al. 2006, 2008).

The morning hours are the most productive and metabolically active period of the day in these hot spring mats. Metametabolomic analyses of the MS mat showed accumulation of metabolites involved in anabolism, including intermediates in amino acid and nucleic acid biosynthesis, during the early morning to midday period (Kim et al. 2015). Peak levels of these nitrogen-rich compounds follow the maximal period of N_2 fixation by *Synechococcus* spp., which occurred between 06:00 and 10:00 in the morning (Kim et al. 2015; Steunou et al. 2008).

Synechococcus spp. conduct photosynthesis and produce glycogen during the day, which is fermented, e.g., to acetate and lactate, during the afternoon and night hours. During peak daylight hours, extreme light and oxygen concentrations induce photorespiration which leads to accumulation of glycolate, a common but toxic by-product of this protection mechanism (Bateson and Ward 1988; Kim et al. 2015). Glycolate is excreted into the extracellular mat environment and thus becomes a resource available to other community members during those hours (Kim et al. 2015). During peak light hours the photosystems of *Synechococcus* spp. are presumably damaged or photoinhibited, leading to cyanobacterial fermentation and the accumulation of lactate in the extracellular matrix of the mat, being available as nutrient for other mat members. Glycolate is probably oxidized to glyoxylate, which has been shown to be photoassimilated by *Roseiflexus* spp. via the 3-hydroxypropionate bi-cycle (Kim et al. 2015; Klatt et al. 2007). Lactate, presumably produced by *Synechococcus* spp. in the afternoon, is also hypothesized to be photoassimilated by *Roseiflexus* spp. as well as other mat community members, which use it as a carbon and energy source (Kim et al. 2015).

Hydrogen is produced as a by-product of nitrogenase activity as well as end product of fermentation. H_2 levels in the mats are low, especially during the day (Revsbech et al. 2016), which presumably is due to its oxidation by chlorophototrophic members of the *Chloroflexi*. Other community members, including the sulfate-reducer *Thermodesulfovibrio yellowstonii*, other chemotrophic members, or even hydrogenotrophic methanogenic archaea, may also consume the hydrogen that is produced (Henry et al. 1994; Klatt et al. 2013; Sandbeck and Ward 1981).

A closed sulfur cycle may exist in the mats. Both sulfate reduction and biogenic sulfide production have been shown to occur, in spite of relatively low sulfate concentrations in the source water (<200 μM; Ball et al. 2004; Dillon et al. 2007). Organisms similar to *Thermodesulfovibrio yellowstonii* were identified as the most abundant sulfate-reducing populations in the upper 1-mm layer of the mat responsible for the measured sulfide production, although additional unidentified *dsrAB* sequences indicate the presence of other sulfate-reducing bacteria in the mats (Dillon et al. 2007; Thiel et al. 2016, 2017). Sulfide is known to act as an electron donor for photoautotrophic *Chloroflexi* (e.g., *Chloroflexus* sp. MS-G; see above) and has been shown to stimulate incorporation of bicarbonate into filamentous *Chloroflexi* in the mats (van der Meer et al. 2005). Genomes of known phototrophic *Chloroflexi* as well as the metagenome of the mat contain *sqr* genes, encoding sulfide:quinone oxidoreductases belonging to the type II family, but none of the *Chloroflexi* strains with sequenced genomes encode *dsr* or *sox* genes (Bryant et al. 2012). In accordance with the absence of *dsr* and *sox* genes, when sulfide is oxidized by *Chloroflexus* sp. or *O. trichoides* cultures, it is deposited as elemental sulfur (polysulfide) outside the cells, often affixed to the cells, and sulfur oxidation to sulfate has not been observed (Bryant et al. 2012; Frigaard and Dahl 2009; Hanada and Pierson 2006; Keppen et al. 2000; Madigan et al. 1974). Oxidation of sulfur to sulfate could possibly occur in the orange-colored undermat by a *Thermocrinis* sp.-like member of the phylum *Aquificae*, leading to a closed sulfur cycle in the mat (Thiel et al., in preparation). Other candidates for sulfur/polysulfide oxidation are *Synechococcus* spp., which may be able to oxidize sulfide quantitatively to thiosulfate like many other cyanobacteria are apparently able to do (De Wit and van Gemerden 1987; Olsen et al. 2015; Rabenstein et al. 1995). This process, which may be temporally correlated with nitrogen fixation, has been shown to stimulate ^{13}C-bicarbonate incorporation into lipid biomarkers (van der Meer et al. 2005). The resulting thiosulfate might ultimately be an alternative source of sulfate to close the sulfur cycle in the mats.

Rare organisms not discussed here, but detected in a 16S rRNA amplicon analysis (Thiel et al. 2016), as well as the α-proteobacterial phototrophs affiliated with these mats, probably do not contribute significantly to community function at 60 °C degrees but might be more important at lower temperatures.

Final Remarks

The chlorophototrophic microbial mats of OS and MS have been intensively studied for the past 50 years, and for the first half of this period, scientists believed that the upper green layer contained members of only two types of chlorophototrophic organisms: *Synechococcus* sp. and *Chloroflexus* sp. The introduction of cultivation-independent, molecular methods based on sequencing of 16S rRNA and other genes roughly doubled the number of organisms identified by revealing the presence of *Roseiflexus* sp. and multiple types of morphologically similar cyanobacteria (e.g., *Synechococcus* spp. A/B-lineage ecotypes, *Thermosynechococcus* sp. C1, and *Synechococcus* sp. C9). Beginning about 10 years ago, cultivation-independent metagenomics, metatranscriptomics, and genomics began to reveal the true complexity of chlorophototrophs and heterotrophs in these mats. Several novel chlorophototrophs, not previously known to science, were identified as described in this chapter. However, it is important to note that several of the chlorophototrophs were not identified initially by metagenomic analysis but were first detected in cultivation studies, once a suitable growth medium was defined by cultivation of *Cab. thermophilum*. Astonishingly, at least 16 different prokaryotic chlorophototrophs coexist and must compete for niches defined by light quality, light quantity, oxygen concentration, and temperature, among other potential niche-defining parameters. With respect to light, specific genes associated with adaptation and acclimation to light intensity have been found (Olsen et al. 2015), but the organisms in the mat utilize different Chls, BChls, carotenoids, and phycobiliproteins to acquire light for photosynthesis.

The experience gained from these efforts to obtain a complete description of the chlorophototrophs in these mats is highly instructive. No single method is sufficient to define the composition of a microbial community, but the combination of modern cultivation-independent methods and classical methods, including fluorescence microscopy, enrichment, and isolation, is a more powerful approach to understanding the composition, physiology, and metabolism of a microbial community. While it is certainly true that meta-omics methods can reveal important information about "who's there" and "what they are doing," these methods cannot substitute for the wealth of information that can be obtained once an organism can be grown axenically. Both in situ and ex situ approaches are important components of the modern microbiologist's toolbox, and the older methods should not be ignored in favor of the currently more fashionable cultivation-independent methods.

Acknowledgements The studies described in this chapter were partly funded by the Division of Chemical Sciences, Geosciences, and Biosciences, Office of Basic Energy Sciences of the Department of Energy through Grant DE-FG02-94ER20137. D. A. B. and D. M. W. additionally acknowledge support from the NASA Exobiology program (NX09AM87G). This work was also partly supported by the US Department of Energy (DOE), Office of Biological and Environmental Research (BER), as part of BER's Genomic Science Program 395 (GSP). This contribution originates from the GSP Foundational Scientific Focus Area (FSFA) at the Pacific Northwest National Laboratory (PNNL) under a subcontract to D.A.B. Some of the nucleotide sequencing was performed as part of a Community Sequencing Program (Project CSP-411) and was performed by the US Department of Energy Joint Genome Institute, which is supported by the Office of Science of the US Department of Energy under Contract No. DE-AC02-05CH11231.

The authors would also like to thank all of the JGI staff members who contributed to obtaining the sequence data. The materials used in this study were collected under permit #YELL-SCI-0129 held by D. M. W. and administered under the authority of Yellowstone National Park. The authors especially thank Christie Hendrix and Stacey Gunther for their advice and assistance.

References

Albuquerque L, Rainey FA, Nobre MF, da Costa MS (2008) *Elioraea tepidiphila* gen. nov., sp. nov., a slightly thermophilic member of the *Alphaproteobacteria*. Int J Syst Evol Microbiol 58:773–778. doi:10.1099/ijs.0.65294-0

Allewalt JP, Bateson MM, Revsbech NP et al (2006) Effect of temperature and light on growth of and photosynthesis by *Synechococcus* isolates typical of those predominating in the Octopus Spring microbial mat community of Yellowstone National Park. Appl Environ Microbiol 72:544–550. doi:10.1128/AEM.72.1.544

Balashov SP, Lanyi JK (2007) Xanthorhodopsin: proton pump with a carotenoid antenna. Cell Mol Life Sci 64:2323–2328. doi:10.1016/j.bbabio.2008.05.005

Balashov SP, Imasheva ES, Boichenko VA et al (2005) Xanthorhodopsin: a proton pump with a light-harvesting carotenoid antenna. Science 309:2061–2064. doi:10.1126/science.1118046

Ball JW, McCleskey RB, Nordstrom DK et al (2004) Water-chemistry data for selected springs, geysers and streams in Yellowstone National Park, Wyoming, 2001-2002. U S Geol Surv Open-File Rep Open-File Report 2004-1316. doi:10.5066/F7M043FS

Bateson MM, Ward DM (1988) Photoexcretion and fate of glycolate in a hot spring cyanobacterial mat. Appl Environ Microbiol 54:1738–1743

Bauld J, Brock TD (1973) Ecological studies of *Chloroflexus*, a gliding photosynthetic bacterium. Arch Mikrobiol 92:267–284. doi:10.1007/BF00409281

Becraft ED, Cohan FM, Kühl M et al (2011) Fine-scale distribution patterns of *Synechococcus* ecological diversity in the microbial mat of Mushroom Spring, Yellowstone National Park. Appl Environ Microbiol 77:7689–7697. doi:10.1128/AEM.05927-11

Becraft ED, Wood JM, Rusch DB et al (2015) The molecular dimension of microbial species: 1. Ecological distinctions among, and homogeneity within, putative ecotypes of *Synechococcus* inhabiting the cyanobacterial mat of Mushroom Spring, Yellowstone National Park. Front Microbiol 6:590. doi:10.3389/fmicb.2015.00590

Béjà O, Aravind L, Koonin EV et al (2000) Bacterial rhodopsin: evidence for a new type of phototrophy in the sea. Science 289:1902–1906. doi:10.1126/science.289.5486.1902

Béjà O, Spudich EN, Spudich JL et al (2001) Proteorhodopsin phototrophy in the ocean. Nature 411:786–789. doi:10.1038/35081051

Berg IA, Keppen OI, Krasil'nikova EN et al (2005) Carbon metabolism of filamentous anoxygenic phototrophic bacteria of the family *Oscillochloridaceae*. Microbiology 74:258–264. doi:10.1007/s11021-005-0060-5

Bhaya D, Grossman AR, Steunou A-S et al (2007) Population level functional diversity in a microbial community revealed by comparative genomic and metagenomic analyses. ISME J 1:703–713. doi:10.1038/ismej.2007.46

Borrego CM, Garcia-Gil J, Cristina XP et al (1998) Occurrence of new bacteriochlorophyll *d* forms in natural populations of green photosynthetic sulfur bacteria. FEMS Microbiol Ecol 26:257–267. doi:10.1111/j.1574-6941.1998.tb00510.x

Bott TL, Brock TD (1969) Bacterial growth rates above 90 degrees C in Yellowstone hot springs. Science 164:1411–1412. doi:10.1126/science.164.3886.1411

Brock TD (1967a) Micro-organisms adapted to high temperatures. Nature 214:882–885. doi:10.1038/214882a0

Brock TD (1967b) Life at high temperatures. Science 158:1012–1019. doi:10.1126/science.158.3804.1012

Brock TD (1972) One hundred years of algal research in Yellowstone National Park. In: Desikachary TV (ed) Taxonomy and biology of blue-green algae. Center for Advanced Study of Botany, Madras, pp 393–405

Brock TD (1978) Thermophilic microorganisms and life at high temperatures. Springer, Berlin. doi:10.1007/978-1-4612-6284-8

Brock TD (1997) The value of basic research: Discovery of *Thermus aquaticus* and other extreme thermophiles. Genetics 146:1207–1210

Brock TD (1998) Early days in Yellowstone microbiology. ASM News 64:137–140

Brock TD, Brock ML (1968) Relationship between environmental temperature and optimum temperature of bacteria along a hot spring thermal gradient. J Appl Microbiol 31:54–58. doi:10.1111/j.1365-2672.1968.tb00340.x

Brock TD, Freeze H (1969) *Thermus aquaticus* gen. nov. and sp. nov., a non-sporulating extreme thermophile. J Bacteriol 98:289–297

Bryant DA, Frigaard N-U (2006) Prokaryotic photosynthesis and phototrophy illuminated. Trends Microbiol 14:488–496. doi:10.1016/j.tim.2006.09.001

Bryant DA, Garcia Costas AM, Maresca JA et al (2007) *Candidatus* Chloracidobacterium thermophilum: an aerobic phototrophic Acidobacterium. Science 317:523–526. doi:10.1126/science.1143236

Bryant DA, Liu Z, Li T et al (2012) Comparative and functional genomics of anoxygenic green bacteria from the taxa *Chlorobi*, *Chloroflexi*, and *Acidobacteria*. In: Burnap R, Vermaas W (eds) Advances in photosynthesis and respiration, vol 33, Functional genomics and evolution of photosynthetic systems. Springer, Dordrecht, pp 47–102. doi:10.1007/978-94-007-1533-2_3

Buchanan BB, Arnon DI (1990) A reverse KREBS cycle in photosynthesis: consensus at last. Photosynth Res 24:47–53. doi:10.1007/BF00032643

Castenholz RW (1969) Thermophilic blue-green algae and the thermal environment. Bacteriol Rev 33:476–504

Chan CS, Chan K-G, Tay Y-L et al (2015) Diversity of thermophiles in a Malaysian hot spring determined using 16S rRNA and shotgun metagenome sequencing. Front Microbiol 6:177. doi:10.3389/fmicb.2015.00177

Chen M (2014) Chlorophyll modifications and their spectral extension in oxygenic photosynthesis. Annu Rev Biochem 83:317–340. doi:10.1146/annurev-biochem-072711-162943

Choi AR, Shi L, Brown LS et al (2014) Cyanobacterial light-driven proton pump, *Gloeobacter* rhodopsin: complementarity between rhodopsin-based energy production and photosynthesis. PLoS One 9, e110643. doi:10.1371/journal.pone.0110643

Copeland J (1936) Yellowstone thermal Myxophyceae. Ann N Y Acad Sci 36:1–232. doi:10.1111/j.1749-6632.1936.tb56976.x

De Wit R, van Gemerden H (1987) Oxidation of sulfide to thiosulfate by *Microcoleus chtonoplastes*. FEMS Microbiol Lett 45:7–13. doi:10.1111/j.1574-6968.1987.tb02332.x

Dillon JG, Fishbain S, Miller SR et al (2007) High rates of sulfate reduction in a low-sulfate hot spring microbial mat are driven by a low level of diversity of sulfate-respiring microorganisms. Appl Environ Microbiol 73:5218–5226. doi:10.1128/AEM.00357-07

Dodsworth JA, Gevorkian J, Despujos F et al (2014) *Thermoflexus hugenholtzii* gen. nov., sp. nov., a thermophilic, microaerophilic, filamentous bacterium representing a novel class in the *Chloroflexi*, *Thermoflexia* classis nov., and description of *Thermoflexaceae* fam. nov. and *Thermoflexales* ord. nov. Int J Syst Evol Microbiol 64:2119–2127. doi:10.1099/ijs.0.055855-0

Doemel WN, Brock TD (1976) Vertical distribution of sulfur species in benthic algal mats. Limnol Oceanogr 21:237–244. doi:10.4319/lo.1976.21.2.0237

Doemel WN, Brock TD (1977) Structure, growth, and decomposition of laminated algal-bacterial mats in alkaline hot springs. Appl Environ Microbiol 34:433–452

Ferris MJ, Ward DM (1997) Seasonal distributions of dominant 16S rRNA-defined populations in a hot spring microbial mat examined by denaturing gradient gel electrophoresis. Appl Environ Microbiol 63:1375–1381

Ferris MJ, Muyzer G, Ward DM (1996a) Denaturing gradient gel electrophoresis profiles of 16S rRNA-defined populations inhabiting a hot spring microbial mat community. Appl Environ Microbiol 62:340–346

Ferris MJ, Ruff-Roberts AL, Kopczynski ED, Bateson MM et al (1996b) Enrichment culture and microscopy conceal diverse thermophilic *Synechococcus* populations in a single hot spring mat habitat. Microbiology 62:1045–1050

Ferris MJ, Kühl M, Wieland A et al (2003) Cyanobacterial ecotypes in different optical microenvironments of a 68°C hot spring mat community revealed by 16S-23S rRNA internal transcribed spacer region variation. Appl Environ Microbiol 69:2893–2898. doi:10.1128/AEM.69.5.2893

Finkel OM, Béjà O, Belkin S (2013) Global abundance of microbial rhodopsins. ISME J 7:448–451. doi:10.1038/ismej.2012.112

Frigaard N-U, Dahl C (2009) Sulfur metabolism in phototrophic sulfur bacteria. In: Poole R (ed) Advances in microbial physiology, vol 54. Academic Press, pp 103–200. doi:10.1016/S0065-2911(08)00002-7

Frigaard N-U, Martinez A, Mincer TJ et al (2006) Proteorhodopsin lateral gene transfer between marine planktonic bacteria and archaea. Nature 439:847–850. doi:10.1038/nature04435

Gaisin VA, Kalashnikov AM, Sukhacheva MV et al (2015) Filamentous anoxygenic phototrophic bacteria from cyanobacterial mats of Alla hot springs (Barguzin Valley, Russia). Extremophiles 19:1–10. doi:10.1007/s00792-015-0777-7

Gan F, Bryant DA (2015) Adaptive and acclimative responses of cyanobacteria to far-red light. Environ Microbiol 17:3450–3465. doi:10.1111/1462-2920.12992

Gan F, Zhang S, Rockwell NC et al (2014) Extensive remodeling of a cyanobacterial photosynthetic apparatus in far-red light. Science 345:1312–1317. doi:10.1126/science.1256963

Gan F, Shen G, Bryant DA (2015) Occurrence of far-red light photoacclimation (FaRLiP) in diverse *Cyanobacteria*. Life (Basel, Switzerland) 5:4–24. doi:10.3390/life5010004

Garcia Costas AMG, Tsukatani Y, Romberger SP et al (2011a) Ultrastructural analysis and identification of envelope proteins of "*Candidatus* Chloracidobacterium thermophilum" chlorosomes. J Bacteriol 193:6701–6711. doi:10.1128/JB.06124-11

Garcia Costas AMG, Tsukatani Y, Rijpstra WIC et al (2011b) Identification of the bacteriochlorophylls, carotenoids, quinones, lipids, and hopanoids of "*Candidatus* Chloracidobacterium thermophilum". J Bacteriol 194:1158–1168. doi:10.1128/JB.06421-11

Garcia Costas AMG, Liu Z, Tomsho LP et al (2012) Complete genome of "*Candidatus* Chloracidobacterium thermophilum", a chlorophyll-based photoheterotroph belonging to the phylum *Acidobacteria*. Environ Microbiol 14:177–190. doi:10.1111/j.1462-2920.2011.02592.x

Gelfand DH, Stoffel S, Lawyer FC et al (1989) Purified thermostable enzyme. Patent US 4889818 A

Gibson J, Pfennig N, Waterbury JB (1984) *Chloroherpeton thalassium* gen. nov. et spec. nov., a non-filamentous, flexing and gliding green sulfur bacterium. Arch Microbiol 138(2):96–101. doi:10.1007/BF00413007

Giovannoni SJ, Bibbs L, Cho J-C et al (2005) Proteorhodopsin in the ubiquitous marine bacterium SAR11. Nature 438:82–85. doi:10.1038/nature04032

Golbeck JH (1993) Shared thematic elements in photochemical reaction centers. Proc Natl Acad Sci U S A 90:1642–1646. doi:10.1073/pnas.90.5.1642

Gomez Maqueo Chew A, Bryant DA (2007) Chlorophyll biosynthesis in bacteria: the origins of structural and functional diversity. Annu Rev Microbiol 61:113–129. doi:10.1146/annurev.micro.61.080706.093242

Gómez-Consarnau L, González JM, Coll-Lladó M et al (2007) Light stimulates growth of proteorhodopsin-containing marine Flavobacteria. Nature 445:210–213. doi:10.1038/nature05381

Gorlenko VM, Bryantseva IA, Kalashnikov AM et al (2014) "*Candidatus* Chloroploca asiatica" gen. nov., sp. nov., a new mesophilic filamentous anoxygenic phototrophic bacterium. Microbiology 83:838–848. doi:10.1134/S0026261714060083

Gregersen LH, Bryant DA, Frigaard N-U (2011) Mechanisms and evolution of oxidative sulfur metabolism in green sulfur bacteria. Front Microbiol 2:116. doi:10.3389/fmicb.2011.00116

Grégoire P, Fardeau M-L, Joseph M et al (2011) Isolation and characterization of *Thermanaerothrix daxensis* gen. nov., sp. nov., a thermophilic anaerobic bacterium pertaining to the phylum *Chloroflexi*, isolated from a deep hot aquifer in the Aquitaine Basin. Syst Appl Microbiol 34:494–497. doi:10.1016/j.syapm.2011.02.004

Gupta RS, Chander P, George S (2013) Phylogenetic framework and molecular signatures for the class Chloroflexi and its different clades; proposal for division of the class *Chloroflexia* class. nov. into the suborder *Chloroflexineae* subord. nov., consisting of *Oscillochloridacea* and the family *Chloroflexaceae* fam. nov., and the suborder *Roseiflexineae* subord. nov., containing the family *Roseiflexaceae* fam. nov. Antonie Van Leeuwenhoek 103:99–119. doi:10.1007/s10482-012-9790-3

Hallenbeck PC, Grogger M, Mraz M, Veverka D (2016) Draft genome sequence of the photoheterotrophic *Chloracidobacterium thermophilum* strain OC1 found in a mat at Ojo Caliente. Genome Announc 4:e01570-15. doi:10.1128/genomeA.01570-15

Hanada S, Pierson BK (2006) The family *Chloroflexaceae*. In: Dworkin M, Falkow S, Rosenberg E, Schleifer KH, Stackebrandt E (eds) The prokaryotes: a handbook on the biology of bacteria. Springer, New York, pp 815–842. doi:10.1007/0-387-30747-8_33

Hanada S, Hiraishi A, Shimada K et al (1995) *Chloroflexus aggregans* sp. nov., a filamentous phototrophic bacterium which forms dense cell aggregates by active gliding movement. Int J Syst Bacteriol 45:676–681. doi:10.1099/00207713-45-4-676

Hanada S, Takaichi S, Matsuura K, Nakamura K (2002) *Roseiflexus castenholzii* gen. nov., sp. nov., a thermophilic, filamentous, photosynthetic bacterium that lacks chlorosomes. Int J Syst Evol Microbiol 52:187–193. doi:10.1099/00207713-52-1-187

Hauruseu D, Koblížek M (2012) Influence of light on carbon utilization in aerobic anoxygenic phototrophs. Appl Environ Microbiol 78:7414–7419. doi:10.1128/AEM.01747-12

Henry EA, Devereux R, Maki JS et al (1994) Characterization of a new thermophilic sulfate-reducing bacterium. *Thermodesulfovibrio yellowstoneii*, gen. nov. and sp. nov.: its phylogenetic relationship to *Thermodesulfobacterium commune* and their origins deep within the bacterial domain. Arch Microbiol 161:62–69. doi:10.1007/BF00248894

Hisada TAH, Okamura K, Hiraishi A (2007) Isolation and characterization of phototrophic purple nonsulfur bacteria from *Chloroflexus* and cyanobacterial mats in hot springs. Microbes Environ 22:405–411. doi:10.1264/jsme2.22.405

Hohmann-Marriott MF, Blankenship RE (2011) Evolution of photosynthesis. Annu Rev Plant Biol 62:515–548. doi:10.1146/annurev-arplant-042110-103811

Holt AS, Hughes DW (1961) Studies of *Chlorobium* chlorophylls. III. *Chlorobium* chlorophyll (650). J Am Chem Soc 83:499–500. doi:10.1021/ja01463a067

Hoogewerf GJ, Jung DO, Madigan MT (2003) Evidence for limited species diversity of bacteriochlorophyll *b*-containing purple nonsulfur anoxygenic phototrophs in freshwater habitats. FEMS Microbiol Lett 218:359–364. doi:10.1016/S0378-1097(02)01195-3

Huber R, Eder W, Heldwein S et al (1998) *Thermocrinis ruber* gen. nov., sp. nov., a pink-filament-forming hyperthermophilic bacterium isolated from Yellowstone National Park. Appl Environ Microbiol 64:3576–3583

Hunter NC, Daldal F, Thurnayer MC, Beatty TJ (eds) (2009) The purple phototrophic bacteria. Advances in Photosynthesis and Respiration, vol 28. Springer, Dordrecht, The Netherlands. ISBN 978-1-4020-8814-8

Imachi H, Sakai S, Lipp JS et al (2014) *Pelolinea submarina* gen. nov., sp. nov., an anaerobic, filamentous bacterium of the phylum *Chloroflexi* isolated from subseafloor sediment. Int J Syst Evol Microbiol 64:812–818. doi:10.1099/ijs.0.057547-0

Imhoff JF (2014) The family *Chromatiaceae*. In: Rosenberg E, DeLong EF, Lory S et al (eds) The prokaryotes: gammaproteobacteria. Springer, Berlin/Heidelberg, pp 151–178. doi:10.1007/978-3-642-38922-1_295

Imhoff JF, Süling J (1996) The phylogenetic relationship among *Ectothiorhodospiraceae*: a reevaluation of their taxonomy on the basis of 16S rDNA analyses. Arch Microbiol 165:106–113. doi:10.1007/s002030050304

Imhoff JF, Trüper HG (1977) *Ectothiorhodospira halochloris* sp. nov., a new extremely halophilic phototrophic bacterium containing bacteriochlorophyll *b*. Arch Microbiol 114:115–121. doi:10.1007/BF00410772

Inoue K, Ono H, Abe-Yoshizumi R et al (2013) A light-driven sodium ion pump in marine bacteria. Nat Commun 4:1678–1687. doi:10.1038/ncomms2689

Jaschke PR, Saer RG, Noll S, Beatty JT (2011) Modification of the genome of *Rhodobacter sphaeroides* and construction of synthetic operons. Methods Enzymol 497:519–538. doi:10.1016/B978-0-12-385075-1.00023-8

Jensen SI, Steunou A-S, Bhaya D et al (2011) *In situ* dynamics of O_2, pH and cyanobacterial transcripts associated with CCM, photosynthesis and detoxification of ROS. ISME J 5:317–328. doi:10.1038/ismej.2010.131

Jiao N, Zhang Y, Zeng Y et al (2007) Distinct distribution pattern of abundance and diversity of aerobic anoxygenic phototrophic bacteria in the global ocean. Environ Microbiol 9:3091–3099. doi:10.1360/02wc0336

Kandori H (2015) Ion-pumping microbial rhodopsins. Front Mol Biosci 2:52. doi:10.3389/fmolb.2015.00052

Kanokratana P, Chanapan S, Pootanakit K, Eurwilaichitr L (2004) Diversity and abundance of bacteria and archaea in the Bor Khlueng hot spring in Thailand. J Basic Microbiol 44:430–444. doi:10.1002/jobm.200410388

Katoh H, Itoh S, Shen JR, Ikeuchi M (2001) Functional analysis of *psbV* and a novel *c*-type cytochrome gene *psbV2* of the thermophilic cyanobacterium *Thermosynechococcus elongatus* strain BP-1. Plant Cell Physiol 42:599–607. doi:10.1093/pcp/pce074

Keppen OI, Baulina OI, Kondratieva EN (1994) *Oscillochloris trichoides* neotype strain DG-6. Photosynth Res 41:29–33. doi:10.1007/BF02184143

Keppen OI, Tourova TP, Kuznetsov BB et al (2000) Proposal of *Oscillochloridaceae* fam. nov. on the basis of a phylogenetic analysis of the filamentous anoxygenic phototrophic bacteria, and emended description of *Oscillochloris* and *Oscillochloris trichoides* in comparison with further new isolates. Int J Syst Evol Microbiol 50:1529–1537. doi:10.1099/00207713-50-4-1529

Kim Y-M, Nowack S, Olsen MT et al (2015) Diel metabolomics analysis of a hot spring chlorophototrophic microbial mat leads to new hypotheses of community member metabolisms. Front Microbiol 6:209. doi:10.3389/fmicb.2015.00209

Kimble LK, Mandelco L, Woese CR, Madigan MT (1995) *Heliobacterium modesticaldum*, sp. nov., a thermophilic heliobacterium of hot springs and volcanic soils. Arch Microbiol 163:259–267. doi:10.1007/BF00393378

Klatt CG, Bryant DA, Ward DM (2007) Comparative genomics provides evidence for the 3-hydroxypropionate autotrophic pathway in filamentous anoxygenic phototrophic bacteria and in hot spring microbial mats. Environ Microbiol 9:2067–2078. doi:10.1111/j.1462-2920.2007.01323.x

Klatt CG, Wood JM, Rusch DB et al (2011) Community ecology of hot spring cyanobacterial mats: predominant populations and their functional potential. ISME J 5:1262–1278. doi:10.1038/ismej.2011.73

Klatt CG, Liu Z, Ludwig M et al (2013) Temporal metatranscriptomic patterning in phototrophic *Chloroflexi* inhabiting a microbial mat in a geothermal spring. ISME J 7:1775–1789. doi:10.1038/ismej.2013.52

Koblížek M (2015) Ecology of aerobic anoxygenic phototrophs in aquatic environments. FEMS Microbiol Rev 39:854–870. doi:10.1093/femsre/fuv032

Kuznetsov BB, Ivanovsky RN, Keppen OI et al (2011) Draft genome sequence of the anoxygenic filamentous phototrophic bacterium *Oscillochloris trichoides* subsp. DG-6. J Bacteriol 193:321–322. doi:10.1128/JB.00931-10

Lanyi JK (2006) Proton transfers in the bacteriorhodopsin photocycle. Biochim Biophys Acta 1757:1012–1018. doi:10.1016/j.bbabio.2005.11.003

Lau MC, Aitchison JC, Pointing SB (2009) Bacterial community composition in thermophilic microbial mats from five hot springs in central Tibet. Extremophiles 13:139–149. doi:10.1007/s00792-008-0205-3

Liu Z, Klatt CG, Wood JM et al (2011) Metatranscriptomic analyses of chlorophototrophs of a hot-spring microbial mat. ISME J 5:1279–1290. doi:10.1038/ismej.2011.37

Liu Z, Klatt CG, Ludwig M et al (2012) "*Candidatus* Thermochlorobacter aerophilum:" an aerobic chlorophotoheterotrophic member of the phylum *Chlorobi* defined by metagenomics and metatranscriptomics. ISME J 6:1869–1882. doi:10.1038/ismej.2012.24

Madigan MT (1984) A novel photosynthetic purple bacterium isolated from a Yellowstone hot spring. Science 225:313–315. doi:10.1126/science.225.4659.313

Madigan MT (1986) *Chromatium tepidum* sp. nov., a thermophilic photosynthetic bacterium of the family *Chromatiaceae*. Int J Syst Evol Microbiol 36:222–227. doi:10.1099/00207713-36-2-222

Madigan MT (2003) Anoxygenic phototrophic bacteria from extreme environments. Photosynth Res 76:157–171. doi:10.1023/A:1024998212684

Madigan MT, Brock TD (1977) Adaptation by hot spring phototrophs to reduced light intensities. Arch Microbiol 113:111–120. doi:10.1007/BF00428590

Madigan MT, Petersen SR, Brock TD (1974) Nutritional studies on *Chloroflexus*, a filamentous photosynthetic, gliding bacterium. Arch Microbiol 100:97–103. doi:10.1007/BF00446309

Maresca JA, Graham JE, Bryant DA (2008) The biochemical basis for structural diversity in the carotenoids of chlorophototrophic bacteria. Photosynth Res 97:121–140. doi:10.1007/s11120-008-9312-3

Maresca JA, Gomez Maqueo Chew A, Ros Ponsatí M et al (2004) The *bchU* gene of *Chlorobium tepidum* encodes the bacteriochlorophyll C-20 methyltransferase. J Bacteriol 186:2558–2566. doi:10.1128/JB.186.9.2558-2566.2004

Martinez-Garcia M, Swan BK, Poulton NJ et al (2011) High-throughput single-cell sequencing identifies photoheterotrophs and chemoautotrophs in freshwater bacterioplankton. ISME J 6:113–123. doi:10.1038/ismej.2011.84

Meeks JC, Castenholz R (1971) Growth and photosynthesis in an extreme thermophile, *Synechococcus lividus* (*Cyanophyta*). Arch Microbiol 78:25–41. doi:10.1007/BF00409086

Melendrez MC, Lange RK, Cohan FM, Ward DM (2011) Influence of molecular resolution on sequence-based discovery of ecological diversity among *Synechococcus* populations in an alkaline siliceous hot spring microbial mat. Appl Environ Microbiol 77:1359–1367. doi:10.1128/AEM.02032-10

Miller SR, Strong AL, Jones KL, Ungerer MC (2009) Bar-coded pyrosequencing reveals shared bacterial community properties along the temperature gradients of two alkaline hot springs in Yellowstone National Park. Appl Environ Microbiol 75:4565–4572. doi:10.1128/AEM.02792-08

Nowack S (2014) Niche character in a temporally varying environment. PhD thesis, Montana State University

Nowack S, Olsen MT, Schaible GA et al (2015) The molecular dimension of microbial species: 2. *Synechococcus* strains representative of putative ecotypes inhabiting different depths in the Mushroom Spring microbial mat exhibit different adaptive and acclimative responses to light. Front Microbiol 6:626. doi:10.3389/fmicb.2015.00626

Nübel U, Bateson MM, Madigan MT et al (2001) Diversity and distribution in hypersaline microbial mats of bacteria related to *Chloroflexus* spp. Appl Environ Microbiol 67:4365–4371. doi:10.1128/AEM.67.9.4365

Nübel U, Bateson MM, Vandieken V et al (2002) Microscopic examination of distribution and phenotypic properties of phylogenetically diverse *Chloroflexaceae*-related bacteria in hot spring microbial mats. Appl Environ Microbiol 68:4593–4603. doi:10.1128/AEM.68.9.4593

Nunoura T, Hirai M, Miyazaki M et al (2013) Isolation and characterization of a thermophilic, obligately anaerobic and heterotrophic marine *Chloroflexi* bacterium from a *Chloroflexi*-dominated microbial community associated with a Japanese shallow hydrothermal system, and proposal for *Thermomarinilinea lacunofontalis* gen. nov., sp. nov. Microbes Environ 28:228–235. doi:10.1264/jsme2.ME12193

Ohashi S, Iemura T, Okada N et al (2010) An overview on chlorophylls and quinones in the photosystem I-type reaction centers. Photosynth Res 104:305–319. doi:10.1007/s11120-010-9530-3

Olsen MT, Nowack S, Wood JM et al (2015) The molecular dimension of microbial species: 3. Comparative genomics of *Synechococcus* strains with different light responses and in situ diel transcription patterns of associated putative ecotypes in the Mushroom Spring microbial mat. Front Microbiol 6:604. doi:10.3389/fmicb.2015.00604

Oren A (2014) The family *Ectothiorhodospiraceae*. In: Rosenberg E, DeLong EF, Lory S et al (eds) The prokaryotes: gammaproteobacteria. Springer, Berlin Heidelberg, Berlin, Heidelberg, pp 199–222

Overmann J (2008) Ecology of phototrophic sulfur bacteria. In: Advances in Photosynthesis and Respiration, vol 27, Hell R, Dahl C, Knaff DB, Leustek T (eds) Sulfur metabolism in phototrophic organisms. Springer, Dordrecht, pp 375–396. doi:10.1007/978-1-4020-6863-8_19

Overmann J, Garcia-Pichel F (2013) The phototrophic way of life. In: Rosenberg E, DeLong EF, Lory S et al (eds) The prokaryotes: prokaryotic communities and ecophysiology. Springer, Berlin Heidelberg, pp 80–136. doi:10.1007/0-387-30742-7_3

Papke RT, Ramsing NB, Bateson MM, Ward DM (2003) Geographical isolation in hot spring cyanobacteria. Environ Microbiol 5:650–659. doi:10.1046/j.1462-2920.2003.00460.x

Pierson BK, Castenholz RW (1971) Bacteriochlorophylls in gliding filamentous prokaryotes from hot springs. Nat New Biol 233:25–27. doi:10.1038/newbio233025a0

Pierson BK, Castenholz RW (1974) A phototrophic gliding filamentous bacterium of hot springs, *Chloroflexus aurantiacus*, gen nov. sp. nov. Arch Microbiol 100:5–24. doi:10.1007/BF00446302

Podosokorskaya OA, Bonch-Osmolovskaya EA, Novikov AA et al (2013) *Ornatilinea apprima* gen. nov., sp. nov., a cellulolytic representative of the class *Anaerolineae*. Int J Syst Evol Microbiol 63:86–92. doi:10.1099/ijs.0.041012-0

Rabenstein A, Rethmeier J, Fischer U (1995) Sulphite as intermediate sulfur compound in anaerobic sulphide oxidation to thiosulphate by marine cyanobacteria. Z Naturforsch 50c:769–774. doi:10.1515/znc-1995-11-1206

Ragon M, Benzerara K, Moreira D et al (2014) 16S rDNA-based analysis reveals cosmopolitan occurrence but limited diversity of two cyanobacterial lineages with contrasted patterns of intracellular carbonate mineralization. Front Microbiol 5:1–11. doi:10.3389/fmicb.2014.00331

Ramsing NB, Ferris MJ, Ward DM (2000) Highly ordered vertical structure of *Synechococcus* populations within the one-millimeter-thick photic zone of a hot spring cyanobacterial mat. Appl Environ Microbiol 66:1038–1049. doi:10.1128/AEM.66.3.1038-1049.2000

Rappé MS, Connon SA, Vergin KL, Giovannoni SJ (2002) Cultivation of the ubiquitous SAR11 marine bacterioplankton clade. Nature 418:630–633. doi:10.1038/nature00917

Resnick S, Madigan MT (1989) Isolation and characterization of a mildly thermophilic nonsulfur purple bacterium containing bacteriochlorophyll *b*. FEMS Microbiol Lett 65:165–170. doi:10.1016/0378-1097(89)90385-6

Revsbech NP, Ward DM (1984) Microelectrode studies of interstitial water chemistry and photosynthetic activity in a hot spring microbial mat. Appl Environ Microbiol 48:270–275

Revsbech, NP, Trampe E, Lichtenberg M et al (2016). *In situ* hydrogen dynamics in a hot spring microbial mat during a diel cycle. Appl Environ Microbiol 82:4209–4217. doi:10.1128/AEM.00710-16

Reysenbach A, Wickham GS, Pace NR (1994) Phylogenetic analysis of the hyperthermophilic pink filament community in Octopus Spring, Yellowstone National Park. Appl Environ Microbiol 60:2113–2119

Reysenbach AL, Voytek M, Mancinelli R (eds) (2001) Thermophiles: biodiversity, ecology, and evolution. Kluwer Academic/Plenum Press, New York. doi:10.1007/978-1-4615-1197-7

Rippka R, Deruelles J, Waterbury JB et al (1979) Generic assignments, strain histories and properties of pure cultures of *Cyanobacteria*. J Gen Microbiol 111:1–61. doi:10.1099/00221287-111-1-1

Ross KA, Feazel LM, Robertson CE et al (2012) Phototrophic phylotypes dominate mesothermal microbial mats associated with hot springs in Yellowstone National Park. Microb Ecol 64:162–170. doi:10.1007/s00248-012-0012-3

Ruff-Roberts AL, Kuenen JG, Ward DM (1994) Distribution of cultivated and uncultivated cyanobacteria and *Chloroflexus*-like bacteria in hot spring microbial mats. Appl Environ Microbiol 60:697–704

Sadekar S, Raymond J, Blankenship RE (2006) Conservation of distantly related membrane proteins: photosynthetic reaction centers share a common structural core. Mol Biol Evol 23:2001–2007. doi:10.1093/molbev/msl079

Saiki R, Gelfand D, Stoffel S et al (1988) Primer-directed enzymatic amplification of DNA with a thermostable DNA polymerase. Science 239:487–491. doi:10.1126/science.2448875

Sandbeck KA, Ward DM (1981) Fate of immediate methane precursors in low-sulfate, hot-spring algal-bacterial mats. Appl Environ Microbiol 41:775–782

Sekiguchi Y, Yamada T, Hanada S et al (2003) *Anaerolinea thermophila* gen. nov., sp. nov. and *Caldilinea aerophila* gen. nov., sp. nov., novel filamentous thermophiles that represent a previously uncultured lineage of the domain *Bacteria* at the subphylum level. Int J Syst Evol Microbiol 53:1843–1851. doi:10.1099/ijs.0.02699-0

Shen G, Gan F, Bryant DA (2016) The siderophilic cyanobacterium *Leptolyngbya* sp. strain JSC-1 acclimates to iron starvation by expressing multiple isiA-family genes. Photosynth Res 128:325–340. doi:10.1007/s11120-016-0257-7

Sidler WA (1994) Phycobilisome and phycobiliprotein structures. In: Bryant DA (ed) Advances in photosynthesis and respiration, vol 1, the molecular biology of cyanobacteria. Springer, Berlin, pp 139–216

Sitz TO, Schmidt RR (1973) Purification of *Synechococcus lividus* by equilibrium centrifugation and its synchronization by differential centrifugation. J Bacteriol 115:43–46

Stanier RY, Cohen-Bazire G (1977) Phototrophic prokaryotes: the cyanobacteria. Ann Rev Microbiol 31:225–274. doi:10.1146/annurev.mi.31.100177.001301

Steensgaard DB, Van Walree CA, Bañeras L et al (1999) Evidence for spatially separate bacteriochlorophyll *c* and bacteriochlorophyll *d* pools within the chlorosomal aggregate of the green sulfur bacterium *Chlorobium limicola*. Photosyn Res 59:231–241. doi:10.1023/A:1006116101525

Steindler L, Schwalbach MS, Smith DP et al (2011) Energy-starved "*Candidatus* Pelagibacter ubique" substitutes light-mediated ATP production for endogenous carbon respiration. PLoS One 6, e19725. doi:10.1371/journal.pone.0019725

Steunou A-S, Bhaya D, Bateson MM et al (2006) *In situ* analysis of nitrogen fixation and metabolic switching in unicellular thermophilic cyanobacteria inhabiting hot spring microbial mats. Proc Natl Acad Sci U S A 103:2398–2403. doi:10.1073/pnas.0507513103

Steunou A-S, Jensen SI, Brecht E et al (2008) Regulation of *nif* gene expression and the energetics of N_2 fixation over the diel cycle in a hot spring microbial mat. ISME J 2:364–378. doi:10.1038/ismej.2007.117

Stevenson AK, Kimble LK, Woese CR, Madigan MT (1997) Characterization of new phototrophic heliobacteria and their habitats. Photosynth Res 53:1–12. doi:10.1023/A:1005802316419

Takaichi S, Tsuji K, Matsuura K, Shimada K (1995) A monocyclic carotenoid glucoside ester is a major carotenoid in the green filamentous bacterium *Chloroflexus aurantiacus*. Plant Cell Physiol 36:773–778

Takaichi S, Maoka T, Yamada M et al (2001) Absence of carotenes and presence of a tertiary methoxy group in a carotenoid from a thermophilic filamentous photosynthetic bacterium *Roseiflexus castenholzii*. Plant Cell Physiol 42:1355–1362. doi:10.1093/pcp/pce172

Tang KH, Feng X, Tang YJ, Blankenship RE (2009) Carbohydrate metabolism and carbon fixation in *Roseobacter denitrificans* OCh114. PLoS One 4, e7233. doi:10.1371/journal.pone.0007233

Tang KH, Tang YJ, Blankenship RE (2011) Carbon metabolic pathways in phototrophic bacteria and their broader evolutionary implications. Front Microbiol 2:165. doi:10.3389/fmicb.2011.00165

Tank M, Bryant DA (2015a) Nutrient requirements and growth physiology of the photoheterotrophic acidobacterium, *Chloracidobacterium thermophilum*. Front Microbiol 06:1–14. doi:10.3389/fmicb.2015.00226

Tank M, Bryant DA (2015b) *Chloracidobacterium thermophilum* gen. nov., sp. nov.: an anoxygenic microaerophilic chlorophotoheterotrophic acidobacterium. Int J Syst Evol Microbiol 65:1426–1430. doi:10.1099/ijs.0.000113

Thiel V, Hamilton TL, Tomsho LP et al (2014) Draft genome sequence of a sulfide-oxidizing, autotrophic filamentous anoxygenic phototrophic bacterium, *Chloroflexus* sp. strain MS-G (*Chloroflexi*). Genome Announc 2:e00872-14. doi:10.1128/genomeA.00872-14

Thiel V, Wood JM, Olsen MT et al (2016) The dark side of the Mushroom Spring microbial mat: life in the shadow of chlorophototrophs. I. Microbial diversity based on 16S rRNA amplicon and metagenomic sequencing. Front Microbiol 7:919. doi:10.3389/fmicb.2016.00919

Thiel V, Hügler M, Ward DM, Bryant DA (2017) The dark side of the Mushroom Spring microbial mat: life in the shadow of chlorophototrophs. II. Metabolic functions of abundant community members predicted from metagenomic analyses. Front. Microbiol., submitted for publication

Tsukatani Y, Romberger SP, Golbeck JH, Bryant DA (2012) Isolation and characterization of homodimeric type-I reaction center complex from *Candidatus* Chloracidobacterium thermophilum, an aerobic chlorophototroph. J Biol Chem 287:5720–5732. doi:10.1074/jbc.M111.323329

Ugalde JA, Podell S, Narasingarao P, Allen EE (2011) Xenorhodopsins, an enigmatic new class of microbial rhodopsins horizontally transferred between archaea and bacteria. Biol Direct 6:52. doi:10.1186/1745-6150-6-52

van der Meer MTJ, Schouten S, Bateson MM et al (2005) Diel variations in carbon metabolism by green nonsulfur-like bacteria in alkaline siliceous hot spring microbial mats from Yellowstone National Park. Appl Environ Microbiol 71:3978–3986

van der Meer MTJ, Schouten S, Sinninghe Damsté JS, Ward DM (2007) Impact of carbon metabolism on 13C signatures of cyanobacteria and green non-sulfur-like bacteria inhabiting a microbial mat from an alkaline siliceous hot spring in Yellowstone National Park (USA). Environ Microbiol 9:482–491. doi:10.1111/j.1462-2920.2006.01165.x

van der Meer MTJ, Klatt CG, Wood J et al (2010) Cultivation and genomic, nutritional, and lipid biomarker characterization of *Roseiflexus* strains closely related to predominant in situ populations inhabiting Yellowstone hot spring microbial mats. J Bacteriol 192:3033–3042. doi:10.1128/JB.01610-09

van Niel CB, Thayer LA (1930) Report on preliminary observations on the microflora in and near the hot springs in Yellowstone National park and their importance for the geological formations. YNP Lib File 7312, Mammoth, WY

Venkata Ramana V, Sasikala C, Takaichi S, Ramana CV (2010) *Roseomonas aestuarii* sp. nov., a bacteriochlorophyll*a* containing alphaproteobacterium isolated from an estuarine habitat of India. Syst Appl Microbiol 33:198–203. doi:10.1016/j.syapm.2009.09.004

Wahlund TM, Madigan MT (1993) Nitrogen fixation by the thermophilic green sulfur bacterium *Chlorobium tepidum*. J Bacteriol 175:474–478

Wahlund TM, Tabita FR (1997) The reductive tricarboxylic acid cycle of carbon dioxide assimilation: initial studies and purification of ATP-citrate lyase from the green sulfur bacterium Chlorobium tepidum. J Bacteriol 179:4859–4867

Wakao N, Yokoi N, Isoyama N et al (1996) Discovery of natural photosynthesis using Zn-containing bacteriochlorophyll in an aerobic bacterium *Acidiphilium rubrum*. Plant Cell Physiol 37:889–893. doi:10.1093/oxfordjournals.pcp.a029029

Ward DM (1998) A natural species concept for prokaryotes. Curr Opin Microbiol 1:271–7. doi:10.1016/S1369-5274(98)80029-5

Ward DM, Weller R, Bateson MM (1990) 16S rRNA sequences reveal numerous uncultured microorganisms in a natural community. Nature 346:183–187. doi:10.1038/346183a0

Ward DM, Bateson MM, Weller R, Ruff-Roberts AL (1992) Ribosomal RNA analysis of microorganisms as they occur in nature. In: Arshall K (ed) Advances in microbial ecology, vol 12. Springer, Boston, pp 219–286. doi:10.1007/978-1-4684-7609-5_5

Ward DM, Ferris MJ, Nold SC, Bateson MM (1998) A natural view of microbial biodiversity within hot spring cyanobacterial mat communities. Microbiol Mol Biol Rev 62:1353–1370

Ward DM, Bateson MM, Ferris MJ et al (2006) Cyanobacterial ecotypes in the microbial mat community of Mushroom Spring (Yellowstone National Park, Wyoming) as species-like units linking microbial community composition, structure and function. Philos Trans R Soc Lond B Biol Sci 361:1997–2008. doi:10.1098/rstb.2006.1919

Ward DM, Castenholz RW, Miller SR (2012a) Ecology of cyanobacteria II: *Cyanobacteria* in geothermal habitats. In: Whitton BA (ed) Ecology of cyanobacteria II. Springer, Dordrecht, pp 39–63. doi:10.1007/978-94-007-3855-3_3

Ward DM, Klatt CG, Wood J et al (2012b) Functional genomics and ecological and evolutionary context: maximizing the value of genomes in systems biology. In: Burnap R, Vermaas W (eds) Advances in photosynthesis and respiration, vol 33, Functional genomics and evolution of photosynthetic systems. Springer, Dordrecht, pp 1–16. doi:10.1007/978-94-007-1533-2

Weller R, Bateson MM, Heimbuch BK et al (1992) Uncultivated cyanobacteria, *Chloroflexus*-like inhabitants, and spirochete-like inhabitants of a hot spring microbial mat. Appl Environ Microbiol 58:3964–3969

Yamada T, Sekiguchi Y, Hanada S et al (2006) *Anaerolinea thermolimosa* sp. nov., *Levilinea saccharolytica* gen. nov., sp. nov. and *Leptolinea tardivitalis* gen. nov., sp. nov., novel filamentous anaerobes, and description of the new classes *Anaerolineae* classis nov. and *Caldilineae* classis nov. in the bacterial phylum *Chloroflexi*. Int J Syst Evol Microbiol 56:1331–1340. doi:10.1099/ijs.0.64169-0

Yamada T, Imachi H, Ohashi A et al (2007) *Bellilinea caldifistulae* gen. nov., sp. nov. and *Longilinea arvoryzae* gen. nov., sp. nov., strictly anaerobic, filamentous bacteria of the phylum *Chloroflexi* isolated from methanogenic propionate-degrading consortia. Int J Syst Evol Microbiol 57:2299–2306. doi:10.1099/ijs.0.65098-0

Yurkov V, Beatty JT (1998) Aerobic anoxygenic phototrophic bacteria. Microbiol Mol Biol Rev 62:695–724

Yurkov VV, Csotonyi JT (2009) New light on aerobic anoxygenic phototrophs. In: Hunter CN, Daldal F, Thurnauer MC, Beatty JT (eds) Advances in photosynthesis and respiration, vol 28, The purple phototrophic bacteria. Springer, Berlin, pp 31–55. doi:10.1007/978-1-4020-8815-5_3

Zeng Y, Feng F, Medová H et al (2014) Functional type 2 photosynthetic reaction centers found in the rare bacterial phylum *Gemmatimonadetes*. Proc Natl Acad Sci U S A 111:7795–7800. doi:10.1073/pnas.1400295111

Zeng Y, Baumbach J, Barbosa EGV et al (2016) Metagenomic evidence for the presence of phototrophic *Gemmatimonadetes* bacteria in diverse environments. Environ Microbiol Rep 8:n/a–n/a. doi:10.1111/1758-2229.12363

The *Chlorobiaceae*, *Chloroflexaceae*, and *Heliobacteriaceae*

Michael T. Madigan, Nicole A. Vander Schaaf, and W. Matthew Sattley

Abstract Here we examine the basic biology of three major groups of "green" anoxygenic phototrophic bacteria: the green sulfur bacteria (*Chlorobiaceae*), the green nonsulfur bacteria (also called the filamentous anoxygenic phototrophs) (*Chloroflexaceae*), and the heliobacteria (*Heliobacteriaceae*). Only organisms that have been grown in laboratory culture are considered. Interestingly, the model organisms for each family are thermophiles: the green sulfur bacterium *Chlorobaculum tepidum*, the filamentous green nonsulfur bacterium *Chloroflexus aurantiacus*, and the hot spring heliobacterium species, *Heliobacterium modesticaldum*. All model green bacteria have had their genomes sequenced, and in the green sulfur bacteria, genome sequences of all recognized species have been completed and compared. Although species in each family are distinct from species in each of the other families in many ways, there are key properties that unite two families to the exclusion of the third. These include the presence of chlorosomes in the green sulfur and green nonsulfur bacteria and the structure of the reaction centers in the green sulfur bacteria and heliobacteria. However, the three families of green-colored bacteria are phylogenetically distinct and thus any similarities are likely the result of horizontal gene transfers.

Keywords Green sulfur bacteria • Chlorosomes • *Chlorobium* • *Chlorobaculum tepidum* • "*Chlorochromatium aggregatum*" *Chlorobiaceae* • Green nonsulfur bacteria • *Chloroflexus aurantiacus* • *Chloroflexaceae* • *Roseiflexus* • Heliobacteria • *Heliobacterium modesticaldum* • *Heliorestis* • *Heliobacteriaceae* • Green sulfur bacteria phylogeny • Green nonsulfur bacteria phylogeny • Heliobacteria phylogeny

---

M.T. Madigan (✉)
Department of Microbiology, Southern Illinois University, Carbondale, IL 62901, USA
e-mail: madigan@siu.edu

N.A.V. Schaaf
Van Andel Research Institute, Grand Rapids, MI 49503, USA

W.M. Sattley
Division of Natural Sciences, Indiana Wesleyan University, Marion, IN 46953, USA

© Springer International Publishing Switzerland 2017
P.C. Hallenbeck (ed.), *Modern Topics in the Phototrophic Prokaryotes*,
DOI 10.1007/978-3-319-46261-5_4

Introduction

Anoxygenic phototrophic bacteria include species of the domain *Bacteria* that fall into six different phyla: the purple bacteria (*Proteobacteria*), green sulfur bacteria (*Chlorobi*), green nonsulfur bacteria—also called the filamentous green bacteria—(*Chloroflexi*), heliobacteria (*Firmicutes*), chloracidobacteria (*Acidobacteria*), and *Gemmatimonas* (*Gemmatimonadetes*). Cyanobacteria are of course phototrophs as well, but they are oxygenic phototrophs and are dealt with elsewhere in this volume.

This chapter will deal exclusively with "green" bacteria. To those in the field of anoxygenic phototrophic bacteria, the term green bacteria typically means the green sulfur and/or green nonsulfur bacteria. But we extend that here to include the heliobacteria, since these organisms share significant photochemical properties with green sulfur bacteria and are also green in color. Although they may be greenish in color, we will not deal here with either the chloracidobacteria (Bryant et al. 2007) or those species of purple bacteria that contain bacteriochlorophyll *b* and because of their complement of carotenoids, are actually green in color (Madigan and Jung 2009). These will be covered in other chapters of this volume.

The *Chlorobiaceae* and *Heliobacteriaceae* are extensive groups, containing a number of genera and species; *Chloroflexaceae*, by contrast, are less diverse, with only a few species obtained in pure culture. Green sulfur and green nonsulfur bacteria share a major property that unites them despite their distinctive phylogenies. Both of these groups contain chlorosomes—the ultimate in light-harvesting structures—that allows these phototrophs to grow at very low light intensities (Oelze and Golecki 1995; Psencík et al. 2009). Species of green sulfur and green nonsulfur bacteria also contain similar antenna pigments: bacteriochlorophylls (Bchls) *c*, *d*, or *e*, and these are packed in dense arrays within their chlorosomes. Heliobacteria, by contrast, lack chlorosomes and any discernable internal photosynthetic membrane system, but instead contain Bchl *g* as well as small amounts of chlorophyll *a* and chlorophyll *a* derivatives in the cytoplasmic membrane (Heinnickel and Golbeck 2007).

The physiologies and habitats of green sulfur bacteria, green nonsulfur bacteria, and the heliobacteria are also a study in contrasts. All of these phototrophs carry out photosynthesis only under anoxic conditions. Green sulfur bacteria and the heliobacteria are strict anaerobes (Asao and Madigan 2010; Imhoff 2014), whereas green nonsulfur bacteria are facultative and capable of respiratory growth in darkness (Pierson and Castenholz 1995). Green sulfur bacteria metabolism is primarily autotrophic. By contrast, heliobacteria are incapable of autotrophy and are dedicated photoheterotrophs when grown in the light; they can also grow chemotrophically in darkness by fermentation (Kimble et al. 1994). Green nonsulfur bacteria can grow as photoautotrophs, photoheterotrophs, or chemotrophs, the latter under highly oxic conditions (Pierson and Castenholz 1995).

Green sulfur bacteria are aquatic phototrophs found in the deepest waters of lakes, inland seas, and other anoxic aquatic habitats where sulfide is present but light levels limit the growth of other phototrophs (Imhoff 2014). Green nonsulfur bacteria are major components of microbial mats that form in the outflows of alkaline hot springs or that develop in shallow marine or hypersaline environments

(Castenholz and Pierson 1995). In contrast to both of these more-or-less aquatic groups, the heliobacteria are soil bacteria, inhabiting anoxic regions of soils where plant exudates likely supply them with the organic matter they need for photoheterotrophic metabolism (Asao and Madigan 2010).

We proceed now to look at these green bacteria with a focus on their diversity and phylogeny, photosystems and physiology, ecology, and genomics. Comprehensive reviews, each focusing on one of these major groups of phototrophs, have been published fairly recently (Asao and Madigan 2010; Hanada 2014; Imhoff 2014; Pierson and Castenholz 1995; Sattley and Swingley 2013) and the interested reader should consult these for more detailed information on any of the organisms covered here.

Chlorobiaceae: The Green Sulfur Bacteria

Diversity and Phylogeny

Green sulfur bacteria are a phylogenetically coherent group of anoxygenic phototrophs that, along with a few chemotrophs, form the phylum *Chlorobi* (Hiras et al. 2015; Imhoff 2014; Imhoff and Thiel 2010). This section will deal only with phototrophic species of *Chlorobi* that constitute the bacterial family *Chlorobiaceae* (Trüper and Pfennig 1992).

The green sulfur bacteria currently consist of four recognized genera: *Chlorobium* (*Chl.*) (Fig. 1), *Chlorobaculum* (*Cba.*), *Prosthecochloris*, and *Chloroherpeton* (*Chp.*). All genera contain several species except for *Chloroherpeton*, which consists of the single species *Chp. thalassium* (Gibson et al. 1984) (Fig. 2). The distribution of antenna Bchls (*c*, *d*, and *e*) shows no specific pattern among green sulfur bacteria, and in some cases, each of three strains of a single species may contain a

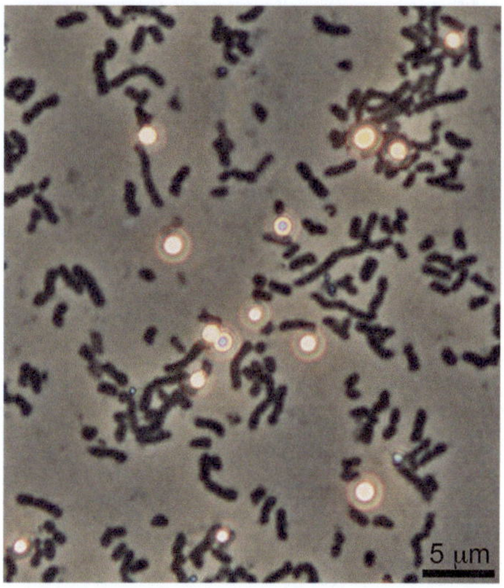

Fig. 1 Phase-contrast color photomicrograph of cells of *Chlorobium* strain PDL showing elemental sulfur globules formed from the oxidation of sulfide

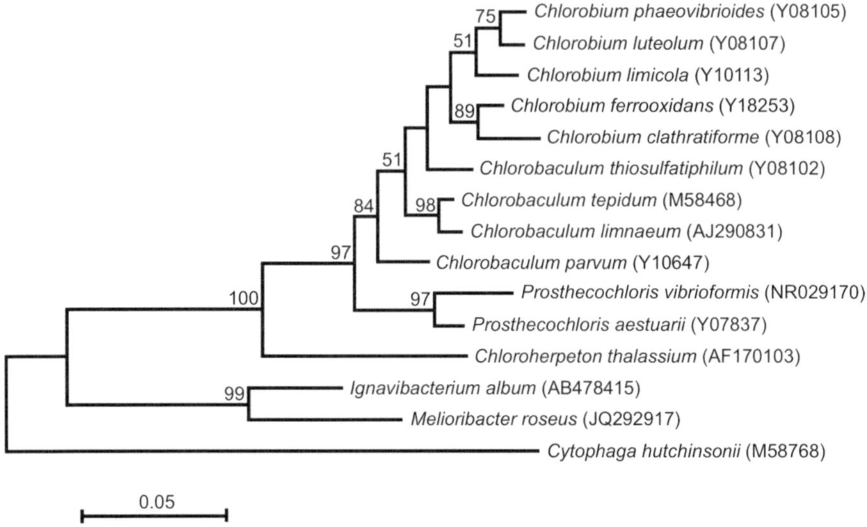

Fig. 2 Phylogenetic tree of cultured free-living species of *Chlorobiaceae* (green sulfur bacteria). This maximum-likelihood phylogenetic tree was created with 1000 bootstrap replications using the Tamura–Nei model and the Nearest-Neighbor-Interchange (NNI) heuristic method. All positions containing gaps and missing data were eliminated, resulting in a total of 1099 nucleotides in the final analysis. The tree was rooted using *Cytophaga hutchinsonii* (phylum *Bacteriodetes*) as the outgroup organism. Genbank accession numbers are listed in parentheses

different antenna pigment. The *Chlorobiaceae* include both freshwater and marine species and the latter require NaCl for growth (Imhoff 2014).

Green sulfur bacteria form a tight phylogenetic unit. Lying basal to the entire group is the marine microbial mat species *Chp. thalassium*. The salt-requiring *Prosthecochloris* species are also deeply branching taxa (Fig. 2). The rod-shaped *Chlorobium* (Fig. 1) and *Chlorobaculum* species are the most derived green sulfur bacteria, with species within each genera grouping fairly tightly. Although by definition, all species of green sulfur bacteria are anoxygenic phototrophs, at least two genera of chemotrophic bacteria—*Ignavibacterium* and *Melioribacter*—group fairly tightly with the *Chlorobi* but lie basal to all phototrophic species (Iino et al. 2010; Kadnikov et al. 2013; Podosokorskaya et al. 2013) (Fig. 2). Interestingly, both of these chemotrophs are moderate thermophiles and anaerobes, as is the phototrophic species *Chlorobaculum tepidum*, suggesting that the green sulfur bacteria may have emerged from thermophilic roots in which *Cba. tepidum*-like organisms are the only thermophilic species of green sulfur bacterium that remain.

Photosystems and Physiology

The major feature of the green sulfur bacterial photosystem is the chlorosome. Chlorosomes (Fig. 3) function as giant antenna systems and contain dense arrays of the light-harvesting Bchls of these phototrophs, Bchls *c*, *d*, or *e* (Oelze and Golecki

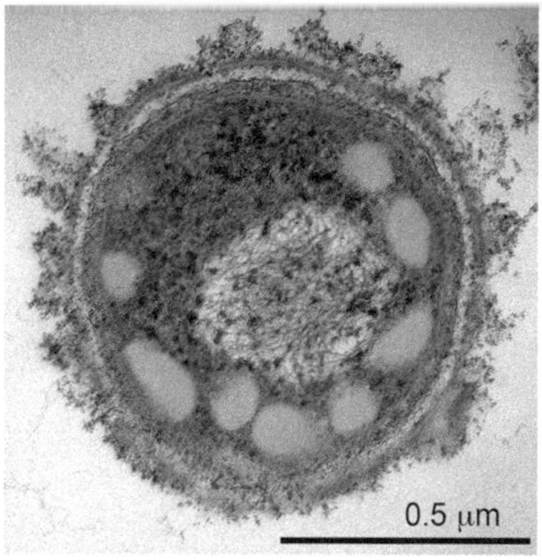

Fig. 3 Transmission electron micrograph of a cross-section of a cell of *Chlorobium* strain PDL. Note the chlorosomes (*light gray*) arranged in the periphery of the cell

1995; Psencík et al. 2009). Light energy absorbed by the antenna pigments is transferred to Bchl *a*, and the latter is present in two locations in the cell. First, Bchl *a* is present in the "Fenna–Matthew–Olson" (FMO) protein, the protein that is located between the chlorosome baseplate and the cytoplasmic membrane and which facilitates energy transmission from the antenna Bchls in the chlorosome to the reaction center embedded in the cytoplasmic membrane. However, in addition to the Bchl *a* in the FMO protein, Bchl *a* is present in the "iron–sulfur" type (type I) photosynthetic reaction center present in the cytoplasmic membrane of green bacteria (Blankenship 2010).

Besides Bchl *a*, the reaction center of green sulfur bacteria also contains small amounts of chlorophyll *a*. When this pigment is reduced by Bchl *a*, its extremely low reduction potential allows it to reduce the iron–sulfur center, also poised at a low reduction potential, and the iron–sulfur center can then reduce ferredoxin (He et al. 2104). The latter is needed to drive key carboxylation reactions in the unique autotrophic pathway of green sulfur bacteria (see later).

Because antenna Bchls are present in such large amounts relative to Bchl *a* in green sulfur bacteria, especially in cultures grown at low light intensities, the color of mass cultures of these phototrophs is typically green (Fig. 4) or brownish green; species containing Bchl *e* are distinctly brown in color. The colors are due to a combination of the antenna pigments, which are all green in color, along with carotenoid pigments. The major carotenoid in green-colored green sulfur bacteria such as *Cba. tepidum* (Fig. 4) is chlorobactene (or chlorobactene derivatives; Takaichi et al. 1997b) and that in brown-colored species is isorenieratene or β-isorenieratene (Imhoff 2014).

Green sulfur bacteria are obligately anaerobic and phototrophic bacteria. Despite numerous attempts to show dark growth, no success has been obtained. Thus, along with *Chloracidobacterium* (Tank and Bryant 2015), green sulfur bacteria are the only other phototrophic bacteria currently known to be unable to grow in darkness; purple sulfur and purple nonsulfur bacteria, green nonsulfur bacteria, heliobacteria, and certain cyanobacteria can all be grown in darkness (Madigan et al. 2015).

Green sulfur bacteria can be grown in pure culture in mineral media containing CO_2 as carbon source and H_2S as photosynthetic electron donor; species of

Fig. 4 Bottle culture of *Chlorobaculum tepidum* strain TLS. When grown in well-buffered media containing both sulfide and thiosulfate as electron donors, yields of 5×10^9 to 10^{10} cells/ml can easily be obtained overnight at 45 °C

Chlorobiaceae require only vitamin B_{12} as a growth factor and grow best around neutral or slightly acidic pH. A few organic compounds such as acetate, pyruvate, and propionate are photoassimilated by green sulfur bacteria and dramatically improve cell yields, but growth is ultimately dependent on a source of reduced sulfur, sulfide being optimal. During growth, sulfide (as well as thiosulfate by those species that can use it) is oxidized to sulfur (Fig. 1), which is then oxidized to sulfate (Gregersen et al. 2011).

Autotrophic growth in green sulfur bacteria is supported not by reactions of the Calvin cycle, the mechanism of autotrophy in phototrophic purple bacteria and cyanobacteria (as well as plants and algae), but instead by a reversal of steps in the citric acid cycle (also called the tricarboxylic acid cycle) referred to as the "reverse TCA cycle" (Evans et al. 1966; Fuchs et al. 1980). Green sulfur bacteria are the only group of anoxygenic phototrophic bacteria in which this pathway has been found to operate.

Ecology

Overviews of the ecology of green sulfur bacteria can be found in the classic reviews of Pfennig (1967), Madigan (1988), and van Gemerden and Mas (1995) (all of these are still well worth a read by anyone just entering this field), and the more recent review of Imhoff (2014). Because they possess chlorosomes (Fig. 3), green sulfur bacteria can thrive in illuminated habitats in which other phototrophs are light-limited. As a result of this and their typically extremely high sulfide tolerance, green sulfur bacteria can form blooms deep in sulfidic lakes or within sulfidic microbial mats where high cell densities greatly attenuate the available light. The ability of green sulfur bacteria to outcompete purple bacteria in light-limited aquatic systems means that they can position themselves deeper in the lake and closer to the source of sulfide emerging from the sediments. Isolates of green bacteria that are present at the lowest light intensities typically contain Bchl *e* as their antenna pigment, such as the population of brown-colored green sulfur bacteria that reside deep in the Black Sea (Manske et al. 2005).

Consortia consisting of green bacteria surrounding and firmly attached to a central motile and chemotrophic bacterium are common in many habitats in which free-living green sulfur bacteria are also present. These symbioses include associations such as "*Chlorochromatium aggregatum*" and "*Pelochromatium roseum*" where the phototrophic component is green or brown, respectively (Overmann 2010). These consortia allow the otherwise nonmotile green sulfur bacteria to move up and down in the water column in response to gradients of light intensity and sulfide.

The consortia consist of a single central bacterial cell named "*Candidatus* Symbiobacter mobilis" surrounded by about 15 cells of the green bacterium (referred to as the "epibiont") *Chlorobium chlorochromatii* (Overmann 2010; Vogl et al. 2006). The epibiont can be isolated and grown in pure culture and in most respects is similar to free-living chlorobia, although it is phylogenetically quite distant from other *Chlorobium* species and thus its taxonomic position is uncertain (Vogl et al. 2006). By contrast to *Chl. chlorochromatii*, the chemotrophic partner of the consortium has undergone widespread gene loss and cannot survive in the absence of its phototrophic partner (Liu et al. 2013).

Whether other benefits for the phototroph (other than motility) occur in these phototrophic consortial arrangements is unclear. However, it is clear that the central cell benefits from nutrients released by the phototroph. Electron micrographs have even suggested that the consortium may be more intimate than originally thought, in that the periplasm regions of the central cell and the epibionts may actually be fused. If true, this would be a direct and very convenient way for the two cells to share nutrients.

Perhaps free-living green bacteria have little need for motility because of their ability to grow at such minimal light intensities. However, it is still curious why no green sulfur bacterium has been found that is flagellated. Flagellar motility is common among anoxygenic phototrophs and is a widespread property of bacteria in general (Madigan et al. 2015). With rampant horizontal gene flow occurring in the bacterial world, why a flagellated green sulfur bacterium has not appeared in any of the many isolates of green bacteria obtained through the years is an interesting ecological conundrum.

Genomics

The rod-shaped and mildly thermophilic green bacterium *Chlorobaculum tepidum* was the first green sulfur bacterium to have its genome sequenced (Eisen et al. 2002). Because of its many desirable properties, in particular its rapid growth and use of thiosulfate as a photosynthetic electron donor, *Cba. tepidum* has emerged as the model species of *Chlorobiaceae*. The fact that this organism is genetically tractable by both conjugation (Azai et al. 2013; Wahlund and Madigan 1995) and transformation (Frigaard and Bryant 2001) has also bolstered its appeal as a laboratory model.

The genome of *Cba. tepidum*, about 2.15 Mbp, is relatively small for a bacterium capable of photoautotrophic growth. Since the complete genome of *C. tepidum* was sequenced, the genomes of many other green sulfur bacteria have been sequenced and analyzed, including species of *Chlorobium*, *Prosthecochloris*, and *Chloroherpeton* (Bryant et al. 2012; Davenport et al. 2010). No major surprises have emerged from the genomes of green sulfur bacteria, as much biochemistry on these phototrophs preceded the genome sequence. As expected, the *Cba. tepidum* genome confirmed the reverse citric acid autotrophic pathway and identified all genes necessary to carry out the processes of nitrogen fixation (Wahlund and Madigan 1995) and sulfur metabolism (Falkenby et al. 2011). The *Cba. tepidum* genome did reveal the genetic foundation for the structure of the chlorosome, the green bacterial photosynthetic reaction center, pigment biosynthesis biochemistry, and other key photosynthetic properties of this phototroph.

An interesting anomaly in the *Cba. tepidum* genome sequence was the relative dearth of genes encoding regulatory functions in this bacterium. Compared with the cyanobacterium *Synechocystis*, *Cba. tepidum* contains roughly eightfold fewer regulatory genes. This could represent selective gene loss due to the habitat of *Cba. tepidum*; the constant nutrient conditions in the hot spring environment in which this phototroph lives (Wahlund et al. 1991) would not require the regulatory machinery of a bacterium living a "feast-or-famine" existence. Thus, the genome of *Cba. tepidum* might have become streamlined to dispense with unused regulatory machinery.

Large scale *in silico* analyses of green bacterial genomes (Bryant et al. 2012; Davenport et al. 2010) have confirmed what 16S rRNA gene phylogenetic analyses first revealed (Fig. 2); with the exception of *Chloroherpeton*, the complement of genes in different genera and species of green sulfur bacteria are quite similar. There are a large number of shared genes (the core genome) that support basic photosynthetic properties in green bacteria, accompanied by a relatively small accessory (pan) genome that supports additional characteristics unique to the habitats of the individual species (Davenport et al. 2010).

One unusual feature of the genomes of *Chlorobium* and *Prosthecochloris* species but not that of *Cba. tepidum* is the presence of a few "giant genes" encoding proteins with a predicted molecular weight of about 1 million (Davenport et al. 2010). These possibly encode adhesin-like proteins that may function to help adhere cells to solid surfaces and/or protect the cell from grazing and other hostile activities. In addition, large scale comparative analyses have revealed that gene synteny in green sulfur bacteria is rather low. That is, although individual genes and operons within the genomes of green sulfur bacteria are quite highly related, there has been significant movement of gene clusters around the chromosomes of different species.

Chloroflexaceae: The Filamentous Green Nonsulfur Bacteria

Diversity and Phylogeny

The family *Chloroflexaceae* (phylum *Chloroflexi*) grew out of the discovery in 1974 of the filamentous, phototrophic, and thermophilic gliding bacterium *Chloroflexus aurantiacus* (Pierson and Castenholz 1974). Because *Chloroflexus* was so different from all known anoxygenic phototrophs, Trüper (1976) formally proposed that a family-level taxon be established to accommodate *Chloroflexus* and organisms related to it. Other genera previously grouped within the *Chloroflexaceae*, such as *Roseiflexus* and *Oscillochloris*, have recently been elevated to their own family status (Gupta et al. 2013). In this section, however, we lump all of these into the *Chloroflexaceae* for convenience and will focus our discussion on the genera *Chloroflexus* and *Roseiflexus* only.

Roseiflexus (Hanada et al. 2002; van der Meer et al. 2010) is a distant relative of *Chloroflexus*, but morphologically, species of these two genera are virtually indistinguishable. Cells of both *Chloroflexus* and *Roseiflexus* are long thin filaments (Fig. 5) that form dense cell masses along with filamentous cyanobacteria in alkaline hot spring or hypersaline microbial mats. Because of their distinct filamentous morphology, these organisms are also called the "filamentous anoxygenic phototrophs" or FAPs for short. Species of the two genera differ fundamentally in their photosynthetic properties, the major difference being the presence of green bacterial antenna Bchls and chlorosomes in species of *Chloroflexus* that are absent from species of *Roseiflexus* (Hanada 2014). Genomic analyses to be discussed later indicate that the core genomes of these two phototrophs are very similar.

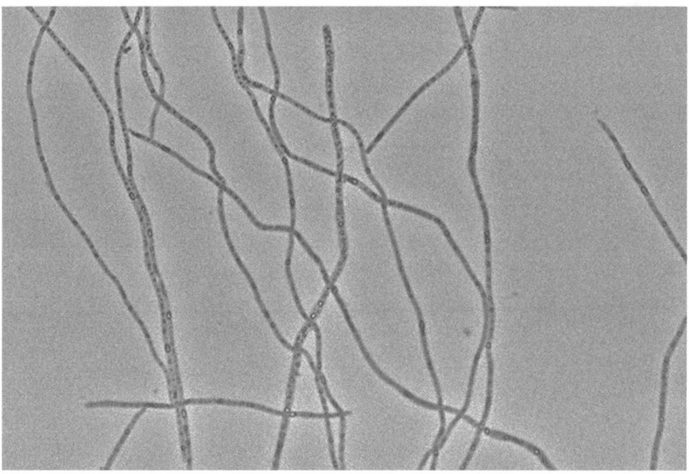

Fig. 5 Phase-contrast photomicrograph of cells of *Roseiflexus* strain RS-1. Cells are about 1 μm in diameter and of indeterminate length

From a phylogenetic perspective, the *Chloroflexaceae* form a tight group of phototrophic species including the genera *Chloroflexus* (*Cfl.*), *Chloronema*, *Oscillochloris*, *Heliothrix*, and *Roseiflexus* (Fig. 6). Of these five genera, only strains of *Chloroflexus* and *Roseiflexus* have been studied in any detail and are the only genera for which pure cultures are available. The phylogenetic tree of *Chloroflexaceae* reveals three main clades: the *Chloroflexus* group, including *Cfl. aurantiacus*, *Cfl. aggregans*, and various strains of *Chloroflexus* related to *Cfl. aurantiacus*; the *Oscillochloris*/*Chloronema* group; and the *Roseiflexus* group (Fig. 6). *Heliothrix* diverges significantly from all other *Chloroflexaceae* and its precise phylogenetic position is unclear since cultures of this organism are not available; its phylogenetic position in Fig. 6 is based on only a partial 16S rRNA sequence. *Roseiflexus* is the most deeply divergent of known phototrophic *Chloroflexi*, and its closest relative is the nonphototrophic filamentous hot spring bacterium *Herpetosiphon* (Fig. 6). The review by Gupta et al. (2013) and Hanada (2014) should be consulted for further information on the current systematics of phototrophic species of *Chloroflexi*.

The discovery of *Chloroflexus* has turned out to be a remarkable story, as aspects of its biology have significantly affected our understanding of the evolution of photosynthesis as well as the diversity of autotrophic metabolism. The fact that species of *Chloroflexus* are major components of the thick microbial biofilms called stro-

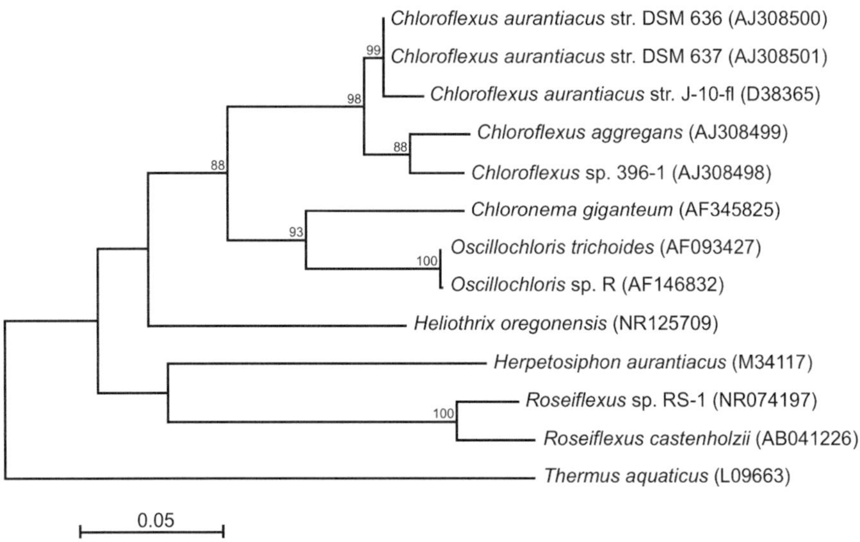

Fig. 6 Phylogenetic tree of cultured species of *Chloroflexaceae* (green nonsulfur bacteria). This maximum-likelihood phylogenetic tree was created with 1000 bootstrap replications using the Tamura–Nei model and the Nearest-Neighbor-Interchange (NNI) heuristic method. All positions containing gaps and missing data were eliminated, resulting in a total of 785 nucleotides in the final analysis (this number of nucleotides was chosen based on the fact that the available *Heliothrix* 16S rRNA gene sequence is only a partial sequence). The tree was rooted using *Thermus aquaticus* (phylum *Deinococcus-Thermus*) as the outgroup organism. Genbank accession numbers are listed in parentheses

matolites (fossilized structures of which have been found in rocks nearly 3.5 billion years old) combined with its deep phylogenetic position on the tree of *Bacteria* and unusual mechanism of autotrophy (Madigan et al. 2015) has led many to believe that *Chloroflexus* has many of the properties of ancient—perhaps even the earliest—phototrophs to evolve on Earth (Blankenship 2010).

Photosystems and Physiology

The 5.3 Mb genome of *Chloroflexus aurantiacus* (Tang et al. 2011) confirmed most of the features of the photosystems and physiology of *Chloroflexus* that had been worked out with this readily cultured phototroph in the 40 years that ensued since its discovery. *Cfl. aurantiacus* is very much a hybrid phototroph, as it contains some features from purple bacteria, green bacteria, and heliobacteria, as well as some features from cyanobacteria; many of these features likely arose in *Cfl. aurantiacus* through horizontal gene transfers from species of these other phototrophic groups (Blankenship 2010). The main components of the *Cfl. aurantiacus* photosystem are its chlorosomes, which contain Bchl *c* (as do those of many green sulfur bacteria), and a quinone-type photosynthetic reaction center (as do all purple bacteria) (Blankenship 2010; Psencík et al. 2009). The reaction center in *Cfl. aurantiacus* is structurally simpler than that of purple bacteria but is clearly of the purple bacterial type (Tang et al. 2011).

Chloroflexus grows both phototrophically (anoxic/light) and aerobically in darkness by respiration (Pierson and Castenholz 1974, 1995). Phototrophic growth includes both photoheterotrophic and photoautotrophic lifestyles. When growing photoautotrophically, *Cfl. aurantiacus* employs the 3-hydroxypropionate pathway for CO_2 fixation, a pathway first discovered in this organism and whose biochemistry was worked out incrementally but has now been solidified (Zarzycki et al. 2009). *Chloroflexus* can also grow fully aerobically by respiring a wide variety of organic compounds (Madigan et al. 1974). The organism can be grown in defined media both phototrophically and chemotrophically but grows best in complex media. Thiamine and biotin are required as growth factors, but such can be replaced with just a small amount (50 mg/l) of yeast extract. The ease by which *C. aurantiacus* can be grown in the laboratory has made it an attractive experimental tool in the field of bacterial photosynthesis.

Roseiflexus differs from *Chloroflexus* primarily by its lack of chlorosomes and their associated green bacterial antenna pigments. *Rof. castenholzii* is readily culturable, and from studies of this species, and to a more limited extent *Roseiflexus* strain RS-1, the parallels between *Chloroflexus* and *Roseiflexus* have become clearer. Both phototrophs can be grown both phototrophically and chemotrophically and both are thermophiles that coexist in alkaline hot spring microbial mats (Hanada et al. 2002; van der Meer et al. 2010). Depending on the strain and the light intensity used to grow the culture, cultures of *Cfl. aurantiacus* can be anywhere from a deep green, to brownish- or yellowish-green, to a deep orange in color; cultures grown with sulfide typically are greenest in color. By contrast, cultures of *Roseiflexus* show little if any greenish tinge and are either reddish-brown or yellow-brown in color. The major carotenoids of *Cfl. aurantiacus* are γ-carotene and β-carotene (Halfen et al. 1972; Pierson and Castenholz 1974), whereas in *Rof. castenholzii*, γ- and β-carotene are absent and instead a variety of keto-carotenes and γ-carotene derivatives are present (Takaichi et al. 2001).

Ecology

Chloroflexaceae are dedicated microbial mat species. These organisms develop in hot spring microbial mats along with various cyanobacteria at temperatures up to about 60 °C (Castenholz and Pierson 1995). In shallow marine and hypersaline systems where microbial mats develop at more moderate temperatures, *Chloroflexus*-like organisms are typically present and tend to form thicker mats than in thermal systems (Pierson and Castenholz 1995; Castenholz and Pierson 1995).

Filamentous anoxygenic phototrophs inhabit both sulfidic mats and mats that contain no detectable sulfide, and thus their metabolic activities in nature probably depend on the physiochemical state of their habitat. In sulfidic mats fed by geothermal sulfide or bacterial sulfate reduction, opportunities for photoautotrophy exist. However, the abundance of cyanobacteria in most *Chloroflexus* mats likely offers ideal opportunities for photoheterotrophy during the day and chemotrophy at night from the organic matter excreted by these oxygenic phototrophs.

Oxygen is a major variable in microbial mats and its consumption and production is a highly dynamic process. Measurements using oxygen microelectrodes show that in hot spring mats, O_2 is produced in the uppermost layers of the mats and can be totally consumed just a few millimeters into the mat (Revsbech and Ward 1984). Sulfide microelectrode measurements in a sulfate-limited Yellowstone microbial mat showed maximal rates of sulfate reduction about 4 mm into the mat and that sulfide-oxidizing bacteria, likely *Chloroflexus* and *Roseiflexus*, are likely responsible for regenerating sulfate from sulfide to maintain the high rates of sulfate reduction. The thermophilic sulfate-reducing bacterium *Thermodesulfovibrio* was the major sulfide producer in this system (Dillon et al. 2007). In marine microbial mats, sulfide is likely to be much more available throughout the mat due to the abundant sulfate present in marine waters (Visscher et al. 1992). However, as in the hot spring habitat, the presence of cyanobacteria in marine mats likely means that oxygen gradients exist and that an abundance of organic matter is available for photoheterotrophic or chemotrophic growth of filamentous anoxygenic phototrophs as well.

A metatranscriptomic study of filamentous green nonsulfur bacteria was carried out in a Yellowstone hot spring mat community to see how *Chloroflexus* and *Roseiflexus* regulate gene expression in response to a diel cycle that affects oxygen levels within the mat. The data showed that at night, when conditions throughout the mat are microoxic to anoxic, these organisms prepare for the next light cycle by synthesizing photopigments and related photosynthetic machinery along with carbon storage polymers such as glycogen and polyhydroxyalkanoates. During the day these polymers are degraded to support photoheterotrophic growth (Klatt et al. 2013). Further, this study confirmed that environmental conditions within these mats are highly dynamic and that the *Chloroflexus* and *Roseiflexus* populations in the mat likely undergo periodic and fairly rapid switches from phototrophy to chemotrophy and vice-versa as physiochemical conditions change on a diel cycle.

Genomics

The complete genome of *Chloroflexus aurantiacus* strain J-10-fl (isolated from a Japanese hot spring and the type strain of the genus and species) has been sequenced (Tang et al. 2011), and draft genome sequences are available for *Roseiflexus castenholzii* and *Roseiflexus* strain RS-1. The J-10-fl sequence of 5,258,541 base pairs encodes 3914 genes, of which 72 % have predicted functions. A draft genome sequence has also been published for a *Chloroflexus* isolate from a Yellowstone hot spring (Thiel et al. 2014).

Despite lacking features such as chlorosomes and the associated antenna features of *Chloroflexus*, the *Roseiflexus* genome is 0.5 Mb larger than that of *Chloroflexus*. However, much of the biochemistry is similar in these two organisms; for example, the complement of genes encoding the 3-hydroxypropionate pathway of autotrophic CO_2 fixation in *Chloroflexus* (Zarzycki et al. 2009) is also present in *Roseiflexus*. Although autotrophic growth in *Roseiflexus* has not been demonstrated, such is likely a matter of simply working with the right strain and getting growth conditions, especially the sulfide concentration, just right, as was the case with *Chloroflexus*. In this connection, autotrophic metabolism in *Chloroflexus*, which was first revealed by simple growth experiments with a single strain over 40 years ago (Madigan and Brock 1975), was only finalized in a biochemical sense rather recently (Zarzycki et al. 2009).

Heliobacteriaceae: The Heliobacteria

Diversity and Phylogeny

The family *Heliobacteriaceae* (phylum *Firmicutes*) consists of 11 isolated, described, and effectively published bacterial species (Madigan et al. 2010; Sattley and Madigan 2014) and one provisional taxon, "*Candidatus* Heliomonas lunata," which has thus far resisted axenic culture (Asao et al. 2012). Unlike all other *Firmicutes*, heliobacteria (a name derived from the Greek "helios," meaning "sun") can grow phototrophically, and they do this using a relatively simple light-harvesting photosystem that contains pigments found in no other phototrophs. Ultrathin sections of heliobacterial cells show that, like other *Firmicutes* but unlike all other

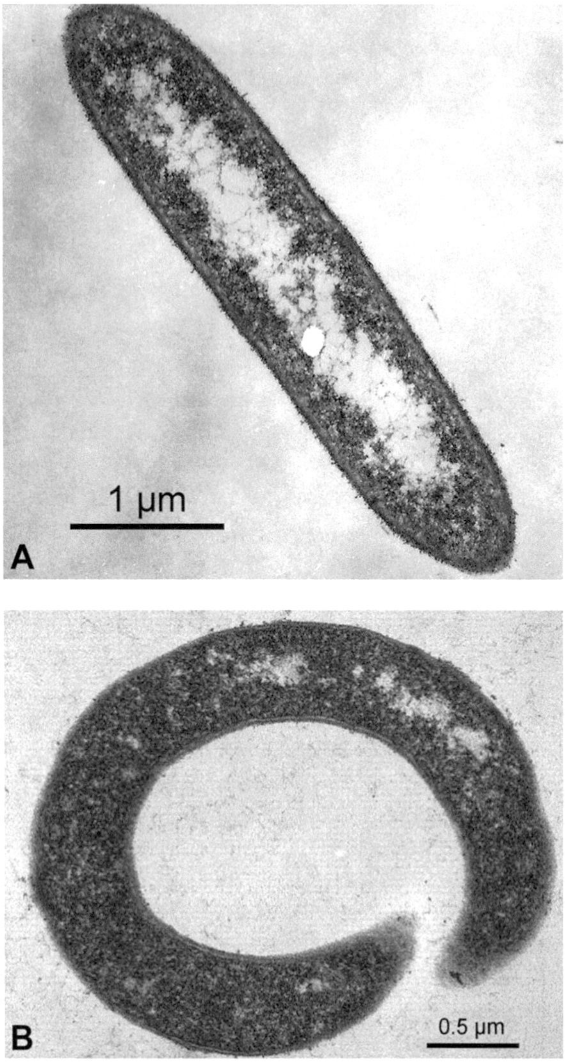

Fig. 7 Transmission electron micrographs of cells of (**a**) *Heliobacterium modesticaldum* and (**b**) *Heliorestis convoluta*. Note the lack of both an outer membrane and specialized, intracytoplasmic light-harvesting structures such as the chlorosomes of green sulfur and green nonsulfur bacteria or the membrane vesicles of purple bacteria

phototrophic bacteria, heliobacteria have a gram-positive cell wall structure characterized by a thick layer of peptidoglycan and no outer membrane (Fig. 7).

Similar to species of *Clostridium*, *Desulfotomaculum*, and *Bacillus*, to which they are closely related, heliobacteria are primarily soil inhabitants that are able to differentiate into heat-resistant endospores that allow the cells to remain viable during inhospitable environmental conditions. Enrichment cultures for heliobacteria can be successfully established from soil samples that have been pasteurized (80 °C for 15 min) before inoculation. This heating process eliminates competing anoxygenic phototrophs, such as purple or green bacteria, but does not affect the viability of heliobacterial endospores. Centrally or subterminally located endospores are often visible in successful primary enrichments for heliobacteria, but for an unknown reason, endospores are rarely observed in highly enriched or pure cultures (Kimble-Long and Madigan 2001; Stevenson et al. 1997).

Heliobacteria are strict anaerobes, and therefore care must be taken to achieve and maintain anoxic culture conditions at all times. Healthy, anoxic cultures of heliobacteria are brownish-green to olive drab in color. By contrast, cultures exposed to oxygen in the presence of light turn emerald green due to irreversible oxidation of the primary photosynthetic pigment, Bchl *g*, and this condition is accompanied by rapid loss of cell viability (Sattley and Madigan 2014; Sattley et al. 2014).

Heliobacteria described to date are either neutrophilic or alkaliphilic. Neutrophilic species inhabit various soils, including those associated with hot springs, and are classified into three genera: *Heliobacterium* (the type genus, consisting of five species), *Heliobacillus* (one species), and *Heliophilum* (one species). Alkaliphilic heliobacteria—isolated from soils and sediments of soda lakes—include species of the genus *Heliorestis* (four total), as well as the candidate taxon "Heliomonas.". Based on 16S ribosomal RNA gene sequence analyses, species of these genera form a monophyletic group of two distinct clades in neighbor-joining phylogenetic trees (Fig. 8). The major clade contains all species of the neutrophilic genera (i.e., *Heliobacterium,*

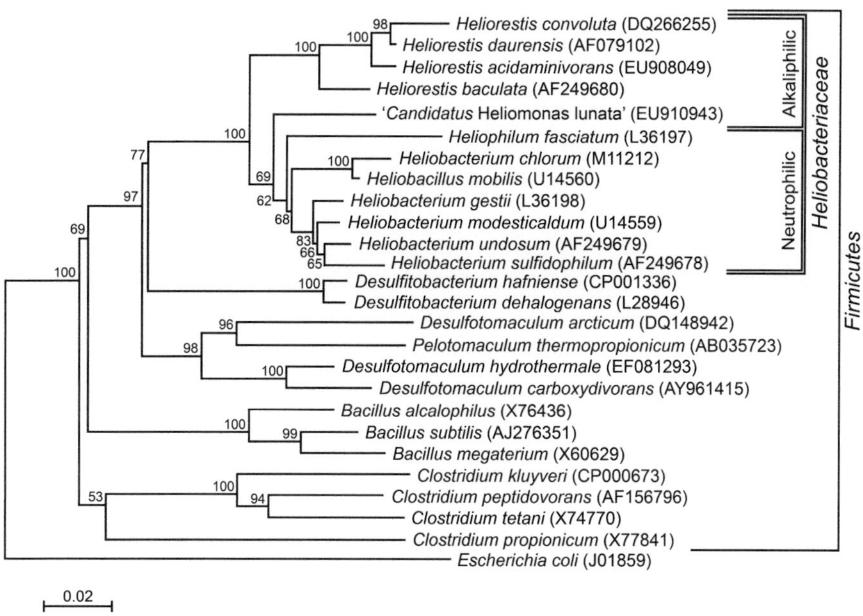

Fig. 8 Phylogenetic tree of cultured heliobacteria and related *Firmicutes*. Only members of the *Heliobacteriaceae* are phototrophic. The tree was generated using the weighted neighbor-joining method in conjunction with the Jukes–Cantor corrected distance model. Bootstrap values (≥50 %) based on 100 replicates are indicated at the nodes. The tree was rooted using *Escherichia coli*, a gammaproteobacterium. GenBank accession numbers indicated in parentheses. Adapted from Sattley and Swingley (2013)

Heliobacillus, and *Heliophilum*) and, interestingly, also the alkaliphile "*Candidatus* Heliomonas lunata." The minor clade contains all four species of the alkaliphilic genus *Heliorestis*. The positioning of "*Candidatus* Heliomonas lunata" at the base of the neutrophilic heliobacterial clade rather than within the *Heliorestis* clade suggests either an alkaliphilic origin for the heliobacteria or the possibility that adaptation to alkaline environments has developed more than once in this group.

With nearly 98 % identical 16S rRNA gene sequences, *Heliobacillus* (*Hba.*) *mobilis* and *Heliobacterium* (*Hbt.*) *chlorum* group more closely together than does *Hbt. chlorum* with the four other species of *Heliobacterium* (*Hbt. gestii*, *Hbt. modesticaldum*, *Hbt. undosum*, and *Hbt. sulfidophilum*), with which it shares <96 % 16S rRNA gene sequence identity. Although their habitats differ significantly, *Hbt. gestii*, isolated from rice paddy soil, forms a tight phylogenetic cluster with *Hbt. modesticaldum*, *Hbt. undosum*, and *Hbt. sulfidophilum*, all of which were isolated from the perimeters of hot springs. The four alkaliphilic *Heliorestis* species (*Heliorestis* [*Hrs.*] *convoluta*, *Hrs. daurensis*, *Hrs. acidaminivorans*, and *Hrs. baculata*) share 16S rRNA gene sequence identities ranging from about 94 to 98 % and form a clade that is also neatly clustered and is clearly resolved from the major clade (Fig. 8). *Heliophilum* (*Hph.*) *fasciatum*, the only representative of the genus, shows the greatest degree of divergence from other heliobacteria. Like *Hbt. gestii*, *Hph. fasciatum* was isolated from rice paddy soils, but it shows <93 % 16S rRNA gene sequence identity to any other heliobacterium.

The different species of heliobacteria exhibit a high degree of morphological variation. Cells of most species of heliobacteria are rod-shaped with diameters of 0.6–1.2 μm and lengths ranging from 2 to 12 μm (Sattley and Madigan 2014). Filaments of about 20 μm are typical for cells of *Hbt. undosum* and *Hrs. daurensis*. Cells of *Hbt. gestii* are spirilla, and cells of *Hrs. convoluta* are unusual tightly coiled rods of varying lengths. Cells of *Hph. fasciatum* are distinctive in that they are tapered rods that aggregate into bundles and coordinate their motility as a single unit (Ormerod et al. 1996). Except for *Hbt. chlorum*, which exhibits gliding motility (Gest and Favinger 1983), and *Hrs. convoluta*, which has an unknown mechanism of motility, all heliobacteria are motile by polar or subpolar flagella, often in a peritrichous arrangement (Sattley and Madigan 2014).

Photosystems and Physiology

Heliobacteria grow optimally under photoheterotrophic conditions using pyruvate as carbon source, but a limited number of other carbon sources, such as lactate, acetate, propionate, and butyrate, are also photoassimilated by various species (Sattley and Madigan 2014; Sattley et al. 2014). In addition, neutrophilic species are capable of dark, chemotrophic growth via pyruvate fermentation (Kimble et al. 1994). Unlike other anaerobic, anoxygenic (non O_2-evolving) phototrophic bacteria, autotrophic growth (CO_2 plus H_2 or H_2S) has not been observed in any species of heliobacteria (Sattley et al. 2014); we discuss the genetic basis for this observation later. Although somewhat limited in usable carbon sources, nutritional versatility in heliobacteria is boosted by the fact that nearly all species tested are strong nitrogen-fixers (Asao and Madigan 2010; Kimble and Madigan 1992). The only reported exception in this regard is "*Candidatus* Heliomonas lunata," in which nitrogen-starved conditions failed to induce nitrogenase activity (Asao et al. 2012).

The defining feature of all heliobacteria is that they synthesize the unique pigment bacteriochlorophyll *g* as their sole bacteriochlorophyll to facilitate phototrophic growth (Brockmann and Lipinski 1983). With peak absorbance in the near infrared (785–790 nm), Bchl *g* allows heliobacteria to exploit wavelengths of light not used by competing phototrophs (Madigan 2006). In addition to Bchl *g*, heliobacteria synthesize small amounts of 8^1-OH-Chlorophyll (Chl) *a* (van de Meent et al. 1991). Both pigments are esterified with farnesol rather than the more typical phytol (Takaichi et al. 1997a, b); see Sattley and Swingley (2013) for proposed biosynthetic mechanisms for these pigments. Heliobacteria also produce unique accessory pigments in the form of C_{30} carotenoids—primarily 4,4′-diaponeurosporene or, in alkaliphilic species, OH-diaponeurosporene glucoside esters; all other phototrophs produce C_{40} carotenoids (Takaichi et al. 2003).

Most phototrophic bacteria concentrate photosynthetic antenna pigments in internal membrane systems, such as the chlorosomes of green bacteria, chromatophores or lamellae in purple phototrophs, or thylakoids in *Cyanobacteria*. In contrast to this, heliobacterial pigments are confined to comparatively small, light-harvesting core complexes bound to the cytoplasmic membrane. The heliobacterial reaction center (HbRC) in these complexes has a type-I (iron–sulfur), homodimeric architecture similar to what is observed in the *Chlorobiaceae* (Oh-Oka 2007), but with no communicating antenna system, the HbRC is the simplest known photosynthetic apparatus (Heinnickel and Golbeck 2007; Sattley et al. 2008).

Although a high-resolution, three-dimensional structure of the HbRC is not yet available, studies of purified RCs from *Hbt. modesticaldum* have shown that each RC homodimer binds the following: 20 Bchl *g*, two Bchl *g*′ (the special pair primary electron donor, P798), two 8^1-OH-Chl *a* (the primary electron acceptor, A_0), ~1.6 menaquinone (the secondary electron acceptor, A_1, primarily consisting of MQ-9), and one 4-4′-diaponeurosporene (Kobayashi et al. 1991; Kondo et al. 2015; Sarrou et al. 2012; Trost and Blankenship 1989). During electron transfer reactions, a proton motive force builds as protons are pumped across the cytoplasmic membrane via a cytochrome (cyt) *bc* complex, allowing for energy conservation and ATP synthesis (Heinnickel and Golbeck 2007; Sattley et al. 2008). Electrons from cyt *bc* are shuttled to the RC through a membrane-bound, diheme cyt c_{553}, which is tethered to the outer leaflet of the cytoplasmic membrane (Oh-Oka et al. 2002). Electrons then pass through the RC and are transferred to a cytoplasmic ferredoxin that provides reducing power for nitrogen fixation and other metabolic processes (Heinnickel and Golbeck 2007; Sattley et al. 2008).

Ecology

Most heliobacteria that have been cultured and isolated thus far are soil microorganisms, and dry rice paddy soils are among the most reliable sources of inocula for establishing successful enrichment cultures (Ormerod et al. 1996;

Stevenson et al. 1997). It is likely that such soils contain significant numbers of heliobacterial endospores and considerably lower numbers of viable purple or green bacteria, which do not produce endospores and would thus be much more vulnerable to desiccation during the dry season (Madigan 2006). As mentioned, pasteurization of such samples prior to inoculation of enrichment culture media allows heliobacterial endospores, if present, to germinate and the resulting vegetative cells to propagate in the absence of competing phototrophs during incubation.

Agricultural soils that support crops other than rice have been much less successful in yielding positive enrichment cultures of heliobacteria. This is possibly because unlike rice paddies, these soils do not routinely experience alternating flooding and desiccation, and therefore the selective advantage that might allow endospore-forming phototrophs to flourish is absent. Alternatively, heliobacteria may establish a mutualistic relationship with rice plants in a manner reminiscent of that between rhizobia and legume plants. Although heliobacteria are free living and do not form root nodules with rice plants, their strong capacity for nitrogen fixation may provide the plants with a supply of usable nitrogen in exchange for organic exudates from the plant roots (Madigan 2006). The transfer of nutrients in the rhizosphere would be easily facilitated during the growing season when the paddy soils are flooded. In addition, anoxic microenvironments in the saturated soil would allow the strictly anaerobic heliobacteria to thrive during this time.

Soils of hot springs and soda lakes have also been shown to be suitable habitats for heliobacteria. *Hbt. undosum*, *Hbt. sulfidophilum*, and *Hbt. modesticaldum* were all isolated from hot spring soils, but only *Hbt. modesticaldum* is thermophilic, with optimal growth occurring at 50–52 °C (Kimble et al. 1995). To date, no heliobacterium has been isolated from hot spring waters, but hot spring microbial mats (containing a mixture of cyanobacteria, *Chloroflexus* and *Roseiflexus* plus many chemotrophs) have yielded additional strains of *Hbt. modesticaldum* (Madigan, unpublished). Similarly, all described isolates of the alkaliphilic genus *Heliorestis* were obtained from shoreline soils of soda lakes. The heliobacterium "*Candidatus* Heliomonas lunata" was enriched from a sediment/water sample of a soda lake and has been maintained in coculture with an unidentified chemotrophic contaminant (Asao et al. 2012). This hints that the diversity of environments suitable for habitation by heliobacteria may be more extensive than first realized.

Genomics

Heliobacteria have now been extensively studied for over 30 years, and the thermophile *Heliobacterium modesticaldum* has proven especially useful for biochemical and genomic studies of heliobacterial photochemistry and physiology (Sattley et al. 2008; Sattley and Blankenship 2010). The *Hbt. modesticaldum* strain Ice1 genome, described by Sattley et al. (2008), consists of a single chromosome of nearly 3.1 Mbp with a G+C content of 57 % and no plasmids. The total number of predicted open reading frames was 3138, with 3000 of these predicted to be protein-encoding genes (see Sattley and Blankenship 2010 for a summary of the protein-encoding genes predicted to play a major role in heliobacterial phototrophy and physiology). In addition, 104 tRNA genes were identified in the *Hbt. modesticaldum* genome, as well as 24 rRNA genes (in eight operons); hypothetical proteins constituted 35 % of the genome (Sattley et al. 2008).

The functional gene category containing the largest number of genes (389, constituting 13 % of the total genetic content) was energy and central intermediary metabolism. Other sizable functional gene categories included cellular processes (e.g., cell division, motility, and sporulation) (273; 9.1 %); protein synthesis, modification, and degradation (247; 8.2 %); and regulatory functions and signal transduction (177; 5.9 %). For a complete list of genetic content in all functional categories, see Sattley et al. (2008).

The obligately heterotrophic metabolism of heliobacteria is unusual for phototrophic bacteria, and the *Hbt. modesticaldum* genome revealed the reason for this autotrophic deficiency. Genes encoding key enzymes for autotrophic pathways found in other phototrophic bacteria, including the Calvin cycle, reverse citric acid cycle, and the hydroxypropionate pathway, were absent in the *Hbt. modesticaldum* genome (Sattley et al. 2008). Interestingly, all of the genes necessary to fix carbon via the reverse citric acid cycle were present except for the gene encoding citrate lyase, an enzyme that splits citrate into oxaloacetate and acetyl-CoA (Sattley et al. 2008). Therefore, if not for the lack of this gene, heliobacteria could presumably fix CO_2 by the same mechanism as green sulfur bacteria (Evans et al. 1966; Fuchs et al. 1980).

The complete genome of *Hbt. modesticaldum* was sequenced several years ago, and other heliobacterial genomes will likely become available in the near future. This should allow for comparative genomic analyses that will elucidate the genetic and evolutionary details of phototrophic metabolisms and other key cellular processes, including nitrogen fixation, carbon and energy metabolism, cell division, chemotaxis, and endosporulation. The relative simplicity of the photosynthetic system in the heliobacteria (Sattley et al. 2008) suggests that this group in particular has an important role to play in unraveling a more detailed understanding of the evolution of photosynthesis in coming years.

Acknowledgements MTM acknowledges current support from the NASA Exobiology Program.

References

Asao M, Madigan MT (2010) Taxonomy, phylogeny, and ecology of the heliobacteria. Photosynth Res 104:103–111

Asao M, Takaichi S, Madigan MT (2012) Amino acid-assimilating phototrophic heliobacteria from soda lake environments: *Heliorestis acidaminivorans* sp. nov. and '*Candidatus* Heliomonas lunata'. Extremophiles 16:585–595

Azai C, Harada J, Oh-oka H (2013) Gene expression system in green sulfur bacteria by conjugative plasmid transfer. PLoS One 8, e82345. doi:10.1371/journal.pone.0082345

Blankenship RE (2010) Early evolution of photosynthesis. Plant Physiol 154:434–438

Brockmann H, Lipinski A (1983) Bacteriochlorophyll *g*. A new bacteriochlorophyll from *Heliobacterium chlorum*. Arch Microbiol 136:17–19

Bryant DA, Costas AMG, Maresca JA et al (2007) *Candidatus* Chloracidobacterium thermophilum: an aerobic phototrophic acidobacterium. Science 317:523–526

Bryant DA, Liu Z, Li T et al (2012) Comparative and functional genomics of anoxygenic green bacteria from the taxa *Chlorobi*, *Chloroflexi*, and *Acidobacteria*. In: Burnap RL, Vermaas WFJ (eds) Functional genomics and evolution of photosynthetic systems. Advances in photosynthesis and respiration, vol 33. Springer, New York, pp 47–102

Castenholz RW, Pierson BK (1995) Ecology of thermophilic anoxygenic phototrophs. In: Blankenship RE, Madigan MT, Bauer CE (eds) Anoxygenic photosynthetic bacteria. Kluwer, Dordrecht, pp 87–103

Davenport C, Ussery DW, Tümmler B (2010) Comparative genomics of green sulfur bacteria. Photosynth Res 104:137–152

Dillon JG, Fishbain S, Miller SR et al (2007) High rates of sulfate reduction in a low-sulfate hot spring microbial mat are driven by a low level of diversity of sulfate-respiring microorganisms. Appl Environ Microbiol 73:5218–5226

Eisen JA, Nelson KE, Paulsen IT et al (2002) The complete genome sequence of *Chlorobium tepidum* TLS, a photosynthetic, anaerobic, green sulfur bacterium. Proc Natl Acad Sci U S A 99:9509–9514

Evans MCW, Buchanan BB, Arnon DI (1966) A new ferredoxin-dependent carbon reduction cycle in a photosynthetic bacterium. Proc Natl Acad Sci U S A 55(4):928–934

Falkenby LG, Szymanska M, Holkenbrink C et al (2011) Quantitative proteomics of *Chlorobaculum tepidum*: insights into the sulfur metabolism of a phototrophic green sulfur bacterium. FEMS Microbiol Lett 323:142–150

Frigaard N-U, Bryant DA (2001) Chromosomal gene inactivation in the green sulfur bacterium *Chlorobium tepidum* by natural transformation. Appl Environ Microbiol 67:2538–2544

Fuchs G, Stupperich E, Eden G (1980) Autotrophic CO_2 fixation in *Chlorobium limicola*. Evidence for the operation of a reductive tricarboxylic acid cycle in growing cells. Arch Microbiol 128:64–71

Gest H, Favinger JL (1983) *Heliobacterium chlorum*, an anoxygenic brownish-green photosynthetic bacterium containing a "new" form of bacteriochlorophyll. Arch Microbiol 136:11–16

Gibson J, Pfennig N, Waterbury JB (1984) *Chloroherpeton thalassium* gen. nov. et spec. nov., a non-filamentous, flexing and gliding green sulfur bacterium. Arch Microbiol 138:96–101

Gregersen LH, Bryant DA, Frigaard N-U (2011) Mechanisms and evolution of oxidative sulfur metabolism in green sulfur bacteria. Front Microbiol. doi:10.3389/fmicb.2011.00116

Gupta RS, Chander P, George S (2013) Phylogenetic framework and molecular signatures for the class *Chloroflexi* and its different clades; proposal for division of the class *Chloroflexi* class. nov. into the suborder *Chloroflexineae* subord. nov., consisting of the emended family *Oscillochloridaceae* and the family *Chloroflexaceae* fam. nov., and the suborder *Roseiflexineae* subord. nov., containing the family *Roseiflexaceae* fam. nov. Antonie van Leeuwenhoek 103:99–119

Halfen LN, Pierson BK, Francis GW (1972) Carotenoids of a gliding organism containing bacteriochlorophylls. Arch Mikrobiol 82:240–246

Hanada S (2014) The Phylum Chloroflexi, the family *Chloroflexaceae*, and the related phototrophic families *Oscillochloridaceae* and *Roseiflexaceae*. In: Rosenberg E et al (eds), The prokaryotes–other major lineages of bacteria and the archaea, pp 515–532. doi:10.1007/978-3-642-38954-2_165

Hanada S, Takaichi S et al (2002) *Roseiflexus castenholzii* gen. nov., sp. nov., a thermophilic, filamentous, photosynthetic bacterium that lacks chlorosomes. Int J Syst Evol Microbiol 52:187–193

He G, Zhang H, King JD et al (2104) Structural analysis of the homodimeric reaction center complex from the photosynthetic green sulfur bacterium *Chlorobaculum tepidum*. Biochemistry 53:4924–4930

Heinnickel M, Golbeck JH (2007) Heliobacterial photosynthesis. Photosynth Res 92:35–53

Hiras J, Wu YW, Eichorst SA et al (2015) Refining the phylum Chlorobi by resolving the phylogeny and metabolic potential of the representative of a deep branching uncultivated lineage. ISME J 10:833–845

Iino T, Mori K, Uchino Y et al (2010) *Ignavibacterium album* gen. nov., sp. nov., a moderately thermophilic anaerobic bacterium isolated from microbial mats at a terrestrial hot spring and proposal of *Ingavibacteria* classis nov., for a novel lineage at the periphery of green sulfur bacteria. Int J Syst Evol Microbiol 60:1376–1382

Imhoff JF (2014) Biology of green sulfur bacteria. In: eLS. John Wiley & Sons Ltd, Chichester. doi:10.1002/9780470015902.a0000458.pub2

Imhoff JF, Thiel V (2010) Taxonomy and phylogeny of *Chlorobiaceae*. Photosynth Res 104:123–136

Kadnikov VV, Mardanov AV, Podosokorskaya OA et al (2013) Genomic analysis of *Melioribacter roseus*, facultatively anaerobic organotrophic bacterium representing a novel deep lineage within *Bacteroidetes/Chlorobi* group. PLoS One 8, e53047

Kimble LK, Stevenson AK, Madigan MT (1994) Chemotrophic growth of heliobacteria in darkness. FEMS Microbiol Lett 115:51–55

Kimble LK, Madigan MT (1992) Nitrogen fixation and nitrogen metabolism in heliobacteria. Arch Microbiol 158:155–161

Kimble LK, Mandelco L, Woese CR et al (1995) *Heliobacterium modesticaldum*, sp. nov., a thermophilic heliobacterium of hot springs and volcanic soils. Arch Microbiol 163:259–267

Kimble-Long LK, Madigan MT (2001) Molecular evidence that the capacity for endosporulation is universal among phototrophic heliobacteria. FEMS Microbiol Lett 199:191–195

Klatt CG, Liu Z, Ludwig M et al (2013) Temporal metatranscriptomic patterning in phototrophic *Chloroflexi* inhabiting a microbial mat in a geothermal spring. ISME J 7:1775–1789. doi:10.1038/ismej.2013.52

Kobayashi M, Watanabe T, Ikegami I et al (1991) Enrichment of bacteriochlorophyll *g′* in membranes of *Heliobacterium chlorum* by ether extraction: unequivocal evidence for its existence in vivo. FEBS Lett 284:129–131

Kondo T, Itoh S, Matsuoka M, Azai C et al (2015) Menaquinone as the secondary electron acceptor in the type I homodimeric photosynthetic reaction center of Heliobacterium modesticaldum. J Phys Chem B 119:8480–8489

Liu Z, Muller J, Li T et al (2013) Genomic analysis reveals key aspects of prokaryotic symbiosis in the phototrophic consortium "*Chlorochromatium aggregatum*". Genome Biol 14:R127

Madigan MT (1988) Microbiology, physiology, and ecology of phototrophic bacteria. In: Zehnder AJB (ed) Biology of anaerobic microorganisms. Wiley, New York, pp 39–111

Madigan MT (2006) The Family *Heliobacteriaceae*. Prokaryotes 4:951–964

Madigan MT, Brock TD (1975) Photosynthetic sulfide oxidation by *Chloroflexus aurantiacus*, a filamentous, photosynthetic gliding bacterium. J Bacteriol 122:782–784

Madigan MT, Jung DO (2009) An overview of purple bacteria: systematics, physiology, and habitats. In: Hunter CN, Daldal F, Thurnauer MC, Beatty JT (eds) The purple phototrophic bacteria. Springer, Dordrecht, pp 1–15

Madigan MT, Petersen SR, Brock TD (1974) Nutritional studies on *Chloroflexus*, a filamentous, photosynthetic gliding bacterium. Arch Microbiol 100:97–103

Madigan MT, Euzéby JP, Asao M (2010) Proposal of *Heliobacteriaceae* fam. nov. Int J Syst Evol Microbiol 60:1709–1710

Madigan MT, Martinko JM, Bender KS et al (2015) Brock biology of microorganisms, 14th edn. Pearson, San Francisco

Manske AK, Glaeser J, Kuypers MMM et al (2005) Physiology and phylogeny of green sulfur bacteria forming a monospecific phototrophic assemblage at a depth of 100 m in the Black Sea. Appl Environ Microbiol 71:8049–8060

Oelze J, Golecki JR (1995) Membranes and chlorosomes of green bacteria: structure, composition and development. In: Blankenship RE, Madigan MT, Bauer CE (eds) Anoxygenic photosynthetic bacteria. Kluwer, Dordrecht, pp 259–278

Oh-Oka H (2007) Type 1 reaction center of photosynthetic heliobacteria. Photochem Photobiol 83:177–186

Oh-Oka H, Iwaki M, Itoh S (2002) Electron donation from membrane-bound cytochrome *c* to the photosynthetic reaction center in whole cells and isolated membranes of *Heliobacterium gestii*. Photosynth Res 71:137–147

Ormerod JG, Kimble LK, Nesbakken T et al (1996) *Heliophilum fasciatum* gen. nov. sp. nov. and *Heliobacterium gestii* sp. nov.: endopore-forming heliobacteria from rice field soils. Arch Microbiol 165:226–234

Overmann J (2010) The phototrophic consortium "Chlorochromatium aggregatum"--a model for bacterial heterologous multicellularity. In: Hallenbeck PC (ed) Recent advances in phototrophic prokaryotes. Springer, Berlin, pp 15–29

Pfennig N (1967) Photosynthetic bacteria. Ann Rev Microbiol 21:285–324

Pierson BK, Castenholz RW (1974) A phototrophic gliding filamentous bacterium of hot springs, *Chloroflexus aurantiacus*, gen. and sp. nov. Arch Microbiol 100:5–24

Pierson BK, Castenholz RW (1995) Taxonomy and physiology of filamentous anoxygenic phototrophs, pp 31–47. In: Hunter CN, Daldal F, Thurnauer MC, Beatty JT (eds) The purple phototrophic bacteria. Springer, Dordrecht, pp 1–15

Podosokorskaya OA, Kadnikov VV, Gavrilov SN et al (2013) Characterization of *Melioribacter roseus* gen. nov., sp. nov., a novel facultatively anaerobic thermophilic cellulolytic bacterium from the class *Ignavibacteria*, and a proposal of a novel bacterial phylum *Ignavibacteriae*. Environ Microbiol. doi:10.1111/1462–2920.12067

Psencík J, Collins AM, Liljeroos L et al (2009) Structure of chlorosomes from the green filamentous bacterium *Chloroflexus aurantiacus*. J Bacteriol 191:6701–6708

Revsbech NP, Ward DM (1984) Microelectrode studies of interstitial water chemistry and photosynthetic activity in a hot spring microbial mat. Appl Environ Microbiol 48:270–275

Sarrou I, Khan Z, Cowgill J et al (2012) Purification of the photosynthetic reaction center from *Heliobacterium modesticaldum*. Photosynth Res 111:291–302

Sattley WM, Blankenship RE (2010) Insights into heliobacterial photosynthesis and physiology from the genome of *Heliobacterium modesticaldum*. Photosynth Res 104:113–122

Sattley WM, Madigan MT (2014) The Family *Heliobacteriaceae*. In: Rosenberg E et al (eds) The prokaryotes–*Firmicutes* and *Tenericutes*. Springer, Dordrecht, pp 185–196. doi:10.1007/9783-642-30120-9_362

Sattley WM, Swingley WD (2013) Properties and evolutionary implications of the heliobacterial genome. In: Beatty JT (ed) Genome evolution of photosynthetic bacteria. Advances in botanical research, vol 66. Academic Press Elsevier, Amsterdam, pp 67–97

Sattley WM, Madigan MT, Swingley WD et al (2008) The genome of *Heliobacterium modesticaldum*, a phototrophic representative of the *Firmicutes* containing the simplest photosynthetic apparatus. J Bacteriol 190:4687–4696

Sattley WM, Asao M, Tang KH et al (2014) Energy conservation in heliobacteria: photosynthesis and central carbon metabolism. In: Hohmann-Marriott MF (ed) The structural basis of biological energy generation. Advances in photosynthesis and respiration, vol 39. Springer, Dordrecht, pp 231–247

Stevenson AK, Kimble LK, Woese CR et al (1997) Characterization of new heliobacteria and their habitats. Photosynth Res 53:1–12

Takaichi S, Inoue K, Akaike M et al (1997a) The major carotenoid in all species of heliobacteria is the C_{30} carotenoid 4,4′-diaponeurosporene, not neurosporene. Arch Microbiol 168:277–281

Takaichi S, Wang ZY, Umetsu M et al (1997b) New carotenoids from the thermophilic green sulfur bacterium *Chlorobium tepidum*: 1′2′-didehydro-γ-carotene, 1′2′-dihydro-chlorobactene, and OH-chlorobactene glucoside ester, and the carotenoid composition of different strains. Arch Microbiol 168:270–276

Takaichi S, Maoka T, Yamada M et al (2001) Absence of carotenes and presence of a tertiary methoxy group in a carotenoid from a thermophilic filamentous photosynthetic bacterium *Roseiflexus castenholzii*. Plant Cell Physiol 42:1355–1362

Takaichi S, Oh-Oka H, Maoka T et al (2003) Novel carotenoid glucoside esters from alkaliphilic heliobacteria. Arch Microbiol 179:95–100

Tang K-H, Barry K, Chertkov O et al (2011) Complete genome sequence of the filamentous anoxygenic phototrophic bacterium *Chloroflexus aurantiacus*. BMC Genomics 12:334

Tank M, Bryant DA (2015) Nutrient requirements and growth physiology of the photoheterotrophic Acidobacterium *Chloroacidobacterium thermophilum*. Front Microbiol. doi:10.3389/fmicb.2015.00226

Thiel V, Hamilton TL, Tomsho LP et al (2014) Draft genome sequence of a sulfide-oxidizing, autotrophic filamentous anoxygenic phototrophic bacterium, *Chloroflexus* sp. strain MS-G (*Chloroflexi*). Genome Announc 2:e0087214. doi:10.1128/genomeA.00872-14

Trost JT, Blankenship RE (1989) Isolation of a photoactive photosynthetic reaction center-core antenna complex from *Heliobacillus mobilis*. Biochemistry 28:9898–9904

Trüper HG (1976) Higher taxa of the phototrophic bacteria: *Chloroflexaceae* fam. nov., a family for the gliding filamentous phototrophic 'green' bacteria. Int J Syst Bacteriol 26:74–75

Trüper HG, Pfennig N (1992) The family Chlorobiaceae. In: Balows A, Trüper HG, Dworkin M, Harder W, Schleifer KH (eds) The prokaryotes, 2nd edn, A handbook on the biology of bacteria. Ecophysiology, isolation, identification, applications. Springer, New York, pp 3583–3592

van de Meent EJ, Kobayashi M, Erkelens C et al (1991) Identification of 8^1-hydroxychlorophyll *a* as a functional reaction center pigment in heliobacteria. Biochim Biophys Acta 1058:356–362

van der Meer MTJ, Klatt C, Wood J et al (2010) Cultivation and genomic, nutritional, and lipid biomarker characterization of *Roseiflexus* strains closely related to predominant *in situ* populations inhabiting Yellowstone hot spring microbial mats. J Bacteriol 192:3033–3042

Van Gemerden H, Mas J (1995) Ecology of phototrophic sulfur bacteria. In: Blankenship RE, Madigan MT, Bauer CE (eds) Anoxygenic photosynthetic bacteria. Kluwer, Dordrecht, pp 49–85

Visscher PT, Prins R, van Gemerden H (1992) Rates of sulfate reduction and thiosulfate consumption in a marine microbial mat. FEMS Microbiol Lett 86:283–293

Vogl K, Glaeser J, Pfannes KR et al (2006) *Chlorobium chlorochromatii* sp nov., a symbiotic green sulfur bacterium isolated from the phototrophic consortium 'Chlorochromatium aggregatum'. Arch Microbiol 185:363–372

Wahlund TM, Madigan MT (1995) Genetic transfer by conjugation in the thermophilic green sulfur bacterium *Chlorobium tepidum*. J Bacteriol 177:81–90

Wahlund TM, Woese CR, Castenholz RW et al (1991) A thermophilic green sulfur bacterium from New Zealand hot springs, *Chlorobium tepidum* sp. nov. Arch Microbiol 156:81–90

Zarzycki J, Brecht V, Muller M et al (2009) Identifying the missing steps of the autotrophic 3-hydroxypropionate CO_2 fixation cycle in *Chloroflexus aurantiacus*. Proc Natl Acad Sci U S A 106:21317–21322

Phototrophic Gemmatimonadetes: A New "Purple" Branch on the Bacterial Tree of Life

Yonghui Zeng and Michal Koblížek

Abstract Photosynthesis first emerged in prokaryotes over three billion years ago and represents one of the most fundamental biological processes on Earth. So far, species capable of performing (bacterio)chlorophyll-based phototrophy have been reported in seven bacterial phyla, i.e., Cyanobacteria, Proteobacteria, Chlorobi, Chloroflexi, Firmicutes, Acidobacteria, and Gemmatimonadetes. Here we review the discovery, physiology, genomic characteristics, environmental distribution, and possible evolutionary origin of the bacterium *Gemmatimonas phototrophica* strain AP64, so far the only phototrophic member of the phylum Gemmatimonadetes. This organism was isolated from a freshwater lake in the Gobi Desert, North China in 2011. It contains fully functional type-2 photosynthetic reaction centers, but they seem to only serve as an auxiliary energy source. Its photosynthesis genes are located in a 42.3 kb long photosynthesis gene cluster which appear to originate from an ancient horizontal gene transfer from a purple phototrophic bacterium. A survey of biomarker genes of phototrophic Gemmatimonadetes bacteria (PGB) in public environmental genomics databases suggests that PGB are widely distributed in diverse environments, including air, river waters/sediment, estuarine waters, lake waters, biofilms, plant surfaces, intertidal sediments, soils, springs, and wastewater treatment plants, but none from marine waters or sediment. PGB make up roughly 0.4–11.9 % of whole phototrophic microbial communities in these habitats. The discovery of PGB presents a strong evidence that genes for anoxygenic phototrophy can be transferred between distant bacterial phyla, providing new insights into the evolution of bacterial photosynthesis.

Y. Zeng (✉)
Nordic Center for Earth Evolution (NordCEE) & Institute of Biology, University of Southern Denmark, Odense 5230, Denmark

Aarhus Institute of Advanced Studies (AIAS), Aarhus University, DK-8000 Aarhus C, Denmark
e-mail: yonghui.sci@gmail.com

M. Koblížek (✉)
Center Algatech, Institute of Microbiology CAS, Třeboň 37981, Czech Republic
e-mail: koblizek@alga.cz

© Springer International Publishing Switzerland 2017

P.C. Hallenbeck (ed.), *Modern Topics in the Phototrophic Prokaryotes*, DOI 10.1007/978-3-319-46261-5_5

Abbreviations

BChl Bacteriochlorophyll
PGB Phototrophic Gemmatimonadetes bacteria
PGC Photosynthesis gene cluster

Introduction

Photosynthesis represents one of the most ancient and fundamental biological processes (Canfield et al. 2006; Hohmann-Marriott and Blankenship 2011). Phototrophic organisms transform solar radiation into metabolic energy which fuels most of the world ecosystems (Falkowski and Raven 2007). It is generally assumed that the earliest phototrophs were anaerobic (living in the absence of free oxygen) anoxygenic (not producing oxygen) prokaryotes (Olson 2006; Hohmann-Marriott and Blankenship 2011). After the rise of oxygenic Cyanobacteria approx. 2.7 Gyr ago, the Earth's atmosphere started to become gradually oxygenated, reaching the present oxygen concentration roughly 0.6 Gyr ago (Nisbet and Sleep 2001). Hence, early anaerobic phototrophs were forced either to adapt to the new oxic conditions or to retreat to anoxic habitats, leading to present phylogenetically and physiologically diverse phototrophic lineages. So far, species employing chlorophyll or bacteriochlorophyll (BChl)-based photosynthetic reaction centers (chlorophototrophs) have been reported in seven bacterial phyla: Cyanobacteria, Proteobacteria, Chlorobi, Chloroflexi, Firmicutes, Acidobacteria, and recently also in Gemmatimonadetes (Fig. 1). While Cyanobacteria, green sulfur bacteria and purple bacteria were already discovered more than 100 years ago (Winogradsky 1888; Schmidle 1901; Nadson 1906; Molisch 1907; reviewed by Gest and Blankenship 2004), green non-sulfur bacteria and heliobacteria were only described during the second half of the 20th century (Pierson and Castenholz 1974; Gest and Favinger 1983). Phototrophic Acidobacteria were first reported from Yellowstone springs in 2007 (Bryant et al. 2007; see also the chapter by Tank et al. in this volume). The last organism representing a novel phylum containing chlorophototrophs is *Gemmatimonas (G.) phototrophica* that belongs to the phylum Gemmatimonadetes (Zeng et al. 2014). Each of these lineages contains a unique apparatus for solar energy conversion differing in the light harvesting complex architecture, pigment composition, and function of their reaction centers (Overmann and Garcia-Pichel 2013; see Table 1 for summaries). In general, photosynthetic reaction centers can be divided into two main groups. FeS-based (type-1) reaction centers are used by green sulfur bacteria (classified into Chlorobi), heliobacteria (phototrophic Firmicutes), and phototrophic Acidobacteria and Gemmatimonadetes. Pheophytin-quinone

Fig. 1 Illustration of the relative phylogenetic relationship of the known seven phyla and their photosystems' features highlighting the novelty of the phototrophic phylum Gemmatimonadetes. The phylogenetic tree was based on the 16S rRNA gene sequences of 29 bacterial phyla with cultured representatives as shown in Zeng et al. (2014) and depicted in an artistic manner without distorting the relative position of each phylum

(type-2) reaction centers are present in green non-sulfur bacteria (Chloroflexi) and purple bacteria (phototrophic Proteobacteria). Oxygenic Cyanobacteria contain both type-1 and type-2 reaction centers.

In this chapter, we review the discovery, physiology, phototrophic properties and genomic characteristics of *G. phototrophica*, and our current understanding of the possible evolutionary origin of its photosynthesis function and the distribution of phototrophic Gemmatimonadetes bacteria (PGB) in the environment.

Table 1 Summary of characteristics of phototrophic Gemmatimonadetes bacteria in comparison with other six bacterial phyla that contain chlorophototrophic members

Phylum	Common name[a]	Reaction centers	Main pigments	Carbon fixation pathway[b]	Oxygen requirement	Year of discovery
Cyanobacteria	Blue-green algae	Type-1, Type-2	Chlorophylls, carotenoids, phycobilins	Reductive pentose phosphate (Calvin–Benson–Bassham) cycle	Aerobic	Nineteenth century
Proteobacteria	Purple bacteria	Type-2	BChl *a*/*b*, carotenoids	Reductive pentose phosphate (Calvin–Benson–Bassham) cycle	Aerobic, semiaerobic, anaerobic	Nineteenth century
Chlorobi	Green sulfur bacteria	Type-1	BChl *a*/*c*/*d*/*e*, carotenoids	Reductive tricarboxylic acid (Arnon–Buchanan) cycle	Anaerobic	1901, 1906
Chloroflexi	Green non-sulfur bacteria	Type-2	BChl *a*/*c*, carotenoids	3-Hydroxypropionate (Fuchs-Holo) bi-cycle	Semiaerobic	1974
Firmicutes	Heliobacteria	Type-1	BChl *g*, carotenoids	*Absent*	Anaerobic	1983
Acidobacteria		Type-1	BChl *a*/*c*, carotenoids	*Absent*	Aerobic, semiaerobic	2007
Gemmatimonadetes		Type-2	BChl *a*, carotenoids	*Absent*	Semiaerobic	2014

[a]Common or historical name used for phototrophic species in the particular phylum
[b]If present

Discovery

Isolation of G. Phototrophica

The phototrophic strain *G. phototrophica* AP64 was isolated during an intense screening of novel anoxygenic phototrophs in freshwater lakes (Zeng et al. 2014). The water sample collected in December 2011 from a freshwater desert lake Tiān é hú (Swan Lake), located in the western Gobi Desert in Inner Mongolia, North China, was inoculated on R2A agar plates (Fig. 2). The screening was performed using a special infra-red (IR) fluorescence imaging system which identified a positive red-pigmented colony (see suppl. material in Zeng et al. 2014). The colony was

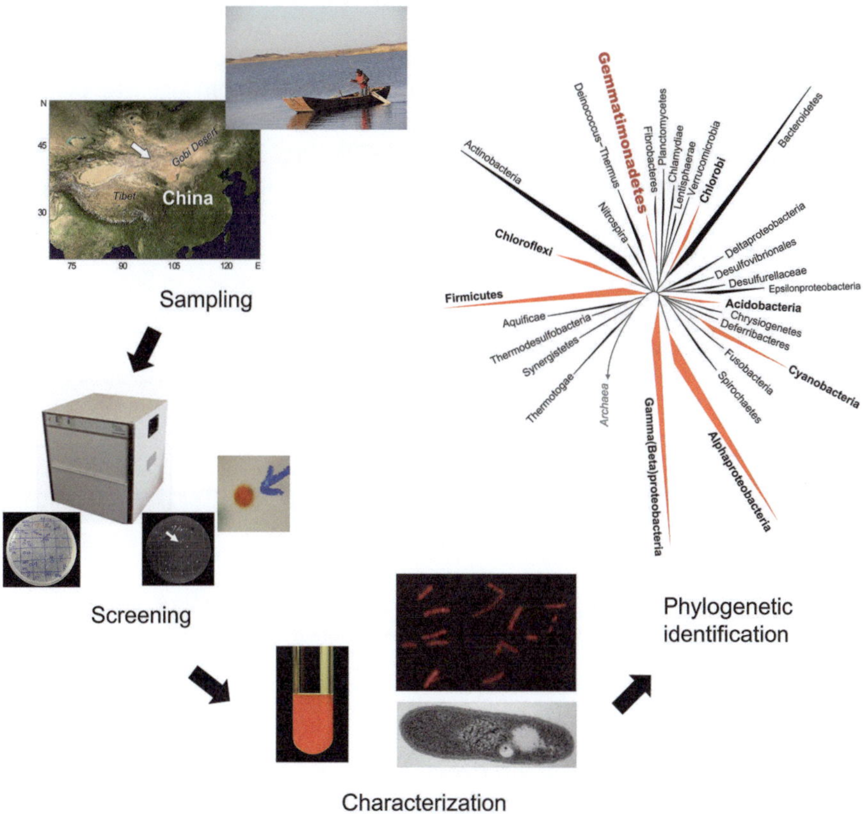

Fig. 2 Illustration of the path of discovery of *Gemmatimonas phototrophica*. (1) Water sample collected at Swan Lake in the western Gobi Desert in Inner Mongolia, North China by Dr. Fuying Feng (in the boat) from Inner Mongolia Agricultural University; (2) The sample was streaked on R2A agar and incubated for approx. 3 weeks. The BChl *a* positive colonies were identified using IR fluorescence imaging system; (3) the pure culture was characterized; and (4) its phylogenetic identity determined from its 16S rRNA sequence

transferred several times onto new agar plates and the obtained isolate was named AP64. The presence of BChl *a* in the cells was confirmed using infra-red epifluorescence microscopy (Fig. 2).

Unexpectedly, the 16S rRNA sequence of the new strain was 96.1 % identical to that of *G. aurantiaca* T-27, the type strain of phylum Gemmatimonadetes (Zhang et al. 2003). Phylum Gemmatimonadetes (Kamagata 2010; Hanada and Sekiguchi 2014) is a sister phylum to Fibrobacteres, but is not closely related to any phyla known to contain chlorophototrophs (see the phylogenetic tree in Fig. 2). Two genera, i.e., *Gemmatimonas* and *Gemmatirosa*, have been proposed in this phylum before with one type species described in each genus, respectively, i.e., *G. aurantiaca* T-27 (Zhang et al. 2003) isolated from a wastewater treatment reactor and *Gemmatirosa kalamazoonensis* (DeBruyn et al. 2013) isolated from organically managed agricultural soil in the USA. The 16S rRNA gene of AP64 shares 96.1 % sequence identity with that of *G. aurantiaca* while shows only 91.2 % identity to that of *Gemmatirosa kalamazoonensis*, suggesting that AP64 and T-27 belong to the same genus whereas *G. kalamazoonensis* represents a different class of the phylum Gemmatimonadetes. This is further confirmed by phylogenetic analysis where AP64 and *G. aurantiaca* form a tight phylogenetic cluster, but distantly related to *Gemmatirosa kalamazoonensis* KBS708 (Fig. 3). Thus, the species represented by this strain was named *G. phototrophica* sp. nov. after its phototrophic lifestyle (Zeng et al. 2015). At the time of writing, a new genus was proposed in phylum Gemmatimonadetes, i.e., *Longimicrobium*, which contains the species *Longimicrobium terrae* isolated from a Mediterranean forest soil in Spain (Pascual et al. 2016). The 16S rRNA genes of *G. phototrophica* and *Longimicrobium terrae* share only 83.6 % sequence identity.

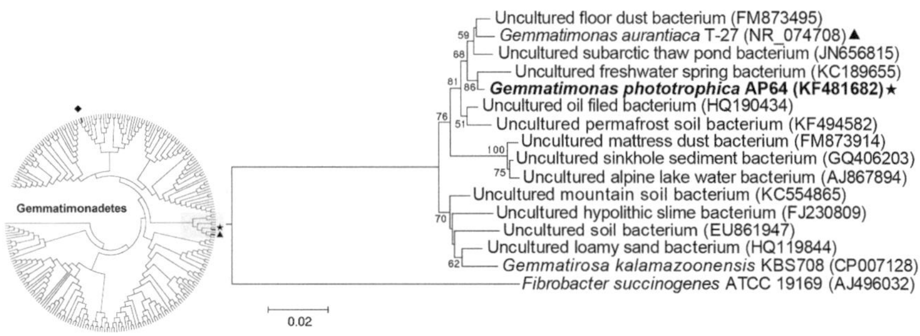

Fig. 3 Phylogenetic tree of representative Gemmatimonadetes 16S rRNA gene sequences. A total of 285 sequences (>1450 bp in length) from environmental clones and pure cultures were multi-aligned and 1130 gap free conserved positions were used for inference of neighbor joining trees. Bootstrap resampling (1000 times) was performed. The 16S rRNA gene sequence of *Fibrobacter succinogenes* was used as the outgroup. Only the topology of the bootstrap consensus tree is shown to illustrate the relative positions of *G. phototrophica* AP64 (*star*), *G. aurantiaca* T-27 (*filled triangle*), and *Gemmatirosa kalamazoonensis* KBS708 (*filled diamond*) on the tree. The branch that shows phylogenetic relationship of *G. phototrophica* and its close relatives is highlighted and zoomed in. Scale bar represents 2 % nucleotide substitution

Morphology and Physiology

Morphology

G. phototrophica cells are rod-shaped, 0.3–0.5 μm in width, and most commonly 1–6 μm in length (Zeng et al. 2015). Occasionally, some cells form up to 12 μm long filaments (Fig. 4a). Similar filaments were also observed in *Gemmatirosa kalamazoonensis* KBS708 (DeBruyn et al. 2013), but they were not reported in *G. aurantiaca* T-27 (Zhang et al. 2003). Structures of gram-negative cell envelope were observed in TEM images (Fig. 4b). Electron-dense small bodies and transparent large vesicle-like structures were present inside cells (Fig. 4b). Cells reproduced by binary fission (Fig. 4b) and often show budding morphology (Fig. 4c), similar to those reported in *G. aurantiaca* T-27 (Zhang et al. 2003) and strain KBS708

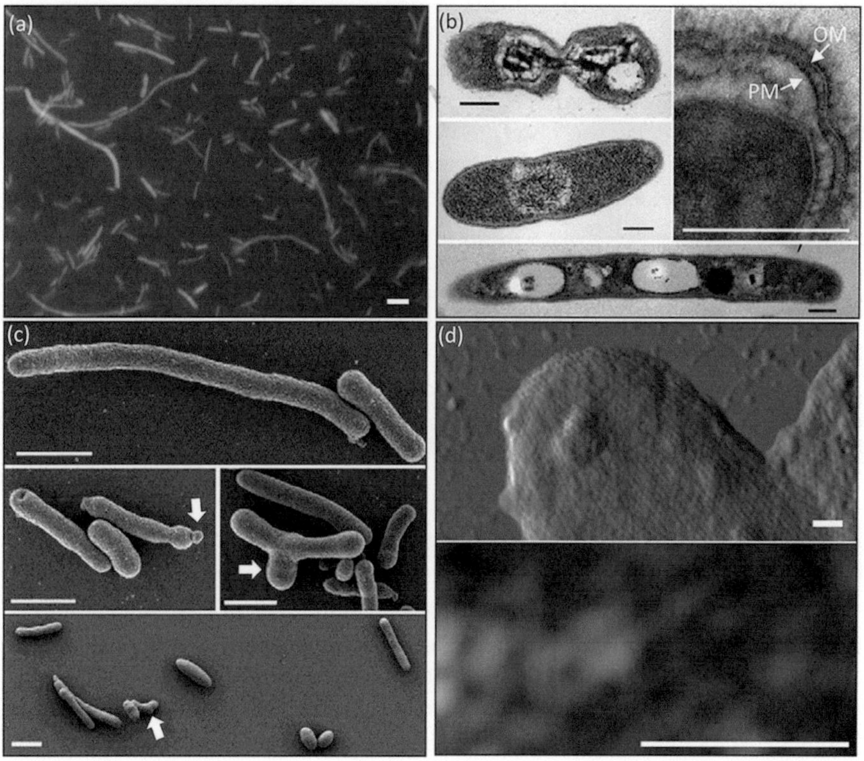

Fig. 4 Microscopic images of *G. phototrophica*'s cells. (**a**) Infra-red (IR) fluorescence image showing rod- and filament-shaped cells. (**b**) TEM images. *OM* outer membrane, *PM* plasma membrane. (**c**) SEM images. Budding structures are marked with *arrows*. (**d**) Atomic force microscopy (AFM) deflection images showing porous surface layer. Bars, 1 μm (**a**, **c**), 200 nm (**b**), 100 nm (**d**) (TEM and SEM imaging by Jason Dean; AFM imaging by David Kaftan). Originally published in Zeng et al. (2015)

(DeBruyn et al. 2013). An ongoing ternary fission through budding was observed (Fig. 4c), indicating its capability of performing various types of cell division. Atomic force microscopy scans showed hexagonal structures (~30 nm in diameter) at cell surface resembling the S-layer (Fig. 4d). Cells were motile, consistent with the fact that its genome contains flagellar biosynthesis genes although flagella were not observed directly by electron microscopy (Zeng et al. 2015).

Cultivation and Physiology

G. phototrophica formed tiny (~0.3 mm), round, smooth, and red-pigmented colonies on agar plates after incubation at 28 °C for 2 weeks under semiaerobic conditions. Interestingly, no growth was observed in the liquid medium. Growth of *G. phototrophica* occurred at 16–30 °C with an optimum temperature of 25–30 °C. The pH range for growth was 6.0–9.0 with an optimum at pH 7.5–8.0. The strain does not require NaCl for growth, but it can tolerate up to 2 g L^{-1} NaCl. Deletion test of components in the R2A$^+$ medium that we used before (Zeng et al. 2014) demonstrated that soluble starch, $MgSO_4$, and ammonium-acetate were not required as no noticeable growth delay was observed in the absence of these nutrients. Similar to *G. aurantiaca* T-27, *G. phototrophica* did not yield comparable biomass on nutrient-rich agar plates (LB, nutrient broth, and double-nutrient-strength R2A$^+$) as on R2A$^+$ agar plates, suggesting that these species do not prefer copiotrophic growth.

The highest biomass yields for *G. phototrophica* were reached at reduced oxygen concentration (9.8–15.2 %), whereas at normal oxygen concentration (21 %) its growth was significantly inhibited (Zeng et al. 2015). No growth was observed under all tested anaerobic conditions in a 4-week incubation. In contrast, *G. aurantiaca* T-27 preferred a fully oxygenic atmosphere and reached a maximum biomass on agar plates after 5–7 days.

Biochemical and Chemotaxonomic Characteristics

Phototrophic *G. phototrophica* AP64 and non-phototrophic *G. aurantiaca* T-27 had in general very similar composition of basic macromolecules as seen from Fourier transform infra-red spectra (Zeng et al. 2015). However, these two species could be differentiated by many specific features such as colony size, pigment composition, optimum pH, optimum salinity and oxygen concentrations for growth, capability of growing in liquid medium, and susceptibility to antibiotics (Zeng et al. 2015). *G. phototrophica* grew well with yeast extract as the sole carbon source. Weak growth with peptone was observed. During a 30-day incubation, no growth was observed with the following compounds as the sole carbon source: casamino acids, sodium succinate, sodium acetate, sodium pyruvate, potato starch, sucrose, L-glutamic acid, L-leucine, L-arginine, L-alanine, L-isoleucine, L-arabinose, D-sorbit, and D-mannitol.

The cells were positive for oxidase and catalase. *G. phototrophica* has a natural resistance to ampicillin, penicillin, paromycin sulfate, polymixin B sulfate, and nystatin, but it was sensitive to neomycin, vancomycin, bacitracin, and gentamycin. The major respiratory quinone is menaquinone-8. The dominant fatty acids are C16:1, C14:1, and C18:1ω9c (Zeng et al. 2015).

Phototrophic Properties

Expression and Functionality of Photosynthetic Apparatus

The expression of photosynthetic reaction centers in *G. phototrophica* was proven by its in vivo absorption spectrum. It displayed two infra-red BChl *a* bands at 819 and 866 nm (Fig. 5a), resembling the absorbance of inner and peripheral light harvesting complexes in some aerobic anoxygenic phototrophic bacteria (Yurkov and Beatty 1998; Selyanin et al. 2016). Carotenoids were responsible for most of the light absorption between 400 and 600 nm, with main absorption peaks at 478, 507, and 542 nm (Fig. 5a). AP64 cells contained 3.5 ± 1.1 (mean ± SD, $n = 5$) mg BChl *a* g^{-1} protein, approx. an order of magnitude less than typical anoxygenic photoautotrophs but similar to BChl *a* levels reported in aerobic anoxygenic phototrophs (Selyanin et al. 2016). Light did not inhibit the BChl *a* biosynthesis. Interestingly, cells that were continuously grown in the dark and subcultured for one year still maintained BChl *a* (Fig. 5a). This indicated that *G. phototrophica* expressed its photosynthetic apparatus constitutively.

G. phototrophica has a complex pigment composition. It contains two forms of BChl *a* esterified either with geranylgeranyl or with phytol side chains (peaks 7 and 9 of Fig. 5c). Based on the chromatography data, we determined that there were 62.1 ± 5.3 BChl *a* molecules per reaction center (mean ± SD, $n = 4$). In addition, *G. phototrophica* contained carotenoids of oscillol series with oscillol 2,2′-dirhamnosides as the major carotenoid (Fig. 5c). These carotenoids probably do not play a role in light harvesting function, but only serve for photoprotection as indicated by the high carotenoid to BChl *a* ratio (5.5 ± 1.7 mol:mol). This notion was supported by the fact that abundant polar carotenoids (peaks 1–4) were also present in the non-phototrophic relative *G. aurantiaca* T-27 (Fig. 5c). In line with the presence of these carotenoids, we identified six carotenogenesis genes (*crtE*, *crtB*, *crtI*, *cruF*, *cruG*, and *crtF*) in the genome of *G. phototrophica*, five of which were also identified in *G. aurantiaca* T-27 except *crtF* (Table 2; Fig. 6). Accordingly, *G. phototrophica* contained small amount of spirilloxanthin which was not identified in *G. aurantiaca*. We speculate that spirilloxanthin is synthesized from didemethylspirilloxanthin in a two step transmethylation reaction catalyzed by *O*-methyltransferase encoded by the *crtF* gene (Fig. 6).

The functionality of *G. phototrophica*'s photosynthetic apparatus was first tested using infra-red kinetic fluorometry. The recorded fluorescence transients (Fig. 5b)

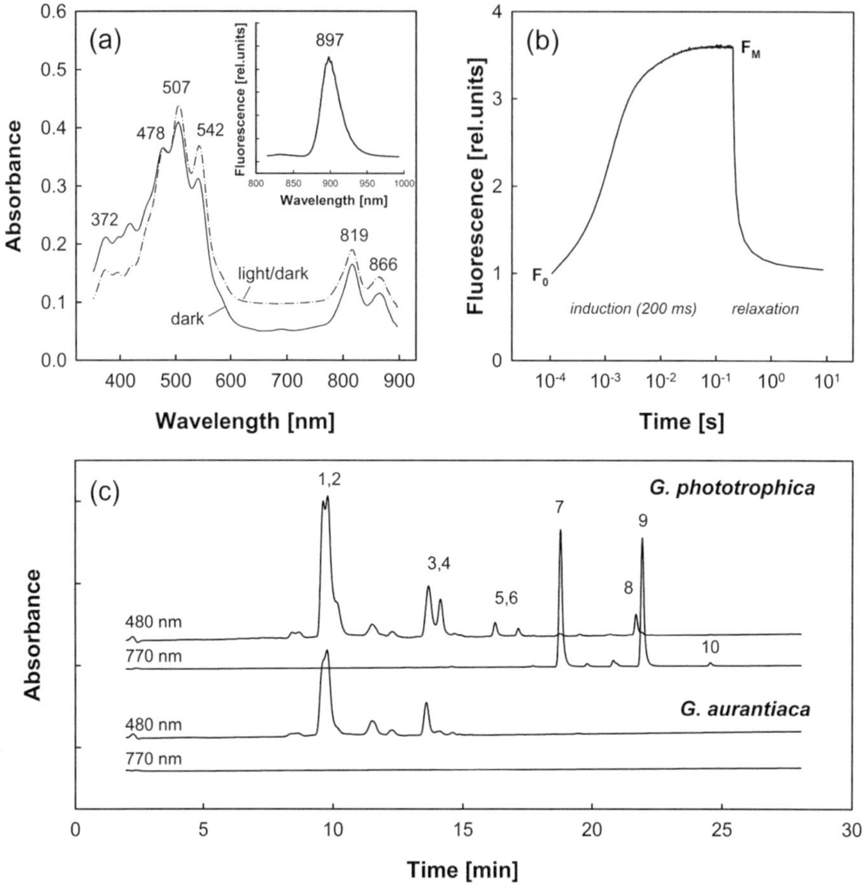

Fig. 5 Photosystem functionality and pigment characterization of *G. phototrophica*. (**a**), in vivo absorption spectrum. The insert shows 77 K fluorescence emission spectrum. *G. phototrophica* cells grew under a 12 h/12 h light/dark cycle (*solid line*) and under continuous darkness (*dotted line*). For cultures under light/dark cycles, the strain was cultured for 2 weeks. For cultures under continuous darkness, colonies were transferred to fresh agar plates every 2 months during a 1-year long experiment. The cells were harvested from the agar plates for analysis (Zeng et al. 2015). (**b**), fluorescence induction and relaxation kinetics recorded by an infra-red fluorometer. (**c**), high performance liquid chromatography (HPLC) elution profile of pigment extracts from *G. phototrophica* and *G. aurantiaca* at 480 nm for carotenoids and at 770 nm for BChl *a* (note that traces are vertically shifted). Numbers above peaks indicate main pigments: 1 and 2, putative (2*S*,2′*S*)-Oscillol 2,2′-di-(α-L-rhamnoside) based on Takaichi et al. (2010); 3–6, unknown carotenoids; 7, BChl a_{GG} (geranylgeranyl); 8, spirilloxanthin; 9, BChl a_P (phytol); 10, bacteriopheophytin a_P

resembled those of purple non-sulfur bacteria and aerobic anoxygenic phototrophs (Koblížek et al. 2010). The recorded F_V/F_M ratios 0.71 ± 0.02 and turnover rates 60.9 ± 18.8 s^{-1} (mean ± SD, *n* = 4) confirmed that *G. phototrophica* contained fully functional type-2 photosynthetic reaction centers connected to an efficient electron transfer chain. Further, the effect of light was tested using respiration assays. Here,

Table 2 Genome characteristics of cultured members of phylum Gemmatimonadetes

	G. phototrophica	*G. aurantiaca*	*Gemmatirosa kalamazoonensis*
NCBI Bioproject ID	PRJNA213561	PRJDA18501	PRJNA194094
Genome size (chromosome, bp)	4,716,552	4,636,964	5,311,527
Plasmids	N.A.	N.A.	1,106,233 bp 1,040,503 bp
GC content	64.4 %	64.3 %	72.6 %
Number of genes	3965	3998	4567
Number of proteins	3388	3892	4513 (+1822 on plasmids)
Number of pseudogenes	523	52	N.A.
Frameshifted genes	12	4	N.A.
tRNA	47	48	48
rRNA	6	3	6
ncRNA	1	3	N.A.
Presence of PGC	Yes	No	No

the exposure of AP64 cells to light caused a 37.5 ± 6.9 % reduction of respiration (see suppl. material in Zeng et al. 2014), which documented that photophosphorylation driven by light energy supplemented oxidative phosphorylation utilizing organic carbon substrates. The capacity of *G. phototrophica* to assimilate CO_2 was tested using radioactive assays. However, only a minimum amount of incorporated radioactivity was observed without significant difference between light and dark treatments. This suggested that this activity in *G. phototrophica* only came from the anaplerotic carboxylation enzymes that were identified in its genome (Zeng et al. 2014). Despite the presence of a light harvesting apparatus, it is not an obligate phototroph as it requires a supply of organic substrates and also grows in the dark. Thus, *G. phototrophica* represents a facultative photoheterotrophic organism, whose ability to harvest light may provide additional energy for its metabolism and improve the economy of carbon utilization, similarly as it was shown in marine photoheterotrophic Proteobacteria (Hauruseu and Koblizek 2012).

Genomics

The genome of *G. phototrophica* AP64 was sequenced by combining Illumina, 454 pyrosequencing, and Sanger chromosome walking technologies. The genome size is 4,716,552 bp with a GC content of 64.4 % (Table 2 and Fig. 7a; GenBank accession number CP011454). The most distinctive characteristic of the genome is the presence of a 42.3-kb-long photosynthesis gene cluster (PGC). It contains 28 genes involved in bacteriochlorophyll biosynthesis, expression and assembly of photosystem, carotenoid biosynthesis, and regulatory functions (Fig. 7a). The clustering of

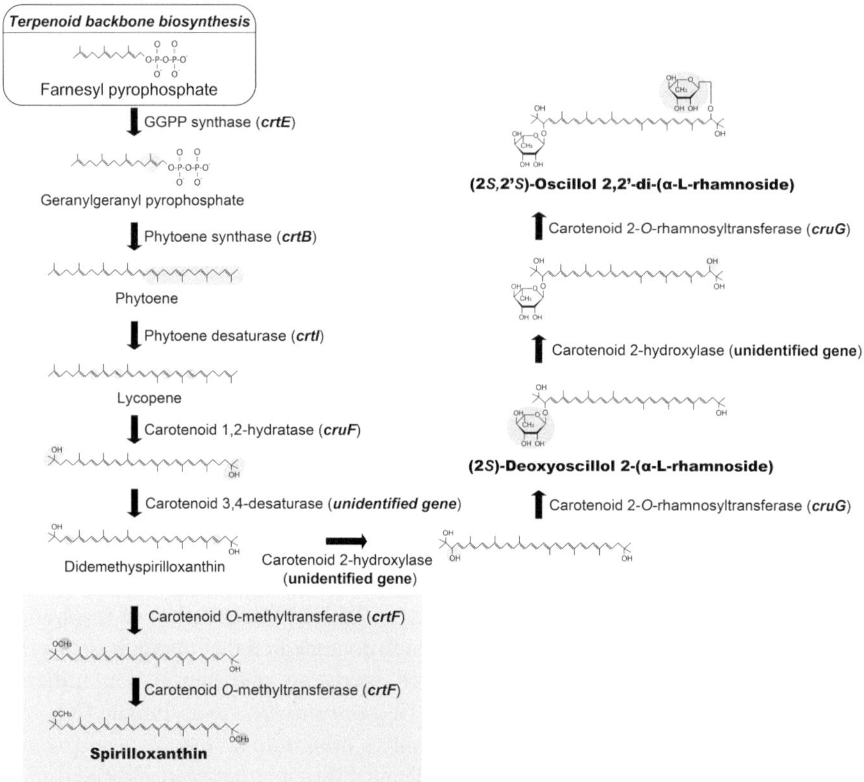

Fig. 6 Putative carotenoid biosynthesis pathway in *G. phototrophica*. The highlighted parts of each compound represent the sites modified by corresponding enzymes. The novel oscillol 2,2′-dirhamnoside biosynthesis pathway which contain two as yet unidentified genes was firstly proposed in *G. aurantiaca* T-27 by Takaichi et al. (2010). The two species share very similar pigment composition and genome content, suggesting that this pathway very likely exists in *G. phototrophica*. In the highlighted box shows the proposed spirilloxanthin biosynthesis steps in *G. phototrophica* catalyzed by the carotenoid *O*-methyltransferase (*crtF*) which is absent in T-27. Gene homologs of *crtE*, *crtB*, *crtI*, *cruF*, *cruG*, and *crtF* have been identified in the genome of *G. phototrophica*, but only *crtF* gene was found inside PGC

photosynthesis genes into PGC represents a typical feature in many purple non-sulfur phototrophic bacteria (Haselkorn et al. 2001; Swingley et al. 2009; Zheng et al. 2011) and in heliobacteria (Xiong et al. 1998), but they are not present in other chlorophototrophic lineages. The PGC arrangement of *G. phototrophica* closely resembles that of Proteobacteria (Fig. 8). The PGC in *G. phototrophica* contained the same conserved gene order *crtF-bchCXYZ* followed by the *pufBALM* operon, which was previously identified in Proteobacteria (Fig. 8). Similarly, the gene arrangement *bchFNBHLM-acsF-lhaA-puhABC* resembled the gene organization found in most Proteobacteria differing only in the position of *acsF* (Fig. 8). The unique position of the *acsF* gene between *bchFNBHLM* and *puhABC* in *G.*

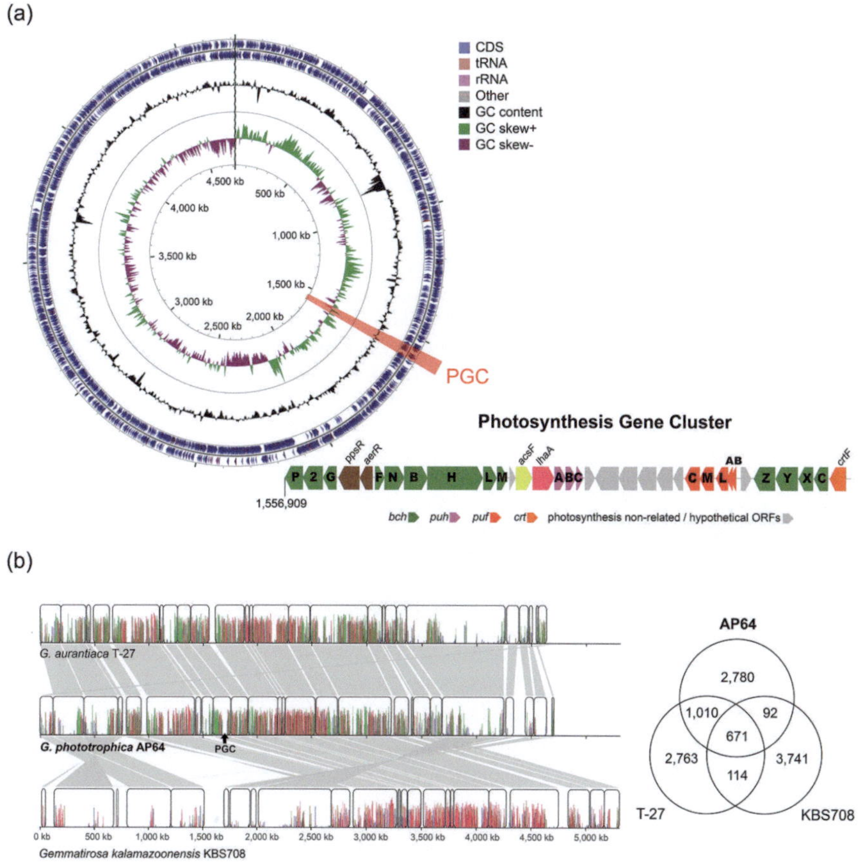

Fig. 7 Genomic features of *G. phototrophica* with the region of photosynthesis gene cluster (PGC) highlighted (**a**) and whole genome sequence comparison of *G. phototrophica* AP64, *G. aurantiaca* T-27, and *Gemmatirosa kalamazoonensis* KBS708 and statistics of their shared and distinct orthologs (**b**). (**a**), *bch* (*green*), bacteriochlorophyll biosynthesis genes; *puh* (*pink*), genes encoding reaction center assembly proteins; *puf* (*red*), genes encoding reaction center proteins; *crt* (*brownish yellow*), carotenoid biosynthesis genes; *gray*, photosynthesis non-related genes or hypothetical ORFs. (**b**), each block represents a region that is presumably homologous and internally free of genomic rearrangement. The height of the similarity profile within blocks was inversely proportional to the average level of conservation over a region of the alignment. Regions that are conserved among all three genomes are shown in *red color*. Regions conserved only among subsets of the genomes have been color coded differently. Fig. 7b originally published in Zeng et al. (2015)

phototrophica was not observed in Proteobacteria, suggesting a different evolution history of photosynthesis in Gemmatimonadetes. The presence of *puf* genes encoding bacterial reaction center subunits suggested the presence of type-2 photosynthetic reaction centers. In addition, the genome contained a complete gene inventory of the bacteriochlorophyll biosynthesis pathway (Table 3).

Fig. 8 Unique gene arrangement in the photosynthesis gene cluster of phototrophic Gemmatimonadetes bacteria in comparison to diverse purple photosynthetic bacteria. Genomes were downloaded from the NCBI WGS database and annotated with the RAST webserver (http://rast.nmpdr.org/). For each category of gene, *bch* (*light green* to *dark green*), bacteriochlorophyll biosynthesis genes; the *acsF* gene encoding the aerobic form of Mg-protoporphyrin IX monomethyl ester oxidative cyclase, which is also involved in the bacteriochlorophyll biosynthesis pathway, is highlighted in *yellow*; *puh* (*pink*), genes encoding reaction center assembly proteins; *puf* (*red*), genes encoding reaction center proteins; *crt* (*brownish yellow*), carotenoid biosynthesis genes. The *crt* genes other than *crtF* which is the only carotenoid biosynthesis gene in the photosynthesis gene cluster of PGB are not shown. Photosynthesis non-related genes or hypothetical ORFs are not shown

We compared genomic similarity and calculated the numbers of shared orthologs among the three type strains *G. phototrophica* AP64, *G. aurantiaca* T-27, and *Gemmatirosa kalamazoonensis* KBS708 that are available in this phylum. Genomes of *G. phototrophica* and *G. aurantiaca* were highly conserved in terms of both genomic similarity and gene synteny with no large genomic rearrangement events detected (Fig. 7b). In contrast, a large portion of genomic regions in *G. phototrophica* and *Gemmatirosa kalamazoonensis* have been subjected to rearrangement and less

Table 3 A complete list of enzymes involved in the bacteriochlorophyll biosynthesis pathway that are predicted to be present in the genome sequence of *G. phototrophica*

Step in the pathway	Enzyme	Gene name	Length (bp)	Inside (●)/ outside (○) PGC
1. L-Glutamate → L-Glutamyl-tRNA	Glutamyl-tRNA synthetase	*gltX*	1464	○
2. →Glutamate-1-semialdehyde	Glutamyl-tRNA reductase	*hemA*	1464	○
3. →5-Amino-levulinate	Glutamate-1-semialdehyde 2,1-aminomutase	*hemL*	1314	○
4. →Porphobilinogen	Porphobilinogen synthase	*hemB*	996	○
5. →Hydroxymethylbilane	Hydroxymethylbilane synthase	*hemC*	999	○
6. →Uroporphyrinogen III	Uroporphyrinogen III synthase	*hemD*	780	○
7. →Coproporphyrinogen III	Uroporphyrinogen III decarboxylase	*hemE*	1095	○
8. →Protoporphyrinogen IX	Coproporphyrinogen oxidase (Aerobic); Coproporphyrinogen dehydrogenase (Anaerobic)	*hemF* *hemN*	996 1365	○ ○
9. →Protoporphyrin IX	Protoporphyrinogen oxidase (Aerobic); Protoporphyrinogen IX dehydrogenase (Anaerobic)	*hemY* *hemG*	1467 435	○ ○
10. →Mg-protoporphyrin IX	Magnesium chelatase subunit I; Magnesium chelatase subunit D; Magnesium chelatase subunit H	*bchI* *bchD* *bchH*	1026 1455 1878 3801	● ● ●
11. →Mg-protoporphyrin IX 13-methyl ester	Mg-protoporphyrin IX methyltransferase	*bchM*	720	●
12. →Divinylprotochlorophyllide	Mg-protoporphyrin IX monomethyl ester oxidative cyclase (Aerobic); Mg-protoporphyrin IX monomethyl ester oxidative cyclase (Anaerobic)	*acsF* *bchE*	1053 1602	● ○
13. →Protochlorophyllide	Divinyl chlorophyllide *a* 8-vinyl-reductase	*bciA*[a]	1284	○

(continued)

Table 3 (continued)

Step in the pathway	Enzyme	Gene name	Length (bp)	Inside (●)/ outside (○) PGC
14. →Chlorophyllide *a*	Protochlorophyllide reductase iron-sulfur ATP-binding protein (Light-independent); Protochlorophyllide reductase subunit N (Light-independent); Protochlorophyllide reductase subunit B (Light-independent)	*bchL* *bchN* *bchB*	969 888 1254 1743	● ● ●
15. →3-Vinyl-bacteriochlorophyllide *a*	Chlorophyllide *a* reductase subunit X; Chlorophyllide *a* reductase subunit Y; Chlorophyllide *a* reductase subunit Z	*bchX* *bchY* *bchZ*	996 1638 1458	● ● ●
16. →3-Hydroxyethyl-bacteriochlorophyllide *a*	3-Vinyl bacteriochlorophyllide *a* hydratase	*bchF*	564	●
17. →Bacteriochlorophyllide *a*	3-Hydroxyethyl-bacteriochlorophyllide *a* dehydrogenase	*bchC*	954	●
18. →Bacteriochlorophyll *a* (esterified with geranylgeraniol)	Bacteriochlorophyll synthase	*bchG*	912	●
19. →Bacteriochlorophyll *a* (esterified with phytol)	Geranylgeranyl reductase	*bchP*	1269	●

The complete genome sequence can be accessed via GenBank accession number CP011454.1
[a]Annotated as *bciA* rather than *bchJ* according to Saunders et al. Biochemistry (2013) 52, 8442–8451 and Canniffe et al. Biochem. J. (2013) 450, 397–405

than 50 % genomic regions were found to be conserved. A similar pattern was seen when comparing the numbers of shared orthologs in each pair of genomes. The unique orthologs (1010) shared by *G. phototrophica* and *G. aurantiaca* were much more than those shared either by *G. phototrophica* and *Gemmatirosa kalamazoonensis* (92) or by *Gemmatirosa kalamazoonensis* and *G. aurantiaca* (114), indicating a close relationship between *G. phototrophica* and *G. aurantiaca* at a genomic level.

The TBLASTN searching in the genomes of *Gemmatirosa kalamazoonensis* and *G. aurantiaca* for homologs of all predicted proteins in *G. phototrophica* was performed to probe the unique genes that occurred in the genome of *G. phototrophica*. A total of 421 ORFs were identified as unique genes in *G. phototrophica* (Zeng et al. 2015), among which a large part (358) were hypothetical genes with unidentified functions. *G. phototrophica* contains cyanophycin synthase and trehalase in its genome, which suggests that it may use cyanophycin and trehalose as an energy storage material (Zeng et al. 2015). Cyanophycin is rich in nitrogen as well as in carbon and its storage inside bacterial cells has a possible relation to enduring long starvation. In addition to serving as an energy source, trehalose is also known as a protective agent that enables bacterial cells to cope with cold and desiccation stresses (Potts 1994). The presence of trehalose perhaps contributes to the capabilities of the members of the phylum Gemmatimonadetes in resisting to environmental stresses, since 16S rRNA gene clone sequences of this phylum have often been found in arid environments.

Evolution

Phylogeny of Photosynthesis Genes in Gemmatimonadetes

The presence of purple photosynthetic reaction centers in *G. phototrophica* challenges our understanding of the evolution of chlorophototrophic species. To elucidate the possible origin of photosynthetic capacity in this organism, the phylogeny of its photosynthesis genes was investigated using genes encoding the initial phase of the bacteriochlorophyll biosynthesis pathway, which are shared by all chlorophototrophic species (Raymond et al. 2002). The enzymes magnesium chelatase, encoded by bchIDH genes (Gibson et al. 1995), and light-independent protochlorophyllide reductase, encoded by bchLNB genes (Fujita et al. 1993), were used to infer the phylogeny of photosynthesis genes in *G. phototrophica* (Fig. 9). The BchIDH and BchLNB trees separated main chlorophototrophic groups and placed *G. phototrophica* AP64 as an early diverging member of the Proteobacterial clade on the BchIDH tree, while inside Proteobacteria on the BchLNB tree. This clearly documented a common origin of the bacteriochlorophyll biosynthesis pathway in *G. phototrophica* and Proteobacteria (Zeng et al. 2014). This conclusion was also supported by the *bchF* phylogeny (Cardona 2016). Similarly, an analysis of reaction center subunit (PufLM) phylogeny placed AP64 onto a branch inside Proteobacteria (Fig. 9), confirming that *G. phototrophica* contained a close homolog of Proteobacterial photosynthetic reaction centers. The similarity of photosynthetic apparatus between *G. phototrophica* and Proteobacteria (Table 1) and their analogous PGC organization as well as the close phylogenetic relationship of their photosynthesis genes strongly suggests that phototrophy in Gemmatimonadetes bacteria originated from an ancient horizontal transfer event of a complete set of photosynthesis genes from a purple photosynthetic bacterium.

It is unclear as to how PGC was transferred between distantly related species. A complete PGC has been reported in a promiscuous self-transmissible R-prime plasmid of *Rhodobacter capsulatus* (Marrs 1981) and in two plasmids of *Roseobacter litoralis* and *Sulfitobacter guttiformis* (Petersen et al. 2012).

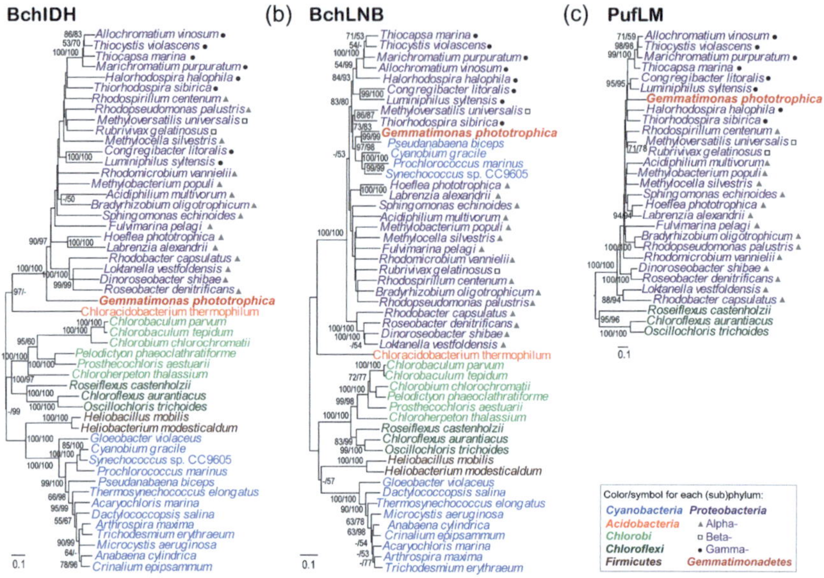

Fig. 9 Phylogenetic analysis of the photosynthesis genes of *G. phototrophica*. (**a**), based on concatenated alignments of magnesium chelatase sequences (BchIDH/ChlIDH, 1810 common amino acid positions). (**b**), based on concatenated alignments of light-independent protochlorophyllide reductase sequences (BchLNB/ChlLNB, 921 common amino acid positions). (**c**), based on concatenated alignments of photosynthetic reaction center subunits' gene sequences (PufLM, 507 common amino acid positions). Maximum likelihood (ML)/neighbor joining (NJ) bootstrap values greater than 50 % are shown on the tree. Scale bars represent changes per position. Originally published in Zeng et al. (2014)

Photosynthesis genes can also be packed into gene transfer agents (Marrs 1974; Lang and Beatty 2000) or bacteriophages (Mann et al. 2003; Lindell et al. 2004; Sharon et al. 2009). Similar mobile elements may have facilitated the acquisition of purple bacterial photosynthetic reaction centers by an ancient counterpart of the heterotrophic Gemmatimonadetes strain T-27. The higher redox potential (~0.5 V) of type-2 reaction centers when compared to type-1 reaction centers may have allowed for an easier incorporation into the respiratory electron transfer chain of this ancient Gemmatimonadetes bacterium. After the adoption of phototrophy as a means of energy production, phototrophic Gemmatimonadetes probably evolved independently, perhaps losing the acquired genes again in some lineages. Yet, *G. phototrophica* represents the first known example of horizontal transfer of a complete photosynthesis gene package between distant bacterial phyla. This gives us a perfect example as to how nature streamlines the genomic arrangement of photosynthesis genes and makes the photosynthesis function transferable between distantly related phototrophic and non-phototrophic bacteria.

Conserved Gene Composition and Arrangement in the Environmental PGCs of Gemmatimonadetes' Origin

Our current knowledge on PGB comes from so far the only culture *G. phototrophica* (Zeng et al., 2014, 2015). To gain more genomic and evolutionary insights into the genomic basis of photosynthesis function in this bacterial group, we performed a reference genome-guided recovery of bacterial genomes from deeply sequenced metagenomes. The complete genome sequence of *G. phototrophica* was used as a reference to search for genomic fragments of its relatives in the three metagenomes OdenseWW, AalborgWW-1, and AalborgWW-2 (Zeng et al. 2016). These metagenomes were chosen because the photosynthetic genes of Gemmatimonadetes' origin have been identified in them and they were deeply sequenced with assembled contigs publicly available. We obtained >1000 positive metagenomic contigs from the metagenomes of OdenseWW and AalborgWW-1 and 2 that were predicted to originate from Gemmatimonadetes-related species, among which we further searched homologs of all genes in the photosynthesis gene cluster (PGC) of *G. phototrophica* as well as genes neighboring the PGC using blastx search.

From the OdenseWW metagenome, ten PGC-related contigs were obtained (for original contig IDs see Fig. 10). Read coverage of these ten contigs fell in a very narrow range from 32.27 to 41.658 (reads per contig) compared to the large range (2.633–26,088.286) for all the contigs in OdenseWW (see suppl. material in Zeng et al. 2016), strongly supporting that these contigs came from the same PGB species. Moreover, these ten PGC-related contigs could be manually joined into a 37.9 kb long PGC based on their 100 % identical overlapping ends (34–62 bases long), but three small gaps remained present inside the *bchN* and *bchD* genes and at the 3′ end of the *pufM* gene (Fig. 10). A similar PGC was assembled from nine metagenomic contigs of AalborgWW-2 with six gaps present inside PGC (Fig. 10).

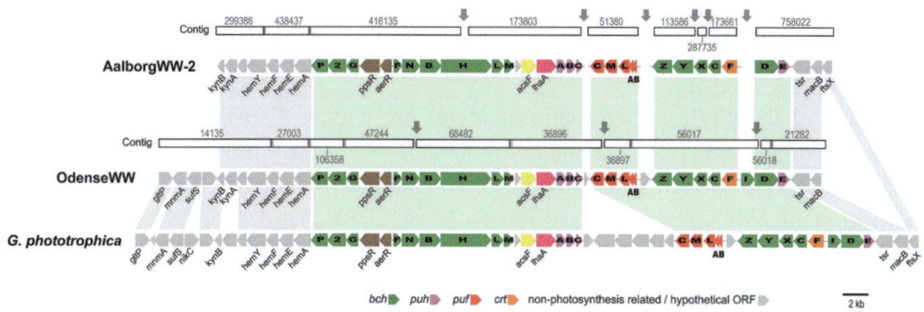

Fig. 10 Comparison of the two photosynthesis gene clusters (PGCs) reconstructed from the Odense wastewater metagenome (OdenseWW) and the Aalborg wastewater metagenome 2 (AalborgWW-2) with the PGC of *G. phototrophica*. Highly similar genomic regions were connected with light colored parallelograms (*gray*, photosynthesis non-related genes; *light green*, photosynthesis related genes). See the legend of Fig. 7 for the description of gene names and colors. Original contig IDs are shown above the assembled PGCs with *arrows* indicating the position of gaps in the assembly. Scale bar, 2 kb in length. Originally published in Zeng et al. (2016)

No PGC fragments were able to be assembled from AalborgWW-1 due to its much shorter average length of Gemmatimonadetes-originated contigs (0.9 kb) than those from OdenseWW (3.1 kb) and AalborgWW-2 (4.8 kb).

Although there are gaps present in both assembled PGCs (OdenseWW-PGC and AalborgWW-2-PGC), no gap appears at the same position of the two PGCs, which indicates that the orientation of each joined contig in OdenseWW-PGC is very likely correct. In AalborgWW-2-PGC, the orientation of *pufBALMC* and *crtF-bchCXYZ* could not be confidentially resolved due to the large gaps present at both ends of the two operons. The composition and arrangement of genes in OdenseWW-PGC and AalborgWW-2-PGC are highly conserved compared to those in *G. phototrophica* (Fig. 10). The amino acid sequence identities between the photosynthesis related ORFs in OdenseWW-PGC and those in *G. phototrophica*'s PGC range from 69 to 95 % (see suppl. material in Zeng et al. 2016). Similarly, the genes flanking the PGC also show high identities (64–90 %) to the counterparts in *G. phototrophica* (see suppl. material in Zeng et al. 2016). All the top blastp hits of genes within and adjacent to OdenseWW-PGC came from the cultured Gemmatimonadetes species (see suppl. material in Zeng et al. 2016), further confirming their Gemmatimonadetes' origin.

The PGC composition and organization seem to be highly conserved among PGB, in contrast to the high diversity in the PGC patterns observed in phototrophic Proteobacteria (Nagashima and Nagashima 2013; Zheng et al. 2011; also Fig. 8). The only difference between OdenseWW-PGC and PGC in *G. phototrophica* is the presence of a 7.2 kb long insert between *puhABC* and *pufBALMC* in *G. phototrophica* which may reflect different evolutionary histories of PGB dwelling on the freshwater Swan Lake in China and PGB residing in Danish wastewater treatment plants.

Environmental Distribution

Methodological Limitations for Detecting Environmental PGB

Members of the phylum Gemmatimonadetes are widely distributed across various natural environments (Hanada and Sekiguchi 2014). Environmental 16S rRNA gene sequence related to Gemmatimonadetes were found in soils (DeBruyn et al. 2011; Portillo et al. 2013), permafrost (Tuorto et al. 2014), rhizospheres (Breidenbach et al. 2015), freshwater lakes and sediments (Gugliandolo et al. 2016; Sheng et al. 2016) activated sludge (Hanada and Sekiguchi 2014; Zhang et al. 2003), deep-sea sediments (Durbin and Teske 2011), gas hydrates, arctic seawater, coastal mobile mud, and marine sponge symbionts (Kamke et al. 2010). High-throughput 16S rRNA gene sequencing indicated that Gemmatimonadetes represent a recognizable fraction of soil microbial communities with relative abundances ranging from 0.2 to 6.5 % (DeBruyn et al. 2011). However, it is impossible to distinguish whether the reported sequences originated from heterotrophic or phototrophic species.

One possible way to distinguish PGB and non-PGB is to detect the presence of their photosynthesis genes. However, caution should be taken when dealing with environmental sequences because most photosynthesis genes found inside the PGC of *G. phototrophica* were closely related to and phylogenetically mixed with those from Proteobacteria. For instance, the commonly used phylogenetic markers—light-independent protochlorophyllide reductase genes (*bchLNB*/*chlLNB*) and genes encoding photosynthetic reaction center subunits (*pufLM*) placed *G. phototrophica* deep into the Proteobacteria branch (Zeng et al. 2014). In contrast, the *acsF* gene encoding the aerobic form of Mg-protoporphyrin IX monomethyl ester oxidative cyclase appears to be a convenient marker gene capable of distinguishing different phototrophic groups (Boldareva-Nuianzina et al. 2013; Zeng et al. 2014). The phylogenetic tree based on AcsF genes clearly separated all known phototrophic phyla except anaerobic phototrophic Firmicutes and Chlorobi into independent clusters, which matched well with the tree based on 16S rRNA genes (Zeng et al. 2014).

Phototrophic Gemmatimonadetes Bacteria Present in Diverse Environments

We searched for Gemmatimonadetes-AcsF like fragments in publicly available metagenome databases including the NCBI's WGS, the JGI's IMG, and the MG-RAST (as of May 30, 2015), with special focus placed on the MG-RAST metagenome webserver (Meyer et al. 2008) which have included over 30,000 public metagenomes. From the WGS and IMG databases, two metagenomes were found to contain Gemmatimonadetes-AcsF like fragments: one from a Danish wastewater treatment bioreactor (OdenseWW) and the other from the Yellowstone Lake (YSLake). In the MG-RAST server, there were 1706 metagenomes reported to contain members of Gemmatimonadetes with relative abundance ranging from below 0.01% up to 2.54% in terms of the number of Gemmatimonadetes-affiliated reads relative to total reads (see suppl. material in Zeng et al. 2016). Based on the source environments' features of metagenomic samples, these Gemmatimonadetes-member-harboring metagenomes could be classified into 20 types of habitats with the top three most sampled ones being soil (536 metagenomes), seawater (231 metagenomes), and wastewater treatment associated habitats (174 metagenomes) (Fig. 11a, b). The Gemmatimonadetes-related sequences were most abundant in soil, wastewater treatment related samples, biofilms, and plant-associated habitats with the largest proportion (2.54%) found in an Arctic tundra permafrost metagenome (MG-RAST ID 4468734.3), while only occasionally found in lakes, rivers, and marine ecosystems (Fig. 11c).

One hundred sixty one metagenomes were identified to contain Gemmatimonadetes-AcsF like sequences with the length of hits ranging from 24 to 361 amino acids (see suppl. material in Zeng et al. 2016 for more details). They came from various types of environments, including air, river waters and sediment,

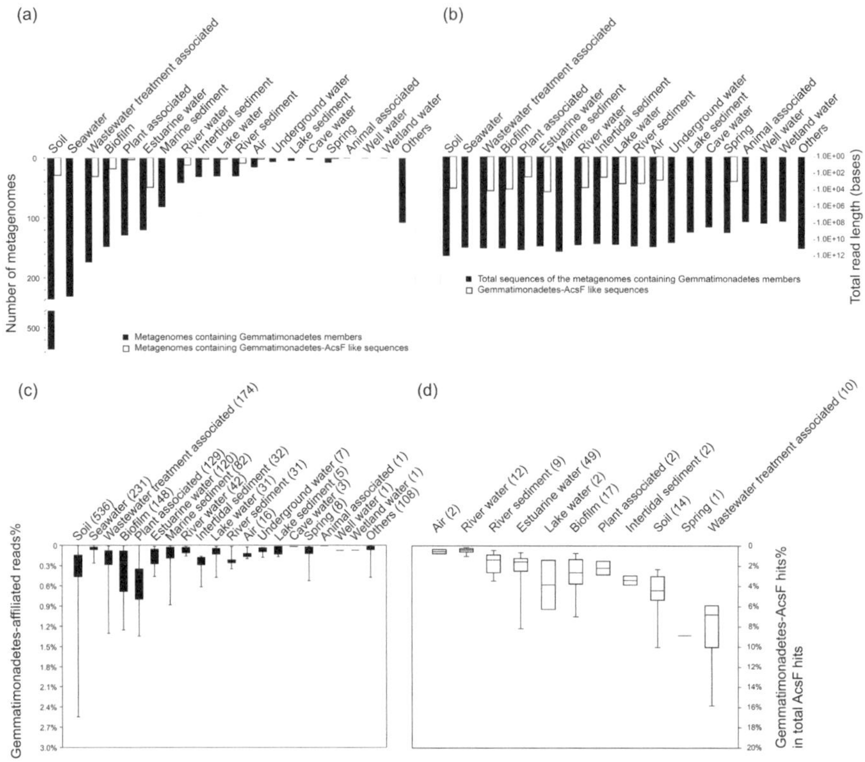

Fig. 11 Summary of the 1706 surveyed metagenomes from the MG-RAST webserver (**a**, **b**) and box plots of the abundance of Gemmatimonadetes-affiliated reads relative to total reads (**c**) and the abundance of Gemmatimonadetes-AcsF hits relative to total AcsF hits (**d**). Metagenomes were classified into 20 categories according to the type of source environments. Only the metagenomes that were reported to contain Gemmatimonadetes members are included. In parentheses are the numbers of metagenome samples. The metagenomes where less than 15 AcsF hits were found were excluded from the calculation of Gemmatimonadetes-AcsF hits' percentage. Originally published in Zeng et al. (2016)

lake waters, estuarine waters, biofilm, plant-associated habitats, intertidal sediment, soil, spring, and wastewater treatment associated samples (Fig. 11d). The highest number of Gemmatimonadetes-AcsF hits was found in a biofilm sample from Dutch costal intertidal sediment (27 hits, MG-RAST ID 4572198.3). Based on percent abundance of Gemmatimonadetes-AcsF hits out of total AcsF hits (Fig. 11d), PGB appear to favor soil and wastewater treatment associated habitats. The reason could be that PGB prefer a low oxygen level in ambient environments as demonstrated in *G. phototrophica* culturing experiments where the normal atmospheric oxygen level greatly inhibited its growth (Zeng et al. 2015); and in soils and wastewater treatment samples, oxygen level gradients are more easily created and thus increase the survival chance of PGB. By contrast, there were no Gemmatimonadetes-AcsF like hits found in the 231 metagenomes from the waters of seas and open oceans where the water columns are generally mixed well and fully oxygenated in the photic zones (Ulloa et al. 2012). Although the data on light availability and oxygen level were lacking in the metadata of the metagenomes surveyed, based on above observations and the physiological data of *G. phototrophica* (Zeng et al. 2015), we speculate that light availability and oxygen level are the two most important factors of shaping the environmental distribution of PGB.

Diverse Phototrophic Gemmatimonadetes Bacteria in the Environment Remain Unexplored

To assess the diversity of environmental PGB we performed a phylogenetic analysis of eight partial and three full-length AcsF sequences retrieved directly from the metagenomic reads or contigs (Fig. 12). These metagenomes came from distinct environments located on three continents, including a bioreactor from a wastewater treatment plant in Odense, Denmark (OdenseWW) (GenBank accession no. APMI00000000, Albertsen et al. 2013), activated sludge from wastewater treatment plants in Aalborg, Denmark (AalborgWW-1 and 2) (MG-RAST project nos. 4487554.3 and 4611649.3, Dueholm et al. 2015) and in Shanghai, China (ShanghaiWW) (MG-RAST project no. 4539352.3, Wang et al. 2014), waters from the Yellowstone Lake, USA (YSLake-SA and MB) (DDBJ accession nos. SRA026894 and SRA026871, Clingenpeel et al. 2011 and Zhou et al. 2015) and from the Lake Gatun in Panama (GatunLake) (MG-RAST project no. 4441590.3, Rusch et al. 2007), soil crust in Nevada, USA (SoilCrust) (MG-RAST project no. 4450752.3, Steven et al. 2012), arable soil in South Korea (KoreanSoil) (MG-RAST project no. 4569532.3, Jung et al. 2014), biofilm from the Stinky Spring in Utah, USA (SulfurSpring, MG-RAST project no. 4528143.3), and surface scum in Lake Taihu in China collected during a cyanobacterial bloom (TaihuBloom) (MG-RAST project no. 4467058.3, Steffen et al. 2012). The translated protein sequences of these 22 AcsF fragments were 75–94 % identical to that of *G. phototrophica* (Zeng et al. 2016). We found that the five *acsF* fragments from YSLake, the six from KoreanSoil, and the three from SulfurSpring could be assembled into single long fragments, respectively, based on the overlapping ends that shared 96.6–99.8 % sequence identities over a length from 118 to 477 bases (see suppl. material in Zeng et al. 2016), indicating that those fragments from the same sample came from the same organism. Phylogenetic analysis of these 11 long AcsF fragments (135–361 amino acids) showed that they formed an independent cluster with *G. phototrophica*, clearly separated from phototrophic Proteobacteria and other phototrophic phyla (Fig. 12), which suggests that these metagenomic gene fragments were very likely of Gemmatimonadetes' origin.

The phylogenetic diversity of PGB appeared to be comparable to that of phototrophic Proteobacteria as reflected by the branches' length on the AcsF tree (Fig. 12). This suggests that PGB in the environment remain largely unexplored and

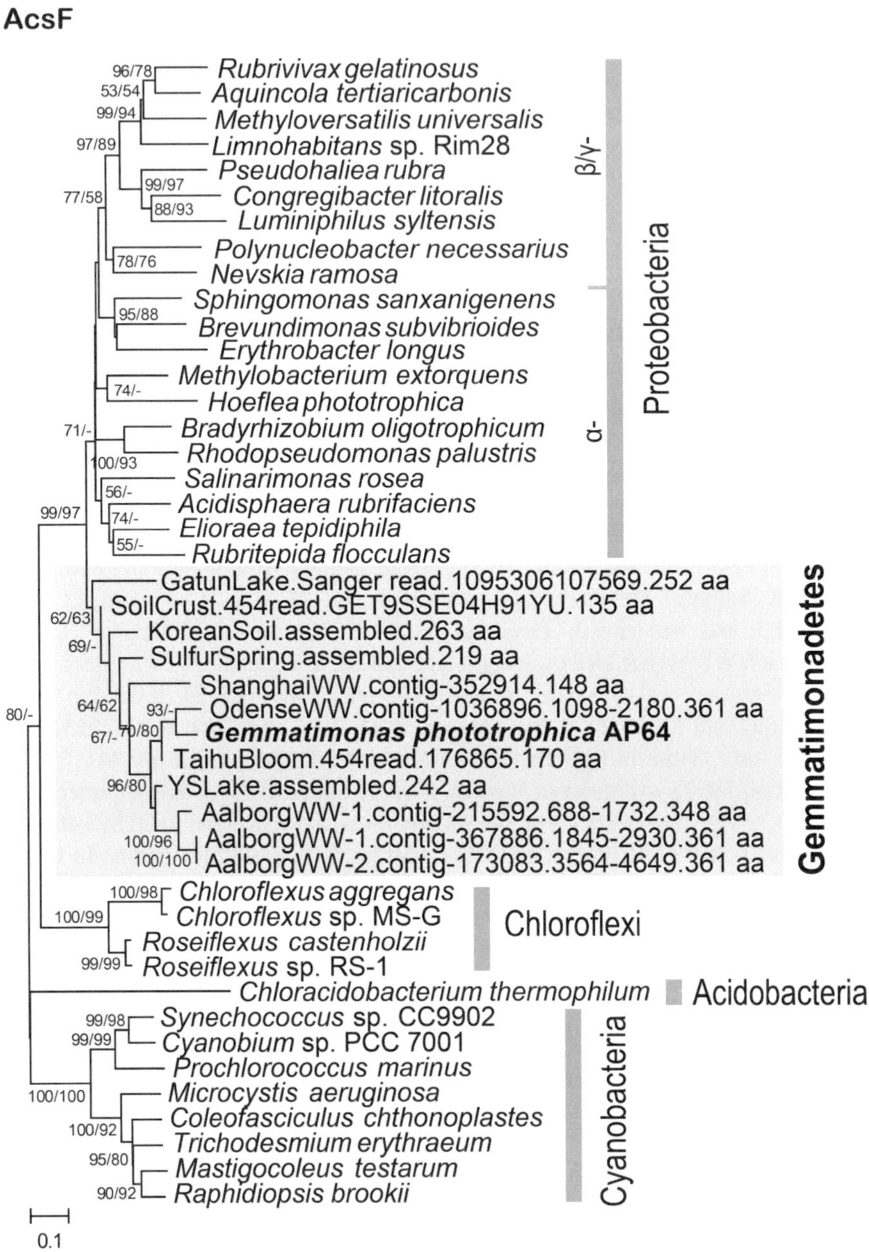

Fig. 12 Phylogenetic analysis of environmental Gemmatimonadetes-AcsF like sequences. Eight partial and three full-length AcsF sequences were retrieved from public metagenomes. In the title of each sequence are shown the abbreviated metagenomes' name, the location of *acsF* gene in a contig or the status of metagenomic fragments, and the length of translated *acsF* protein sequences. See text for sample name abbreviations. "Assembled" means that the original reads were manually assembled into a continuous longer contig. Both neighbor joining and maximum likelihood trees were inferred. A bootstrapping analysis of 1000 resamplings was performed. Values above 50 % are shown. Scale bar represents a 10 % amino acid substitution rate. Originally published in Zeng et al. (2016)

awaits further discoveries. We calculated the relative abundance of PGB in total aerobic phototrophic communities by classifying each AcsF hit into a phototrophic (sub)phylum based on its blastp results against NCBI nr database and pooling hits from the same type of environment prior to calculation. Among the 11 types of environment where Gemmatimonadetes-AcsF like hits were found (Fig. 13), Alphaproteobacterial AcsF dominated the types Air (60.6 %), Lake water (55.7 %), Biofilm (48.1 %), Plant associated (68.6 %), Intertidal sediment (38.3 %), Soil (67.8 %), and Wastewater treatment associated (51.8 %), while Betaproteobacterial AcsF dominated River water (74.4 %), River sediment (61.9 %), and Estuarine water (62.5 %). The Spring type which contained only one metagenome was dominated by cyanobacterial AcsF (58.8 %), followed by Alphaproteobacteria (29.4 %) and Gemmatimonadetes (8.8 %). The highest proportion of PGB was found in

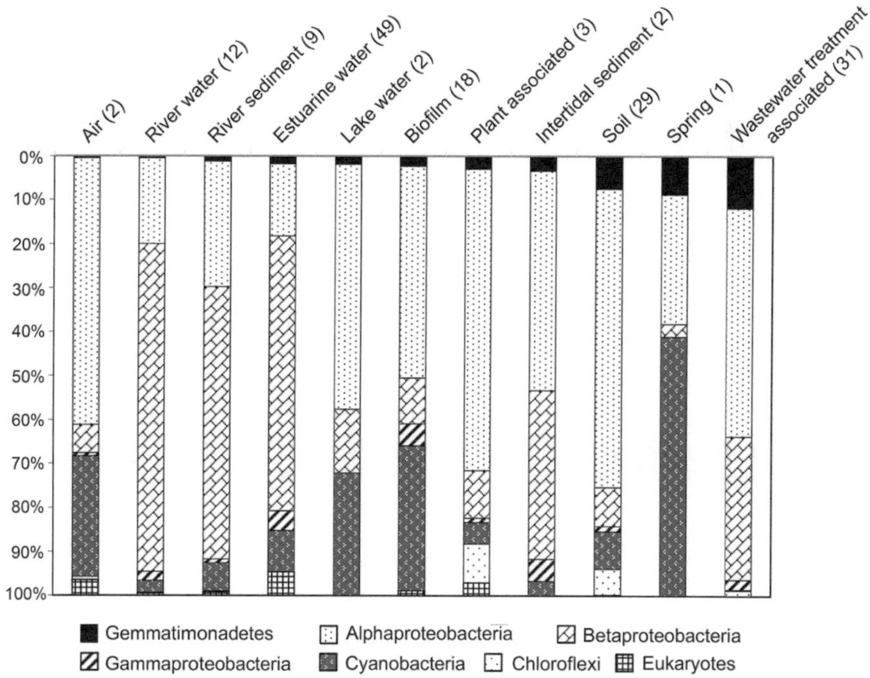

Fig. 13 Putative taxonomic distribution of all AcsF hits in the 158 MG-RAST metagenomes where Gemmatimonadetes-AcsF like hits were found. The AcsF hits were pooled according to the environmental categories of their source metagenomes. Eleven categories were applied, i.e., Air (2/417, total Gemmatimonadetes-AcsF hits/total AcsF hits), River water (30/6104), River sediment (12/1000), Estuarine water (85/4711), Lake water (3/158), Biofilm (47/2034), Plant associated (3/102), Intertidal sediment (2/60), Soil (42/556), Spring (3/34), and Wastewater treatment associated (59/494). The taxonomy of each hit was determined based on its top blastp hits against the NCBI nr database. A strict e-value rule was set for determining a Gemmatimonadetes-AcsF like sequence. In parentheses are the numbers of metagenome samples investigated. Originally published in Zeng et al. (2016)

wastewater treatment associated samples (11.9 %), while air metagenomes contained the lowest number of Gemmatimonadetes-AcsF like hits (0.4 %).

Our analysis of public metagenomic data demonstrated that these organisms turned out to be widely distributed in the environment and exhibited a high genetic diversity. PGB roughly make up 0–11.9 % of total prokaryotic phototrophs. In certain environments like soils, plant-associated habitats, and springs, the abundance of PGB was comparable to that of phototrophic Betaproteobacteria (Imhoff 2006), indicating the importance of this recently found phototrophic group in these habitats. The presence of PGB's marker genes in geographically dispersed and non-equivalent environments suggests that phototrophy may be a common strategy in Gemmatimonadetes rather than an unusual event occurring only in a specific environment.

Perspectives

Physiology and ecology of PGB remain largely unexplored due to the lack of cultured representatives and direct environmental survey using biomarker genes such as AcsF. The limited amount of physiological data from *G. phototrophica* suggests that PGB prefer low oxygen conditions for growth (Zeng et al. 2015). What still needs to be explored are characteristics of photosynthesis apparatus in PGB. The presence of two absorption bands in the IR part of *G. phototrophica* spectrum resembles the LH1 and LH2 complexes in phototrophic Proteobacteria. On the other hand, the *G. phototrophica* genome does not contain LH2 genes. This may indicate an unusual architecture of its photosynthetic complexes. Another open question that remains is the apparent lack of regulation of photosynthetic apparatus expression. *G. phototrophica* expresses its photosynthetic complexes even under constant darkness, at the same level as in the light/dark cycle. However, how light and oxygen regulate the expression of their photosynthesis genes is unknown. So far it is not clear whether the expression is constitutive, but the organization of photosynthesis genes in the PGC may indicate that they are under some regulatory control.

It is also unclear what evolutionary and physiological advantages PGB can gain by assembling such metabolically expensive complex (Kirchman and Hanson 2013). A plausible explanation is that the additional energy source helps PGB cope with unfavorable environmental conditions, e.g., nutrient depletion (Zeng et al. 2014). More PGB cultures from different environments and more physiological experiments are needed to test this hypothesis.

The search of AcsF gene fragments of Gemmatimonadetes' origin in public metagenomic databases provided us with more insights into the distribution and diversity of this bacterial group. Genes of PGB were identified in a number of terrestrial and freshwater aquatic environments. Taking into account the large variability of sequencing depth among the surveyed metagenomes, we speculate that PGB may be present in even more environments which were so far not covered with sufficient sequencing depth. There is also another possibility that can cause such underestimation, which is that some PGB may contain unknown types of AcsF gene equivalents and thus were not identified by the methods that we used.

The database searching provided strong evidence for the widespread presence of phototrophic Gemmatimonadetes in the environment. An even more convincing piece of evidence came from the successful assembly of two almost complete PGCs of Gemmatimonadetes' origin from two Danish wastewater metagenomes. All three Gemmatimonadetes-PGCs identified so far, i.e., PGC of *G. phototrophica*, OdenseWW-PGC, and AalborgWW-2-PGC, are highly conserved in gene composition and arrangement. It should be noted that despite that the source environments of the OdenseWW metagenome and *G. phototrophica* strain AP64 are located on two different continents, they are phylogenetically closely related as seen from the AcsF tree (Fig. 12) and thus may represent only a single type of PGC pattern in the Gemmatimonadetes phylum. The conservation of PGC gene content and arrangement could be much lower among the PGB from the GatunLake, SoilCrust, KoreanSoil, and SulfurSpring environments that are more distantly related to *G. phototrophica*. More PGC data from other Gemmatimonadetes bacteria are needed to address if there are variations in Gemmatimonadetes' PGCs.

This molecular evidence strongly demonstrates that we have ample opportunities to isolate these organisms from environments and have a good chance to answer the questions regarding the genomic diversity, physiological potentials, and ecological significance of those as-yet-uncultured PGB. Understanding the genomic basis and variations of their photosynthesis function will eventually enable us to freely edit and engineer their photosystems, which will open many potential applications in biotechnology.

Acknowledgement The main content of this chapter is a compilation of three research papers we published before (Zeng et al. 2014, 2015, 2016). We thank Fuying Feng, Hana Medová, Jason Dean, Vadim Selyanin, Martin Lukes, David Kaftan, and Jan Baumbach and his group members in University of Southern Denmark for their help in different periods along our journey of researching this novel phototrophic bacterial group. Y.Z.'s postdoctoral research in University of Southern Denmark was supported by the ERC advanced grant "OXYGEN." M.K. would like to acknowledge the support from Czech projects GAČR P501/12/G055 and Algatech Plus LO1416.

References

Albertsen M, Hugenholtz P, Skarshewski A, Nielsen KL, Tyson GW, Nielsen PH (2013) Genome sequences of rare, uncultured bacteria obtained by differential coverage binning of multiple metagenomes. Nat Biotechnol 31(6):533–538

Boldareva-Nuianzina EN, Blahova Z, Sobotka R, Koblížek M (2013) Distribution and origin of oxygen-dependent and oxygen-independent forms of Mg-protoporphyrin monomethylester cyclase among phototrophic proteobacteria. Appl Environ Microbiol 79(8):2596–2604

Breidenbach B, Pump J, Dumont MG (2015) Microbial community structure in the rhizosphere of rice plants. Front Microbiol 6:1537

Bryant DA, Costas AMG, Maresca JA, Chew AGM, Klatt CG, Bateson MM, Tallon LJ, Hostetler J, Nelson WC, Heidelberg JF, Ward DM (2007) Candidatus chloracidobacterium thermophilum: An aerobic phototrophic acidobacterium. Science 317(5837):523–526

Canfield DE, Rosing MT, Bjerrum C (2006) Early anaerobic metabolisms. Philos Trans R Soc B 361(1474):1819–1834
Canniffe DP, Jackson PJ, Hollingshead S, Dickman MJ, Hunter CN (2013) Identification of an 8-vinyl reductase involved in bacteriochlorophyll biosynthesis in *Rhodobacter sphaeroides* and evidence for the existence of a third distinct class of the enzyme. Biochem J 450(2):397–405
Cardona T (2016) Origin of bacteriochlorophyll *a* and the early diversification of photosynthesis. PLoS One 11:e0151250
Clingenpeel S, Macur RE, Kan J, Inskeep WP, Lovalvo D, Varley J, Mathur E, Nealson K, Gorby Y, Jiang H, LaFracois T, McDermott TR (2011) Yellowstone lake: high-energy geochemistry and rich bacterial diversity. Environ Microbiol 13(8):2172–2185
DeBruyn JM, Nixon LT, Fawaz MN, Johnson AM, Radosevich M (2011) Global biogeography and quantitative seasonal dynamics of Gemmatimonadetes in soil. Appl Environ Microbiol 77(17):6295–6300
DeBruyn JM, Fawaz MN, Peacock AD, Dunlap JR, Nixon LT, Cooper KE, Radosevich M (2013) *Gemmatirosa kalamazoonesis* gen. nov., sp nov., a member of the rarely-cultivated bacterial phylum Gemmatimonadetes. J Gen Appl Microbiol 59(4):305–312
Dueholm MS, Albertsen M, Stokholm-Bjerregaard M, McIlroy SJ, Karst SM, Nielsen PH (2015) Complete genome sequence of the bacterium Aalborg_AAW-1, representing a novel family within the candidate phylum SR1. Genome Announc 3(3)
Durbin AM, Teske A (2011) Microbial diversity and stratification of South Pacific abyssal marine sediments. Environ Microbiol 13:3219–3234
Falkowski PG, Raven JA (2007) Aquatic photosynthesis, 2nd edn. Princeton Univ Press, Princeton, p 484
Fujita Y, Matsumoto H, Takahashi Y, Matsubara H (1993) Identification of a *nifDK*-like gene (ORF467) involved in the biosynthesis of chlorophyll in the cyanobacterium *Plectonema boryanum*. Plant Cell Physiol 34(2):305–314
Gest H, Blankenship RE (2004) Time line of discoveries: anoxygenic bacterial photosynthesis. Photosynth Res 80(1–3):59–70
Gest H, Favinger JL (1983) *Heliobacterium chlorum*, an anoxygenic brownish-green photosynthetic bacterium containing a "new" form of bacteriochlorophyll. Arch Microbiol 136(1):11–16
Gibson LCD, Willows RD, Kannangara CG, Von Wettstein D, Hunter CN (1995) Magnesium-protoporphyrin chelatase of *Rhodobacter sphaeroides*: reconstitution of activity by combining the products of the bchH,-I, and-D genes expressed in *Escherichia coli*. Proc Natl Acad Sci U S A 92(6):1941–1944
Gugliandolo C, Michaud L, Lo Giudice A, Lentini V, Rochera C, Camacho A, Maugeri TL (2016) Prokaryotic community in lacustrine sediments of Byers Peninsula (Livingston Island, Maritime Antarctica). Microb Ecol 71:387–400
Hanada S, Sekiguchi Y (2014) The phylum Gemmatimonadetes. In: Rosenberg E, DeLong EF, Lory S, Stackebrandt E, Thompson F (eds) The prokaryotes - other major lineages of bacteria and the archaea. Springer, Berlin, pp 677–681
Haselkorn R, Lapidus A, Kogan Y, Vlcek C, Paces J, Paces V, Ulbrich P, Pecenkova T, Rebrekov D, Milgram A, Mazur M, Cox R, Kyrpides N, Ivanova N, Kapatral V, Los T, Lykidis A, Mikhailova N, Reznik G, Vasieva O, Fonstein M (2001) The *Rhodobacter capsulatus* genome. Photosynth Res 70(1):43–52
Hauruseu D, Koblizek M (2012) Influence of light on carbon utilization in aerobic anoxygenic phototrophs. Appl Environ Microbiol 78(20):7414–7419
Hohmann-Marriott MF, Blankenship RE (2011) Evolution of photosynthesis. Annu Rev Plant Biol 62:515–548
Imhoff JF (2006) The phototrophic beta-proteobacteria. In: The prokaryotes. Springer, pp 593–601
Jung J, Choi S, Hong H, Sung J-S, Park W (2014) Effect of red clay on diesel bioremediation and soil bacterial community. Microb Ecol 68(2):314–323
Kamagata Y (2010) Phylum XXI. Gemmatimonadetes Zhang, Sekiguchi, Hanada, Hugenholtz, Kim, Kamagata and Nakamura 2003, 1161VP. In: Krieg NR, Staley JT, Brown DR et al. (eds) Bergey's manual of systematic bacteriology, vol IV, 2 edn. Springer, New York, pp 781–784

Kamke J, Taylor MW, Schmitt S (2010) Activity profiles for marine sponge-associated bacteria obtained by 16S rRNA vs 16S rRNA gene comparisons. ISME J 4:498–508

Kirchman DL, Hanson TE (2013) Bioenergetics of photoheterotrophic bacteria in the oceans. Environ Microbiol Rep 5(2):188–199

Koblížek M, Mlčoušková J, Kolber Z, Kopecký J (2010) On the photosynthetic properties of marine bacterium COL2P belonging to *Roseobacter* clade. Arch Microbiol 192(1):41–49

Lang AS, Beatty JT (2000) Genetic analysis of a bacterial genetic exchange element: the gene transfer agent of *Rhodobacter capsulatus*. Proc Natl Acad Sci U S A 97(2):859–864

Lindell D, Sullivan MB, Johnson ZI, Tolonen AC, Rohwer F, Chisholm SW (2004) Transfer of photosynthesis genes to and from *Prochlorococcus* viruses. Proc Natl Acad Sci U S A 101(30):11013–11018

Mann NH, Cook A, Millard A, Bailey S, Clokie M (2003) Marine ecosystems: bacterial photosynthesis genes in a virus. Nature 424(6950):741

Marrs B (1974) Genetic recombination in *Rhodopseudomonas capsulata*. Proc Natl Acad Sci U S A 71(3):971–973

Marrs B (1981) Mobilization of the genes for photosynthesis from *Rhodopseudomonas capsulata* by a promiscuous plasmid. J Bacteriol 146(3):1003–1012

Meyer F, Paarmann D, D'Souza M, Olson R, Glass EM, Kubal M, Paczian T, Rodriguez A, Stevens R, Wilke A, Wilkening J, Edwards RAC (2008) The metagenomics RAST server - a public resource for the automatic phylogenetic and functional analysis of metagenomes. BMC Bioinforma 9:386

Molisch H (1907) Die Purpurbakterien nach neuen Untersuchungen. Gustav Fischer, Jena, Germany

Nadson GA (1906) The morphology of inferior algae. III. *Chlorobium limicola* Nads., the green chlorophyll bearing microbe. Bull Jard Bot St Petersp 6:190

Nagashima S, Nagashima KV (2013) Comparison of photosynthesis gene clusters retrieved from total genome sequences of purple bacteria. In: Beatty JT (ed) Genome evolution of photosynthetic bacteria. Elsevier B.V., San Diego, pp 151–178

Nisbet EG, Sleep NH (2001) The habitat and nature of early life. Nature 409(6823):1083–1091

Olson JM (2006) Photosynthesis in the Archean Era. Photosynt Res 88:109–117

Overmann J, Garcia-Pichel F (2013) The phototrophic way of life. In: Rosenberg E, DeLong EF, Lory S, Stackebrandt E, Thompson F (eds) The prokaryotes: prokaryotic communities and ecophysiology, 4th edn. Springer, Berlin, pp 203–257

Pascual J, García-López M, Bills GF, Genilloud O (2016) *Longimicrobium terrae* gen. nov., sp. nov., a novel oligotrophic bacterium of the underrepresented phylum Gemmatimonadetes isolated through a system of miniaturized diffusion chambers. Int J Syst Evol Microbiol. doi:10.1099/ijsem.0.000974

Petersen J, Brinkmann H, Bunk B, Michael V, Paeuker O, Pradella S (2012) Think pink: photosynthesis, plasmids and the *Roseobacter* clade. Environ Microbiol 14(10):2661–2672

Pierson BK, Castenholz Rw (1974) A phototrophic gliding filamentous bacterium of hot springs, *Chloroflexus aurantiacus*, gen. and sp. nov. Arch Microbiol 100(1):5–24

Portillo MC, Leff JW, Lauber CL, Fierer N (2013) Cell size distributions of soil bacterial and archaeal taxa. Appl Environ Microbiol 79:7610–7617

Potts M (1994) Desiccation tolerance of prokaryotes. Microbiol Rev 58(4):755–805

Raymond J, Zhaxybayeva O, Gogarten JP, Gerdes SY, Blankenship RE (2002) Whole-genome analysis of photosynthetic prokaryotes. Science 298(5598):1616–1620

Rusch DB, Halpern AL, Sutton G, Heidelberg KB, Williamson S, Yooseph S, Wu D, Eisen JA, Hoffman JM, Remington K, Beeson K, Tran B, Smith H, Baden-Tillson H, Stewart C, Thorpe J, Freeman J, Andrews-Pfannkoch C, Venter JE, Li K, Kravitz S, Heidelberg JF, Utterback T, Rogers Y-H, Falcon LI, Souza V, Bonilla-Rosso G, Eguiarte LE, Karl DM, Sathyendranath S, Platt T, Bermingham E, Gallardo V, Tamayo-Castillo G, Ferrari MR, Strausberg RL, Nealson K, Friedman R, Frazier M, Venter JC (2007) The sorcerer II global ocean sampling expedition: northwest Atlantic through eastern tropical pacific. PLoS Biol 5(3):398–431

Saunders AH, Golbeck JH, Bryant DA (2013) Characterization of BciB: a ferredoxin-dependent 8-vinyl-protochlorophyllide reductase from the green sulfur bacterium *Chloroherpeton thalassium*. Biochemistry 52(47):8442–8451
Schmidle W (1901) Neue Algen aus dem Gebiete des Oberrheins. Beih Bot Zentralbl 10:179–180
Selyanin V, Hauruseu D, Koblížek M (2016) The variability of light-harvesting complexes in aerobic anoxygenic phototrophs. Photosynth Res 128(1):35–43
Sharon I, Alperovitch A, Rohwer F, Haynes M, Glaser F, Atamna-Ismaeel N, Pinter RY, Partensky F, Koonin EV, Wolf YI, Nelson N, Beja O (2009) Photosystem I gene cassettes are present in marine virus genomes. Nature 461(7261):258–262
Sheng P, Yu Y, Zhang G, Huang J, He L, Ding J (2016) Bacterial diversity and distribution in seven different estuarine sediments of Poyang Lake, China. Environ Earth Sci 75:479
Steffen MM, Li Z, Effler TC, Hauser LJ, Boyer GL, Wilhelm SW (2012) Comparative metagenomics of toxic freshwater cyanobacteria bloom communities on two continents. PLoS One 7(8)
Steven B, Gallegos-Graves LV, Yeager CM, Belnap J, Evans RD, Kuske CR (2012) Dryland biological soil crust cyanobacteria show unexpected decreases in abundance under long-term elevated CO_2. Environ Microbiol 14(12):3247–3258
Swingley WD, Blankenship RE, Raymond J (2009) Evolutionary relationships among purple photosynthetic bacteria and the origin of proteobacterial photosynthetic systems. In: Hunter CN, Daldal F, Thurnauer MC, Beatty JT (eds) Advances in photosynthesis and respiration, the purple phototrophic bacteria, vol 28. Springer, Dordrecht, pp 17–29
Takaichi S, Maoka T, Takasaki K, Hanada S (2010) Carotenoids of *Gemmatimonas aurantiaca* (Gemmatimonadetes): identification of a novel carotenoid, deoxyoscillol 2-rhamnoside, and proposed biosynthetic pathway of oscillol 2, 2'-dirhamnoside. Microbiology 156(3):757–763
Tuorto SJ, Darias P, McGuinness LR, Panikov N, Zhang TJ, Haggblom MM, Kerkhof LJ (2014) Bacterial genome replication at subzero temperatures in permafrost. ISME J 8:139–149
Ulloa O, Canfield DE, DeLong EF, Letelier RM, Stewart FJ (2012) Microbial oceanography of anoxic oxygen minimum zones. Proc Natl Acad Sci U S A 109(40):15996–16003
Wang Z, Guo F, Liu L, Zhang T (2014) Evidence of carbon fixation pathway in a bacterium from candidate phylum SBR1093 revealed with genomic analysis. PLoS One 9(10), e109571
Winogradsky S (1888) Zur morphologie und Physiologie der Schwefelbakterien. Arthur Felix, Leipzig
Xiong J, Inoue K, Bauer CE (1998) Tracking molecular evolution of photosynthesis by characterization of a major photosynthesis gene cluster from *Heliobacillus mobilis*. Proc Natl Acad Sci U S A 95(25):14851–14856
Yurkov VV, Beatty JT (1998) Aerobic anoxygenic phototrophic bacteria. Microbiol Mol Biol Rev 62(3):695–724
Zeng Y, Feng F, Medova H, Dean J, Koblizek M (2014) Functional type 2 photosynthetic reaction centers found in the rare bacterial phylum Gemmatimonadetes. Proc Natl Acad Sci U S A 111(21):7795–7800
Zeng Y, Selyanin V, Lukes M, Dean J, Kaftan D, Feng F, Koblížek M (2015) Characterization of the microaerophilic, bacteriochlorophyll *a*-containing bacterium *Gemmatimonas phototrophica* sp nov., and emended descriptions of the genus *Gemmatimonas* and *Gemmatimonas aurantiaca*. Int J Syst Evol Microbiol 65:2410–2419
Zeng Y, Baumbach J, Vieira Barbosa EG, Azevedo V, Zhang C, Koblížek M (2016) Metagenomic evidence for the presence of phototrophic Gemmatimonadetes bacteria in diverse environments. Environ Microbiol Rep 8(1):139–149
Zhang H, Sekiguchi Y, Hanada S, Hugenholtz P, Kim H, Kamagata Y, Nakamura K (2003) *Gemmatimonas aurantiaca* gen. nov., sp nov., a gram-negative, aerobic, polyphosphate-accumulating micro-organism, the first cultured representative of the new bacterial phylum Gemmatimonadetes phyl. nov. Int J Syst Evol Microbiol 53:1155–1163
Zheng Q, Zhang R, Koblizek M, Boldareva EN, Yurkov V, Yan S, Jiao N (2011) Diverse arrangement of photosynthetic gene clusters in aerobic anoxygenic phototrophic bacteria. PLoS One 6(9):e25050
Zhou J, Sun D, Childers A, McDermott TR, Wang Y, Liles MR (2015) Three novel virophage genomes discovered from yellowstone lake metagenomes. J Virol 89(2):1278–1285

Aerobic Anoxygenic Phototrophs: Four Decades of Mystery

Vladimir Yurkov and Elizabeth Hughes

Abstract The aerobic anoxygenic phototrophs (AAP) are an important group of bacteria making up large proportions of bacterial communities in both marine and freshwater systems. They thrive in the extreme conditions of hot springs, hypersaline spring systems, and hydrothermal vents and in the presence of high concentrations of toxic metal(loid) oxides. They likely evolved from the purple non-sulfur bacteria, to fill an oxygenated environmental niche, carrying out oxygen-dependant anoxygenic photosynthesis. Investigations into the ecological significance of AAP are in their infancy, although some speculations have now been proposed. Additionally, modern studies are beginning to touch the paradox that is bacteriochlorophyll *a* synthesis in the presence of oxygen as well as the role of abundant carotenoids in AAP. The presence of numerous AAP in every environment tested, in addition to their unique photosynthetic arrangement, are mysteries that have garnered much attention among scientists since their discovery.

Introduction

The aerobic anoxygenic phototrophs (AAP) are a diverse group of bacteria, which produce bacteriochlorophyll (BChl) *a* under oxygenated conditions. The first discovered species, *Roseobacter denitrificans* and *Erythrobacter longus*, were isolated from marine environments almost 40 years ago (Shiba et al. 1979; Shiba and Simidu 1982; Shiba 1991). Soon after that *Sandaracinobacter sibiricus*, *Erythromonas ursincola*, *Erythromicrobium hydrolyticum*, *Erythromicrobium ramosum*, *Erythromicrobium ezovicum*, and *Roseococcus thiosulfatophilus* were found in freshwater hot temperature springs (Yurkov and Gorlenko 1990, 1992a, b; Yurkov et al. 1993a, 1994, 1997). Since then research into AAP has exploded. They have been detected in high abundance in every habitat tested, including oceans, seas, freshwater lakes (Fig. 1a), and river systems around the world, as well as in extreme settings like the phyllosphere (Stiefel et al. 2013), hypersaline spring systems

V. Yurkov (✉) • E. Hughes
Department of Microbiology, University of Manitoba, Winnipeg, MB, Canada
e-mail: vyurkov@umanitoba.ca; umhughe4@myumanitoba.ca

© Springer International Publishing Switzerland 2017

P.C. Hallenbeck (ed.), *Modern Topics in the Phototrophic Prokaryotes*,
DOI 10.1007/978-3-319-46261-5_6

(Fig. 1b), hot springs (Fig. 1c), meromictic lakes, hydrothermal vents (Fig. 1d), mine tailings (Fig. 1e), and soil crusts (Fig. 1f) (Rathgeber et al. 2004; Yurkov and Csotonyi 2009; Yurkov and Hughes 2013).

Unlike other anoxygenic phototrophs, AAP require oxygen for both growth and photosynthetic electron transport. They are obligate heterotrophs able to supplement their energy generation with photosynthesis, which physiologically makes them photoheterotrophs and gives them a competitive advantage against other heterotrophs. The main light-harvesting pigment is BChl *a*, which is incorporated into reaction

Fig. 1 Diversity of AAP habitats: (**a**) Lake Winnipeg, Manitoba, Canada; (**b**) East German Creek system, Northern Manitoba, Canada; (**c**) Sulfur Mountain hot springs, Banff, Alberta, Canada; (**d**) hydrothermal vent, Juan de Fuca Ridge, Pacific Ocean; (**e**) Central Mine, Nopiming Provincial Park, Canada; and (**f**) soil crusts, Manitoba, Canada

centers (RC) and light-harvesting (LH) complexes, resembling those of the purple non-sulfur bacteria (PNSB). However, while they do carry out anoxygenic photosynthesis in a similar manner as the PNSB, they differ in that it is not coupled to carbon fixation, as all AAP lack the key enzyme of the Calvin cycle, RUBISCO (Yurkov and Beatty 1998; Yurkov and Csotonyi 2009; Yurkov and Hughes 2013). Minimal levels of carbon fixation can occur through anapleurotic reactions in the TCA cycle, but it is not sufficient to support bacterial growth (Yurkov and Hughes 2013).

Phylogenetically, AAP can be most closely related to PNSB or to non-phototrophs among the α-, β-, and γ-proteobacteria. This raises questions into how the AAP evolved. When the atmosphere first became oxygenated 2.5 GYa ago through the action of oxygenic phototrophs such as cyanobacteria, it opened up new niches for life (Beatty 2005). At this point, did the AAP evolve from PNSB as selection pressure forced them to move into these newly formed habitats in contact with oxygen? Conversely, were the AAP originally aerobic heterotrophs that gained a photosynthetic gene cluster (PGC) through horizontal transfer from the PNSB? Or is it a combination of both those events? Have they remained closely related to the PNSB because they evolved from them by losing photosynthetic genes upon coming into contact with oxygen and then regained them later through horizontal gene transfer? There are speculations to support all these options (Yurkov and Csotonyi 2009; Yurkov and Hughes 2013), and in fact, the answer may be a combination of all proposals. This is only one of many mysteries still surrounding the AAP four decades post-discovery.

Despite intensive research, there are many unanswered questions about the AAP. Our chapter will explore the riddles, hypotheses, and new discoveries about the adaptations of the BChl *a* synthetic pathway in the presence of both light and oxygen, the role of carotenoids in light harvesting and photoprotection, the ecological significance of AAP in their natural habitats, as well as novel approaches to determine diversity and abundance in aquatic environments. It is primarily a review of scientific reports that appeared since 2013. For previous analyses of AAP, we recommend reading Yurkov and Beatty (1998), Rathgeber et al. (2004), Yurkov (2006), Yurkov and Csotonyi (2009), and Yurkov and Hughes (2013).

Diversity, Abundance, and Ecological Significance

Distribution and Enumeration

Investigations of AAP abundance and distribution have always been somewhat complicated. There is no defined medium that can be used to select specifically for AAP in culture-dependant approaches, as they grow best on complex carbon sources, allowing for a wide range of heterotrophs to thrive alongside them. Therefore, AAP must instead be identified among other colonies by their bright pigmentation, ranging from yellow, orange, and red to pink and purple. They should then be tested spectrophotometrically for the presence of BChl *a* (Yurkov and Csotonyi 2009; Yurkov and Hughes 2013). Enumeration in such manners can result in an underestimation of abundance as a large percentage of microbes do not grow in a laboratory setting. Unfortunately, culture-independent sequencing methods are not accurate either, as there are no genes found to date that are exclusive to AAP (Yurkov and Hughes 2013). Some studies have tried to enumerate AAP in natural samples using detection of the *pufL* and *pufM* genes, which encode subunits of the RC (Yurkov and Hughes 2013). However, this can result in an overestimation of AAP abundance as the genes are too closely related to those of PNSB, making it impossible to distinguish between the two physiologically different groups with any certainty. If *pufLM* genes do not allow for an accurate study of abundance and distribution, are there any other genes restricted to the AAP that have not yet been identified? Full genome sequences of cultured strains may provide us with that answer.

Recently, the proposed solution to this conundrum has been to use infrared epifluorescence to detect BChl *a* in uncultured bacteria from both marine and freshwater habitats (Garcia-Chaves et al. 2016). While BChl *a* is the same light-harvesting pigment synthesized by PNSB, they are assumed to not be producing it in the presence of oxygen. Therefore, sample collection is limited to the aerated portion of the water column to ensure there is sufficient oxygen present and therefore any BChl *a* found must be from an aerobic bacterium (Mašín et al. 2012; Garcia-Chaves et al. 2016). This might be particularly effective in well-stratified lakes, where an aerated upper layer and an anaerobic bottom portion are separated by a sharp thermocline. Unfortunately, this method is also not completely accurate as several PNSB, for example, *Rhodobacter capsulatus*, can produce some BChl *a* in oxygenated environments (Hebermehl and Klug 1998).

Investigations in both marine and freshwater environments using infrared epifluorescence have revealed a great morphological diversity and abundance of AAP as well as their growth patterns. There was a clear seasonal variation with higher bacterial counts in summer and lower in winter (Čuperová et al. 2013; Ferrera et al. 2014; Fauteux et al. 2015) and also a clear negative correlation with distance from shore in marine habitats including the Delaware coast and estuary (Stegman et al. 2014), the Northeast Pacific and Arctic Ocean (Boeuf et al. 2013), and close to various islands in the Pacific Ocean (Ritchie and Johnson 2012), with AAP abundance decreasing as sampling sites became distanced from the shoreline. Apparently AAP make up a significant portion of the bacterial community in marine environments (10–14 % of total bacteria in the Northeast Pacific and Arctic Oceans (Boeuf et al. 2013) and 3.5–7.9 % in the Uwa Sea (Sato-Takabe et al. 2015)). The data agree with previous studies, showing that AAP compose roughly 10 % of the bacterial community in the open ocean (Rathgeber et al. 2004; Yurkov and Csotonyi 2009; Yurkov and Hughes 2013). While most of the early distribution analysis of AAP was done in the open ocean, recently there has been increased interest in freshwater systems, where AAP represent as high as 29 % of all prokaryotes in Austrian alpine lakes (Čuperová et al. 2013) and 2–12 % in lakes throughout Germany, Finland, and Poland. Oligotrophic and mesotrophic lakes show higher presence of AAP over eutrophic lakes (Mašín et al. 2012), though earlier studies have suggested that enumeration can be even higher, up to 30 % in eutrophic habitats, such as freshwater hot temperature spring mats, which commonly play home to AAP (Yurkov and Beatty 1998; Yurkov and Hughes 2013).

Throughout all tested habitats, there is a broad taxonomic diversity of AAP species. However, in some environments, patterns are being recognized in the distribution of genera. For instance, in higher alpine lakes in Austria, almost all AAP belonged to the *Sphingomonadaceae* family. This could be because members of the family, such as the genus *Sphingomonas*, resist high levels of UV-B radiation (Cŭperová et al. 2013). This may be a beneficial trait in regions where the water is very clear and the surrounding area has very little forest shade to protect the lake occupants from the damaging effects of intensive sunlight (Cŭperová et al. 2013). Conversely, *Citromicrobium* relatives have been found to dominate in the aerated upper twilight zone in the Pacific Ocean, suggesting they may be able to use very dim light or specific wavelengths that can pass through the water column for photosynthesis, allowing them to occupy a specific niche that other AAP cannot (Zheng et al. 2015). Although this is one published speculation, in our opinion it has little merit, as all known AAP are able to grow indefinitely as heterotrophs without using light for photosynthesis. We propose that in this case, competition may be too high in illuminated spaces causing the *Citromicrobium* relatives to move into the upper twilight or dark zones, where they can effectively grow purely heterotrophically.

While some current methods allow observation of the patterns discussed above and achieve relatively accurate bacterial counts, culturing methods cannot be completely replaced. Only pure laboratory cultures of AAP allow us to study them in detail to define if our theoretical speculations are true and to discover novel physiological and biochemical traits that AAP possess.

Phylogenetic Conundrums

All AAP, like their PNSB counterparts, belong to the proteobacteria. The vast majority of taxonomically classified strains are part of the α-proteobacteria with one species, *Roseatales depolymerans*, belonging to the β-proteobacteria, and two species, *Congregibacter litoralis* and *Chromocurvus halotolerans*, to the γ-proteobacteria, though many culture-independent studies have shown that there are far more β- and γ-proteobacterial AAP present in nature, with the β-proteobacteria dominating freshwater environments and γ-proteobacteria being prominent in saline systems (Ritchie and Johnson 2012; Yurkov and Hughes 2013; Zheng et al. 2015; Lehours and Jeanthon 2015). One notable exception is found in the Northeast Pacific and Arctic Ocean, where β-proteobacteria were unexpectedly reported in high numbers when they are not usually observed in abundance in the open ocean (Boeuf et al. 2013).

Since our latest review (Yurkov and Hughes 2013), several novel genera and species have been described (Table 1): *Roseibacula alcaliphilum* isolated from Lake Doroninskoe, which is a meromictic soda lake in Siberia (Nuyanzina-Boldareva and

Table 1 Determinative characteristics of novel AAP described since 2012[a]

Species	Habitat, site of isolation	Phylogenetic affiliation[b]	Color	Cell shape and size	In vivo BChl peaks
Roseibacula alcaliphilum[c] (Nuyanzina-Boldareva and Gorlenko 2014)	Lake Doroninskoe (meromictic soda lake) surface water	α-Proteobacteria, *Roseinatronobacter* and *Rhodobaca* (96 %)	Pink	Oval, 0.5–1.0 × 1.5–1.7 μm	810, 874 nm
Citrimicrobium luteum (Jung et al. 2014)	Gut of sea cucumber *Stichopus japonicus*	α-Proteobacteria, *Citromicrobium bathyomarinum* (98.4 %)	Yellow	Coccoid to ovoid rods, 0.8–1.0 × 1.0–1.5 μm	ND[d]
Humitalea rosea (Margesin and Zhang 2013)	Petroleum hydrocarbon-contaminated soil, Italy	α-Proteobacteria, *Roseococcus* and *Rubritepida* (<92.8 %)	Pale pink	Short rods, 0.9 × 1.4–1.7 μm	764 nm (In vitro), In vivo ND
Blastomonas aquatica (Xiao et al. 2015)	Lake water on the Tibetan plateau	α-Proteobacteria, *Blastomonas natatoria* (99.0 %)	Yellow to orange brown	Ovoid to rod	866 nm
Gemmatimonas phototrophica (Zeng et al. 2015)	Freshwater Swan Lake, Gobi Desert	Gemmatimonadetes, *Gemmatimonas aurantiaca* (96.1 %)	Red	Rods, 0.3–0.5 × 1–6 μm	819, 866 nm

[a]New species described before 2012 are summarized in Yurkov and Hughes (2013), Yurkov and Csotonyi (2009), Rathgeber et al. (2004), and Yurkov and Beatty (1998)
[b]Proteobacterial subclass, nearest 16S rRNA relative and phylogenetic distance to nearest relative
[c]Names in bold indicate novel genera
[d]ND indicates not determined

Gorlenko 2014), and *Citrimicrobium luteum*, which is from the gut of a sea cucumber (Jung et al. 2014). Petroleum hydrocarbon-contaminated soil in Italy revealed the AAP *Humitalea rosea* (Margesin and Zhang 2013). *Blastomonas aquatica* was discovered in fresh to brackish water in Lakes Peng Co and Namtso on the Tibetan Plateau, China (Xiao et al. 2015). *Gemmatimonas phototrophica* is an interesting new species, which grows best under limited oxygen concentrations (Zeng et al. 2015). It is similar to *Roseicyclus mahoneyensis*, strain ML6, which is photosynthetically most active under microaerophilic conditions, though heterotrophic growth remains optimal aerobically (Rathgeber et al. 2012). Physiologically *G. phototrophica* groups better with AAP rather than PNSB, as it cannot grow in fully anaerobic conditions and produces BChl *a* in the dark. However, light does not inhibit its synthesis (Zeng et al. 2015), a trait which has also been observed in AAP strain EG13, isolated from a hypersaline spring system in Manitoba, Canada (Csotonyi et al. 2015). Another novel aspect is that it does not belong to the proteobacteria; it is part of the *Gemmatimonadetes* and not the *Proteobacteria*, which begs the question whether it should truly be recognized as AAP.

Recently, Koblížek considered a new definition of AAP (Koblížek 2015), which suggests to include all Bchl *a*-producing bacteria which, in laboratory conditions, grow primarily aerobically. This would allow the inclusion of the phototrophic methylotrophs and rhizobia as well as unusual species such as *G. phototrophica*. It was suggested that because they are aerobic BChl *a*-containing bacteria, they should be considered AAP despite other metabolic, taxonomical, and phylogenetic differences. Additionally, as more and more strains are undergoing full genome sequencing, it is being found that some species, which were presumed to be aerobic non-phototrophic heterotrophs, have all the necessary genes for fully operational photosynthetic pigment-protein complexes and photosynthetic electron transfer (Koblížek 2015). Having never produced BChl *a* and photosynthetic complexes in a controlled laboratory setting suggests that they are simply not being grown under appropriate conditions to express photosynthesis genes. One example is the obligately aerobic β-proteobacterium *Aquincola tertiaricarbonis*, strain L108, which was originally reported to not produce BChl *a* aerobically (Lechner et al. 2007). However, it does in fact synthesize photosynthetic protein complexes with bound BChl *a* when faced with a sudden decrease in organic carbon availability (Rohwerder et al. 2013). Hence, this species can now be re-categorized as an AAP. Such previously published misleading taxonomical descriptions open up the question of how many other AAP are masquerading as non-phototrophic heterotrophs in laboratory studies. Also, what growth conditions are best for inducing BChl *a* production in facultative phototrophs? Will patterns emerge or will every species behave differently? At this time we do not have the answers; however, studies on heterotrophs that have been sequenced to reveal the presence of a full PGC may help bring to light some solutions.

Morphology and Its Link to Ecological Significance

The AAP are morphologically highly diverse. They come in many forms: rods (both short and long), cocci, vibroid, almost cyclic as in *R. mahoneyensis* (Fig. 2a) (Rathgeber et al. 2005), and pleomorphic as found in *E. ramosum* (Fig. 2b), *Citromicrobium bathyomarinum* (Fig. 2c), and *Porphyrobacter meromictius* (Yurkov and Csotonyi 2009; Yurkov and Hughes 2013). They can form highly complex clusters that resemble corals or rosettes (Fig. 2d) (Yurkov and Hughes 2013). The only morphology that has not yet been observed is classic spirilloid, though studies in the Sargasso Sea indicate they may exist, but simply have not yet successfully been cultured in the laboratory (Sieracki et al. 2006).

Interestingly, on average, AAP cells in marine and freshwater habitats are larger than other members of their respective bacterial communities (Stegman et al. 2014; Fauteux et al. 2015; Garcia-Chaves et al. 2015; Sato-Takabe et al. 2015). In one study that analyzed samples from 43 lakes throughout Québec, Canada, the size of AAP cells was positively correlated with the amount of dissolved organic carbon (Fauteux et al. 2015). This size difference sets AAP apart from other heterotrophs in aquatic systems and has led to some interesting theories on their possible ecological significance and role.

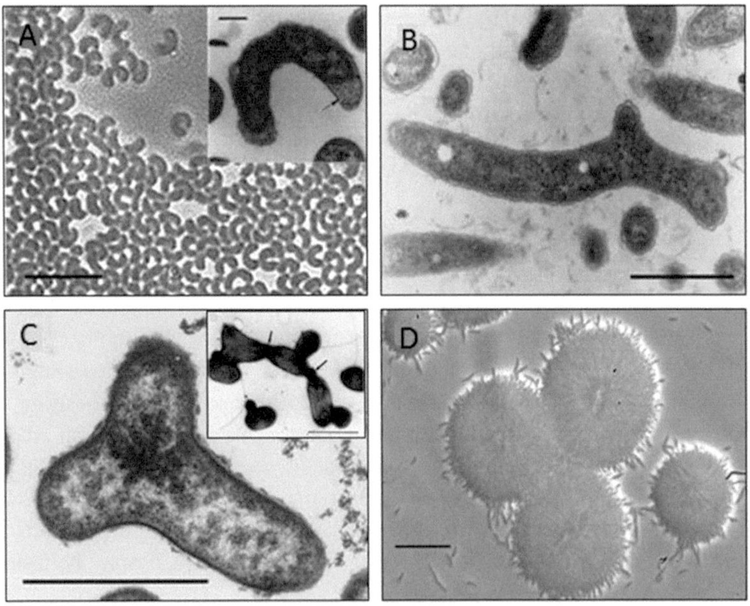

Fig. 2 Morphological variations among AAP. (**a**) *Roseicyclus mahoneyensis* with some almost cyclic cells (bar 5 µm, inset bar 0.25 µm); (**b**) branching *Erythromicrobium ramosum* (bar 1 µm); (**c**) *Citromicrobium bathyomarinum* Y-shaped cell (bar 1 µm), inset pleomorphic cells connected by membranous material (indicated by *arrows*, bar 1 µm); and (**d**) strain BL7 clusters (bar 2.5 µm)

As AAP are larger than the average non-phototrophic heterotrophic bacterium in communities, they are preferentially grazed upon by protists and zooplankton, placing them at the very bottom of the food chain in freshwater and marine ecosystems (Stegman et al. 2014; Garcia-Chaves et al. 2015; Mašín et al. 2012; Sato-Takabe et al. 2015). Zooplankton grazing in particular allows AAP energy to be passed to higher trophic levels (Garcia-Chaves et al. 2015). AAP also have a higher growth rate (1.5–3× higher) than some other heterotrophic bacteria (Stegman et al. 2014; Garcia-Chaves et al. 2015), possibly because they can profit by generating higher levels of energy more quickly due to their photosynthetic capacity. This allows the consumption of higher amounts of dissolved organic carbon at a faster rate before AAP are consumed by zooplankton, making a disproportionately large contribution to carbon cycling and total biomass considering their natural abundance (Stegman et al. 2014; Garcia-Chaves et al. 2015; Sato-Takabe et al. 2015).

When an organism at any trophic level is being grazed upon, it develops some defensive strategies (Mašín et al. 2012). It was proposed that AAP are no exception in this regard. In 27 studied lakes throughout Germany, Poland, and Finland, AAP were often found attached to particles that are too large to be consumed by zooplankton (Mašín et al. 2012). For now, it remains a speculation whether or not this is truly a defense mechanism. In our opinion, it is more likely AAP are simply attached to these particles as they are rich in organics, which serve as a good food source for bacteria rather than to be using them as an escape mechanism from predating zooplankton. In general, the ecological role of AAP in the majority of habitats is very poorly investigated. It still requires a lot of time and research effort to draw a clear image of their significance.

Carbon Utilization and Metabolic Pathways

Carbon Metabolic Pathways

It was established long ago that AAP are obligate heterotrophs as they lack the key enzyme RUBISCO of the Calvin cycle (Yurkov and Beatty 1998; Yurkov and Csotonyi 2009; Yurkov and Hughes 2013). Recent genome sequence data from a number of species including *Sandarakinorhabdus* sp. AAP62 (Zeng et al. 2013a), *Blastomonas* sp. AAP53 (Zeng et al. 2013b), *Porphyrobacter* sp. AAP82 (Li et al. 2013), and *Sphingomonas* sp. FukuSWIS1 (Salka et al. 2014) have only confirmed the absence of this cycle. While incapable of autotrophy, AAP, like many other heterotrophs, fix a minimal amount of CO_2 via anapleurotic reactions (Fig. 3) (Yurkov and Hughes 2013), though it is at insufficient levels to support growth.

What AAP do excel at is breaking down highly complex organics due to the activity of a broad range of proteins and enzymes. This makes them indispensable in carbon cycling in both saline and freshwater systems, as naturally available carbon is generally very complex both structurally and chemically for digestion by other microbes (Fauteux et al. 2015).

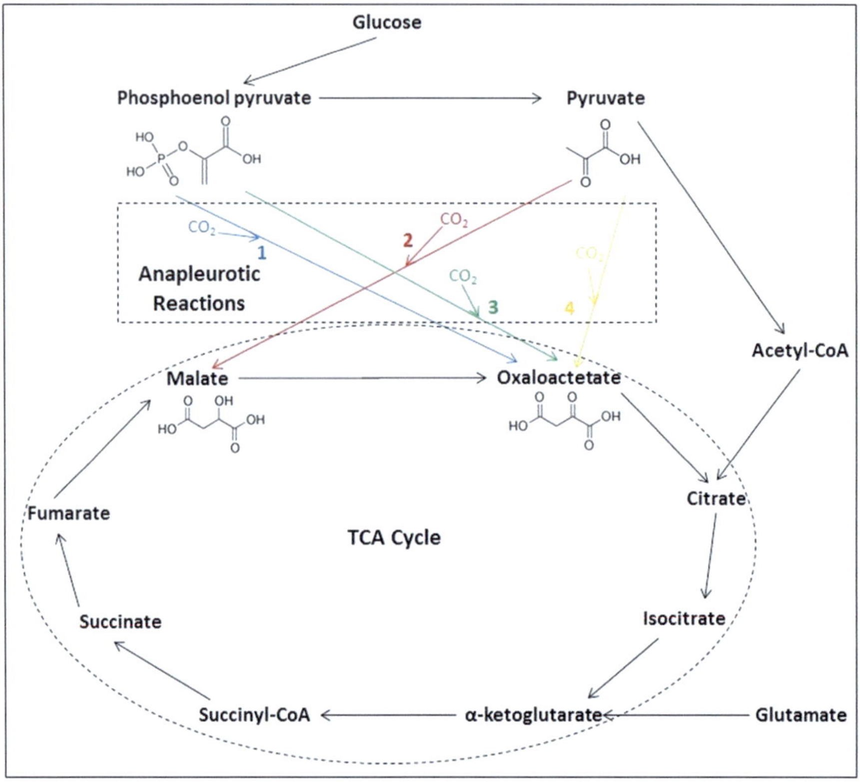

Fig. 3 Fixation of CO_2 through anapleurotic reactions encoded by the enzymes (1, *blue*) phosphoenolpyruvate carboxykinase, (2, *red*) malic enzyme, (3, *green*) phosphoenolpyruvate carboxylase, and (4, *yellow*) pyruvate carboxylase replenishes the TCA cycle. Glucose feeds into the TCA cycle through pyruvate and therefore anapleurotic reactions or acetyl-CoA. Glutamate replenishes the TCA cycle through α-ketoglutarate

Interestingly, when grown on defined medium, the carbon source can have an impact on levels of photosynthesis and anapleurotic reactions in *Erythrobacter* sp. NAP1 (Hauruseu and Koblížek 2012). Four commonly used compounds were tested: glutamate, pyruvate, acetate, and glucose. When provided with glutamate, growth was inhibited by high light intensities, though greater amounts of biomass were produced with a moderate amount of light. Additionally, almost none of the incorporated carbon came from anapleurotic reactions (1 %). Conversely, light intensity made no difference when medium contained pyruvate instead of glutamate. However, 4 % (dark) and up to 11 % (light) of incorporated carbon originated from anapleurotic reactions (Hauruseu and Koblížek 2012). It seems to be species specific, as these values are lower than the 10–15 % anapleurotic CO_2 fixation carried out by *R. denitrificans* (Hauruseu and Koblížek 2012). It was discovered long ago that photosynthetic activity allows AAP to increase their productivity (Yurkov and Gemerden 1993), and now it is confirmed that the organic carbon source also makes a difference. For instance, light exposure allowed 30 % biomass increases, when grown in media containing pyruvate; however, photosynthesis in the presence of glucose resulted in 49.1 % increase (Hauruseu and Koblížek 2012).

It is possible that anapleurotic reactions are used by AAP simply to replenish the TCA cycle. Anapleurotic reactions use pyruvate or phosphoenolpyruvate as a substrate (Fig. 3); hence, more carbon fixation occurs in pyruvate- and glucose-grown cultures. Glutamate, however, can be easily converted to α-ketoglutarate to feed into the TCA cycle (Fig. 3) (Hauruseu and Koblížek 2012). It is also conceivable that this explanation ties into why light intensity makes a difference with glutamate and not with pyruvate. With glutamate, higher light intensities provide more energy than is required by the cell, which in the end causes downregulation of metabolism. However, anapleurotic reactions using pyruvate as a substrate require a large amount of energy, so the possibility of excess is avoided and metabolism is not downregulated (Hauruseu and Koblížek 2012).

Mechanisms of Toxic Heavy Metal(loid) Oxide Resistance and Reduction

As mentioned above, AAP are commonly enumerated in high numbers in extreme environments. While it remains uncertain how and why they have evolved mechanisms to allow growth in such a wide range of extremes, their capability to be very comfortable in the presence of toxic heavy metal(loid) oxides has been researched. These oxides include tellurite, tellurate, selenite, selenate, metavanadate, and orthovanadate (Yurkov and Csotonyi 2003). AAP show resistance to tellurite, for example, of up to 2000 μg/ml, a highly significant number considering that many bacterial species are killed by only 1 μg/ml (Maltman and Yurkov 2015). How is such great resistance to metal(loid) oxides achieved?

One strategy of resistance by AAP is to use toxic compounds in metabolic processes or for energy production. Strains EG13 and EG8 are vanadiphilic, meaning vanadium oxides, which usually kill most life in very small concentrations, actually enhances the bacterial growth capabilities (Csotonyi et al. 2015). Similarly, *E. ursincola*, KR99, and *E. ramosum*, E5, increase biomass and ATP production when grown in the presence of tellurite (Maltman and Yurkov 2014, 2015). Alternately, instead of benefitting from the oxides, some species can simply tolerate them. For instance, *C. bathyomarinum*, JF1, experiences a lag phase when tellurite is added to the medium to adapt to the shock and then resumes normal biomass production (Maltman and Yurkov 2014). *R. thiosulfatophilus*, RB3; *E. ezovicum*, E1; *S. sibiricus*, RB 16-17; *E. hydrolyticum*, E4; and *Erythrobacter litoralis*, T4, all have decreased biomass and ATP levels in the presence of high concentrations of tellurite, suggesting it has a toxic effect on the cells (Maltman and Yurkov 2015).

A search was also carried out for a tellurite-specific reductase in strains KR99, E5, RB3, E1, RB 16-17, E4, T4, and JF1. Cells were fractionated to determine in what compartment of the cell the tellurite reductase was localized. Additionally, protein synthesis was halted with antibiotics before cells were introduced into the medium containing tellurite to check whether the enzyme was constitutively expressed or de novo synthesis was required. Reductase activity in strain E4 was only possible in fully intact cells and de novo protein synthesis was required. T4 and JF1 also had a requirement for de novo synthesis and needed an intact cytoplasmic membrane as was supported by activity in spheroplast lysate fractions (Maltman and Yurkov 2014). In the other tested strains, a constitutively expressed membrane-bound tellurite reductase was present, suggesting the function of a different enzyme (Maltman and Yurkov 2015). Future research will answer how other AAP resist such high levels of toxic compounds and in some cases have evolved to benefit from them.

One of the most common observations with high resistance to toxic metal(loid) oxide strains is that they reduce them to less toxic elemental forms, followed by color changes in the culture. Reduction of tellurite and tellurate to tellurium results in a black appearance, selenite and selenate conversion to selenium brings a red color, and vanadate being reduced to vanadium appears as bluish coloration (Yurkov and Csotonyi 2003; Maltman and Yurkov 2015). As oxides are reduced, the elements accumulate inside or are released outside the cells, as seen by electron microscopy of *E. ramosum* (Fig. 4a), *E. litoralis* (Fig. 4b), and *R. thiosulphatophilus* (Fig. 4c). In the future this could be an indispensable ability to remove toxic compounds for bioremediation in industrial processes, as well as offer a great potential for biometallurgy by collecting concentrated pure tellurium or other accumulated elements as they are quite rare in the biosphere.

Because AAP are so highly resistant to tellurite, in particular, the question was asked whether tellurite resistance is tied to photosynthesis. BChl *a* and carotenoids were monitored in *E. ramosum*, E5; *E. ursincola*, KR99; *C. bathyomarinum*, JF1; *E. litoralis*, T4; and *Erythrobacter* relative strain EG15 in the presence and absence of tellurite (Csotonyi et al. 2014). BChl *a* increased when cultures were grown with tellurite in strains KR99, JF1, T4, and EG15, indicating this pigment may play the role of an antioxidant, but decreased in strain E5, showing oxidative stress could be exceeding the cell's ability to cope. However, when carbon availability was decreased, BChl *a* levels increased in E5 despite the presence of tellurite (Csotonyi et al. 2014). The same pattern holds true for cellular carotenoids. They were increased in KR99, T4, and EG15 and decreased in E5, though under substrate limitation the carotenoids of E5 did increase. The only strain that broke the pattern is JF1, as its carotenoids decreased, while BChl *a* was augmented. Notably, in strain T4, zeaxanthin, spirilloxanthin, β-carotene, and erythroxanthin sulfate were elevated, and bacteriorubixanthinal levels declined (Csotonyi et al. 2014). Speculations into the above results suggested three hypotheses. First, tellurite may act on the promoters to induce transcription of the entire PGC, even if carotenoids are the only pigments involved in oxidative defense. Second, BChl *a* may act in conjunction with carotenoids as antioxidants. Third, resistance to tellurite may be quite energetically costly, causing the upregulation of the photosynthetic apparatus to supply the cells with as much additional energy as possible (Csotonyi et al. 2014). Ongoing experiments will help to tell which strategy is actually used by AAP cells.

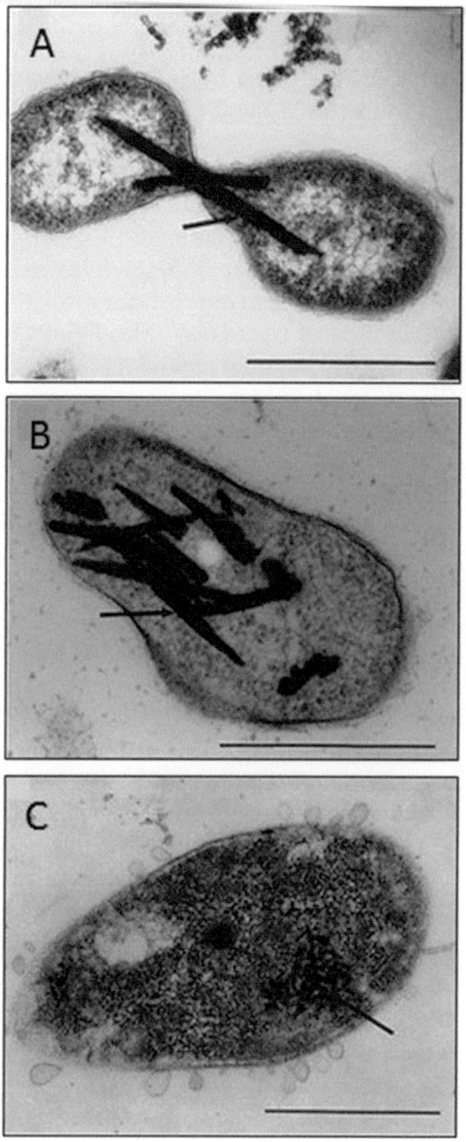

Fig. 4 Electron microscopy showing intracellular accumulation of tellurium (indicated by *arrows*) as a product of tellurite reduction. (**a**) Tellurium crystals interfering with cell division in *E. ramosum*, (**b**) tellurium accumulated in *E. litoralis*, and (**c**) smaller tellurium crystals accumulated in *R. thiosulfatophilus*. Bars, 0.5 mm

Photosynthetic Pigments and Their Significance

The Enigma of Aerobic Bacteriochlorophyll a *Synthesis*

The process of BChl *a* synthesis in the presence of both light and oxygen results in the formation of triplet BChl and singlet oxygen, which exerts high levels of oxidative stress on the cells (Yurkov and Hughes 2013). This is a barrier that all anoxygenic phototrophs have had to overcome. Some phototrophs, for instance, PNSB, produce BChl *a* primarily under anaerobic conditions, decreasing the possibility of oxidative stress. Contrarily, AAP produce their photosynthetic pigments mainly in the dark, so they can then be used up during periods of light. Also AAP keep the number of photosynthetic units to a minimum to limit the toxic effect and have much higher levels of carotenoids to protect the cells from damage. One notable exception to the rule is strain EG13, which synthesizes BChl *a* with either illumination or not (Csotonyi et al. 2015). It has a very good strategy to deal with oxidative stress, possibly tied to its resistance to toxic vanadium oxides as described in section "Mechanisms of Toxic Heavy Metal(loid) Oxide Resistance and Reduction," though continuing research is required for a better understanding of the exact mechanism (Csotonyi et al. 2015).

Genomic and proteomic studies have also been conducted to define the differences in BChl *a* synthesis in cells of AAP versus PNSB and chlorophyll *a* synthesis in oxygenic phototrophs such as cyanobacteria, as all synthetic pathways start similarly. One major difference is in the genes encoding the enzyme magnesium-protoporphyrin IX monomethylester cyclase (Boldareva-Nuianzina et al. 2013). There are two options possible. The gene *acsF*, which uses an O_2 molecule to catalyze the reaction, and the other, *bchE*, instead take an oxygen atom from H_2O (Fig. 5). The genome sequences of 53 phototrophic proteobacteria were compared. Purple sulfur bacteria contain only the *bchE* gene, PNSB all have both forms, and tested AAP all have *acsF*; however, several also have *bchE* (Boldareva-Nuianzina et al. 2013). This delivers a conundrum, as it was previously expected that anaerobic bacteria would have the oxygen-independent form and aerobic phototrophs, the oxygen-dependant type only. Speculatively, *acsF* was gained very early in evolution, as it has a highly conserved position within the PGC: in the *puh* operon in α- and β-proteobacteria and on the end of the PGC in γ-proteobacteria. Hypothetically, it was passed from cyanobacteria to PNSB

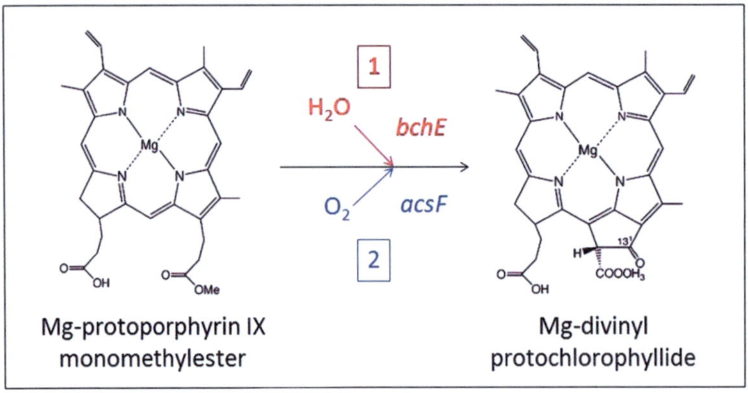

Fig. 5 Conversion of Mg-protoporphyrin IX monomethylester to Mg-divinyl protochlorophyllide using two different forms of the enzyme Mg-protoporphyrin IX monomethylester cyclase. The first form (1, *red*) is encoded by the gene *bchE* to use an oxygen atom from H_2O. The second (2, *blue*) is encoded by the gene *acsF* to use atmospheric O_2

through horizontal gene transfer, allowing PNSB to proliferate into oxygenated environments and initiate the evolutionary process to become AAP (Boldareva-Nuianzina et al. 2013). Likely, the gene was acquired very early in evolution, since a phylogenetic tree based on *acsF* sequences grouped all studied strains the same way as 16S rRNA gene sequences would. Therefore, the gene probably evolved in conjunction with the evolution of each species, rather than being a separate entity recently gained. This may not be true of *bchE*, which appears to be highly variable in position and sequence, as compared to the conserved 16S rRNA gene sequences (Boldareva-Nuianzina et al. 2013).

Another known step in BChl *a* synthesis is the conversion of protochlorophyllide to chlorophyllide by the enzyme protochlorophyllide reductase (Kaschner et al. 2014). There are two forms of this reductase: an oxygen-sensitive dark operating type, which is likely used by anoxygenic phototrophs, and an oxygen-insensitive light-dependant form, possibly employed by oxygenic phototrophs such as cyanobacteria and algae (Kaschner et al. 2014). However, the AAP *Dinoroseobacter shibae*, strain DFL12T, possesses both (Kaschner et al. 2014). Why would an anoxygenic phototroph use an enzyme for BChl *a* synthesis that requires both light and oxygen in conjunction as synthesis of the pigment in this way results in oxidative stress? This is just another unanswered question. It seems that as we explore deeper into AAP BChl *a* synthesis, we just come up with more questions, riddles, and mysteries.

Carotenoids and Their Role in AAP

In comparison to PNSB, AAP have a very high cellular presence of carotenoids, resulting in the vibrant colors of their cultures, such as red, orange, yellow, pink, and purple (Rathgeber et al. 2004; Yurkov and Csotonyi 2009; Yurkov and Hughes 2013). The roles of cellular carotenoids are still debatable, although two are proposed. The first considers their use as accessory light-harvesting pigments in photosynthesis, which may allow the cells to harvest light in a wider range of wavelengths, maximizing the efficiency of photosynthetic energy generation. The second suggests that carotenoids are protecting cells against the damaging effects of light, permitting BChl *a* synthesis aerobically. Realistically, the answer might be a combination of both, with some carotenoids bound to the photosynthetic RC to aid in light harvesting and others acting as protectors against the oxidative stress.

One clear example of an organism that likely uses carotenoids in light harvesting is *Roseobacter*, strain OBYS0001, which has spheroidenone as its dominant pigment (Sato-Takabe et al. 2012, 2014). Higher levels of spheroidenone were produced under substrate-deficient conditions. Apparently, pigments may provide some benefit to starving cells. Combined with the knowledge that this carotenoid absorbs green light and that under substrate deficiency the photochemical efficiency was increased compared to in substrate-replete conditions, the cells may be collecting green light to increase their photosynthetic energy generation. Hence cells can survive, when other heterotrophs would perish (Sato-Takabe et al. 2012). Energy transfer from spheroidenone to BChl *a* was also confirmed by Šlouf et al. (2013) in *R. denitrificans*.

Zeaxanthin, one of the major pigments in *Erythrobacter* sp. NAP1, is not being used for energy transfer (Šlouf et al. 2013), suggesting that its main role is in protection against the aggressive triplet BChl *a* and singlet oxygen (Šlouf et al. 2013). However, the organism also has significant levels of bacteriorubixanthinal, which was shown to harvest light energy and pass it to BChl *a* (Šlouf et al. 2013). A similar composition of carotenoids was previously reported in *E. longus* and *E. litoralis*, though *Erythromicrobium* species such as *E. ramosum* have zeaxanthin only as a minor component of their carotenoid complement with high levels of bacteriorubixanthinal, erythroxanthin sulfate, and β-carotenes (Yurkov et al. 1993b; Yurkov and Beatty 1998). *R. thiosulfatophilus* is different as 95 % of its carotenoid complement is the C_{30} carotenedioate and its diglucosylester (Yurkov et al. 1993a; Yurkov and Beatty 1998), both of which participate in light harvesting.

One AAP species can have numerous different carotenoids, in some cases up to 20 or more, with each one presumably playing a different role. Whether all species have carotenoids that aid in light harvesting, or if the majority are used for protection, or if there are other metabolic functions to carry out is still unresolved. The only way to answer is to continue culture-dependent studies of carotenoids in AAP.

Photosynthetic Apparatus and Electron Transport Chain

New Discoveries on Reaction Centers and Light-Harvesting Complexes

The membrane-bound pigment-protein complexes making up the photosynthetic apparatus in AAP are almost identical to those of the PNSB. BChl *a* is the primary light-harvesting pigment bound into a RC and LH1 complex, with some species also employing a peripheral LH2. The RC can be observed spectrophotometrically by producing absorption peaks at 800 nm and 860–870 nm due to the incorporated BChl *a* as well as a peak at 750–760 nm of bacteriopheophytin (BPheo), a BChl precursor, also bound to the RC proteins. It is a complex composed of three protein subunits (L, H, and M) with four bound BChl *a* molecules, two BPheo, two ubiquinones, a nonheme iron, and carotenoids (Yurkov and Beatty 1998; Yurkov and Csotonyi 2009; Yurkov and Hughes 2013). There can be anywhere between 140 and 1800 RC per cell, a full order of magnitude less than what is produced by the PNSB (Selyanin et al. 2015). The small number of photosynthetic units is seemingly to keep the oxidative stress in the cells (caused by the formation of triplet BChl *a* and singlet oxygen) to a minimum, while still taking advantage of as much of the energy generation photosynthesis allows for.

The LH1 complex can also be observed spectrophotometrically typically as an absorption peak at 870 nm, though this can be somewhat shifted: at 879 nm in *C.*

halotolerans and *Charonomicrobium ambiphototrophicum* or as far as 855 nm in *R. thiosulfatophilus* (Yurkov and Csotonyi 2009; Yurkov and Hughes 2013). These shifts are caused by varying protein environments of BChl *a*. Structurally, the LH1 complex is a ring of 16 heterodimeric alpha and beta subunits with approximately 30 BChl *a* molecules and bound photosynthetic carotenoids (Yurkov and Beatty 1998; Tang et al. 2010).

The LH2 complex is found in only some AAP species and, when present, allows for more efficient photosynthetic activity. It has a similar ring structure as the LH1, but is composed of only 8–9 heterodimeric alpha/beta subunits (Tang et al. 2010). Spectrophotometrically two peaks are present: one at around 805 nm and the other at 850 nm. However, there can also be shifts in the peaks, indicating differences in the protein settings may exist from species to species. Absorbance maxima have been seen, for instance, at 832 nm in *E. ramosum* or 835 nm in *Porphyrobacter dokdonensis* (Yurkov and Hughes 2013). Additionally, in rare cases such as in *R. mahoneyensis*, the LH2 complex has a monomodal peak, absorbing at 805 nm, instead of the typical dual peaks (Rathgeber et al. 2005). This unusual monomodal complex has only been found in two other genera, *Roseobacter* and *Rubrimonas* (Shiba 1991; Suzuki et al. 1999).

The genes encoding the LH1 complex are located alongside those for the RC L and M subunits in the *puf* operon, while the RC H subunit is placed in the *puh* operon. Alpha and beta portions of the LH2 complex are placed in the *puc* operon (Zheng et al. 2011; Yurkov and Hughes 2013). Encoding the LH2 on a separate operon from the RC-LH1 complex permits independent gene regulation. This means that cells may turn on or off the genes for the LH2 without entirely eliminating photosynthetic activity as has been observed in *Citromicrobium litoralis* (Spring et al. 2009).

Recent studies have shown that exposure to light can decrease cellular respiration by about 25 %. Obviously, AAP can replace oxidative phosphorylation with photophosphorylation to some extent (Hauruseu and Koblížek 2012). Also, light intensity can have an effect on the size of the RC-LH1 complex. *Roseobacter litoralis* under high light intensity had more RC-LH1 complexes than under low light. The complexes were smaller in cells grown in the light (39 ± 3 BChl *a* molecules) compared to grown in the dark (115 ± 30 BChl *a* molecules) (Selyanin et al. 2015). It's likely that the cultures grown under very low illumination had larger antennae to maximize light harvesting, and when grown under intensive light, they are smaller to reduce damage to the RC and cells from the high illumination (Selyanin et al. 2015). It would be interesting to study if this pattern holds true for other AAP species, both marine and freshwater.

The Photosynthetic Electron Transport Chain

Once light energy has been harvested by BChl *a*, it has to be funneled through the LH complexes into the RC, so the light energy can be converted into chemical energy (ATP) for further use in different metabolic processes. This is where the cyclic photosynthetic electron transport chain comes in. Once BChl *a* is excited, the energy is passed to the primary electron acceptor, which is a quinone (Q_A). It is then passed to cytochromes (cyt), first the membrane-bound cytbc_1 complex and then the cyt*c*, which is either soluble in the periplasmic space or tightly bound to the RC. The electron is then cycled back to the beginning of the electron transport chain (Yurkov and Hughes 2013).

While the chain described above is highly similar in AAP and PNSB, there are a couple of key differences. The first, and possibly most important, is that in AAP the Q_A redox midpoint potential is quite high (in the range of +5 to +150 mV, depending on the species) (Yurkov and Beatty 1998; Rathgeber et al. 2012) compared to the quinone in PNSB, which is always negative. This may be among the major reasons why AAP can only carry out photosynthesis in the presence of oxygen. The Q_A would be over-reduced under anaerobic conditions, making it incapable of accepting more electrons, so the cyclic electron transport comes to a grinding halt (Rathgeber et al. 2004; Yurkov and Csotonyi 2009; Yurkov and Hughes 2013). Second, PNSB have an alternative quinol oxidase pathway to keep the Q_A in the appropriate redox state (Yurkov and Csotonyi 2009; Yurkov and Hughes 2013). If the AAP had such pathways, they could possibly photosynthesize anaerobically. There are species, *R. mahoneyensis*, strain ML6, and *C. ambiphototrophicum*, EG17, in particular, that developed ways around this problem as they are AAP/PNSB intermediates (Rathgeber et al. 2005, 2012; Csotonyi et al. 2011). ML6 is most active photosynthetically under microaerophilic conditions, and EG17 produces the same photosynthetic pigment-protein complexes in both the presence and absence of oxygen. Future experimentation on intermediate species may help to find the principle difference between AAP and PNSB photosynthesis.

Further Riddles and Concluding Remarks

While investigations of AAP have made great leaps in the past several years, there is still so much we do not understand, and every discovery we make delivers new questions that have yet to be answered. We still discuss without much certainty whether AAP evolved from PNSB or the heterotrophs, as well as how and why they evolved coping strategies for so many extreme conditions. Based on the wide range of recently discovered primarily aerobic bacteria that are capable of BChl *a* production and oxygen-dependent anoxygenic photosynthesis, it might be the right time for a more inclusive definition of the term AAP. Modern ecological studies are at last revealing some potential roles AAP may play in aquatic systems; however, we must continue our experimentations to make more definitive remarks. Additionally, we still have not conclusively agreed on what sets the AAP photosynthetic electron transport chain and BChl *a* synthetic pathways apart from those of the PNSB, which allows them to function aerobically and restricts anaerobic flow of photosynthetic reactions. These are only a few of the many mysteries left to solve. Continued research is necessary to understand such an abundant, diverse, and fascinating group of bacteria.

Acknowledgments This research was supported by an NSERC Canada Discovery Grant and GETS funds from the University of Manitoba held by Dr. V. Yurkov.

References

Beatty JT (2005) On the natural selection and evolution of the aerobic phototrophic bacteria. In: Govindjee JT, Beatty HG, Allan JF (eds) Discoveries in photosynthesis. Springer, pp 1099–1104

Boeuf D, Cottrell MT, Kirchman DL, Lebaron P, Jeanthon C (2013) Summer community structure of aerobic anoxygenic phototrophic bacteria in the western arctic ocean. FEMS Microbiol Ecol. doi:10.1111/1574-6941.12130

Boldareva-Nuianzina EN, Bláhová Z, Sobotka R, Koblížek M (2013) Distribution and origin of oxygen-dependent and oxygen-independent forms of mg-protoporphyrin monomethylester cyclase among phototrophic proteobacteria. Appl Environ Microbiol 79(8):2596–2604

Csotonyi JT, Stackebrandt E, Swiderski J, Schumann P, Yurkov V (2011) An alphaproteobacterium capable of both aerobic and anaerobic anoxygenic photosynthesis but incapable of photoautotrophy: *Charonomicrobium ambiphototrophicum*, gen. nov., sp. nov. Photosynth Res 107:257–268

Csotonyi JT, Maltman C, Yurkov V (2014) Influence of tellurite on synthesis of bacteriochlorophyll and carotenoids in aerobic anoxygenic phototrophic bacteria. Trends Photochem Photobiol 16:1–17

Csotonyi JT, Maltman C, Swiderski J, Stackebrandt E, Yurkov V (2015) Extremely 'Vanadiphilic' multiple metal-resistant and halophilic aerobic anoxygenic phototrophs, strains EG13 and EG8, from hypersaline springs in Canada. Extremophiles 19:127–134

Cŭperová Z, Holzer E, Salka I, Sommaruga R, Koblížek M (2013) Temporal changes and altitudinal distribution of aerobic anoxygenic phototrophs in mountain lakes. Appl Environ Microbiol 79(20):6439–6446

Fauteux L, Cottrell M, Kirchman DL, Borrego CM, Garcia-Chaves MC, del Giorgio PA (2015) Patterns in abundance, cell size and pigment content of aerobic anoxygenic phototrophic bacteria along environmental gradients in Northern Lakes. PLoS One. doi:10.1371/journal.pone.0124035

Ferrera I, Borrego CM, Salazar G, Gasol JM (2014) Marked seasonality of aerobic anoxygenic phototrophic bacteria in the coastal NW Mediterranean sea as revealed by cell abundance, pigment concentration and pyrosequencing of *pufM* gene. Environ Microbiol 16(9):2953–2965

Garcia-Chaves MC, Cottrell DL, Derry AM, Bogard MJ, del Giorgio PA (2015) Major contribution of both zooplankton and protists to the top-down regulation of freshwater aerobic anoxygenic phototrophic bacteria. Aquat Microb Ecol 76:71–83. doi:10.3354/ame01770

Garcia-Chaves M, Cottrell MT, Kirchman DL, Ruiz-González C, del Giorgio PA (2016) Single-cell activity of freshwater aerobic anoxygenic phototrophic bacteria and their contribution to biomass production. ISME. doi:10.1038/ismej.2015.242

Hauruseu D, Koblížek M (2012) Influence of light on carbon utilization in aerobic anoxygenic phototrophs. Appl Environ Microbiol 78(20):7414–7419

Hebermehl M, Klug G (1998) Effect of oxygen on translation and posttranslational steps in expression of photosynthesis genes in *Rhodobacter capsulatus*. J Bacteriol 180(15):3983–3987

Jung HJ, Cha IT, Yim KJ, Song HS, Cho K, Kim D, Lee HW, Lee JK, Seo MJ, Roh SW, Lee SJ (2014) *Citrimicrobium luteum* gen. nov., sp. nov., aerobic anoxygenic phototrophic bacterium isolated from the gut of a sea cucumber *Stichopus japonicas*. J Microbiol 52(10):819–824

Kaschner M, Loeschcke A, Krause J, Minh BQ, Heck A, Endres S, Svensson V, Wirtz A, von Haeseler A, Jaeger KE, Drepper T, Krauss U (2014) Discover of the first light-dependent protochlorophyllide oxidoreductase in anoxygenic phototrophic bacteria. Mol Microbiol 93(5):1066–1078

Koblížek M (2015) Ecology of aerobic anoxygenic phototrophs in aquatic environments. FEMS Microbiol Rev. doi:10.1093/femsre/fuv032

Lechner U, Brodkorb D, Geyer R, Hause G, Härtig C, Auling G, Fayolle-Guichard F, Piveteau P, Müller R, Rohwerder T (2007) *Aquincola tertiaricrbonis* gen. nov., sp. nov., a tertiary butyl moiety-degrading bacterium. IJSEM 57:1295–1303

Lehours AC, Jeanthon C (2015) The hydrological context determines the beta-diversity of aerobic anoxygenic phototrophic bacteria in European arctic seas but does not favor endemism. Front Microbiol 6(638)

Li X, Koblížek M, Feng F, Li Y, Jian J, Zeng Y (2013) Whole-genome sequence of a freshwater aerobic anoxygenic phototroph, *Porphyrobacter* sp. Strain AAP82, isolated from the Huguangyan Maar Lake in Southern China. Genome Announc 1(2):e0007213

Maltman C, Yurkov V (2014) The impact of tellurite on highly resistant marine bacteria and strategies for its reduction. Int J Environ Eng Nat Res 1(3):109–119

Maltman C, Yurkov V (2015) The effect of tellurite on highly resistant freshwater aerobic anoxygenic phototrophs and their strategies for reduction. Microorganisms 3:826–838

Margesin R, Zhang DC (2013) *Humitalea rosea* gen. nov., sp. nov., an aerobic bacteriochlorophyll-containing bacterium of the family *Acetobacteraceae* isolated from soil. IJSEM 63:1411–1416

Mašín M, Čuperová Z, Hojerová E, Sallka I, Grossart HP, Koblížek M (2012) Distribution of aerobic anoxygenic phototrophic bacteria in glacial lakes of Northern Europe. Aquat Microbial Ecol 66:77–86

Nuyanzina-Boldareva EN, Gorlenko VM (2014) *Roseibacula alcaliphilum* gen. nov. sp. nov., a new alkaliphilic aerobic anoxygenic phototrophis bacterium from a meromictic soda lake doroninskoe (East Siberia, Russia). Microbiology 83(4):381–390

Rathgeber C, Beatty JT, Yurkov V (2004) Aerobic phototrophic bacteria: new evidence for the diversity, ecological importance and applied potential of this previously overlooked group. Photosynth Res 81:113–128

Rathgeber C, Yurkova N, Stackebrandt E, Schumann P, Beatty JT, Yurkov V (2005) *Rosiecyclus mahoneyensis* gen. nov., sp. nov., an aerobic phototrophic bacterium isolated from a meromictic lake. IJSEM 55:1597–1603

Rathgeber C, Alric J, Hughes E, Vermeglio A, Yurkov V (2012) The photosynthetic apparatus and photoinduced electron transfer in the aerobic phototrophic bacteria *Roseicyclus mahoneyensis* and *Porphyrobacter meromictius*. Photosynth Res 110(3):193–203

Ritchie AE, Johnson ZI (2012) Abundance and genetic diversity of aerobic anoxygenic phototrophic bacteria of coastal regions of the Pacific Ocean. Appl Environ Microbiol 78:2858–2866

Rohwerder T, Müller RH, Weichler T, Schuster J, Hübschmann T, Müller S, Harms H (2013) Cultivation of *Aquincola tertiaricarbonis* L108 on the fuel oxygenate tert-butyl alcohol induces aerobic anoxygenic photosynthesis at extremely low feeding rates. Microbiology 159:2180–2190

Salka I, Srivastava A, Allgaier M, Grossart HP (2014) The draft genome sequence of *Sphingomonas* sp. Strain FukuSWIS1, obtained from acidic lake grosse fuchskuhle, indicates photoheterotrophy and a potential for humic matter degradation. Genome Announc 2(6)

Sato-Takabe Y, Hamasaki K, Suzuki K (2012) Photosynthetic characteristics of marine aerobic anoxygenic phototrophic bacteria *Roseobacter* and *Erythrobacter* strains. Arch Microbiol 194:331–341

Sato-Takabe Y, Hamasaki K, Suzuki K (2014) Photosynthetic competence of the marine aerobic anoxygenic phototrophic bacterium *Roseobacter* sp. under organic substrate limitation. Microbes Environ 29(1):100–103

Sato-Takabe Y, Suzuki S, Shishikura R, Hamasaki K, Tada Y, Kataoka T, Yokokawa T, Yoshie N, Suzuki S (2015) Spatial distribution and cell size of aerobic anoxygenic phototrophic bacteria in the Uwa Sea, Japan. J Oceanogr 71:151–159

Selyanin V, Hauruseu D, Koblížek M (2015) The variability of light harvesting complexes in aerobic anoxygenic phototrophs. Photosynth Res. doi:10.1007/s11120-015-0197-7

Shiba T (1991) *Roseobacter litoralis* gen. nov., sp. nov., and *Roseobacter denitrificans* sp. nov., aerobic pink-pigmented bacteria which contain bacteriochlorophyll *a*. Syst Appl Microbiol 14(2):140–145
Shiba T, Simidu U (1982) *Erythrobacter longus* gen. nov., sp. nov., an aerobic bacterium which contains bacteriochlorophyll *a*. Int J Syst Bacteriol 32:211–217
Shiba T, Simidu U, Taga N (1979) Distribution of aerobic bacteria which contain bacteriochlorophyll *a*. Appl Environ Microbiol 38:43–45
Sieracki ME, Gilg IC, Their IC, Poulton NJ, Goericke R (2006) Distribution of planktonic aerobic photoheterotrophic bacteria in the northwest Atlantic. Limnol Oceanogr 51:38–46
Šlouf V, Fuciman M, Dulebo A, Kaftan D, Koblížek M, Frank HA, Polívka T (2013) Carotenoid charge transfer states and their role in energy transfer processes in LH1-RC complexes from aerobic anoxygenic phototrophs. J Phys Chem 117:10987–10999
Spring S, Lunsdorf H, Fuchs BM, Tindall BJ (2009) The photosynthetic apparatus and it regulation in the aerobic gammaproteobacterium *Congregibacter litoralis* gen. nov., sp. nov. PLoS One 4(3), e4866
Stegman MR, Cottrell MT, Kirchman DL (2014) Leucine incorporation by aerobic anoxygenic phototrophic bacteria in the Delaware estuary. ISME J 8:2339–2348
Stiefel P, Zambelli T, Vorholt JA (2013) Isolation of optically targeted single bacteria by application of fluidic force microscopy to aerobic anoxygenic phototrophs from the phyllosphere. Appl Environ Microbiol 79(16):4895–4905
Suzuki T, Muroga Y, Takahama M, Nishimura Y (1999) Roseivivax halodurans gen. nov., sp. nov. and Roseivivax halotolerans sp. nov., aerobic bacteriochlorophyll-containing bacteria isolated from a saline lake. Int J Syst Bact 49:629–634
Tang K, Zong R, Zhang F, Xiao N, Jiao N (2010) Characterization of the photosynthetic apparatus and proteome of *Roseobacter denitrificans*. Curr Microbiol 60:124–133
Xiao N, Liu Y, Liu X, Gu Z, Jiao N, Liu H, Zhou Y, Shen L (2015) *Blastomonas aquatica* sp. nov., a bacteriochlorophyll-containing bacterium isolated from lake water. IJSEM 65:1653–1658
Yurkov V (2006) Aerobic phototrophic proteobacteria. In: Dworkin M, Falkow S, Rosenberg F, Schleifer KH, Stackebrandt E (eds) Prokaryotes, vol 5, 3rd edn. Springer, New York, pp 562–584
Yurkov V, Beatty JT (1998) Aerobic anoxygenic phototrophic bacteria. Microbiol Mol Biol Rev 62:695–724
Yurkov V, Csotonyi J (2003) Aerobic anoxygenic phototrophs and heavy metalloid reducers from extreme environments. Recent Res Dev Bacteriol 1:247–300
Yurkov V, Csotonyi J (2009) New light on aerobic anoxygenic photosynthesis. In: Hunter CN, Daldal F, Thurnauer MC, Beatty JT (eds) The purple phototrophic bacteria. Springer Science+Business Media B.V., pp 31–55
Yurkov VV, Gemerden H (1993) Impact of light/dark regimen on growth rate, biomass formation and bacteriochlorophyll synthesis in *Erythromicrobium hydrolyticum*. Arch Microbiol 159:84–89
Yurkov VV, Gorlenko VM (1990) *Erythrobacter sibiricus* sp. nov., a new freshwater aerobic bacterial species containing bacteriochlorophyll *a*. Microbiol (New York) 59(1):120–126
Yurkov VV, Gorlenko VM (1992a) A new genus of freshwater aerobic bacteriochlorophyll a-containing bacteria, *Roseococcus* gen. nov. Microbiol (New York) 60(5):902–907
Yurkov VV, Gorlenko VM (1992b) New species of aerobic bacteria from the genus *Erythromicrobium* containing bacteriochlorophyll a. Microbiol (New York) 61(2):248–255
Yurkov V, Hughes E (2013) Genes associated with the peculiar phenotypes of the aerobic anoxygenic phototrophs. In: Beatty JT, Jacquot JP, Gadal P (eds) Genome evolution of photosynthetic bacteria, vol 66. Elsevier, pp 327–358
Yurkov V, Gad'on N, Drews G (1993a) The major part of polar carotenoids of the aerobic bacteria *Roseococcus thiosulfatophilus*, RB3 and *Erythromicrobium ramosum*, E5 is not bound to the bacteriochlorophyll a complexes of the photosynthetic apparatus. Arch Microbiol 160:372–376

Yurkov VV, Gorlenko VM, Kompantseva EI (1993b) A new type of freshwater aerobic orange colored bacterium *Erythromicrobium* gen. nov., containing bacteriochlorophyll a. Microbiol (New York) 61(2):256–260

Yurkov V, Stackebrandt E, Holmes A, Fuerst JA, Hugenholtz P, Golecki J, Gorlenko VM, Kompantseva EI, Drews G (1994) Phylogenetic positions of novel aerobic, bacteriochlorophyll *a*-containing bacteria and description of *Roseococcus thiosulfatophilus* gen. nov., sp. nov., *Erythromicrobium ramosum* gen nov., sp. nov., and *Erythrobacter litoralis* sp. nov. Int J Syst Bacteriol 44:427–434

Yurkov V, Stackebrandt E, Buss O, Verméglio A, Gorlenko V, Beatty JT (1997) Reorganization of the genus *Erythromicrobium*: description of "*Erythromicrobium sibiricum*" as *Sandaracinobacter sibiricus* gen. nov., sp. nov., and of "*Erythromicrobium ursincola*" as *Erythromonas ursincola* gen. nov., sp. nov. Int J Syst Bacteriol 47:1172–1178

Zeng Y, Feng F, Liu Y, Li Y, Koblížek M (2013a) Genome sequences and photosynthesis gene cluster composition of a freshwater aerobic anoxygenic phototroph, *Sandarakinorhabdus* sp. Strain AAP62, isolated from the Shahu Lake in Ningxia, China. Genome Announc 1(1)

Zeng Y, Koblížek M, Feng F, Liu Y, Wu Z, Jian J (2013b) Whole-genome sequencing of an aerobic anoxygenic phototroph, *Blastomonas* sp. Strain AAP53, isolated from a freshwater desert lake in inner Mongolia. Genome Announc 1(2)

Zeng Y, Selyanin V, Lukeš M, Dean J, Kaftan D, Feng F, Koblížek M (2015) Characterization of the microaerophilic, bacteriochlorophyll *a*-containing bacterium *Gemmatimonas phototrophica* sp. nov., and emended descriptions of the genus *Gemmatimonas* and *Gemmatimonas aurantiaca*. IJSEM 65:2410–2419

Zheng Q, Zhang R, Koblížek M, Boldareva EN, Yurkov V, Yan S, Jiao N (2011) Diverse arrangement of photosynthetic gene clusters in aerobic anoxygenic phototrophic bacteria. PLoS ONE 6(9):e25050. doi:10.1371/journal.pone.0025050

Zheng Q, Liu Y, Steindler L, Jiao N (2015) Pyrosequencing analysis of aerobic anoxygenic phototrophic bacterial community structure in the oligotrophic Western Pacific Ocean. FEMS Microbiol Lett. doi:10.1093/femsle/fnv034

Part II
Environment and Ecology

Desert Crusts

Patrick C. Hallenbeck

Abstract A significant fraction of the world's land surface is covered by arid and semiarid land. Desert crusts, microbial communities formed from cyanobacteria, algae, fungi, and bacteria, are important ecosystems that stabilize and enrich desert soils. Cyanobacteria are key players, often providing physical cohesion, primary production, and life-supporting nitrogen fixation. Here the overall structure of crusts, the important microbial partners, and the microbial diversity that is present are discussed. Some of the special features of these communities and the individual organisms are their tolerances to desiccation and to high levels of UV radiation. Mechanisms to survive long periods of severe dehydration include the synthesis of large amounts of trehalose and the excretion of copious quantities of unique exopolysaccharides. Adaptations that allow survival in spite of high levels of UV radiation include mobility and the synthesis of natural sunscreens, scytonemin, and mycosporines. Much remains to be learned about these ubiquitous microbial consortia, whose functionalities and interrelationships are beginning to be probed at the molecular level. This chapter reviews the general microbial aspects of desert crusts and gives a special emphasis to the involvement of cyanobacteria.

Introduction

Cyanobacteria have had a major influence on the earth for at least the last 2.95 billion years (Planavsky et al. 2014). These organisms are of great ecological importance as they have expanded over time to colonize, in either free-living or symbiotic forms, most of the ecological niches available on the earth (Bolhuis et al. 2014; Makhalanyane et al. 2015a; Moreira et al. 2013; Sukenik et al. 2015). One ecological niche of significant interest is biological soil crusts, in particular desert crusts, found in arid and hyperarid regions. Dry lands make up 41.3 % of the land surface

P.C. Hallenbeck (✉)
Department of Biology, Life Sciences Research Center, United States Air Force Academy, 2355 Faculty Drive, Colorado Springs, CO 80840, USA

Département de microbiologie, infectiologie et immunologie, Université de Montréal, Montréal, Québec, Canada
e-mail: hallenbeckassociates@gmail.com

© Springer International Publishing Switzerland 2017
P.C. Hallenbeck (ed.), *Modern Topics in the Phototrophic Prokaryotes*,
DOI 10.1007/978-3-319-46261-5_7

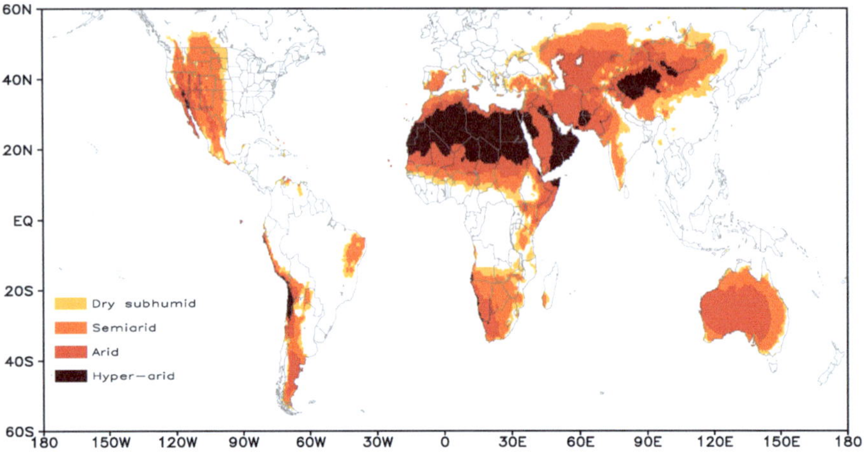

Fig. 1 Global distribution of drylands. Reproduced from Feng and Fu (2013) under the Creative Commons license

with semidesert (arid 10.6 %) and desert (hyperarid 6.6 %) being a significant fraction of that (Fig. 1). Arid and semiarid regions are expanding on an appreciable scale due to land degradation, deterioration in water supplies, and climate changes. Concerns about the lands worldwide that are increasingly threatened with deterioration led the United Nations General Assembly to declare the United Nations Decade for Deserts and the Fight Against Desertification (January 2010 to December 2020) in an effort to promote action to protect the world's dry lands and to prevent further desertification (United Nations 2016).

Biological desert crusts, an assemblage of cyanobacteria, fungi, bacteria, and green algae, cover much of the semiarid and arid regions (Fig. 2). The communities found in these crusts, one type of biological soil crust, are mixtures of algae, cyanobacteria, bacteria, fungi, lichens, and mosses whose exact composition varies with soil type, age, and local environmental factors such as shading, temperatures, rainfall, wind, etc. (Janatkova et al. 2013; Johansen 1993; Li et al. 2013, 2014; Pointing and Belnap 2012; Makhalanyane et al. 2015b). These consortia carry out a variety of ecological functions related to water retention and soil conditioning; fixing nitrogen and carbon dioxide and releasing mineral-bound nutrients such as phosphate, permitting eventually the proliferation of higher life forms (Belnap 2003). Not only is there academic interest in studying desert crust development and the crust-associated communities, but, as the world experiences increased anthropogenic desertification, there is a practical need to understand how to recolonize these lands (Wang et al. 2009; Xu et al. 2013).

The ecological niche where these organisms are found is characterized by high levels of radiation due to incoming solar and reflected light. Soil surface temperatures vary widely, from −20 °C to over 70 °C depending upon time of day and season. Precipitation is sparse and infrequent, and the water content of these crusts can drop to 5 % or less (dry weight basis). Therefore, the key to survival for desert crust

Fig. 2 Mature desert crust from the Colorado Plateau

organisms is their ability to cope with extreme dehydration. Although a great deal is known about desert crusts at the macroscopic level (Wu et al. 2013a; Rosentreter et al. 2007; Makhalanyane et al. 2015b), relatively little information is available at the molecular level. Photoautotrophic organisms are of particular interest and importance in terms of overall desert crust ecology.

While the vast majority of desert crust studies have examined "hot" deserts, as well as what have been called "cold" deserts, such as parts of the Colorado Plateau, very few studies have been done on truly cold deserts, the dry valleys of Antarctica. Initial reports of nearly "sterile" environments were in fact due to the inability to culture any microorganisms from these soils. However, when examined by modern molecular methods, a wide variety of taxonomically unique taxa have been uncovered, suggesting a rich and diverse community despite the inhospitable conditions and extremely oligotrophic soils (Cary et al. 2010). The recent application of GeoChip technology, a technique that would also be extremely useful in the study of more "normal" desert soils, has revealed a diverse functional ecology with genes for various major metabolic pathways, autotrophic and heterotrophic carbon metabolism and diazotrophy (Chan et al. 2012, 2013). Moreover, an examination of the different niche environments available, open soil, hypolithic (growth underneath translucent rocks, i.e., quartz), chasmoendolithic (growth in cracks or fissures), and cryptoendolithic (growth in pores), revealed significant differences in the functional ecologies of the different communities (Chan et al. 2012, 2013). While this review, given the focus of this volume, will highlight the roles played by cyanobacteria, general concepts and advances in knowledge about desert crusts in general and about desiccation resistance will also be covered in order to provide the appropriate context.

A variety of cyanobacteria have been identified worldwide in desert crusts, both nitrogen fixing and nonnitrogen fixing. Nitrogen-fixing cyanobacteria are of specific interest for their key role in supplying fixed nitrogen to the desert crust community (Yeager et al. 2007, 2012). Some members have been studied for their ability to carry out hydrotaxis (Pringault and Garcia-Pichel 2004), suggesting that they possess an interesting cellular system for transducing signals related to water availability. Their response to dehydration/rehydration has been studied (Rajeev et al. 2013), and the genome of a ubiquitous member of this family found in desert crusts worldwide has become available (Starkenburg et al. 2011).

General Considerations

Desert crusts, complex microbial communities commonly found in hot and cold desert environments, have long been attributed with a variety of ecological and environmental roles. In particular, a number of early studies credited them with reducing erosion (Johansen 1993). In addition, they undoubtedly act to increase nitrogen inputs, estimated to be of the order of 25 kg N ha^{-1} y^{-1}, through active nitrogen fixation by cyanobacteria as well as heterotrophic bacteria, and act to increase the organic carbon content of the soils they cover, with rates of between 6 and 23 kg C ha^{-1} y^{-1} being estimated for semiarid areas of Southwestern USA. In essence, one function of these communities is to "terraform" and condition the soil for subsequent colonization by other ecosystem communities. Given the large surface areas of the world that are implicated, cryptogamic crusts make a significant overall contribution to global nutrient fluxes, fixing large amounts of atmospheric CO_2 (>2.6 pg C y^{-1}) and N_2 (>49 Tg y^{-1}) (Elbert et al. 2012).

Fixed nitrogen input into the soil through the action of cyanobacteria and heterotrophic bacteria may be of particular importance. Although highly variable on a daily basis, since no fixation apparently occurs in the dry state, or when temperatures are either too high or low, N_2 fixation by cyanobacterial-dominated crusts of the Colorado Plateau, especially those containing the lichen *Collema* (photobiont *Nostoc* sp.), has been estimated to input fixed nitrogen at rates of 9–13 kg ha^{-1} $year^{-1}$ (Belnap 2002). The desert soil crust N cycle appears to lack appreciable anammox or anaerobic denitrification activities, thus suggesting that the major part of the fixed N might find its way into the surrounding environment (Strauss et al. 2012). One study found that mature crusts were much more effective at N_2 fixation (tenfold) than poorly developed crusts and that in all crusts examined *nifH* sequences were mainly (78–100 %) cyanobacterial in origin (Yeager et al. 2004). In fact, rates of both carbon and nitrogen fixation are much higher in late successional crusts than in early ones, strongly suggesting that disturbance of mature crusts, which returns them to early successional stages, has very large impacts (reduction) on primary productivity and nitrogen fixation (Housman et al. 2006).

Cyanobacteria are considered to be the primary producers in most desert crusts, and recent evidence points to the possibility of a significant amount of excretion of fixed carbon compounds with subsequent metabolite "sharing" and cross feeding (Baran et al. 2015). This study used *Microcoleus vaginatus*, a cyanobacterium dominating early crusts (see below), and showed that a broad range of exometabolites were excreted. Additionally, when paired with likely heterotrophic bacterial partners, extensive cross feeding was demonstrated. This suggests that desert crust microbial communities, in addition to being physically associated, may be tied together metabolically. Thus, during early crust formation, when the nonnitrogen fixer *Microcoleus vaginatus* dominates, metabolite excretion may support the heterotrophic bacteria responsible for the fixed nitrogen input into this ecosystem (Pepe-Ranney et al. 2016).

Although not specifically studied, it is very likely that different organisms in these communities solubilize phosphate and trace minerals, thus enriching the soil for growth of the crust and for potential future successional communities. In addition, the filamentous bacteria and fungi that are present undoubtedly contribute to soil stabilization through various mechanisms (Belnap and Gardner 1993; Pointing and Belnap 2012; Ogut et al. 2010; Tao et al. 2008; Pérez et al. 2007; de Oliveira Mendes et al. 2014).

Diversity

General Considerations

As in all of microbial ecology, one debate concerning desert crust diversity has centered around biogeographic issues. Is distribution affected by allopatric speciation or is as suggested by Beijerinck, "everything is everywhere but it is the environment that selects" (O'Malley 2008). Although most past studies have been at the local scale, some recent studies have begun to examine this issue at the continental or even worldwide scale.

As yet, there is no real consensus on this issue. One continental-wide (North America) study of biological soil crusts showed that, at least on the phylum level, no evident biogeographic pattern could be observed (Garcia-Pichel et al. 2013). On the other hand, when a single organism important in desert crust formation, *Microcoleus vaginatus*, was examined on a worldwide scale, appreciable diversity was found and apparent differences in continental distribution was evident (Dvořák et al. 2012). Similarly, another study examined this issue by analyzing the phylogenetics of the ubiquitous desert cyanobacterium *Chroococcidiopsis* using massively parallel pyrosequencing of samples from a variety of desert locations (Bahl et al. 2011). Evidence was presented for divergence in these samples dating to around 2.5 Ga, or at least as far back as the onset of global aridity (~1.8 Ga) at which time two hot desert clades and one cold desert clade were established. The fact that such phylogenies can be established strongly suggests that, in general, there is a lack of dispersal between different habitats, i.e., hot desert into cold, or even within climatically similar deserts. Obviously, these local communities will be structured and shaped by a variety of local variations in factors that are significant in promoting growth; pH, temperature, water activity, fixed nitrogen, minerals, and salinity (Angel et al. 2010; Büdel et al. 2009; Demergasso et al. 2004; Fierer et al. 2012a; Garcia-Pichel et al. 2013; Hagemann et al. 2015; Li et al. 2013, 2014; Schmidt et al. 2012; Stomeo et al. 2012). There is even evidence for local biohistory influencing crust outcome (Lan et al. 2015; Steven et al. 2015).

Green Algae

Unicellular green algae have also been found to be components of desert crusts from different parts of the world. Since they are relatively indistinguishable morphologically, the true diversity of this class of organisms to be found in desert crusts was not appreciated until molecular techniques were used. However, more recent studies have shown that at least five different green algal classes are implicated in different desert crust communities which have also been shown to contain many previously undescribed taxa (Lewis and Flechtner 2002; Cardon et al. 2008). Thus, adaptation to this environment appears to have caused a large radiation in diversity, not only genetic but also in terms of physiological adaptation, for example, different photophysiologies and variations in desiccation tolerance. This translates into very significant DNA sequence variations, especially in comparison with the database sequences derived from green algae isolated from aquatic environments (Lewis and Lewis 2005).

Thus, phylogenetic studies demonstrate that desert lineages are distinct from aquatic ones and give evidence for at least 14 separate transitions from aquatic to terrestrial life. Survival under these conditions requires adaptations not normally seen in the aquatic green algae. For example, *Chlorella ohadii*, newly isolated from desert sand crusts, shows remarkable insensitivity to high light intensities, with photosynthetic oxygen evolution unaffected by exposure to 3500 μmol photons m^{-2} s^{-1}

(Treves et al. 2013). On the other hand, at the same time, there was a significant drop (90 %) in variable fluorescence (F_v), strongly suggesting the intervention of a mechanism to dissipate excess light energy. Similar observations have been made with the cyanobacterium *Microcoleus* (see below) (Ohad et al. 2010).

Bacteria

The bacterial population of cryptogamic crusts of the Colorado Plateau (Southwestern USA) have been reported to be dominated by *Actinobacteria* (Garcia-Pichel et al. 2003), whereas this phylum appears to be less abundant in at least some other hot deserts (Abed et al. 2010). A metagenomic analysis has shown that in general both hot and cold desert microbiomes have a significantly different phylogenetic composition than microbiomes from other soil types. Some of this of course might be due to differences in soil pH, previously shown to be a very important factor controlling microbial soil diversity (Fierer and Jackson 2006).

On the other hand, in specific cases, other factors as well might control species composition as seen by an examination of gypsum-containing soils which were shown to have higher amounts of *Actinobacteria* and *Proteobacteria* when compared with shale and sandstone soils, which, on the other hand, had greater amounts of *Cyanobacteria* (Steven et al. 2013). Another study carrying out a PCR-based survey of environmental 16S rRNA genes of the Sonoran Desert found, in order of their importance, members of the *Cyanobacteria*, *Proteobacteria*, *Actinobacteria*, and *Acidobacteria*, with minor amounts of *Bacteroidetes*, *Chloroflexi*, and *Gemmatimonadetes* (Nagy et al. 2005). A very similar distribution was found for desert crusts of the Colorado Plateau where *Microcoleus vaginatus* and *M. steenstrupi* were dominant among the phototrophs, and *Actinobacteria* accounted for 12 %, β-*Proteobacteria* and *Bacteroidetes* each accounted for 10 % of the 16S rRNA recovered (Gundlapally and Garcia-Pichel 2006). Although little studied, it is not surprising that bacterial populations have been found to be stratified over millimeter scales in these desert crusts (Garcia-Pichel et al. 2003).

A recent study using a combination of different pretreatments and isolation strategies showed that it was possible to recover a large number of bacterial isolates from early successional stage desert crust, 105 phylotypes, a significant fraction of the diversity detected by metagenomics (da Rocha et al. 2015). In fact, the distribution of isolates, *Actinomycetes*>*Proteobacteria*>*Firmicutes*>*Bacteroidetes*, roughly mirrors what has been previously found using culture-independent methods. This work sets the stage for detailed physiological analysis of these isolates, important in the desert crust community structure and relatively distant from known and well-studied bacteria.

Nevertheless, a variety of studies have shown that the desert microbiomes have relatively sparse phylogenetic and functional diversity as compared to microbiomes from other environments and, compared to them, are enriched in genes coding for dormancy and osmoregulation while having genes associated with nutrient cycling and catabolism of complex carbon compounds at lower relative abundance (Fierer et al. 2012b). The same study noted that desert soils are also much less rich in genes for antibiotic biosynthesis, suggesting that competitive interactions are of less importance in this environment. Nevertheless, a variety of studies have shown that desert soils typically can be found to contain several different genera, *Actinobacteria*, *Bacteroidetes*, and *Proteobacteria*, with many of the isolates often being novel species (Makhalanyane et al. 2015b).

Fungi

Fungi in desert crusts have been relatively little studied and mostly through cultivation studies, which nevertheless have shown an appreciable degree of diversity. Of course a large number of fungi, principally *Ascomycetes* and, less frequently, *Basidiomycetes*, are found as the mycobiont partners in desert crust lichens. Aside from these, many of the free-living fungi appear to belong to a group of highly pigmented "black," or dematiaceous fungi, with rigid walls and high contents of melanin, carotenoids, and mycosporines, making them highly resistant to environmental stresses (Sterflinger et al. 2012). Among these are often found fungi like *Cladosporium*, *Stachybotrys*, and *Pleospora* and sometimes *Aspergillus* and *Eurotium.* As might be imagined, thermophilic and thermotolerant types can be readily isolated from hot desert crust samples. Fungi are thought to be ecologically important in the desert crust ecosystems potentially contributing to UV resistance, soil stabilization, and solubilization of phosphorous and minerals (Makhalanyane et al. 2015). In this regard, the role and importance of the fungal mycobiont partner in lichens is evident, but the nature and significance of the contribution of free-living microcolonial "black" fungi remains to be determined (Sterflinger et al. 2012). A recent molecular study using rRNA gene fingerprinting demonstrated that desert crusts of the Southwestern USA contain a considerable amount of diversity with the majority of the fungi belonging to either *Ascomycota* or *Pleosporales* (Bates et al. 2012). In addition, evidence was obtained for geographic specificity of assemblages.

Archaea

Relatively little is known about archaeal populations of desert crusts. At least one metagenomics survey has suggested that they are present as a relatively minor fraction, ~6 %, in hot deserts, which nevertheless represents a greater abundance than found in other soil types examined (Fierer and Jackson 2006). The principle group detected belonged within the relatively recently described *Thaumarchaeota*, well known for their involvement in the nitrogen cycle in many habitats.

Cyanobacteria

General Distribution of Cyanobacteria in Desert Crusts

Cyanobacteria have long been recognized as essential components of desert crusts, and early studies demonstrated that these crusts were dominated by filamentous cyanobacteria, with *Microcoleus*, *Phormidium*, *Plectonema*, *Schizothrix*, *Nostoc*, *Tolypothrix*, and *Scytonema* being described as common genera found worldwide in hot and cold deserts (Johansen 1993). As noted above for bacteria and in common with many other biofilms, cyanobacteria have been found to be stratified with desert crusts, with maximal photosynthetic activity being observed one millimeter below the surface (Raanan et al. 2016a). The dominance of cyanobacteria in this environment implies that they play the role of keystone species, probably through their abilities to appropriately modify the physical matrix, to provide fixed carbon as primary producers and, in most cases, to provide fixed nitrogen.

Although it has been long supposed that one of the significant roles played by cyanobacteria was to establish and maintain crust integrity, in realty there has been little direct proof for this hypothesis. One relatively recent study has now shown this to be the case by testing the ability of four different filamentous cyanobacteria, previous isolated from desert crusts, *Microcoleus vaginatus*, *Phormidium tenue*, *Scytonema javanicum*, and *Nostoc* sp., to consolidate sand (Hu et al. 2002). Erosion was then studied using a wind tunnel, permitting the researchers to test the effects of a number of variables.

One study, by combining phylogenetic and morphological analysis, was able to define six distinct clusters of cyanobacteria observed in four biological crust types from the Colorado Plateau (Garcia-Pichel et al. 2001). One of these did not appear to have a cultivated representative strain with which to compare, and another, although "*Phormidium*-like," contained only sequences from desert crust cyanobacteria and was quite distant from *Phormidium* species. A new name, *Xeronema*, was proposed for this cluster. However, this seems like a poor choice since it is already the name of a genus of flowering plants. Nonetheless, DGGE analysis suggests that members of this group are ubiquitous and therefore certainly deserving of further study.

Arid grasslands in some areas of the Southwestern USA have been shown to contain three types of crust even though they are growing on the same red sandstone soil substratum: cyanobacterial crust, lichen-dominated crust, and moss-dominated crust (Redfield et al. 2002). The cyanobacterial diversity in the three types of crusts was analyzed using terminal restriction fragment length polymorphism (TRF or T-RFLP) and 16S rDNA sequence analysis. The results showed that the cyanobacterial crust, while containing other genera, was dominated by strains of *Microcoleus vaginatus*, which also the most abundant cyanobacterial species in the moss crust. On the other hand, this organism was only a minor component of lichen-based crusts where strains related to *Chroococcidiopsis* and *Oscillatoria* were in the majority.

Similarly, a study of biological soil crusts in the Gurbantunggut Desert, China, showed that diverse morphotypes and phylotypes can exist in the same desert depending upon local conditions: position on a sand dune, type of soil texture, and available phosphate (Zhang et al. 2011). In most cases, filamentous cyanobacteria predominated in this study. It was study found that the upper layers of the crusts were dominated by *Microcoleus vaginatus*, with appreciable amounts of *Oscillatoria aeruginosa*, *Synechococcus parvus*, *Oscillatoria tenuis*, *Chlorococcum humicola*, *Navicula* sp., and *Hantzschia amphioxys*. At least in these samples, morphotype diversity was largely influenced by phosphate concentrations.

Hypolithic and Endolithic Cyanobacteria

Cyanobacteria are in the majority in hypolithic communities observed worldwide, representing 47–96 % of the bacterial phylotypes recovered (Caruso et al. 2011). Rocks supporting hypolithic growth include quartz, prehnite, and agate. There is some evidence that hypoliths from warm deserts are mainly coccoid *Pleurocapsales* of the genus *Chroococcidiopsis*, whereas the hypoliths from extreme cold and polar deserts are more likely to be filamentous oscillatorian morphotypes (Chan et al. 2012). The contribution of cyanobacteria to the overall ecology of hypolithic communities is probably key due to their autotrophic carbon fixation. In many hyperarid desert hypolithic communities, cyanobacteria may well be the only primary producers.

A study of hypoliths found in deserts of Northern Australia found that up to one-third of the OTUs recovered grouped within *Chroococcidiopsis* and that photosynthetic activity required a minimum soil moisture content of 15 % (Tracy et al. 2010). As well, the most common cyanobacterial 16S rRNA sequences recovered from hypoliths from the hyperarid Atacama Desert were related to *Chroococcidiopsis* (Warren-Rhodes et al. 2006). Not surprisingly, this study found that abundance, diversity, and the residence time of organic carbon decreased significantly with decreasing annual rainfall. However, even under extremely low rainfall conditions (<1 mm/year), hypolithic communities can be found if local conditions, such as fog along the coast, permit (Azúa-Bustos et al. 2011). Even under these conditions, DNA analysis showed that strains related to *Chroococcidiopsis* appeared to be in the majority with some strains related to other well-known desert inhabitants, *Microcoleus*, *Nostoc*, and *Scytonema*, also being recovered.

A more recent, more complete study of organisms of this hyperarid region, part of a continuing effort where more than 480 cyanobacterial cultures have been isolated, found a variety of species richness and distribution patterns depending upon sampling site, with it being impossible to culture any cyanobacteria form the majority of the sites (88) examined (Patzelt et al. 2014). Nevertheless, molecular analysis demonstrated a surprising diversity with a variety of filamentous forms belonging to Nostocophycideae, Synechococcophycideae, and Oscillatoriophycideae, along with the coccoid *Chroococcidiopsis*.

Endolithic environments, where microorganisms can grow inside translucent rocks, such as halites, gypsum, and carbonates, are very related microhabitats. These minerals might help trap some moisture as well as provide some light at short distances (~mm) beneath the surface while at the same time largely blocking UV radiation and greatly attenuating the intense solar visible light. A variety of bacteria and a unique cyanobacterium, related to *Halothece*, have been recovered from halite samples (Robinson et al. 2015). Cyanobacteria associated with both halites and gypsum have been reported to make large quantities of carotenoids and scytonemin, a UV-protectant compound (see below). The known hypolith, *Chroococcidiopsis*, has been found to be an active colonizer of gypsum (Wierzchos et al. 2015).

Desiccation Tolerance

Obviously an important attribute of many desert crust microorganisms is the ability to survive extended periods of desiccation. Desiccation tolerance, also referred to as anhydrobiosis, is as yet poorly understood (Potts 2001). It is likely that a number of different strategies and mechanisms are involved, from means for maintaining protein stability in the absence of structurally important water of hydration, to repair processes involved in responses to UV damage and reactive oxygen species (ROS) (Billi and Potts 2000, 2002).

For a thorough discussion of the relevant biophysical properties of water and the biological structural and functional considerations related to the consequences of its loss or reduction through various drying processes, see the discussion by Potts (1994). Different cellular components and structures are either directly or indirectly sensitive to dehydration. For one thing, membrane integrity and fluidity are affected since membranes are held together by the surface tension of water and drying can affect the membrane transition temperature (T_m) with potentially disastrous and lethal effects on permeability. How desiccation-tolerant microbes escape this fate is not known, perhaps they have a different polar/nonpolar nature of their membranes. Likewise, proteins depend upon bound water to maintain their three-dimensional structure, and removal of intracellular water below the levels required to maintain a surface layer covering cellular proteins can cause irreversible denaturation of key proteins. On the other hand, cellular DNA is probably indirectly affected with the impossibility of carrying out repair during the desiccated dormant state.

The water content of a "normal" microbial cell is around 70 %, whereas desiccated cells can reach 3–10 % by weight (Potts 1994). Coping with such a loss of water and surviving intact pose special problems. For example, the driest state, 3 %, is well below the amount of water required to fully cloak cellular proteins in a monolayer of water, 30–40 %. It seems that one mechanism of surmounting this problem is the synthesis of large amounts of sucrose or trehalose, and some desert crust organisms can accumulate large amounts of these compounds, up to 20 % by weight. However, it would seem that, although these same compounds are used by some organisms when challenged osmotically, through what is called "preferential exclusion," a different mechanism applies when these compounds are made as protectants against desiccation.

In fact, a wide variety of compatible solutes are made by a range of microbes, and these compounds have been suggested to serve a variety of purposes, from desiccation tolerance and osmoregulation to reserves of carbon, nitrogen, and energy (Welsh 2000). "Preferential exclusion" by so-called compatible solutes, sugars, amino acids, polyols, etc., act to stabilize proteins by forcing the remaining water to form shells around the proteins, hydrating them. However, this cannot be the mechanism of action of sucrose or trehalose in anhydrobiosis since, as noted above, desiccated cells do not even contain enough water to fully hydrate their complement of proteins. Instead, these polyhydroxy compounds are thought to act through a replacement mechanisms, serving to "hydrate" cellular proteins through the formation of the requisite hydrogen bonds. In fact, by the same mechanism, trehalose could stabilize the cellular membrane if it was present on both sides. Finally, these compounds could act in another manner, through the formation of aqueous glasses, a property of some solutes at low enough water activity or low enough temperature, thus avoiding the complete dehydration of the microbial cell at temperatures below the "glass" melting point, 90 °C for trehalose/water (Potts 1994, 2001).

Similarly, some desiccation-resistant cyanobacteria produce large amounts of extracellular polysaccharide which presumably offers resistance. For example, up to 60 % of the dry weight of a colony of *Nostoc commune* can be formed of the glycan matrix, a novel polysaccharide composed of a 1-4-linked xylogalactoglucan backbone decorated with D-ribofuranose and 3-*O*-[(*R*)-1-carboxyethyl]-D-glucuronic acid (nosturonic acid) groups (Helm et al. 2000). The unusual properties of this compound and its abundance may be put to use, either through the formation of a biological glass, or, otherwise, to protect the cells from desiccation.

Relatively little is known about how the special challenges faced by photosynthetic organisms are met. One of the challenges for photosynthetic microorganisms like cyanobacteria is how to survive in a desiccated state where metabolism is necessarily dormant, but the photosynthetic pigments, chlorophyll, and phycobiliproteins are still potentially capable of absorbing photons and generating high-energy states whose dissipation can be quite deleterious. One response is to synthesize chromophores which can safely absorb potentially damaging light (see below). However, some organisms, like *Leptolyngbya*, lack this capacity and hence must rely on another strategy for survival.

One study has suggested that this organism is spared from excessive photochemical reactions in the dry state by small changes in the thylakoid membranes which appear sufficient to totally quench the absorbed light energy (Bar-Eyal et al. 2015). Another study showed that over 50 % of photosystem II (PSII) activity could be recovered within 5 min of rehydration of a desert crust sample that had been kept in the dry state for over 1 month (Harel et al. 2004). As well, this study suggested that one survival mechanism of the cyanobacteria within the crust is the unusual ability to increase the rate of PSII repair with light intensity and time of exposure, resulting in only very low levels of photoinhibition under high light intensity.

A study of the tolerance of desert green algae to desiccation examined the recovery of the quantum yield of photosynthesis (F_v/F_m) upon rehydration and found that desert algae showed remarkably different recovery abilities than their aquatic relatives. Cells incubated over extended periods of time in darkness after desiccation were more likely to recover than cells which had been illuminated during this time period (Gray et al. 2007). As with the isolated green algae, reactivation of photosystem II in lichen soil crusts exposed to repeated dehydration/rehydration cycles occurs over a remarkably short time period, several hours (Wu et al. 2013b). An early study examined the responses of several different *Nostoc*, *N. flagelliforme*, *N. commune*, and *Nostoc* sp., rewetting after 2 years of dryness (Scherer et al. 1984). Rapid rewetting was noted with respiration recovering the fastest (30 min), followed by photosynthesis (~7 h), and, much later (~140 h), nitrogen fixation.

Dynamics of Resuscitation

Of course, there is a great deal of interest in understanding how the individual organisms and entire community structure responds to cycles of desiccation and rehydration. Nevertheless, this is a difficult study to make in any great detail and was not really amenable to examination until the advent of modern molecular tools. One recent study followed the dynamic changes in microbial community structure following rewetting by analyzing rRNA using a stable isotope approach and H_2O^{18} (Angel and Conrad 2013). Cyanobacteria were found to be the dominant group two weeks after rehydration and incubation in the light.

In order to attempt to mimic natural cycles, but under controlled conditions, another study employed a chamber which allowed desiccation to take place at controllable rates (Raanan et al. 2016b). A strain of *Leptolyngbya ohadii* isolated from the desert crust obtained from the Negev desert was used and examined under conditions, light cycle and drying rate, that corresponded to local meteorological data. Just as observed with the whole crust, recovery was strongly dependent upon the previous rate of dehydration, with slower rates favoring recovery (Raanan et al. 2016b). Insight into the molecular programming leading to responses to desiccation and rehydration by *Microcoleus vaginatus* were gained by carrying out an in situ whole-genome transcriptional time course (Rajeev et al. 2013). This analysis shows that the onset of desiccation triggers the induction of genes for C and N storage and response to osmotic, oxidative, and photooxidative stresses. Hydration led to the immediate induction of genes for DNA repair and regulatory processes, with photosynthesis being reestablished within one hour of wetting.

Production of Exopolysaccharides

In general, cyanobacteria as a group have the capacity to synthesize and excrete a wide variety of extracellular polysaccharides (De Philippis and Vincenzini 1998; Pereira et al. 2009; Rossi and De Philippis 2015). Most polysaccharides synthesized (80 %) contain six to ten different monosaccharides, about 90 % have one or more uronic acids, and as well most contain noncarbohydrate groups such as sulfate, peptides, acetyl, or pyruvoyl groups (De Philippis et al. 2001). Although this has been little exploited, many have interesting properties that suggest biotechnological uses as emulsifying or thickening agents, for cation absorption, etc.

For terrestrial cyanobacteria, some of which are found in desert crusts, exopolysaccharides have been proposed to serve a variety of functions: the stabilization of the physical matrix upon which they are growing, the sequestration of metal cations, as a matrix to absorb excreted UV-protectant compounds, protection against desiccation, and, perhaps, the basis for gliding motility. In fact, the rapid rehydration of *Nostoc* colonies upon rewetting to give macroscopic gelatinous masses leads to the naming of this genus in the Middle Ages, before the invention of the microscope (Potts 1997)!

Besides stabilizing the physical structure of the growth matrix and providing a store of fixed carbon to support growth of heterotrophs (Mager and Thomas 2011), cyanobacterial exopolysaccharides improve its hydraulic conductivity (Rossi et al. 2012). The cohesive ability necessary for stabilization of fine sand particles seems to be a function of the specific composition of the polysaccharide (Hu et al. 2003). Simple amendment of desert soil with cyanobacterial polysaccharide appears to enhance shrub growth (Xu et al. 2013). Interestingly, given the rampant desertification that is underway in many parts of the world, cyanobacteria and their excreted polysaccharides may play an important role in restoration efforts through the use of induced biological crusts. Recent research indicates that these cyanobacterial compounds impart many beneficial properties in terms of stabilization and improved hydrological behavior (Chen et al. 2014; Colica et al. 2014, 2015).

Protection Against High Levels of UV Radiation

Desert crust cyanobacteria might be thought to have high levels of resistance to radiation. Even the aquatic *Anabaena* sp. PCC7120 has been found to be relatively resistant to gamma radiation, with 50 % survival at a dose of 5.4 kGy (Singh et al. 2013). The same study also found that this strain showed desiccation tolerance with 50 % survival after 6 days. Another more pertinent study has shown that cyanobacteria of relevance to desert crusts, *Microcoleus vaginatus*, *Nostoc* sp., and *Scytonema javanicum*, are relatively resistant to UV-B radiation (Chen et al. 2013).

Of course, desert crust microorganisms are by nature exposed to relatively high levels of UV radiation, and cyanobacteria from this environment have been shown to have a number of mechanisms for evading or combating UV exposure: avoidance, defense, and repair (Ehling-Schulz and Scherer 1999). Most filamentous cyanobacteria possess gliding motility and therefore in principle can escape too high light intensities through downward migration. In this regard, it is interesting to note that UV-B has been shown to induce migration of the marine cyanobacterium *Microcoleus chthonoplastes* in a microbial mat, implying that it contains an as yet to be described sensor capable of perceiving light of these wavelengths (Bebout and Garcia-Pichel 1995). In addition, UV damage is avoided in at least one cyanobacterium through the replacement of the normal D1 protein, integral part of the PSII reaction center, with a UV-B inducible isozyme (Campbell et al. 1998). Additionally, changes in PSII may enable some desert crust cyanobacteria to survive high light intensities in general (Ohad et al. 2010).

Many desert crust cyanobacteria have been shown to produce UV-absorbing compounds, natural "sunscreens," as protection against UV-induced photodamage.

The classic example is scytonemin, a unique UV-A absorbing pigment, which has a dimeric indole-phenolic structure and is produced only by terrestrial cyanobacteria (Garcia-Pichel and Castenholz 1991; Gao and Garcia-Pichel 2011; Proteau et al. 1993). This specialized pigment is made by a unique, specialized biosynthetic pathway, one known to be possessed by a number of cyanobacteria (Sorrels et al. 2009).

In addition to scytonemin, many cyanobacteria, as well as other lower organisms, are capable of synthesizing water-soluble mycosporine amino acids (MAAs), amino acids, and amino alcohols linked to a cyclohexanone chromophore, which absorbs between 310 and 360 nm. Indeed, studies have shown that MAAs protect cyanobacteria against harmful UV radiation (Ehling-Schulz et al. 1997). One study, using *Nostoc commune* isolated from desert crust, showed that exposure to UV-A and UV-B increased the synthesis of scytonemin and a UV-absorbing mycosporine along with extracellular polysaccharide (Ehling-Schulz et al. 1997). There were differences in induction patterns depending upon whether UV-B or UV-A was supplied, suggesting that synthesis is regulated by different UV photoreceptors.

Recently, microarrays were used to examine the global expression response of *Nostoc punctiforme* to UV-A exposure (Soule et al. 2013). Roughly 10 % of the genes (573/6903) were affected, with upregulation of 473 and downregulation of 100. Notably the downregulated genes included those for photosynthetic pigment biosynthesis, while the upregulated genes included genes encoding scytonemin biosynthesis as well as antioxidant enzymes (catalase, superoxide dismutase, etc.). Strikingly, almost half the upregulated genes could not be assigned to functional categories, demonstrating that much remains to be learned about this response. The extremely desiccation- and radiation-resistant *Chroococcidiopsis* can even survive the intense UV radiation of outer space for more than 1 year (Cockell et al. 2011).

Nitrogen Fixation

One study carried out on desert crusts of the Colorado Plateau combined a *nifH* (encodes Fe-protein of nitrogenase) environmental survey with the isolation of representative nitrogen-fixing cyanobacteria and concluded that 89 % of the recovered *nifH* sequences were from cyanobacteria of the *Nostocales* order (Yeager et al. 2007). This suggests that the majority of nitrogen fixation maybe carried out by cyanobacteria with only a minor role for heterotrophic nitrogen fixation, in contrast to some earlier proposals (Billings et al. 2003; Johnson et al. 2005). The isolated strains were morphologically identified as *Nostoc commune* and *Scytonema hyalinum* and strains belonging to *Tolypothrix* and *Spirirestis*, but as noted elsewhere here final taxonomic designation should rely on more detailed molecular analysis.

Specific Desert Crust Cyanobacteria

Much remains to be discovered in terms of specific taxa of cyanobacteria associated with desert crusts, and it is relatively easy to isolate new species and even new genera (Řeháková et al. 2007). Nevertheless, years of standard microbiological

Table 1 Cyanobacterial morphotypes found in desert crusts of the Colorado Plateau[a]

Morphotype	Type	Relative abundance			
		Sandy soil	Shale	Gypsum	Silt
Ensheathed filaments	*M. vaginatus*	++++	++	n.d.	++++
Ensheathed filaments	*Schizothrix* sp.	++	++	+	++
Single filaments	*Phormidium* sp.	+	n.d.	++	+
Heterocystous filamentous	*Scytonema* sp.	+	++	++	++
Heterocystous filamentous	*Nostoc* sp.	n.d.	n.d.	+++	+++
Heterocystous filamentous	*Chlorogloeopsis* sp.	n.d.	n.d.	n.d.	+

[a]Adapted from Garcia-Pichel et al. (2001)

work has enabled the isolation and characterization of some of the major cyanobacterial players in desert crust communities. Some of the relevant details are given briefly in what follows. Typical morphotypes of cyanobacteria found in the desert of the Colorado Plateau are given in Table 1.

Nostocales

A variety of organisms falling in this order can often be isolated from desert crusts, in particular mature ones. In one study, two new *Nostoc* species, *N. indistinguendum* sp. nov. and *N. desertorum*, sp. nov. as well as a new genus and species within this order, *Mojavia pulchra*, were identified (Řeháková et al. 2007). A great deal of evidence, particularly on the molecular level, supports these assignments and also suggests that various isolates previously identified as known terrestrial species isolated from humid soils based on morphological considerations have been misidentified. Strains often identified, based on old morphological criteria, as *Scytonema*, are commonly observed in and isolated from desert crusts (Fig. 3a). There is some

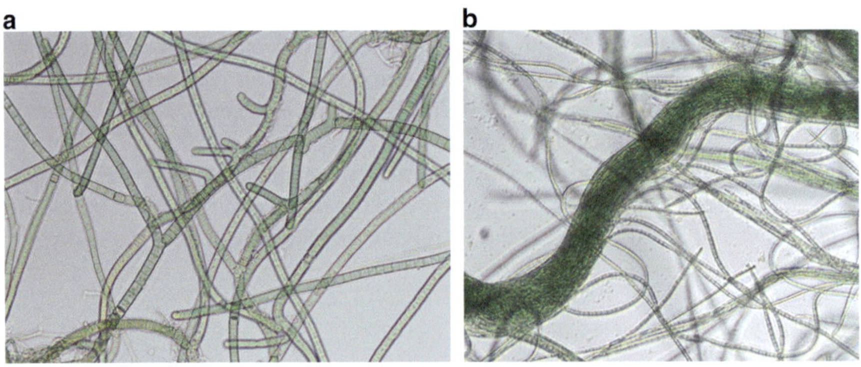

Fig. 3 Desert crust cyanobacteria. Typical isolates from Colorado Plateau desert crusts are shown. (**a**) *Scytonema* sp. (**b**) Ensheathed trichomes of *Microcoleus vaginatus*

suggestion that this genus is in fact polyphyletic, and a new taxonomy for these strains will probably emerge in the near future with the application of molecular approaches. Already a number of draft genome sequences of strains isolated from diverse habitats have recently become available.

Microcoleus

It has long been assumed that cyanobacteria have, as with other free-living microorganisms, a ubiquitous occurrence. This has certainly been true for *Microcoleus*, a genus within the *Oscillatoriales* order with reports of the isolation of *M. vaginatus* from sites spread around the globe. One classical distinguishing morphological feature is that the filaments are often found as ensheathed bundles of trichomes (Fig. 3b). An early application of molecular analysis of 16S rRNA and 16S rRNA–18S rRNA internal transcribed spacer (ITS) sequences showed that many strains of *M. vaginatus* isolated from desert soils in the Southwestern USA were in fact *M. steenstrupii* and that there was enough divergence within these to suggest that this group contains several cryptic species (Boyer et al. 2002). Some variation even within the much more closely related “true” *M. vaginatus* strains was also noted.

The degree of identity of *M. vaginatus* from various locales and the possible role of geographic barriers, which could lead to some kind of allopatric speciation, were investigated in a relatively recent study using the 16S rRNA and 16S–23S rRNA ITS sequences of *M. vaginatus* isolates from three different continents (Dvořák et al. 2012). Analysis showed that a broad genetic diversity was present and that strains isolated in Europe were a separate lineage from strains isolated in North America and Asia. Thus, geographic isolation, at least on the continental scale, can lead to differential evolution of local strains over the more than 4 million years since the estimated divergence. Thus, seemingly cosmopolitan strains can possess cryptic divergence when examined at the molecular level (Boyer et al. 2002; Dvořák et al. 2012).

A metabolomics study has shown that *Microcoleus vaginatus* PCC 9802 is capable of producing mercaptohistidine betaine (ergothioneine) and, more interestingly, unlike nine other cyanobacteria examined, a series of unusual oligosaccharides, possibly based on seven carbon sugar alcohols (Baran et al. 2013). These compounds may be important in promoting gliding motility in this organism, known to be capable of vertical migration in desert soils (Garcia-Pichel and Pringault 2001; Garcia-Pichel et al. 2001).

These cyanobacteria normally have a very characteristic sheath encasing multiple filaments (Fig. 3b). Microscopic studies have shown that this sheath, by virtue of its polysaccharide content, binds soil particles and, most likely, cationic minerals. In addition, electron micrographs show that filaments are completely encased in sheaths when dry, suggesting that they might also confer desiccation tolerance to this organism (Belnap and Gardner 1993). *Desertifilum tharense*, a novel cyanobacterium related to *Microcoleus*, but with more than 5 % 16S rRNA sequence divergence, has been described from crusts of the Thar Desert in India (Dadheech et al. 2012). On the basis of their 16S rRNA sequence, this is a distinct lineage while sharing some morphological features with *Microcoleus* and *Phormidium*.

Pseudoanabaenales

This is an order that was recently split off from *Oscillatoriales* and contains the polyphyletic *Leptolyngbya*. Within the *Leptolyngbya* lie organisms that have been isolated from desert crusts from a variety of locations and which have been recently recognized as a new genus, *Nodosilinea* (Perkerson et al. 2011). Another group has identified a number of new *Oculatella* species in this order, *O. atacamensis* and *O. mojavensis*, isolated from different desert habitats (Osorio-Santos et al. 2014).

Gloeocapsopsis

Recently a novel *Gloeocapsopsis* (order *Chroococcales*), a hypolith isolated from the Atacama Desert, was described. This cyanobacterium appears to be highly tolerant to desiccation with 69 % survival after 2 years at 0.4 a_w (Azua-Bustos et al. 2014). Upon desiccation, this organism was found to increase the synthesis and its content of sucrose and trehalose, compatible solutes. There is some indication that desiccation tolerance requires the capacity to repair DNA, and thus there is potential overlap with radiation resistance.

Chroococcidiopsis

Indeed, when *Chroococcidiopsis* strains (order *Chroococcidiopsidales*), important members of desert crusts from hyperarid areas and known for their robust desiccation tolerance, were examined for radiation resistance, there was 35–80 % survival of doses of 2.5 kGy (Billi et al. 2000). As noted above, representatives of this genus have been isolated from a large number of hyperarid habitats. They are characterized as being spherical, unicellular cells, commonly found in groups or clumps where they can be covered by a common envelope or sheath.

Responses of Desert Crusts to Climate Change

Although not enough is known presently about the ecophysiology of desert crusts to be able to predict with any certitude how they might respond to the different challenges of a climate-induced change in the desert environments in which they are found, several studies have recently suggested several possible scenarios. Analysis of a large-scale, long-term (10 years) experiment where a desert area as kept under elevated CO_2 shows that, in contrast to expectations, cyanobacterial biomass actually decreased under prolonged exposure to elevated CO_2 and that this decrease was due to a reduction along multiple lineages as there was no apparent reduction in taxonomic richness (Steven et al. 2012a, b). Additionally, total microbial biomass was the same between elevated and ambient CO_2 samples, suggesting that the decline in cyanobacterial biomass was accompanied by an equal increase in other groups. A study of the biogeographic distribution of two *Microcoleus* species, *M. vaginatus* and *M. steenstrupii*, found that *M. vaginatus* was more prevalent at coder sites and *M. steenstrupii* more prevalent at hotter sites, suggesting that *M. vaginatus* will be replaced by *M. steenstrupii* as the earth warms (Garcia-Pichel et al. 2013).

Biotechnological Aspects

Although relatively little exploited until now, the different organisms found in desert crusts can be imagined to have the capacity to make a number of compounds of potential biotechnological interest. At some point therefore, there may be an interest in the cultivation of specific organisms in order to carry out large-scale production of particular compounds. Recently, a specific type of photobioreactor has been proposed for this end (Kuhne et al. 2014). In this case, the cyanobacterium *Trichocoleus sociatus* was grown in an immersed fashion instead of the normal submerged mode, leading to a 35 % increase in growth rate and a sevenfold increased production of product, extracellular polymeric substances (EPS).

Microcoleus vaginatus has been reported to produce an unusual mixture of four normal and a relatively low concentrations, more than 60 alkanes, with the dominant compounds being heptadecane (12 %), 7-methylheptadecane (7.8 %), hexadecanoic acid (6.5 %), (Z)-9-hexadecenoic acid (5.6 %), 4-ethyl-2,2,6,6-tetramethylheptane (2.8 %), (Z)-9-octadecenoic acid (2.8 %), and 4-methyl-5-propylnonane (2.7 %) (Dembitsky et al. 2001).

Conclusion

Desert crusts, with their keystone cyanobacteria, are important ecological communities covering relatively significant areas of land under what can be considered largely inhospitable climatic conditions. They have evolved special adaptations permitting survival and proliferation under adverse conditions. A fair amount is already known about diversity and biogeography of some of the various crust organisms and about some of the desiccation protection and UV protection mechanisms involved. The future should bring much more detailed information about the important physical and metabolic interactions between the key microbial partners and the details of the molecular mechanisms of adaptation to desiccation and rehydration, including the cellular programs involved. Finally, given the cryptic metabolic capacities involved, it is likely that desert crust organisms represent a largely untapped resource of compounds of potential biotechnological interest.

Acknowledgments This work was carried out under a CRADA cooperative agreement (FA7000-16-2-0006) between USAF Academy and Hallenbeck Associates (P.C. Hallenbeck). The views expressed here are those of the authors and do not reflect the official policy or position of the US Air Force, the Department of Defense, or the US Government.

References

Abed RMM, Al Kharusi S, Schramm A, Robinson MD (2010) Bacterial diversity, pigments and nitrogen fixation of biological desert crusts from the Sultanate of Oman. Fems Microbiol Ecol 72(3):418–428

Angel R, Conrad R (2013) Elucidating the microbial resuscitation cascade in biological soil crusts following a simulated rain event. Environ Microbiol 15(10):2799–2815

Angel R, Soares MI, Ungar ED, Gillor O (2010) Biogeography of soil archaea and bacteria along a steep precipitation gradient. ISME J 4:553–563

Azúa-Bustos A, González-Silva C, Mancilla RA, Salas L, Gómez-Silva B, McKay CP, Vicuña R (2011) Hypolithic cyanobacteria supported mainly by Fog in the coastal range of the Atacama desert. Microb Ecol 61(3):568–581

Azua-Bustos A, Zúñiga J, Arenas-Fajardo C, Orellana M, Salas L, Rafael V (2014) Gloeocapsopsis AAB1, an extremely desiccation-tolerant cyanobacterium isolated from the Atacama Desert. Extremophiles 18(1):61–74

Bahl J, Lau MCY, Smith GJD, Vijaykrishna D, Cary SC, Lacap DC, Lee CK, Papke RT, Warren-Rhodes KA, Wong FKY, McKay CP, Pointing SB (2011) Ancient origins determine global biogeography of hot and cold desert cyanobacteria. Nat Commun 2:163

Baran R, Ivanova NN, Jose N, Garcia-Pichel F, Kyrpides NC, Gugger M, Northen TR (2013) Functional genomics of novel secondary metabolites from diverse cyanobacteria using untargeted metabolomics. Mar Drugs 11:3617–3631

Baran R, Brodie EL, Mayberry-Lewis J, Hummel E, Da Rocha UN, Chakraborty R, Bowen BP, Karaoz U, Cadillo-Quiroz H, Garcia-Pichel F, Northen TR (2015) Exometabolite niche partitioning among sympatric soil bacteria. Nat Commun 6

Bar-Eyal L, Eisenberg I, Faust A, Raanan H, Nevo R, Rappaport F, Krieger-Liszkay A, Setif P, Thurotte A, Reich Z, Kaplan A, Ohad I, Paltiel Y, Keren N (2015) An easily reversible structural change underlies mechanisms enabling desert crust cyanobacteria to survive desiccation. Biochimica Et Biophysica Acta-Bioenergetics 1847(10):1267–1273

Bates ST, Nash TH III, Garcia-Pichel F (2012) Patterns of diversity for fungal assemblages of biological soil crusts from the southwestern United States. Mycologia 104(2):353–361

Bebout BM, Garcia-Pichel F (1995) UV-B induced vertical migrations of cyanobacteria in a microbial mat. Appl Environ Microbiol 61:4215–4222

Belnap J (2002) Nitrogen fixation in biological soil crusts from southeast Utah, USA. Biol Fertil Soils 35:128–135

Belnap J (2003) The world at your feet: desert biological soil crusts. Front Ecol Environ 1(4):181–189

Belnap J, Gardner JS (1993) Soil microstructure in soils of the Colorado Plateau: the role of the cyanobacterium *Microcoleus vaginatus*. Great Basin Nat 53(1):40–47

Billi D, Potts M (2000) Life without water: responses of prokaryotes to desiccation. In: Storey KB, Storey JM (eds) Environmental stressors and gene responses. Elsevier Science BV, Amsterdam, pp 181–192

Billi D, Potts M (2002) Life and death of dried prokaryotes. Res Microbiol 153(1):7–12

Billi D, Friedmann EI, Hofer KG, Caiola MG, Ocampo-Friedmann R (2000) Ionizing-radiation resistance in the desiccation-tolerant cyanobacterium *Chroococcidiopsis*. Appl Environ Microbiol 66(4):1489–1492

Billings SA, Schaeffer SM, Evans RD (2003) Nitrogen fixation by biological soil crusts and heterotrophic bacteria in an intact Mojave Desert ecosystem with elevated CO2 added soil carbon. Soil Biol Biochem 35:643–649

Bolhuis H, Cretoiu MS, Stal LJ (2014) Molecular ecology of microbial mats. Fems Microbiol Ecol 90:335–350

Boyer SL, Johansen JR, Flechtner VR, Howard GL (2002) Phylogeny and genetic variance in terrestrial microcoleus (cyanophyceae) species based on sequence analysis of the 16S rRNA gene and associated 16S-23S ITS region 1. J Phycol 38(6):1222–1235

Büdel B, Darienko T, Deutschewitz K, Dojani S, Friedl T, Mohr KI, Salisch M, Reisser W, Weber B (2009) Southern African biological soil crusts are ubiquitous and highly diverse in drylands, being restricted by rainfall frequency. Microb Ecol 57(2):229–247

Campbell D, Eriksson MJ, Oquist G, Gustafsson P, Clarke AK (1998) The cyanobacterium synechococcus resists UV-B by exchanging photosystem II reaction-center D1 proteins. Proc Natl Acad Sci U S A 95:364–369

Cardon ZG, Gray DW, Lewis LA (2008) The green algal underground: evolutionary secrets of desert cells. BioScience 58(2):114–122

Caruso T, Chan Y, Lacap DC, McKay CP, Pointing SB (2011) Stochastic and deterministic processes interact to determine global biogeography of arid soil bacteria. ISME J 5:1406–1413

Cary SC, McDonald IR, Barrett JE, Cowan DA (2010) On the rocks: the microbiology of Antarctic Dry Valley soils. Nat Rev Micro 8(2):129–138

Chan Y, Lacap DC, Lau MCY, Ha KY, Warren-Rhodes KA, Cockell CS, Cowan DA, McKay CP, Pointing SB (2012) Hypolithic microbial communities: between a rock and a hard place. Environ Microbiol 14:2272–2282

Chan Y, van Nostrand JD, Zhou J, Pointig SB, Farrell RL (2013) Functional ecology of an Antarctic Dry Valley. Proc Natl Acad Sci U S A 110:8990–8995

Chen L, Deng S, De Philippis R, Tian W, Wu H, Wang J (2013) UV-B resistance as a criterion for the selection of desert microalgae to be utilized for inoculating desert soils. J Appl Phycol 25(4):1009–1015

Chen L, Rossi F, Deng S, Liu Y, Wang G, Adessi A, De Philippis R (2014) Macromolecular and chemical features of the excreted extracellular polysaccharides in induced biological soil crusts of different ages. Soil Biol Biochem 78:1–9

Cockell CS, Rettberg P, Rabbow E, Olsson-Francis K (2011) Exposure of phototrophs to 548 days in low Earth orbit: microbial selection pressures in outer space and on early earth. ISME J 5(10):1671–1682

Colica G, Li H, Rossi F, Li D, Liu Y, De Philippis R (2014) Microbial secreted exopolysaccharides affect the hydrological behavior of induced biological soil crusts in desert sandy soils. Soil Biol Biochem 68:62–70

Colica G, Li H, Rossi F, De Philippis R, Liu Y (2015) Differentiation of the characteristics of excreted extracellular polysaccharides reveals the heterogeneous primary succession of induced biological soil crusts. J Appl Phycol 27(5):1935–1944

da Rocha UN, Cadillo-Quiroz H, Karaoz U, Rajeev L, Klitgord N, Dunn S, Viet T, Buenrostro M, Bowen BP, Garcia-Pichel F, Mukhopadhyay A, Northen TR, Brodie EL (2015) Isolation of a significant fraction of non-phototroph diversity from a desert biological Soil crust. Front Microbiol 6

Dadheech PK, Abed RMM, Mahmoud H, Mohan MK, Krienitz L (2012) Polyphasic characterization of cyanobacteria isolated from desert crusts, and the description of *Desertifilum tharense* gen. et sp. nov. (Oscillatoriales). Phycologi 51(3):260–270

de Oliveira Mendes G, Moreira de Freitas AL, Liparini Pereira O, Ribeiro da Silva I, Bojkov Vassilev N, Dutra Costa M (2014) Mechanisms of phosphate solubilization by fungal isolates when exposed to different P sources. Ann Microbiol 64(1):239–249

De Philippis R, Vincenzini M (1998) Exocellular polysaccharides from cyanobacteria and their possible applications. FEMS Microbiol Rev 22:151–175

De Philippis R, Sili C, Paperi R, Vincenzini M (2001) Exopolysaccharide-producing cyanobacteria and their possible exploitation: a review. J Appl Phycol 13(4):293–299

Dembitsky VM, Dor I, Shkrob I, Aki M (2001) Branched alkanes and other apolar compounds produced by the cyanobacterium *Microcoleus vaginatus* from the Negev desert. Russ J Bioorg Chem 27:110–119

Demergasso C, Casamayor EO, Chong G, Galleguillos P, Escudero L, Pedros-Alio C (2004) Distribution of prokaryotic genetic diversity in athalassohaline lakes of the Atacama Desert, Northern Chile. FEMS Microbiol Ecol 48:57–69

Dvořák P, Hašler P, Poulíčková A (2012) Phylogeography of the *Microcoleus vaginatus* (Cyanobacteria) from three continents – a spatial and temporal characterization. PLoS One 7(6):e40153. doi:10.1371/journal.pone.0040153

Ehling-Schulz M, Bilger W, Scherer S (1997) UV-B-induced synthesis of photoprotective pigments and extracellular polysaccharides in the terrestrial cyanobacterium *Nostoc commune*. J Bacteriol 179:1940–1945

Ehling-Schulz M, Scherer S (1999) UV protection in cyanobacteria. Eur J Phycol 34(4):329–338

Elbert W, Weber B, Burrows S, Steinkamp J, Budel B, Andreae MO, Poschl U (2012) Contribution of cryptogamic covers to the global cycles of carbon and nitrogen. Nat Geosci 5(7):459–462

Feng S, Fu Q (2013) Expansion of global drylands under a warming climate. Atmos Chem Phys 13:10081–10094. doi:10.5194/acp-13-10081-2013

Fierer N, Jackson RB (2006) The diversity and biogeography of soil bacterial communities. Proc Natl Acad Sci U S A 103(3):626–631

Fierer N, Leff JW, Adams BJ, Nielsen UN, Bates ST, Lauber CL, Owens S, Gilbert JA, Wall DH, Caporaso JG (2012a) Cross-biome metagenomic analyses of soil microbial communities and their functional attributes. Proc Natl Acad Sci 109(52):21390–21395

Fierer N, Lauber CL, Ramirez KS, Zaneveld J, Bradford MA, Knight R (2012b) Comparative metagenomic, phylogenetic and physiological analyses of soil microbial communities across nitrogen gradients. ISME J 6:1007–1017

Gao Q, Garcia-Pichel F (2011) Microbial ultraviolet sunscreens. Nat Rev Micro 9(11):791–802

Garcia-Pichel F, Castenholz RW (1991) Characterization and biological implications of scytonemin, a cyanobacterial sheath pigment. J Phycol 27(3):395–409

Garcia-Pichel F, Pringault O (2001) Microbiology: Cyanobacteria track water in desert soils. Nature 413(6854):380–381

Garcia-Pichel F, López-Cortés A, Nübel U (2001) Phylogenetic and morphological diversity of cyanobacteria in soil desert crusts from the Colorado Plateau. Appl Environ Microbiol 67(4):1902–1910

Garcia-Pichel F, Johnson SL, Youngkin D, Belnap J (2003) Small-scale vertical distribution of bacterial biomass and diversity in biological soil crusts from arid lands in the Colorado Plateau. Microb Ecol 46:312–321

Garcia-Pichel F, Loza V, Marusenko Y, Mateo P, Potrafka RM (2013) Temperature drives the continental-scale distribution of key microbes in topsoil communities. Science 340(6140):1574–1577

Gray DW, Lewis LA, Cardon ZG (2007) Photosynthetic recovery following desiccation of desert green algae (Chlorophyta) and their aquatic relatives. Plant Cell Environ 30(10):1240–1255

Gundlapally SR, Garcia-Pichel F (2006) The community and phylogenetic diversity of biological soil crusts in the Colorado Plateau studied by molecular fingerprinting and intensive cultivation. Microb Ecol 52(2):345–357

Hagemann M, Henneberg M, Felde VJMNL, Drahorad SL, Berkowicz SM, Felix-Henningsen P, Kaplan A (2015) Cyanobacterial diversity in biological soil crusts along a precipitation gradient, Northwest Negev Desert, Israel. Microb Ecol 70(1):219–230

Harel Y, Ohad I, Kaplan A (2004) Activation of photosynthesis and resistance to photoinhibition in cyanobacteria within biological desert crust. Plant Physiol 136(2):3070–3079

Helm RF, Huang Z, Edwards D, Leeson H, Peery W, Potts M (2000) Structural characterization of the released polysaccharide of desiccation-tolerant nostoc commune DRH-1. J Bacteriol 182(4):974–982

Housman DC, Powers HH, Collins AD, Belnap J (2006) Carbon and nitrogen fixation differ between successional stages of biological soil crusts in the Colorado Plateau and Chihuahuan Desert. J Arid Environ 66(4):620–634

Hu CX, Liu YD, Song LR, Zhang DK (2002) Effect of desert soil algae on the stabilization of fine sands. J Appl Phycol 14(4):281–292

Hu C, Liu Y, Paulsen BS, Petersen D, Klaveness D (2003) Extracellular carbohydrate polymers from five desert soil algae with different cohesion in the stabilization of fine sand grain. Carbohydr Polym 54:33–42

Janatkova K, Rehakova K, Dolezal J, Simek M, Chlumska Z, Dvorsky M, Kopecky M (2013) Community structure of soil phototrophs along environmental gradients in arid Himalaya. Environ Microbiol 15(9):2505–2516

Johansen JR (1993) Cryptogamic crusts of semiarid and arid lands of North America. J Phycol 29(2):140–147

Johnson SL, Budinoff CR, Belnap J, Garcia-Pichel F (2005) Relevance of ammonium oxidation in biological soil crust communities. Environ Microbiol 7:1–12

Kuhne S, Strieth D, Lakatos M, Muffler K, Ulber R (2014) A new photobioreactor concept enabling the production of desiccation induced biotechnological products using terrestrial cyanobacteria. J Biotechnol 192:28–33

Lan S, Wu L, Zhang D, Hu C (2015) Analysis of environmental factors determining development and succession in biological soil crusts. Sci Total Environ 538:492–499

Lewis LA, Flechtner VA (2002) Green algae (Chlorophyta) of desert microbiotic crusts: diversity of North American taxa. Taxon 51:443–451

Lewis LA, Lewis PO (2005) Unearthing the molecular phylodiversity of desert soil green algae (Chlorophyta). Syst Biol 54(6):936–947

Li K, Liu R, Zhang H, Yun J (2013) The diversity and abundance of bacteria and oxygenic phototrophs in saline biological desert crusts in Xinjiang, Northwest China. Microb Ecol 66(1):40–48

Li H, Rao B, Wang G, Shen S, Li D, Hu C, Liu Y (2014) Spatial heterogeneity of cyanobacteria-inoculated sand dunes significantly influences artificial biological soil crusts in the Hopq Desert (China). Environ Earth Sci 71(1):245–253

Mager DM, Thomas AD (2011) Extracellular polysaccharides from cyanobacterial soil crusts: a review of their role in dryland soil processes. J Arid Environ 75:91–97

Makhalanyane TP, Valverde A, Velazquez D, Gunnigle E, Van Goethem MW, Quesada A, Cowan DA (2015a) Ecology and biogeochemistry of cyanobacteria in soils, permafrost, aquatic and cryptic polar habitats. Biodivers Conserv 24:819–840

Makhalanyane TP, Valverde A, Gunnigle E, Frossard A, Ramond J-B, Cowan DA (2015b) Microbial ecology of hot desert edaphic systems. Fems Microbiol Rev 39(2):203–221

Moreira C, Vasconcelos V, Antunes A (2013) Phylogeny and biogeography of cyanobacteria and their produced toxins. Mar Drugs 11:4350–4369

Nagy ML, Perez A, Garcia-Pichel F (2005) The prokaryotic diversity of biological soil crusts in the Sonoran Desert (Organ Pipe Cactus National Monument, AZ). Fems Microbiol Ecol 54(2):233–245

O'Malley MA (2008) 'Everything is everywhere: but the environment selects': ubiquitous distribution and ecological determinism in microbial biogeography. Stud Hist Philos Sci Part C Stud Hist Philos Biol Biomed Sci 39(3):314–325

Ogut M, Er F, Kandemir N (2010) Phosphate solubilization potentials of soil acinetobacter strains. Biol Fertil Soils 46(7):707–715

Ohad I, Raanan H, Keren N, Tchernov D, Kaplan A (2010) Light-induced changes within photosystem II protects microcoleus sp. in biological desert sand crusts against excess light. PLoS ONE 5(6):e11000. doi:10.1371/journal.pone.0011000

Ohad I, Berg A, Berkowicz SM, Kaplan A, Keren N (2011) Photoinactivation of photosystem II: is there more than one way to skin a cat? Physiol Plant 142:79–86

Osorio-Santos K, Pietrasiak N, Bohunick M, Miscoe LH, Kováčik L, Martin MP, Johansen JR (2014) Seven new species of Oculatella (Pseudanabaenales, Cyanobacteria): taxonomically recognizing cryptic diversification. Eur J Phycol 49(4):450–470

Patzelt DJ, Hodač L, Friedl T, Pietrasiak N, Johansen JR (2014) Biodiversity of soil cyanobacteria in the hyper-arid Atacama Desert, Chile. J Phycol 50(4):698–710

Pepe-Ranney C, Koechli C, Potrafka R, Andam C, Eggleston E, Garcia-Pichel F, Buckley DH (2016) Non-cyanobacterial diazotrophs mediate dinitrogen fixation in biological soil crusts during early crust formation. ISME J 10(2):287–298

Pereira S, Zille A, Micheletti E, Moradas-Ferreira P, De Philippis R, Tamagnini P (2009) Complexity of cyanobacterial exopolysaccharides: composition, structures, inducing factors and putative genes involved in their biosynthesis and assembly. FEMS Microbiol Rev 33:917–941

Pérez E, Sulbarán M, Ball MM, Yarzábal LA (2007) Isolation and characterization of mineral phosphate-solubilizing bacteria naturally colonizing a limonitic crust in the south-eastern venezuelan region. Soil Biol Biochem 39(11):2905–2914
Perkerson RB III, Johansen JR, Kovácik L, Brand J, Kaštovský J, Casamatta DA (2011) A unique pseudoanabaenalean (Cyanobacteria) genus nodosilinea gen.nov. based on morphological and molecular data. J Phycol 47(6):1397–1412
Planavsky NJ, Asael D, Hofmann A, Reinhard CT, Lalonde SV, Knudsen A, Wang X, Ossa FO, Pecoits E, Smith AJB, Beukes NJ, Bekker A, Johnson TM, Konhauser KO, Lyons TW, Rouxel OJ (2014) Evidence for oxygenic photosynthesis half a billion years before the great oxidation event. Nat Geosci 7:283–286
Pointing SB, Belnap J (2012) Microbial colonization and controls in dryland systems. Nat Rev Microbiol 10(8):551–562
Potts M (1994) Desiccation tolerance of prokaryotes. Microbiol Rev 58:755–805
Potts M (1997) Etymological basis for the genus name *Nostoc* (cyanobacteria). Int J Syst Bact 47:584
Potts M (2001) Desiccation tolerance: A simple process? Trends Microbiol, 9553–559
Pringault O, Garcia-Pichel F (2004) Hydrotaxis of cyanobacteria in desert crusts. Microb Ecol 47(4):366–373
Proteau PJ, Gerwick WH, Garcia-Pichel F, Castenholz R (1993) The structure of scytonemin, an ultraviolet sunscreen pigment from the sheaths of cyanobacteria. Experientia 49(9):825–829
Raanan H, Felde VJMNL, Peth S, Drahorad S, Ionescu D, Eshkol G, Treves H, Felix-Henningsen P, Berkowicz SM, Keren N, Horn R, Hagemann M, Kaplan A (2016a) Three-dimensional structure and cyanobacterial activity within a desert biological soil crust. Environ Microbiol 18:372–383
Raanan H, Oren N, Treves H, Berkowicz SM, Hagemann M, Pade N, Keren N, Kaplan A (2016b) Simulated soil crust conditions in a chamber system provide new insights on cyanobacterial acclimation to desiccation. Environ Microbiol 18:414–426
Rajeev L, da Rocha UN, Klitgord N, Luning EG, Fortney J, Axen SD, Shih PM, Bouskill NJ, Bowen BP, Kerfeld CA, Garcia-Pichel F, Brodie EL, Northen TR, Mukhopadhyay A (2013) Dynamic cyanobacterial response to hydration and dehydration in a desert biological soil crust. ISME J 7(11):2178–2191
Redfield E, Barns SM, Belnap J, Daane LL, Kuske CR (2002) Comparative diversity and composition of cyanobacteria in three predominant soil crusts of the Colorado Plateau. FEMS Microbiol Ecol 40(1):55–63
Řeháková K, Johansen JR, Casamatta DA, Xuesong L, Vincent J (2007) Morphological and molecular characterization of selected desert soil cyanobacteria: three species new to science including *Mojavia pulchra* gen. et sp. Nov. Phycologia 46(5):481–502
Robinson CK, Wierzchos J, Black C, Crits-Christoph A, Ma B, Ravel J, Ascaso C, Artieda O, Valea S, Roldán M, Gómez-Silva B, DiRuggiero J (2015) Microbial diversity and the presence of algae in halite endolithic communities are correlated to atmospheric moisture in the hyperarid zone of the Atacama Desert. Environ Microbiol 17:299–315
Rosentreter R, Bowker M, Belnap, J (2007) A field guide to biological soil crusts of western U.S. drylands. U.S. Government Printing Office, Denver, Colorado
Rossi F, De Philippis R (2015) Role of cyanobacterial exopolysaccharides in phototrophic biofilms and in complex microbial mats. Life 5(2):1218–1238
Rossi F, Potrafka RM, Garcia-Pichel F, De Philippis R (2012) The role of the exopolysaccharides in enhancing hydraulic conductivity of biological soil crusts. Soil Biol Biochem 46:33–40
Scherer S, Ernst A, Chen T-W, Böger P (1984) Rewetting of drought-resistant blue-green algae: time course of water uptake and reappearance of respiration, photosynthesis, and nitrogen fixation. Oecologia 62(3):418–423
Schmidt SK, Nemergut DR, Todd BT, Lynch RC, Darcy JL, Cleveland CC, King AJ (2012) A simple method for determining limiting nutrients for photosynthetic crusts. Plant Ecol Divers 5(4):513–519

Singh H, Anurag K, Apte SK (2013) High radiation and desiccation tolerance of nitrogen-fixing cultures of the cyanobacterium *Anabaena* sp. strain PCC 7120 emanated for the genome/proteome repair capabilities. Photosynth Res 118:71–81

Sorrels CM, Proteau PJ, Gerwick WH (2009) Organization, evolution, and expression analysis of the biosynthetic gene cluster for scytonemin, a cyanobacterial UV-absorbing pigment. Appl Environ Microbiol 75:4861–4869

Soule T, Gao Q, Stout V, Garcia-Pichel F (2013) The global response of *Nostoc punctiforme* ATCC 29133 to UVA stress, assessed in a temporal DNA microarray study. Photochem Photobiol 89(2):415–423

Starkenburg SR, Reitenga KG, Freitas T, Johnson S, Chain PSG, Garcia-Pichel F, Kuske CR (2011) Genome of the cyanobacterium microcoleus vaginatus FGP-2, a photosynthetic ecosystem engineer of arid land soil biocrusts worldwide. J Bacteriol 193(17):4569–4570

Sterflinger K, Tesei D, Zakharova K (2012) Fungi in hot and cold deserts with particular reference to microcolonial fungi. Fungal Ecol 5:453–462

Steven B, Gallegos-Graves LV, Starkenburg SR, Chain PS, Kuske CR (2012a) Targeted and shotgun metagenomic approaches provide different descriptions of dryland soil microbial communities in a manipulated field study. Environ Microbiol Rep 4(2):248–256

Steven B, Gallegos-Graves LV, Yeager CM, Belnap J, Evans RD, Kuske CR (2012b) Dryland biological soil crust cyanobacteria show unexpected decreases in abundance under long-term elevated CO2. Environ Microbiol 14(12):3247–3258

Steven B, Gallegos-Graves LV, Belnap J, Kuske CR (2013) Dryland soil microbial communities display spatial biogeographic patterns associated with soil depth and soil parent material. FEMS Microbiol Ecol 86:101–113

Steven B, Kuske CR, Gallegos-Graves LV, Reed SC, Belnap J (2015) Climate change and physical disturbance manipulations result in distinct biological soil crust communities. Appl Environ Microbiol 81(21):7448–7459

Stomeo F, Makhalanyane TP, Valverde A, Pointing SB, Stevens MI, Cary CS, Tuffin MI, Cowan DA (2012) Abiotic factors influence microbial diversity in permanently cold soil horizons of a maritime-associated Antarctic Dry Valley. FEMS Microbiol Ecol 82:326–340

Strauss SL, Day TA, Garcia-Pichel F (2012) Nitrogen cycling in desert biological soil crusts across biogeographic regions in the Southwestern United States. Biogeochemistry 108(1-3):171–182

Sukenik A, Quesada A, Salmaso N (2015) Global expansion of toxic and non-toxic cyanobacteria: effect on ecosystem functioning. Biodivers Conserv 24:889–908

Tao G-C, Tian S-J, Cai M-Y, Xie G-H (2008) Phosphate-solubilizing and -mineralizing abilities of bacteria isolated from soils. Pedosphere 18(4):515–523

Tracy CR, Streten-Joyce C, Dalton R, Nussear KE, Gibb KS, Christian KA (2010) Microclimate and limits to photosynthesis in a diverse community of hypolithic cyanobacteria in northern Australia. Environ Microbiol 12:592–607

Treves H, Raanan H, Finkel OM, Berkowicz SM, Keren N, Shotland Y, Kaplan A (2013) A newly isolated *Chlorella* sp from desert sand crusts exhibits a unique resistance to excess light intensity. FEMS Microbiol Ecol 86(3):373–380

United Nations. http://www.un.org/en/events/desertification_decade/background.shtml. Accessed 28 June 2016

Wang W, Liu Y, Li D, Hu C, Rao B (2009) Feasibility of cyanobacterial inoculation for biological soil crusts formation in desert area. Soil Biol Biochem 41(5):926–929

Warren-Rhodes KA, Rhodes KL, Pointing SB, Ewing SA, Lacap DC, Gómez-Silva B, Amundson R, Friedmann EI, McKay CP (2006) Hypolithic cyanobacteria, dry limit of photosynthesis, and microbial ecology in the hyperarid Atacama Desert. Microb Ecol 52(3):389–398

Welsh DT (2000) Ecological significance of compatible solute accumulation by micro-organisms: from single cells to global climate. FEMS Microbiol Rev 24:263–290

Wierzchos J, DiRuggiero J, Vitek P, Artieda O, Souza-Egipsy V, Skaloud P, Tisza M, Davila AF, Vilchez C, Garbayo I, Ascaso C (2015) Adaptation strategies of endolithic chlorophototrophs to survive the hyperarid and extreme solar radiation environment of the Atacama Desert. Front Microbiol 6

Wu L, Zhang G, Lan S, Zhang D, Hu C (2013a) Microstructures and photosynthetic diurnal changes in the different types of lichen soil crusts. Eur J Soil Biol 59:48–53
Wu L, Lan S, Zhang D, Hu C (2013b) Functional reactivation of photosystem II in lichen soil crusts after long-term desiccation. Plant and Soil 369(1-2):177–186
Xu Y, Rossi F, Colica G, Deng S, De Philippis R, Chen L (2013) Use of cyanobacterial polysaccharides to promote shrub performances in desert soils: a potential approach for the restoration of desertified areas. Biol Fertil Soils 49(2):143–152
Yeager CM, Kornosky JL, Housman DC, Grote EE, Belnap J, Kuske CR (2004) Diazotrophic community structure and function in two successional stages of biological soil crusts from the Colorado Plateau and Chihuahuan Desert. Appl Environ Microbiol 70:973–983
Yeager CM, Kornosky JL, Morgan RE, Cain EC, Garcia-Pichel F, Housman DC, Belnap J, Kuske CR (2007) Three distinct clades of cultured heterocystous cyanobacteria constitute the dominant N-2-fixing members of biological soil crusts of the Colorado Plateau, USA. Fems Microbiol Ecol 60(1):85–97
Yeager CM, Kuske CR, Carney TD, Johnson SL, Ticknor LO, Belnap J (2012) Response of biological soil crust diazotrophs to season, altered summer precipitation, and year-round increased temperature in an arid grassland of the Colorado Plateau, USA. Front Microbiol 3
Zhang B, Zhang Y, Downing A, Niu Y (2011) Distribution and composition of cyanobacteria and microalgae associated with biological soil crusts in the Gurbantunggut Desert, China. Arid Land Res Manag 25(3):275–293

Cyanobacteria in Terrestrial Symbiotic Systems

Jouko Rikkinen

Abstract Filamentous cyanobacteria are important primary producers and N_2 fixers in many terrestrial environments. As reduced nitrogen is often limiting, some thalloid liverworts (Marchantiophyta), hornworts (Anthocerophyta), the water fern *Azolla* (Salviniales), cycads (Cycadophyta), and the angiosperm *Gunnera* (Gunnerales) have evolved the ability to establish stable and structurally well-defined symbioses with N_2-fixing cyanobacteria. Also a wide diversity of lichen-forming fungi have cyanobacteria as photosynthetic symbionts or as N_2-fixing symbionts. Cyanolichen symbioses have evolved independently in different fungal lineages, and evolution has often resulted in convergent morphologies in distantly related groups. DNA techniques have provided a wealth of new information on the diversity of symbiotic cyanobacteria and their hosts. The fact that many plants and fungi engage in many different symbioses simultaneously underlines the probable significance of diffuse evolutionary relationships between different symbiotic systems, including cyanobacterial and mycorrhizal associations. This review introduces the reader to recent research on symbiotic cyanobacteria in terrestrial ecosystems and shortly describes the astonishing range of diversity in these ecologically important associations.

Introduction

Mutually beneficial symbiotic interactions are an inherent feature of most ecological communities. Nitrogen is essential for growth of land plants, but the availability of reduced nitrogen in the soil is often limiting. Diazotrophic bacteria are able to convert atmospheric dinitrogen (N_2) to ammonia (NH_3) that can be utilized by plants, and consequently, numerous land plants form an association with N_2-fixing bacteria. These interactions include the morphologically and physiologically highly coevolved symbioses between legumes and rhizobia, between actinorhizal plants

J. Rikkinen (✉)
Finnish Museum of Natural History, Botany Unit, University of Helsinki,
P.O. Box 7, Helsinki FIN-00014, Finland
e-mail: jouko.rikkinen@helsinki.fi

© Springer International Publishing Switzerland 2017

P.C. Hallenbeck (ed.), *Modern Topics in the Phototrophic Prokaryotes*,
DOI 10.1007/978-3-319-46261-5_8

and *Frankia*, and a plethora of more casual associations between plants and prokaryotic diazotrophs (Bothe et al. 2010; Santi et al. 2013).

Filamentous cyanobacteria are important N_2 fixers in aquatic ecosystems but also in many terrestrial environments (Cornelissen et al. 2007; Menge and Hedin 2009; Elbert et al. 2012; Lindo et al. 2013; Rousk et al. 2013; Arróniz-Crespo et al. 2014). However, only a restricted and highly paraphyletic assemblage of land plants establish well-defined symbioses with cyanobacteria. These include two genera of thalloid liverworts (Marchantiophyta), all hornworts (Anthocerophyta), one genus of ferns (*Azolla*, Salviniales), all cycads (Cycadophyta), and one isolated genus of angiosperms (*Gunnera*, Gunnerales). Within the symbiotic structures produced by plants, the symbiotic cyanobacteria (cyanobionts) are sustained by sugars provided by the host plant (Adams and Duggan 2008; Meeks 2009; Ekman et al. 2013). Most plant–cyanobacterium symbioses are facultative in the sense that both partners can be cultured separately, but plants in the wild invariably associate with cyanobacterial symbionts.

Nitrogen is also essential for fungal growth and not surprisingly also many fungi associate with N_2-fixing prokaryotes, including diazotrophic cyanobacteria (Schneider et al. 2016). In addition, a wide diversity of lichen-forming fungi have cyanobacteria as photosynthetic symbionts (photobionts). Cyanolichen symbioses have evolved independently in many fungal lineages, and convergent evolution has often resulted in similar thallus morphologies in distantly related groups of lichen-forming fungi (Rikkinen 2002).

This review introduces the reader to recent research on terrestrial symbiotic cyanobacteria and shortly describes the range of variation in their associations with land plants and fungi. Many if not most plants and fungi engage in multiple symbioses simultaneously. This provides vast possibilities for indirect effects between different symbionts, as well as the potential for each symbiont to modify the benefits and costs the other symbionts impose on their hosts. Thus, special attention is given to possible diffuse connections between cyanobacterial and other symbioses, including the diverse mycorrhizal associations of early land plants.

Symbiotic Cyanobacteria

While cyanobacteria are diverse and sometimes abundant in many terrestrial ecosystems, only a small minority of them establish well-defined symbioses with eukaryote hosts. *Nostoc* is by far the most common genus of cyanobacteria associated with land plants and fungi (Fig. 1). Also several other nostocalean and stigonematalean genera (Howard-Azzeh et al. 2014), including *Calothrix*, *Dichothrix*, *Stigonema*, and the recently re-circumscribed *Rhizonema* (Lücking et al. 2009, 2013a; Dal-Forno et al. 2013), include symbiotic taxa. Lichen-forming ascomycetes of the predominately tropical order Lichenomycetes seem to mainly associate with *Gloeocapsa*, *Anacystis*, and other non-nostocalean cyanobacteria (Tschermak-Woess 1988; Voytsekhovich et al. 2011a, b). However, DNA studies on the biological diversity in these symbioses have only barely begun (Ortiz-Álvarez et al. 2015).

Fig. 1 Lichen-symbiotic and free-living terrestrial cyanobacteria. **A** Foliose bipartite cyanolichen (*Leptogium* sp., Peltigerales) with cyanobionts (*Nostoc*) inside thin gelatinous thallus. The *brown disks* are apothecia of the fungal symbiont. **B–D** Morphological variation of *Nostoc* cyanobionts in three specimens of epiphytic *Leptogium* photographed through the thin upper cortex of hydrated thalli. Note the large heterocysts in **B** and **C**. The small translucent bodies between cyanobacterial trichomes are fungal hyphae in optical cross section. **E–F** Morphological variation in free-living colony-forming *Nostoc*. **G** Foliose bipartite cyanolichen (*Coccocarpia* sp., Peltigerales) with *Rhizonema* cyanobionts in the photobiont layer. Scale bars **A**, **E–G** 5 mm; **B–D** 5 μm

When in symbiosis, the phenotype of cyanobacteria is often modified, complicating direct comparisons between symbiotic cyanobacteria and their aposymbiotic relatives (Bergman and Hällbom 1982; Schüßler 2012; Ran et al. 2010). However, DNA methods can now be used to accurately identify symbiotic cyanobacteria both from fresh biological material (Rikkinen 2013) and dry herbarium specimens (Wright et al. 2001; Palinska et al. 2006). At the generic level, the standard DNA method used for identifying symbiotic cyanobacteria is 16S rRNA gene sequencing. However, due to many unresolved problems in cyanobacterial taxonomy, the symbiotic taxa cannot presently be identified to bacterial species (Oren and Garrity 2014; Pinevich 2015).

Symbiotic *Nostoc* can benefit their hosts either by providing fixed nitrogen, as in all plant symbioses and some cyanolichens, or by serving as a source of both carbon and nitrogen, as in most cyanolichens. Several studies have shown that lichen symbiontic *Nostoc* genotypes are closely related to plant symbiotic and free-living forms of the same genus (Paulsrud 2002; Rikkinen 2004; Papaefthimiou et al. 2008a; Yamada et al. 2012). They have also indicated that many fungal hosts are highly selective with respect to their cyanobionts. On the other hand, some lichen-forming fungi are known to associate with many different cyanobiont genotypes and frequently share them with other fungal species (Kaasalainen et al. 2012; O'Brien et al. 2013; Magain and Sérusiaux 2014; see Rikkinen 2013 for recent review). Bryophytes and cycads do not seem to be quite as selective in their cyanobiont choice and can often associate with several different *Nostoc* genotypes (Rikkinen and Virtanen 2008; Gehringer et al. 2010; Yamada et al. 2012).

As a whole, attempts to characterize the cyanobacterial symbiont have so far been made for a minute fraction of all known terrestrial symbioses. However, the complete genomic sequence of the symbiotically competent cyanobacterium *Nostoc punctiforme* (strain ATCC 29133 or PCC 73102), originally isolated from the coralloid roots of the cycad *Macrozamia*, has been determined (Meeks et al. 2001). Also the genome of the cyanobiont of the water fern *Azolla* ("*Nostoc azollae*" 0708) has been sequenced (Vigil-Stenman et al. 2015), and genomic sequences of several lichen-associated *Nostoc* genotypes are expected to become available soon (Grube et al. 2014). Metagenome sequencing of cyanolichens is also providing new insights into the genetic diversity of cyanobacterial symbioses (Kampa et al. 2013; Sigurbjörnsdóttir et al. 2015).

Cyanobacterial Adaptations to Symbiosis

Species of *Nostoc* form multicellular filaments in which, especially under conditions of combined nitrogen deprivation, some cells differentiate into heterocysts (Fig. 1), which have a thickened cell wall and provide a micro-aerobic environment for the functioning of the highly oxygen-sensitive enzyme nitrogenase, which is essential for the conversion of N_2 to NH_3 (Dodds et al. 1995; Flores and Herrero 2010). The vegetative cells and heterocysts of *Nostoc* are mutually interdependent;

the latter import photosynthates from vegetative cells and provide nitrogen in return. Multicellularity evolved in cyanobacteria already 2.5 billion years ago (Tomitani et al. 2006; Schirrmeister et al. 2011, 2013, 2015) and has since been lost and regained several times in different cyanobacterial lineages (Flores and Herrero 2010; Claessen et al. 2014).

The vegetative cells of *Nostoc* can also differentiate into thick-walled resting stages (akinetes) and/or hormogonia which are short, small-celled filaments that move by gliding (Meeks et al. 2002). Hormogonia are attracted to root extracts and to certain sugars, including glucose, and can penetrate into tissues and cells of plant and fungal hosts (Meeks and Elhai 2002; Adams and Duggan 2008; Ekman et al. 2013). They differentiate in response to a range of signals, including the still uncharacterized hormogonium-inducing factor (HIF) that is released by plants under nitrogen starvation (Campbell and Meeks 1989; Nilsson et al. 2006). For more on the role and mechanisms of cyanobacterial motility in plant infection, see Adams and Duggan (2008) and Adams et al. (2013).

When inside the plant host, hormogonium production is repressed by hormogonium-repressing factors. Liaimer et al. (2015) reported that the nonribosomal peptide nostopeptolide produced by symbiotic *Nostoc* can function either as a hormogonium-repressing factor or as a chemoattractant, depending on its extracellular concentration. Splitt and Risser (2016) found that plant-derived sucrose or sucrose analogs can both repress hormogonia and induce the production of a polysaccharide sheath that seems to play a role in the establishment and maintenance of the symbiotic state in *Nostoc*. The signaling between the plant and the cyanobiont may also involve arabinogalactan proteins (Jackson et al. 2012; Adams et al. 2013).

Inside the host plant, the symbiotic cyanobacterium can undergo drastic morphological and physiological changes. The vegetative cells may be enlarged and show irregularities of shape. The rate of CO_2 fixation tends to be reduced, whereas N_2 fixation is stimulated and ammonium assimilation downregulated (Adams and Duggan 2008; Adams et al. 2013). In plant symbioses and some cyanolichens, the frequency of heterocysts increases, often reaching 30–40 %, which is several times higher than that typically found in free-living *Nostoc* (Meeks and Elhai 2002). The doubling time of symbiotic *Nostoc* may be slowed down to a fraction of that in free-living cyanobacteria, which ensures that the cyanobiont does not outgrow its host. For details on genetic aspects of plant–cyanobiont interactions, the reader is referred to Adams et al. (2013).

In exchange for fixed nitrogen, the plant-symbiotic cyanobacteria extract a carbon cost from their hosts. The high rate of N_2 fixation cannot be maintained by the limited photosynthetic capacity of the cyanobionts themselves and must be supported by reduced carbon from the host. Symbiotic cyanobacteria are facultative heterotrophs and the sugars most frequently assimilated are fructose, glucose, and sucrose. From a carbon budget perspective, the overall costs to the host for maintaining cyanobacterial N_2 fixation and respiration have not been determined, but they may be considerable. Especially in situations where soil nitrogen is easily obtained, the cost of maintaining cyanobionts may outweigh the benefits and explain why cyanobacterial symbioses are not more common among land plants (Ekman et al. 2013).

The cell surfaces of cyanobacteria are covered with complex carbohydrates which act as barriers against different types of stress and can also act as specific recognition factors (Kehr and Dittmann 2015). Recognition of compatible cyanobiont cells is believed to be performed by specific lectins produced and secreted by the plant or fungal host. Lectins have been identified from the fungal hosts of several cyanolichen species (Manoharan et al. 2012; Miao et al. 2012), and a glycosylated arginase acting as a fungal lectin has been found to bind to *Nostoc* from cyanolichens to function in the recruitment and adhesion of cyanobiont cells to the hyphal surface (Díaz et al. 2009, 2011; Vivas et al. 2010).

Vertical transmission of cyanobionts from one host generation to the next has evolved independently in many different lineages of cyanolichens (Rikkinen 2002), in thalloid liverworts (Rikkinen and Virtanen 2008), and in the cyanobacterium–*Azolla* symbiosis (Adams et al. 2013). The production of specialized symbiotic propagules helps to maintain pairwise symbiotic interactions over time and promotes the likelihood of coevolution between specific partners. The cyanobiont of *Azolla* spends its entire life cycle within the host plant and due to genomic erosion is now unable to grow aposymbiotically (Ran et al. 2010; Vigil-Stenman et al. 2015). Also some lichen-symbiotic *Nostoc* genotypes are "unculturable" and may well have lost their ability to grow outside lichen thalli (Rikkinen 2013).

Kaasalainen et al. (2012) related the high diversity of microcystin variants and corresponding genes in lichen cyanobionts to the evolutionary effects of symbiotic dispersal. When packaged into propagules of symbiotically dispersing lichens, the cyanobionts invariably experience extreme genetic bottlenecks. On the other hand, the close association with one or more fungal hosts likely promotes the selection for traits different from those typically experienced by nonsymbiotic cyanobacteria. As a whole, the recurrent bottlenecks and other populations shaping effects may have been instrumental for the evolution of the present genetic and chemical diversity in lichen cyanobionts.

Cyanobacteria produce a wide range of secondary metabolites, including many toxic compounds (Calteau et al. 2014; Dittmann et al. 2015). The cyanobionts of some cyanolichens produce hepatotoxic microcystins and nodularin in symbiosis (Oksanen et al. 2004; Kaasalainen et al. 2009, 2012, 2013), and nodularin was recently also found from the roots of cycads (Gehringer et al. 2012). These nonribosomal peptides are familiar from cyanobacterial blooms in aquatic ecosystems, where they have caused animal poisonings around the world and also pose a threat to human health. It is also possible that some lichen-symbiotic cyanobacteria can contribute to the defense against their hosts and/or parasitic fungi by producing hassalladins or other mycotoxic compounds (Vestola et al. 2014; Shishido et al. 2015).

Cyanolichens

'Lichens are traditionally defined as ecologically obligate symbioses between fungal hosts and symbiotic green algae and/or cyanobacteria. The term "cyanolichen" is used of lichens with a cyanobacterial symbiont (cyanobiont), either as the sole photosynthetic partner (photobiont) or as a N_2-fixing symbiont in addition to a primary eukaryotic photobiont. Lichen symbioses do not have independent scientific names; all partners of the symbiosis have their own names and the name of the intact "lichen species" refers to the fungal partner alone. Cyanolichen symbioses are ecologically obligate in the sense that the mycobionts of most cyanolichens cannot be cultured without the appropriate cyanobionts, and most of the cyanobionts do not appear to establish aposymbiotic populations outside lichen thalli.

Symbioses between different fungi and cyanobacteria have evolved independently on many occasions, and in some lineages, the symbiosis seems to have been lost. Numerous cyanolichen species, possibly even the majority, have not yet been found and described (Lumbsch et al. 2011; Lücking et al. 2014; Moncada et al. 2014). A lichen-symbiotic lifestyle appears to have evolved at least ten times in the Ascomycota and five times in the Basidiomycota, with a vast majority of the ca. 18,000 currently accepted species of lichen-forming fungi belonging to the ascomycetes (Lücking et al. 2014). Only about 10 % of the currently known species of mycobionts establish lichen symbioses with cyanobacteria, with the overwhelming majority associating with green algae (Chlorophyta), especially certain groups of trebouxiophytes and Trentepohliales or rarely with other types of eukaryotic algae (Rikkinen 2002, 2015).

Among the fungi, cyanobacteria establish symbiotic associations mainly with ascomycetes (Ascomycota). Glomeromycetes (Glomeromycota) and basidiomycetes (Basidiomycota) are common partners in mycorrhizal symbioses, but associations between them and cyanobacteria are rare. Prominent exceptions include basidiomycetes of the *Dictyonema* clade (Dal-Forno et al. 2013; Lücking et al. 2014) and the enigmatic glomeromycete *Geosiphon* (Schüßler 2012), which does not well fit into the commonly used definition of a "lichen-forming fungus." The polyphyletic chytrids (Chytridiomycota) include many parasites of cyanobacteria (Gerphagnon et al. 2013; Rohrlack et al. 2015), but no mutualistic forms have been described.

The Lichen Thallus

Within the lichen thallus, the cyanobiont provides sugar (as glucose) and/or fixed atmospheric dinitrogen (as ammonium) to the mycobiont. The fungus, in turn, provides the cyanobionts water, carbon dioxide, and stable environment that is relatively well buffered against environmental extremes and grazing invertebrates.

On the basis of their general habit, cyanolichens have traditionally been grouped into foliose, fruticose, and crustose species (Fig. 2). This division is obviously artificial and convergent forms have repeatedly evolved in different fungal lineages. The thalli of most cyanolichens are foliose, i.e., their dorsiventral thalli are flat and lobate and mainly grow horizontally. Fruticose cyanolichens produce shrubby, often upright thalli with cylindrical lobes that may be attached to the substrate only by a narrow base. Crustose cyanolichens produce relatively undifferentiated thalli that often grow very tightly attached to or even partly immersed into the substrate.

Fig. 2 Cyanolichen growth forms. **A** Foliose bipartite cyanolichen (*Peltigera praetextata*, Peltigerales) with cyanobionts (*Nostoc*) in a layer below the upper cortex of the stratified thallus. **B** Foliose cephalodiate lichen (*Peltigera leucophlebia*, Peltigerales) with cyanobionts (*Nostoc*) in cephalodia on the upper surface of the *green* algal thallus. **C** Fruticose cephalodiate lichen (*Stereocaulon* sp., Lecanorales) with cyanobionts in large saccate cephalodia. **D** Crustose bipartite cyanolichen (*Collema* sp., Peltigerales). **E** Gelatinous bipartite cyanolichen (*Leptogium saturninum*, Peltigerales). Scale bars 5 mm

The thalli of most lichens are clearly stratified and have a protective upper cortex, a well-defined photobiont layer immediately below, and a loose medulla that effectively facilitates gas exchange. The lower surface may have rhizines which attach the lichen to substrate, help to maintain airspace below the thallus, and contribute to water absorption. The effective exchange of metabolites obviously requires an intimate connection between the mycobiont and photobionts. In most cyanolichens, thin-walled mycobiont hyphae penetrate the gelatinous sheaths of cyanobionts but are not in direct contact with the cyanobacterial cell wall. For more details of the anatomy of lichen thalli, the reader is referred to the recent review by Honegger (2012).

The light-absorbing fungal pigments and refractive structures in the cortex influence the quantity and quality of light that reaches the photobiont layer (Bjerke et al. 2005; Wu et al. 2014). The structural organization of the thallus also influences water relations; while thick thalli need a relatively long time to fill their internal water storage, the fully hydrated thalli can also continue photosynthesis for extended periods while drying (Gauslaa 2014; Merinero et al. 2014). In most cyanolichens, the structural characteristics of the thallus can be modified in response to the prevailing environment. The genetic regulation of lichen thallus formation and metabolite exchange is still almost unknown, but it must involve extensive molecular crosstalk between the partners (Chua et al. 2012; Kampa et al. 2013; Wang et al. 2014).

The cyanobionts of gelatinous cyanolichens or so-called jelly lichens (Oksanen et al. 2002; Otálora et al. 2013, 2014) are not organized into a distinct photobiont layer but occur more freely within the medulla (Fig. 2). The extensive gelatinous sheaths of the cyanobionts can absorb large quantities of water, and consequently, gelatinous cyanolichens typically exhibit major changes in thallus dimensions during wetting and drying and are in this respect similar to the gelatinous colonies of free-living *Nostoc* (Fig. 1).

Based on symbiont composition, cyanolichens have traditionally been divided into two artificial groups: bipartite cyanolichens and tripartite or cephalodiate lichens (Fig. 2). In bipartite cyanolichens, the cyanobionts typically form a photobiont layer immediately below the upper cortex. In cephalodiate lichens, the fungus produces a compound thallus with green algal and cyanobacterial symbionts (Fig. 3). In most cases, the green algal symbionts form a photobiont layer, while the cyanobionts are restricted to cephalodia, i.e., specific structures housing cyanobionts in an otherwise green algal lichen thallus (Cornejo and Scheidegger 2013). Lichen cyanobionts can deliver both sugar and fixed nitrogen to their fungal partners, and the relative importance of these two activities differs in bipartite and cephalodiate cyanolichens. In cephalodiate lichens, the green algal photobionts typically produce most photosynthate and the cephalodial cyanobionts mainly fix nitrogen (Green et al. 2008; Nash 2008). Concurrently, the cephalodial cyanobionts tend to show higher heterocyst frequencies and higher rates of N_2 fixation than those of bipartite cyanolichens. As a whole, cephalodia occur in hundreds of lichen species, often hidden in the medulla or on the lower surface of the thallus, reflecting the primary role of cephalodial cyanobionts in N_2 fixation.

Cephalodiate cyanolichens can either have one cyanobacterial genotype in all cephalodia of a single lichen thallus (Paulsrud et al. 1998, 2001) or house different cyanobacterial genotypes in different cephalodia (Paulsrud et al. 2000; Kaasalainen et al. 2009), and some tripartite lichens have the two different photobionts in what appears to be the same photobiont layer (Henskens et al. 2012). It is possible that some mycobionts can respond to habitat variability and environmental change by selecting between different cyanobiont genotypes with different environmental optima, in a similar way as has been described from some green algal lichens (Casano et al. 2011). Some cyanolichen mycobionts can produce distinct thallus morphologies in symbiosis with compatible green algae and cyanobacteria, respectively (Fig. 3). The disparate morphs can either combine to form one chimeric thallus or

Fig. 3 Cephalodia and photosymbiodemes. **A** Cephalodia with *Nostoc* cyanobionts on upper surface of *Peltigera aphthosa* (Peltigerales). **B** The cephalodia are delimited by a continuous fungal cortex. **C** Poorly delimited cephalodia with *Nostoc* cyanobionts on the stems of *Stereocaulon* sp. (Lecanorales). **D** Large cephalodia with *Nostoc* cyanobionts on the upper surface of *Placopsis* sp. (Trapeliales). **E** Photosymbiodeme of *Pseudocyphellaria* sp. (Peltigerales). *Green* algal lobes with *Nostoc* cyanobionts in internal cephalodia developing from bipartite thallus of the same lichen. **F** Independent *Dendriscocaulon* cyanomorph of *Sticta* sp. (Peltigerales). **G** Photosymbiodeme of *Nephroma arcticum* (Peltigerales). *Green* algal lobes with *Nostoc* cyanobionts in internal cephalodia developing from bipartite thallus of the same lichen. Scale bars 5 mm

grow separately. Chimeric lichens with green algae and cyanobacteria as photobionts in different parts of one thallus are called photosymbiodemes, and the two types of free-living morphs are called chloromorphs (chlorosymbiodemes) and cyanomorphs (cyanosymbiodemes), respectively. Many previously unknown photosymbiodemes have recently been described from different cyanolichen lineages (Aptroot and Schumm 2009; Magain et al. 2012; Magain and Sérusiaux 2014). They can potentially be used for studying the influence of photobiont choice on thallus morphogenesis and metabolism under identical conditions of growth and host association (Schelensog et al. 2000).

While lichens are usually perceived as pairwise interactions between one fungus and one or two photobionts, they regularly also incorporate a plethora of associated microorganisms. Lichen thalli host diverse and apparently specialized communities of non-phototrophic bacteria (Grube et al. 2009; Hodkinson and Lutzoni 2009; Hodkinson et al. 2012; Sigurbjörnsdóttir et al. 2014; Aschenbrenner et al. 2016). Also the diversity of associated microfungi is remarkable; nearly 2000 species of lichenicolous fungi have been described and thousands of additional species are believed to await description (Arnold et al. 2009; U'Ren et al. 2012; Werth et al. 2013). Many lichenicolous fungi grow only on cyanolichens (Fig. 5), either as host-specific parasites or as broad-spectrum pathogens, saprotrophs, or commensals (Lawrey and Diederich 2016). In addition, some green algal lichens regularly start their development on cyanolichen thalli (Dal-forno et al. 2013), while others associate with colonies of "free-living" cyanobacteria, presumably in order to access fixed nitrogen (Rikkinen 2002).

Ascolichens and Basidiolichens

Among the Ascomycota, cyanobacteria form symbioses especially with Lecanoromycetes (Miadlikowska et al. 2014) and Lichinomycetes (Schultz et al. 2001). An overwhelming majority of cyanolichen mycobionts belong to Lecanoromycetes (Fig. 4). Species that form bipartite symbioses with cyanobacteria are largely restricted to the Peltigerales, while cephalodiate taxa occur more sporadically among different lecanoromycete orders (Miadlikowska et al. 2014). Most fungi in the Peltigerales (e.g., Collemataceae, Nephromataceae, Pannariaceae, and Peltigeraceae) associate with symbiotic *Nostoc*, while others (e.g., Coccocarpiaceae) associate with other nostocalean and/or stigonematalean cyanobionts (Lücking et al. 2009, 2013b; Ekman et al. 2014; Spirbille et al. 2014; Zúñiga et al. 2015, see Rikkinen 2013 for more detailed review).

All fungi in the predominately tropical ascomycete class Lichinomycetes are lichen symbiotic, and most of them seem to establish bipartite symbioses with *Gloeocapsa*, *Anacystis*, and other non-nostocalean cyanobacteria (Tschermak-Woess 1988; Schultz et al. 2001; Voytsekhovich et al. 2011a, b). However, Ortiz-Álvarez et al. (2015) reported that the cyanobionts of two European species of *Lichina* (Lichinomycetes) were closely related to marine and freshwater strains of *Rivularia* (Nostocales).

Fig. 4 Ascolichens and basidiolichens. **A** Fertile specimen of bipartite ascolichen *Fuscopannaria* sp. (Peltigerales) with apothecia on the upper surface of the thallus. **B** Bipartite ascolichen *Peltigera* sp. (Peltigerales) with apothecia on upper surface of upright thallus lobes. **C** Bipartite ascolichen *Peltigera* sp. (Peltigerales) producing symbiotic propagules (soredia) in well-delimited structures (soralia) on upper surface of the thallus. **D** Bipartite basidiolichen *Dictyonema* s.lat. (Agaricales) with *Rhizonema* cyanobionts in corticoid basidiomata. Scale bars 5 mm

While lichenization is generally rare among basidiomycetes, it has clearly evolved several times in unrelated groups. Seven currently accepted genera form structurally well-delimited symbioses with green algae or cyanobacteria. In addition, several other groups of basidiomycetes include lichenicolous species, some of which grow on cyanolichens (Diederich 1996; Lawrey and Diederich 2016). The major lineages of Basidiomycota that contain lichen-forming species also exhibit saprotrophic, pathogenic, mycorrhizal, and lichenicolous taxa (Lawrey et al. 2009). For example, some species of *Athelia* (Atheliales) associate with cyanobacteria, while others are saprophytes or grow on coccomyxoid green algae or as lichen parasites (Oberwinkler 2012).

Recent studies have revealed that especially the *Dictyonema* clade (Fig. 4) in the Hygrophoraceae (Agaricales) includes a large number of hitherto undescribed species (Lücking et al. 2013a, 2014). All species of this clade are lichen-forming (Dal-Forno et al. 2013, 2016). The basal clade of *Dictyonema* s. lat. includes fungi with appressed-filamentous thalli and a simple hyphal sheath around the cyanobacterial photobiont filaments (Lawrey et al. 2009). *Dictyonema* s. str. is a paraphyletic grade of species that form appressed to shelflike filamentous thalli with the cyanobiont filaments surrounded by jigsaw puzzle-shaped hyphal sheaths. *Acantholichen*, *Corella*, and *Cora* are nested within *Dictyonema* s.lat and have coiled cyanobacterial filaments forming clusters, wrapped within a dense hyphal sheath formed by jigsaw puzzle-shaped cells of the mycobiont, appearing unicellular or pseudo-colonial. *Acantholichen* is sister to *Corella*, both in turn forming a sister clade to *Cora* (Lawrey et al. 2009; Dal-Forno et al. 2013). The Neotropical genus *Acantholichen* includes microsquamulose species with dark blue to gray thalli, with spiny apical cells on both thallus surfaces, giving them a coarsely white-pruinose appearance (Jørgensen 1998).

The cyanobionts of the *Dictyonema* and related basidiomycetes were previously identified as *Scytonema* or *Chroococcus*, depending on whether their filaments were short coiled or not (Tschermak-Woess 1988). Lücking et al. (2009) revealed that they form a lineage among nostocalean cyanobacteria clearly distinct from *Scytonema* species and named them *Rhizonema* (Lücking et al. 2009). In *Dictyonema*, *Cora*, and *Acantholichen*, the cyanobionts are enveloped by fungal hyphae and penetrated by haustoria. Thus, like in parasitic chytrids and some cyanophilous ascomycetes, the basidiomycete hosts actually penetrate the cell walls of cyanobacteria in search of nutrition (Oberwinkler 1984, 2012).

Lawrey et al. (2009) pointed out that the high concentration and diversity of lichen-forming taxa in the Hygrophoraceae suggest a predisposition toward lichen symbiosis in this family, even though nutritional modes in the lineage are not markedly different from many other families in the Agaricales, including mostly saprotrophic and ectomycorrhizal fungi. However, the family included many bryophilous taxa which associate either with the plants directly or with photosynthetic microbial communities containing complex mixtures of eukaryotic algae, cyanobacteria, and moss protonemata on water-retaining peat or well-decayed wood. This may indicate a tendency to switch from a saprotrophic to a lichen-symbiotic mode of nutrition in bryophilous species of this family (Lawrey et al. 2009).

Cyanolichen Ecology

Cyanolichens are found in many types of terrestrial environments ranging from tropical rain forests and semideserts to arctic tundra (Fig. 5). Their diversity and abundance are highest in relatively humid climates. Epiphytic species do particularly well in the moist and cool conditions of higher elevations in tropical mountains and in maritime regions of higher latitudes. They are often well represented in the epiphyte communities of old-growth boreal and temperate forests where they intercept and help to retain atmospheric moisture, sequester nutrients, and provide

Fig. 5 Ecology of cyanolichens. **A** Lichens are important winter feed for reindeer which graze heavily on terricolous *green* algal species. However, the ruminants avoid eating cyanolichens, possibly due to toxic compounds produced by lichen cyanobionts. **B** Mollusk grazing limits the distribution of some cyanolichens in boreal rain forests and other humid habitats. **C** Selective feeding by unidentified beetle of *Nostoc* cyanobionts from cephalodia of *Nephroma arcticum* (Peltigerales). **D** Unidentified lichenicolous fungus parasitizing the photobiont layer of *Peltigera* sp. (Peltigerales). **E** Secondary substances produced by many lichen-symbiotic fungi produce can play a role in defense against herbivores. This *Pseudocyphellaria* species produces *yellow* vulpinic acid and other pulvinic acid derivatives in the medulla, and the toxic compound also accumulates in the symbiotic propagules (soredia) of the lichen. Scale bars 2 mm

habitat and food for many invertebrates. Many epiphytic species thrive in microhabitats that combine moderate light intensities and ample moisture with periodic drying events.

In closed canopy forests, epiphytic cyanolichens are often most abundant in the lower and mid canopy sections, where the quantity and quality of light are moderated by the overlying canopy. Cyanolichens are generally less diverse in arid climates, but can be important components of biological soil crusts in semideserts, where they help to stabilize the soil and contribute to its fertility. The N_2-fixing cyanobionts of cyanolichens contribute significant amounts of nitrogen to the ecosystem (Gavazov et al. 2010; Elbert et al. 2012). Flexibility between alternative nitrogen fixation pathways in *Nostoc* can be significant for cyanolichens that grow on nutrient-poor substrates and get most of their mineral nutrition through aerial deposition (Darnajoux et al. 2014; Hodkinson et al. 2014). Epiphytic cyanolichens are highly susceptible to the adverse effects of air pollution and can be used as indicators of clean air. In addition, many species have been hard hit by logging and are now more or less restricted to old-growth forests and can be used as indicators of long habitat continuity. For more information on the habitat ecology of cyanolichens, see Rikkinen (2015).

Many cyanolichen species facilitate the reproduction and simultaneous dispersal of their symbiotic consortium by producing symbiotic propagules. However, the mycobionts of most cyanolichen species only produce fungal spores and must thus reestablish their symbiotic association at each reproductive cycle. A compatible partner can be obtained either from a population of free-living cyanobionts or from the thallus or symbiotic diaspore of another cyanolichen. As appropriate cyanobionts are not likely to be ubiquitously distributed, the local availability of cyanobionts can explain many patterns of cyanolichen species occurrence, and shared symbiont specificity may lead to facilitative interactions between different fungal hosts. For example, some spore-dispersed mycobionts appear to be facilitated by the prior establishment of other cyanolichen species that produce symbiotic propagules (Rikkinen et al. 2002; Rikkinen 2003, 2013; Fedrowitz et al. 2011, 2012; Belinchón et al. 2014; Dal Grande et al. 2014) or by bryophytes that house appropriate cyanobacteria (Cornejo and Scheidegger 2016).

Geosiphon pyriformis–Nostoc Symbiosis

Geosiphon pyriforme (Geosiphonaceae, Archaeosporales) is the only species of Glomeromycota that is known to form a well-defined symbiosis with cyanobacteria (Schüßler 2012). While this unique association has only been found from a few sites in central Europe, there is full reason to believe that the symbiosis type is ancient and that other similar associations still exist and await discovery. The *Geosiphon pyriformis–Nostoc* symbiosis seems to represent a "living fossil" of multibiont associations that once connected the cyanobacterial and mycorrhizal symbioses of early bryophytes and contributed to the evolution of early terrestrial ecosystems.

The Glomeromycota is a monophyletic group of fungi that was previously included as an order in the former Zygomycetes. Details in the taxonomic history of the lineage were recently reviewed by Redecker and Schüßler (2014). Glomeromycotan fungi establish arbuscular mycorrhizal (AM) symbioses with land plants including most vascular plants and many hornworts and liverworts. The symbiotic fungi colonize plant roots but also efficiently penetrate the soil and improve plant water and mineral nutrient uptake in return for plant-assimilated carbon (van der Heijden et al. 2015).

Geosiphon pyriformis is unique among the presently known glomeromycetan fungi not only in associating with symbiotic cyanobacteria but also because it has not yet been shown to form mycorrhizae with bryophytes or vascular plants. The substrate ecology and establishment of the *Geosiphon pyriformis–Nostoc* symbiosis were described by Mollenhauer and Mollenhauer (1988) who managed to maintain the symbiontic consortium in the laboratory. The fungus grows among free-living cyanobacteria and bryophytes in the topsoil of humid, nutrient-poor sites and regularly occurs in close association with moss protonemata. The *Nostoc* symbionts are housed endosymbiotically within small multinucleate and vacuolated fungal bladders on the soil surface, and the fungal hyphae can produce new vesicles when a specific primordial stage of *Nostoc* is present. The cyanobacteria are incorporated by the fungus near the hyphal tip, which then swells and develops into a pear-shaped aboveground bladder up to 2 mm long. Each bladder results from a single incorporation event and different bladders formed by the mycelium can thus house different *Nostoc* genotypes. Within the terminal bladders, the cyanobionts proliferate and provide photosynthate and fixed nitrogen to the host. At times the fungus forms large resting spores similar to those of related glomeromycotan fungi.

Geosiphon bladders are essentially large, multikaryotic cells with a photosynthetically active region in the exposed apex and a more restricted storage region at the base, partly embedded in the soil and attached to the underground mycelium. All cyanobionts of each bladder are in one membrane-bound cellular compartment (symbiosome). There is a rudimentary fungal cell wall between the symbiosome membrane (fungal plasma membrane) and the enclosed *Nostoc* cells, basically similar to the thin cell walls of arbuscules of the AM fungi within colonized plant cells. For more details on the anatomy and ultrastructure, see Schüßler (2012), and for genetic aspects, see Schüßler et al. (2006, 2007) and Ellerbeck et al. (2013).

The endosymbiotic cyanobacterium isolated by Mollenhauer was originally identified as *Nostoc punctiforme*. There is a level of specificity in the *Geosiphon pyriformis–Nostoc* association as some genotypes of *Nostoc* are incorporated by the fungus, while in others the development of the cyanobacterium ceases at an early stage or it is not incorporated at all (Schüßler 2012). The cyanobiont can be readily isolated and cultivated without the fungal partner. The vegetative cells of the cyanobiont within symbiotic bladders are larger than the cells of typical free-living *Nostoc*, apparently because of the high osmotic pressure inside the bladder, but their ultrastructure is not markedly modified. Also the heterocyst frequency of cyanobionts within the bladder is similar to that of free-living *Nostoc*, indicating that the cyanobionts play a principal role in photosynthesis and not in N_2 fixation (Kluge et al. 1991, 1992). The cyanobionts within *Geosiphon* bladders get water, carbon dioxide, and all inorganic nutrients except nitrogen from the fungal host. The fungus may also offer protection against heavy metals and other abiotic stress factors (Scheloske et al. 2001; Wojtczak and Janik 2016).

While *Geosiphon pyriforme* is the only AM fungus known to have endosymbiotic cyanobacteria, many other AM fungi host biotrophic endobacteria in their cytoplasm (Naumann et al. 2010; Torres-Cortésa et al. 2015) and also symbiotic bladders of *Geosiphon* house such endosymbionts. The *Mollicutes* (mycoplasma-related endobacteria, MRE) are related to *Mycoplasma* and *Phytoplasma* species which are biotrophic parasites in animals and plants. The coccoid bacteria appear to possess a Gram-positive cell wall and are not surrounded by host membrane (Torres-Cortésa et al. 2015).

MRE are common in the mycelia and spores of AMF and are thus also indirectly associated with plants. Recently they have also been identified from species of *Endogone* (Mucoromycotina), a fungal genus that also includes plant mycorrhizal species, implying that the symbiosis between MRE and fungi would predate the divergence between glomeromycetes and mucoromycetes. They are presumed to play some role in the AM symbiosis, but the nature of their role is unknown and it thus remains unclear whether they actually are mutualistic associates or parasites of their fungal hosts (Desirò et al. 2013b, 2015; Naito et al. 2015; Toomer et al. 2015).

Bryophyte–Cyanobacterium Symbioses

Bryophytes (liverworts, mosses, and hornworts) are generally recognized as the oldest living land plants (Shaw et al. 2011; Ligrone et al. 2012). While the exact relationships between the three lineages continue to remain ambiguous (Wickett et al. 2014), the recently most widely accepted view has placed liverworts as sister to all other embryophytes (Qiu et al. 2006). Unfortunately the fossil record is extremely fragmentary for all bryophytes and is of limited value for resolving the earliest divergences and radiations (Taylor et al. 2009). Spore fossils predate plant megafossils by 40 million years and give some clues to the pioneering events in land colonization some 470 Mya (Edwards et al. 2014; Brown et al. 2015).

Among the extant bryophytes, only the hornworts and some liverworts establish structurally well-defined symbioses with cyanobacteria. The cyanobionts are usually *Nostoc*, although also other cyanobacterial genera have been mentioned as symbionts of some bryophyte species (West and Adams 1997; Costa et al. 2001; Rikkinen and Virtanen 2008).

Liverwort Symbioses

Liverworts include ca. 5000 extant species in three classes. Among these, the Haplomitriopsida, including Treubiales with two genera and the monotypic Haplomitriales, is sister to all other taxa. The Blasiales with two monotypic genera is sister to the remaining complex thalloid liverworts, the Marchantiopsida, with ca.

340 extant species (Wahrmund et al. 2008). All the remaining liverworts belong to the diverse Jungermanniopsida, which includes both leafy and simple thalloid forms. Villarreal et al. (2016) estimated that the complex thalloid liverworts diverged 295 Mya (250–365 Mya) and the Marchantiidae (excluding Blasiales) 262 Mya (226–327 Mya).

Blasia pusilla and *Cavicularia densa* (Blasiales) are the only extant liverworts that form morphologically well-defined symbioses with cyanobacteria. The *Nostoc* cyanobionts of the thalloid liverworts occupy spherical auricles that develop on the underside of the thallus. The auricle starts its development from a three-celled mucilage hair, and *Nostoc* hormogonia enter the auricle when it is a small, dome-shaped structure. Concomitant growth of cyanobiont and auricle expansion eventually results in a larger, more or less globular structure that is easily visible even with the naked eye.

Auricles of *Blasia pusilla* have two slime papillae, with one (the inner slime papilla) partly filling the young auricle cavity and the other (the outer slime papilla) arising from the thallus adjacent to the auricle (Fig. 6). The cyanobacteria enter the auricles as hormogonia, lose their motility, and differentiate heterocysts. As a cyanobiont colony grows, filamentous protrusions derived from the inner slime papilla grow into the cyanobiont colony and increase the surface area of contact between the cyanobiont and the bryophyte, thus enhancing nutrient exchange between the host and symbiont (Duckett et al. 1977; Renzaglia 1982; Renzaglia et al. 2000; Kimura and Nakano 1990).

Blasia and *Cavicularia* reproduce asexually by producing gemmae which facilitate the simultaneous dispersal of the host and *Nostoc* cyanobionts. *Blasia* produces ovoid gemmae in bottle-shaped receptacles and stellate gemmae on the upper surface of the tips of thallus lobes (Fig. 6). *Cavicularia* produces both types of gemmae in crescent-shaped receptacles forming close to the tips of mature thallus lobes (Fig. 7). The ovoid gemmae of both liverwort species are nonsymbiotic and only associate with cyanobacteria when they have dispersed and develop into thallus primordia. However, the stellate gemmae essentially represent miniature thalli, equipped with two symbiotic auricles, which are regularly infected by symbiotic *Nostoc*, while the propagules are attached to the parent gametophyte (Rikkinen and Virtanen 2008).

During gametophyte growth, the symbiosis is continuously reestablished as young auricles are infected by *Nostoc* hormogonia. Concurrently, several different *Nostoc* genotypes are often present in single thalli. However, some *Nostoc* genotypes are dominant and widespread and typically shared by most bryophytes within a given locality (West and Adams 1997; Costa et al. 2001; Rikkinen and

Fig. 6 (continued) a recently infected young auricle with *Nostoc* propagating between the inner slime papilla (x) and the auricle wall (cells with chloroplasts). **L** Branched transfer cells (t) and detached *Nostoc* cells revealed by squashing a developing auricle. **M** Young basidioma of *Blasiphalia pseudogrisella* (Hymenochaetales) attached to rhizoids of *Blasia pusilla*. Several bottle-shaped receptacles (r) releasing ovoid, non-symbiotic gemmae have formed on the upper surface of the liverwort. Scale bars **A–D**, M 2 mm; **E–J** 10 μm; **E–J** 20 μm

Fig. 6 Cyanobacterial symbiosis of the liverwort *Blasia pusilla* (Blasiales). **A** Young gametophyte with stellate gemmae on upper surfaces of thallus lobes. Some cyanobacterial auricles on the lower surface of the liverwort can be seen as *dark spots* through the thin marginal sections of thallus lobes. **B** *Blasia pusilla* is a pioneer colonizer of clayey soils. **C** The stellate gemmae are essentially miniature thalli and tend to be infected by symbiotic *Nostoc* while being still attached to the parent gametophyte. **D** Mature auricle with symbiotic *Nostoc* on the lower surface of the gametophyte. **E–H** Series of optical cross sections through a young uninfected auricle in the meristematic region of stellate gemma; *n*=*Nostoc* hormogonia and trichomes outside the auricle, *a* auricle, *o* outer slime papilla, *x* inner slime papilla. **I** *Nostoc* trichome with terminal heterocyst in contact with outer slime papilla. **J** Optical cross section through young uninfected auricle showing the outer slime papilla (o) and inner slime papilla (x). The *Nostoc* hormogonia infect the auricle through the minute opening seen just left of the inner slime papilla. **K** Optical cross section (from above) of

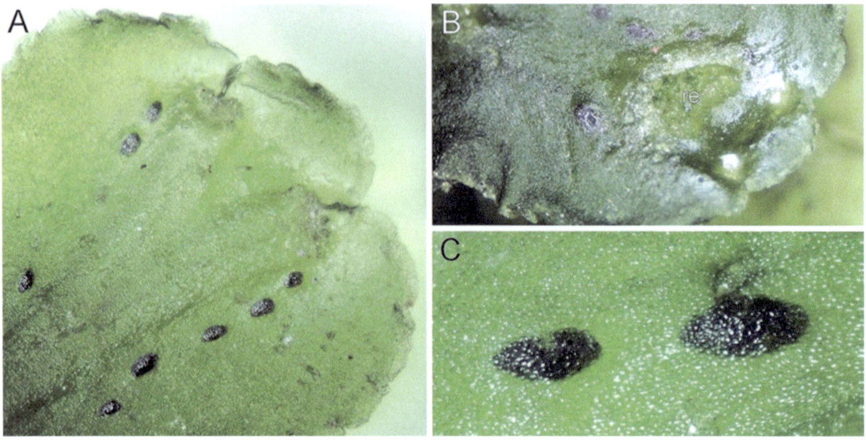

Fig. 7 Cyanobacterial symbiosis of the liverwort *Cavicularia densa* (Blasiales). **A** Tip of thallus lobe with cyanobacterial auricles on the lower surface visible as *dark spots* on the upper surface. **B** Mature receptacle (re) producing ovoid and stellate gemmae closely associated with *Nostoc* colonies of mature auricles. **C** Close-up of two mature auricles as seen through the upper surface of the thallus. Scale bars 2 mm

Virtanen 2008). The primary cyanobionts of *Blasia* and *Cavicularia* are closely related and the same *Nostoc* genotypes have also been found from terricolous cyanolichens (Rikkinen 2004, 2009, 2013; Rikkinen and Virtanen 2008; Papaefthimiou et al. 2008b).

The stellate gemmae of *Blasia* and *Cavicularia* can promote the persistence of specific cyanobacterial associations over the critical dispersal phase. During subsequent gametophyte development, new auricles are continuously formed. At this stage, some of the infecting cyanobacteria can represent suboptimal cyanobacteria from the soil (West and Adams 1997). This is not a serious problem, as only a few auricles infected by appropriate *Nostoc* genotypes are needed to support thallus growth. Hence, the cyanobacterial diversity of bryophyte thalli may more reflect difficulties in avoiding accidental infections, than the lack of symbiont preference (Rikkinen and Virtanen 2008).

Some leafy liverworts (Jungermanniopsida) including *Frullania* and *Porella* species regularly have cyanobacteria in the watersacs of their leaves (Dalton and Chatfield 1987). Cornejo and Scheidegger (2016) reported that *Frullania asagrayana* and the rare and endangered epiphytic cyanolichen *Erioderma pedicellatum* share identical *Rhizonema* cyanobionts. The liverwort is the principal substrate of *Erioderma* during its juvenile stage, and *Rhizonema* within the watersacs of liverwort leaves seems to facilitate the establishment and growth of the rare cyanolichen. Since *Rhizonema* has not been found "free-living," the availability and distribution of cyanobionts likely represent a critical limit to the establishment and recruitment of ascospore-dispersed cyanolichen species (Cornejo and Scheidegger 2016). The watersacs on some liverworts are also involved in other ecological interactions, including zoophagy of protozoa (Barthlott et al. 2000).

The basidiomycete *Blasiphalia pseudogrisella* (Hymenochaetales) forms appressoria on the rhizoids of *Blasia* (Fig. 6). The germinating spores of the fungus have also been observed to infect slime papillae in the meristematic region of liverwort gemmae. The fungus belongs to the *Rickenella* clade with other species that associate with bryophytes or green algae or are predacious on nematodes. *Blasiphalia* seems to only associate with *Blasia* and may be dispersed via infected gemmae of the liverwort host (Redhead 1981; Larsson et al. 2006).

In *Blasia* and *Cavicularia*, the slime papillae of symbiotic auricles play a crucial role in symbiosis by attracting hormogonia and developing into transfer cells that mediate metabolite exchange between the cyanobiont and the plant host. Also the slime papillae of other liverworts may be involved in symbiotic interactions with soil microbes (Fig. 8). Such associations have not been described in any detail, but casual observations reveal that the mucilage produced by slime papillae commonly supports a diversity of bacteria and filamentous fungi and some of them also infect the papillae. It seems likely that some of the associated bacteria are diazotrophic or in other ways beneficial to the plant.

In this context, one must emphasize that the production of slime papillae is an apomorphic feature that characterizes all extant liverworts. Slime papillae are ephemeral structures and typically produced in specific localities in growing liverworts where there is an obvious need for fixed nitrogen and/or other growth promoters, such as at the tips of growing thalli or leaves, in gemmiferous regions, around

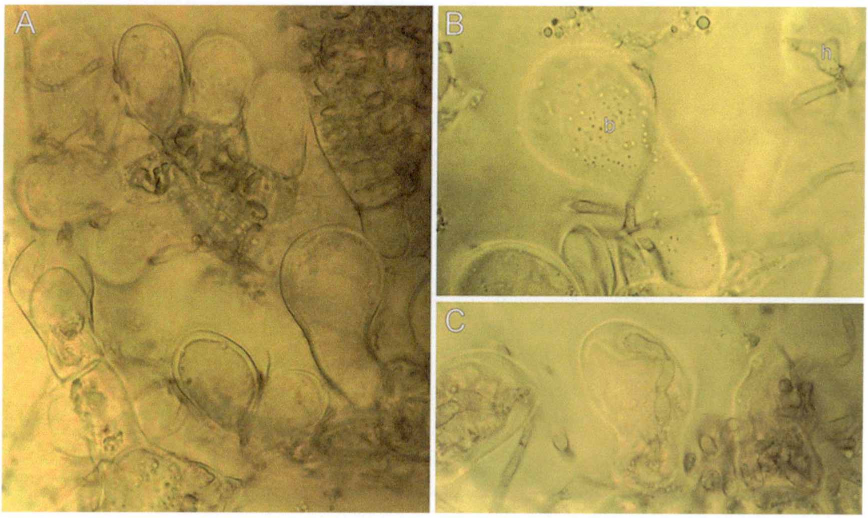

Fig. 8 Interactions between slime papillae of leafy liverwort (*Odontoschisma elongatum*, Cephaloziaceae) with bacteria and fungi. **A** Young uninfected slime papillae at margins and surface of reduced underleaves near the growing tip of the plant; the fragile slime papillae are rarely preserved in dry herbarium material. **B** Bacteria (b) and fungal hyphae attached to the surface of young slime papilla; eventually the fungus infects the papilla with haustoria (h). **C** At later stages the slime papillae become heavily infected by fungal hyphae and are eventually shed away. Scale bars 10 μm

gametangia, etc. They are also typically produced in locations where moisture conditions are favorable for prolonged microbial activity, like in between rhizoids on the lower surfaces of liverwort thalli, around water reservoirs of lobed leaves and watersacs, etc. Without roots and a vascular system, liverworts cannot effectively transport nutrients like vascular plants. Fixing nitrogen where and when it is needed and then shedding the slime papillae may be a very widespread strategy among this ancient lineage of minute land plants.

Among thalloid liverworts, associations with glomeromycete fungi are widespread in both the complex (Marchantiopsida) and simple (Jungermanniopsida, Metzgeriidae) lineages. However, these fungi appear to be absent from *Blasia* and *Cavicularia* as well as several derived lineages. The intracellular endophytes typically infect thalloid liverworts through the rhizoids and colonize particular taxon-specific regions of the thalli where they form trunk hyphae with numerous short-lived, regularly dichotomizing arbuscular side branches (Pressel et al. 2010). The swollen rhizoids in many families of leafy liverworts (e.g., Cephaloziaceae, Calypogeiaceae) can be packed with hyphae of the ascomycete *Rhizoscyphus ericae* (Leotiomycetes), which also forms ericoid mycorrhizas with vascular plants. Pressel et al. (2008) pointed out that the origins of the ascomycete associations in liverworts may date back to more than 250 Mya. Thus, they long predate the ericoid mycorrhizas which arose only 106–114 Mya.

Hornwort Symbioses

Hornworts (Anthocerophyta) include about 220 currently accepted species in 12 genera (Villarreal et al. 2015). Several recent phylogenies have placed hornworts as the closest extant relatives of the tracheophytes (Groth-Malonek et al. 2005; Qiu et al. 2006; Chang and Graham 2011), but lately this has again been challenged (Cox et al. 2014; Wickett et al. 2014). The monospecific *Leiosporoceros* is sister to all other hornworts and is currently placed in a separate class Leiosporocerotopsida (Villarreal et al. 2015). Among many other unique features, many hornworts have a chloroplast with a central pyrenoid and a carbon concentration mechanism unknown from other land plants (Li et al. 2009; Xue et al. 2010; Villarreal et al. 2013). They are widely distributed in moist temperate and tropical habitats, mainly as pioneer colonizers of nutrient-poor substrates (Fig. 9). The tropics have the highest diversity of species per area, particularly tropical Asia and the Neotropics (Villarreal et al. 2010; Villarreal and Renner 2014).

Cyanobacterial symbiosis with *Nostoc* is an universal feature in hornwort gametophytes. The association is established via apically derived clefts on the lower surface of the thallus. The cavities are connected to the thallus surface via stomata-like pores. When the *Nostoc* cyanobiont enters a young cleft, the middle lamella between hornwort cells separates to form a schizogenous space. As the cavity forms, it is filled with mucilage produced by surrounding plant cells (Rodgers and Stewart 1977; Renzaglia 1978; Adams 2002). Most hornworts produce mucilage

Fig. 9 Cyanobacterial symbiosis of hornworts (Anthocerophyta). **A** *Anthoceros agrestis* (Anthocerotales) is a pioneer colonizer of clayey soils. The horn-shaped sporophytes grow from archegonia embedded in the upper surface of the gametophyte. **B** Mature hornwort sporophytes split into two halves lengthwise, releasing the spores. **C** *Nostoc* colonies housed in mucilage clefts can be seen as *dark spots* on the lower surface of thallus. Scale bars 2 mm

clefts continuously during thallus growth and each of them becomes individually infected by *Nostoc* hormogonia. The presence of the cyanobiont stimulates enlargement of the symbiotic clefts and noninfected cavities do not develop further (Rodgers and Stewarts 1977). The mature cyanobiont colonies are usually small and more or less globular. However, in mature *Leiosporoceros* thalli, the *Nostoc* cyanobionts form long and branching strands that run parallel to the main axis of the thallus (Villarreal and Renzaglia 2006). Also young thalli of *Leiosporoceros* have mucilage clefts at the thallus apex and *Nostoc* invasion seems to take place through them. Later no more clefts are produced and the established cyanobacterial strands start to elongate in synchrony with thallus growth and branching (Villarreal and Renzaglia 2006).

As most hornworts continuously produce new ventral mucilage clefts near the thallus apex throughout their life span, it is likely that multiple cyanobacterial genotypes will invade different clefts of a single hornwort thallus. The presence of multiple cyanobacterial genotypes within a single hornwort thallus have indeed been confirmed using molecular markers (West and Adams 1997; Costa et al. 2001). Even young *Leiosporoceros* thalli typically have many lobes, each with several mucilage clefts, and this can lead to cyanobiont variability within single mature thalli (Villarreal and Renzaglia 2006). The development of integrated cyanobiont network in *Leiosporoceros* eliminates the need for multiple cyanobacterial invasions during thallus growth. The canals also allow extensive surface contact for exchange between the partners along the length of the strand. A single invasion of the hornwort thallus can be adaptive as it eliminates the need to constantly attract new, appropriate symbionts. On the other hand, multiple invasions could more effectively ensure successful colonization and allow for more flexibility in cyanobiont choice (Villarreal and Renzaglia 2006).

Hornworts establish mycorrhizal symbioses with both Glomeromycota and Mucoromycotina fungi, often simultaneously (Ligrone 1988; Schüßler 2000; Desirò et al. 2013a; Bidartondo et al. 2011). Interestingly, the principal mode of fungal entry in the hornworts is via mucilage clefts and not through rhizoids as in most liverworts. Desirò et al. (2013a) noted that the fungal hyphae occur in close association with symbiotic *Nostoc* within the mucilage, possibly indicating a relationship resembling that of *Geosiphon*. They also found that the more abundant the cyanobacteria, the less likely the hornworts were to harbor mycorrhizal fungi. For example, the thalli of *Leiosporoceros* were found to be fungus-free (Desirò et al. 2013a).

Moss Symbioses

The mosses (Bryophyta) are the largest group of bryophytes consisting of some 10,000 species in three or four distinct clades usually recognized as classes. Although the order of diversification is still controversial, it is clear that the peat mosses (Sphagnopsida) with four genera and the monogeneric Takakiopsida are sister to all other mosses, followed by the Andreaeopsida with two genera and then the remainder of more advanced orders (Cox et al. 2010).

Many mosses regularly house epiphytic cyanobacterial colonies on their surfaces, especially in sheltered spots between leaves and rhizoids (Fig. 10). While most of these associations are undoubtedly facultative, they definitely play an important role in N_2 fixation, especially in boreal and arctic and in some temperate ecosystems (Solheim et al. 1996; Zackrisson et al. 2004; Gavazov et al. 2010; Lindo and Whiteley 2011; Turetsky et al. 2012; Rousk et al. 2013; Arróniz-Crespo et al. 2014). Feather mosses in boreal forests commonly house heterocystous cyanobacteria (e.g., *Nostoc*, *Calothrix*, *Stigonema*) in addition to other N_2-fixing bacteria (Houle et al. 2006; Cornelissen et al. 2007; Ininbergs et al. 2011). In peat bogs, cyanobacteria are often present in the dead and water-filled hyaline cells in the leaves and stems of *Sphagnum*. With an elevated pH, the hyaline cells may represent a favorable microhabitat for cyanobacteria in acidic peatlands (Berg et al. 2013; Kostka et al. 2016). Lindo et al. (2013) concluded that bryophyte–cyanobacterium associations carry out several important functions in nitrogen-limited boreal and arctic ecosystems through their production of recalcitrant litter, thermal protection of soils, and role as the primary source of nitrogen through N_2 fixation. The composition of nitrogen-fixing moss-associated cyanobacterial communities differs between moss species, sites, and seasons, but the exact patterns remain largely unexplored (Ininbergs et al. 2011; Bay et al. 2013; Leppänen et al. 2013; Warshan et al. 2016).

While many specialized fungi grow on mosses (Döbbeler 1997; Davey and Currah 2006; Kauserud et al. 2008) and many of them appear to be species-specific (Higgins et al. 2007; Stenroos et al. 2010), mutualistic symbioses between mosses and fungi have not been found (Pressel et al. 2010). Thus, all fungal associates of mosses seem to be either parasites or saprotrophs, and mosses represent the only division of land plants that does not seem to establish mycorrhizal symbioses with

Fig. 10 Cyanobacterial symbioses of mosses (Bryophyta). **A** *Sphagnum riparium* (Sphagnales) and other peat mosses can house cyanobacteria in the water-filled hyaline cells of their leaves and stems. **B** Moss-associated cyanobacteria play an important role in biological N_2 fixation in arctic and boreal regions but also in montane rain forests. **C** *Hylocomium splendens* (Hypnales) and other feather mosses commonly have heterocystous cyanobacteria and other N_2-fixing prokaryotes in sheltered spots between their leaves. Scale bars 2 mm

Glomeromycota. This is quite peculiar considering the early divergence and wide extant diversity of mosses (Liu et al. 2011; Taylor et al. 2015a).

Azolla Symbiosis

The genus *Azolla* (Salviniales, Pteridophyta) contains seven extant species of small aquatic ferns (Evrard and Van Hove 2004; Metzgar et al. 2007; Pereira et al. 2011). Several species are widely distributed in tropical and warm temperate regions (Fig. 11). The genus has an extensive fossil history starting from the Cretaceous based principally on isolated megaspores (Hall 1974; Collison 1980; Taylor et al. 2009).

Fig. 11 Cyanobacterial symbiosis of the water fern *Azolla* (Salviniales). **A** *Azolla filiculoides* is native to warm temperate and tropical parts of the Americas and Australasia and has become invasive in several regions including western Europe and Africa. Scale bar 5 mm. **B** *Azolla rubra* is widespread throughout the Pacific and in southern Asia. In New Zealand, it is common in shallow water bodies such as ponds, lake margins, and slow-flowing streams and also grows in swamps on muddy ground. **C** The introduced *Azolla filiculoides* (*Azolla* section *Azolla*) and the larger native *Azolla nilotica* (*Azolla* section *Rhizosperma*) growing together on the muddy shore of Lake Naivasha, Kenya

Azolla has branching rhizomes with adventitious roots and alternate bilobed leaves (Fig. 11). The transparent ventral lobe of the lead helps the plant to float, whereas the dorsal lobe is photosynthetic and contains a cavity with symbiotic cyanobacteria. In mature *Azolla* plants, the cyanobionts are located in the periphery of the leaf cavity in mucilage between internal and external envelopes (Adams et al. 2013). The adaxial epidermis of the leaf cavity contains a pore with an opening that is larger in younger leaves. The growth of the cyanobionts is coordinated with the growth of the host plant. The apical meristem of each branch contains a colony of undifferentiated cyanobionts. Cyanobacteria from the colony are introduced into the leaf primordium before the development of the leaf cavity is complete. During aging, the cyanobionts show decreases in cell division, increases in the size of vegetative cells, and an increase in heterocyst frequency up to 20–30 % in mature leaves (Adams et al. 2013). Simple hair cells are involved in the transport of sugars (sucrose) from the photosynthetic mesophyll cells of *Azolla* to the leaf cavity. Also, primary branched hair cells with transfer cell morphology may be involved in nutrient transfer. Ekman et al. (2008) studied the proteomics of the *Azolla* cyanobiont and found that processes related to energy production, nitrogen and carbon metabolism, and stress-related functions were upregulated compared with free-living cyanobacteria, whereas photosynthesis and metabolic turnover rates were downregulated. In addition to the primary cyanobiont, the leaf cavities of *Azolla* regularly house diverse communities of other bacteria which may also play a role in the symbiosis (Nierzwicki-Bauer and Aulfinger 1991; Lechno-Yossef and Nierzwicki-Bauer 2002; Zheng et al. 2008).

The *Azolla* cyanobionts retain their association with the fern throughout its life cycle. The vertical transfer of the cyanobiont to new generations is facilitated by the female reproductive structure (megasporocarp) of the heterosporous fern. During colonization of the megasporocarp, cyanobiont hormogonia enter through pores at the top of the indusium and then differentiate into akinetes in a synchronized manner. The details of this process are described in Zheng et al. (2009). Papaefthimiou et al. (2008b) and Sood et al. (2008) studied cyanobacterial diversity in *Azolla* and found different cyanobacterial genotypes from different species and also diversity within a single *Azolla* species. The cyanobacteria differed most markedly between members of the two sections of the genus *Azolla*.

Ran et al. (2010) provided evidence that there has been selective streamlining of the primary cyanobiont genome which has resulted in an organism devoted to nitrogen fixation and devoid of autonomous growth. This has led to marked loss of function within gene categories for basic metabolic processes such as glycolysis, replication, and nutrient uptake. Phylogenetic analysis grouped the *Azolla* cyanobiont (*Nostoc azollae* 0708) with *Raphidiopsis* and *Cylindrospermopsis*, which have the smallest known genomes among multicellular cyanobacteria (Stucken et al. 2010). Other molecular characteristics indicate a close phylogenetic relationship to *Nostoc* and *Anabaena* strains (Baker et al. 2003; Papaefthimiou et al. 2008b; Ran et al. 2010). For more information on the identity of the *Azolla*-associated cyanobacteria, the reader is referred to recent reviews by Adams et al. (2013) and Pereira and Vasconcelos (2014).

The *Azolla* symbiosis is of considerable economic importance (Brouwer et al. 2014). The symbiotic plants are used as a nitrogen-rich biofertilizer especially in rice paddies and/or fertilization of fields. Owing to its high protein content, the plants can also be used as a fodder for pigs, ducks, and other domestic animals. Finally, due to its rapid growth and floating lifestyle, *Azolla* can also be used to remove nitrates, phosphorous, and heavy metals from polluted water. On the other hand, the mat-forming and easily spreading plants have also become noxious weeds in many tropical and temperate regions (Fig. 11).

Cycad Symbioses

Cycads (Cycadales) are an ancient lineage of evergreen, palmlike gymnosperms that can be traced back to the Paleozoic (Fig. 12). The oldest cycad fossils are from the Pennsylvanian and the general morphology of the group has remained relatively unchanged since then (Taylor et al. 2009). Cycads and *Ginkgo* diverged in the late Carboniferous or early Permian, and the most recent common ancestor of the living cycads lived in the late Permian. They diversified and became more abundant in the Triassic when primitive seed-plant floras were replaced by conifers and cycads after the Permian–Triassic mass extinction. Cycads reached a diversity peak during the Jurassic and remained relatively stable in terms of diversity during the Cretaceous when the extant lineages were established. Although all extant cycad lineages are

Fig. 12 Cyanobacterial symbiosis of cycads (Cycadales). **A** Over 100 species of *Cycas* (Cycadaceae) are widely distributed in southern Asia and Australasia. **B** *Nostoc* cyanobionts of all cycads are housed within the cortex of specialized coralloid roots

ancient, the modern species do not seem to be much older than 12 million years (Nagalingum et al. 2011; Salas-Leiva et al. 2013; Xi et al. 2013; Condamine et al. 2015; Silvestro et al. 2015).

Cycads are insect pollinated and their early diversification may have been linked to the cycad insect pollinator of the order Thysanoptera originating and diversifying in the late Permian. While some cycad–insect interactions may be ancient, this does not seem to be the case for all associations between cycads and beetles (Schneider et al. 2002; Terry et al. 2007; Downie et al. 2008; Peñalver et al. 2012). Dinosaurs have been proposed as key dispersers of cycad seeds during the Mesozoic, and temporal variation in cycad diversity and abundance has been linked to faunal changes. However, when assessing the fossil evidence, Butler et al. (2009) could not find unequivocal support for coevolutionary interactions between cycads and herbivorous dinosaurs.

Extant cycads include ca. 300 species that are classified in 12 genera in three families, widely but patchily distributed in tropical and subtropical regions of the Americas, Africa, Southeast Asia, and Australia (Wang and Ran 2014). Most cycads have a stout trunk with a large crown of tough leaves and can vary in height from 20 cm to almost 20 m at maturity and a thick taproot that can extend many meters into the soil. They also produce lateral roots, some of which develop into specialized coralloid roots that house the cyanobacterial symbionts (Fig. 12). The coralloid roots grow sideways or upward toward the soil surface. The cycad–cyanobacterium symbiosis is still the only known example of a naturally occurring plant root–cyanobacterium symbiosis in plants, and the ability of many cycads to thrive in nutrient-poor soils is generally attributed to the cyanobacterial symbiont (Lindblad 1990, 2009; Costa et al. 1999; Adams et al. 2013).

Symbiotic cyanobacteria are not present in precoralloid roots of cycad seedlings, but their presence is required for further development into coralloid roots (Lindblad 2009; Adams et al. 2013). Precoralloid roots likely release chemicals that induce hormogonium formation in and act as chemoattractants for symbiotic cyanobacteria (Ow et al. 1999). The invasion of plant tissue is thought to occur through apical lenticels and/or injured epidermal cells, and also other soil microbes may be involved in the process (Nathanielsz and Staff 1975; Lobakova et al. 2003; Lindblad 2009). Once inside the root, the cyanobacteria migrate inward through the outer cortex and eventually establish a well-defined zone between the inner and outer cortices of the coralloid root. Specialized plant cells within the cyanobacterial zone facilitate the transfer of nutrients between the symbiotic partners (Nathanielsz and Staff 1975; Ahern and Staff 1994; Lindblad 2009; Adams et al. 2013).

As the coralloid roots of most cycad species are deep beneath the soil surface, the cyanobionts live in complete darkness. However, they retain a full photosynthetic apparatus, associated pigments, and carbon-fixing potential (Lindblad et al. 1985). Nitrogenase activity is several folds higher in cycad-symbiotic *Nostoc* than in free-living forms and increases with increasing heterocyst frequency, until reaching a maximum at heterocyst frequencies of around 25–35 % (Lindblad et al. 1991). There is a developmental gradient from comparatively low heterocyst frequency in the growing tips of the coralloid roots to very high (up to 46 %) in the older parts of the roots. Many heterocysts in the older parts are inactive and this explains the decrease in nitrogenase activity of aging coralloid roots (see Lindblad 2009 and Adams et al. 2013 for further details).

The cyanobionts of cycads are usually *Nostoc*, but also *Calothrix* has sometimes been found. Cycads can house multiple *Nostoc* genotypes in single plants as well as in single roots, and there seems to be little specificity between cycad species and their cyanobionts (Lindblad et al. 1989; Grobbelaar et al. 1987; Costa et al. 1999, 2004; Zheng et al. 2002; Gehringer et al. 2010; Thajuddi et al. 2010; Yamada et al. 2012). Most symbiotic *Nostoc* genotypes reported from cycads are not identical to those typically found in thalloid bryophytes and/or lichen-forming fungi and are more closely related to some free-living *Nostoc* genotypes (Rikkinen 2004; Rikkinen and Virtanen 2008; Gehringer et al. 2010). However, the cyanobionts within cycad roots may differ markedly from those found in the surrounding soil (Cuddy et al. 2012).

Associations Between Angiosperms and Cyanobacteria

As pointed out by Osborne and Bergman (2009), structurally well-defined symbioses between angiosperms and nitrogen-fixing bacteria are relatively uncommon in nature. The *Gunnera–Nostoc* symbiosis remains the only known symbiosis between angiosperms and cyanobacteria. It differs markedly from other plant–cyanobacterium associations in being a true endosymbiosis with *Nostoc* cyanobionts housed within the cells of the host plant.

Gunnera Symbiosis

All species of *Gunnera* establish endosymbiotic associations with *Nostoc* (Bergman 2002; Osborne and Sprent 2002; Khamar et al. 2010; Fernández-Martínez et al. 2013). The genus includes ca. 60 species of small, medium, or sometimes huge herbs with spiral leaves that have wide blades with prominent palmate venation and toothed margins (Fig. 13). They also have stipule-like "scale leaves" and relatively large and branched inflorescences bearing numerous small and inconspicuous, almost exclusively wind-pollinated flowers. Most species are perennial and some can reach a considerable age; the only annual species is sister to the rest of the genus (Wanntorp et al. 2001, 2003; Wanntorp and Ronse De Craene 2005; Rutishauser et al. 2004).

The order Gunnerales occupies an isolated position among the core eudicots and the divergence from the stem lineage occurred already ca. 115 Mya (Soltis et al. 2003; Magallón and Castillo 2009; Vekemans et al. 2014). In addition to the genus *Gunnera* (Gunneraceae), the lineage only includes the genus *Myrothamnus* (Myrothamnaceae) with two species of small xerophytic shrubs in southern Africa and Madagascar (Apg 2016). The natural range of extant *Gunnera* species includes South America, Africa, and the Australasian region, with some species reaching Hawaii and southern Mexico in the north (Wanntorp and Wanntorp 2003). This range indicates that the plants were widely distributed in Gondwana before its final breakup (Wanntorp and Wanntorp 2003).

Gunnera also has a long fossil history based on its characteristic pollen going back to the Early Cretaceous. Fossil pollen has been identified from all continents of the Southern Hemisphere but also from India and North America, as well as sediments in the Indian and South Atlantic Oceans (Wanntorp et al. 2004a, b). Several

Fig. 13 Cyanobacterial symbiosis of *Gunnera* (Gunnerales). **A** *Gunnera manicata* is originally native to Brazil and Columbia but now widely grown as a garden ornamental in temperate and tropical regions. **B** The growing tips of the rhizomes of mature *Gunnera* plants are well protected by scale leaves

species are cultivated as garden ornamentals, and especially *Gunnera tinctoria*, originally a native of Chile, has naturalized and become invasive in several regions with humid temperate climates, including Ireland, the UK, the Azores, the USA, and New Zealand (Osborne and Sprent 2002; Gioria and Osborne 2009, 2013; Fennell et al. 2013, 2014).

The *Nostoc* cyanobionts of *Gunnera* are maintained in the stem cortex of the rhizome of larger species and the stolon of smaller species (Bergman 2002; Osborne and Sprent 2002; Khamar et al. 2010). Special glands are formed on the stem immediately below the leaves. The gland is formed by up to nine papillae surrounding one central papilla. The papillae secrete thick mucilage which induces hormogonium differentiation in *Nostoc*. Between the papillae, and leading into the stem tissue, are deep channels through which the mucilage is released. The mucilage attracts hormogonia between the papillae and further into the interior of the gland where appropriate *Nostoc* genotypes are taken into plant cells (Söderbäck and Bergman 1993; Johansson and Bergman 1994; Rasmussen et al. 1994, 1996; Uheda and Silvester 2001).

Within *Gunnera* cells, the *Nostoc* hormogonia develop into filaments with heterocysts. The cyanobiont proliferates and the infected plant cells divide repeatedly to form internal colonies within the stem cortex. The glands are the only known entry point for *Nostoc* hormogonia into the cortical tissue of *Gunnera*. Along the rhizomes, the cyanobionts occur as well-defined colonies and show different stages of development, indicating that they were each formed through successive and separate infection processes involving different glands. Within the plant cells, the cyanobionts are surrounded by the host cell plasmalemma and the membrane acts as the interface through which the exchange of metabolites takes place. Although only young glands can incorporate *Nostoc* hormogonia, new glands continue to develop at the base of each new leaf when the stem grows. Gland development in *Gunnera* also takes place in the absence of symbiotic cyanobacteria, but nitrogen limitation seems to be a prerequisite for their development (Bergman 2002; Wang et al. 2004; Chiu et al. 2005; Osborne and Bergman 2009; Khamar et al. 2010; Adams et al. 2013).

Molecular studies have confirmed that the cyanobionts of *Gunnera* belong to *Nostoc* and are closely related to some symbiotic and free-living genotypes of the genus (Rasmussen and Svenning 2001; Svenning et al. 2005; Papaefthimiou et al. 2008a). There is considerable phenotypic and genotypic variation among the cyanobionts (Bergman et al. 1992). Several different *Nostoc* genotypes may sometimes be present in a single *Gunnera* plant (Nilsson et al. 2000), but in most cases, each plant has housed only one *Nostoc* genotype (Guevara et al. 2002). Fernández-Martínez et al. (2013) found no genetic variability among *Nostoc* cyanobionts within single *Gunnera* plants, while the cyanobionts of neighboring plants could be markedly different. The *Nostoc* genotypes were closely related to those found in terricolous cyanolichens and bryophytes. *Nostoc* isolates from cycads and bryophytes also readily invade *Gunnera* cells and vice versa (Adams et al. 2013).

The evolutionary origin of the *Gunnera* glands is a mystery. The bright red glands form at right angles to successive petioles and are clearly visible at the developing cotyledon (for images, see Osborne and Bergman 2009 and Adams et al. 2013). The gland originates internally from the lateral meristem and eventually ruptures the epidermis, exposing the surface of epidermal cells. Osborne and Bergman (2009) pointed out that at later stages of development, the glands are often closely associated with adventitious root formation. The glands in young seedlings are initiated prior to shoot emergence and are evident as a dome-shaped structure with red (anthocyanin) pigmentation. A similar pigmentation is also present in the tips of developing adventitious roots, suggesting that the glands might evolve as highly modified adventitious roots. In any case, the specialized morphology and sugar accumulation in the glands of *Gunnera* seedlings are crucial for early stages in the establishment of the *Nostoc* endosymbiosis.

Khamar et al. (2010) investigated changes in the carbohydrate metabolism during *Gunnera* gland development and discovered that plant cells within the mature glands accumulated high levels of soluble sugars (including high concentration of glucose and fructose and lower concentration of sucrose) prior to the arrival of cyanobacteria and that the various sugars affected the *Nostoc* cyanobionts' ability to form motile hormogonia. They concluded that in a nitrogen-limiting environment, *Gunnera* glands secrete mucilage with minute of soluble sugars, which attract *Nostoc* to the gland surface but does not interfere with hormogonium formation. The hormogonia penetrate into the gland and enter cells at the base of the gland presumably because they are attracted by the high levels of glucose and fructose inside the cells, while within the plant cells, further hormogonia are not formed because of the high levels of soluble sugars (Khamar et al. 2010).

The accumulation of soluble sugars in the cells of mature but still noninfected *Gunnera* gland indicates that the nonsugar-secreting structure could have originally evolved from a sugar-secreting gland functionally similar but not necessarily homologous to extrafloral (EF) nectaries of other eudicots (Marazzi et al. 2013). Also the conspicuous red pigmentation of the glands in *Gunnera* seedlings might play a role in attracting flying insects capable of vectoring symbiotically compatible *Nostoc* genotypes from the decomposing old rhizomes of well-established *Gunnera* colonies to the germinating seedlings. While the mucilage secreted by the glands largely consists of nondigestible carbohydrates (Rasmussen et al. 1994), with barely detectible levels of soluble sugars (Khamar et al. 2010), the gland tissue itself is rich in soluble sugars that may well attract insects. The possible role of insect vectors in the dispersal ecology of the *Nostoc–Gunnera* symbiosis should be elucidated in field studies within the natural range of *Gunnera* species.

It is possible that the original function of *Gunnera* glands could have been comparable to those of EF nectaries that are produced on the leaves, stems, and virtually any other parts of thousands of plant species. These structures have been described for nearly 4000 vascular plant species representing nearly 750 genera in over 100 families (Weber and Keeler 2013). Their primary function is to attract aggressive and often mutualistic insects, particularly ants which, while foraging for nectar, protect the host plant from herbivorous animals (Heil et al. 2004; Villamil et al. 2013; Weber and Agrawal 2014). Besides ants, the EF nectaries of different plants attract a wide variety of other arthropods including many Diptera. Some of them also provide visual cues for foraging arthropods and also the emission of odors can attract mutualistic insects (Koptur 1992).

Weber and Keeler (2013) estimated that extrafloral nectaries have evolved independently at least 457 times in vascular plants. Even among the ferns, extrasoral nectaries have arisen independently several times, and they first evolved long before floral nectar and long before ants appeared in the fossil record (Marazzi et al. 2012; Koptur et al. 2013). Therefore, their original function could not have involved ants or other social hymenopterans. Whatever the primary role, the sugar secreted has impacted the diversification of several groups on arthropods and fungi. For example, several lineages of sooty molds have specialized to use plant exudates and insect honey dew for nutrition (Rikkinen et al. 2003; Schmidt et al. 2014).

According to the EF nectary hypothesis, the secretions of ancestral *Gunnera* glands would have been involved in an indirect anti-herbivore defense by attracting and hence increasing the presence of putative plant defenders on the plant stem. Thus, the mutualism between the ancestral *Gunnera* species and the putative (facultative) insect associates would represent the original state, whereas the association between the plant and *Nostoc* would have derived later and then evolved into an ecologically obligate cyanobacterial symbiosis.

Interestingly, Campbell et al. (2013) reported that some epiphytic cyanolichens on conifers may utilize glucose excreted by the EF nectaries of neighboring angiosperms. They proposed that the exogenous source of carbon could help to sustain the lichen mycobionts during drought-induced periods of inactivity of their cyanobacterial symbionts.

Other Angiosperms

Many angiosperms regularly house epiphytic cyanobacteria on their stems, leaves, or root surfaces. While most of these associations are probably accidental and of no particular significance to the plants, in some cases, N_2 fixation of associated cyanobacteria might be significant. For example, cyanobacteria including *Nostoc* and *Calothrix* are commonly seen on the exposed roots of tropical epiphytic orchids (Fig. 14), and sometimes cyanobacteria and associated fungi can form sheaths around aerial roots (Tsavkelova et al. 2003). As the cyanobacteria are capable of N_2 fixation, they may benefit the plant and/or the endomycorrhizal fungi that live within the specialized velamen of the aerial roots (Teixeira da Silva et al. 2015).

Establishment of artificial associations between nitrogen-fixing microorganisms and agricultural crop plants could potentially reduce the demand for chemically produced nitrogen fertilizers (Gusev et al. 2002; Prasanna et al. 2009). *Nostoc* hormogonia are known to be attracted to root extracts of many nitrogen-deprived vascular plants (Fig. 14), including rice and wheat (Gantar et al. 1995; Nilsson et al. 2002, 2005). Floating cyanobacterial communities on the soil–water surface are widely used in agriculture as nitrogen-supplementing biofertilizers in rice paddy fields, especially in Southeast Asia, and nitrogen accretions of 25–30 kg N/ha have been attributed to their N_2 fixation (Prasanna et al. 2009).

Fig. 14 Cyanobacterial associations of tracheophytes. **A** Heterocystous cyanobacteria and other N_2-fixing prokaryotes often grow on and within the aerial roots of tropical orchids, with presumed benefits to the plants or their mycorrhizal fungi. **B** *Nostoc* hormogonia are attracted to nitrogen-deprived vascular plants. Here lichen-symbiotic *Nostoc* isolated from a *Peltigera* species and cultured on nitrogen-deficient medium (Z8) form an artificial association with a somatic embryo plant of Scots pine (*Pinus sylvestris*)

Early Evolution of Terrestrial Symbiotic Systems

Through endosymbiosis, certain cyanobacteria evolved into the plastids of photoautotrophic eukaryotes some 900 million years ago (Deusch et al. 2008; Dagan et al. 2013; Shih and Matzke 2013; Ochoa de Alda et al. 2014; Ruhfel et al. 2014; Ku et al. 2015). Already much earlier cyanobacteria evolved oxygenic photosynthesis and changed the Earth's atmosphere from anoxic to oxic. As nitrogenase is highly sensitive to oxygen, most N_2-fixing prokaryotes became confined to anoxic environmental niches. Cyanobacteria themselves evolved several strategies to protect nitrogenase from oxygen, including a temporal separation of oxygenic photosynthesis and N_2 fixation in unicellular forms and the differentiation of a specialized cell, the heterocyst, in some filamentous forms. The sensitivity of nitrogenase to oxygen may also have enhanced the early evolution of cyanobacterial symbioses, as many symbiotic structures represent micro-aerobic environments that help to protect cyanobacterial cells from atmospheric oxygen (Rikkinen 2002).

It is possible that the evolution of some cyanobacterial symbioses correlated temporally with geological periods of increased oxygen in the atmosphere. High atmospheric oxygen levels (over 30 %) were experienced during the Carboniferous, and this may have seriously interfered with nitrogen fixation and given adaptive value to any structures and mechanisms that helped to isolate N_2-fixing cells from the atmosphere. Hence the early evolution of some symbiotic structures, like the coralloid roots of cycads, could parallel the development of arthropod and amphibian gigantism, which was directly facilitated by the hyperoxic atmosphere (Payne et al. 2009, 2011, 2012). Unfortunately, no fossilized roots of early cycads have yet been identified (Taylor et al. 2009).

The mycelia of mycorrhizal fungi associate with soil bacteria, many of which are capable of N_2 fixation (Scheublin et al. 2010; Valverde et al. 2016). Such associations probably experienced a triumph in the late Permian, when soil bacteria, following the atmospheric transition to hypoxia, became more efficient in nitrogen fixation. While giant amphibians and arthropods disappeared, the cyanobacterial symbioses did not and the reduction in atmospheric O_2 levels did not directly threaten them. However, the competitive edge and ecological significance of cyanobacterial N_2 fixation may have started to dwindle when diffuse rhizosphere associations with soil bacteria became a more viable option for nitrogen-starved plants (Rikkinen 2002). Rhizosphere associations between fungi and bacteria must have promoted the evolution of mycorrhizal symbioses and the subsequent diversification of seed plants. The radiation of modern ferns, bryophytes, fungal endophytes, and lichens was in turn facilitated by the diversification of seed plants (Schneider et al. 2004; Schuettpelz and Pryer 2009; Feldberg et al. 2014; Li et al. 2014; Lóriga et al. 2014; Sundue et al. 2016).

Thalloid liverworts, hornworts, and cycads together represent a rather conspicuous proportion of all extant plant lineages that can be reliably traced back to pre-Permian times. Also *Azolla* and *Gunnera* can be traced back to the Cretaceous (Taylor et al. 2009). Also some lineages of extant cyanobacterial lichens are probably ancient. The oldest fossil of a stratified cyanolichen is from the Lower Devonian (Honegger et al. 2012). While very few Paleozoic or Mesozoic lichen fossils have so far been found, many amber fossils demonstrate that many extant genera of lichen-forming fungi had already evolved by the Tertiary (Rikkinen and Poinar 2002, 2008; Kaasalainen et al. 2015; Lumbsch and Rikkinen 2016). The initial diversification of the Pezizomycotina (Ascomycota) occurred already in the Ordovician, followed by repeated splits of lineages throughout the Phanerozoic (Beimforde et al. 2014).

The earliest fungal symbioses may have in many respects resembled some extant crustose cyanolichens, especially salt-tolerant forms that live on rocky ocean shores. Littoral habitats along ancient shorelines brought cyanobacteria, green algae, and fungi into close contact under conditions where there were both a need and opportunity for establishing new symbiotic interactions (Rikkinen 2002; Edwards and Kenrick 2015). The earliest lichen-like symbioses probably evolved long before the initial evolution of mycorrhizal symbioses, the subsequent rise of vascular plants, and the later diversification of parasitic and saprophytic fungi. Accordingly, some modern cyanolichens may preserve biological features that closely mirror those that evolved in similar cyanobacterial symbioses at early stages of terrestrial evolution.

The fact that most vascular plants, liverworts, and hornworts include species that form symbioses with AM fungi indicates that also these symbioses evolved during very early stages in the evolution of land plants. The AM fungi have a definite fossil record extending to the early Devonian ca. 400 Mya, which is before early vascular plants had acquired roots (Kenrick and Strullu-Derrien 2014; Strullu-Derrien et al. 2014), and the oldest fossil spores of possible AM fungi date back to the Ordovician ca. 460 Mya (Taylor et al. 2015b). Molecular data indicate that glomeromycetes have always been symbiotic and suggest that the first mycorrhizae may have involved mucoromycetes, the sister lineage glomeromycetes (Bidartondo et al. 2011; Strullu-Derrien et al. 2014). The mycobiont of an early lichen-like symbiosis from the Early Devonian with nonseptate hyphae may have belonged to one of these groups (Taylor et al. 1997, 2015b), and already the common ancestor of Glomeromycota and Mucoromycota was probably capable of establishing symbioses with cyanobacteria and/or early photosynthetic eukaryotes (Delaux et al. 2013, 2015; Field et al. 2015a, b; Tang et al. 2016).

The *Geosiphon pyriformis–Nostoc* endosymbiosis may represent one living example of early symbioses between glomeromycetes, cyanobacteria, and bryophytes (Schüßler and Walker 2011; Schüßler 2012). Such symbioses could be seen as "evolutionary precursors" to more modern types of AM symbioses (Schüßler and Walker 2011; Selosse et al. 2015; Wojtczak and Janik 2016). It is noteworthy that a symbiotic cyanobacterium has been identified from the cortex of *Aglaophyton major*, one of the earliest early land plants in the Devonian (Taylor and Krings 2005). Somewhat resembling the situation seen in modern hornworts, the filamentous cyanobacterium seems to have entered its host via stomatal pores, colonized the substomatal chambers, and then spread throughout the outer cortical tissue, where it has been preserved in fossil specimens.

Concluding Remarks

Symbiotic interactions between cyanobacteria, plants, and fungi have evolved repeatedly during the course of evolution. A trend toward versatile and diffuse associations with non-photosynthetic bacteria may have reduced the relative importance of cyanobacterial symbioses during fungal and vascular plant evolution. However, cyanolichens are still an important component of biodiversity in many terrestrial ecosystems and have provided unique opportunities for the diversification of both symbiotic cyanobacteria and fungi. During their evolution, many lichen-forming fungi may have established associations with plant-symbiotic *Nostoc* genotypes and/or vice versa, leading to repeated cyanobiont switches between terricolous lichens, thalloid bryophytes, cycads, and/or *Gunnera*. Likely these cyanobacterial symbioses have also had many diffuse effects on mycorrhizal and plant–insect mutualisms. Clearly we are only beginning to unravel the complex network of biological interactions and evolutionary processes in which symbiotic cyanobacteria have evolved. DNA techniques offer accurate methods for studying the genetic diversity of symbiotic cyanobacteria in situ and for making comparisons between lichen cyanobionts and their nonsymbiotic relatives. In the future, we can expect many interesting new findings from previously ignored cyanobiont lineages and unexplored ecological settings. They can add significantly to what we already know about symbiotic cyanobacteria and their diverse hosts.

References

Adams DG (2002) Cyanobacteria in symbiosis with hornworts and liverworts. In: Rai AN, Bergman B, Rasmussen U (eds) Cyanobacteria in symbiosis. Kluwer, Dordrecht, pp 117–135

Adams DG, Duggan PS (2008) Cyanobacteria-bryophyte symbioses. J Exp Bot 59:1047–1058. doi:10.1093/jxb/ern005

Adams D, Bergman B, Nierzwicki-Bauer S et al (2013) Cyanobacterial-plant symbioses. In: Rosenberg E, DeLong E, Lory S et al (eds) The prokaryotes. Springer, Berlin, pp 359–400

Ahern CP, Staff IA (1994) Symbiosis in cycads: the origin and development of coralloid roots in *Macrozamia communis* (Cycadaceae). Am J Bot 81:1559–1570

Apg IV (2016) An update of the Angiosperm Phylogeny Group classification for the orders and families of flowering plants: APG IV. Bot J Linn Soc 181:1–80

Aptroot A, Schumm F (2009) Chimeras occur on the pantropical lichinomycete *Phyllopeltula corticola*. Lichenologist 42:307–310. doi:10.1017/S0024282909990351

Arnold AE, Miadlikowska J, Higgins KL et al (2009) A phylogenetic estimation of trophic transition networks for ascomycetous fungi: are lichens cradles of symbiotrophic fungal diversification? Syst Biol 58:283–297. doi:10.1093/sysbio/syp001

Arróniz-Crespo M, Pérez-Ortega S, De Los RA et al (2014) Bryophyte-cyanobacteria associations during primary succession in recently deglaciated areas of Tierra del Fuego (Chile). PLoS One 9(5):e96081. doi:10.1371/journal.pone.0096081

Aschenbrenner IA, Cernava T, Berg G et al (2016) Understanding microbial multi-species symbioses. Front Microbiol 7:180. doi:10.3389/fmicb.2016.00180

Baker JA, Entsch B, McKay DB (2003) The cyanobiont in an *Azolla* fern is neither *Anabaena* nor *Nostoc*. FEMS Microbiol Lett 229:43–47. doi:10.1016/S0378-1097(03)00784-5

Barthlott W, Fischer E, Frahm J-P (2000) First experimental evidence for zoophagy in the hepatic *Colura*. Plant Biol 2:93–97. doi:10.1055/s-2000-9150

Bay G, Nahar N, Oubre M, Whitehouse MJ et al (2013) Boreal feather mosses secrete chemical signals to gain nitrogen. New Phytol 200:54–60. doi:10.1111/nph.12403

Beimforde C, Feldberg K, Nylinder S et al (2014) Estimating the Phanerozoic history of the Ascomycota lineages: combining fossil and molecular data. Mol Phylogenet Evol 78:386–398. doi:10.1016/j.ympev.2014.04.024

Belinchón R, Yahr R, Ellis CJ (2014) Interactions among species with contrasting dispersal modes explain distributions for epiphytic lichens. Ecography 37:001–007. doi:10.1111/ecog.01258

Berg A, Danielsson A, Svensson BH (2013) Transfer of fixed-N from N_2-fixing cyanobacteria associated with the moss *Sphagnum riparium* results in enhanced growth of the moss. Plant Soil 362:271–278. doi:10.1007/s11104-012-1278-4

Bergman B (2002) *Nostoc-Gunnera* symbiosis. In: Rai AN, Bergman B, Rasmussen U (eds) Cyanobacteria in symbiosis. Kluwer, Dordrecht, pp 207–232

Bergman B, Hällbom L (1982) *Nostoc* of *Peltigera canina* when lichenized and isolated. Can J Bot 60:2092–2098

Bergman B, Johansson C, Söderbäck E (1992) The *Nostoc-Gunnera* symbiosis. New Phytol 122:379–400

Bidartondo MI, Read DJ, Trappe JM et al (2011) The dawn of symbiosis between plants and fungi. Biol Lett 7:574–577. doi:10.1098/rsbl

Bjerke JW, Gwynn-Jones D, Callaghan TV (2005) Effects of enhanced UV-B radiation in the field on the concentration of phenolics and chlorophyll fluorescence in two boreal and arctic-alpine lichens. Environ Exp Bot 53:139–149. doi:10.1016/j.envexpbot.2004.03.009

Bothe H, Schmitz O, Yates MG et al (2010) Nitrogen fixation and hydrogen metabolism in cyanobacteria. Microbiol Mol Biol Rev 74:529–551. doi:10.1128/MMBR.00033-10

Brouwer P, Bräutigam A, Külahoglu C et al (2014) *Azolla* domestication towards a biobased economy? New Phytol 202:1069–1082. doi:10.1111/nph.12708

Brown RC, Lemmon BE, Shimamura M et al (2015) Spores of relictual bryophytes: diverse adaptations to life on land. Rev Palaeobot Palynol 216:1–17. doi:10.1016/j.revpalbo.2015.01.004

Butler RJ, Barrett PM, Kenrick P et al (2009) Testing co-evolutionary hypotheses over geological timescales: interactions between Mesozoic non-avian dinosaurs and cycads. Biol Rev Camb Philos Soc 84:73–89. doi:10.1111/j.1469-185X.2008.00065.x

Calteau A, Fewer DP, Latifi A et al (2014) Phylum-wide comparative genomics unravel the diversity of secondary metabolism in Cyanobacteria. BMC Genomics 15:977. doi:10.1186/1471-2164-15-977

Campbell EL, Meeks JC (1989) Characteristics of hormogonia formation by symbiotic *Nostoc* spp. in response to the presence of *Anthoceros punctatus* or its extracellular products. Appl Environ Microbiol 55:125–131

Campbell J, Bengtson P, Fredeen AL et al (2013) Does exogenous carbon extend the realized niche of canopy lichens? Evidence from sub-boreal forests in British Columbia. Ecology 94:1186–1195. doi:10.2307/23435960

Casano LM, del Campo EM, García-Breijo FJ et al (2011) Two *Trebouxia* algae with different physiological performances are ever-present in lichen thalli of *Ramalina farinacea*. Coexistence versus competition? Environ Microbiol 13:806–818. doi:10.1111/j.1462-2920.2010.02386.x

Chang Y, Graham SW (2011) Inferring the higher-order phylogeny of mosses (Bryophyta) and relatives using a large, multigene plastid data set. Am J Bot 98:839–849. doi:10.3732/ajb.0900384

Chiu WL, Peters GA, Levieille G et al (2005) Nitrogen deprivation stimulates symbiotic gland development in *Gunnera manicata*. Plant Physiol 139:224–230. doi:10.1104/pp.105.064931

Chua JPS, Wallace EJS, Yardley JA et al (2012) Gene expression indicates a zone of heterocyst differentiation within the thallus of the cyanolichen *Pseudocyphellaria crocata*. New Phytol 196:862–872. doi:10.1111/j.1469-8137.2012.04272.x

Claessen D, Rozen DE, Kuipers OP et al (2014) Bacterial solutions to multicellularity: a tale of biofilms, filaments and fruiting bodies. Nat Rev Microbiol 12:115–124. doi:10.1038/nrmicro3178

Collison ME (1980) A new multiple-floated *Azolla* from the Eocene of Britain with a brief review of the genus. Paleontology 23:213–229

Condamine FL, Nagalingum NS, Marshall CR, Morlon H (2015) Origin and diversification of living cycads: a cautionary tale on the impact of the branching process prior in Bayesian molecular dating. BMC Evol Biol 15:65. doi:10.1186/s12862-015-0347-8

Cornejo C, Scheidegger C (2013) Morphological aspects associated with repair and regeneration in *Lobaria pulmonaria* and *L. amplissima* (Peltigerales, Ascomycota). Lichenologist 45:285–289. doi:10.1017/S0024282912000813

Cornejo C, Scheidegger C (2016) Cyanobacterial gardens: the liverwort *Frullania asagrayana* acts as a reservoir of lichen photobionts. Environ Microbiol Rep. doi:10.1111/1758-2229.12386 [Epub ahead of print]

Cornelissen JHC, Lang MH, Soudzilovskaia NA et al (2007) Comparative cryptogam ecology: a review of bryophyte and lichen traits that drive biogeochemistry. Ann Bot 99:987–1001. doi:10.1093/aob/mcm030

Costa JL, Paulsrud P, Lindblad P (1999) Cyanobiont diversity within coralloid roots of selected cycad species. FEMS Microbiol Ecol 28:85–91. doi:10.1016/S0168-6496(98)00095-6

Costa J-L, Paulsrud P, Rikkinen J et al (2001) Genetic dversity of *Nostoc* symbionts endophytically associated with two bryophyte species. Appl Environ Microbiol 67:4393–4396. doi:10.1128/AEM.67.9.4393-4396.2001

Costa JL, Romero EM, Lindblad P (2004) Sequence based data supports a single *Nostoc* strain in individual coralloid roots of cycads. FEMS Microbiol Ecol 49:481–487. doi:10.1016/j.femsec.2004.05.001

Cox CJ, Goffinet B, Wickett NJ et al (2010) Moss diversity: a molecular phylogenetic analysis of genera. Phytotaxa 9:175–195

Cox CJ, Li B, Foster PG et al (2014) Conflicting phylogenies for early land plants are caused by composition biases among synonymous substitutions. Syst Biol 63:272–279. doi:10.3732/ajb.0900384

Cuddy WS, Neilan BA, Gehringer MM (2012) Comparative analysis of cyanobacteria in the rhizosphere and as endosymbionts of cycads in drought-affected soils. FEMS Microbiol Ecol 80:204–215. doi:10.1111/j.1574-6941.2011.01288.x

Dagan T, Roettger M, Stucken K et al (2013) Genomes of Stigonematalean cyanobacteria (subsection V) and the evolution of oxygenic photosynthesis from prokaryotes to plastids. Genome Biol Evol 5:31–44. doi:10.1093/gbe/evs117

Dal Grande F, Beck A, Cornejo C et al (2014) Molecular phylogeny and symbiotic selectivity of the green algal genus *Dictyochloropsis* s.l. (Trebouxiophyceae): a polyphyletic and widespread group forming photobiont-mediated guilds in the lichen family Lobariaceae. New Phytol 202:447–455. doi:10.1111/nph.12678

Dal-Forno M, Lawrey JD, Sikaroodi M et al (2013) Starting from scratch, evolution of the lichen thallus in the basidiolichens *Dictyonema* (Agaricales, Hygrophoraceae). Fungal Biol 117:584–598. doi:10.1016/j.funbio.2013.05.006

Dal-Forno M, Lücking R, Bungartz F et al (2016) From one to six: unrecognized species diversity in the genus *Acantholichen* (lichenized Basidiomycota: Hygrophoraceae). Mycologia 108:38–55. doi:10.3852/15-060

Dalton DA, Chatfield JM (1987) A new nitrogen-fixing cyanophyte–hepatic association: *Nostoc* and *Porella*. Am J Bot 72:781–784

Darnajoux R, Constantin J, Miadlikowksa J et al (2014) Is vanadium a biometal for boreal cyanolichens? New Phytol 202:765–771. doi:10.1111/nph.12777

Davey ML, Currah RS (2006) Interactions between mosses (Bryophyta) and fungi. Botany 84:1509–1519. doi:10.1139/B10-004

Delaux PM, Séjalon-Delmas N, Bécard G et al (2013) Evolution of the plant-microbe symbiotic 'toolkit'. Trends Plant Sci 18:298–304. doi:10.1016/j.tplants.2013.01.008

Delaux PM, Radhakrishnan J et al (2015) Algal ancestor of land plants was preadapted for symbiosis. Proc Natl Acad Sci U S A 112:13390–13395. doi:10.1073/pnas.1515426112

Desirò A, Duckett JG, Pressel S et al (2013a) Fungal symbioses in hornworts: a chequered history. Proc R Soc B 280:20130207. doi:10.1098/rspb.2013.0207

Desirò A, Naumann M, Epis S et al (2013b) *Mollicutes*-related endobacteria thrive inside liverwort-associated arbuscular mycorrhizal fungi. Environ Microbiol 15:822–836. doi:10.1111/j.1462-2920.2012.02833.x

Desirò A, Faccio A, Kaech A et al (2015) *Endogone*, one of the oldest plant-associated fungi, host unique *Mollicutes*-related endobacteria. New Phytol 205:1464–1472. doi:10.1111/nph.13136

Deusch O, Landan G, Roettger M et al (2008) Genes of cyanobacterial origin in plant nuclear genomes point to a heterocyst-forming plastid ancestor. Mol Biol Evol 25:748–761. doi:10.1093/molbev/msn022

Díaz EM, Sacristan M, Legaz ME et al (2009) Isolation and characterization of a cyanobacterium-binding protein and its cell wall receptor in the lichen *Peltigera canina*. Plant Signal Behav 4:598–603. doi:10.4161/psb.4.7.9164

Díaz EM, Vicente-Manzanares M, Sacristan M et al (2011) Fungal lectin of *Peltigera canina* induces chemotropism of compatible *Nostoc* cells by constriction-relaxation pulses of cyanobiont cytoskeleton. Plant Signal Behav 6:1525–1536. doi:10.4161/psb.6.10.16687

Diederich P (1996) The lichenicolous heterobasidiomycetes. Bibl Lichenol 61:1–198

Dittmann E, Gugger M, Sivonen K et al (2015) Natural product biosynthetic diversity and comparative genomics of the cyanobacteria. Trends Microbiol 23(10):642–652. doi:10.1016/j.tim.2015.07.008

Döbbeler P (1997) Biodiversity of bryophilous ascomycetes. Biodivers Conserv 6:721–738. doi:10.1023/A:1018370304090

Dodds WK, Gudder DA, Mollenhauer D (1995) The ecology of *Nostoc*. J Phycol 31:2–18

Downie DA, Donaldson JS, Oberprieler RG (2008) Molecular systematics and evolution in an African cycad-weevil interaction: *Amorphocerini* (Coleoptera: Curculionidae: Molytinae) weevils on *Encephalartos*. Mol Phylogenet Evol 47:102–116. doi:10.1016/j.ympev.2008.01.023

Duckett JG, Prasad AKSK, Davies DA et al (1977) A cytological analysis of the *Nostoc*–bryophyte relationship. New Phytol 79:349–362. doi:10.1111/j.1469-8137.1977.tb02215.x

Edwards D, Kenrick P (2015) The early evolution of land plants, from fossils to genomics: a commentary on Lang (1937) 'On the plant-remains from the Downtonian of England and Wales'. Philos Trans R Soc Lond B Biol Sci 370. doi:10.1098/rstb.2014.0343

Edwards D, Morris JL, Richardson JB et al (2014) Cryptospores and cryptophytes reveal hidden diversity in early land floras. New Phytol 202:50–78. doi:10.1111/nph.12645

Ekman M, Tollbäck P, Bergman B (2008) Proteomic analysis of the cyanobacterium of the *Azolla* symbiosis: identity, adaptation, and NifH modification. J Exp Bot 59:1023–1034. doi:10.1093/jxb/erm282

Ekman M, Picossi S, Campbell EL et al (2013) A *Nostoc punctiforme* sugar transporter necessary to establish a cyanobacterium-plant symbiosis. Plant Physiol 161:1984–1992. doi:10.1104/pp.112.213116

Ekman S, Wedin M, Lindblom L et al (2014) Extended phylogeny and a revised generic classification of the Pannariaceae (Peltigerales, Ascomycota). Lichenologist 46:627–656. doi:10.1017/S002428291400019X

Elbert W, Weber B, Burrows S et al (2012) Contribution of cryptogamic covers to the global cycles of carbon and nitrogen. Nat Geosci 5:459–462. doi:10.1038/ngeo1486

Ellerbeck M, Schüßler A, Brucker D et al (2013) Characterization of three ammonium transporters of the glomeromycotan fungus *Geosiphon pyriformis*. Eukaryot Cell 12:1554–1562. doi:10.1128/EC.00139-13

Evrard C, Van Hove C (2004) Taxonomy of the American *Azolla* species (Azollaceae): a critical review. Syst Geogr Plant 74:301–318

Fedrowitz K, Kaasalainen U, Rikkinen J (2011) Genotype variability of *Nostoc* symbionts in three epiphytic *Nephroma* species in a boreal forest landscape. Bryologist 114:220–230. doi:10.1639/0007-2745-114.1.220

Fedrowitz K, Kaasalainen U, Rikkinen J (2012) Geographic mosaic of symbiont selectivity in a genus of epiphytic cyanolichens. Ecol Evol 2:2291–2303. doi:10.1002/ece3.343

Feldberg K, Schneider H2, Stadler T et al (2014) Epiphytic leafy liverworts diversified in angiosperm-dominated forests. Sci Rep 4:5974. doi:10.1038/srep05974

Fennell M, Murphy JE, Gallagher T et al (2013) Simulating the effects of climate change on the distribution of an invasive plant, using a high resolution, local scale, mechanistic approach: challenges and insights. Glob Chang Biol 19:1262–1274. doi:10.1111/gcb.12102

Fennell M, Gallagher T, Vintro LL et al (2014) Using soil seed banks to assess temporal patterns of genetic variation in invasive plant populations. Ecol Evol 4:1648–1658. doi:10.1002/ece3.1043

Fernández-Martínez MA, de Los RA, Sancho LG et al (2013) Diversity of endosymbiotic *Nostoc* in *Gunnera magellanica* from Tierra del Fuego, Chile. Microb Ecol 66:335–350. doi:10.1007/s00248-013-0223-2

Field KJ, Rimington WR, Bidartondo MI (2015a) First evidence of mutualism between ancient plant lineages (Haplomitriopsida liverworts) and Mucoromycotina fungi and its response to simulated Palaeozoic changes in atmospheric CO_2. New Phytol 205:743–756. doi:10.1111/nph.13024

Field KJ, Rimington WR, Bidartondo MI (2015b) Functional analysis of liverworts in dual symbiosis with Glomeromycota and Mucoromycotina fungi under a simulated Palaeozoic CO_2 decline. ISME J. doi:10.1038/ismej.2015.204

Flores E, Herrero A (2010) Compartmentalized function through cell differentiation in filamentous cyanobacteria. Nat Rev Microbiol 8:39–50. doi:10.1038/nrmicro2242

Gantar M, Rowell P, Kerby NW, Sutherland IW (1995) Role of extracellular polysaccharides in the colonization of wheat (*Triticum vulgare* L.) roots by N2-fixing cyanobacteria. Biol Fert Soils 19:41–48. doi:10.1007/BF00336345

Gauslaa Y (2014) Rain, dew, and humid air as drivers of morphology, function and spatial distribution in epiphytic lichens. Lichenologist 46:1–16. doi:10.1017/S0024282913000753

Gavazov KS, Soudzilovskaia NA, van Logtestijn RSP et al (2010) Isotopic analysis of cyanobacterial nitrogen fixation associated with subarctic lichen and bryophyte species. Plant Soil 333:507–517. doi:10.1007/s11104-010-0374-6

Gehringer MM, Pengelly JJ, Cuddy WS et al (2010) Host selection of symbiotic cyanobacteria in 31 species of the Australian cycad genus: *Macrozamia* (Zamiaceae). Mol Plant-Microbe Interact 23:811–822. doi:10.1094/MPMI-23-6-0811

Gehringer MM, Adler L, Roberts AA et al (2012) Nodularin, a cyanobacterial toxin, is synthesized in planta by symbiotic *Nostoc* sp. ISME J 6:1834–1847. doi:10.1038/ismej.2012.25
Gentili F, Nilsson M-C, Zackrisson O et al (2005) Physiological and molecular diversity of feather moss associative N_2-fixing cyanobacteria. J Exp Bot 56:3121–3127. doi:10.1093/jxb/eri309
Gerphagnon M, Latour D, Colombet J, Sime-Ngando T (2013) Fungal parasitism: life cycle, dynamics and impact on cyanobacterial blooms. PLoS One 8(4), e60894. doi:10.1371/journal.pone.0060894
Gioria M, Osborne BA (2009) The impact of *Gunnera tinctoria* (Molina) Mirbel invasions on soil seed bank communities. J Plant Ecol 2:153–167. doi:10.1093/jpe/rtp013
Gioria M, Osborne BA (2013) Biological flora of the British Isles: *Gunnera tinctoria*. J Ecol 101:243–264. doi:10.1111/1365-2745.12022
Green TGA, Nash TH III, Lange OL (2008) Physiological ecology of carbon dioxide exchange. In: Nash TH III (ed) Lichen biology, 2nd edn. Cambridge University Press, Cambridge, pp 152–181
Grobbelaar N, Scott WE, Hattingh W, Marshall J (1987) The identification of the coralloid endophytes of the southern African cycads and the ability of the isolates to fix dinitrogen. S Afr J Bot 53:111–118
Groth-Malonek M, Pruchner D, Grewe F et al (2005) Ancestors of trans-splicing mitochondrial introns support serial sister group relationships of hornworts and mosses with vascular plants. Mol Biol Evol 22:117–125. doi:10.1093/molbev/msh259
Grube M, Cardinale M, de Castro JV et al (2009) Species-specific structural and functional diversity of bacterial communities in lichen symbioses. ISME J 3:1105–1115. doi:10.1038/ismej.2009.63
Grube M, Berg G, Andrésson ÓS et al (2014) Lichen genomics: prospects and progress. In: Martin F (ed) The ecological genetics of fungi. Wiley, pp 191–212. doi:10.1002/9781118735893.ch9
Guevara R, Armesto JJ, Caru M (2002) Genetic diversity of *Nostoc* microsymbionts from *Gunnera tinctoria* revealed by PCR-STRR fingerprinting. Microb Ecol 44:127–136. doi:10.1007/s00248-002-1019-y
Gusev MV, Baulina OI, Gorelova OA, Lobakova ES, Korzhenevskaya TG (2002) Artificial cyanobacterial–plant symbioses In: Rai AN, Bergman B, Rasmussen U (eds) Cyanobacteria in Symbiosis. Kluwer, Dordrecht, pp 253–312. doi:10.1007/0-306-48005-0_14
Hall JW (1974) Cretaceous salviniaceae. Ann Miss Bot Gard 61:354–367
Heil M, Greiner S, Meimberg H (2004) Evolutionary change from induced to constitutive expression of an indirect plant resistance. Nature 430:205–208. doi:10.1038/nature02703
Henskens FL, Green TGA, Wilkins A (2012) Cyanolichens can have both cyanobacteria and green algae in a common layer as major contributors to photosynthesis. Ann Bot 110:555–563. doi:10.1093/aob/mcs108
Higgins KL, Arnold AE, Miadlikowska J et al (2007) Phylogenetic relationships, host affinity, and geographic structure of boreal and arctic endophytes from three major plant lineages. Mol Phylogenet Evol 42:543–555. doi:10.1016/j.ympev.2006.07.012
Hodkinson BP, Lutzoni F (2009) A microbiotic survey of lichen-associated bacteria reveals a new lineage from the Rhizobiales. Symbiosis 49:163–180. doi:10.1007/s13199-009-0049-3
Hodkinson BP, Gottel NR, Schadt CW et al (2012) Photoautotrophic symbiont and geography are major factors affecting highly structured and diverse bacterial communities in the lichen microbiome. Environ Microbiol 14:147–161. doi:10.1111/j.1462-2920.2011.02560.x
Hodkinson BP, Allen JL, Forrest L et al (2014) Lichen-symbiotic cyanobacteria associated with *Peltigera* have an alternative vanadium-dependent nitrogen fixation system. Eur J Phycol 49:11–19. doi:10.1080/09670262.2013.873143
Honegger R (2012) The symbiotic phenotype of lichen-forming ascomycetes and their endo- and epibionts. In: Hock B (ed) The Mycota IX—Fungal associations, 2nd edn. Springer, Berlin. doi:10.1007/978-3-642-30826-0_15
Honegger R, Edwards D, Axe L (2012) The earliest records of internally stratified cyanobacterial and algal lichens from the lower Devonian of the Welsh Borderland. New Phytol 197:264–275. doi:10.1111/nph.12009

Houle D, Gauthier SB, Paguet S et al (2006) Identification of two genera of N_2-fixing cyanobacteria growing on three feather moss species in Boreal forests of Quebec, Canada. Can J Bot 84:1025–1029. doi:10.1139/b06-059

Howard-Azzeh M, Shamseer L, Schellhorn HE et al (2014) Phylogenetic analysis and molecular signatures defining a monophyletic clade of heterocystous cyanobacteria and identifying its closest relatives. Photosynth Res 122:171–185. doi:10.1007/s11120-014-0020-x

Ininbergs K, Bay G, Rasmussen U et al (2011) Composition and diversity of nifH genes of nitrogen-fixing cyanobacteria associated with boreal forest feather mosses. New Phytol 192:507–517. doi:10.1111/j.1469-8137.2011.03809.x

Jackson O, Taylor O, Adams DG et al (2012) Arabinogalactan proteins occur in the free-living cyanobacterium genus *Nostoc* and in plant-*Nostoc* symbioses. Mol Plant Microbe Interact 25:1338–1349. doi:10.1094/MPMI-04-12-0095-R

Johansson C, Bergman B (1994) Reconstitution of the *Gunnera manicata* linde symbiosis: cyanobacterial specificity. New Phytol 126:643–652. doi:10.1111/j.1469-8137.1994.tb02960.x

Jørgensen PM (1998) *Acantholichen pannarioides*, a new basidiolichen from South America. Bryologist 101:444–447. doi:10.2307/3244185

Kaasalainen U, Jokela J, Fewer DP et al (2009) Microcystin production in the tripartite cyanolichen *Peltigera leucophlebia*. Mol Plant Microb Interact 22:695–702. doi:10.1094/MPMI-22-6-0695

Kaasalainen U, Fewer DP, Jokela J et al (2012) Cyanobacteria produce a high variety of hepatotoxic peptides in lichen symbiosis. Proc Natl Acad Sci U S A 109:5886–5891. doi:10.1073/pnas.1200279109

Kaasalainen U, Fewer DP, Jokela J et al (2013) Lichen species identity and diversity of cyanobacterial toxins in symbiosis. New Phytol 198:647–651. doi:10.1111/nph.12215

Kaasalainen U, Heinrichs J, Krings M et al (2015) Alectorioid morphologies in paleogene lichens: new evidence and re-evaluation of the fossil *Alectoria succini* Mägdefrau. PLoS One 10(6), e0129526. doi:10.1371/journal.pone.0129526

Kampa A, Gagunashvili AN, Gulder TA et al (2013) Metagenomic natural product discovery in lichen provides evidence for a family of biosynthetic pathways in diverse symbioses. Proc Natl Acad Sci U S A 110:E3129–E3137. doi:10.1073/pnas.1305867110

Kauserud H, Mathiesen C, Ohlson M (2008) High diversity of fungi associated with living parts of boreal forest bryophytes. Botany 86:1326–1333. doi:10.1139/B08-102

Kehr JC, Dittmann E (2015) Biosynthesis and function of extracellular glycans in cyanobacteria. Life 5:164–180. doi:10.3390/life5010164

Kenrick P, Strullu-Derrien C (2014) The origin and early evolution of roots. Plant Physiol 166:570–580. doi:10.1104/pp.114.244517

Khamar HJ, Breathwaite EK, Prasse CE et al (2010) Multiple roles of soluble sugars in the establishment of *Gunnera-Nostoc* endosymbiosis. Plant Physiol 154:1381–1389. doi:10.1104/pp.110.162529

Kimura J, Nakano T (1990) Reconstitution of a *Blasia-Nostoc* symbiotic association under axenic conditions. Nova Hedwig 50:191–200

Kluge M, Mollenhauer D, Mollenhauer R (1991) Photosynthetic carbon assimilation in *Geosiphon pyriforme* (Kützing) F.v. Wettstein, an endosymbiotic association of fungus and cyanobacterium. Planta 185:311–315. doi:10.1007/BF00201049

Kluge M, Mollenhauer D, Mollenhauer R et al (1992) *Geosiphon pyriforme*, an endosymbiotic consortium of a fungus and a cyanobacterium (*Nostoc*), fixes nitrogen. Bot Acta 105:343–344. doi:10.1111/j.1438-8677.1992.tb00309.x

Koptur S (1992) Extrafloral nectary-mediated interactions between insects and plants. In: Bernays E (ed) Insect–plant interactions, vol IV. CRC Press, Boca Raton, pp 81–129

Koptur S, Palacios-Rios M, Díaz_Castelazo et al (2013) Nectar secretion on fern fronds associated with lower levels of herbivore damage: field experiments with a widespread epiphyte of Mexican cloud forest remnants. Ann Bot 111:1277–1283. doi:10.1093/aob/mct063

Kostka JE, Weston DJ, Glass JB (2016) The *Sphagnum* microbiome: new insights from an ancient plant lineage. New Phytol. doi:10.1111/nph.13993. [Epub ahead of print]

Ku C, Nelson-Sathi S, Roettger M et al (2015) Endosymbiotic origin and differential loss of eukaryotic genes. Nature 524:427–432. doi:10.1038/nature14963

Larsson KH, Parmasto E, Fischer M et al (2006) Hymenochaetales: a molecular phylogeny for the hymenochaetoid clade. Mycologia 98:926–936. doi:10.3852/mycologia.98.6.926

Lawrey JD, Diederich P (2016) Lichenicolous fungi–worldwide checklist, including isolated cultures and sequences available. http://www.lichenicolous.net [5/27/2016]. Accessed 5 June 2016

Lawrey JD, Lücking R, Sipman HJM et al (2009) High concentration of basidiolichens in a single family of agaricoid mushrooms (Basidiomycota: Agaricales: Hygrophoraceae). Mycol Res 113:1154–1171. doi:10.1016/j.mycres.2009.07.016

Lechno-Yossef S, Nierzwicki-Bauer SA (2002) *Azolla-Anabaena* symbiosis. In: Rai AN, Bergman B, Rasmussen U (eds) Cyanobacteria in symbiosis. Kluwer, Dordrecht, pp 153–178

Leppänen SM, Salemaa M, Smolander A et al (2013) Nitrogen fixation and methanotrophy in forest mosses along a N deposition gradient. Environ Exp Bot 90:62–69. doi:10.1016/j.envexpbot.2012.12.006

Li L, Wang B, Liu Y, Qiu YL (2009) The complete mitochondrial genome sequence of the hornwort *Megaceros aenigmaticus* shows a mixed mode of conservative yet dynamic evolution in early land plant mitochondrial genomes. J Mol Evol 68:665–678. doi:10.1007/s00239-009-9240-7

Li FW, Villarreal JC, Kelly S et al (2014) Horizontal transfer of an adaptive chimeric photoreceptor from bryophytes to ferns. Proc Natl Acad Sci U S A 111:6672–6677. doi:10.1073/pnas.1319929111

Liaimer A, Helfrich EJ, Hinrichs K et al (2015) Nostopeptolide plays a governing role during cellular differentiation of the symbiotic cyanobacterium *Nostoc punctiforme*. Proc Natl Acad Sci U S A 112:1862–1867. doi:10.1073/pnas.1419543112

Ligrone R (1988) Ultrastructure of a fungal endophyte in *Phaeoceros laevis* (L.) Prosc. (Anthocerophyta). Bot Gaz 149:92–100. doi:10.1086/337695

Ligrone R, Duckett JG, Renzaglia KS (2012) Major transitions in the evolution of early land plants: a bryological perspective. Ann Bot 109:851–871. doi:10.1093/aob/mcs017

Lindblad P (1990) Nitrogen and carbon metabolism in coralloid roots of cycads. Mem NY Bot Gard 57:104–113

Lindblad P (2009) Cyanobacteria in symbiosis with cycads. In: Pawlowski K (ed) Prokaryotic symbionts in plants. Springer, Berlin, pp 225–233. doi:10.1007/7171_118

Lindblad P, Bergman B, Hofsten AV et al (1985) The cyanobacterium-*Zamia* symbiosis: an ultrastructural study. New Phytol 101:707–716. doi:10.1111/j.1469-8137.1985.tb02876.x

Lindblad P, Haselkorn R, Bergman B et al (1989) Comparison of DNA restriction fragment length polymorphisms of *Nostoc* strains in and from cycads. Arch Microbiol 152:20–24. doi:10.1007/BF00447006

Lindblad P, Atkins CA, Pate JS (1991) N-2-fixation by freshly isolated *Nostoc* from coralloid roots of the cycad *Macrozamia riedlei* (Fisch. ex Gaud.) Gardn. Plant Physiol 95:753–759

Lindo Z, Whiteley JA (2011) Old trees contribute bio-available nitrogen through canopy bryophytes. Plant Soil 342:141–148. doi:10.1007/s11104-010-0678-6

Lindo Z, Nilsson MC, Gundale MJ (2013) Bryophyte-cyanobacteria associations as regulators of the northern latitude carbon balance in response to global change. Glob Chang Biol 19:2022–2035. doi:10.1111/gcb.12175

Liu Y, Xue JY, Wang B, Li L, Qiu YL (2011) The mitochondrial genomes of the early land plants *Treubia lacunosa* and *Anomodon rugelii*: dynamic and conservative evolution. PLoS One 6(10), e25836. doi:10.1371/journal.pone.0025836

Lobakova ES, Orazova MK, Dobrovol'skaya TG (2003) Microbial complexes occurring on the apogeotropic roots and in the rhizosphere of cycad plants. Microbiology 72:628–633

Lóriga J, Schmidt AR, Moran RC, Feldberg K, Schneider H, Heinrichs J (2014) The first fossil of a bolbitidoid fern belongs to the early-divergent lineages of *Elaphoglossum* (Dryopteridaceae). Am J Bot 101:1466–1475. doi:10.3732/ajb.1400262

Lücking R, Lawrey JD, Sikaroodi M et al (2009) Do lichens domesticate photobionts like farmers domesticate crops? Evidence from a previously unrecognized lineage of filamentous cyanobacteria. Am J Bot 96:1409–1418. doi:10.3732/ajb.0800258

Lücking R, Barrie FR, Genney D (2013a) *Dictyonema coppinsii*, a new name for the European species known as *Dictyonema interruptum* (Basidiomycota: Agaricales: Hygrophoraceae), with a validation of its photobiont *Rhizonema* (Cyanoprokaryota: Nostocales: Rhizonemataceae). Lichenologist 46:261–267. doi:10.1017/S0024282913000352

Lücking R, Dal-Forno M, Lawrey JD et al (2013b) Ten new species of lichenized Basidiomycota in the genera *Dictyonema* and *Cora* (Agaricales: Hygrophoraceae), with a key to all accepted genera and species in the *Dictyonema* clade. Phytotaxa 139:1–38. doi:10.11646/phytotaxa.139.1.1

Lücking R, Dal-Forno M, Sikaroodi M (2014) A single macrolichen constitutes hundreds of unrecognized species. Proc Natl Acad Sci U S A 111:11091–11096. doi:10.1073/pnas.1403517111

Lumbsch T, Rikkinen J. (2016) Evolution of lichens. In: Dighton J, White Jr JF, Oudemans P (eds) The fungal community: its organization and role in the ecosystem, 4th edn. CRC Press (in press)

Lumbsch HT, Ahti T, Altermann S et al (2011) One hundred new species of lichenized fungi: a signature of undiscovered global diversity. Phytotaxa 18:1–127

Magain N, Sérusiaux E (2014) Do photobiont switch and cephalodia emancipation act as evolutionary drivers in the lichen symbiosis? A case study in the Pannariaceae (Peltigerales). PLoS One 9(2):e89876. doi:10.1371/journal.pone.0089876

Magain N, Goffinet B, Sérusiaux E (2012) Further photomorphs in the lichen family Lobariaceae from Reunion (Mascarene archipelago) with notes on the phylogeny of *Dendriscocaulon* cyanomorphs. Bryologist 115:243–254. doi:10.1639/0007-2745-115.2.243

Magallón SS, Castillo AA (2009) Angiosperm diversification through time. Am J Bot 96:349–365

Manoharan SS, Miao VPW, Andrésson ÓS (2012) LEC-2, a highly variable lectin in the lichen *Peltigera membranacea*. Symbiosis 58:91–98. doi:10.1007/s13199-012-0206-y

Marazzi B, Ané C, Simon MF et al (2012) Locating evolutionary precursors on a phylogenetic tree. Evolution 66:3918–3930. doi:10.1111/j.1558-5646.2012.01720.x

Marazzi B, Bronstein JL, Koptur S (2013) The diversity, ecology and evolution of extrafloral nectaries: current perspectives and future challenges. Ann Bot 111:1243–1250. doi:10.1093/aob/mct109

Meeks JC (2009) Physiological adaptations in nitrogen-fixing *Nostoc*-plant symbiotic associations. In: Pawlowski K (ed) Prokaryotic symbionts in plants. Springer, Berlin, pp 181–205. doi:10.1007/7171_2007_101

Meeks JC, Elhai J (2002) Regulation of cellular differentiation in filamentous cyanobacteria in free-living and plant-associated symbiotic growth states. Microbiol Mol Biol Rev 66:94–121. doi:10.1128/MMBR.66.1.94-121.2002

Meeks JC, Elhai J, Thiel T et al (2001) An overview of the genome of *Nostoc punctiforme*, a multicellular, symbiotic cyanobacterium. Photosynth Res 70:85–106. doi:10.1023/A:1013840025518

Meeks JC, Campbell EL, Summers ML et al (2002) Cellular differentiation in the cyanobacterium *Nostoc punctiforme*. Arch Microbiol 178:395–403. doi:10.1007/s00203-002-0476-5

Menge DN, Hedin LO (2009) Nitrogen fixation in different biogeochemical niches along a 120 000-year chronosequence in New Zealand. Ecology 90:2190–2201

Merinero S, Hilmo O, Gauslaa Y (2014) Size is a main driver for hydration traits in cyano- and cephalolichens of boreal rainforest canopies. Fungal Ecol 7:59–66. doi:10.1016/j.funeco.2013.12.001

Metzgar J, Schneider H, Pryer K (2007) Phylogeny and divergence time estimates for the fern genus *Azolla* (Salviniaceae). Int J Plant Sci 168:1045–1053. doi:10.1086/519007

Miadlikowska J, Kauff F, Högnabba F et al (2014) A multigene phylogenetic synthesis for the class Lecanoromycetes (Ascomycota): 1307 fungi representing 1139 infrageneric taxa, 312 genera and 66 families. Mol Phylogenet Evol 79:132–168. doi:10.1016/j.ympev.2014.04.003

Miao VPW, Manoharan SS, Snæbjarnarson V et al (2012) Expression of lec-1, a mycobiont gene encoding a galectin-like protein in the lichen *Peltigera membranacea*. Symbiosis 57:23–31. doi:10.1007/s13199-012-0175-1

Mollenhauer D, Mollenhauer R (1988) *Geosiphon* cultures ahead. Endocyt Cell Res 5:69–73
Moncada B, Ready B, Lücking R (2014) A phylogenetic revision of Hawaiian *Pseudocyphellaria* sensu lato (lichenized Ascomycota: Lobariaceae) reveals eight new species and a high degree of inferred endemism. Bryologist 117:119–160. doi:10.1639/0007-2745-117.2.119
Nagalingum NS, Marshall CR, Quental TB et al (2011) Recent synchronous radiation of a living fossil. Science 334:796–799. doi:10.1126/science.1209926
Naito M, Morton JB, Pawlowska TE (2015) Minimal genomes of mycoplasma-related endobacteria are plastic and contain host-derived genes for sustained life within Glomeromycota. Proc Natl Acad Sci U S A 112:7791–7796. doi:10.1073/pnas.1501676112
Nash TH III (2008) Nitrogen, its metabolism and potential contribution to ecosystems. In: Nash TH III (ed) Lichen biology, 2nd edn. Cambridge University Press, Cambridge, pp 216–233
Nathanielsz CP, Staff IA (1975) A mode of entry of blue-green algae into the apogeotrophic roots of *Macrozamia communis*. Am J Bot 62:232–235
Naumann M, Schüßler A, Bonfante P et al (2010) The obligate endobacteria of arbuscular mycorrhizal fungi are ancient heritable components related to the *Mollicutes*. ISME J 4:862–871. doi:10.1038/ismej.2010.21
Nierzwicki-Bauer SA, Aulfinger H (1991) Occurrence and ultrastructural characterization of bacteria in association with and isolated from *Azolla caroliniana*. Appl Environ Microbiol 57:3629–3636
Nilsson M, Bergman B, Rasmussen U (2000) Cyanobacterial diversity in geographically related and distant host plants of the genus *Gunnera*. Arch Microbiol 173:97–102. doi:10.1007/s002039900113
Nilsson M, Bhattacharya J, Rai AN, Bergman B (2002) Colonization of roots of rice (*Oryza sativa*) by symbiotic *Nostoc* strains. New Phytol 156:517–525. doi:10.1046/j.1469-8137.2002.00534.x
Nilsson M, Rasmussen U, Bergman B (2005) Competition among symbiotic cyanobacterial *Nostoc* strains forming artificial associations with rice (*Oryza sativa*). FEMS Microbiol Lett 245:139–144. doi:10.1016/j.femsle.2005.03.010
Nilsson M, Rasmussen U, Bergman B (2006) Cyanobacterial chemotaxis to extracts of host and nonhost plants. FEMS Microbiol Ecol 55:382–390. doi:10.1111/j.1574-6941.2005.00043.x
O'Brien HE, Miadlikowska J, Lutzoni F (2013) Assessing population structure and host specialization in lichenized cyanobacteria. New Phytol 198:557–566. doi:10.1111/nph.12165
Oberwinkler F (1984) Fungus-alga interactions in basidiolichens. Nova Hedw 79:739–774
Oberwinkler F (2012) Basidiolichens. In: Hock B (ed) The Mycota vol IX, fungal associations, 2nd edn. Springer, Berlin, pp 341–362
Ochoa de Alda JA, Esteban R, Diago ML et al (2014) The plastid ancestor originated among one of the major cyanobacterial lineages. Nat Commun 5:4937. doi:10.1038/ncomms5937
Oksanen I, Lohtander K, Paulsrud P et al (2002) A molecular approach to cyanobacterial diversity in a rock-pool community involving gelatinous lichens and free-living *Nostoc* colonies. Ann Bot Fennici 39:93–99
Oksanen I, Jokela J, Fewer DP et al (2004) Discovery of rare and highly toxic microcystins from lichen associated cyanobacterium *Nostoc* sp. strain IO-102-I. Appl Environ Microbiol 70:5756–5763. doi:10.1128/AEM.70.10.5756-5763.2004
Oren A, Garrity GM (2014) Proposal to change general consideration 5 and principle 2 of the international code of nomenclature of prokaryotes. Int J Syst Evol Microbiol 64:309–310. doi:10.1099/ijs.0.059568-0
Ortiz-Álvarez R, de Los RA, Fernández-Mendoza F et al (2015) Ecological specialization of two photobiont-specific maritime cyanolichen species of the genus *Lichina*. PLoS One 10(7), e0132718. doi:10.1371/journal.pone.0132718
Osborne B, Bergman B (2009) Why does *Gunnera* do it and other angiosperms don't? An evolutionary perspective on the *Gunnera-Nostoc* symbiosis. In: Pawlowski K (ed) Prokaryotic symbionts in plants. Springer, Berlin, pp 207–224. doi:10.1007/7171_2007_116
Osborne BA, Sprent JI (2002) Ecology of *Nostoc-Gunnera* symbiosis. In: Rai AN, Bergman B, Rasmussen U (eds) Cyanobacteria in symbiosis. Kluwer, Dordrecht, pp 233–251. doi:10.1007/0-306-48005-0_13

Otálora MA, Aragoń G, Martinez I, Wedin M (2013) Cardinal characters on a slippery slope—a reevaluation of phylogeny, character evolution, and evolutionary rates in the jelly lichens (Collemataceae s. str). Mol Phylogenet Evol 68:185–198. doi:10.1016/j.ympev.2013.04.004
Otálora MA, Jørgensen PM, Wedin M (2014) A revised generic classification of the jelly lichens, collemataceae. Fungal Div 64:275–293. doi:10.1007/s13225-013-0266-1
Ow MC, Gantar M, Elhai J (1999) Reconstitution of a cycad-cyanobacterial association. Symbiosis 27:125–134
Palinska KA, Thomasius CF, Marquardt J et al (2006) Phylogenetic evaluation of cyanobacteria preserved as historic herbarium exsiccata. Int J Syst Evol Microbiol 56:2253–2263. doi:10.1099/ijs.0.64417-0
Papaefthimiou D, Hrouzek P, Mugnai MA et al (2008a) Differential patterns of evolution and distribution of the symbiotic behaviour in nostocacean cyanobacteria. Int J Syst Evol Microbiol 58:553–564. doi:10.1099/ijs.0.65312-0
Papaefthimiou D, Van Hove C, Lejeune A et al (2008b) Diversity and host specificity of *Azolla* cyanobionts. J Phycol 44:60–70. doi:10.1111/j.1529-8817.2007.00448.x
Paulsrud P (2002) The *Nostoc* symbiont of lichens: diversity, specificity and cellular modifications. PhD Thesis, Uppsala University, Sweden
Paulsrud P, Rikkinen J, Lindblad P (1998) Cyanobiont specificity in some *Nostoc*-containing lichens and in a *Peltigera aphthosa* photosymbiodeme. New Phytol 139:517–524. doi:10.1046/j.1469-8137.1998.00220.x
Paulsrud P, Rikkinen J, Lindblad P (2000) Spatial patterns of photobiont diversity in some *Nostoc*-containing lichens. New Phytol 146:291–299. doi:10.1046/j.1469-8137.2000.00647.x
Paulsrud P, Rikkinen J, Lindblad P (2001) Field investigations on cyanobacterial specificity in *Peltigera aphthosa*. New Phytol 152:117–123. doi:10.1046/j.0028-646x.2001.00234.x
Payne JL, Boyer AG, Brown JH et al (2009) Two-phase increase in the maximum size of life over 3.5 billion years reflects biological innovation and environmental opportunity. Proc Natl Acad Sci U S A 106:24–27. doi:10.1073/pnas.0806314106
Payne JL, McClain CR, Boyer AG et al (2011) The evolutionary consequences of oxygenic photosynthesis: a body size perspective. Photosynth Res 107:37–57. doi:10.1007/s11120-010-9593-1
Payne JL, Groves JR, Jost AB et al (2012) Late paleozoic fusulinoidean gigantism driven by atmospheric hyperoxia. Evolution 66:2929–2939. doi:10.1111/j.1558-5646.2012.01626.x
Peñalver E, Labandeira CC, Barrón E et al (2012) Thrips pollination of mesozoic gymnosperms. Proc Natl Acad Sci U S A 109:8623–8628. doi:10.1073/pnas.1120499109
Pereira AL, Vasconcelos V (2014) Classification and phylogeny of the cyanobiont *Anabaena azollae* strasburger: an answered question? Int J Syst Evol Microbiol 64:1830–1840. doi:10.1099/ijs.0.059238-0
Pereira AL, Martins M, Oliveira MM et al (2011) Morphological and genetic diversity of the family Azollaceae inferred from vegetative characters and RAPD markers. Plant Syst Evol 297:213–226. doi:10.1007/s00606-011-0509-0
Pinevich AV (2015) Proposal to consistently apply the International Code of Nomenclature of Prokaryotes (ICNP) to names of the oxygenic photosynthetic bacteria (cyanobacteria), including those validly published under the International Code of Botanical Nomenclature (ICBN)/International Code of Nomenclature for algae, fungi and plants (ICN), and proposal to change Principle 2 of the ICNP. Int J Syst Evol Microbiol 65:1070–1074. doi:10.1099/ijs.0.000034
Prasanna R, Jaiswal P, Nayak S et al (2009) Cyanobacterial diversity in the rhizosphere of rice and its ecological significance. Indian J Microbiol 49:89–97. doi:10.1007/s12088-009-0009-x
Pressel S, Ligrone R, Duckett JG (2008) The ascomycete *Rhizoscyphus ericae* elicits a range of host responses in the rhizoids of leafy liverworts; an experimental and cytological analysis. Fieldiana 47:59–72. doi:10.3158/0015-0746-47.1.59
Pressel S, Bidartondo MI, Ligrone R et al (2010) Fungal symbioses in bryophytes: new insights in the twenty first century. Phytotaxa 9:238–253
Qiu Y, Li L, Wang B et al (2006) The deepest divergences in land plants inferred from phylogenomic evidence. Proc Natl Acad Sci U S A 103:15511–15516. doi:10.1073/pnas.0603335103

Ran L, Larsson J, Vigil-Stenman T et al (2010) Genome erosion in nitrogen-fixing vertically transmitted endosymbiotic multicellular cyanobacterium. PLoS One 5, e11486. doi:10.1371/journal.pone.0011486

Rasmussen U, Svenning MM (2001) Characterization by genotypic methods of symbiotic *Nostoc* strains isolated from five species of *Gunnera*. Arch Microbiol 176:204–210. doi:10.1007/s002030100313

Rasmussen U, Johansson C, Bergman B (1994) Early communication in the *Gunnera-Nostoc* symbiosis: plant-induced cell differentiation and protein synthesis in the cyanobacterium. MPMI 7:696–702

Rasmussen U, Johansson C, Renglin A et al (1996) A molecular characterization of the *Gunnera-Nostoc* symbiosis: comparison with *Rhizobium*- and *Agrobacterium*-plant interactions. New Phytol 133:391–398. doi:10.1111/j.1469-8137.1996.tb01906.x

Redecker D, Schüßler (2014) Glomeromycota. In: McLaughlin DJ, Spatafora JW (eds) The Mycota VII part A, systematics and evolution, 2nd edn. Springer, Berlin, pp 251–269

Redhead SA (1981) Parasitism of bryophytes by agarics. Can J Bot 59:63–67

Renzaglia KS (1978) A comparative morphology and developmental anatomy of the Anthocerotophyta. J Hattori Bot Lab 44:31–90

Renzaglia KS (1982) A comparative developmental investigation of the gametophyte generation in the Metzgeriales (Hepatophyta). Bryophyt Bibl 24:1–238

Renzaglia KS, Duff RJ, Nickrent DL et al (2000) Vegetative and reproductive innovations of early land plants: implications for a unified phylogeny. Trans R Soc Lond 355:769–793. doi:10.1098/rstb.2000.0615

Rikkinen J (2002) Cyanolichens: an evolutionary overview. In: Rai AN, Bergman B, Rasmussen U (eds) Cyanobacteria in symbiosis. Kluwer, Dordrecht, pp 73–96

Rikkinen J (2003) Ecological and evolutionary role of photobiont-mediated guilds in lichens. Symbiosis 34:99–110

Rikkinen J (2004) Ordination analysis of $tRNA^{Leu}$ (UAA) intron sequences in lichen-forming Nostoc strains and other cyanobacteria. Symb Bot Upsal 34:377–391

Rikkinen J (2009) Relations between cyanobacterial symbionts in lichens and plants. In: Pawlowski K (ed) Prokaryotic symbionts in plants. Springer, Berlin, pp 265–270. doi:10.1007/7171_2007_102

Rikkinen J (2013) Molecular studies on cyanobacterial diversity in lichen symbioses. MycoKeys 6:3–32. doi:10.3897/mycokeys.6.3869

Rikkinen J (2015) Cyanolichens. Biodivers Conserv 24:973–993. doi:10.1007/s10531-015-0906-8

Rikkinen J, Poinar G (2002) Fossilised Anzia (Lecanorales, lichen-forming Ascomycota) from European tertiary amber. Mycol Res 106:984–990. doi:10.1017\S0953756202005907

Rikkinen J, Poinar G (2008) A new species of *Phyllopsora* (Lecanorales, lichenforming Ascomycota) from Dominican amber, with remarks on the fossil history of lichens. J Exp Bot 59:1007–1011. doi:10.1093/jxb/ern004

Rikkinen J, Virtanen V (2008) Genetic diversity in cyanobacterial symbionts of thalloid bryophytes. J Exp Bot 59:1013–1021. doi:10.1093/jxb/ern003

Rikkinen J, Oksanen I, Lohtander K (2002) Lichen guilds share related cyanobacterial symbionts. Science 297:357. doi:10.1126/science.1072961

Rikkinen J, Dörfelt H, Schmidt AR et al (2003) Sooty moulds from European tertiary amber, with notes on the systematic position of *Rosaria* ('Cyanobacteria'). Mycol Res 107:251–256. doi:10.1017/S0953756203007330

Rodgers GA, Stewart WD (1977) The cyanophyte-hepatic symbiosis. I. Morphology and physiology. New Phytol 78:441–458. doi:10.1111/j.1469-8137.1977.tb04851.x

Rohrlack T, Haande S, Molversmyr Å et al (2015) Environmental conditions determine the course and outcome of phytoplankton chytridiomycosis. PLoS One 10(12):e0145559. doi:10.1371/journal.pone.0145559

Rousk K, Jones DL, Deluca TH (2013) Moss-cyanobacteria associations as biogenic sources of nitrogen in boreal forest ecosystems. Front Microbiol 4:150. doi:10.3389/fmicb.2013.00150

Ruhfel BR, Gitzendanner MA, Soltis PS et al (2014) From algae to angiosperms-inferring the phylogeny of green plants (Viridiplantae) from 360 plastid genomes. BMC Evol Biol 14:23. doi:10.1186/1471-2148-14-23

Rutishauser R, Wanntorp L, Pfeifer E (2004) *Gunnera herteri* – developmental morphology of a dwarf from Uruguay and S Brazil (Gunneraceae). Plant Syst Evol 248:219–241. doi:10.1007/s00606-004-0182-7

Salas-Leiva DE, Meerow AW, Calonje M et al (2013) Phylogeny of the cycads based on multiple single-copy nuclear genes: congruence of concatenated parsimony, likelihood and species tree inference methods. Ann Bot 112:1263–1278. doi:10.1093/aob/mct192

Santi C, Bogusz D, Franche C (2013) Biological nitrogen fixation in non-legume plants. Ann Bot 111:743–767. doi:10.1093/aob/mct048

Schelensog M, Schroeter B, Green TGA (2000) Water dependent photosynthetic activity of lichens from New Zealand: differences in the green algal and the cyanobacterial thallus parts of photosymbiodemes. Bibl Lichenol 75:149–160

Scheloske S, Maetz M, Schüßler A (2001) Heavy metal uptake of *Geosiphon pyriforme*. Nucl Instrum Method B 181:659–663. doi:10.1016/S0168-583X(01)00534-1

Scheublin TR, Sanders IR, Keel C et al (2010) Characterisation of microbial communities colonising the hyphal surfaces of arbuscular mycorrhizal fungi. ISME J 4:752–763. doi:10.1038/ismej.2010.5

Schirrmeister BE, Antonelli A, Bagheri HC (2011) The origin of multicellularity in cyanobacteria. BMC Evol Biol 11:45. doi:10.1186/1471-2148-11-45

Schirrmeister BE, de Vos JM, Antonelli A (2013) Evolution of multicellularity coincided with increased diversification of cyanobacteria and the great oxidation event. Proc Natl Acad Sci U S A 110:1791–1796. doi:10.1073/pnas.1209927110

Schirrmeister BE, Gugger M, Donoghue PC (2015) Cyanobacteria and the great oxidation event: evidence from genes and fossils. Palaeontology 58:769–785. doi:10.1111/pala.12178

Schmidt AR, Beimforde C, Seyfullah LJ et al (2014) Amber fossils of sooty moulds. Rev Palaeobot Palynol 200:53–64. doi:10.1016/j.revpalbo.2013.07.002

Schneider D, Wink M, Sporer F et al (2002) Cycads: their evolution, toxins, herbivores and insect pollinators. Naturwissenschaften 89:281–294

Schneider H, Schuettpelz E, Pryer KM et al (2004) Ferns diversified in the shadow of angiosperms. Nature 428:553–557. doi:10.1038/nature02361

Schneider K, Resl P, Spribille T (2016) Escape from the cryptic species trap: lichen evolution on both sides of a cyanobacterial acquisition event. doi:10.1111/mec.13636

Schuettpelz E, Pryer KM (2009) Evidence for a cenozoic radiation of ferns in an angiosperm-dominated canopy. Proc Natl Acad Sci U S A 106:11200–11205. doi:10.1073/pnas.0811136106

Schultz M, Arendholz WR, Büdel B (2001) Origin and evolution of the lichenized ascomycete order Lichinales: monophyly and systematic relationships inferred from ascus, fruiting body and SSU rDNA evolution. Plant Biol 3:116–123

Schüßler A (2000) *Glomus claroideum* forms an arbuscular mycorrhiza-like symbiosis with the hornwort *Anthoceros punctatus*. Mycorrhiza 10:15–21. doi:10.1007/s005720050282

Schüßler A (2012) The *Geosiphon–Nostoc* endosymbiosis and its role as a model for arbuscular mycorrhiza research. In: Hock B (ed) The Mycota IX—fungal associations, 2nd edn. Springer, Berlin. doi:10.1007/978-3-642-30826-0_5

Schüßler A, Walker C (2011) Evolution of the ‘plant-symbiotic’ fungal phylum, Glomeromycota. In: Pöggeler S, Wöstemeyer J (eds) The Mycota VII—evolution of fungi and fungal-like organisms. Springer, Berlin, pp 163–185. doi:10.1007/978-3-642-19974-5_7

Schüßler A, Martin H, Cohen D et al (2006) Characterization of a carbohydrate transporter from symbiotic glomeromycotan fungi. Nature 444:933–936. doi:10.1038/nature05364

Schüßler A, Martin H, Cohen D et al (2007) Arbuscular mycorrhiza: studies on the *Geosiphon* symbiosis lead to the characterization of the first glomeromycotean sugar transporter. Plant Signal Behav 2:431–434. doi:10.4161/psb.2.5.4465

Selosse MA, Strullu-Derrien C, Martin FM et al (2015) Plants, fungi and oomycetes: a 400-million year affair that shapes the biosphere. New Phytol 206:501–506. doi:10.1111/nph.13371

Shaw AJ, Szövényi P, Shaw B (2011) Bryophyte diversity and evolution: windows into the early evolution of land plants. Am J Bot 98:352–369. doi:10.3732/ajb.1000316

Shih PM, Matzke NJ (2013) Primary endosymbiosis events date to the later proterozoic with cross-calibrated phylogenetic dating of duplicated ATPase proteins. Proc Natl Acad Sci U S A 110:12355–12360. doi:10.1073/pnas.1305813110

Shishido TK, Humisto A, Jokela J et al (2015) Antifungal compounds from cyanobacteria. Mar Drugs 13:2124–2140. doi:10.3390/md13042124

Sigurbjörnsdóttir MA, Heiðmarsson S, Jónsdóttir AR et al (2014) Novel bacteria associated with arctic seashore lichens have potential roles in nutrient scavenging. Can J Microbiol 60:307–317. doi:10.1139/cjm-2013-0888

Sigurbjörnsdóttir MA, Andrésson ÓS, Vilhelmsson O (2015) Analysis of the *Peltigera membranacea* metagenome indicates that lichen-associated bacteria are involved in phosphate solubilization. Microbiology 161:989–996. doi:10.1099/mic.0.000069

Silvestro D, Cascales-Miñana B, Bacon CD, Antonelli A (2015) Revisiting the origin and diversification of vascular plants through a comprehensive bayesian analysis of the fossil record. New Phytol 207:425–436. doi:10.1111/nph.13247

Söderbäck E, Bergman B (1993) The *Nostoc-Gunnera* symbiosis: carbon fixation and translocation. Physiol Plant 89:125–132. doi:10.1111/j.1399-3054.1993.tb01795.x

Solheim B, Endal A, Vigstad H (1996) Nitrogen fixation in arctic vegetation and soils from Svalbard, Norway. Polar Biol 16:35–40. doi:10.1007/BF02388733

Soltis DE, Senters AE, Zanis MJ et al (2003) Gunnerales are sister to other core eudicots: implications for the evolution of pentamery. Am J Bot 90:461–470. doi:10.3732/ajb.90.3.461

Sood A, Prasanna R, Prasanna BM, Singh PK (2008) Genetic diversity among and within cultured cyanobionts of diverse species of *Azolla*. Folia Microbiol 53:35–43. doi:10.1007/s12223-008-0005-2

Spirbille T, Tønsberg T, Stebentheiner E et al (2014) Reassessing evolutionary relationships in the filamentous cyanolichen genus *Spilonema* (Peltigerales, Lecanoromycetes). Lichenologist 46:373–388. doi:10.1017/S0024282913000601

Splitt SD, Risser DD (2016) The non-metabolizable sucrose analog sucralose is a potent inhibitor of hormogonium differentiation in the filamentous cyanobacterium *Nostoc punctiforme*. Arch Microbiol 198:137–147. doi:10.1007/s00203-015-1171-7

Stenroos S, Laukka T, Huhtinen S et al (2010) Multiple origins of symbioses between ascomycetes and bryophytes suggested by a five-gene phylogeny. Cladistics 26:281–300. doi:10.1111/j.1096-0031.2009.00284.x

Strullu-Derrien C, Kenrick P, Pressel S et al (2014) Fungal associations in *Horneophyton ligneri* from the Rhynie Chert (c. 407 million year old) closely resemble those in extant lower land plants: novel insights into ancestral plant-fungus symbioses. New Phytol 203:964–979. doi:10.1111/nph.12805

Stucken K, John U, Cembella A et al (2010) The smallest known genomes of multicellular and toxic cyanobacteria: comparison, minimal gene sets for linked traits and the evolutionary implications. PLoS One 5, e9235. doi:10.1371/journal.pone.0009235

Sundue MA, Testo WL, Ranker TA (2016) Morphological innovation, ecological opportunity, and the radiation of a major vascular epiphyte lineage. Evolution 69:2482–2495. doi:10.1111/evo.12749

Svenning MM, Eriksson T, Rasmussen U (2005) Phylogeny of symbiotic cyanobacteria within the genus *Nostoc* based on 16S rDNA sequence analyses. Arch Microbiol 183:19–26. doi:10.1007/s00203-004-0740-y

Tang N, San Clemente H, Roy S (2016) A survey of the gene repertoire of *Gigaspora rosea* unravels conserved features among glomeromycota for obligate biotrophy. Front Microbiol 7:233. doi:10.3389/fmicb.2016.00233

Taylor TN, Krings M (2005) Fossil microorganisms and land plants: associations and interactions. Symbiosis 40:119–135

Taylor TN, Hass H, Kerp H (1997) A cyanolichen from the lower devonian rhynie chert. Am J Bot 84:992–1004

Taylor NT, Taylor EL, Krings M (2009) Paleobotany. The biology and evolution of fossil plants, 2nd edn. Academic, London
Taylor NT, Krings M, Taylor EL (2015a) Fossil fungi. Academic, London
Taylor ZN, Rice DW, Palmer JD (2015b) The complete moss mitochondrial genome in the angiosperm *Amborella* is a chimera derived from two moss whole-genome transfers. PLoS One 10(11), e0137532. doi:10.1371/journal.pone.0137532
Teixeira da Silva JA, Tsavkelova EA, Zeng S et al (2015) Symbiotic in vitro seed propagation of *Dendrobium*: fungal and bacterial partners and their influence on plant growth and development. Planta 242:1–22. doi:10.1007/s00425-015-2301-9
Terry I, Walter GH, Moore C et al (2007) Odor-mediated push-pull pollination in cycads. Science 318:70. doi:10.1126/science.1145147
Thajuddi N, Muralithara G, Sundaramoorth M et al (2010) Morphological and genetic diversity of symbiotic cyanobacteria from cycads. J Basic Microbiol 50:254–265. doi:10.1002/jobm.200900343
Tomitani A, Knoll AH, Cavanaugh et al (2006) The evolutionary diversification of cyanobacteria: molecular-phylogenetic and paleontological perspectives. Proc Natl Acad Sci USA 103:5442–5447. doi:10.1073/pnas.0600999103
Toomer KH, Chen X, Naito M et al (2015) Molecular evolution patterns reveal life history features of mycoplasma-related endobacteria associated with arbuscular mycorrhizal fungi. Mol Ecol 24:3485–3500. doi:10.1111/mec.13250
Torres-Cortésa G, Ghignoneb S, Bonfantec P et al (2015) Mosaic genome of endobacteria in arbuscular mycorrhizal fungi: transkingdom gene transfer in an ancient mycoplasma-fungus association. Proc Natl Acad Sci U S A 112:7785–7790. doi:10.1073/pnas.1501540112
Tsavkelova EA, Lobakova ES, Kolomeĭtseva GL et al (2003) Localization of associative cyanobacteria on the roots of epiphytic orchids. Microbiology 72:86–91. doi:10.1023/A:1022286225013
Tschermak-Woess E (1988) The algal partner. In: Galun M (ed) CRC handbook of lichenology, vol I. CRC Press, Boca Raton, pp 39–92
Turetsky MR, Bond-Lamberty B, Euskirchen E et al (2012) The resilience and functional role of moss in boreal and arctic ecosystems. New Phytol 196:49–67. doi:10.1111/j.1469-8137.2012.04254.x
U'Ren JM, Lutzoni F, Miadlikowska J et al (2012) Host- and geographic structure of endophytic and endolichenic fungi at a continental scale. Am J Bot 99:898–914. doi:10.3732/ajb.1100459
Uheda E, Silvester WB (2001) The role of papillae during the infection process in the *Gunnera – Nostoc* symbiosis. Plant Cell Physiol 42:780–783. doi:10.1093/pcp/pce097
Valverde A, De Maayer P, Oberholster T et al (2016) Specific microbial communities associate with the rhizosphere of *Welwitschia mirabilis*, a living fossil. PLoS One 11, e0153353. doi:10.1371/journal.pone.0153353
van der Heijden MG, Martin FM, Selosse MA et al (2015) Mycorrhizal ecology and evolution: the past, the present, and the future. New Phytol 205:1406–1423. doi:10.1111/nph.13288
Vekemans D, Proost S, Vanneste K et al (2014) Gamma paleohexaploidy in the stem lineage of core eudicots: significance for MADS-box gene and species diversification. Mol Biol Evol 29:3793–3806. doi:10.1093/molbev/mss183
Vestola J, Shishido TK, Jokela J et al (2014) Hassallidins, antifungal glycolipopeptides, are widespread among cyanobacteria and are the end-product of a nonribosomal pathway. Proc Natl Acad Sci U S A 111:E1909–E1917. doi:10.1073/pnas.1320913111
Vigil-Stenman T, Larsson J, Nylander JA et al (2015) Local hopping mobile DNA implicated in pseudogene formation and reductive evolution in an obligate cyanobacteria-plant symbiosis. BMC Genomics 16:193. doi:10.1186/s12864-015-1386-7
Villamil N, Márquez-Guzmán J, Boege K (2013) Understanding ontogenetic trajectories of indirect defence: ecological and anatomical constraints in the production of extrafloral nectaries. Ann Bot 112:701–709. doi:10.1093/aob/mct005
Villarreal JC, Renner SS (2014) A review of molecular-clock calibrations and substitution rates in liverworts, mosses, and hornworts, and a timeframe for a taxonomically cleaned-up genus *Nothoceros*. Mol Phylogenet Evol 78:25–35. doi:10.1016/j.ympev.2014.04.014

Villarreal AJC, Renzaglia KS (2006) Structure and development of *Nostoc* strands in *Leiosporoceros dussii* (Anthocerotophyta): a novel symbiosis in land plants. Am J Bot 93:693–705. doi:10.3732/ajb.93.5.693

Villarreal JC, Cargill DC, Hagborg A et al (2010) A synthesis of hornwort diversity: patterns, causes and future work. Phytotaxa 9:150–166. doi:10.11646/phytotaxa.9.1.8

Villarreal JC, Forrest LL, Wickett N, Goffinet B (2013) The plastid genome of the hornwort *Nothoceros aenigmaticus* (Dendrocerotaceae): phylogenetic signal in inverted repeat expansion, pseudogenization, and intron gain. Am J Bot 100:467–477. doi:10.3732/ajb.1200429

Villarreal JC, Cusimano N, Renner SS (2015) Biogeography and diversification rates in hornworts: the limitations of diversification modeling. Taxon 64:229–238. doi:10.12705/642.7

Villarreal AJC, Crandall-Stotler BJ, Hart ML et al (2016) Divergence times and the evolution of morphological complexity in an early land plant lineage (Marchantiopsida) with a slow molecular rate. New Phytol 209:1734–1746. doi:10.1111/nph.13716

Vivas M, Sacristan M, Legaz ME et al (2010) The cell recognition model in chlorolichens involving a fungal lectin binding to an algal ligand can be extended to cyanolichens. Plant Biol 12:615–621. doi:10.1111/j.1438-8677.2009.00250.x

Voytsekhovich A, Mikhailyuk TI, Darienko TM (2011a) Lichen photobionts. 1: biodiversity, ecophysiology and co-evolution with the mycobiont. Algologia 21:3–26

Voytsekhovich A, Mikhailyuk TI, Darienko TM (2011b) Lichen photobionts. 2: origin and correlation with mycobiont. Algologia 21:151–177

Wahrmund U, Groth-Malonek M, Knoop V (2008) Tracing plant mitochondrial DNA evolution: rearrangements of the ancient mitochondrial gene cluster trnA-trnT-nad7 in liverwort phylogeny. J Mol Evol 66:621–629. doi:10.1007/s00239-008-9114-4

Wang XQ, Ran JH (2014) Evolution and biogeography of gymnosperms. Mol Phylogenet Evol 75:24–40. doi:10.1016/j.ympev.2014.02.005

Wang CM, Ekman M, Bergman B (2004) Expression of cyanobacterial genes involved in heterocyst differentiation and dinitrogen fixation along a plant symbiosis development profile. Mol Plant Microbe Interact 17:436–443. doi:10.1094/MPMI.2004.17.4.436

Wang YY, Liu B, Zhang XY et al (2014) Genome characteristics reveal the impact of lichenization on lichen-forming fungus *Endocarpon pusillum* Hedwig (Verrucariales, Ascomycota). BMC Genomics 15:34. doi:10.1186/1471-2164-15-34

Wanntorp L, Ronse De Craene LP (2005) The *Gunnera* flower: key to eudicot diversification or response to pollination mode? Int J Plant Sci 166:945–953

Wanntorp L, Wanntorp H-E (2003) The biogeography of *Gunnera* L.: vicariance and dispersal. J Biogeogr 30:979–987. doi:10.1046/j.1365-2699.2003.00895.x

Wanntorp L, Wanntorp H-E, Oxelman B et al (2001) Phylogeny of *Gunnera*. Plant Syst Evol 226:85–107. doi:10.1007/s006060170075

Wanntorp L, Wanntorp H-E, Rutishauser R (2003) On the homology of the scales in *Gunnera* (Gunneraceae). Bot J Linnean Soc 142:301–308. doi:10.1046/j.1095-8339.2003.00185.x

Wanntorp L, Dettmann ME, Jarzen DM (2004a) Tracking the mesozoic distribution of *Gunnera*: comparison with the fossil pollen species *Tricolpites reticulatus* cookson. Rev Palaeobot Palynol 132:163–174. doi:10.1016/j.revpalbo.2004.06.001

Wanntorp L, Praglowski J, Grafström E (2004b) New insights into the pollen morphology of the genus *Gunnera* (Gunneraceae). Grana 43:15–21

Warshan D, Bay G, Nahar N et al (2016) Seasonal variation in nifH abundance and expression of cyanobacterial communities associated with boreal feather mosses. ISME J. doi:10.1038/ismej.2016.17 [Epub ahead of print]

Weber MG, Agrawal AA (2014) Defense mutualisms enhance plant diversification. Proc Natl Acad Sci U S A 111:16442–16447. doi:10.1073/pnas.1413253111

Weber MG, Keeler KH (2013) The phylogenetic distribution of extrafloral nectaries in plants. Ann Bot 111:1251–1261. doi:10.1093/aob/mcs225

Werth S, Millanes AM, Wedin M et al (2013) Lichenicolous fungi show population subdivision by host species but do not share population history with their hosts. Fungal Biol 117:71–84. doi:10.1016/j.funbio.2012

West N, Adams DG (1997) Phenotypic and genotypic comparison of symbiotic and free-living cyanobacteria from a single field site. Appl Environ Microbiol 63:4479–4484

Wickett NJ, Mirarab S, Nguyen N (2014) Phylotranscriptomic analysis of the origin and early diversification of land plants. Proc Natl Acad Sci U S A 111:E4859–E4868. doi:10.1073/pnas.1323926111

Wojtczak G, Janik P (2016) Phytoremediation with *Geosiphon*-like symbiosis? Environ Sci Pollut Res Int 23:5992–5994. doi:10.1007/s11356-016-6135-1

Wolfe BE, Husband BC, Klironomos JN (2005) Effects of a belowground mutualism on an aboveground mutualism. Ecol Lett 8:218–223. doi:10.1111/j.1461-0248.2004.00716.x

Wright D, Prickett T, Heim RF et al (2001) Form species *Nostoc commune* (Cyanobacteria). Int J Syst Evol Microbiol 51:1839–1852. doi:10.1099/00207713-51-5-1839

Wu L, Zhang G, Lan S et al (2014) Longitudinal photosynthetic gradient in crust lichens' thalli. Microb Ecol 67:888–896. doi:10.1007/s00248-014-0366-9

Xi Z, Rest JS, Davis CC (2013) Phylogenomics and coalescent analyses resolve extant seed plant relationships. PLoS One 8(11):e80870. doi:10.1371/journal.pone.0080870

Xue JY, Liu Y, Li L, Wang B, Qiu YL (2010) The complete mitochondrial genome sequence of the hornwort *Phaeoceros laevis*: retention of many ancient pseudogenes and conservative evolution of mitochondrial genomes in hornworts. Curr Genet 56:53–61. doi:10.1007/s00294-009-0279-1

Yamada S, Ohkubo S, Miyashita et al (2012) Genetic diversity of symbiotic cyanobacteria in *Cycas revoluta* (Cycadaceae). FEMS Microbiol Ecol 81:696–706. doi:10.1111/j.1574-6941.2012.01403.x

Zackrisson O, DeLuca TH, Nilsson M-C et al (2004) Nitrogen fixation increases with successional age in boreal forests. Ecology 85:3327–3334. doi:10.1890/04-0461

Zheng WW, Song TY, Bao XD et al (2002) High cyanobacterial diversity in coralloid roots of cycads revealed by PCR fingerprinting. FEMS Microbiol Ecol 40:215–222. doi:10.1111/j.1574-6941.2002.tb00954.x

Zheng S-P, Chen B, Guan X et al (2008) Diversity analysis of endophytic bacteria within *Azolla microphylla* using PCR-DGGE and electron microscopy. Chin J Agric Biotechnol 16:508–514. doi:10.1017/S1479236208002441

Zheng W, Bergman B, Chen B et al (2009) Cellular responses in the cyanobacterial symbiont during its vertical transfer between plant generations in the *Azolla microphylla* symbiosis. New Phytol 181:53–61. doi:10.1111/j.1469-8137.2008.02644.x

Zúñiga C, Leiva D, Ramírez-Fernández L et al (2015) Phylogenetic diversity of *Peltigera* cyanolichens and their photobionts in southern Chile and Antarctica. Microbes Environ 30:172–179. doi:10.1264/jsme2.ME14156

Phototrophic Microbial Mats

Lucas J. Stal, Henk Bolhuis, and Mariana Silvia Cretoiu

Abstract Microbial mats are structured, small-scale microbial ecosystems, and similar as biofilms cover a substratum like a tissue. A general characteristic of a microbial mat is the steep physicochemical gradients that are the result of the metabolic activities of the mat microorganisms. Virtually every microbial mat is formed through autotrophic metabolism and through the fixation of atmospheric dinitrogen. Chemoautotrophic organisms fuel these processes in the absence of light. In illuminated environments photoautotrophic organisms are the driving force and these mats are subject of this chapter. In the vast majority of cases, primary production by the oxygenic phototrophic cyanobacteria is the basis of a diverse community that forms a living entity with a macroscopic habitus. This entity has its own physiology that is the result of interaction, communication, cooperation, and competition of the individual functional groups of microorganisms. Organic matter is remineralized and in sulfur-dominated environments sulfate-reducing bacteria are responsible for end-oxidation that leads to the production of sulfide, which is used by anoxygenic photoautotrophic bacteria. Aerobic and anaerobic anoxygenic phototrophic bacteria and proteorhodopsin-containing bacteria are important as secondary producers and take care of the decomposition of organic matter in a process that is aided by light.

L.J. Stal (✉)
Department of Marine Microbiology and Biogeochemistry, Royal Netherlands Institute for Sea Research and Utrecht University, Yerseke, The Netherlands

Department of Aquatic Microbiology, University of Amsterdam, Amsterdam, The Netherlands
e-mail: lucas.stal@nioz.nl

H. Bolhuis • M.S. Cretoiu
Department of Marine Microbiology and Biogeochemistry, Royal Netherlands Institute for Sea Research and Utrecht University, Yerseke, The Netherlands

© Springer International Publishing Switzerland 2017
P.C. Hallenbeck (ed.), *Modern Topics in the Phototrophic Prokaryotes*,
DOI 10.1007/978-3-319-46261-5_9

Introduction

Microbial mats are benthic communities of different functional groups of microorganisms that are organized in a vertically stratified way, determined by the physicochemical gradients that are the result of the microbial community's metabolic activities (Jørgensen et al. 1983). Microbial mats are often found in environments that largely exclude grazing organisms that would otherwise prevent the accumulation of a large density of standing stock biomass that characterizes microbial mats (Fenchel 1998). A microbial mat is recognized as a macroscopic phenomenon with a characteristic habitus (Fig. 1). A microbial mat can be considered as a living entity formed by the component microorganisms that together perform the physiological functions of the whole.

Microbial mats are found in extreme environments such as hypersaline ponds, hot springs, hot- and cold dry deserts, and coastal intertidal environments (Stal 2012). The latter does perhaps not strike as an extreme environment such as the other. However, intertidal sediments experience large changes in environmental conditions. Intertidal microbial mats are more or less regularly inundated by the sea, experience desiccation, and are subject to huge salinity changes (evaporation, rain), large temperature fluctuations, nutrient depletions, and oxygen levels change from supersaturation to anoxic conditions that coincide with high levels of toxic sulfide (Stal et al. 1985).

In illuminated environments most microbial mats are built by cyanobacteria, oxygenic photoautotrophic bacteria that colonize bare sediments and form the basis of the microbial mat through the primary production of organic matter synthesized from the photosynthetic fixed carbon dioxide. Such mats are therefore often also referred to as cyanobacterial mats or phototrophic microbial mats. The photosynthetic produced organic matter serves as a substrate for a variety of functionally different microorganisms (Stal 2001). Therefore, cyanobacteria represent only part of the community of which purple and colorless sulfur bacteria and sulfate-reducing bacteria may account for 40 % of the bacterial community (Visscher

Fig. 1 A coastal microbial mat and a cross section of it showing the vertical stratified communities of phototrophic microorganisms

and Van Gemerden 1993). Cyanobacteria have even been reported to be a minor component of a hypersaline microbial mat (Ley et al. 2006) and of mats growing on the beaches of the Orkney islands, which were dominated by anoxygenic phototrophs (Wieland et al. 2003).

Microbial mats are exceptionally productive ecosystems in which cyanobacteria are the main primary producers, although eukaryotic microalgae such as diatoms and anoxygenic phototrophic bacteria may contribute to the primary production as well in addition to chemosynthetic production by sulfate-reducing and methanogenic bacteria (Fourçans et al. 2004; Wieland et al. 2003). In the Ebro Delta the production of organic carbon as high as ~200 g C m^{-2} y^{-1} was reported (Urmeneta et al. 1998). The highest daily productions reported in microbial mats in the coastal North Sea and in the tropical hypersaline Solar Lake were 5–6 g C m^{-2} d^{-1} (Stal 2012), which probably would give similar yearly productions as in the Ebro Delta mats. Microbial mats possess high cell densities. For example, in the hypersaline evaporitic microbial mats of Guerrero Negro (Baja California, Mexico) cell densities of 4.2×10^9 cm^{-3} have been reported (Jahnke et al. 2014).

Microbial mats are also characterized by high diversity. For instance, the microbial mats of the hypersaline lagoons at Guerrero Negro generated more than 1500 16S rRNA sequences representing over 750 species (Ley et al. 2006). Bolhuis and Stal (2011) found an even higher diversity in coastal microbial mats of 2000–4000 bacteria and 100–300 archaea and concluded that these mats were among the most diverse ecosystems known. Bolhuis et al. (2014) showed that coastal microbial mats are more diverse than mats from the more extreme hypersaline and hot springs. The latter were the least diverse and high temperature is apparently a strong selective parameter.

Laminated microbial mats have been considered as the modern counterparts of Archean and Proterozoic stromatolites (Hamilton et al. 2016). Stromatolites are lithified fossil remains of microbial mats that developed in shallow coastal waters and that show remarkable morphological similarities with certain modern microbial mats.

Fairy Rings

The microbial mats developing on the bottom of the shallow evaporation ponds in Guerande salterns show a phenomenon that has been termed “fairy rings” and the origin have been subject of speculation (Gerdes et al. 1993) (Fig. 2). These discoid structures are known from the fossil record and been interpreted as originating from medusas (coelenterate) and known as *Cyclomedusa*. The fairy rings forming in the microbial mats of the Guerande salterns are extant concentric ring-shaped structures and are most likely of microbial origin. Similar phenomena were discovered in coastal microbial mats and were possibly caused by an infection with the fungus *Emericellopsis* sp. (Carreira et al. 2015). It was hypothesized that this fungus infects the mat-forming cyanobacteria and kill them and subsequently migrate outwards

Fig. 2 "Fairy rings" in microbial mats of the Guerande salterns (France)

from the point of first infection. In the centrum cyanobacteria would re-colonize the sediment, hence forming rings of cyanobacteria followed by rings of infection. This is an interesting hypothesis. Fungi form an important but neglected component of microbial mats as are other microbial eukarya.

The Vertical Distribution and Migration of Microbial Mat Organisms

The oxygenic photosynthetic activity causes an oxygen supersaturation in the layer of cyanobacteria, which is in principle the sunlight exposed (green, blue-green, or dark green) top layer of the microbial mat. Sometimes a biofilm of epipelic diatoms covers the cyanobacterial mat and in that case these oxygenic phototrophic eukaryotes form a brown layer on top of the mat. In other cases, the cyanobacterial mat is hidden under a layer of recently deposited sand. In both cases, the generally low light adapted cyanobacteria are protected from excessive high irradiance. This "sunglass" effect may also be achieved when a top layer of sheath material is left behind from dead cyanobacteria. These sheaths are composed of recalcitrant polymeric substances and may contain the UV-protection pigment scytonemin or other sunscreen pigments such as mycosporine-like amino acids (Garcia-Pichel and Castenholz 1991, 1993).

Hence, while the cyanobacteria at the utmost surface may be photoinhibited, deeper in the mat they will find their optimum light conditions, while even deeper the cyanobacteria may become light limited or even live in permanent darkness

(Lassen et al. 1992; Wieland et al. 2003). Not only the total irradiance decreases strongly in the cyanobacterial mat but also the photopigments of the cyanobacteria higher in the mat will absorb the corresponding wavelengths, leaving the wavelengths that cannot be absorbed by cyanobacteria. Sunlight follows a diurnal sinus and on top of that fluctuations of light happen due to partly overcast. While microbial mats often develop in sedimentary environments and sediment deposits on top of the mat, motile cyanobacteria may constantly migrate and positioning themselves optimally in the ever changing light gradient in the mat (Dillon et al. 2009).

Motile cyanobacteria migrate by gliding movement and the direction and speed is guided by light (phototaxis, photokinesis, and photophobic response), including UV light (Bebout and Garcia-Pichel 1995), although there are also reports of chemotaxis (e.g., towards CO_2, bicarbonate, and O_2 (Malin and Walsby 1985); fructose and sulfide (Richardson and Castenholz 1989); water (hydrotaxis) (Pringault and Garcia-Pichel 2004); salt (halotaxis) (Kohls et al. 2010), and inhibition of motility by sulfide (Richardson and Castenholz 1987). Immotile cyanobacteria (many of the benthic unicellular cyanobacteria are immotile) will follow the light gradient by growing faster at optimum (average) light intensity.

Anoxygenic phototrophic bacteria migrate in response to the fluctuating and shifting vertical gradients of O_2 and pH (Fourçans et al. 2006). These authors used the microscale depth distribution of *pufM*, a gene that codes for one of the subunits of the photosynthetic reaction center of anoxygenic phototrophic bacteria, and found that the distribution of the biomass of filamentous cyanobacteria was the most important factor determining the depth distribution of the anoxygenic phototrophic organisms. This day and night migration is an efficient behavior to cope with the fluxes of oxygen and sulfide.

At the basis of the cyanobacterial mat there is insufficient light of the spectral quality needed to support oxygenic photosynthesis. Consequently, cyanobacteria at the basis of the mat will be unable to sustain metabolism and eventually die and decompose, providing substrate for chemosynthetic microorganisms. The thickness of the cyanobacterial layer is thus a function of the light penetrating the mat, which is a function of the incident light intensity and the sun angle (Ramsing et al. 2000). Also the properties of the sediment play a role. Quartz and carbonate sand scatter light more than silt and mud particles and the proportional composition of these sediment types determine the attenuation of light (Watermann et al. 1999).

The Physiology of Phototrophic Microbial Mats

Obviously, the evolution of oxygen occurs only when daylight is present. Initially, the oxygen diffuses out of the mat at a lower rate than its photosynthetic production, which results in supersaturation until diffusion out is in equilibrium with oxygen evolution. When light ceases, diffusion becomes higher than oxygen production and the level of dissolved oxygen goes down (Revsbech et al. 1983). In the dark, cyanobacteria switch to respiration to cover maintenance energy requirements. Dark energy generation may also cover the demands of metabolic processes such as the fixation of dinitrogen (see below) and sometimes even may support growth (Stal and Moezelaar 1997). When respiration by the cyanobacteria and other microorganisms in the mat exceeds diffusion of oxygen from the overlying water or atmosphere, the mat turns anoxic. This can often happen within minutes because of the high demand of the dense standing stock of microorganisms (Villbrandt et al. 1990).

Hence, in the dark the microbial mat is anoxic almost to the very surface (any oxygen diffusing into the mat is scavenged immediately). Anaerobic processes are therefore dominating the mat's metabolism. This includes the cyanobacteria that no longer can rely on aerobic respiration of the storage compounds that were synthesized during the daytime. Mat-forming cyanobacteria are known to be able to ferment their intracellular storage carbohydrate and produce low-molecular organic compounds such as acetate, lactate, and ethanol, in addition to hydrogen and CO_2 gas and formic acid (Stal and Moezelaar 1997). These fermentation products are excellent substrates for sulfate-reducing bacteria that form sulfide.

Sulfide serves as electron donor for anoxygenic phototrophic bacteria that oxidize it back to sulfate, via elemental sulfur or other sulfur compounds with intermediate oxidation states (van Gemerden 1993). The anaerobic anoxygenic purple sulfur bacteria form a layer below the cyanobacteria. Sometimes these communities are separated by a layer of iron oxides that could represent a buffer between the aerobic and permanent anaerobic layers (see below) (Stal 2001). Green sulfur bacteria sometimes form a layer beneath the purple sulfur bacteria (Pierson et al. 1987). Both groups of anoxygenic bacteria are obligate or facultative (some purple sulfur bacteria) anaerobes. Purple sulfur bacteria use far red light, which is not used by the cyanobacteria and consequently is not attenuated and penetrates the cyanobacterial layer. Green sulfur bacteria photosynthesize at very low light intensities (as low as 0.015 μmol photons m^{-2} s^{-1}, which equals moon light) (Manske et al. 2005). Cyanobacteria may also exhibit an anoxygenic phototrophic mode of metabolism or even may even grow in this mode (see below) (Cohen et al. 1975).

Microbial mats may produce H_2 in a variety of ways. Most importantly, H_2 is a product of fermentation of organic matter, including the reserve compounds in phototrophic organisms (Stal and Moezelaar 1997; Burow et al. 2012) but a second good candidate is the nitrogen-fixing enzyme complex nitrogenase that evolves at least one H_2 for every N_2 reduced to ammonia (Kim and Rees 1994). Under saturation of nitrogenase with N_2 may increase the H_2 evolution accordingly. Non-heterocystous filamentous mat-forming cyanobacteria such as *Lyngbya* and *Microcoleus* are important for the export of H_2 (Burow et al. 2012; Hoffmann et al. 2015). It has been speculated that the escape of microbial mat produced H_2 might have been an important process that led to the oxidation of the primitive Earth (Hoehler et al. 2001). While in anoxic marine sediments the H_2 is efficiently used by sulfate-reducing bacteria this seems not to be the case in microbial mats (Hoehler et al. 2002). H_2 production in mats is light controlled and follows the daily light cycle. Methanogens rather than sulfate-reducing bacteria take advantage and consume the H_2.

Sulfate-reducing bacteria are in majority obligate anaerobic organisms although there are some that withstand up to atmospheric levels of oxygen and a few seem even to respire low levels of oxygen (Canfield and Des Marais 1991; Sigalevich et al. 2000). These microorganisms are not pigmented but the sulfide that they produce precipitates with iron to the black iron sulfide (which may further react with elemental sulfur to the gray pyrite). The black horizon below the varicolored photosynthetic layers is often denoted as the layer of sulfate reduction. This is, however, not necessarily true and sulfate-reducing bacteria as well as sulfate reduction are encountered throughout the microbial mat. Colorless sulfur-oxidizing bacteria thrive at the interface of oxygen and sulfide, which is moving up and down in a day–night cycle. Hence, although these bacteria are abundantly present, they also do not form a distinct vertical stratification (Visscher et al. 1992) (although among these groups different species or ecotypes may be vertically stratified, Risatti et al. 1994). Besides the said bacteria a plethora of other functional groups of microorganisms occurs in microbial mats, making them among the most diverse microbial ecosystems known (Bolhuis and Stal 2011).

Sulfide is produced by sulfate-reducing bacteria or by bacteria that reduce any inorganic sulfur compound with a redox state between sulfate (+6) and sulfide (−2). Particularly, the reduction of elemental sulfur to sulfide is important in microbial mats. The reduction of elemental sulfur to sulfide (sulfur respiration, disproportionation, or as electron sink in fermentation) and its subsequent oxidation back to elemental sulfur by anoxygenic photosynthesis in microbial mats is also known as the mini sulfur cycle (van Gemerden 1993).

Sulfide is an extremely toxic compound, especially for any organism with electron transport chains. Sulfide is inhibitory for oxygenic phototrophs, including cyanobacteria the growth of which may cease even at moderately concentrations of sulfide, and as a result the metabolic activity of the entire ecosystem may come to a halt. However, in microbial mats this effect is counteracted by colorless and purple sulfur bacteria. Colorless sulfur bacteria have a high affinity for sulfide but generally need oxygen to oxidize it. Some are capable of using nitrate as terminal electron acceptor (denitrification). But nitrate is in very low supply in microbial mats and oxygen is available only during daytime in the top layer of the mat. Under low supply of oxygen colorless sulfur bacteria oxidize sulfide incompletely and produce substrates that can be used by the purple sulfur bacteria such as elemental sulfur, polysulfide, or thiosulfate. For the oxidation of these substrates purple sulfur bacteria need light and anaerobic conditions. Hence, only during the daytime it is possible to keep the sulfide concentration low. During the dark, sulfide is produced by a plethora of different microorganisms, including purple sulfur bacteria, cyanobacteria, and, of course, the sulfate-reducing bacteria. The latter group of organisms receives its substrate from the cyanobacteria, purple sulfur bacteria, and many other chemoorganotrophic bacteria that produce the low-molecular compounds from fermentation (Stal and Moezelaar 1997). Purple sulfur bacteria such as *Thiocapsa* and *Chromatium* are responsible for the fact that very little H_2S produced in microbial mats reaches the atmosphere.

Another mechanism that keeps the sulfide concentration low is through its reaction with iron, which is abundantly available in coastal microbial mats. Ferrous iron forms FeS with sulfide that precipitates and gives the typical black color to the deeper anoxic part of the microbial mat. Sulfide reduces oxidized (ferric) iron to ferrous iron and sulfur, which subsequently reacts with sulfide to FeS, which may further react with sulfur to form pyrite (FeS_2). Iron sulfide and pyrite are insoluble and stable minerals that undergo slow chemical or biological oxidation. Pyrite can be formed directly from polysulfide and this reaction may be faster than the reaction between FeS and elemental sulfur and may be either chemical or biological (Howarth 1979).

N_2 Fixation

The fixation of dinitrogen (N_2) is an important process in microbial mats. It provides an infinite source of one of the quantitatively most important elements for organisms.

All microbial mats that have been investigated for dinitrogen fixation appeared indeed to be diazotrophic (Severin and Stal 2010). Microbial mats are dense communities containing a high amount of biomass in which nitrogen is quantitatively the second most important element (after carbon), amounting ~10 % of the dry weight. The synthesis and maintenance of this biomass demands therefore sufficient nitrogen, which is provided by the biological fixation of atmospheric dinitrogen. There is an ongoing debate about which microorganisms are actually responsible for this fixation of N_2.

Only certain Bacteria and a few Archaea that are capable of synthesizing nitrogenase are able to fix dinitrogen (N_2) (Kim and Rees 1994). Nitrogenase is an enzyme complex of dinitrogenase reductase (Fe-enzyme) and dinitrogenase (Fe-Mo-enzyme). The former reduces the latter using reduced ferredoxin as electron acceptor and 2 ATP per electron transferred. Reduced dinitrogenase reduces N_2 to $2NH_3$ and H_2. Hence, the fixation of N_2 comes at a considerable energetic cost. Another important feature is that nitrogenase is sensitive to oxygen and functions only under near anoxic conditions (Gallon 1992). These properties limit the ability and ecological feasibility of using N_2 as source of nitrogen to organisms that (1) possess the genetic potential to synthesize functional nitrogenase, (2) provide sufficient energy and low-potential reducing equivalents, and (3) live in an anoxic environment or otherwise possess a strategy that protects nitrogenase from oxygen inactivation. No eukaryotic organism is known to possess the capacity of dinitrogen fixation except when in symbiosis with a diazotrophic bacterium (established symbioses of Eukarya with diazotrophic Archaea are not known) (Rai et al. 2000).

Measurements of nitrogenase activity in microbial mats show different patterns during a 24 h day, depending on the type of mat and its development, including the time of the year, the actual microbial composition, and the prevailing conditions. Villbrandt et al. (1990) monitored nitrogenase activity in two intertidal coastal mats in different stage of development. In summer (June) the mature, well-developed microbial mat, two sharp peaks of nitrogenase activity were observed, one at sunset and one at sunrise. Later in the year (September) nitrogenase activity was much lower and confined to the day. The developing (immature) mat also showed these activity peaks in summer but less pronounced because most nitrogenase activity was confined to the dark. In September nitrogenase activity was low and confined to the morning hours. The interpretation of these remarkable patterns was as follows. The mature mat turned anoxic at night, while during the daytime the mat is supersaturated with oxygen due to the oxygenic photosynthetic activity of the cyanobacteria in the mat. Such conditions are not supportive for the energy-demanding N_2 fixation. However, at sunset, photosynthesis ceases and the oxygen concentration drops, while low light is still available. At sunrise, the mat is still anoxic, but light is available. The non-heterocystous diazotrophic cyanobacteria use these windows of opportunity to fix N_2. The developing mat is characterized by low standing stock biomass and these mats do not turn anoxic at night. This allows aerobic respiration to cover the energy demand of the diazotrophs, although also in the developing mat sunset and sunrise seemed still optimum for N_2 fixation. This pattern was experimentally proven in an axenic culture of the mat-forming cyanobacterium *Lyngbya aestuari* (at the time assigned as *Oscillatoria* sp.) that was cultured under an alternating light-dark, aerobic–anaerobic cycle (Stal and Heyer 1987).

An investigation of the composition of *nifH*, one of the structural genes coding for the iron protein of nitrogenase, showed that approximately half of the sequences could be attributed to cyanobacteria (non-heterocystous filamentous and unicellular species), while 25 % belonged to Gammaproteobacteria (mainly purple sulfur bacteria) and Deltaproteobacteria (mainly sulfate-reducing bacteria). The other 25 % were unassigned (Severin et al. 2010). However, when the expression of the gene was considered, cyanobacteria contributed 60–80 %, while Gammaproteobacteria contributed 20–40 %. Only in the morning hours expression of *nifH* of Gammaproteobacteria was more important than that of cyanobacteria (Fig. 3). This seems logic because of the accumulation of sulfide during the night and the availability of light in the morning would stimulate the photosynthetic activity of this group of organisms.

The Case of *Microcoleus* (*Coleofasciculus*) *chthonoplastes*

Microcoleus chthonoplastes (re-assigned as *Coleofasciculus chthonoplastes* by Siegesmund et al. 2008) is a filamentous cyanobacterium belonging to Section 3 (Rippka et al. 1979), meaning that it does not differentiate cells. *Microcoleus* means "small sheath" (*Coleofasciculus* means "small sheathed bundle") and *chthonoplastes* means the "creator of the subterranean" or better the "maker of soil." This name refers to the organisms' capability of forming microbial mats, basically an organic film covering an inorganic surface. *M. chthonoplastes* is a cosmopolitan cyanobacterium that is characterized by long trichomes that are surrounded in a common

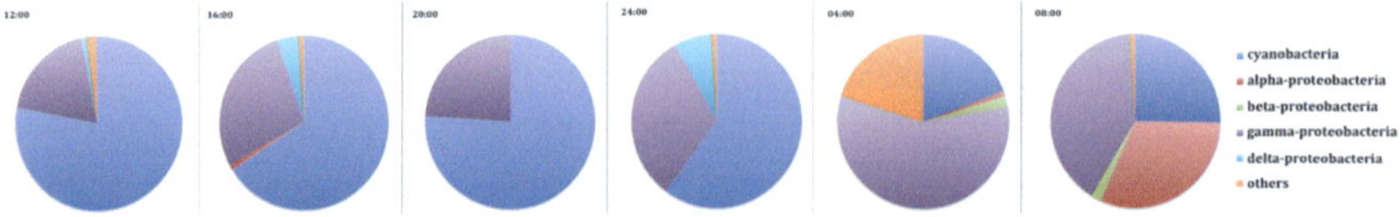

Fig. 3 Expression of nifH of different groups of microorganisms during a 24 h day

polysaccharide sheath (Fig. 4). This sheath may contain many tens of trichomes and the sheath may be divided into compartments. The sheath may be lost when the organism is isolated and cultured in the laboratory. This has led to misidentification of morphological similar, opportunistic isolates from microbial mats as *M. chthonoplastes* (Garcia-Pichel et al. 1996; Siegesmund et al. 2008). This cyanobacterium is known as a mat-builder worldwide.

The function of the sheath of *M. chthonoplastes* has been debated and is unknown but may be to protect the organism from desiccation. Microbial mats in intertidal sediments are often exposed to cycles of periodic desiccation. The sheath would retain water and in that way allows the organism to survive long periods of drought. It has been observed that upon hydration of a dehydrated bundle of *M. chthonoplastes*, the trichomes moved quickly out of the sheath (Campbell 1979). There may be other advantages of the sheath such as protection from grazing and from other environmental threads such as toxic chemicals (heavy metals, sulfide), while the many bundles in the mat would entangle with each other, trapping the sediment and in this way make a strong network that increases the erosion threshold of the sediment (Yallop et al. 1994). This would lead to a stable environment and provide a matrix in which the microbial mat community can develop protected from physical and biological forces.

For a long time, *M. chthonoplastes* mats are known to fix atmospheric nitrogen. However, the identity of the diazotrophic organism has been and still is a matter of debate (Zehr et al. 1995; Steppe et al. 1996). Because many cyanobacteria are capable of nitrogen fixation, *M. chthonoplastes* was obviously a candidate. Initially, it was conceived that the sheathed bundle is an adaptation to protect the oxygen-sensitive nitrogenase, but this has never proven, nor was a convincing conceptual model presented that explained how this could work (Potts et al. 1978). There have been reports of isolation of N_2-fixing *M. chthonoplastes*. Pearson et al. (1979) reported aerobic nitrogen fixation in *M. chthonoplastes*, but this isolate was later re-identified as *Symploca* sp. (Fredriksson et al. 1998). Diazotrophic *M. chthonoplastes* "strain 11," which was isolated from a coastal microbial mat from one of the Wadden Sea barrier islands, was re-identified as *Geitlerinema* sp. (Bolhuis et al. 2010). Villbrandt and Stal (unpublished results) screened all isolates of true *M. chthonoplastes* that were analyzed by Garcia-Pichel et al. (1996) and were unable to induce nitrogenase activity in any of them, not even under anoxic and anoxygenic conditions.

PCR amplification of *nifH* from *M. chthonoplastes* mats did not reveal cyanobacterial sequences but mainly sequences related to Gamma- and Deltaproteobacteria (Zehr et al. 1995; Olson et al. 1999; Omoregie et al. 2004). This led to the assumption that cyanobacteria in these mats were unimportant as diazotrophs and that N_2 fixation was mainly attributed to anoxygenic phototrophic bacteria and sulfate-reducing bacteria. It was thought that N_2 fixation was possible because of a joint venture between cyanobacteria and chemotrophic bacteria and that the former provide the organic substrate for the latter, exchanging fixed nitrogen in return (Steppe et al. 1996). Indeed, other bacteria can be found tightly associated with the *M. chthonoplastes* polysaccharide sheath or even within it (D'Amelio et al. 1987). These bacteria included members of the sulfate-reducing Desulfovibrionaceae (Deltaproteobacteria).

Bolhuis et al. (2010) discovered that the genome of *M. chthonoplastes* contained a complete *nif* cluster, but that most of it, including the structural genes *nifHDK*, was identical to those of Deltaproteobacteria or *Chlorobia*. These authors proved that this cluster must have been obtained by *M. chthonoplastes* through horizontal gene transfer, possibly from a sulfate-reducing bacterium. They also reported the same *nifHDK* genes in (true) *M. chthonoplastes* strains, isolated by different research groups from microbial mats from various parts of the world. This indicated that the transfer of this gene cluster must have occurred early in the evolution of *M. chthonoplastes*. This discovery explained the earlier observations of a lack of cyanobacterial *nif* genes in mats of *M. chthonoplastes*.

Induction of nitrogenase activity *M. chthonoplastes* failed and also transcription of *nifHDK* was not detected, even when anaerobic conditions were applied. Many anaerobically N_2-fixing cyanobacteria induce nitrogenase only when first starved for nitrogen under normal oxygenic photoautotrophic conditions until bleaching

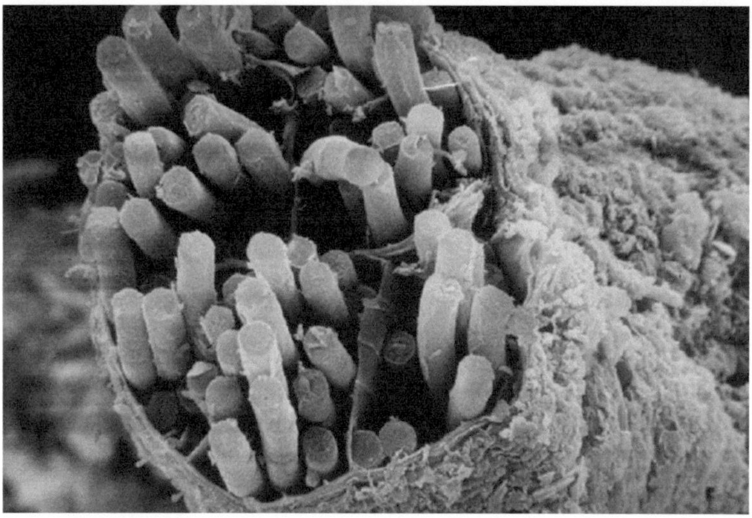

Fig. 4 *Microcoleus (Coleofasciculus) chthonoplastes*

commences (which is a sign that the phycobiliproteins are starting to degrade). Subsequently, the culture is made anoxic and the inhibitor of oxygenic photosynthesis (PS II), DCMU, is added in order to assure continued anaerobic conditions, after which nitrogenase activity appears in 3–12 h (Rippka and Waterbury 1977). However, this procedure did not lead to the induction of nitrogenase activity or transcription of the *nif* genes in *M. chthonoplastes*. Remarkably, using primers specific for the *M. chthonoplastes nifHDK*, transcription was found in the mat (together with nitrogenase activity, although this activity could not be unequivocally attributed to *M. chthonoplastes*) (Bolhuis et al. 2010). Hence, obviously the chosen conditions in the laboratory did not allow nitrogenase induction. Preliminary results indicate that growing *M. chthonoplastes* on sand (hence, benthic rather than as a liquid culture) when it also forms again the bundles as under natural conditions in the mat, or under low O_2 levels (5 %) induced *nif* gene transcription and revealed nitrogenase activity (Bolhuis and Stal, unpublished results).

Anoxygenic Photosynthesis by Cyanobacteria

While oxygenic photosynthesis uses water as the electron donor, anoxygenic photosynthetic CO_2 fixation relies on other electron donors such as reduced sulfur compounds, organic acids, hydrogen, nitrite, or ferrous iron (Padan 1979). Cyanobacteria thriving in microbial mats or sulfureta (environments dominated by sulfur) are known to perform photosystem I-dependent anoxygenic photosynthesis using sulfide as electron donor (van Gemerden 1993). Other electron donors for anoxygenic photosynthesis in cyanobacteria are not known. Cyanobacteria oxidize sulfide to elemental sulfur or thiosulfate, but in fact not much is known about anoxygenic photosynthesis of cyanobacteria and it its role in microbial mats.

In some cases cyanobacteria are the predominant organisms responsible for the oxidation of sulfide in microbial mats such as those formed in sulfidic streams (Klatt et al. 2016). These authors observed that the observed anoxygenic photosynthesis and oxidation of sulfide was insensitive to DCMU, a potent and specific inhibitor of the oxygenic photosystem II, while in the dark the concentration of sulfide in the mat increased, indicating that sulfide oxidation was light-dependent. The depth-integrated rate of sulfide oxidation was 1.5 μmol m^{-2} s^{-1} and cyanobacteria were the only significant sink of sulfide.

Sulfide inhibits the oxygenic photosystem II, which is required to induce anoxygenic photosynthesis in cyanobacteria capable of it such as some strains of *Geitlerinema* (Solar Lake *Oscillatoria limnetica*) that grow anoxygenic and oxidize sulfide to sulfur (Cohen et al. 1975). The Solar Lake cyanobacteria are permanently exposed to high concentrations (several mM) of sulfide. The mat-forming *M. chthonoplastes* strain 11 (now re-identified as *Geitlerinema*, SAG31.92, and CCY9619) simultaneously performs oxygenic and anoxygenic photosynthesis and therefore tolerates only moderately levels of sulfide (0.15 mM). *M. chthonoplastes* mats are in general not experiencing high concentrations of free sulfide. One reason that *M. chthonoplastes* does not tolerate high concentrations of sulfide is that it requires low amounts of oxygen as a nutrient (de Wit et al. 1988). Growth of *M. chthonoplastes* occurred at higher sulfide concentrations (<1 mM), provided that oxygen was supplied. *M. chthonoplastes* strain 11 oxidizes sulfide to thiosulfate (de Wit and van Gemerden 1987a). The mat-forming *Oscillatoria limosa* strain 23 (re-assigned as *Lyngbya aestuarii* PCC8106 and CCY9616) is unable to perform sulfide-dependent anoxygenic photosynthetic CO_2 fixation when photosystem II is inhibited by DCMU (Stal 1995). However, CO_2 fixation tripled in the presence of 0.5 mM sulfide and even at 1 mM sulfide CO_2 fixation was still 74 % of the control. *L. aestuarii* exhibited sulfide- and light-dependent uptake of O_2. The oxidation of sulfide in the light may both lower the O_2 concentration, which would increase the carboxylation reaction of RubisCO as well as detoxify sulfide (Stal 1995). This physiology is obviously beneficially for mat-forming cyanobacteria.

Free sulfide occurs in a pH-dependent equilibrium of the gaseous H_2S and the ions HS^- and S^{2-}. Between pH 7 and pH 9 virtually all sulfide is present as HS^-. Above pH 9 it is mostly S^{2-} and below pH 7 it is mostly H_2S. Since H_2S is a gas it can diffuse into the cell whereas the sulfide ions need to be actively transported. Sulfide is therefore more toxic at lower pH. The cyanobacterial layer of the mat is characterized by a high pH (up to pH 10) resulting from the photosynthetic fixation of CO_2 in the light, but in the dark the pH may drop to 7. The high pH will limit the access to sulfide for photosynthesis in the light. However, cyanobacteria capable of anoxygenic photosynthesis are apparently capable of uptake of the sulfide ion. More importantly, the sediments of many marine environments where microbial mats develop are rich in iron and sulfide will react with it and precipitates as FeS (iron sulfide) or FeS_2 (pyrite). Ferric iron will be reduced to ferrous iron by sulfide producing FeS and elemental sulfur, which may further react to pyrite. The reaction with iron will largely eliminate free sulfide in the mat (Wieland et al. 2005). Hence, in microbial mats exposed to high pH and iron, anoxygenic photosynthesis by cyanobacteria may be negligible most of the time but episodically, when conditions have changed, the resistance of mat-forming cyanobacteria against sulfide and the capacity of using it in an anoxygenic photosynthetic mode may give these organisms an essential advantage that allow them to thrive in microbial mats (Klatt et al. 2016).

Mat-forming cyanobacteria such as *M. chthonoplastes* may accumulate iron on the polysaccharide sheath (Stal 2001). In contact with sulfide, the iron-loaded sheath of *M. chthonoplastes* turns black from FeS. It has been conceived that this may actually protect the cyanobacterium from toxic free sulfide as well as from supersaturating oxygen concentrations. At night when the sulfide concentration increases and may reach the cyanobacterial mat, the sulfide reacts with iron to ferrous sulfide. During the day the photosynthetic evolved oxygen reacts with the ferrous iron and oxidizes it to ferric iron. This will keep the oxygen concentration low and allows efficient CO_2 fixation by minimizing the oxygenation activity of the CO_2-fixing enzyme RubisCO. Many coastal microbial mats show a layer of oxidized iron immediately beneath the cyanobacterial mat and it has been conceived that this may similarly serve as a buffer separating the oxygenated top layer from the deeper sulfide-rich layer. However, besides this the iron layer may also be the result of the activity of anoxygenic phototrophic bacteria using ferrous iron as electron donor.

Anoxygenic Phototrophic Bacteria

Purple sulfur bacteria are a notable component of coastal microbial mats provided that sufficient sulfide is produced (Wieland et al. 2005). These anoxygenic phototrophic bacteria harvest light in the far-red part of the (photosynthetic active radiation) light spectrum that is not absorbed by the cyanobacteria and is not much attenuated by the sediment. Moreover, they exhibit some tolerance to oxygen (and some species are even capable of using it) and therefore very well suited to live in the proximity of cyanobacteria (Overmann and Garcia-Pichel 2006; Meyer et al. 2011). Rarely, green sulfur bacteria appear in coastal microbial mats, but they have been reported as a visible layer below the purple sulfur bacteria (Pierson et al. 1987). Green sulfur bacteria have very low light requirements and have a higher affinity towards sulfide than purple sulfur bacteria that allow them to compete with the latter when sulfide is short in supply. They are also much less tolerant to oxygen and therefore there localization in the deeper part of the mat fit these requirements perfectly (van Gemerden 1984; Pringault et al. 1999; Hamilton et al. 2016).

Microbial mats not always possess a clearly visible layer of purple sulfur bacteria and de Wit et al. (1995) explained this phenomenon using a model that described the competition between colorless and purple sulfur bacteria. These authors showed that either colorless sulfur bacteria dominate or that they co-exist with purple sulfur bacteria that in that case are by far in majority. The question whether colorless sulfur bacteria or purple sulfur bacteria are responsible for the removal of sulfide in the microbial mat depends very much on the availability of oxygen. Under anoxic conditions the purple sulfur bacteria will use up all sulfide and under excess oxygen colorless sulfur bacteria will do the job, even though purple sulfur bacteria such as *T. roseopersicina* are capable of doing the same, but lack the high affinity for sulfide of colorless sulfur bacteria such as *Thiobacillus* (*Acidithiobacillus*). While the latter removes oxygen, it allows the purple sulfur bacteria to synthesize their photopigments and using the sulfur intermediates produced by the colorless sulfur bacteria. This forms the perfect conditions for a co-existence of the two functionally different groups of microorganisms in the mat (van Gemerden 1993).

Disproportionation of reduced sulfur compounds may be important in microbial mats. This is essentially a fermentation of inorganic compounds during which part of the substrate is oxidized (serves as an electron donor) and another part of the same substrate is reduced (serves as an electron acceptor). Examples are among others elemental sulfur (including polysulfides), sulfite, and thiosulfate (Bak and Pfennig 1987; Canfield and Thamdrup 1996).

The role of anoxygenic phototrophic bacteria in microbial mats has been and still is neglected. Purple sulfur bacteria may also be important because they can contribute importantly to the physical stabilization of the matrix of the mat (Grant and Gust 1987). The predominant purple sulfur bacteria in coastal microbial mats are the Gammaproteobacteria *Thiocapsa roseopersicina*, *Thiocystis violacea*, and *Allochromatium vinosum* (Imhoff 2001). For instance, *T. roseopersicina*, which can be easily cultured and can reach abundances of 10^6–10^7 cells cm^{-3} (van Gemerden et al. 1989a). The success of this organism in coastal microbial mats is attributed to its high metabolic versatility and flexibility and its capacity to acclimate to a wide range of salt concentrations. Although *T. roseopersicina* is basically an anaerobic organism, it has a wide tolerance to O_2 and is even capable of aerobic growth in the dark (de Wit and van Gemerden 1987b, 1990). The latter property may, however, be of minor relevance in most microbial mats, because aerobic conditions in darkness do not occur frequently, except perhaps in "inverted" mats when anoxygenic purple sulfur bacteria appear at the surface (van Gemerden et al. 1989b). Vacuolated bacteria such as *Thiocapsa rosea* and *Lamprobacter modestohalophilus*, and non-vacuolated bacteria such as *Marichromatium gracile* and *Ectothiorhodospira* sp. have been isolated from mats in the Ebro Delta (Martinez-Alonso et al. 2005; Villanueva et al. 2010). There was a high diversity of members of the Chromatiaceae in the microbial mats of the Ebro Delta and many novel species of purple sulfur bacteria were present in these mats (Martinez-Alonso et al. 2005).

From the purple layers of hypersaline microbial mats other members of the family Chromatiaceae such as *Halochromatium salexigens*, *H. glycolicum*, and *Halothiocapsa halophila* have been isolated (Caumette et al. 1988, 1991, 1997).

Because green sulfur bacteria are obligate anaerobic organisms and sensitive to oxygen they are rarely encountered in microbial mats where oxygen may periodically reach the entire photic zone (which may be a few millimeters). However, green sulfur bacteria also grow at very low irradiances (Manske et al. 2005) and therefore may proliferate below the layer of purple sulfur bacteria, provided that continuous anoxia is guaranteed (Nicholson et al. 1987; Pierson et al. 1987). Green non-sulfur bacteria such as *Chloroflexus* and *Roseiflexus* are anoxygenic phototrophs that are an important component in anoxygenic hypersaline and some thermophilic microbial mats (van der Meer et al. 2005). They contribute importantly to gross primary production and community respiration and they depend strongly on light in the near infrared region (Polerecky et al. 2007).

Purple sulfur bacteria are anaerobic anoxygenic phototrophic organisms that use sulfide or other reduced forms of sulfur as electron donor with which they fix CO_2. Other substrates include are polysulfides, FeS (Visscher et al. 1990), and DMS (Visscher and van Gemerden 1991). Sulfide is oxidized to elemental sulfur, which is stored intracellularly (de Wit and van Gemerden 1987b). This process is usually fast because it involves only one oxidation step (two electrons) so that the resulting elemental sulfur is unavailable for other organisms and is further oxidized to sulfate when sulfide is no longer available. The depletion of sulfide may increase the risk that the organisms become exposed to oxygen. Purple sulfur bacteria thriving in these microbial mats therefore resist moderate concentrations of oxygen. The most common purple sulfur bacterium in microbial mats, *Thiocapsa roseopersicina*, is even capable of growing chemotrophically using sulfide or thiosulfate and oxygen (de Wit and van Gemerden 1987b). A large proportion of the electrons are now used to generate energy for CO_2 fixation and the yield will be considerably lower when compared to photoautotrophic growth, when light serves as source of energy and all electrons are available for CO_2 fixation. Under alternating oxic–anoxic conditions *T. roseopersicina* uses its chemotrophic and phototrophic growth potential simultaneously (Schaub and van Gemerden 1994). The synthesis of photopigments cannot take place in the presence of oxygen, which would be the case during the day. However, under an alternating light-oxic and dark-anoxic regime, which simulates the actual situation in a microbial mat, synthesis of bacteriochlorophyll and the carotenoid spirilloxanthin in *T. roseopersicina* occurs during the dark at the expense of glycogen that is stored during the day. Although photopigments cannot be synthesized during the day, *T. roseopersicina* could grow photosynthetically using the photopigments synthesized anaerobically in the dark (de Wit and van Gemerden 1990). This property is probably the reason for the success of *T. roseopersicina* in microbial mats because under the fluctuating conditions of oxygen and light it outcompetes purple sulfur bacteria that otherwise possess a much higher affinity for sulfide.

Polysulfides are chains of elemental sulfur with a negatively charged sulfur atom at each end (S_x^{2-}) and are formed by a chemical reaction of elemental sulfur with sulfide. Polysulfide is soluble and is more toxic than sulfide. Polysulfide is stabile at high pH and plays therefore an important role in microbial mats. Both green and purple sulfur bacteria oxidize polysulfide to sulfate (Visscher et al. 1990). Initially it was assumed that this reaction occurs only in microbial mats with low concentrations of iron so that free sulfide may be present (Jørgensen and Cohen 1977). However, high concentrations of polysulfides have been reported to occur in a cyanobacterial mat, indicating that this compound may be relevant for the sulfur cycle in marine microbial mats (van Gemerden 1993). Although polysulfides are an order of magnitude more toxic for most organisms than sulfide it may serve as the form of elemental sulfur that is transported into cells (Steudel et al. 1990). *Thiocapsa roseopersicina*, for instance, is capable of anoxygenic photosynthesis at the expense of polysulfide (Visscher et al. 1990).

Thiosulfate is formed as an intermediate in photosynthetic sulfide oxidation by green and purple sulfur bacteria (Steinmetz and Fischer 1982; Steudel et al. 1990) and can be used as electron donor by most colorless and photosynthetic sulfur bacteria. Oxidation of thiosulfate may result in the formation of tetrathionate (Then and Trüper 1981) or being split into sulfate or sulfite and elemental sulfur, which are subsequently further oxidized to sulfate (van Gemerden 1993).

Iron Photosynthesis

Ehrenreich and Widdel (1994) discovered iron-dependent anoxygenic photosynthesis by purple bacteria related to *Chromatium* and *Rhodobacter* (Gamma- and Alphaproteobacteria, respectively). The authors proposed that these microorganisms may have been responsible for the formation of Proterozoic banded-iron formations that were formed 1.8–2.5 billion years before present. Another explanation could be the chemical or chemotrophic biological aerobic oxidation of iron using the oxygen evolved by cyanobacteria as seen in hot spring microbial mats (Pierson et al. 1999). Aerobic chemotrophic oxidation of iron under near neutral pH may be through the activities of organisms such as *Gallionella* and *Leptothrix* (Emerson et al. 2010). *Chlorobium ferrooxidans* is a member of the green sulfur bacteria and carries out ferrous–iron-dependent anoxygenic photosynthesis but is unable to use sulfide or other reduced sulfur compounds (Heising et al. 1999). Anoxygenic iron-dependent photosynthesis by green sulfur bacteria belonging to the *Chlorobiaceae* was found at great depth in an iron-rich, sulfide-poor lake (Crowe et al. 2008). No reports have been published showing iron photosynthesis in microbial mats, but based on the current knowledge there is no reason why this should not occur.

Purple Non-Sulfur Bacteria

Purple non-sulfur bacteria are phylogenetically dispersed throughout the subclass of Alpha- and Betaproteobacteria (Lee et al. 2005) and are important and often dominant organisms in microbial mats (Guyoneaud et al. 1996; Ranchou-Peyruse et al. 2004). Purple non-sulfur bacteria distinguish from purple sulfur bacteria that they do not oxidize sulfide to elemental sulfur (as intermediate). These organisms are also more versatile than the mostly obligate photoautotrophic purple sulfur bacteria. Purple non-sulfur bacteria can be photoorganoheterotrophic, using light as the source of energy and organic matter as electron donor and carbon source, but they can also use organic matter as source of energy and live chemoorganoheterotrophic. Organic substrates include volatile fatty acids, lactate, pyruvate, malate, fumarate, succinate, glycerol, and certain amino acids (Guyoneaud et al. 2002). Some species can also live photolithotrophic using sulfide, sulfite, or thiosulfate as electron donors (Imhoff 2001; Guyoneaud et al. 2002; Caumette et al. 2007). Purple non-sulfur bacteria may be facultative anaerobes and may respire aerobically. Typical genera in microbial mats include *Rhodobacter*, *Rhodobium*, *Rhodovibrio*, *Rhodothalassium*, *Rhodospira*, *Rhodospirillum*, *Rhodocysta*, and *Phaeospirillum* (Imhoff et al. 1998; Glaeser and Overmann 1999; Guyoneaud et al. 2002).

Aerobic Anoxygenic Phototrophic Bacteria

Aerobic anoxygenic phototrophs (AAP) are bacteria belonging to the Alpha-, Beta-, and Gammaproteobacteria (Yurkov and Beatty 1998; Koblížek 2015). AAPs are facultative photoheterotrophs that possess a photosynthetic reaction center containing bacteriochlorophyll *a*, with which they harvest light energy but are unable to fix CO_2 except small amounts through anaplerotic routes (Tang et al. 2009). AAPs are basically secondary producers that recycle organic matter. AAPs have a wide distribution and occur in almost any illuminated environment, both freely suspended in freshwater and marine aquatic environments as well as in aggregates and attached to seaweeds and in biofilms. It is therefore not surprising that AAPs have been found in microbial mats and stromatolites (Shiba et al. 1979; Shiba 1991; Nishimura et al. 1994). AAPs have been isolated from alkaline cyanobacterial mats (Yurkov and Gorlenko 1992; Yurkov and van Gemerden 1993; Yurkov et al. 1994) and soil crusts (Csotonyi et al. 2010). The clone library of *pufM* from a cyanobacterial mat revealed many aerobic anoxygenic phototrophic bacteria although isolation and cultivation revealed that anaerobic purple sulfur bacteria formed the majority (Ranchou-Peyruse et al. 2006). Also in other phototrophic mats aerobic and anaerobic anoxygenic phototrophic bacteria co-exist (Thiel et al. 2010). However, very little is known about the role of AAPs in microbial mats, which should not surprise considering the fact that hardly anything is known about the ecology of these microorganisms in any environment. AAPs are often found associated with algal cultures where they probably live from organic matter exuded by the alga and perhaps also supply essential growth factors to the alga, while having access to light (Brauer et al. 2015; Koblížek 2015). This mutual relationship may also exist between AAPs and cyanobacteria in microbial mats. The effect of light on these organisms has been and still is disputed. Obviously, AAPs must have an advantage for having a photosynthetic reaction center, otherwise it would have been lost and it would not explain their wide distribution and phylogenetic diversity. The effect of light may be small but sufficiently important to guarantee a long-term advantage that may not easily be reproduced in laboratory cultures. Another important environmental parameter that affects AAPs is temperature. However, it is unclear whether AAPs require a higher temperature for growth or that the effect of temperature is indirect. In aquatic environments where the positive effect of higher temperature was reported it is likely that the elevated temperature caused a stratification of the water column, which would keep the AAPs in the illuminated surface layer due to the absence of mixing. This would, however, not apply to microbial mats.

Acknowledgments The research has received funding from the European Union Seventh Framework Programme (FP7/2007-2013) under grant agreement n° 311975. This publication reflects the views only of the authors, and the European Union cannot be held responsible for any use, which may be made of the information contained therein.

References

Bak F, Pfennig N (1987) Chemolithotrophic growth of *Desulfovibrio sulfodismutans* sp. nov. by disproportionation of inorganic sulfur compounds. Arch Microbiol 147:184–189

Bebout BM, Garcia-Pichel F (1995) UV B-induced vertical migrations of cyanobacteria in a microbial mat. Appl Environ Microbiol 61:4215–4222

Bolhuis H, Stal LJ (2011) Analysis of bacterial and archaeal diversity in coastal microbial mats using massive parallel 16S rRNA gene tag sequencing. ISME J 5:1701–1712

Bolhuis H, Cretoiu MS, Stal LJ (2014) Molecular ecology of microbial mats. FEMS Microbiol Ecol 90:335–350

Bolhuis H, Severin I, Confurius-Guns V, Wollenzien UIA, Stal LJ (2010) Horizontal transfer of the nitrogen fixation gene cluster in the cyanobacterium Microcoleus chthonoplastes. The ISME J 4:121–130

Brauer VS, Stomp M, Bouvier T, Fouilland E, Leboulanger C, Confurius-Guns V, Weissing FJ, Stal LJ, Huisman J (2015) Competition and facilitation between the marine nitrogen-fixing cyanobacterium *Cyanothece* and its associated bacterial community. Front Microbiol 5:795

Burow LC, Woebken D, Bebout BM, McMurdie PJ, Singer SW, Pett-Ridge J, Prufert-Bebout L, Spormann AM, Weber PK, Hoehler TM (2012) Hydrogen production in photosynthetic microbial mats in the Elkhorn Slough estuary, Monterey Bay. ISME J 6:863–874

Campbell SE (1979) Soil stabilization by a prokaryotic desert crust: implications for Precambrian land biota. Orig Life 9:335–348

Canfield DE, Des Marais DJ (1991) Aerobic sulfate reduction in microbial mats. Science 251:1471–1473

Canfield DE, Thamdrup B (1996) Fate of elemental sulfur in an intertidal sediment. FEMS Microbiol Ecol 19:95–103

Carreira C, Staal M, Falkoski D, de Vries RP, Middelboe M, Brussaard CPD (2015) Disruption of photoautotrophic intertidal mats by filamentous fungi. Environ Microbiol 17:2910–2921

Caumette P (1988) Characterization of *Chromatium salexigens* sp. nov., a halophilic *Chromatiaceae* isolated from Mediterranean Salinas. Syst Appl Microbiol 10:284–292

Caumette P, Baulaigue R, Matheron R (1991) *Thiocapsa halophila* sp. nov., a new halophilic phototrophic purple sulfur bacterium. Arch Microbiol 155:170–176

Caumette P, Imhoff JF, Suling J, Matheron R (1997) *Chromatium glycolicum* sp. nov., a moderately halophilic purple sulfur bacterium that uses glycolate as substrate. Arch Microbiol 167:11–18

Caumette P, Guyoneaud R, Duran R, Cravo-Laureau C, Matheron R (2007) *Rhodobium pfennigii* sp. nov., a phototrophic purple non-sulfur bacterium with unusual bacteriochlorophyll a antennae, isolated from a brackish microbial mat on Rangiroa atoll, French Polynesia. Int J Syst Evol Microbiol 57:1250–1255

Cohen Y, Padan E, Shilo M (1975) Facultative anoxygenic photosynthesis in the cyanobacterium *Oscillatoria limnetica*. J Bacteriol 123:855–861

Crowe SA, Jones C, Katsev S, Magen C, O'Neill AH, Sturm A, Canfield DE, Haffner GD, Mucci A, Sundby B, Fowle DA (2008) Photoferrotrophs thrive in an Archean ocean analogue. Proc Natl Acad Sci 105:15938–15943

Csotonyi J, Swiderski J, Stackebrandt E, Yurkov V (2010) A new environment for aerobic anoxygenic phototrophic bacteria: biological soil crusts. Environ Microbiol Rep 2:651–656

D'Amelio ED, Cohen Y, Des Marais DJ (1987) Association of a new type of gliding, filamentous, purple phototrophic bacterium inside bundles of *Microcoleus chthonoplastes* in hypersaline cyanobacterial mats. Arch Microbiol 147:213–220

de Wit R, van Gemerden H (1987a) Oxidation of sulfide to thiosulfate by *Microcoleus chthonoplastes*. FEMS Microbiol Ecol 45:7–13

de Wit R, van Gemerden H (1987b) Chemolithotrophic growth of the phototrophic sulfur bacterium *Thiocapsa roseopersicina*. FEMS Microbiol Ecol 45:117–126

de Wit R, van Gemerden H (1990) Growth of the phototrophic purple sulfur bacterium *Thiocapsa roseopersicina* under oxic/anoxic regimens in the light. FEMS Microbiol Ecol 73:69–76

de Wit R, van Boekel WHM, van Gemerden H (1988) Growth of the cyanobacterium *Microcoleus chthonoplastes* on sulfide. FEMS Microbiol Ecol 53:203–209

de Wit R, van den Ende FP, van Gemerden H (1995) Mathematical simulation of the interactions among cyanobacteria, purple sulfur bacteria and chemotrophic sulfur bacteria in microbial mat communities. FEMS Microbiol Ecol 17:117–135

Dillon JG, Miller S, Bebout B, Hullar M, Pinel N, Stahl DA (2009) Spatial and temporal variability in a stratified hypersaline microbial mat. FEMS Microbiol Ecol 68:46–58

Ehrenreich A, Widdel F (1994) Anaerobic oxidation of ferrous iron by purple bacteria, a new type of phototrophic metabolism. Appl Environ Microbiol 60:4517–4526
Emerson D, Fleming EJ, McBeth JM (2010) Iron-oxidizing bacteria: an environmental and genomic perspective. Annu Rev Microbiol 64:561–583
Fenchel T (1998) Formation of laminated cyanobacterial mats in the absence of benthic fauna. Aquat Microb Ecol 14:235–240
Fourçans A, García de Oteyza T, Wieland A, Solé A, Diestra E, van Bleijswijk J, Grimalt JO, Kühl M, Esteve I, Muyzer G, Caumette P, Duran R (2004) Characterization of functional bacterial groups in a hypersaline microbial mat community (Salins-de-Giraud, Camargue, France). FEMS Microbiol Ecol 51:55–70
Fourçans A, Solé A, Diestra E, Ranchou-Peyruse A, Esteve I (2006) Vertical migration of phototrophic bacterial populations in a hypersaline microbial mat from Salins-de-Giraud (Camargue, France). FEMS Microbiol Ecol 57:367–377
Fredriksson C, Malin G, Siddiqui PJA, Bergman B (1998) Aerobic nitrogen fixation is confined to a subset of cells in the non-heterocystous cyanobacterium *Symploca* PCC 8002. New Phytol 140:531–538
Gallon JR (1992) Reconciling the incompatible: N_2 fixation and O_2. New Phytol 122:571–609
Garcia-Pichel F, Castenholz RW (1991) Characterization and biological implications of scytonemin, a cyanobacterial sheath pigment. J Phycol 27:395–409
Garcia-Pichel F, Castenholz RW (1993) Occurrence of UV-absorbing, mycosporine-like compounds among cyanobacterial isolates and an estimate of their screening capacity. Appl Environ Microbiol 59:163–169
Garcia-Pichel F, Prufert-Bebout L, Muyzer G (1996) Phenotypic and phylogenetic analyses show *Microcoleus chthonoplastes* to be a cosmopolitan cyanobacterium. Appl Environ Microbiol 62:3284–3291
Gerdes G, Claes M, Dnajtschik-Piewak K, Riege H, Krumbein WE, Reineck HE (1993) Contribution of microbial mats to sedimentary surface structures. Facies 29:61–74
Glaeser J, Overmann J (1999) Selective enrichment and characterization of *Roseospirillum parvum*, gen. nov. and sp. nov., a new purple nonsulfur bacterium with unusual light absorption properties. Arch Microbiol 171:405–416
Grant J, Gust G (1987) Prediction of coastal sediment stability from photopigment context of mats of purple sulphur bacteria. Nature 330:244–246
Guyoneaud R, Matheron R, Baulaigue R, Podeur K, Hirschler A, Caumette P (1996) Anoxygenic phototrophic bacteria in eutrophic coastal lagoons of the French Mediterranean and Atlantic Coasts (Prevost Lagoon, Arcachon Bay, Certes fishponds). Hydrobiologia 329:33–43
Guyoneaud R, Mouné S, Eatock C, Bothorel V, Hirschler-Rea A, Willison J, Duran R, Liesack W, Herbert R, Matheron R, Caumette P (2002) Characterization of three spiral-shaped purple nonsulfur bacteria isolated from coastal lagoon sediments, saline sulfur springs, and microbial mats: emended description of the genus *Roseospira* and description of *Roseospira marina* sp. nov., *Roseospira navarrensis* sp. nov., and *Roseospira thiosulfatophila* sp. nov. Arch Microbiol 178:315–324
Hamilton TL, Bryant DA, Macalady JL (2016) The role of biology in planetary evolution: cyanobacterial primary production in low-oxygen Proterozoic oceans. Environ Microbiol 18:325–340
Heising S, Richter L, Ludwig W, Schink B (1999) *Chlorobium ferrooxidans* sp. nov., a phototrophic green sulfur bacterium that oxidizes ferrous iron in coculture with a "*Geospirillum*" sp. strain. Arch Microbiol 172:116–124
Hoehler TM, Bebout BM, Des Marais DJ (2001) The role of microbial mats in the production of reduced gases on the early Earth. Nature 412:324–327
Hoehler TM, Albert DB, Alperin MJ, Bebout BM, Martens CS, Des Marais DJ (2002) Comparative ecology of H_2 cycling in sedimentary and phototrophic ecosystems. Antonie Van Leeuwenhoek 81:575–585
Hoffmann D, Maldonado J, Wojciechowski MF, Garcia-Pichel F (2015) Hydrogen export from intertidal cyanobacterial mats: sources, fluxes and the influence of community composition. Environ Microbiol 17:3738–3753

Howarth RW (1979) Pyrite: its rapid formation in a salt marsh and its importance in ecosystem metabolism. Science 203:49–50

Imhoff JF (2001) True marine and halophilic anoxygenic phototrophic bacteria. Arch Microbiol 176:243–254

Imhoff JF, Petri R, Süling J (1998) Reclassification of species of the spiral-shaped phototrophic purple non-sulfur bacteria of the a-Proteobacteria: description of the new genera *Phaeospirillum* gen. nov., *Rhodovibrio* gen. nov., *Rhodothalassium* gen. nov. and *Roseospira* gen. nov. as well as transfer of *Rhodospirillum fulvum* to *Phaeospirillum fulvum* comb. nov., of *Rhodospirillum molischianum* to *Phaeospirillum molischianum* comb. nov., of *Rhodospirillum salinarum* to *Rhodovibrio salinarum* comb. nov., of *Rhodospirillum sodomense* to *Rhodovibrio sodomensis* comb. nov., of *Rhodospirillum salexigens* to *Rhodothalassium salexigens* comb. nov. and of *Rhodospirillum mediosalinum* to *Roseospira mediosalina* comb. nov. Int J Syst Bacteriol 48:793–798

Jahnke LL, Turk-Kubo KA, Parenteau MN, Green SJ, Kubo MDY, Vogel M, Summons RE, Des Marais DJ (2014) Molecular and lipid biomarker analysis of a gypsum-hosted endoevaporitic microbial community. Geobiology 12:62–82

Jørgensen BB, Cohen Y (1977) Solar Lake (Sinai). 5. The sulfur cycle of the benthic cyanobacterial mats. Limnol Oceanogr 22:657–666

Jørgensen BB, Revsbech NP, Cohen Y (1983) Photosynthesis and structure of benthic microbial mats: micro-electrode and SEM studies of four cyanobacterial communities. Limnol Oceanogr 28:1075–1093

Kim J, Rees DC (1994) Nitrogenase and biological nitrogen fixation. Biochemistry 33:389–397

Klatt JM, Meyer S, Häusler S, Macalady JL, de Beer D, Polerecky L (2016) Structure and function of natural sulphide-oxidizing microbial mats under dynamic input of light and chemical energy. ISME J 10:921–933

Koblížek M (2015) Ecology of aerobic anoxygenic phototrophs in aquatic environments. FEMS Microbiol Rev 39:854–870

Kohls K, Abed RMM, Polerecky L, Weber M, de Beer D (2010) Halotaxis of cyanobacteria in an intertidal hypersaline microbial mat. Halotaxis of cyanobacteria in an intertidal hypersaline microbial mat. Environ Microbiol 12:567–575

Lassen C, Ploug H, Jørgensen BB (1992) A fibre-optic scalar irradiance microsensor - application for spectral light measurements in sediments. FEMS Microbiol Ecol 86:247–254

Lee K-B, Liu C-T, Anzai Y, Kim H, Aono T, Oyaizu H (2005) The hierarchical system of the '*Alphaproteobacteria*': description of *Hyphomonadaceae* fam. nov., *Xanthobacteraceae* fam. nov. and *Erythrobacteraceae* fam. nov. Int J Syst Evol Microbiol 55:1907–1919

Ley RE, Harris JK, Wilcox J, Spear JR, Millern SR, Bebout BM, Maresca JA, Bryant DA, Sogin ML, Pace NR (2006) Unexpected diversity and complexity of the Guerrero Negro hypersaline microbial mat. Appl Environ Microbiol 72:3685–3695

Malin G, Walsby AE (1985) Chemotaxis of a cyanobacterium on concentration gradients of carbondioxide, bicarbonate and oxygen. J Gen Microbiol 131:2643–2652

Manske AK, Glaeser J, Kuypers MMM, Overmann J (2005) Physiology and phylogeny of green sulfur bacteria forming a monospecific phototrophic assemblage at a depth of 100 meters in the Black Sea. Appl Environ Microbiol 71:8049–8060

Martinez-Alonso M, van Bleijswijk J, Gaju N, Muyzer G (2005) Diversity of anoxygenic phototrophic sulfur bacteria in the microbial mats of the Ebro delta: a combined morphological and molecular approach. FEMS Microbiol Ecol 52:339–350

Meyer KM, Macalady JL, Fulton JM, Kump LR, Schaperdoth I, Freeman KH (2011) Carotenoid biomarkers as an imperfect reflection of the anoxygenic phototrophic community in meromictic Fayetteville Green Lake. Geobiology 9:321–329

Nicholson JAM, Stolz JF, Pierson BK (1987) Structure of a microbial mat at Great Sippewissett Marsh, Cape Cod, Massachusetts. FEMS Microbiol Ecol 45:343–364

Nishimura Y, Muroga Y, Saito S, Shiba T, Takamiya K-I, Shioi Y (1994) DNA relatedness and chemotaxonomic feature of aerobic bacteriochlorophyll-containing bacteria isolated from coasts of Australia. J Gen Appl Microbiol 40:287–296

Olson JB, Litaker RW, Paerl HW (1999) Ubiquity of heterotrophic diazotrophs in marine microbial mats. Aquat Microb Ecol 19:29–36
Omoregie EO, Crumbliss LL, Bebout BM, Zehr JP (2004) Determination of nitrogen-fixing phylotypes in *Lyngbya* sp. and *Microcoleus chthonoplastes* cyanobacterial mats from Guerrero Negro, Baja California, Mexico. Appl Environ Microbiol 70:2119–2128
Overmann J, Garcia-Pichel F (2006) The phototrophic way of life. In: Dworkin M, Falkow S, Rosenberg E, Schleifer KH (eds) The prokaryotes, vol 2. Springer, Heidelberg, pp 32–85
Padan E (1979) Facultative anoxygenic photosynthesis in cyanobacteria. Annu Rev Plant Physiol 30:27–40
Pearson HW, Howsley R, Kjeldsen CK, Walsby AE (1979) Aerobic nitrogenase activity associated with a non-heterocystous filamentous cyanobacterium. FEMS Microbiol Lett 5:163–169
Pierson BK, Oesterle A, Murphy GL (1987) Pigments, light penetration, and photosynthetic activity in the multi-layered microbial mats of Great Sippewissett salt marsh, Massachusetts. FEMS Microbiol Ecol 45:365–376
Pierson BK, Parenteau MN, Griffin BM (1999) Phototrophs in high-iron-concentration microbial mats: physiological ecology of phototrophs in an iron-depositing hot spring. Appl Environ Microbiol 65:5474–5483
Polerecky L, Bachar A, Schoon R, Grinstein M, Jørgensen BB, de Beer D, Jonkers HM (2007) Contribution of *Chloroflexus* respiration to oxygen cycling in a hypersaline microbial mat from Lake Chiprana, Spain. Environ Microbiol 9:2007–2024
Potts M, Krumbein WE, Metzger J (1978) Nitrogen fixation rates in anaerobic sediments determined by acetylene reduction, a new ^{15}N field assay, and simultaneous total N ^{15}N determination. In: Krumbein WE (ed) Environmental biogeochemistry and geomicrobiology vol 3: methods, metals and assessment. Ann Arbor Science, Ann Arbor Michigan, pp 753–769
Pringault O, Garcia-Pichel F (2004) Hydrotaxis of cyanobacteria in desert crusts. Microbial Ecol 47:366–373. Appl Environ Microbiol 66:1038–1049
Pringault O, Epping E, Guyoneaud R, Khalili A, Kühl M (1999) Dynamics of anoxygenic photosynthesis in an experimental green sulphur bacteria biofilm. Environ Microbiol 1:295–305
Rai AN, Söderbäck E, Bergman B (2000) Cyanobacterium-plant symbioses. New Phytol 147:449–481
Ramsing NB, Ferris MJ, Ward DM (2000) Highly ordered vertical structure of *Synechococcus* populations within the one-millimeter-thick photic zone of a hot spring cyanobacterial mat. Appl Environ Microbiol 66:1038–1049
Ranchou-Peyruse A, Moppert X, Hourcade E, Hernandez G, Caumette P, Guyoneaud R (2004) Characterization of brackish anaerobic bacteria involved in hydrocarbon degradation: a combination of molecular and culture-based approaches. Ophelia 58:255–262
Ranchou-Peyruse A, Herbert R, Caumette P, Guyoneaud R (2006) Comparison of cultivation-dependent and molecular methods for studying the diversity of anoxygenic purple phototrophs in sediments of an eutrophic brackish lagoon. Environ Microbiol 8:1590–1599
Revsbech NP, Jørgensen BB, Blackburn TH, Cohen Y (1983) Microelectrode studies of the photosynthesis and O_2, H_2S and pH profiles of a microbial mat. Limnol Oceanogr 28:1062–1074
Richardson LL, Castenholz RW (1987) Diel vertical movements of the cyanobacterium *Oscillatoria terebriformis* in a sulfide-rich hot spring microbial mat. Appl Environ Microbiol 53:2142–2150
Richardson LL, Castenholz RW (1989) Chemokinetic motility responses of the cyanobacterium *Oscillatoria terebriformis*. Appl Environ Microbiol 55:261–263
Rippka R, Waterbury JB (1977) The synthesis of nitrogenase by non-heterocystous cyanobacteria. FEMS Microbiol Lett 2:83–86
Rippka R, Deruelles J, Waterbury JB, Herdman M, Stanier RY (1979) Generic assignments, strain histories and properties of pure cultures of cyanobacteria. J Gen Microbiol 111:1–61
Risatti JB, Capman WC, Stahl DA (1994) Community structure of a microbial mat: the phylogenetic dimension. Proc Natl Acad Sci U S A 91:10173–10177
Schaub BEM, van Gemerden H (1994) Simultaneous phototrophic and chemotrophic growth in the purple sulfur bacterium *Thiocapsa roseopersicina* M1. FEMS Microbiol Ecol 13:185–195

Severin I, Stal LJ (2010) Diazotrophic microbial mats. In: Seckbach J, Oren A (eds) Microbial mats. Modern and ancient microorganisms in stratified systems. Springer, Heidelberg, pp 321–339

Severin I, Acinas SG, Stal LJ (2010) Diversity of nitrogen-fixing bacteria in cyanobacterial mats. FEMS Microbiol Ecol 73:514–525

Shiba T (1991) *Roseobacter litoralis* gen. nov., sp. nov., and *Roseobacter denitrificans* sp. nov., aerobic pink-pigmented bacteria which contain bacteriochlorophyll *a*. Syst Appl Microbiol 14:140–145

Shiba T, Simidu U, Taga N (1979) Distribution of aerobic bacteria which contain bacteriochlorophyll *a*. Appl Environ Microbiol 38:43–45

Siegesmund M, Johansen JR, Karsten U, Friedl T (2008) *Coleofasciculus* gen. nov. (Cyanobacteria): morphological and molecular criteria for revision of the genus *Microcoleus* Gomont. J Phycol 44:1572–1585

Sigalevich P, Baev MV, Teske A, Cohen Y (2000) Sulfate reduction and possible aerobic metabolism of the sulfate-reducing bacterium *Desulfovibrio oxyclinae* in a chemostat coculture with *Marinobacter* sp. strain MB under exposure to increasing oxygen concentrations. Appl Environ Microbiol 66:5013–5018

Stal LJ (1995) Physiological ecology of cyanobacteria in microbial mats and other communities. New Phytol 131:1–32

Stal LJ (2001) Coastal microbial mats: the physiology of a small-scale ecosystem. S Afr J Bot 67:399–410

Stal LJ (2012) Microbial mats and stromatolites. In: Whitton BA (ed) The ecology of cyanobacteria. Springer, Dordrecht, pp 61–120

Stal LJ, Heyer H (1987) Dark anaerobic nitrogen fixation (acetylene reduction) in the cyanobacterium *Oscillatoria* sp. FEMS Microbiol Ecol 45:227–232

Stal LJ, Moezelaar R (1997) Fermentation in cyanobacteria. FEMS Microbiol Rev 21:179–211

Stal LJ, van Gemerden H, Krumbein WE (1985) Structure and development of a benthic marine microbial mat. FEMS Microbiol Ecol 31:111–125

Steinmetz MA, Fischer U (1982) Cytochromes, rubredoxin and sulfur metabolism of the non-thiosulfate-utilizing green sulfur bacterium *Pelodictyon luteolum*. Arch Microbiol 132:204–210

Steppe TF, Olson JB, Paerl HW, Litaker RW, Belnap J (1996) Consortial N_2 fixation: a strategy for meeting nitrogen requirements of marine and terrestrial cyanobacterial mats. FEMS Microbiol Ecol 21:149–156

Steudel R, Holdt G, Visscher PT, van Gemerden H (1990) Search for polythionates in cultures of *Chromatium vinosum* after sulfide incubation. Arch Microbiol 153:432–437

Tang K-H, Feng X, Tang YJ, Blankenship RE (2009) Carbohydrate metabolism and carbon fixation in *Roseobacter denitrificans* OCh114. PLoS One 4(10):e7233

Then J, Trüper HG (1981) The role of thiosulfate in sulfur metabolism of *Rhodopseudomonas globiformis*. Arch Microbiol 130:143–146

Thiel V, Tank M, Neulinger SC, Gehrmann L, Dorador C, Imhoff JF (2010) Unique communities of anoxygenic phototrophic bacteria in saline lakes of Salar de Atacama (Chile): evidence for a new phylogenetic lineage of phototrophic *Gammaproteobacteria* from *pufLM* gene analyses. FEMS Microbiol Ecol 74:510–522

Urmeneta J, Alcoba Ó, Razquín E, Tarroja E, Navarrete A, Guerrero R (1998) Oxygenic photosynthesis and respiratory activity in microbial mats of the Ebro delta, Spain, by oxygen exchange method. Curr Microbiol 37:151–155

van der Meer MTJ, Schouten S, Bateson MM, Nübel U, Wieland A, Kühl M, De Leeuw JW, Sinninghe Damste JS, Ward DM (2005) Diel variations in carbon metabolism by green nonsulfur-like bacteria in alkaline siliceous hot spring microbial mats from Yellowstone national park. Appl Environ Microbiol 71:3978–3986

van Gemerden H (1984) The sulfide affinity of phototrophic bacteria in relation to the location of elemental sulfur. Arch Microbiol 139:289–294

van Gemerden H (1993) Microbial mats: a joint venture. Mar Geol 113:3–25

van Gemerden H, de Wit R, Tughan CS, Herbert RA (1989a) Development of mass blooms of *Thiocapsa roseopersicina* on sheltered beaches on the Orkney Islands. FEMS Microbiol Lett 62:111–118 van Gemerden H, Tughan CS, de Wit R, Herbert RA (1989b) Laminated microbial ecosystems on sheltered beaches in Scapa Flow, Orkney Islands. FEMS Microbiol Ecol 62:87–102

Villanueva L, del Campo J, Guerrero R, Geyer R (2010) Intact phospholipid and quinone biomarkers to assess microbial diversity and redox state in microbial mats. Microb Ecol 60:226–238

Villbrandt M, Stal LJ, Krumbein WE (1990) Interactions between nitrogen fixation and oxygenic photosynthesis in a marine cyanobacterial mat. FEMS Microbiol Ecol 74:59–72

Visscher PT, van Gemerden H (1991) Production and consumption of dimethylsulfoniopropionate in marine microbial mats. Appl Environ Microbiol 57:3237–3242

Visscher PT, van Gemerden H (1993) Sulfur cycling in laminated marine microbial ecosystems. In: Oremland RS (ed) Biogeochemistry of global change: radiatively active trace gases. Chapman and Hall, New York, pp 672–690

Visscher PT, Nijburg JW, van Gemerden H (1990) Polysulfide utilization by *Thiocapsa roseopersicina*. Arch Microbiol 155:75–81

Visscher PT, Prins RA, van Gemerden H (1992) Rates of sulfate reduction and thiosulfate consumption in a marine microbial mat. FEMS Microbiol Ecol 86:283–293

Watermann F, Hillebrand H, Gerdes G, Krumbein WE, Sommer U (1999) Competition between benthic cyanobacteria and diatoms as influenced by different grain sizes and temperatures. Mar Ecol Prog Ser 187:77–87

Wieland A, Kühl M, McGowan L, Fourçans A, Duran R, Caumette P, Garcia de Oteyza T, Grimalt JO, Solé A, Diestra E, Herbert RA (2003) Microbial mats on the Orkney Islands revisited: microenvironment and microbial community composition. Microb Ecol 46:371–390

Wieland A, Zopfi J, Benthien M, Kühl M (2005) Biogeochemistry of an iron-rich hypersaline microbial mat (Camargue, France). Microb Ecol 49:34–49

Yallop ML, de Winder B, Paterson DM, Stal LJ (1994) Comparative structure, primary production and biogenic stabilization of cohesive and non-cohesive marine sediments inhabited by microphytobenthos. Estuar Coast Shelf Sci 39:565–582

Yurkov VV, Beatty JT (1998) Aerobic anoxygenic phototrophic bacteria. Microbiol Mol Biol Rev 62:695–724

Yurkov V, Gorlenko VM (1992) Ecophysiological peculiarities of phototrophic microbial communities of Bolsherechensky thermal springs. Microbiology 61:115–122

Yurkov VV, van Gemerden H (1993) Abundance and salt tolerance of obligately aerobic, phototrophic bacteria in a marine microbial mat. Neth J Sea Res 31:57–62

Yurkov V, Krasilnikova EN, Gorlenko VM (1994) Thiosulfate metabolism in aerobic bacteriochlorophyll-a containing bacteria. Microbiology 63:181–188

Zehr JP, Mellon M, Braun S, Litaker W, Steppe T, Paerl HW (1995) Diversity of heterotrophic nitrogen fixation genes in a marine cyanobacterial mat. Appl Environ Microbiol 61:2527–2532

Part III
Bioremediation, Secondary Metabolites and Other Applied Aspects

Photosynthetic Purple Non Sulfur Bacteria in Hydrogen Producing Systems: New Approaches in the Use of Well Known and Innovative Substrates

Alessandra Adessi, Elisa Corneli, and Roberto De Philippis

Abstract During the last few years, progress has been made in developing cleaner and more efficient bioenergy producing systems. Innovative processes and novel substrates were assessed at lab scale, in order to investigate and promote a sustainable development of photobiological hydrogen production. Recent and innovative processes and the use of novel substrates are discussed in this chapter. The main focus is on photofermentation systems conducted on biomass derived substrates, as these are considered to be the applicative goal of hydrogen production. Indeed, it is also present a short *excursus* on some synthetic media, investigated as interesting opportunities for enlarging applicability of the hydrogen technology. The number of new findings here reported demonstrates that it is worth continuing the efforts for increasing the knowledge on the photofermentation process for H_2 production, in particular owing to the need of reducing the use of fossil fuels for mitigating the emissions of GHG in the atmosphere.

Keywords Biohydrogen • Purple non sulfur bacteria • Biomass • Waste disposal • Energy conversion

A. Adessi
Institute of Chemistry of Organometallic Compounds (ICCOM), CNR, Florence, Italy

E. Corneli
Institute of Life Sciences, Scuola Superiore Sant'Anna, Pisa, Italy

R. De Philippis (✉)
Institute of Chemistry of Organometallic Compounds (ICCOM), CNR, Florence, Italy

Department of Agrifood Production and Environmental Sciences, University of Florence, Florence, Italy
e-mail: roberto.dephilippis@unifi.it

© Springer International Publishing Switzerland 2017
P.C. Hallenbeck (ed.), *Modern Topics in the Phototrophic Prokaryotes*,
DOI 10.1007/978-3-319-46261-5_10

Abbreviations

DW	Dry weight
OMWW	Olive mill waste waters
PHA	Polyhydroxyalkanoate
PHB	Poly-β-hydroxybutyrate
PNSB	Purple non sulfur bacteria
VFAs	Short chain volatile fatty acids
VS	Volatile solids

Bioenergies and Biohydrogen

Bioenergy production and use is rising in many countries to diversify energy sources and to promote environmental quality, mitigation of climate change, energy security, and economic growth, including the development of rural economies (Weiland 2010; El Bassam 2010; Appels et al. 2011; IRENA 2013). Bioenergy derives from the conversion of biomass, where biomass may be used directly as fuel or processed into liquids and gases (IRENA 2013) and according to Directive 2009/28/EC of the European Parliament and the Council of 23 April 2009 on the promotion of the use of energy from renewable sources, the term biomass means "the biodegradable fraction of products, waste and residues of biological origin from agriculture (including vegetal and animal substances), forestry and related industries including fisheries and aquaculture, as well as the biodegradable fraction of industrial and municipal waste."

Concerning policy target for energy from renewable sources in transport, the development of biofuels plays a fundamental role (El Bassam 2010). The importance of producing biofuels is linked to the need of reducing fossil fuel extraction and consumption, with the aim of decreasing the rise of atmospheric CO_2 and in general to decrease fuel impact on global climate change (El Bassam 2010; Frigon and Guiot 2010; Appels et al. 2011). Biofuels are commonly separated into three different groups according to their level of development and the feedstocks used, though there is no universally agreed definition (IEA 2009; El Bassam 2010). In general, 1st generation biofuels include mature technologies for the production of bioethanol from sugar and starch crops, biodiesel from oil crops and animal fats, and biomethane from the anaerobic digestion of wet biomass; 2nd generation biofuels include several biofuels, such as bioethanol and biodiesel, produced from conventional technologies, but using novel feedstocks, like alternative starch, oil and sugar crops, or lignocellulosic materials (e.g., straw, wood, and grass); 3rd generation or advanced biofuels are at the earlier stages of research and development (e.g., biofuels from algae, hydrogen from biomass reforming, and biohydrogen) (IEA 2009; El Bassam 2010). The 2nd and 3rd generation biofuels are more sustainable, with biomass at lower costs and lower greenhouse gas emissions than 1st generation ones, avoiding the replacement of food and forage production by energy crops (IEA 2009).

In general, the increased bioenergy use can lead to increased demand for biomass, thus to a possible competition for land currently used for food production (IEA 2009; Appels et al. 2011). Several aspects are related to this concept, on one hand, the increasing global population (nine billions in the 2050, according to UN estimations) results in an increase of food and animal feed demand, on the other hand, the use of croplands and forests for energy crops production could be detrimental to biodiversity and to soil and water resources (IEA 2009). On these grounds, government policies and industrial efforts need to be directed to achieve bioenergy potential targets in the longer term, making sure of increasing biomass yield levels, global food production, promoting the technology development, the diffusion of best sustainable agricultural practices, and a sustainable use of residues and wastes for bioenergy, which present limited environmental risks and impacts and need to be encouraged and promoted globally (IEA 2009; Appels et al. 2011).

Hydrogen is a promising energy carrier, is the most abundant element in the universe, and represents a clean and renewable biofuel, with high conversion efficiency (Holladay et al. 2009; Christopher and Dimitrios 2012). Currently, about 96 % of hydrogen is synthesized from fossil fuels, in particular from methane reforming, and the remaining percentage is produced by water electrolysis and can be used directly in internal combustion engines or in fuel cells, after appropriate purification, without a direct combustion (Holladay et al. 2009; Christopher and Dimitrios 2012; Adessi and De Philippis 2014). Hydrogen can play an important role in decarbonizing the transport sector in the long-term period, as there is no CO_2 emission during its combustion, and it can be derived from many renewable sources including biomass and water. However, the deployment of hydrogen vehicles and a related fueling infrastructure is still missing or inadequate for a successful market application in the contest of hydrogen economy (Holladay et al. 2009; IEA 2009). Comparing with anaerobic digestion, which is classified within the biochemical conversion processes as a robust and widely applied technology, the biological hydrogen production is a technology in progress (Frigon and Guiot 2010). Biohydrogen production processes can be classified into different groups, as follows: biophotolysis of water by microalgae and cyanobacteria; photodecomposition of organic compounds by photosynthetic bacteria, i.e., photofermentation; dark fermentation of organic compounds by anaerobic or facultative anaerobic bacteria; and bioelectrohydrogenesis (Das and Veziroglu 2008; Hallenbeck et al. 2009; Holladay et al. 2009). Combined systems can be created with the dark fermentation yielding biohydrogen as first stage followed by the second stage of either anaerobic digestion, yielding biomethane, or photofermentation, yielding biohydrogen (Hallenbeck and Ghosh 2009; Adessi et al. 2012a; Argun and Kargi 2011; Gómez et al. 2011; Hay et al. 2013).

This chapter will be focused on photofermentation as either a single stage process, or in dark fermentation/photofermentation systems both sequential (two stage) and combined (co-cultures), using novel biomass derived substrates.

Photofermentation by Purple Non Sulfur Bacteria

Photofermentation is carried out by purple non sulfur bacteria (PNSB), that are anoxygenic phototrophic bacteria, converting substrates to hydrogen, carbon dioxide, and microbial biomass (Heiniger et al. 2012; Adessi and De Philippis 2014; Hallenbeck and Liu 2016). During hydrogen biosynthesis, nitrogenase enzyme reduces molecular nitrogen and protons to ammonia and hydrogen (Eq. 1) (Heiniger et al. 2012; Adessi and De Philippis 2014; Hallenbeck and Liu 2016):

$$N_2 + 8H^+ + 8e^- + 16ATP \rightarrow 2NH_3 + H_2 + 16ADP \quad (1)$$

This reaction is energy demanding, requiring ATP produced through cyclic photophosphorylation in absence of oxygen with artificial or solar light as the energy source, and the reducing power from the catabolism of carbon compounds in the tricarboxylic acid (TCA) cycle, preferably low-molecular weight organic acids that can easily enter the TCA cycle (Fig. 1) (Adessi and De Philippis 2014; Hallenbeck and Liu 2016). Thus, photobiological hydrogen production using PNSB depends mainly on nitrogen fixation, ATP production, and carbon sources catabolism (Keskin et al. 2011; Adessi and De Philippis 2014). In absence of molecular nitrogen, nitrogenase can dissipate the reducing equivalents coming from other metabolic processes producing hydrogen (Heiniger et al. 2012; Adessi

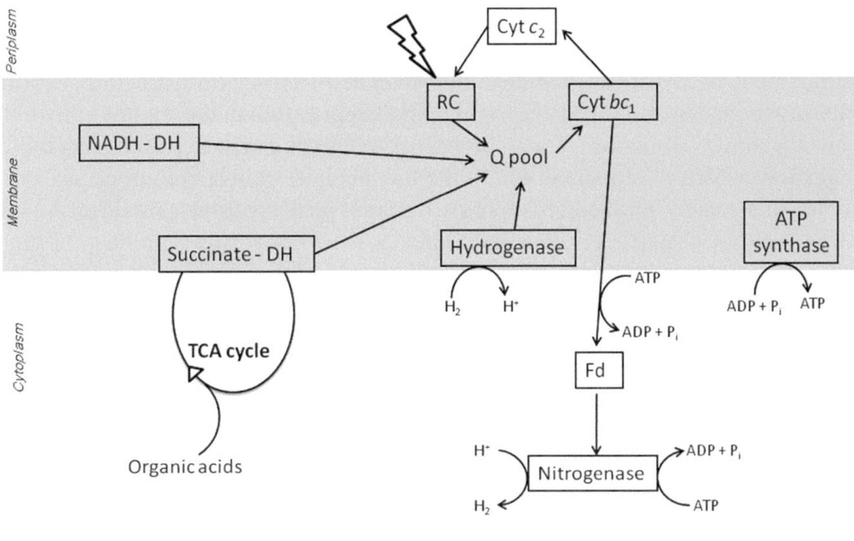

Fig. 1 Main processes related to hydrogen production, under photoheterotrophic growth in non-nitrogen fixing conditions: anoxygenic photosynthesis, ATP synthesis, TCA cycle, hydrogenase, and nitrogenase activities. The *straight black arrows* indicate the electron flow. The *lightning symbol* indicates light excitation. *Cyt* bc_1 cytochrome bc_1 complex, *Cyt* c_2 cytochrome c_2, *Fd* ferredoxin, *RC* Reaction Center, *Succinate – DH* succinate dehydrogenase, *NADH-DH* NADH dehydrogenase. (Image from Adessi and De Philippis 2012)

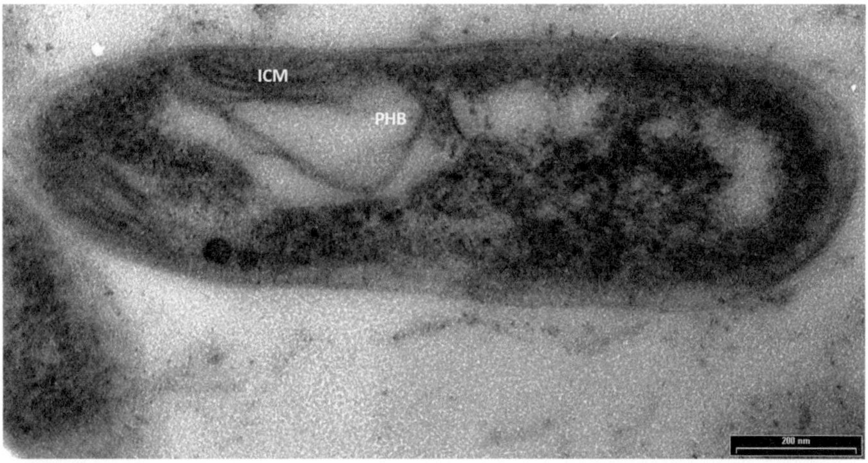

Fig. 2 Electron micrograph of *Rhodopseudomonas palustris* 42OL. Whole cell containing PHB granules, longitudinal section. *PHB* poly-β-hydroxybutyrate granules, *ICM* intra-cytoplasmic membranes

and De Philippis 2014). Otherwise, in cells with active nitrogenase, hydrogen can represent an electron donor, oxidized by the uptake hydrogenase enzyme (Keskin et al. 2011; Adessi and De Philippis 2013) (Fig. 1). It has to be stressed that even if the process is anaerobic, a microanaerobic nitrogenase activity was found in some PNSB with low oxygen concentration (Hallenbeck and Liu 2016). In general, PNSB are a diverse group of anoxygenic phototrophic bacteria with a versatile metabolism, which are able to use a variety of organic acids and sugars, depending on the species selected (Argun and Kargi 2011; Eroğlu et al. 2014). The most studied PNSB species for photobiological hydrogen production are *Rhodobacter sphaeroides*, *Rhodopseudomonas palustris* (Fig. 2), *Rhodobacter capsulatus*, *Rhodobacter sulfidophilus*, and *Rhodospirillum rubrum* (Argun and Kargi 2011). PNSB are capable of producing polyhydroxybutyrate (PHB), that is a member of polyhydroxyalkanoate (PHA) family and represents a biodegradable polymer, that can be used for the production of biodegradable plastics (Keskin et al. 2011; Wu et al. 2012; Adessi and De Philippis 2014). PHB is a carbon storage polymer (visible in Fig. 2) that can be used as carbon and energy source during starvation, since it has low solubility, high molecular weight, and inert nature, causing negligible osmotic pressure on cell (Wu et al. 2012). In PNSB, PHB biosynthesis represents a competitive reductive reaction compared to the nitrogenase activity, which makes it undesirable in hydrogen producing systems (Hustede et al. 1993; Vincenzini et al. 1997; Redwood et al. 2009; Wu et al. 2012).

Generally, the advantages of the photobiological hydrogen production mainly concern the high substrate to hydrogen conversion yields, the possibility to use a wide spectrum of sunlight, the absence of oxygen-evolving reactions, and the possibility of coupling the process with other kinds of fermentation, like the combined system with the dark fermentation (Keskin et al. 2011; Adessi and De Philippis 2014). PNSB can also directly use organic acids or sugars in the photofermentation process composing a single stage system, even if until now only a few studies have assessed the use of sugars as substrates for this purpose (Argun and Kargi 2011; Hallenbeck and Liu 2016) (see "Single Stage Photofermentation Processes"). Otherwise, PNSB can utilize organic acids coming from the dark fermentation, in combined processes: (a) sequential or two stage system (see "Sequential Dark/Photofermentation Processes (Two Stage Systems)"); (b) combined processes forming a co-culturing system (see "Combined Dark/Photofermentation Processes (Co-Cultures)").

For what concerns process parameters, strict control of environmental conditions is essential for efficient hydrogen production. Optimal pH and temperature ranges were reported to be 6.8–7.5 and 30–35 °C, respectively (Argun and Kargi 2011; Eroğlu et al. 2014). Generally, the ammonium in the medium is used for growing until the ammonium concentration decreases under the inhibition threshold for nitrogenase, which is around 2.5 mM (45.1 mg L^{-1}), so H_2 production can start and cellular growth almost stops (Argun and Kargi 2011; Adessi et al. 2012b). Ammonium may be naturally present in the starting substrate, like it frequently happens in wastewaters, but also generated by protein degradation during dark fermentation step (Gómez et al. 2011). Optimum volatile fatty acids (VFA) concentrations were reported to be between 1800 and 2500 mg L^{-1} (Argun and Kargi 2011). The photosynthetic efficiency, also known as light conversion efficiency, is an important and commonly used indicator of the photofermentation and it is defined as the efficiency on the basis of the hydrogen-related energy produced per unit of light energy absorbed (Adessi and De Philippis 2014). It can vary from 0.2 to 9.3 % depending on several factors, like the quality and quantity of light, the biological parameters, like pigment composition, quantum requirements and PNSB strain metabolism, and the kind of substrates used for the fermentation (Argun and Kargi 2011; Adessi and De Philippis 2014). Substrate to hydrogen conversion is another important indicator of the photofermentation, as the catabolism of carbon sources provides electrons in the photosynthesis process (Adessi and De Philippis 2014). This parameter represents the ratio between the moles of hydrogen produced and the moles theoretically obtainable if all the substrate consumed was converted to CO_2 and H_2 (Adessi and De Philippis 2014). Substrate to hydrogen conversion is affected by the C/N in the medium, because with high C/N values the nitrogenase activity is enhanced and with low C/N values the cell growth occurs, instead of hydrogen production (Keskin et al. 2011; Adessi and De Philippis 2014). Also the PHB production competes with the conversion of substrates to hydrogen, as it uses carbon and reducing power coming from carbon sources metabolism (Keskin et al. 2011; Adessi and De Philippis 2014).

Main problems to be faced in photofermentation are: (a) to avoid ammonia inhibition, (b) to increase VFAs availability, (c) to allow an effective light distribution through fermentation medium, and (d) to avoid the metabolic shift from H_2 production to PHB synthesis (Argun and Kargi 2011; Keskin et al. 2011). Rate and yields of H_2 production could be enhanced by metabolic engineering aimed at: (a) blocking competing pathways in order to increase the electron flux from substrate to hydrogen (e.g., by inducing PHB-defection); (b) inactivating uptake hydrogenase; (c) reducing pigment content and enhancing electron flow to nitrogenase; (d) deregulating nitrogenase in order to induce a low sensitivity to ammonium concentration; and (e) enhancing nitrogenase activity (Adessi and De Philippis 2014; Hallenbeck and Liu 2016).

Low light conversion efficiency, high energy demand, low turnover of nitrogenase, and high cost of hydrogen impermeable photobioreactors are some of the main critical issues to be addressed for making economically sustainable the production of hydrogen by photofermentation processes (Keskin et al. 2011; Adessi and De Philippis 2014; Hallenbeck and Liu 2016). In general, light distribution should be as uniform as possible, in particular when using as substrate complex biomasses, since they may interfere with the adsorption spectra of pigments or may contain particles that shade the cells. Moreover, feedstocks characterized by low C/N values favor cell growth instead of H_2 production (Keskin et al. 2011; Adessi and De Philippis 2014). Design, scale-up, and optimization of photobioreactors are fundamental issues, since the cultivation system must be closed in order to maintain anaerobic conditions and to prevent hydrogen dispersion, and requires high illuminated surfaces, efficient mixing and gas exchange system, and temperature control (Adessi and De Philippis 2014). Systems for efficient solid–liquid–gas phase separation, for a higher gas recovery efficiency have been identified in cell immobilization (Tsygankov and Kosourov 2014), but only few studies have been carried out on large scale, or on biomass derived substrates. This issue will be discussed in "Immobilized Systems."

Novel Fermentation Systems

In recent years, innovative processes and novel substrates were tested at lab scale in order to investigate and promote a sustainable development of the photobiological hydrogen production. The various photofermentation processes, that have been designed up to now, and the most investigated novel substrates are schematically represented in Fig. 3.

Recent and innovative processes will be discussed in "Innovative Processes." This section mainly focuses on photofermentation systems conducted on biomass derived substrates, as these are considered to be the applicative goal of hydrogen production. Afterwards, novel substrates will be described in "Novel Substrates," first focusing on the most studied biomass derived substrates ("Biomass Derived Substrates") and then reporting an *excursus* on some synthetic media, investigated as interesting opportunities for enlarging applicability of the hydrogen technology ("Synthetic Substrates").

Innovative Processes

Single Stage Photofermentation Processes

A single stage system is composed by the sole photofermentation stage, where PNSB use substrates, containing mainly organic acids or sugars (Argun and Kargi 2011; Hallenbeck and Liu 2016). Only few and recent studies reported the single stage

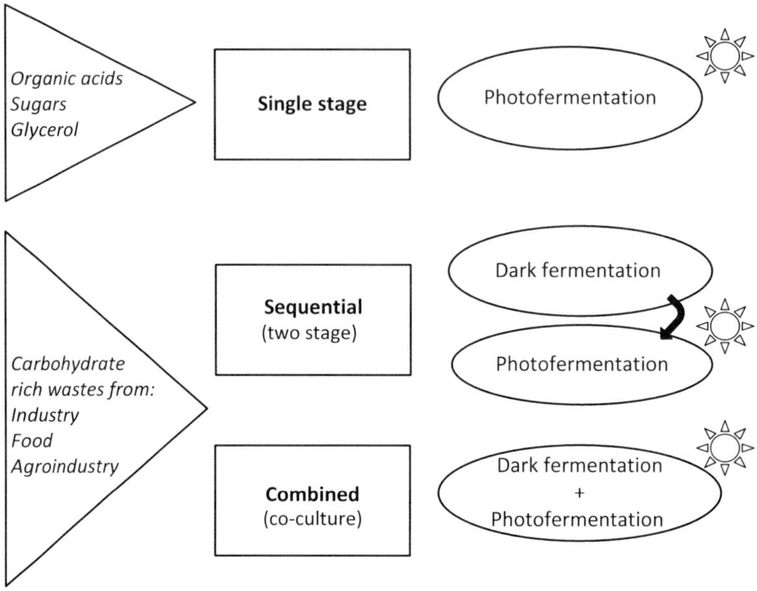

Fig. 3 Novel fermentation systems for a sustainable development of the photobiological hydrogen production. Novel substrates can be used in the following innovative processes: (**a**) single stage photofermentation (Sect. "Single Stage Photofermentation Processes," Table 1); (**b**) sequential two stage fermentation systems (Sect. "Sequential Dark/Photofermentation Processes (Two Stage Systems)," Table 2); and (**c**) combined systems or co-cultures (Sect. "Combined Dark/Photofermentation Processes (Co-Cultures)," Table 3)

photofermentation of sugars, in most cases glucose, while several single stage studies have been carried out using various types of wastewaters, such as effluents from the food and agro-industry, or industrial wastes, like crude glycerol (Hay et al. 2013; Hallenbeck and Liu 2016). One of the first works on this issue was carried out by Zürrer and Bachofen (1981), who reported the hydrogen potential of lactate and lactate-containing wastes for photofermentation processes carried out by the PNSB *Rs. rubrum*.

In Table 1, recent studies on single stage photofermentation using substrates rich in organic acids are reported. Several works investigated the photofermentation of olive mill wastewaters (OMWW) with different operational conditions. Eroğlu et al. (2006) carried out the single stage photofermentation of OMWW with *Rb. sphaeroides* OU001, reporting a higher potential with a clay pretreatment (35.0 L (L medium)$^{-1}$), than with the row substrate (8.0 L (L medium)$^{-1}$). Eroğlu et al. (2008b) reported a single stage assay using a temperature controlled flat plate solar bioreactor, in order to assess the photobiological hydrogen production from *Rb. sphaeroides* O.U.001 in outdoor conditions, and they obtained a H_2 production of 11.4 L (L medium)$^{-1}$. In another work, Eroğlu et al. (2008a) reported that the photofermentative hydrogen production with OMWW was doubled by using the clay pretreated effluent (31.5 L (L medium)$^{-1}$), comparing with the not pretreated one. The effectiveness of using the clay treatment as the optimal method for a fast and low-cost

Table 1 Recent studies on single stage photofermentation from biomass derived substrates

Substrate	Reactor operation mode	Photofermentation inoculum	Temperature (°C)	Hydrogen potential (see notes)	Reference
Olive mill wastewater (4 % in water)	Batch	*Rb. sphaeroides* OU001	n.r.	8.0[a]	Eroğlu et al. (2006)
(Pretreated; 4 % in water)	Batch	*Rb. sphaeroides* OU001	n.r.	35.0[a]	Eroğlu et al. (2006)
(4 % in water)	Batch	*Rb. sphaeroides* O.U.001	32	11.4[a]	Eroğlu et al. (2008b)
(Pretreated; 4 % in water)	Batch	*Rb. sphaeroides* O.U.001	n.r.	31.5[a]	Eroğlu et al. (2008a)
(Pretreated; 4 % in water)	Batch	*Rb. sphaeroides* O.U.001	n.r.	2.1–31.5[a]	Eroğlu et al. (2009)
(2 % in water)	Batch	*Rb. sphaeroides* O.U.001	32	0.05[a]	Eroglu et al. (2010)
(Pretreated; 30 % in water)	Batch	*Rp. palustris* 42OL	30	5.28[b]	Pintucci et al. (2013)
(Pretreated; 30 % in water)	Batch	*Rp. palustris* 6A	30	13.5[c]	Pintucci et al. (2015)
Diary wastewater (40 % in water)	Batch	*Rb. sphaeroides* OU001	28	1.97[a]	Seifert et al. (2010a)
Hydrolyzed wheat starch	Batch	*Rb. sphaeroides* RV	30	1.23[d]	Kapdan et al. (2009)
Beet molasses	Batch	*Rb. capsulatus* JP91	30	10.5[e]	Keskin and Hallenbeck (2012)
Black strap molasses	Batch	*Rb. capsulatus* JP91	30	8.0[e]	Keskin and Hallenbeck (2012)
Hydrolyzed bagasse + yeast extract	Batch	*Rhodobium marinum* NBRC 100434	30	2.67[a]	Anam et al. (2012)
Soy sauce wastewater + yeast extract (10 % in water)	Batch	*Rd. marinum* (Sanur)	30	0.55[a]	Anam et al. (2012)
Brewery wastewaters (10 % in water)	Batch	*Rb. sphaeroides* OU001	28	2.24[a]	Seifert et al. (2010b)
Crude glycerol	Batch	*Rp. palustris* CGA009	30	4.0[f]	Sabourin-Provost and Hallenbeck (2009)

(continued)

Table 1 (continued)

Substrate	Reactor operation mode	Photofermentation inoculum	Temperature (°C)	Hydrogen potential (see notes)	Reference
	Batch	*Rp. palustris* CGA009	30	6.1[f]	Ghosh et al. (2012a)
	Batch	*Rp. palustris* CGA009	30	6.69[f]	Ghosh et al. (2012b)
	Batch	*Rp. palustris* NCIMB 11774	n.r.	6.0[f]	Pott et al. (2013)

n.r. = not reported
[a]mL g dry weight^{-1} h^{-1}
[b]mol mol glycerol^{-1}
[c]L L medium^{-1}
[d]mol mol glucose^{-1}
[e]mol mol sucrose^{-1}
[f]mL L^{-1} h^{-1}

treatment of OMWW was afterwards confirmed by Eroglu et al. (2010). The investigation on batch cultures grown under continuous light or light/dark diurnal cycles with OMWW as substrate gave a similar hydrogen production between the two conditions tested (0.05 L (L medium)$^{-1}$), but the light/dark diurnal cycles condition showed a pronounced lag in biomass and hydrogen accumulation (Eroglu et al. 2010). Different irradiances for hydrogen production using dephenolized OMWW from *Rp. palustris* 42OL were investigated by Pintucci et al. (2013), who found that the higher was the irradiance, the higher were the hydrogen yield and rate. Pintucci et al. (2015) investigated different culture mixing using dephenolized OMWW from *Rp. palustris* 6A and reported the highest hydrogen production using an impeller equipped with five turbines.

Some experiments were performed in order to investigate different starting concentrations of the substrate, such as the photofermentation assays reported by Seifert et al. (2010a, b), who investigated the hydrogen production from *Rb. sphaeroides* O.U.001, using different concentrations of dairy and brewery wastewaters. Kapdan et al. (2009) analyzed the effects of initial sugar concentration on hydrogen yield with hydrolyzed wheat starch, reporting the best photofermentation performance at 5 g L^{-1}, equal to 1.23 mol (mol glucose)$^{-1}$. Ghosh et al. (2012a) investigated the effects of nitrogen source and different concentrations of crude glycerol on hydrogen production using *Rp. palustris* CGA009. They reported that at 20 mM of glycerol and 4 mM of glutamate the highest hydrogen yield was obtained, equal to 6.1 mol H_2 (mol crude glycerol)$^{-1}$, a yield of 87 % of the theoretical. In another study, Ghosh et al. (2012b) investigated the interactive effects among several process parameters: light intensity and concentrations of crude glycerol and glutamate on the stoichiometric conversion of crude glycerol to hydrogen. They observed the optimal conditions with a light intensity of 175 W m^{-2}, 30 mM of glycerol, and 4.5 mM of glutamate, resulting in 6.69 mol (mol crude glycerol)$^{-1}$, a yield 96 % of the theoretical.

Table 2 Recent studies on sequential systems of dark fermentation followed by the photofermentation

Substrate	Reactor operation mode	Dark fermentation inoculum 1st stage	Photofermentation inoculum 2nd stage	Temperature (°C)	Hydrogen potential 1st stage (see notes)	Hydrogen potential 2nd stage (see notes)	Reference
Olive mill wastewater	Batch	Sewage sludge	*Rb. sphaeroides* O.U.001	35 (1st stage) 32 (2nd stage)	n.i.	29[a]	Eroğlu et al. (2006)
Ensiled olive pomace	Batch	Heat-shocked anaerobic sludge	*Rp. palustris* CGA676	37 (1st stage) 28 (2nd stage)	<0.1 [a]	n.d.	Corneli et al. 2016a, submitted; Adessi et al. 2016a
Sugar beet molasses	Batch	*C. saccharolyticus* DSM 8903	*Rb. capsulatus* hup$^-$ (YO3)	72 (1st stage) 30–33 (2nd stage)	4.2 [b]	9.5[b]	Özgür et al. (2010)
Sugarcane vinasse	Anaerobic fluidized bed reactor (1st stage) Batch (2nd stage)	Heat-shocked granular sludge	Phototrophic hydrogen-producing consortium	30	0.34 [c]	5.5[c]	Lazaro et al. (2015)
Cassava	Batch	Heat-shocked cattle dung	*Rb. sphaeroides* ZX-5	37 (1st stage) 30 (2nd stage)	199[d]	611[d]	Zong et al. (2009)
Cassava starch	Batch	Heat-shocked anaerobic sludge	*Rp. palustris*	35 (1st stage) 30 (2nd stage)	240.4[d]	131.9[d]	Su et al. (2009)
	Batch (immobilized 2nd stage)	Mixed anaerobic bacteria (mainly *Clostridium* species)	Mixed photosynthetic bacteria (mainly *Rp. palustris*)	30	2.53[e]	3.54[e]	Cheng et al. (2011b)

(continued)

Table 2 (continued)

Substrate	Reactor operation mode	Dark fermentation inoculum 1st stage	Photofermentation inoculum 2nd stage	Temperature (°C)	Hydrogen potential 1st stage (see notes)	Hydrogen potential 2nd stage (see notes)	Reference
Chlorella pyrenoidosa and cassava starch	Batch	Hydrogen-producing bacteria (mainly *C. butyricum*)	Photosynthetic bacteria (mainly *Rp. palustris*)	35 (1st stage) 30 (2nd stage)	276.2[a]	388[a]	Xia et al. (2014)
Corn stalk	Batch	Heat-shocked cow dung	*Rb. sphaeroides* HY01	35	192.9[a]	401.5[a]	Yang et al. (2015)
Ensiled *Zea mays*	Batch	Heat-shocked anaerobic sludge	*Rp. palustris* CGA676	37 (1st stage) 28 (2nd stage)	13.8[a]	228.7[a]	Corneli et al. 2016a, submitted; Adessi et al. 2016a
Potato	Batch	Microbial consortium	*Rb. capsulatus* B10	37 (1st stage) 28 (2nd stage)	0.7[e]	4.9[e]	Laurinavichene et al. (2010)
Rice straw	Batch by (immobilized 2nd stage)	Heat-shocked anaerobic sludge	Mixed photosynthetic bacteria	35 (1st stage) 30 (2nd stage)	108–155[a]	217–328[a]	Cheng et al. (2011a)
Food waste	Batch	Heat-shocked cattle dung	*Rb. sphaeroides* ZX-5	37 (1st stage) 30 (2nd stage)	220[d]	541[d]	Zong et al. (2009)
Vegetable waste	Batch	Autochthonous chemoheterotrophic microflora	*Rp. palustris* CGA676	r.t. (1st stage) 30 (2nd stage)	n.i.	3.9–9.6[f]	Adessi et al. (2012b)
Water hyacinth	Batch (immobilized 2nd stage)	Heat-shocked anaerobic sludge	*Rp. palustris*	35 (1st stage) 30 (2nd stage)	73.5[a]	522.6[a]	Su et al. (2010)

Substrate	Reactor operation mode	Dark fermentation inoculum 1st stage	Photofermentation inoculum 2nd stage	Temperature (°C)	Hydrogen potential 1st stage (see notes)	Hydrogen potential 2nd stage (see notes)	Reference
Arthrospira platensis	Batch	Mixed anaerobic bacteria (mainly *Clostridium* species)	Mixed photosynthetic bacteria (mainly *Rp. palustris*)	35 (1st stage) 30 (2nd stage)	96.6[d]	337.0[d]	Cheng et al. (2012)
Ensiled *Arundo donax*	Batch	Heat-shocked anaerobic sludge	*Rp. palustris* CGA676	37 (1st stage) 28 (2nd stage)	<0.1[a]	7.5[a]	Corneli et al. 2016a, submitted; Adessi et al. 2016a
Ground wheat solution	Batch	Heat-shocked anaerobic sludge	*Rb. sphaeroides* RV	30	1.87[e]	2.68[e]	Argun and Kargi (2010a)
Wheat bran	Batch	Heat-shocked anaerobic sludge	*Rp. palustris* CGA676	37 (1st stage) 28 (2nd stage)	18.9[a]	463.0[a]	Corneli et al. 2016a, submitted; Adessi et al. 2016a
Crude glycerol	Batch	*Klebsiella* sp. TR17	*Rp. palustris TN1*	40 °C	5.74[g]	0.68[g]	Chookaew et al. (2015)
Seawater (+glucose)	Batch	*Thermotoga neapolitana* (capnophilic lactic fermentation)	*Rp. palustris* 42OL	72 (1st stage) 28 (2nd stage)	2.6[e]	6.8[e]	Dipasquale et al. (2015)

n.i. = not investigated
r.t. = room temperature
[a]NL $kgVS^{-1}$
[b]mol $molsucrose^{-1}$
[c]mmol L^{-1}
[d]mL g^{-1}
[e]mol $molglucose^{-1}$
[f]mL L^{-1} h^{-1}
[g]mmol g COD^{-1}

Table 3 Recent studies on co-culture system of dark fermentation and photofermentation

Substrate	Reactor operation mode	Dark fermentation inoculum	Photofermentation inoculum	Temperature (°C)	Hydrogen potential (see notes)	Reference
Sugarcane distillery effluent	Continuous	*C. freundii* 01, *E. aerogenes* E10	*Rp. palustris* P2	n.c.	0.53[a]	Vatsala et al. (2008)
Ground wheat solution	Batch	Heat-shocked anaerobic sludge	*Rb. sphaeroides* NRRL B-1727; DSMZ-158; RV, *Rp. palustris* DSMZ-127	30	156.8[b]	Argun et al. (2009a)
	Batch	Heat-shocked anaerobic sludge	*Rb. sphaeroides* NRRL B-1727; DSMZ-158; RV, *Rp. palustris* DSMZ-127	30	176[b]	Argun et al. (2009b)
	Continuous	*C. beijerinkii* DSMZ-791	*Rb. sphaeroides* RV	30	90[b]	Argun and Kargi (2010b)
	Batch	Heat-shocked anaerobic sludge	*Rb. sphaeroides* RV	30	218[b]	Argun and Kargi (2010c)
	Batch	Heat-shocked anaerobic sludge	*Rb. sphaeroides* NRRL	30	0.36[c]	Ozmihci and Kargi (2010)
Distillers wheat grains	Batch	*E. coli* BW 25113	*Rb. sphaeroides* MDC6521	30	5.16[d]	Sargsyan et al. (2016)
Calophyllum inophyllum oil cake	Batch	*E. aerogenes* MTCC 8558	*Rb. sphaeroides* MTCC 9765	30	7.95[e]	Arumugam et al. (2014)

n.c. not controlled
[a] kg 100 m^{-3} h^{-1}
[b] mL g $starch^{-1}$
[c] mol mol $glucose^{-1}$
[d] mmol L^{-1} $days^{-1}$
[e] L L^{-1}

Sequential Dark/Photofermentation Processes (Two Stage Systems)

The two stage system is composed by a first stage of dark fermentation, which is followed by a second stage of photofermentation in separated reactors. During dark fermentation, heterotrophic bacteria convert organic substrates, mainly carbohydrate-rich materials, into organic products, hydrogen, and carbon dioxide (Holladay et al. 2009; Abo-Hashesh and Hallenbeck 2012; Ghimire et al. 2015). The key enzyme of the process is hydrogenase, that in anaerobic condition reduces protons to hydrogen, neutralizing the electrons coming from the organic compounds oxidation (Argun and Kargi 2011; Ghimire et al. 2015). Dark fermentation effluents are characterized by the presence of large amounts of volatile fatty acids and lactate. Since the main fermentation end products of dark fermentation are acetic and butyric acids, being the latter in excess with respect to the former, the process is also called acetate/butyrate-type fermentation (Hawkes et al. 2007; Argun and Kargi 2011; Abo-Hashesh and Hallenbeck 2012; Ghimire et al. 2015). The pathways leading to the synthesis of H_2 and to the formation of these two acids allow the highest theoretical conversion of glucose to H_2 compared those producing other acids. In particular, the conversion of 1 mol of glucose to acetic acid involves the production of 4 mol of H_2, while the conversion of glucose to butyric acid involves the production of 2 mol of H_2 (Hawkes et al. 2007; Argun and Kargi 2011; Abo-Hashesh and Hallenbeck 2012; Ghimire et al. 2015). In order to have a high H_2 yield, dark fermentation processes need to be carried out under anaerobiosis and with a low partial pressure of hydrogen (Guo et al. 2010; Abo-Hashesh and Hallenbeck 2012). Strict anaerobic conditions are necessary if the inoculum is composed by strict anaerobic species, such as those belonging to the *Clostridium* genus. Otherwise, oxygen can be present in traces if the inoculum is composed of facultative anaerobic species, such as those belonging to the *Enterobacter* genus, or mixed cultures (Bartacek et al. 2007; Guo et al. 2010; Abo-Hashesh and Hallenbeck 2012). When the inoculum is a non-sterile culture, e.g., digestate of biogas plants or sewage sludge, a pretreatment (e.g., heat shock, acid or alkaline treatment, aeration, and sonication) is recommended in order to reduce the activity of H_2 consumer microorganisms such as homoacetogens, methanogens, nitrate, and sulfate reducing bacteria (Guo et al. 2010; Argun and Kargi 2011; Abo-Hashesh and Hallenbeck 2012; Ghimire et al. 2015).

The oxidation of organic compounds for hydrogen production requires a low hydrogen partial pressure (Guo et al. 2010; Abo-Hashesh and Hallenbeck 2012). Among the techniques used to decrease the hydrogen concentration in order to increase the hydrogen yield of the system, the most frequently used are the agitation of the medium, the insufflation of molecular nitrogen, and hydrogen stripping (Guo et al. 2010; Abo-Hashesh and Hallenbeck 2012). The pH value is another important factor, which can affect the hydrogen yield, the metabolic products, and the structure of the microbial community (Cappai et al. 2014; Guo et al. 2010). Better process performances can be achieved at pH values ranging between 5 and 6 for food wastes, while a neutral pH (7–7.5) is advisable for plant residues and for livestock wastes (Cappai et al. 2014; Guo et al. 2010). In particular, pH 5 is the minimum value that most heterotrophic bacteria can tolerate (Abo-Hashesh and Hallenbeck 2012). By literature, studies on hydrogen production by dark fermentation in batch, without pH correction and using sucrose as a substrate, appear to be quite frequently carried out (Argun and Kargi 2011; Guo et al. 2010). Typical dark fermentation processes were carried out under mesophilic conditions. However, it was shown that thermophilic conditions lead to higher hydrogen yields due to: (a) higher rates of substrate decomposition; (b) better hydrolysis of recalcitrant molecules such as lignocellulosic constituents of vegetable residues, and (c) faster metabolic activity of H_2 producing thermophilic bacteria. Moreover, high temperatures also inhibit growth and activity of hydrogen consuming microorganisms (Guo et al. 2010; Argun and Kargi 2011; Abo-Hashesh and Hallenbeck 2012). However, thermophilic conditions imply a higher energy consumption than mesophilic conditions (Argun and Kargi 2011; Guo et al. 2010), even if it is possible to use the widely available hot waters deriving from the cooling systems of many industrial processes.

In the sequential system, the effluents derived from the dark fermentation processes are subsequently used as substrate for the photofermentation stage. The investigation of renewable substrates for the sequential systems started some years ago and one of the first study was carried out by Fascetti et al. (1998), reporting the photosynthetic hydrogen production using *Rb. sphaeroides* RV on acidogenic fermentation effluents of municipal solids wastes, mainly consisting of fruit and vegetables wastes. Recent studies on the use of biomass derived substrates such as energy crops, crop residues, agroindustrial and industrial residues in dark/photofermentation sequential systems are reported in Table 2. Heat treated inocula for the dark fermentation stage and mesophilic conditions are usually used; as inocula, pure cultures of *Rb. sphaeroides* or *Rp. palustris* were most frequently used. Concerning OMWW, they are rich in organic acids and, as above reported, can be used in one stage photofermentation. However, since the light penetration is difficult due to the dark color of OMWW, Eroglu et al. (2010) first reported that the dark fermentation step implies a positive effect on the subsequent photofermentation, enhancing the efficiency of the process.

Various operational conditions have been tested in order to maximize hydrogen yield and to optimize the process (Table 1). For example, Su et al. (2009) investigated different starting raw cassava starch concentrations, from 10 to 25 g L^{-1}, using heat-shocked anaerobic sludge and *Rp. palustris* as inocula of the first and the second stages, respectively, and obtained the maximum hydrogen yield of 240.4 mL (g starch)$^{-1}$ and of 131.9 mL (g starch)$^{-1}$, in the dark fermentation and the photofermentation stages, respectively, using a starch concentration of 10 g L^{-1}. Another study assessed the effect of the starting concentration of the substrate, Laurinavichene et al. (2010) reported that a starting concentration of potato homogenate of 400 g L^{-1} allowed to obtain an overall maximum hydrogen production equal to 5.6 mol (mol glucose)$^{-1}$, using a microbial consortium for the first stage and *Rb. capsulatus* B10 for the second one. Also Cheng et al. (2011a) reported different starting substrate concentrations, obtaining the highest overall hydrogen yield of 463 mL (gVS)$^{-1}$ with a concentration of microwave-assisted alkali pretreated rice straw of 50 g L^{-1}. Su et al. (2010) reported the highest overall hydrogen yield in a sequential system of 596.1 mL (gVS)$^{-1}$, using pretreated water hyacinth at a concentration of 10 g L^{-1}.

Concerning the kind of light source for the photobiological hydrogen production, Argun and Kargi (2010a) found that halogen lamp was the most suitable light source for the photofermentation of dark fermentation effluents of ground wheat solution, yielding the highest cumulative hydrogen production of 2.68 mol (mol glucose)$^{-1}$.

Cheng et al. (2012) assessed the ammonium concentration effect on the photobiological hydrogen production and they reported that the reduced content of ammonium, from 2.2 to 2.7 mM, in the dark fermentation effluent of *Arthrospira platensis*, enhanced the hydrogen potential from 96.6 to 337.0 mL (g DW)$^{-1}$. Also Adessi et al. (2012b) investigated the ammonium concentration using in the photofermentation stage the mutant strain *Rp. palustris* CGA676, which constitutively expresses nitrogenase genes, reporting the highest hydrogen production rate of 9.6 mL L^{-1} h^{-1} in the 3-fold diluted medium containing 2.03 mM of ammonium. The same PNSB strain was used by Corneli et al. (2016a, b, submitted) in a process, reported in the patent filed F1.S0061.12.IT.1 (Adessi et al. 2016a), aimed at assessing the photofermentative hydrogen potential of the effluents of the dark fermentation of ensiled maize, ensiled giant reed, ensiled olive pomace, and wheat bran. Under the conditions tested, the highest performance of the strain was observed in the presence of maize and wheat bran (228.7 and 463.0 NL (kgVS)$^{-1}$, respectively).

Xia et al. (2014) investigated the effect of different C/N based on different ratios of *Chlorella pyrenoidosa* and cassava starch in a codigestion. Their higher dark and photofermentative hydrogen potential was equal to 664.2 mL (gVS)$^{-1}$ at a C/N of 25.3. Chookaew et al. (2015) focused on the optimal conditions for hydrogen production from *Rp. palustris* TN1 on dark fermentation effluent of crude glycerol and they found that the fivefold diluted effluent, without the supplement of yeast extract and $NaHCO_3$ and 2 mM glutamate corresponded to the optimum condition, with a cumulative hydrogen production of 0.68 mmol g COD^{-1}.

A novel and high yielding experiment was reported by Dipasquale et al. (2015); they carried out the photofermentation using capnophilic lactic fermentation effluents of seawater supplemented with glucose utilized by *Thermotoga neapolitana* and they reported an overall hydrogen potential of 9.4 mol molglucose^{-1}.

Combined Dark/Photofermentation Processes (Co-Cultures)

The co-culture system is composed by dark and photofermentative bacteria in a coupled fermentation for biohydrogen production, both fermentations taking place simultaneously in the same bioreactor (Keskin et al. 2011; Adessi et al. 2012a; Eroğlu et al. 2014; Pachapur et al. 2015a; Hallenbeck and Liu 2016). Co-culturing is considered advantageous in comparison with sequential fermentation due to: (a) a possible reduction in the fermentation time; (b) an increase in hydrogen production yields, rate, and substrate conversion efficiencies; and (c) the elimination of some operations needed for using the effluents of dark fermentation to feed light-dependent fermentation (e.g., H adjustment, medium sterilization or dilution), being the process carried out in only one bioreactor (Keskin et al. 2011; Adessi et al. 2012a; Eroğlu et al. 2014; Singh and Wahid 2015; Hallenbeck and Liu 2016).

On the other hand, the main drawback concerns the differences in nutrients requirements and growth conditions (e.g., growth rate and acid-resistant capacity) that the two different types of microorganisms might have (Adessi et al. 2012a; Zagrodnik and Laniecki 2015; Sargsyan et al. 2016). The possible accumulation of organic acids and ammonium, and the decrease of pH can negatively affect the overall process, which can also be affected by a decrease in light penetration due to suspended solids and cell growth (Singh and Wahid 2015; Zagrodnik and Laniecki 2015). The rate of volatile fatty acids production by dark fermentative bacteria can be higher than the utilization rate by photofermentative bacteria and thus the growth of the former can be limited by the decrease of the pH value (Liu et al. 2010; Singh and Wahid 2015). Thus, the system should be carefully controlled in media composition, in environmental conditions and in bacteria ratio in order to promote the growth and the activity of all the bacterial species in the co-culture (Liu et al. 2010; Adessi et al. 2012a; Singh and Wahid 2015; Zagrodnik and Laniecki 2015; Sargsyan et al. 2016). By the literature, only few studies reported the co-culture system for dark and photofermentation. Typically, studies tested co-cultures of pure bacterial strains, being glucose the most frequently studied substrate (Liu et al. 2010; Adessi et al. 2012a; Zagrodnik and Laniecki 2015). However, recent investigations on biomass derived substrates reported some co-culture assays, some of them using pure bacterial cultures, while others used mixed cultures, such as heat-shocked anaerobic sludge for dark fermentation, and selected bacterial consortia for photofermentation (Table 3). In general, these studies reported mesophilic batch tests assessing the potential of simple carbohydrate-rich substrates. The comparison among the different studies on complex substrates is quite difficult, because of the different operational conditions adopted and the units of measure of the results.

Several studies carried out a co-culture system using ground wheat solution as substrate in mesophilic batch tests (Argun et al. 2009a, b; Argun and Kargi 2010c; Ozmihci and Kargi 2010) (Table 3). Argun et al. (2009a) assessed the effects of the substrate and of cell concentration on hydrogen production, using heat-shocked anaerobic sludge and a photofermentative bacterial consortium, as inocula, and they reported the highest hydrogen potential equal to 156.8 mL (g starch)$^{-1}$ with a biomass to substrate ratio of 0.22 g cells (g substrate)$^{-1}$. A similar result (176 mL (g starch)$^{-1}$) was obtained by Argun et al. (2009b) with the dark to light biomass ratio of 1/7. Argun and Kargi (2010c) investigated different light sources, intensities, and illumination regime and reported the highest hydrogen potential (218 mL (g starch)$^{-1}$) using a halogen lamp. Ozmihci and Kargi (2010) compared different mixed cultures for hydrogen production and reported the highest performance (0.36 mol (mol glucose)$^{-1}$) using heat-shocked anaerobic sludge and *Rb. sphaeroides* NRRL, as inocula.

Also Argun and Kargi (2010b) reported a co-culture study using ground wheat solution as substrate, but in a continuous system. They used *Clostridium beijerinkii* DSMZ-791 and *Rb. sphaeroides* RV, as dark and photofermentative strains, respectively, and they reported the highest hydrogen potential of 90 mL (g starch)$^{-1}$ at hydraulic residence time of 6 days.

Other studies reported hydrogen potentials of co-culture systems using different kinds of biomass derived substrates (Table 3). Vatsala et al. (2008) evaluated the hydrogen production of a mixture of pure cultures (*Citrobacter freundii* 01, *Enterobacter aerogenes* E10, and *Rp. palustris* P2) using sugarcane distillery effluent in a mesophilic 100 m^3 scale reactor and they reported a hydrogen potential of 0.53 kg 100 m^{-3} h^{-1}. In another study, *Calophyllum inophyllum* oil cake was investigated as substrate in a co-culture system of *E. aerogenes* and *Rb. sphaeroides* under dark and photofermentative conditions in a mesophilic batch test and the hydrogen potential was 7.95 L L^{-1} (Arumugam et al. 2014). Sargsyan et al. (2016) reported a co-culture batch system using distillers wheat grains with mixed cultures of *Escherichia coli*, as dark fermentative bacteria, and *Rb. sphaeroides*, as photofermentative bacteria. Their maximal rate of H_2 production was 5.16 mmol L^{-1} day^{-1} in the twofold diluted medium.

Immobilized Systems

Immobilized whole cell techniques represent a reliable approach to dark and photofermentation for the enhancement of continuous hydrogen production, compared to suspended cell systems, since they are more efficient in solid/liquid/gas separation and can be operated at high dilution rates without the risk of biomass washout (Chang et al. 2002).

Methods of immobilization can be either natural or artificial: natural immobilization refers to the spontaneous or enhanced formation of biofilm and granules; oppositely, artificial cell entrapments assume the use of matrices or substrates for attachment, entrapment, or encapsulation of microorganisms. Immobilized cells, and in particular cells in biofilms, are usually characterized by enhanced resistance to the presence of toxic components or other extreme culture conditions as compared to cells in suspension, due to the diffusion barrier constituted by the matrix (Tsygankov and Kosourov 2014).

In every immobilized system there is a natural separation of solid, liquid, and gaseous phases; this not only facilitates gas recovery, but also the repeated use of biomass. The separation of phases makes immobilized cultures have higher volumetric rates of hydrogen production, compared to suspended cell systems (Tsygankov and Kosourov 2014). However, in some cases the yield of hydrogen production can still be lower than that of the suspended cell systems. This could be due to low substrate conversion efficiency, or mass transfer limitations arising from the matrix barrier. In order to obtain higher yields of hydrogen production in immobilization process, it is necessary to develop new immobilized materials for cell entrapment (Singh and Wahid 2015).

At the present time, a number of materials for immobilization are under investigation for the immobilization of PNSB, such as latex (Gosse et al. 2007), carbon fibers (Xie et al. 2012), or a mixture of different immobilizing materials (Wang et al. 2010, 2012). Biofilm reactors are starting to be studied intensively as well (Tian et al. 2010; Zhang et al. 2010; Guo et al. 2011).

However, at present all the immobilized systems have been studied under lab conditions, in batch, with relatively small-volume photobioreactors and mostly with synthetic substrates.

Only a limited number of studies have been carried out on this specific topic, recently, and are indicated in Table 2. Su et al. (2010) reported a study on immobilized *Rp. palustris*, in the second stage of a sequential system degrading water hyacinth biomass. Pure *Rp. palustris* cultures were immobilized in alginate granules of 3–4 mm of diameter. The authors tested four different water hyacinth concentrations: the volumetric hydrogen production increased with increasing the amount of water hyacinth, while the yields decreased. This confirmed that the phase separation allows a very good separation and recovery of the gas produces, but immobilization in alginate beads may interfere with optimal nutrient exchange when the concentration of substrate increases, thus giving lower yields.

Cheng et al. (2011a, b) used the same immobilization system, but for entrapping mixed photosynthetic bacteria. A preliminary compared study on immobilized *Rp. palustris* and immobilized mixed photosynthetic bacteria showed increased volumetric production and yields (20 % and 24 % yield increase, respectively) as compared to suspended cell systems (Cheng et al. 2011b), but this part of the study was conducted on synthetic media containing acetate. Synthetic media surely allow a faster diffusion than complex waste derived substrates. However, the yields reported for the conversion of rice straw (Cheng et al. 2011a) showed the best result, 328 mL H_2 (g TVS)$^{-1}$, with the highest substrate concentration tested. Cassava starch (Cheng et al. 2011b) resulted to be a very interesting substrate, yielding 3.54 mol of H_2 per mole of initial glucose.

The number of recent studies about the use of immobilized systems with biomass derived substrates is very poor, unfortunately. Moreover, they all report very standard immobilization matrixes and in batch processes. As above mentioned, the main advantages in using immobilized cells are the stability of the process and the possibility of carrying out continuous feeding. This feature would be the best solution in particular when working with wastes, whose organic matter content needs to be reduced.

Novel Substrates

Biomass Derived Substrates

One interesting feature of PNSB is their capability to use, for the production of H_2, biomass derived substrates, like residues deriving from industrial or agricultural processes, that are, in many cases, available in large amounts. However, the sustainability of fermentative process mainly depends on the kind of substrate employed as carbon source (Bartacek et al. 2007; Frigon and Guiot 2010).

A large portion of possible wastes for energy recovery is composed of food and agricultural wastes. Depending on their composition, those wastes have to be treated before using them for photofermentation. Indeed, most frequently it is indicated a previous fermentation step, either hydrogenogenic or not as reported in Table 2.

Several substrates have been proposed and studied for two stage sequential systems, varying among energy crops, crop residues, biodegradable residues and byproducts produced by the livestock and agroindustrial sectors, food waste, and organic fraction of municipal solid waste (Frigon and Guiot 2010; Weiland 2010; Appels et al. 2011). The use of substrates rich in fermentescible sugars or of complex matrix deeply affects the overall efficiency of the process, because of the different physico-chemical properties: simple sugars typically lead to obtain higher energy potential and production rate than complex organic materials (Frigon and Guiot 2010). Nevertheless, often the highest is the biodegradability of the biomass, the highest is the environmental cost (Bartacek et al. 2007; Frigon and Guiot 2010). The use of conventional arable crops, like maize and sorghum, for energy use need careful consideration of land availability and food demand, while, at least in the medium term, lignocellulosic crops (both herbaceous and woody) provide environmental advantages, since they can be produced on marginal and degraded lands, requiring lower technical input (Lewandowski et al. 2003; Angelini et al. 2009; Frigon and Guiot 2010). An interesting opportunity for the bioenergy supply chain is represented by the perennial grasses, like giant reed, switchgrass, and miscanthus, that are high yielding no-food crops with good adaptability to marginal areas (Lewandowski et al. 2003; Angelini et al. 2009; Dragoni et al. 2015). They have resistance to drought stress and to pathogens and phytophagous insects, are good competitor against weeds, and can be used for phytoremediation (Lewandowski et al. 2003; Angelini et al. 2009). Only recently the fermentation product deriving from the dark fermentation of the above-mentioned no-food crops has been used for photofermentation (Adessi et al. 2016a; Corneli et al. 2016a, b submitted).

Another interesting opportunity consists in the use of agroindustrial residues, that are renewable, abundant, economic, and no land-demanding (Schievano et al. 2009; Guo et al. 2010). A bioenergy valorization implies no additional costs for other treatments or disposals (Schievano et al. 2009; Guo et al. 2010). Agroindustrial systems produce abundant and diverse feedstocks, such as lignocellulosic materials, crop residues, vegetable oils, animal fats, protein-rich waste, pre-digested wastewater sludges, animal slurries and manures, waste paper, and household waste, that can contribute to the biomass demand for bioenergy supply chain (Schievano et al. 2009). Bioenergy use of these substrates needs to properly face and respect other uses of residues, like the use of crop and agroindustrial residues as animal feed (Nonhebel 2007), in order to be not in conflict with food production. One of the main issues about energy crops (i.e., crop and agroindustrial residues) is the lignocellulosic content, characterized by its low biodegradability (Frigon and Guiot 2010).

Some agroindustrial and food residues are already rich in organic acids and can be used in one stage photofermentation, such as olive mill and dairy wastewaters (Table 1). However, since the light penetration may result difficult due to the dark color of the medium, especially for olive mill wastewaters, Eroğlu et al. (2006) reported that pretreatments such as dark fermentation (Table 2) can enhance hydrogen production in the photofermentation stage, not only by increasing the amount of readily available organic acids, but also by color depletion.

Carbohydrate-rich substrates such as molasses, hydrolyzed wheat starch, hydrolyzed bagasse, or substrates composed of a mixture of acids and sugars, such as soy sauce or dairy wastewaters, have been used as well for direct photofermentation, since some PNSB species are able to convert sugars to hydrogen with interesting rates (Table 1).

In general, renewable resources constitute an abundant and low-cost material for biohydrogen production and their use is fundamental for large-scale sustainable application. Hence, investigations on novel substrates in order to enlarge the knowledge on photobiological hydrogen production are needed.

Among them, crude glycerol was studied as a possible industrial waste to be used for further energy recovery. Indeed, the current technology for biodiesel production (a base-catalyzed trans-esterification of oils) produces 1 kg of crude glycerol per 10 L of biodiesel, thus the glycerol fraction has become a waste disposal problem (Ghosh et al. 2012a; Johnson and Taconi 2007). A certain number of research papers have been published recently on the topic (Sects. "Single Stage Photofermentation Processes and Sequential Dark/Photofermentation Processes (Two Stage Systems)," and Tables 1 and 2), most of them reporting single stage photofermentation processes using *Rp. palustris* (Sabourin-Provost and Hallenbeck 2009; Ghosh et al. 2012a, b, c; Pott et al. 2013), giving conversion efficiencies ranging from 75 to 100 %. Chookaew et al. (2015) reported a dark/photofermentation sequential process conducted by *Klebsiella* sp. and *Rp. palustris*, giving a much lower conversion, namely 10.4 % of the theoretical yield.

Besides the chemical characteristics of all the substrates mentioned earlier, biomass storage is essential for the sustainability of the overall fermentation technology. Up to now, ensiling is the common way of storage for biomasses, and is widely used, for example, in anaerobic digestion plants (Weiland 2010; Dragoni et al. 2015). Wet feedstocks (25–35 % of total solids) can be ensiled with the purpose to maintain and use them in time, as in the livestock industry (Weiland 2010; Dragoni et al. 2015). In the case of using substrates for photofermentation, ensiling could be an opportunity for degrading the fermentescible substrates to organic acids thus enhancing both fermentation and photofermentation (Corneli et al. 2016a; Corneli et al. 2016a, b submitted). Indeed, during ensiling, after a short initial aerobic phase, the fermentation starts under anaerobic conditions with the production of lactic acid by lactic acid bacteria, with the consequent decrease of pH to about 4.0 (Weiland 2010). With this level of acidity, in few days, the growth of undesirable microorganisms, such as enterobacteria, clostridia, and yeasts, which consume nutrients and energy, is inhibited and subsequently the process settles (Weiland 2010; Herrmann et al. 2011). Thus, anaerobic conditions together with a rapid production of lactic acid allow a good conservation of the biomass, in terms of both nutrients and energy (Weiland 2010; Herrmann et al. 2011) also giving an excellent substrate for photofermentation with PNSB.

Synthetic Substrates

The use of synthetic media for hydrogen production processes is important, since it gives the possibility to investigate the behavior of the microorganisms in a controlled system, where the culture medium is completely defined. Thus, synthetic substrates are used for research studies on very innovative culturing systems, for the characterization of new or engineered strains, or for exploring new metabolic routes. However, in this section only the synthetic substrates investigated for broadening the range of the biomass derived substrates utilizable for photofermentation will be discussed. Indeed, part of the research on substrates investigates the possibility of expanding the medium composition combinations, thus increasing the range of applicability of the hydrogen production process.

A few recent research papers were focused on the possibility of using glucose for direct photofermentation, in order to skip the dark fermentation step when using sugar-containing waste substrates. The feasibility of the process majorly stands in the fact that the hydrogen yield obtained (mol H_2 mol glucose^{-1}) has to be higher or comparable to the ones obtained by combined dark/photofermentation systems. Indeed, from the initial low values of 3.3 mol H_2 (mol glucose)$^{-1}$ (Abo-Hashesh et al. 2013), an increase to 5.5 mol H_2 (mol glucose)$^{-1}$ (Ghosh et al. 2012c) and finally to 9.0 mol H_2 (mol glucose)$^{-1}$ (Abo-Hashesh et al. 2013) was obtained with a hup$^-$ strain of *Rb. capsulatus*. Recently, an interesting study carried out on a mixture of sugars and acids (namely, glucose, xylose, and acetate, that are the main products in palm oil hydrolysates) using *Rb. sphaeroides* S10, reported a 45 % substrate-to-hydrogen conversion efficiency when the substrate was composed of 5 mM glucose, 18 mM xylose, and 7 mM acetate (Pattanamanee et al. 2015). Another recent study was conducted with a mutant strain of the marine organism *Rhodovulum sulfidophilum*, giving a yield of 7.1 mol H_2 (mol glucose)$^{-1}$ (Cai and Wang 2014). These authors used in their experiments marine water, which opens a completely different scenario for possible low-cost substrates. Indeed, the application of seawater for bacterial fermentative production is of increasing interest (Maeda et al. 2000), since large-scale cultivation systems need to be sustainable in terms of water resources. Hence, brackish water, wastewater, and seawater seem the most appropriate for large-scale culturing. Most of hydrogen production studies on salt containing media were conducted using marine photosynthetic bacteria such as *Rhodobium marinum* (Ike et al. 1997), *Rhodovulum sulfidophilum* P5 (Cai and Wang 2012, 2013, 2014), as well as a marine mixed phototrophic bacterial consortium (Cai and Wang 2012). Recently, a study on the use of freshwater *Rp. palustris* was conducted on a substrate derived from threefold diluted seawater (Dipasquale et al. 2015). In order to further increase the possible combinations of processes that can be carried out Adessi et al. (2016b) investigated a range of salt concentrations that can be suitable for hydrogen production with the same freshwater *Rp. palustris* strain, up to 3.9 % salt content. Thus, the possibility of using *Rp. palustris*, that is extremely versatile in terms of carbon sources utilization (Larimer et al. 2004; Adessi et al. 2016c), also on salt containing substrates, would enhance the applicability of the hydrogen production process and the prospect of its cost-reduction.

Another interesting substrate with a direct applicability is glycerol. As above mentioned, crude glycerol is produced as a side product of the biodiesel manufacturing industry. Direct photofermentation of pure glycerol was studied prior to the use of crude glycerol by Sabourin-Provost and Hallenbeck (2009) and by Pott et al. (2013), giving a conversion of glycerol to hydrogen of 75 % and 80–85 %, respectively.

Ethanol has recently reached some attention in photofermentation. The possibility of using substrates containing small percentages of ethanol is an opportunity for increasing the number of wastes that can be used. Recently, Kim et al. (2014) demonstrated that ethanol can increase lactate utilization in *Rb. sphaeroides*, acting as an enhancer for addressing reducing power to nitrogenase: the yield was increased from 1.5–2.2 mol H_2 (mol lactate)$^{-1}$, by adding 0.2 % of ethanol. It was first demonstrated that hydrogen could be produced from ethanol as the sole substrate by *Rhodopseudomonas* sp. (Fuji et al. 1983). This was brought up again by Liu et al. (2015) with a culture of *Rp. palustris* grown in presence of ethanol up to 2 %; this culture gave a yield of 2 mol H_2 (mol ethanol)$^{-1}$ (i.e., 33 %).

Conclusions

In recent years, progress has been made in developing cleaner and more efficient bioenergy producing systems. In order for bioenergies to become increasingly competitive with other energy sources, logistics and infrastructures must be addressed and further technological innovation should lead to more efficient and cleaner conversion of a diverse range of feedstocks, in the view of promoting clean biofuel production, and the subsequent decarbonization of energy sources and fuels.

Facing the increasing relevance of the hydrogen economy, improving biohydrogen production yields, investigating novel substrates, and developing the technology at plant scale represent imperative tasks and photofermentation systems may be considered as alternatives capable of attaining these goals. The investigation on abundant and low-cost renewable biomass derived substrates is of relevant importance, trying to find sustainable feedstocks for novel fermentation systems, and this chapter showed how novel substrates can be suitable for biohydrogen producing applications, with limited pretreatments.

The photobiological hydrogen production is a technology in progress, which can be classified as single stage process and as dark/photofermentation systems both sequential (two stage) and combined (co-cultures). Furthermore, the cell immobilization techniques could enhance the continuous hydrogen production, compared to suspended cell systems.

However, at the current state, the design of suitable and efficient photobioreactors is still to be achieved and no cost-effective approaches have been developed yet both for dark fermentation, where biogas plant-like technology could be used, and for photofermentation, where new and efficient photobioreactors need to be realized at plant scale. Investigations should focus on several issues, such as (a) a longer retention time for the low biodegradable substrates; (b) the research of novel, robust, and versatile inocula; (c) the codigestion of biomasses in order to balance the C/N ratio; and (d) the study of fermentation systems using continuous culture, with the view to translate the process from lab to plant scale. Many aspects have still to be optimized, but in the last few years the number of new findings demonstrates that it is worth continuing the efforts for increasing the knowledge on the photofermentation process for H_2 production, in particular owing to the need of reducing the use of fossil fuels for mitigating the emissions of GHG in the atmosphere.

Acknowledgements AA and RDP acknowledge CNR (Italian National Research Council) (EFOR project), and Ente Cassa di Risparmio di Firenze (Project HYDROLAB^2) for funding their researches cited in this review. RDP would also like to mention the contribution given to his activities by the participation in the IEA-HIA (International Energy Agency—Hydrogen Implementation Agreement), Annex 34. EC acknowledges Scuola Superiore Sant'Anna of Pisa for supporting and funding her PhD research project.

References

Abo-Hashesh M, Hallenbeck PC (2012) Fermentative hydrogen production. In: Hallenbeck PC (ed) Microbial technologies in advanced biofuels production. Springer, pp 77–92

Abo-Hashesh M, Desaunay N, Hallenbeck PC (2013) High yield single stage conversion of glucose to hydrogen by photofermentation with continuous cultures of *Rhodobacter capsulatus* JP91. Bioresour Technol 128:513–517

Adessi A, De Philippis R (2013) Purple bacteria: electron acceptors and donors. In: Lennarz WJ, Lane MD (eds) The encyclopedia of biological chemistry, vol 3. Academic, Elsevier, Waltham, MA, pp 693–699

Adessi A, De Philippis R (2014) Photosynthesis and hydrogen production in purple non sulfur bacteria: fundamental and applied aspects. In: Zannoni D, De Philippis R (eds) Microbial bioenergy: hydrogen production, vol 38. Springer, pp 269–290

Adessi A, De Philippis R, Hallenbeck PC (2012a) Combined systems for maximum substrate conversion. In: Hallenbeck PC (ed) Microbial technologies in advanced biofuels production. Springer, pp 107–126

Adessi A, McKinlay JB, Harwood CS et al (2012b) A *Rhodopseudomonas palustris* nifA* mutant produces H_2 from-containing vegetable wastes. Int J Hydrogen Energy 37:15893–15900

Adessi A, De Philippis R, Corneli E et al (2016a) Produzione fotobiologica poliidrossibutirrato da biomasse con batteri rossi non sulfurei. Italian patent F1 S0061 12 IT 1, filed March 17, 2016

Adessi A, Concato M, Sanchini A et al (2016b) Hydrogen production under salt stress conditions by a freshwater *Rhodopseudomonas palustris* strain. Appl Microb Biotechnol 100:2917–2926

Adessi A, Spini G, Presta L et al (2016c) Draft genome sequence and overview of the purple non sulfur bacterium *Rhodopseudomonas palustris* 42OL. Stand Genomic Sci 11:1

Anam K, Habibi MS, Harwati TU et al (2012) Photofermentative hydrogen production using *Rhodobium marinum* from bagasse and soy sauce wastewater. Int J Hydrogen Energy 37:15436–15442

Angelini LG, Ceccarini L, Nassi o Di Nasso N et al (2009) Comparison of *Arundo donax* L. and *Miscanthus x giganteus* in a long-term field experiment in Central Italy: analysis of productive characteristics and energy balance. Biomass Bioenerg 33:635–643

Appels L, Lauwers J, Degrève J et al (2011) Anaerobic digestion in global bio-energy production: potential and research challenges. Renew Sust Energ Rev 15:4295–4301

Argun H, Kargi F (2010a) Photo-fermentative hydrogen gas production from dark fermentation effluent of ground wheat solution: effects of light source and light intensity. Int J Hydrogen Energy 35:1595–1603

Argun H, Kargi F (2010b) Bio-hydrogen production from ground wheat starch by continuous combined fermentation using annular-hybrid bioreactor. Int J Hydrogen Energy 35:6170–6178
Argun H, Kargi F (2010c) Effects of light source, intensity and lighting regime on bio-hydrogen production from ground wheat starch by combined dark and photo-fermentations Int J Hydrogen Energy 35:1604–1612
Argun H, Kargi F (2011) Bio-hydrogen production by different operational modes of dark and photo-fermentation: an overview. Int J Hydrogen Energy 36:7443–7459
Argun H, Kargi F, Kapdan IK (2009a) Effects of the substrate and cell concentration on bio-hydrogen production from ground wheat by combined dark and photo-fermentation. Int J Hydrogen Energy 34:6181–6188
Argun H, Kargi F, Kapdan IK (2009b) Hydrogen production by combined dark and light fermentation of ground wheat solution. Int J Hydrogen Energy 34:4305–4311
Arumugam A, Sandhya M, Ponnusami V (2014) Biohydrogen and polyhydroxyalkanoate co-production by *Enterobacter aerogenes* and *Rhodobacter sphaeroides* from *Calophyllum inophyllum* oil cake. Bioresour Technol 164:170–176
Bartacek J, Zabranska J, Lens PN (2007) Developments and constraints in fermentative hydrogen production. Biofuel Bioprod Bior 1:201–214
Cai J, Wang G (2012) Hydrogen production by a marine photosynthetic bacterium, *Rhodovulum sulfidophilum* P5, isolated from a shrimp pond. Int J Hydrogen Energy 37:15070–15080
Cai J, Wang G (2013) Screening and hydrogen-producing characters of a highly efficient H 2-producing mutant of *Rhodovulum sulfidophilum* P5. Bioresour Technol 142:18–25
Cai J, Wang G (2014) Hydrogen production from glucose by a mutant strain of *Rhodovulum sulfidophilum* P5 in single-stage photofermentation. Int J Hydrogen Energy 39:20979–20986
Cappai G, De Gioannis G, Friargiu M et al (2014) An experimental study on fermentative H_2 production from food waste as affected by pH. Waste Manage 34:1510–1519
Chang JS, Lee KS, Lin PJ (2002) Biohydrogen production with fixed-bed bioreactors. Int J Hydrogen Energy 27:1167–1174
Cheng J, Su H, Zhou J et al (2011a) Microwave-assisted alkali pretreatment of rice straw to promote enzymatic hydrolysis and hydrogen production in dark-and photo-fermentation. Int J Hydrogen Energy 36:2093–2101
Cheng J, Su H, Zhou J et al (2011b) Hydrogen production by mixed bacteria through dark and photo fermentation. Int J Hydrogen Energy 36:450–457
Cheng J, Xia A, Liu Y et al (2012) Combination of dark-and photo-fermentation to improve hydrogen production from *Arthrospira platensis* wet biomass with ammonium removal by zeolite. Int J Hydrogen Energy 37:13330–13337
Chookaew T, Sompong O, Prasertsan P (2015) Biohydrogen production from crude glycerol by two stage of dark and photo fermentation. Int J Hydrogen Energy 40:7433–7438
Christopher K, Dimitrios R (2012) A review on exergy comparison of hydrogen production methods from renewable energy sources. Energ Environ Sci 5:6640–6651
Corneli E, Adessi A, Dragoni F et al (2016 submitted) Agroindustrial residues and energy crops for the production of H_2 and PHB via photofermentation. Biores Technol (under review)
Corneli E, Dragoni F, Adessi A et al (2016b) Energy conversion of biomass crops and agroindustrial residues by combined biohydrogen/biomethane system and anaerobic digestion. Bioresour Technol 211:509–518
Das D, Veziroglu N (2008) Advances in biological hydrogen production processes. Int J Hydrogen Energy 33:6046–6057
Dipasquale L, Adessi A, d'Ippolito G et al (2015) Introducing capnophilic lactic fermentation in a combined dark-photo fermentation process: a route to unparalleled H_2 yields. Appl Microb Biotechnol 99:1001–1010
Dragoni F, Ragaglini G, Corneli E et al (2015) Giant reed (*Arundo donax* L.) for biogas production: land use saving and nitrogen utilisation efficiency compared with arable crops. Ital J Agron 10:192–201
El Bassam N (2010) Handbook of bioenergy crops: a complete reference to species, development and applications. Earthscan, London/Washington, DC

Eroğlu E, Eroğlu İ, Gündüz U et al (2006) Biological hydrogen production from olive mill wastewater with two-stage processes. Int J Hydrogen Energy 31:1527–1535
Eroğlu E, Eroğlu İ, Gündüz U et al (2008a) Effect of clay pretreatment on photofermentative hydrogen production from olive mill wastewater. Bioresour Technol 99:6799–6808
Eroğlu İ, Tabanoğlu A, Gündüz U et al (2008b) Hydrogen production by *Rhodobacter sphaeroides* OU 001 in a flat plate solar bioreactor. Int J Hydrogen Energy 33:531–541
Eroğlu E, Eroğlu İ, Gündüz U et al (2009) Treatment of olive mill wastewater by different physicochemical methods and utilization of their liquid effluents for biological hydrogen production. Biomass Bioenerg 33:701–705
Eroglu E, Gunduz U, Yucel M et al (2010) Photosynthetic bacterial growth and productivity under continuous illumination or diurnal cycles with olive mill wastewater as feedstock. Int J Hydrogen Energy 35:5293–5300
Eroğlu I, Özgür E, Eroğlu E et al (2014) Applications of photofermentative hydrogen production. In: Zannoni D, De Philippis R (eds) Microbial bioenergy: hydrogen production, vol 38. Springer, pp 237–267
Fascetti E, D'addario E, Todini O et al (1998) Photosynthetic hydrogen evolution with volatile organic acids derived from the fermentation of source selected municipal solid wastes. Int J Hydrogen Energy 23:753–760
Frigon JC, Guiot SR (2010) Biomethane production from starch and lignocellulosic crops: a comparative review. Biofuel Bioprod Bior 4:447–458
Fuji T, Nakazawa A, Sumi N et al (1983) Utilization of alcohols by *Rhodopseudomonas* sp No 7 isolated from n-propanol–enrichment cultures. Agr Biol Chem Tokio 47:2747–2753
Ghimire A, Frunzo L, Pirozzi F et al (2015) A review on dark fermentative biohydrogen production from organic biomass: process parameters and use of by-products. Appl Energ 144:73–95
Ghosh D, Tourigny A, Hallenbeck PC (2012a) Near stoichiometric reforming of biodiesel derived crude glycerol to hydrogen by photofermentation. Int J Hydrogen Energy 37:2273–2277
Ghosh D, Sobro IF, Hallenbeck PC (2012b) Stoichiometric conversion of biodiesel derived crude glycerol to hydrogen: response surface methodology study of the effects of light intensity and crude glycerol and glutamate concentration. Bioresour Technol 106:154–160
Ghosh D, Sobro IF, Hallenbeck PC (2012c) Optimization of the hydrogen yield from single-stage photofermentation of glucose by *Rhodobacter capsulatus* JP91 using response surface methodology. Bioresour Technol 123:199–206
Gómez X, Fernández C, Fierro J et al (2011) Hydrogen production: two stage processes for waste degradation. Bioresour Technol 102:8621–8627
Gosse JL, Engel BJ, Rey FE et al (2007) Hydrogen production by photoreactive nanoporous latex coatings of nongrowing *Rhodopseudomonas palustris* CGA009. Biotechnol Progr 23:124–130
Guo XM, Trably E, Latrille E et al (2010) Hydrogen production from agricultural waste by dark fermentation: a review. Int J Hydrogen Energy 35:10660–10673
Guo CL, Zhu X, Liao Q et al (2011) Enhancement of photo-hydrogen production in a biofilm photobioreactor using optical fiber with additional rough surface. Bioresour Technol 102:8507–8513
Hallenbeck PC, Ghosh D (2009) Advances in fermentative biohydrogen production: the way forward? Trends Biotechnol 27:287–297
Hallenbeck PC, Liu Y (2016) Recent advances in hydrogen production by photosynthetic bacteria. Int J Hydrogen Energy 41:4446–4454
Hallenbeck PC, Ghosh D, Skonieczny MT et al (2009) Microbiological and engineering aspects of biohydrogen production. Indian J Microbiol 49:48–59
Hawkes FR, Hussy I, Kyazze G et al (2007) Continuous dark fermentative hydrogen production by mesophilic microflora: principles and progress. Int J Hydrogen Energy 32:172–184
Hay JXW, Wu TY, Juan JC (2013) Biohydrogen production through photo fermentation or dark fermentation using waste as a substrate: overview, economics, and future prospects of hydrogen usage. Biofuel Bioprod Bior 7:334–352
Heiniger EK, Oda Y, Samanta SK et al (2012) How posttranslational modification of nitrogenase is circumvented in *Rhodopseudomonas palustris* strains that produce hydrogen gas constitutively. Appl Environ Microbiol 78:1023–1032

Herrmann C, Heiermann M, Idler C (2011) Effects of ensiling, silage additives and storage period on methane formation of biogas crops. Bioresour Technol 102:5153–5161
Holladay JD, Hu J, King DL et al (2009) An overview of hydrogen production technologies. Catal Today 139:244–260
Hustede E, Steinbüchel A, Schlegel HG (1993) Relationship between the photoproduction of hydrogen and the accumulation of PHB in non-sulphur purple bacteria. Appl Microb Biotechnol 39:87–93
IEA (International Energy Agency) (2009) Bioenergy–a sustainable and reliable energy source. International Energy Agency Bioenergy, Paris, France
Ike A, Toda N, Tsuji N et al (1997) Hydrogen photoproduction from CO_2-fixing microalgal biomass: application of halotolerant photosynthetic bacteria. J Ferment Bioeng 84:606–609
IRENA (2013) Statistical issues: bioenergy and distributed renewable energy. http://www.irena.org/DocumentDownloads/Publications/Statistical%20issues_bioenergy_and_distributed%20renewable%20_energy.pdf
Johnson DT, Taconi KA (2007) The glycerin glut: options for the value-added conversion of crude glycerol resulting from biodiesel production. Environ Progr 26:338–348
Kapdan IK, Kargi F, Oztekin R et al (2009) Bio-hydrogen production from acid hydrolyzed wheat starch by photo-fermentation using different *Rhodobacter* sp. Int J Hydrogen Energy 34:2201–2207
Keskin T, Hallenbeck PC (2012) Hydrogen production from sugar industry wastes using single-stage photofermentation. Bioresour Technol 112:131–136
Keskin T, Abo-Hashesh M, Hallenbeck PC (2011) Photofermentative hydrogen production from wastes. Bioresour Technol 102:8557–8568
Kim DH, Lee JH, Kang S et al (2014) Enhanced photo-fermentative H_2 production using *Rhodobacter sphaeroides* by ethanol addition and analysis of soluble microbial products. Biotechnol Biofuel 7:79
Larimer FW, Chain P, Hauser L et al (2004) Complete genome sequence of the metabolically versatile photosynthetic bacterium *Rhodopseudomonas palustris*. Nat Biotechnol 22:55–61
Laurinavichene TV, Belokopytov BF, Laurinavichius KS et al (2010) Towards the integration of dark-and photo-fermentative waste treatment. 3. Potato as substrate for sequential dark fermentation and light-driven H_2 production. Int J Hydrogen Energy 35:8536–8543
Lazaro CZ, Varesche MBA, Silva EL (2015) Sequential fermentative and phototrophic system for hydrogen production: an approach for Brazilian alcohol distillery wastewater. Int J Hydrogen Energy 40:9642–9655
Lewandowski I, Scurlock JM, Lindvall E et al (2003) The development and current status of perennial rhizomatous grasses as energy crops in the US and Europe. Biomass Bioenerg 25:335–361
Liu BF, Ren NQ, Tang J et al (2010) Bio-hydrogen production by mixed culture of photo-and dark-fermentation bacteria. Int J Hydrogen Energy 35:2858–2862
Liu Y, Ghosh D, Hallenbeck PC (2015) Biological reformation of ethanol to hydrogen by *Rhodopseudomonas palustris* CGA009. Bioresour Technol 176:189–195
Maeda, I., Mizoguchi, T., Miura, Y., Yagi, K., Shioji, N., & Miyasaka, H. (2000). Influence of sulfate-reducing bacteria on outdoor hydrogen production by photosynthetic bacterium with seawater. Current microbiology, 40(3), 210–213
Nonhebel S (2007) Energy from agricultural residues and consequences for land requirements for food production. Agr Syst 94:586–592
Özgür E, Mars AE, Peksel B et al (2010) Biohydrogen production from beet molasses by sequential dark and photofermentation. Int J Hydrogen Energy 35:511–517
Ozmihci S, Kargi F (2010) Comparison of different mixed cultures for bio-hydrogen production from ground wheat starch by combined dark and light fermentation. J Ind Microbiol Biot 37:341–347
Pachapur VL, Sarma SJ, Brar SK et al (2015) Co-culture strategies for increased biohydrogen production. Int J Energ Res 39:1479–1504

Pattanamanee W, Chisti Y, Choorit W (2015) Photofermentive hydrogen production by *Rhodobacter sphaeroides* S10 using mixed organic carbon: effects of the mixture composition. Appl Energ 157:245–254

Pintucci C, Giovannelli A, Traversi ML et al (2013) Fresh olive mill waste deprived of polyphenols as feedstock for hydrogen photo-production by means of *Rhodopseudomonas palustris* 42OL. Renew Energ 51:358–363

Pintucci C, Padovani G, Giovannelli A et al (2015) Hydrogen photo-evolution by *Rhodopseudomonas palustris* 6A using pre-treated olive mill wastewater and a synthetic medium containing sugars. Energ Conver Manage 90:499–505

Pott RW, Howe CJ, Dennis JS (2013) Photofermentation of crude glycerol from biodiesel using *Rhodopseudomonas palustris*: comparison with organic acids and the identification of inhibitory compounds. Bioresour Technol 130:725–730

Redwood MD, Paterson-Beedle M, Macaskie LE (2009) Integrating dark and light bio-hydrogen production strategies: towards the hydrogen economy. Rev Environ Sci Biotechnol 8:149–185

Sabourin-Provost G, Hallenbeck PC (2009) High yield conversion of a crude glycerol fraction from biodiesel production to hydrogen by photofermentation. Bioresour Technol 100:3513–3517

Sargsyan H, Trchounian K, Gabrielyan L et al (2016) Novel approach of ethanol waste utilization: biohydrogen production by mixed cultures of dark-and photo-fermentative bacteria using distillers grains. Int J Hydrogen Energy 4:2377–2382

Schievano A, D'Imporzano G, Adani F (2009) Substituting energy crops with organic wastes and agro-industrial residues for biogas production. J Environ Manage 90:2537–2541

Seifert K, Waligorska M, Laniecki M (2010a) Hydrogen generation in photobiological process from dairy wastewater. Int J Hydrogen Energy 35:9624–9629

Seifert K, Waligorska M, Laniecki M (2010b) Brewery wastewaters in photobiological hydrogen generation in presence of *Rhodobacter sphaeroides* OU 001. Int J Hydrogen Energy 35:4085–4091

Singh L, Wahid ZA (2015) Methods for enhancing bio-hydrogen production from biological process: a review. J Ind Eng Chem 21:70–80

Su H, Cheng J, Zhou J et al (2009) Improving hydrogen production from cassava starch by combination of dark and photo fermentation. Int J Hydrogen Energy 34:1780–1786

Su H, Cheng J, Zhou J et al (2010) Hydrogen production from water hyacinth through dark-and photo-fermentation. Int J Hydrogen Energy 35:8929–8937

Tian X, Liao Q, Zhu X et al (2010) Characteristics of a biofilm photobioreactor as applied to photo-hydrogen production. Bioresour Technol 101:977–983

Tsygankov A, Kosourov S (2014) Immobilization of photosynthetic microorganisms for efficient hydrogen production. In: Zannoni D, De Philippis R (eds) Microbial bioenergy: hydrogen production, vol 38. Springer, pp 321–347

Vatsala TM, Raj SM, Manimaran A (2008) A pilot-scale study of biohydrogen production from distillery effluent using defined bacterial co-culture. Int J Hydrogen Energy 33:5404–5415

Vincenzini M, Marchini A, Ena A et al (1997) H and poly-β-hydroxybutyrate, two alternative chemicals from purple non sulfur bacteria. Biotechnol Lett 19:759–762

Wang YZ, Liao Q, Zhu X et al (2010) Characteristics of hydrogen production and substrate consumption of *Rhodopseudomonas palustris* CQK 01 in an immobilized-cell photobioreactor. Bioresour Technol 101:4034–4041

Wang X, Yang G, Feng Y et al (2012) Optimizing feeding composition and carbon–nitrogen ratios for improved methane yield during anaerobic co-digestion of dairy, chicken manure and wheat straw. Bioresour Technol 120:78–83

Weiland P (2010) Biogas production: current state and perspectives. Appl Microb Biotechnol 85:849–860

Wu SC, Liou SZ, Lee CM (2012) Correlation between bio-hydrogen production and polyhydroxybutyrate (PHB) synthesis by *Rhodopseudomonas palustris* WP3-5. Bioresour Technol 113:44–50

Xia A, Cheng J, Ding L et al (2014) Enhancement of energy production efficiency from mixed biomass of *Chlorella pyrenoidosa* and cassava starch through combined hydrogen fermentation and methanogenesis. Appl Energ 120:23–30

Xie GJ, Liu BF, Ding J et al (2012) Enhanced photo-H_2 production by *Rhodopseudomonas faecalis* RLD-53 immobilization on activated carbon fibers. Biomass Bioenerg 44:122–129

Yang H, Shi B, Ma H et al (2015) Enhanced hydrogen production from cornstalk by dark-and photo-fermentation with diluted alkali-cellulase two-step hydrolysis. Int J Hydrogen Energy 40:12193–12200

Zagrodnik R, Laniecki M (2015) The role of pH control on biohydrogen production by single stage hybrid dark-and photo-fermentation. Bioresour Technol 194:187–195

Zhang C, Zhu X, Liao Q et al (2010) Performance of a groove-type photobioreactor for hydrogen production by immobilized photosynthetic bacteria. Int J Hydrogen Energy 35:5284–5292

Zong W, Yu R, Zhang P et al (2009) Efficient hydrogen gas production from cassava and food waste by a two-step process of dark fermentation and photo-fermentation. Biomass Bioenerg 33:1458–1463

Zürrer H, Bachofen R (1981) Hydrogen production from lactate and lactate-containing wastes by the photosynthetic bacterium *Rhodospirillum rubrum*. Stud Environ Sci 9:31–36

Engineering Cyanobacteria for Biofuel Production

Rui Miao, Adam Wegelius, Claudia Durall, Feiyan Liang, Namita Khanna, and Peter Lindblad

Abstract Fast depletion of petroleum resources and environmental concerns due to rapidly increasing fossil fuel related CO_2 emissions have prompted scientists to find more sustainable and environmental friendly fuel alternatives. Considering this, algal conversion of CO_2 to biofuels has received increased attention in recent times. In particular, cyanobacteria have been considered as promising candidates for biofuel production considering their fast growth rate, ability to fix carbon dioxide, and their genetic tractability. In parallel to the advancements in synthetic biology and genetic engineering, several proofs of concept studies have emerged demonstrating the ability of cyanobacteria to produce different kinds of biofuels including alcohols, hydrogen, and fatty acid derived biofuels. However, presently their low titer values impede their commercial success. In this perspective, we review the recent publications on engineering cyanobacteria for biofuel production discuss the challenges and scope of improvements for advancing cyanobacterial fuel production.

Keywords Alcohols • Alka(e)nes • Biofuels • Cyanobacteria • Carbon concentrating mechanism • Fatty acid alkyl ester • Hydrogen • PEPC

Introduction

Fossil fuel limitation is a worldwide problem with implications on energy resources, global peace, and environmental perspectives. The consequence of such a limitation has created an urgent demand for new, sustainable energy carriers. To circumvent this problem, the International Energy Agency has targeted to fulfil more than 1/4th of the global transportation fuel demand with biofuels by 2050 (Fairley 2011). Currently, bioethanol production from biomass fermentation, biodiesel production from transesterification, and biogas production from anaerobic digestion are the

R. Miao • A. Wegelius • C. Durall • F. Liang • N. Khanna • P. Lindblad (✉)
Microbial Chemistry, Department of Chemistry - Ångström, Uppsala University, Box 523, SE-75120 Uppsala, Sweden
e-mail: peter.lindblad@kemi.uu.se

© Springer International Publishing Switzerland 2017

P.C. Hallenbeck (ed.), *Modern Topics in the Phototrophic Prokaryotes*,
DOI 10.1007/978-3-319-46261-5_11

most commercialized conventional biofuel processes (Guo et al. 2015). However, these conventional biofuels have several disadvantages such as requirement of arable land for cultivation of biomass, low energy content, and difficulties in storage and long distance transportation. Use of photosynthetic microorganisms to produce biofuels offers a tangible solution to these problems.

Photosynthetic microorganisms, such as cyanobacteria, evolved on Earth billions years ago. They contributed to the formation of the early atmosphere and were the key components in the evolution of eukaryotic photosynthesis (Herrero and Flores 2008). Much of our fundamental understanding of photosynthesis comes from these organisms. However, only more recently have they been considered for biofuel production. Simultaneously, with the development of genetic engineering, a high number of advanced biofuels have been obtained from engineered cyanobacteria, such as ethanol from pyruvate pathway, higher alcohols via the keto-acid and the Ehrlich pathways, terpene-based fuels from the mevalonate pathway, fatty acids and alkanes from fatty acid biosynthesis pathways, and hydrogen from dark and photofermentation pathways (Zhang et al. 2011; Khanna and Lindblad 2015; Tan et al. 2011; Wang et al. 2013) (Fig. 1). Despite the success achieved in lab-scale, it still remains a challenge to commercialize these biofuel processes on an industrial-scale. Low titer values in large-scale bioreactors have been attributed mainly to their low photosynthetic productivity compared to their sugar-based heterotrophic counterparts (e.g., *Escherichia coli* and *Saccharomyces cerevisiae*) (Savakis and Hellingwerf 2015). However, a recent study carried out to determine the maximum theoretical yields of a few selected chemicals in *E.coli* and the cyanobacterium *Synechocystis* PCC 6803. Results showed that autotrophically grown *Synechocystis* PCC 6803 had higher theoretical potential as compared to heterotrophically grown *E. coli* (Nogales et al. 2012). The study thus suggests that photosynthetic metabolism in cyanobacteria may not be the sole inherent reason for the lower productivity of the

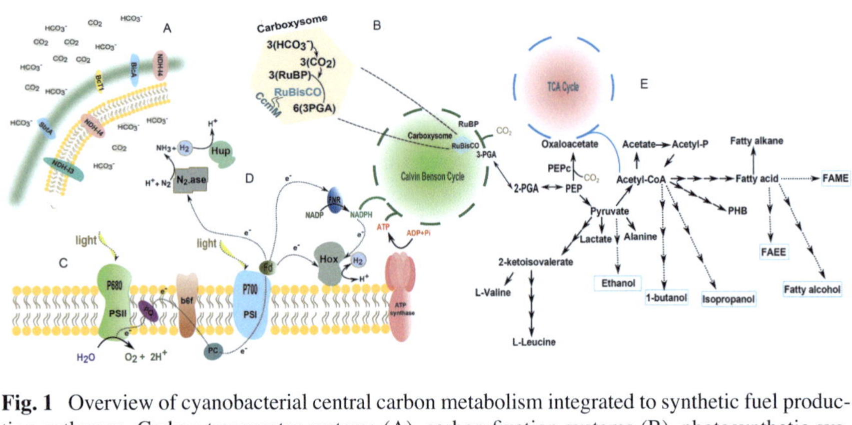

Fig. 1 Overview of cyanobacterial central carbon metabolism integrated to synthetic fuel production pathways. Carbon transporter systems (A), carbon fixation systems (B), photosynthetic systems (C), hydrogen production systems (D), pyruvate and acetyl-CoA-based metabolic systems (E). *Dotted lines* in E represent the introduced heterogenous pathways for different biofuels discussed in this chapter. The chemicals in *blue boxes* represent the potential biofuel molecules

targeted chemicals. Conducting metabolic optimization at systems level appears relevant to maximize productivity.

Engineering of Cyanobacterial Chassis for Optimal Biofuel Production

Over the last decade, several proof of concept studies on producing different biofuel compounds in cyanobacteria have been demonstrated (Ducat et al. 2011). However, the titers achieved in cyanobacteria are lower in comparison to *E coli*. The chassis thus needs to be optimized to make it cost competitive. It has been suggested that besides optimizing the production pathways, improving the robustness and photosynthetic efficiency of the host organism may also contribute towards the development of a suitable chassis for higher biofuel productivity. So far, primary focus has been laid on optimization of the light harvesting capacity, enhancement of carbon assimilation, optimization of carbon flux rerouting, modification of the regulatory systems, enhancement of genetic stability, and improvement of tolerance towards the targeted biofuels. Below, each of these approaches is discussed in more details.

Optimization of Light Harvesting Capacity

Light is one of the key resources and also the primary limitation for increasing biomass accumulation and metabolite production in cyanobacteria (Ducat et al. 2011). Though, compared to plants, cyanobacteria have a much higher capacity to convert captured solar energy into biomass, their light harvesting efficiency still remains a question (Dismukes et al. 2008). Photosystem II (PSII) and Photosystem I (PSI) catalyze the central reactions of energy conversion in cyanobacteria. Both photosystems comprise of internal antenna systems consisting of chlorophylls and carotenoids that capture light. In addition, both are functionally connected to the peripheral antenna complexes, the large membrane associated phycobilisomes (PBS), used to increase the absorption cross-section of the light-capturing system.

The light harvesting system in cyanobacteria is plagued by two potential drawbacks. First is the presence of large light harvesting complexes, which primarily may be attributed to their evolutionary adaptation to low light. The large antennas ensure enough absorption of incident light for photosynthesis under low light conditions; however, the same becomes a burden under increased light intensity. In fact, after reaching the maximum photosynthetic rate, additional light does not result in higher photosynthesis, on the contrary, it leads to decreased photosynthetic rates and photo-damage. As a consequence, CO_2 fixation, oxygen evolution, and hydrogen photo-production become limited due to low electron transfer within the photosynthetic apparatus.

To date, very few studies have been performed to reduce the antenna sizes in cyanobacteria. A phycocyanin (PC)-deficient strain of *Synechocystis* PCC 6714 was developed using chemical mutagenesis. It contained one third of the phycocyanin content compared to wild type cells but showed 1.5 times higher photosynthetic productivity under high light conditions (Nakajima and Ueda 1997). However, the potential of this deficient strain as a biofuel producer has not yet been evaluated. Moreover, a series of truncated mutants have also been examined in *Synechocystis* PCC 6803. Three phycobilisome antenna-deficient strains, (i) a core-membrane linker protein (LCM) deficient strain (Shen et al. 1993), (ii) a PC-deficient strain, and (iii) a completely PBS-deficient strain (Ajlani and Vernotte 1998), were developed to compare their influence on photosynthetic electron transport and responses to different light conditions (Bernát et al. 2009). All the above strains achieved higher PSII/PSI ratios and showed higher linear electron transport activities compared to wild type cells. This result indicates that PBS antenna-deficiency can be applied in developing efficient host strains to produce biofuels such as hydrogen. Furthermore, another PBS-deficient study showed higher oxygen evolution and biomass accumulation in *Synechocystis* PCC 6803 strain grown under normal CO_2 condition. However, the biomass accumulation preponderance disappeared under CO_2 saturated conditions (Lea-Smith et al. 2014).

A second drawback of the light harvesting system in cyanobacteria is their narrow absorption spectrum range. Most of the oxygen-evolving photosynthetic organisms can utilize only the spectrum of photosynthetically active radiation, ranging from 400 to 700 nm (Blankenship and Chen 2013). However, few photosynthetic microorganisms can use the spectrum above 700 nm due to their evolvement of chlorophyll d and f. Recently, these two chlorophylls have gained much attention in oxygenic photosynthesis solar spectrum expansion research (Chen et al. 2010). Considering this, it would be interesting to genetically introduce different chlorophylls into the inherent photosynthetic systems with the aim to enhance their light harvesting capacity and light dependent photosynthetic reactions (Chen et al. 2004; Miyashita et al. 1997). This subsequently may lead to supply of more energy and carbon skeletons towards biofuel production.

Enhancement of Carbon Assimilation

In cyanobacteria, photosynthesis is responsible to generate high energy molecules, NADPH and ATP, and these molecules are used in the Calvin–Benson–Bassham (CBB) cycle to produce glyceraldehyde-3-phosphate (G3P), which can further synthesize the primary nutrient resources in photosynthetic organisms.

The CBB cycle is the primary carbon fixation pathway in nature. It widely occurs in photoautotrophic organisms and can be divided into three stages: carboxylation, reduction, and regeneration. It is believed that carboxylation and Ribulose-1,5-biphosphate (RuBP) regeneration are the two key stages of CBB cycle-based CO_2 assimilation under certain conditions. Besides the CBB cycle, other carbon fixing enzymes such as phosphoenolpyruvate carboxylase (PEPc) also play supplementary roles in providing carbon skeletons in photosynthetic organisms (Luinenburg and Coleman 1990).

Considering carbon assimilation, the availability of sufficient carbon source in water can be a potential problem for cyanobacteria, owing to the rather low CO_2 diffusion capacity between air and water. To overcome this constraint, cyanobacteria have evolved a carbon concentrating mechanism (CCM) that comprises of three parts: CO_2/HCO_3^- transporters, carboxysomes, and the carbonic anhydrase located in the carboxysomes. Over the last few years, our understanding of the carbon fixation mechanism and CCM has increased significantly (Badger and Price 2003; Durall and Lindblad 2015). In this chapter, we discuss the components of carbon assimilation in some details, and also provide some recent examples of genetic/metabolic engineering that have been carried out to improve carbon fixation efficiencies.

Carbon Fixation in CBB Cycle

RuBP Carboxylation

In the CBB cycle, RuBP carboxylation is carried out by Ribulose-1,5-bisphosphate carboxylase/oxygenase (RuBisCO), a bi-substrate enzyme located in the carboxysome, that catalyzes RuBP and CO_2 into two 3-phosphoglycerate (3PGA) molecules that are later directed into pyruvate metabolism (Calvin and Benson 1948). RuBisCO also catalyzes RuBP and O_2 to form one 3PGA molecule and one 2-phosphoglycolate (2PGA) molecule. 2PGA is a known toxic intermediate and can be routed through the photorespiration pathway that results in the loss of previously fixed CO_2.

Unfortunately, RuBisCO does not have a strong ability to distinguish CO_2 from O_2 as substrate, which leads to decreased carboxylation efficiency in the presence of O_2. Furthermore, RuBisCO has a low substrate affinity which means it requires high concentration of substrates. However, evolution bears evidence to the fact that the ability to fix CO_2 was not always a problem. When cyanobacteria evolved RuBisCO billions of years ago, the CO_2 concentration in the atmosphere was many folds higher leading to higher rate of carboxylation. However, along with the increase of O_2 level, the carboxylation efficiency of RuBisCO faced a setback, and the low CO_2 level in the atmosphere became a rate-limiting step (Badger and Price 2003). As a consequence, plants evolved to devote up to half of their leaf-soluble protein and 25 % of their cellular nitrogen to generate enough RuBisCO to achieve acceptable rates of photosynthesis (Iwaki et al. 2006).

A number of studies have been carried out in cyanobacteria to engineer or overexpress RuBisCO to improve their carbon fixation capacity in an effort to increase both their photosynthetic efficiency and productivity of carbon-based compounds (Table 1). This was demonstrated in an engineered free fatty acid (FFA) producing *Synechococcus* PCC 7002 strain where the cells showed more than three-fold increase in FFA production when coupled with overexpression of the RuBisCO subunits from *Synechococcus elongatus* PCC 7942 (Ruffing 2014). In another similar study, overexpression of RuBisCO from *Synechococcus* PCC 6301 significantly

Table 1 Cyanobacterial strains with engineered Ribulose-1,5-bisphosphate carboxylase/oxygenase (RuBisCO) or fructose-1,6-/sedoheptulose-1,7-bisphosphatase (FBPase)

Host	Overexpressed enzyme	Resource	Mutagenesis	Main effect	Reference
Synechococcus PCC 7002	RuBisCO large and small subunits	*Synechococcus elongatus* PCC 7942	–	Threefold increase of free fatty acids (FFA) production	(Ruffing 2014)
Synechococcus elongatus PCC 7942	RuBisCO large and small subunits	*Synechococcus elongatus* PCC 7942	–	Unchanged oxygen evolution and FFA production	(Ruffing 2013)
Synechococcus elongatus PCC 7942	RuBisCO large and small subunits	*Synechococcus elongatus* PCC 6301	–	1.4-fold increase in total RuBisCO activity; unchanged photosynthetic O_2 production; 2-fold higher isobutyraldehyde production	(Atsumi et al. 2009)
Synechococcus elongatus PCC 7942	RuBisCO large and small subunits	*Allochromatium vinosum*	–	1.5- to 4-fold increase of RuBisCO activity	(Iwaki et al. 2006)
Anabaena PCC 7120	Chloroplast FBPase	*Triticum aestivum* (wheat)		1.4-fold higher of FBPase activity; increased growth and photosynthesis	(Ma et al. 2005)
Synechocystis PCC 6803	–	–	T65A/S or G404A (1P-binding site) of RuBisCO	Reduced K_{cat} (CO_2); one third of V_c; 3-fold increase of RuBisCO content	(Marcus et al. 2011)

Synechocystis PCC 6803	–	–	H298A (latch pocket) or H327Q (RuBP 5P-binding site) of RuBisCO	Increased K_m (RuBP); reduced catalytic turnover; minor to moderate effect on photosynthetic rate; unchanged photoautotrophic growth	(Marcus et al. 2005)
Synechocystis PCC 6803	–	–	C172A, C192A, C172A + C192A (adjacent to catalytic sites)	Unchanged growth rate; decreased photosynthesis	(Marcus et al. 2003)
Synechocystis PCC 6803			C247A (linking two large subunits) of RuBisCO	Unchanged growth rate; decreased photosynthesis	(Marcus et al. 2003)
Synechococcus elongates PCC 6301	–	–	A340N (in loop 6) of RuBisCO	Increased $S_{C/O}$ by 9 % and *Vc* by 19 %	(Madgwick et al. 1998)

enhanced the production of isobutyraldehyde in an engineered *Synechococcus elongatus* PCC 7942 strain though the oxygen evolution remained the same (Atsumi et al. 2009). In the same host, the total RuBisCO activity was also increased by overexpressing heterogeneous RuBisCO from *Allochromatium vinosum* (Iwaki et al. 2006). The overexpression driven by the *Synechococcus elongatus* PCC 7942 *psbAI* promoter resulted in 2.5- to 4-fold increase in total RuBisCO activity while overexpression driven by *Synechocystis* PCC 6803 *psbAII* promoter showed a 1.5–2 times higher RuBisCO activity. Along with the increase of RuBisCO activity, both engineered strains showed higher photosynthesis compared to wild type cells. This indicates that RuBisCO is one of the limiting factors to increase photosynthesis rate in *Synechococcus elongatus* PCC 7942.

RuBP Regeneration

RuBP regeneration is another key part of CBB cycle that involves eight enzymes in the entire catalytic process. Three molecules of RuBP are generated from five G3P molecules in this regeneration step. According to modeling and experimental results in plant, in addition to RuBisCO, three enzymes in the RuBP regeneration pathway, i) sedoheptulose 1,7-bisphosphatase (SBPase), ii) fructose 1,6-biphosphate aldolase (FBA), and iii) transketolase, excert major control on directing carbon flux in the cells (Zhu et al. 2007). Similar results were recently reported in *Synechocystis* PCC 6803 where increased expression of RuBisCO, FBA, SBPase (bifunctional protein fructose-1,6-/sedoheptulose-1,7-bisphosphatase), or TK accumulated more dry cell weight under 100 μmol photons m^{-2} s^{-1} light intensity. Moreover, RuBisCO, FBA, and SBPase engineered strains grew faster and had higher maximal oxygen evolution rate as compared to the control strains (Liang and Lindblad 2016).

SBPase regulates the rate of photosynthesis by affecting the RuBP regeneration capacity. In plants, SBPase catalyzes the irreversible dephosphorylation of sedoheptulose 1,7-bisphosphate (SBP) to sedoheptulose-7-phosphate (S7P). In cyanobacteria, this function is carried out by a bifunctional enzyme, fructose-1,6-/sedoheptulose-1,7-bisphosphatase (FBP/SBP-ase), which dephosphorylates both fructose-1,6-biphosphate and sedoheptulose-1,7-biphosphate. Cyanobacterial FBP/SBP-ase has no homology with either FBPase or SBP-ase from plants (Tamoi et al. 1996). However, overexpressing a cyanobacterial FBP/SBP-ase in an algae *Euglena gracilis* resulted in enhanced biomass and wax ester production (Ogawa et al. 2015) and engineered tobacco plant with cyanobacterial FBP/SBP-ase gained higher photosynthesis rate, higher carbohydrate accumulation, and higher growth rate (Miyagawa et al. 2001). Vice-versa, heterologous expression of wheat chloroplastic FBPase in the cyanobacterium *Anabaena* PCC 7120 also resulted in increased photosynthesis (Ma et al. 2005). To our knowledge, this is the only available report demonstrating a higher photosynthetic rate in cyanobacteria by overexpressing an alternate enzyme of the carbon metabolism, other than RuBisCO.

Another key enzyme in RuBP regeneration is fructose 1,6-biphosphate aldolase (FBA), which catalyzes the reversible aldol condensation of dihydroxyacetone phosphate (DHAP) and G3P. FBA is involved in three metabolic pathways: CBB cycle, glycolysis, and gluconeogenesis. Together with SBPase, FBA is also considered as a potential target to improve photosynthesis. There are two types of FBAs, FBA-I and FBA-II, based on their different catalyzing mechanisms. FBA-I (FbaI) forms a Schiff base with its substrates while FBA-II (FbaA) needs divalent ions as cofactor to obtain activity. Cyanobacteria mainly possess FBA-II even though small portions of FBA-I have been detected in *Synechocystis* PCC 6803 (Nakahara et al. 2003). Unfortunately, the overexpression of this enzyme has not been reported in cyanobacteria. A study in potato plant reported significant metabolic impacts by down regulating FBA. The reduction of FBA inhibited photosynthesis, altered levels of sugars and starch, and retarded the growth of plant (Haake et al. 1998). In contrast, the up-regulation of FBA in tobacco resulted in enhanced growth and photosynthesis (Uematsu et al. 2012).

A third candidate to improve RuBP regeneration is transketolase (TK). It catalyzes the reversible transfer of a two-carbon ketol group from S7P to G3P for the generation of xylulose-5-phosphate (Xu5P) and ribose-5-phosphate (R5P) or the transfer of a ketol group from F6P to G3P for the generation of Xu5P and erythrose-4-phosphate (E4P) (Schenk et al. 1998). The importance of TK in photosynthesis and carbon metabolism has been demonstrated in the tobacco plant. TK overexpressed tobacco showed a chlorotic phenotype and thiamine auxotrophy. On the other hand, it accumulated more starch, fructose, and glucose (Khozaei et al. 2015). As expected, upon supplementation of thiamin this phenotype disappeared. In contrast, down regulating TK expression in tobacco decreased RuBP regeneration and photosynthesis (Henkes et al. 2001). These results demonstrated that there are crucial responses on the TK activity level in the cells. It would be interesting to expand these studies into cyanobacteria in an attempt to increase their photosynthetic potentials.

Carbon Fixation from Phosphoenolpyruvate Carboxylases

Besides RuBisCO, phosphoenolpyruvate carboxylase (PEPc) is another important secondary carbon fixing enzyme prevalent in various organisms such as cyanobacteria, algae, C4, and CAM plants (Lepiniec et al. 1993; Rivoal et al. 1996; Chollet et al. 1996). The enzyme is known to catalyze an irreversible conversion of phosphoenolpyruvate to oxaloacetate and inorganic phosphate in the presence of bicarbonate and Mg^{2+} (Chen et al. 2002; Owttrim and Colman 1986; Rivoal et al. 1998). So far, only one form of the enzyme has been identified in cyanobacteria together with two other carboxylases, carbamoyl phosphate synthetase (CPS) and acetyl-CoA carboxylase (Owttrim and Colman 1986). CPS is involved in the formation of pyrimidine and arginine (Kasahara and Obmori 1997) whereas acetyl-CoA carboxylase serves as the first enzyme of the fatty acid biosynthesis pathway (Kothe et al. 2009). Due to its function in carbon fixation, modulation of PEPc expression has been touted as another route to improve the photosynthetic efficiency in cyanobacteria as well as in higher plants. Indeed, up-regulation of the endogenous PEPc in *Anabaena* 7120 resulted in higher enzyme activity and the carbon fixation capacity of the engineered strain (Jia et al. 2014). In addition, down regulation of PEPc showed increased tolerance to low temperature, acidic or alkaline pH, and high salinity (Jia et al. 2014).

Recently, in an interesting modeling study, several synthetic metabolic carbon fixation pathways that were hypothesized to be more efficient in carbon fixation compared to any of the existing native pathways (Bar-Even et al. 2010). The modeling results suggested that synthetic lactate MOG pathway and alanine MOG pathway that both use PEPc are the best alternatives for achieving maximal carbon assimulation. Lately it was experimentally verified that the overexpression of the native PEPc in *Synechocystis* PCC 6803 demonstrated faster growth under low light intensity (3 μmole photons m^{-2} s^{-1}) as compared to wild type cells. This was further validated by higher in vitro PEPc activities and in vivo protein levels in the cells overexpressing PEPc (Durall et al. 2016).

Carbon Concentrating Mechanism

CO_2/HCO_3^- Transporters

So far, five different carbon transporters have been identified in cyanobacteria based on the common laboratory strains *Synechococcus* PCC 7942 and *Synechocystis* PCC 6803 (Durall et al. 2015; Price et al. 2002). Two are classified as CO_2 uptake transporters while the other three are collectively grouped as HCO_3^- uptake transporters. They are located in either the plasma membrane or the thylakoid membrane in cyanobacteria (Price et al. 2002; Price 2011; Mi et al. 1995). All these uptake systems serve to accumulate HCO_3^- in the cytosol of the cell, which is subsequently used by the carbonic anhydrase, located in the carboxysomes, to increase the CO_2 concentration around RuBisCO. The five different transporters are briefly described below.

1. BCT1 is a high HCO_3^- affinity carbon transporter, located in the plasma membrane. So far, its presence is limited to only freshwater and brackish strains of β-carboxysome-cyanobacteria (explained later) and as such has only rarely been found in some α-carboxysome-cyanobacteria (explained later) (Price 2011; Woodger et al. 2003). However, physiologically, the transporter has only been characterized in *Synechococcus elongatus* PCC 7942 where it is encoded by *cmpABCD* operon that belongs to the bacterial ATP-binding cassette (ABC) transporter family (Price et al. 2007; Omata et al. 1999). The transcription of the *cmpABCD* operon is induced by severe inorganic carbon limitation and is regulated by light.
2. BicA is a plasma membrane localized Na^+ dependent low HCO_3^- affinity transporter that belongs to the SulP/SLC26 protein family (Price 2011). In some cyanobacteria, the expression of this transporter increases under low inorganic carbon levels (Price et al. 2007). Suggestions to improve CO_2 concentrating mechanism in cyanobacteria have considered overexpression of this transporter. Recently, in one such study, improvement in biomass accumulation was recorded when additional copies of the endogenous bicarbonate transporter were intro-

duced into *Synechocystis* PCC 6803 under the control of an inducible promoter. As expected, the engineered strain grew almost twice as fast as the wild type cells (Kamennaya et al. 2015).

3. SbtA is another Na^+ dependent, high HCO_3^- affinity transporter, located in the plasma membrane that is induced by low inorganic carbon concentrations (Price 2011; Shibata et al. 2002). It is commonly found in β-carboxysome-cyanobacteria and its homologues may also be prevalent in some α-carboxysome-cyanobacteria (Price 2011). It acts as an Na^+/ HCO_3^- symporter with a relatively low flux rate (Price 2011; Shibata et al. 2002). SbtA is colocated with SbtB, and the function of the later has remained elusive for several years. Recently, it was shown that SbtB from cyanobacteria is repressed by the native SbtA activity in *E. coli*. It was thus speculated that this protein may regulate the activity of the transporter in darkness.

   The second group of transporters (NDH-$I_{3/4}$) is responsible for CO_2-uptake and bears homology to the plastoquinone oxidoreductase NADPH dehydrogenase (NDH-1) respiratory complexes. These transporters have largely been found in β-cyanobacteria while α-cyanobacteria may possess either one or none of them. Since CO_2 can passively transport across the plasma membrane, the importance of these transporters has mainly been associated with the hydration of CO_2 to HCO_3^- inside the cell. This conversion is achieved by the catalytic activity of Chp (for CO$_2$ hydration proteins), which differ (ChpY or ChpX) depending on the type of transporter (Price et al. 2002; Mi et al. 1995).

4. NDH-I_3 is a CO_2-uptake transporter induced by low levels of inorganic carbon. They are localized on the thylakoid membranes. For this transporter, the hydration activity of CO_2 to HCO_3^- is achieved by ChpY (Woodger et al. 2003).
5. NDH-I_4 is another CO_2-uptake transporter, expressed constitutively in the cell. So far, its intracellular location remains elusive. For this transporter, the hydration activity of CO_2 to HCO_3^- is achieved by ChpX (Price et al. 2002; Maeda et al. 2002).

From a genetic perspective, to improve CO_2-uptake, the above discussed transporters may be further evaluated for their role and possibility of transgenic overexpression in engineered cyanobacterial cells.

Carbonic Anhydrase

Carbonic anhydrase locally provides CO_2 to RuBisCO by catalyzing the reversible conversion between CO_2 and bicarbonate in carboxysome. There are two types of carboxysomes: α-carboxysome and β-carboxysome (Heinhorst et al. 2014). In α-carboxysomes, the carboxysomal carbonic anhydrase (CsoSCA), RuBisCO, and the carboxysome structure protein (CsoS2) are included in the carbonic anhydrase rich layer interior of the carboxysome shell. Similarly, in β-carboxysomes, carbonic anhydrase (CcaA), RuBisCO, and the carboxysome structure protein (CcmM58 and CcmN) also form a carbonic anhydrase rich layer attached to the carboxysome shell.

Carbonic anhydrase exists ubiquitously in nature and can be classified into five classes α, β, γ, δ, and ε, based on their amino acid sequences. Cyanobacteria mainly have β and γ type carbonic anhydrases. β-carbonic anhydrase is associated with the carboxysomal shell (CcaA in β-carboxysomes and CsoSCA in α-carboxysomes) to support the elevation of CO_2 concentration around RuBisCO (Espie and Kimber 2011; Kaplan et al. 2008). This was demonstrated in a carbonic anhydrase (*icfA*) deletion *Synechococcus* PCC 7942 strain, which showed higher CO_2 requirement and less CO_2 fixation compared to wild type cells (Fukuzawa et al. 1992). This indicates a crucial need of carbonic anhydrase in photosynthetic CO_2 fixation in cyanobacteria.

Overexpressing carbonic anhydrase in *Synechococcus elongatus* PCC 7942 resulted in an increased level of carbonic anhydrase in the cytosol without any increase in the RuBisCO catalyzed carbon fixation activity (Price and Badger 1989). The study suggested the importance of localization of carbonic anhydrase within the carboxysomes and moreover suggested the requirement of co-expression with CcmM. These findings led to a related investigation to overexpress CcmM in *Synechococcus elongatus* PCC 7942 which achieved higher expression of carbonic anhydrase, however, RuBisCO activity in these cells remained unchanged (Long et al. 2010).

Optimization of Carbon Flux Rerouting

Glycogen is one of the major photosynthetic sinks in cyanobacteria. The synthesis of glycogen is catalyzed by glucose-1-phosphate adenylyltransferase (GlgC), glycogen synthase, and 1,4-alpha glucan branching enzyme (Suzuki et al. 2010). This storage metabolism is a major source of fixed carbon and reducing power. Decoupling of the glycogen synthesis could redirect the carbon source to alternate biosynthetic processes (Carrieri et al. 2015). This was illustrated in a *Synechococcus elongatus* 7942 *glgC* deletion strain that could redirect its carbon flux and reducing power towards the production of a non-native metabolite, isobutanol (Li et al. 2014). Under high light condition the *glgC* deletion strain showed growth impairment. However, induction of the isobutanol pathway rescued the growth defects. This study suggested redox imbalance due to accumulation of reduced NAD(P)H, under high light conditions, as the most likely reason for the observation. Subsequent induction of isobutanol pathway facilitated the cells into an alternate route to release the redox stress.

Modification of Regulatory System and Enhancement of Genetic Stability

To obtain an optimal host strain, its metabolism and regulatory systems must be well-characterized under a wide variety of relevant conditions. Natively in cyanobacteria, regulation of gene expression using RNA regulatory mechanisms may have evolved to overcome their highly dynamic environmental and diurnal lifestyles (Berla et al. 2013).

Many of the highly expressed regulatory RNAs belong to the family of unique non-coding RNAs (ncRNAs), which are found in nearly all sequenced cyanobacteria (Gierga et al. 2009; Mitschke et al. 2011). Their prevalence clearly suggests their importance in cyanobacterial metabolic regulations. However, to date, very few of them have been functionally characterized. A study of the *Synechocystis* PCC 6803 Hox-hydrogenase operon, which encodes all the proteins required for the pentameric bidirectional NiFe-hydrogenase (HoxEFUYH), demonstrated that this operon is flanked by two ncRNAs. One of them, *syR1*, is strongly up-regulated under high light while the other, *ncr0700*, was maximally expressed in darkness (Mitschke et al. 2011). Overexpression of the endogenous SyRl driven by an inducible promoter led to a severe phenotype with reduced pigmentation. These results suggest an intrinsic control of ncRNAs in regulating the dark/light adaptation in cyanobacteria. The antisense transcriptional regulation may thus be a hurdle in overexpressing endogenous genes in cyanobacteria. The study thus implores the need for a deeper understanding of these regulators, especially the ones associated with photosynthesis, carbon transport, and nutrient assimilation, in order to increase the production of, e.g., biofuels in cyanobacteria.

There are a desirable number of reports on genetic engineering in cyanobacteria. However, the genetic instability of cyanobacteria has only been discussed briefly (Jones 2014), even though it has become a non-negligible obstacle which may be referred to as genetic instability, single gene mutation, or wild type "revenge". In a study of an ethylene-forming enzyme expressing *Synechococcus* strain, the engineered population lost the capability to produce ethylene due to an unaware curtail of this overexpressed gene sequence (Takahama et al. 2003). A similar phenomenon was discovered when a synthetic isopropanol pathway was introduced into a *Synechococcus* strain (Kusakabe et al. 2013). There was an obstinate single nucleotide mutation in one of the genes in this pathway which resulted in reduced productivity of isopropanol. All of these instabilities can occur in the early stage of genetic engineering period as well as during the extended cultivation period (Dienst et al. 2014). From an industrial perspective, it can be excruciating to have an unstable strain in a larger photobioreactor system.

Transcriptomic responses have been observed as a consequence of engineered cyanobacterial cells producing special biofuels (Liu et al. 2012; Qiao et al. 2012; Tian et al. 2013; Zhu et al. 2013). According to these results, the genetic instability may be caused by the genes encoding DNA repair and mutation mechanisms under stressful conditions. Unfortunately, there is no simple technique to solve this problem. One possibility may be the development of highly inducible promoters for the transcription of the, e.g., introduced biofuel producing pathway.

Although cyanobacteria have been studies for decades, we still lack fully regulated promoters (Huang et al. 2010; Camsund and Lindblad 2014). A tightly regulated promoter induced by anhydrotetracycline was developed but it's not practical to use, especially in industrial applications due to the high cost and light sensitivity of the inducer (Huang and Lindblad 2013). A light regulated promoter would be a perfect choice. One green-light-regulated promoter has been developed by modifying the native *cpcG2* promoter with an insertion of a Shine-Dalgarno-like sequence (Abe et al. 2013).

Improvement of Tolerance Towards Target Biofuel Chemicals

To be able to use cyanobacteria as a robust chassis for high rate biofuel production, its tolerance capacity to the stress caused by the production of the fuels should be considerably addressed. Many chemical products are toxic to cyanobacteria even at low concentrations. They decrease the growth rate of the host cells, limiting their production potential and the possibility for industrial applications. Systematic studies have been conducted to address the responses of cyanobacteria to solvents such as ethanol (Qiao et al. 2012), n-butanol (Tian et al. 2013), and hexane (Liu et al. 2012). It has been suggested that the photosynthetic microorganisms are more susceptible to the detrimental effects of solvents. This is because of their extremely high sensitivity to the redox state of key molecules and their intricate organization of the membrane-bound photosynthetic apparatus (Anfelt et al. 2013).

Understanding the stress responses will help to build more robust systems for future biofuel applications. In theory, improving tolerance to products should allow for higher titers. The stress responses induced by ethanol, butanol, hexane, and lactic acid have been extensively investigated via transcriptomics, metabolomics, and proteomics. The results suggested heat-shock proteins, oxidative stress related proteins, transporters, along with modification of cell membranes, and the proteins involved in common stress responses should be highlighted along with some metabolites such as 3-phosphoglycerate, glycine, and urea (Wang et al. 2013; Dienst et al. 2014; Liu et al. 2012; Qiao et al. 2012; Tian et al. 2013; Zhu et al. 2013; Anfelt et al. 2013; Borirak et al. 2015; Jin et al. 2014).

The development of mass cultivation in saline water for biotechnological applications is highly desirable. However, it can be challenging owing to the presence of high NaCl concentrations that can impair the growth and metabolism of the chosen biological chassis. For instance, a 50 % growth inhibition was observed in the freshwater cyanobacterium *Synechocystis* PCC 6803 cultivated in a medium containing 4 % (m/v) NaCl (Qiao et al. 2013). However, recently, it was demonstrated that *Synechocystis* PCC 6803 can be grown in an artificial seawater medium supplemented with nitrogen and phosphorus sources (Iijima et al. 2015). The study thus opens up the possibility of cultivating freshwater cyanobacteria using seawater in the future.

Not all compounds can fuse out of cell membranes. Therefore, secretion mechanisms may be needed to get rid of unwanted or even harmful compounds. Many microorganisms have native transporters to export waste products but cyanobacteria lack many of those (Niederholtmeyer et al. 2010). Engineered *Synechococcus elongatus* PCC 7942 strains with introduced invertase, glucose facilitator, and lactate transporter all showed increased productivity for each of these substances (Niederholtmeyer et al. 2010).

Cyanobacteria Metabolic Engineering for Different Biofuel Production

Hydrogen Production in Cyanobacteria

Cyanobacteria have gained attention as promising candidates for microbial hydrogen production. Indeed, a large number of cyanobacteria are capable of molecular hydrogen evolution and this is a rare feature among oxygenic photoautotrophs.

Three enzymes are known to be directly involved in cyanobacterial hydrogen metabolism: uptake hydrogenase (Hup-hydrogenase), nitrogenase, and bidirectional Hox-hydrogenase (Hox-hydrogenase) (Tamagnini et al. 2002). Some cyanobacteria strains possess all the three enzymes; some lack the Hox-hydrogenase, while still others possess only the Hox-hydrogenase. Nitrogenase plays a key-role in the fixation of atmospheric nitrogen by diazotrophs and catalyzes proton reduction in a by-process. Hup-hydrogenase is found in close association to nitrogenase, as it catalyzes the oxidation of hydrogen produced during nitrogen fixation. The bidirectional Hox-hydrogenase, found both in diazotrophic and non-diazotrophic cyanobacteria, can catalyze both hydrogen oxidation and proton reduction.

Metabolic Engineering of Hup-Hydrogenase for Hydrogen Production

Cyanobacterial Hup-hydrogenases are oxygen sensitive enzymes that belong to the NiFe-hydrogenase family. They catalyze the irreversible oxidation of hydrogen under physiological conditions. The enzyme consists of at least two subunits, small and large, encoded by *hupS* and *hupL*, respectively. The presence of a third, membrane anchoring, subunit has been suggested, but not yet been identified (Tamagnini et al. 2007). The large catalytic subunit contains a NiFe-complex active site, while the small catalytic subunit contains a [4Fe-4S]-cluster that can transfer electrons from the catalytic core. The main role of the enzyme is considered to be the reutilization of reduction equivalents wasted on hydrogen production by nitrogenase, effectively lowering the cost of nitrogen fixation. This recycling is beneficial for the cell by enabling the generation of ATP through the oxyhydrogen reaction, which is also permitting the removal of oxygen from the proximity of nitrogenase; and by supplying electrons to different cell functions (Tamagnini et al. 2002, 2007; Bothe et al. 2010; Khetkorn et al. 2013). In heterocystous cyanobacteria, the active Hup-hydrogenase is found limited to the heterocyst although both transcription and translation have been confirmed in vegetative cells in some strains (Tamagnini et al. 2007; Bothe et al. 2010; Khetkorn et al. 2013). Due to the strong bias towards hydrogen oxidation and exclusive participation in hydrogen uptake reactions, the Hup-hydrogenase is not of any interest for large-scale photobiological hydrogen production. However, it was reported recently that the Hup-hydrogenase in *Nostoc punctiforme* ATCC 29133 was successfully turned towards hydrogen production by modifying the proximal FeS-cluster of the small subunit (HupS) using site directed mutagenesis (Raleiras et al. 2016).

As the function of Hup-hydrogenase is the oxidation of molecular hydrogen produced by nitrogenase, a natural strategy to enhance hydrogen evolution from the nitrogen fixing strains is the disruption or removal of this enzyme (Table 2) (Khanna and Lindblad 2015; Khetkorn et al. 2013). This has been carried out successfully in several heterocystous cyanobacteria with the help of a rapid devolpment of molecular tools and an increase of complete cyanobacterial genome sequences. Hup-hydrogenase deficient strains of *Anabaena variabilis* (Happe et al. 2000), *Anabaena* PCC 7120 (Lindberg et al. 2011; Masukawa et al. 2002), *Nostoc punctiforme* ATCC 29133 (Lindberg et al. 2002), *Nostoc* PCC 7422 (Yoshino et al. 2006), and *Anabaena siamensis* TISTR 8012 (Khetkorn et al. 2012) have all been reported to produce hydrogen at significantly higher rates than their respective wild type cells (Table 2). More recently, the ΔHupW strain of *Anabaena* PCC 7120, in

Table 2 Nitrogenase-based hydrogen evolution from engineered filamentous cyanobacteria

Strain	H_2 evolution μmol mg chl a^{-1} h^{-1}	Condition	Reference
Anabaena variabilis ATCC 29413 ΔHupSL	135	Ar, 100 μmol photons m^{-2} s^{-1}	(Happe et al. 2000)
Nostoc. punctiforme ATCC 29133 ΔHupL	14	Air, light	(Lindberg et al. 2002)
Anabaena PCC 7120 ΔHupL ΔHoxW	53	Ar, 10 W m^{-2}	(Masukawa et al. 2002)
Nostoc PCC 7422 ΔHupL	100	Ar + 5%CO2, 70 μmol photons m^{-2} s^{-1}	(Yoshino et al. 2006)
Anabaena siamensis TISTR 8012 ΔHupS	29.7	Ar, 200 μ μmol photons m^{-2} s^{-1}	(Khetkorn et al. 2012)
Anabaena PCC 7120 ΔHupW	3.3	Air, 10 W m^{-2}	(Lindberg et al. 2011)
Anabaena PCC 7120 ΔHupW	850	Air–Ar cycles, 44 μmol photons m^{-2} s^{-1}, 3 L bioreactor	(Nyberg et al. 2015)
Anabaena PCC 7120 ΔHupW	710	20 % Ar/80 % N_2— Ar cycles, 44 μmol photons m^{-2} s^{-1}, 3 L bioreactor	(Nyberg et al. 2015)
Anabaena PCC 7120 (ΔHup) sc-R248H	19	Air, 60–70 μmol photons m^{-2} s^{-1}	(Masukawa et al. 2010)
Anabaena PCC 7120 (ΔHup) sc-R248H	16	Ar, 60–70 μmol photons m^{-2} s^{-1}	(Masukawa et al. 2010)
Anabaena PCC 7120 (ΔHup) dc-R248H	29	N_2, 60–65 μmol photons m^{-2} s^{-1}	(Masukawa et al. 2014)
Anabaena PCC 7120 (ΔHup) dc-R248H	29	Ar, 60–65 μmol photons m^{-2} s^{-1}	(Masukawa et al. 2014)
Anabaena PCC 7120 (ΔHup) dc-Q193S	29	N_2, 60–65 μmol photons m^{-2} s^{-1}	(Masukawa et al. 2014)
Anabaena PCC 7120 (ΔHup) dc-Q193S	35	Ar, 60–65 μmol photons m^{-2} s^{-1}	(Masukawa et al. 2014)

which *hupW*, encoding a putative hydrogenase-specific protease, is inactivated, was grown and examined for hydrogen production in a 5.0 L flat panel photobioreactor with 3 L culture volume. This strain was able to turn light energy into hydrogen gas with an impressive energy conversion efficiency of 4.0 % under altering Ar/N_2 (20/80 %) and Ar (100 %) atmospheres, respectively (Nyberg et al. 2015).

Metabolic Engineering of Nitrogenase for Hydrogen Production

The process of N_2-fixation is widely spread within the cyanobacterial phylum and has been found in unicellular, filamentous heterocystous, and filamentous non-heterocystous strains. The key player in this process is a metalloprotein-complex called nitrogenase. This enzyme catalyzes the breaking of N_2 bonds, that later leads to the formation of ammonia. During N_2-fixation, H_2 is produced as a by-product. There are three known nitrogenases in nature: Mo-nitrogenase, V-nitrogenase, and Fe-nitrogenase. They are classified by the kind of metal atoms close to the active site and only the two former have been found in cyanobacteria (Bothe et al. 2010). Mo-nitrogenase, harbouring a molybdenum atom in the active site, is the most common nitrogenase and it is present in all nitrogen fixing cyanobacteria (Bothe et al. 2010). V-nitrogenase utilizes a vanadium atom in the active site and it is commonly viewed as a backup nitrogenase under Mo-starved conditions (Bothe et al. 2010). Mo-nitrogenase is a multi-protein complex that comprises of two proteins: a dinitrogenase (MoFe-protein) and a dinitrogenase reductase (Fe-protein). The MoFe-protein is the catalyst for reducing the N_2 bonds while the Fe-protein relays electrons from the donor to the MoFe-protein. The MoFe-protein is a $\alpha_2\beta_2$ heterotetramer whose subunits are encoded by *nifD* and *nifK* and Fe-protein is a homodimer encoded by *nifH*.

The nitrogenase complex is an oxygen-intolerant enzyme and requires low levels of molecular oxygen to sustain N_2-fixing activity. Thus the enzyme cannot be active in the presence of oxygenic photosynthesis, where high levels of oxygen are continuously produced by water splitting at photosystem II. Thus different physiological variants of cyanobacteria strains have evolved to facilitate the N_2-fixation. Unicellular and non-heterocystous filamentous variants have the N_2-fixing machinery and the water splitting reaction center within the same cell making spatial separation of the two processes impossible. Most commonly, the separation of the two incompatible reactions in these strains is temporal, in many cases following a diurnal cycle (Bothe et al. 2010), e.g. during the dark period, the energy for N_2-fixation is provided by oxidative respiration. The most fascinating solution for the requirement of strictly low oxygen level is the development of specialized cells in filamentous heterocystous cyanobacteria. In these strains, a subset of the cells in the filaments undergo a drastic differentiation to become heterocysts. These heterocysts provide a semi-anaerobic environment for the nitrogenase to perform nitrogen fixation under oxygenic photosynthesis. The key features of these specialized cells are the thick glycolipid and polysaccharide cell envelope that obstruct the entrance of

O_2 from outside, the down regulation of O_2 evolution from PSII, and the high respiration level to quickly devour any oxygen that enters or evolves within the cell.

It has been suggested that electron flux redirection from N_2 reduction to H_2 reduction would greatly improve the feasibility of nitrogenase-based hydrogen production for large-scale applications (Khetkorn et al. 2013). Replacing the atmosphere gas N_2 with Ar-gas can enhance hydrogen production from nitrogenase (Barney et al. 2004; Benemann and Weare 1974). However, this approach has been deemed too costly for large-scale hydrogen production.

Another more sophisticated and potentially more feasible strategy is the alteration of nitrogenase by mutagenesis. This has been tried with some success on several occasions over the last years. In one such approach, the supply of N_2 to the active site of the MoFe-protein of Nif2 in *Anabaena variabilis* was limited by reducing the size of the presumed gas channel using mutagenesis. This resulted in a 4-fold increase of H_2 production in an N_2 atmosphere for the first hours after combined nitrogen deprivation (Weyman et al. 2010). This H_2 evolution from the engineered strain in N_2 was shown to be as high as from wild type cells under Ar. The engineered strain also showed similar H_2 evolution in Ar as wild type cells, suggesting that the mutation may be limiting the size of the gas channel, preventing N_2 from entering into the active site. However, it should be noted that the *A. variabilis* Nif2, as opposed to other nitrogenases in the same organism, is expressed in vegetative cells immediately after diazotrophic induction. It is thus lacking the protection and anaerobic conditions that are provided by the heterocysts, and the hydrogen production needs to be preceded by anaerobic condition and chemical inhibition of PSII.

Another strategy involves mutagenesis of amino acids directly or indirectly involved in the N_2 reduction catalysis. Several amino acid modifications made in close proximity to the active site in MoFe-protein in a Hup-hydrogenase deficient *Anabaena* PCC 7120 strain, resulted in enhanced H_2 production when cultured under N_2 complemented with 5 % CO_2. The most promising strain, with R284H mutation in the α subunit, had both higher H_2 evolution rate and twelve times higher hydrogen accumulation than the reference strains in 7 days (Masukawa et al. 2010). Interestingly, H_2 production from the R248H strain was higher both in air and in N_2 compared to in Ar, indicating a positive effect of N_2 on proton reduction by this nitrogenase variant. These features make this strain more feasible for large-scale hydrogen production by keeping high production rate with reduced cost on cultivation gases. On the other hand, R248H was unable to grow diazotrophically due to the introduced changes in the nitrogenase active site. This could be the disadvantageous for long-term hydrogen production because the cultures eventually have to be supplemented with a nitrogen source. This makes the cultivation more expensive and demanding.

More recently, double and single crossover variants of the R248H strain and one other promising strain from the same study, bearing the Q193S mutation, were shown to produce notable amounts of hydrogen for up to three weeks under 5 % CO_2 supplemented N_2 atmosphere with regular gas exchange (Masukawa et al. 2014). The production rates and the product accumulation of these engineered strains matched or exceeded those obtained when cultivated under Ar. Also, the

Q193S double crossover variant showed clear signs of diazotrophic growth, making it even more promising for prolonged hydrogen production.

Metabolic Engineering of Hox-Hydrogenase for Hydrogen Production

The cyanobacterial Hox-hydrogenase is a heteropentameric NiFe-hydrogenase encoded by *hoxEFUYH*. Like Hup-hydrogenase, Hox-hydrogenase belongs to the NiFe-hydrogenase family. Prevalence of the enzyme has been recorded in both filamentous and unicellular cyanobacteria, and in both N_2-fixing and non-N_2-fixing strains. It is composed of two protein complexes, a catalytic-complex where the large and small catalytic subunits are encoded by *hoxH* and *hoxY* and a diaphorase-complex encoded by *hoxEFU*. Functionally, Hox-hydrogenase catalyzes both hydrogen production and hydrogen uptake reactions.

Expression of Hox-hydrogenase is a complex process that involves the assembly of active site, FeS clusters as well as post-translational activation. This requires a system of maturation factors encoded by the *hyp*-operon. The process is well studied in *E. coli*, but has not yet been investigated in depth in cyanobacteria. The physiological function of Hox-hydrogenase has been debated for a long time (Khanna and Lindblad 2015; Carrieri et al. 2011), but it is generally suggested that it works as a valve to take care of the overabundance of electrons either from fermentation metabolism or from photosynthesis. During less reductive conditions, it recycles the electrons by catalyzing hydrogen oxidation.

The Hox-hydrogenase catalyzes the proton reduction and hydrogen oxidation in an ATP-independent manner and thus holds promise of being energetically efficient. However the bidirectional characteristics of this enzyme, as well as its sensitivity to oxygen, require tightly controlled anoxic conditions and the constant removal of produced hydrogen to limit the unwanted reverse reaction. Despite these drawbacks, this enzyme is interesting for hydrogen production applications as it is the only known cyanobacterial hydrogenase that can catalyze proton reduction in any detectable rates. Besides, Hox-hydrogenase has a slight bias towards hydrogen production over hydrogen uptake (Mclntosh et al. 2011), and is only temporarily inactivated by oxygen, and can be quickly reactivated under anaerobic, reductive conditions (Mclntosh et al. 2011; Sigfridsson et al. 2015; Vignais and Billoud 2007; Vincent et al. 2007).

For a long time it has been widely accepted that the electron donor of Hox-hydrogenase is NAD(P)H, releasing electrons to the FeS clusters of the diaphorase subunits (Aubert-Jousset et al. 2011). More recent studies suggest that ferredoxin and flavodoxin can donate electrons to the hydrogenase as well (Khanna and Lindblad 2015; Gutekunst et al. 2013). It has been shown that Hox-hydrogenase takes the role of an electron sink, stores reducing power in the form of hydrogen during photosynthesis (Appel et al. 2000) and aids the redox balance under fermentative conditions (Stal and Moezelaar 2006).

The Hox-hydrogenase is widely spread and highly conserved within the cyanobacterial phylum (Khanna et al. 2016). This indicates a crucial contribution of

Hox-hydrogenase to the cell fitness. Several studies of Hox-hydrogenase deletion *Synechocystis* strains have been conducted over the years to bring clarity to the exact physiological role of the enzyme, and many propose a function during changing or very specific growth conditions (Aubert-Jousset et al. 2011; Appel et al. 2000; Pinto et al. 2011; Eckert et al. 2012). It was lately discovered that a *hoxE-H*-deficient *Synechocystis* strain exhibited severely reduced growth rate in prolonged darkness compared to wild type cells, while the growth rate in continuous light or in 12 h dark/light cycles remained unchanged (De Rosa et al. 2015). Interestingly, this pattern was reported both under aerobic and anaerobic conditions, indicating that H_2 production and uptake by the catalytic-complex do not facilitate growth in darkness. This was further demonstrated by *hypA* and *hypB* (maturation genes, known to be responsible for the insertion of Ni to the catalytic site) deletion strains, having no hydrogenase moiety but an intact *HoxEFU* diaphorase-complex. The deletion strains manifested growth rates similar to wild type cells when cultivated under prolonged light or in darkness (De Rosa et al. 2015). Large-scale proteomics analysis revealed that the expression levels of several subunits of NDH-1 protein-complex were reduced under conditions of prolonged darkness. This suggested that the diaphorase-complex of Hox-hydrogenase may contribute to the regulation of electron transport chain under conditions when both photosynthetic and respiratory pathways were down regulated, thereby allowing *Synechocystis* to grow in the absence of a light source. This may explain both the homology between Hox-hydrogenase and NDH-1 and the constitutive expression of *hox* and *hyp* under aerobic conditions.

One of the major obstacles for biological light driven hydrogen production from the Hox-hydrogenase is the oxygen sensitivity of the enzyme. To date, the exact nature of O_2 intolerance in Hox-hydrogenase remains elusive (Ghirardi 2015). Despite this, some interesting strategies to improve O_2 tolerance of cyanobacterial Hox-hydrogenases have been investigated, though with limited success.

In nature, there are several examples of NiFe-hydrogenases that are less oxygen sensitive than the Hox-hydrogenase in cyanobacteria. An obvious strategy is the heterologous expression of such a hydrogenase in a cyanobacterial host. Unfortunately, the maturation systems of those NiFe-hydrogenases may be specific (English et al. 2009), and this may be a major technical obstacle for heterologous expression of functional proteins. In one of the very few successful attempts, a NiFe-hydrogenase from *Alteromonas macleodii* "Deep ecotype" was expressed together with accessory genes in the cyanobacterium *Synechococcus elongatus* PCC 7942, and reported to exhibit in vivo hydrogenase activity (Weyman et al. 2011). Although interesting and important from a technical point of view, the NiFe-hydrogenase in question shows only very limited tolerance against O_2.

Based on extensive research of the NiFe-hydrogenase oxygen tolerance in the gram-negative proteobacterium *Desulfovibrio fructosovorans* (Dementin et al. 2009), the oxygen tolerance of the *Synechocystis* Hox-hydrogenase was successfully improved by altering the gas diffusion channels to limit the accessibility of the active site (Cano et al. 2014). Further, the same mutation that rendered the increased oxygen tolerance also exhibited a notable change in the bias towards hydrogen production in vivo, however it is still low. This first report of an engineered cyanobacterial Hox-hydrogenase with sustained O_2 tolerance is promising for future applications.

In addition to the problem of oxygen sensitivity, is the low photon conversion efficiency preventing a large-scale light driven hydrogen production from Hox-hydrogenase. Only a very limited fraction of the reductant power provided by photosynthesis is used in hydrogen production even under optimal conditions. The major part of the electrons flux is directed towards CO_2 fixation via the CBB cycle. Thus, an interesting design to improve the electron flux towards hydrogen production is to physically couple a Hox-hydrogenase to PSI in order to facilitate a direct electron transfer from photosynthesis to proton reduction.

The validity of such an approach was demonstrated when an oxygen tolerant NiFe-hydrogenase from *Ralstonia eutropha* was chemically fused to the PSI subunit PsaE from the cyanobacterium *Thermosynechococcus elongatus*, assembled together with PSI, and catalyzed light driven hydrogen production in vitro (Ihara et al. 2006). The hydrogenase-PsaE fusion was shown to self-assemble in solution together with PsaE-free PSI, and was able to receive electrons directly from PSI. It was suggested that this approach might also be interesting for increasing in vivo hydrogen production efficiency. However it held the drawback of being strongly inhibited by ferredoxin and ferredoxin-$NADP^+$ reductase (FNR). To address this and to further increase the electron flux from PSI to Hox-hydrogenase, the PsaE subunits were chemically linked to cytochrome C3 from *Desulfovibrio vulgaris*, known to relay electrons between redox partners, instead of hydrogenase. Structural modeling suggested that the addition of cytochrome C3 to the design would rearrange the Ferredoxin binding site of PSI, possibly increasing the electron flow towards hydrogenase. This design was indeed capable of light driven hydrogen production in vitro from *Desulfovibrio vulgaris* NiFe-hydrogenase in the presence of ferredoxin and FNR (Ihara et al. 2006).

To use this strategy to enhance photobiological hydrogen production from living cyanobacteria, a complete design with the Cytochrome C3-PSI fusion has to be synthesized and assembled in vivo and this is not a trivial task. Even though, this may still be a way forward since it possibly can improve light driven hydrogen production only by expressing a Cytochrome C3-PsaE fusion protein. It has to be noted that this strategy has not been tried with any cyanobacterial hydrogenase, and with the challenges on heterologous expression of NiFe-hydrogenases, this remains an uncertainty for the usefulness of this technique.

Production of Alcohols in Cyanobacteria

Besides hydrogen production, cyanobacteria have been used to produce numbers of carbon-based compounds, such as ethanol, butanol, isoprene, fatty acid, and alkane. The advantage to produce carbon-based compounds in cyanobacteria is that a virtuous carbon cycle can be established in the eco-system because of the thriving growth of cyanobacteria in exceeding CO_2 condition (Quintana et al. 2011). Most of the investigations have been carried out using two appealing cyanobacterial strains: *Synechococcus elongatus* PCC 7942 and *Synechocystis* PCC 6803. In this section, the production of several different alcohols in the model strains is introduced as examples for a better understanding of metabolic engineering strategies in cyanobacteria (Table 3).

Metabolic Engineering for Ethanol Production

Ethanol has been used as transportation engine fuel for decades. Nowadays, more than 95 % of the gasoline used in the USA contains at least 5 % ethanol. Owing to the development of flexible fuel vehicles, the demand for ethanol has dramatically

Table 3 Ethanol, isopropanol, and 1-butanol production in cyanobacteria

Organism	Overexpressed genes	Host modification	Promoter	Titer	Reference
Ethanol production					
Synechococcus elongatus PCC 7942	*pdc; adh*	–	P*rbcL*	0.23 g L^{-1} in 28 days	(Deng and Coleman 1999)
Synechocystis PCC 6803	*pdc; adh*	–	P*psbA2*	0.55 g L^{-1} in 6 days	(Dexter and Fu 2009)
Synechocystis PCC 6803	*pdc; slr1192*	Deletion of PHB pathway	P*rbcL*	5.5 g L^{-1} in 26 days	(Gao et al. 2012)
Isopropanol production					
Synechococcus elongatus PCC 7942	*thl; adc; atoAD; adh*	–	P_L*lac*O_1	0.027 g L^{-1} in 9 days	(Kusakabe et al. 2013)
Synechococcus elongatus PCC 7942	*thl; adc; atoAD; adh*	–	P_L*lac*O_1	0.146 g L^{-1} in 10 days	(Hirokawa et al. 2015)
1-butanol production					
Synechococcus elongatus PCC 7942	*atoB; hbd; crt; ter; adhE2*	–	P*trc*; P_L*lac*O_1	0.015 g L^{-1} in 7 days	(Lan and Liao 2011)
Synechococcus elongatus PCC 7942	*nphT7; phaB; phaJ; ter; bldh; yqhD*	–	P*trc*; P_L*lac*O_1	0.03 g L^{-1} in 17 days	(Lan and Liao 2012)
Synechococcus elongatus PCC 7942	*nphT7; phaB; phaJ; ter; pduP; yqhD*	–	P*trc*; P_L*lac*O_1	0.4 g L^{-1} in 12 days	(Lan et al. 2013)
Synechocystis PCC 6803	*slr1193; slr1994; phaJ; ter; adhE2; xfpk*	Deletion of PHB pathway	P*phaAB*; P*psbA2;*P*trc*	0.037 g L^{-1} in 8 days	(Anfelt et al. 2015)

increased over the years. The largest ethanol producer and exporter is the USA, with a reported production of 14,300 million US gallons in 2014. Brazil also plays an important role by contributing 25 % of the world's ethanol production in 2014 (Beiter 2014). Both countries have devoted abundant agricultural land to fulfil their demand for ethanol production via traditional biomass fermentation methods. To resolve the issue of land use between edible crop and fuel production, cyanobacteria may be selected as a potential photoautotrophic platform for ethanol production. Synthetic pathways for ethanol production in both *Synechococcus elongatus* PCC 7942 and *Synechocystis* PCC 6803 have been developed (Fig. 2).

Ethanol can be naturally produced in various microorganisms. *Zymomonas mobilis* is one of those native ethanol producers. It uses two enzymes, pyruvate decarboxylase encoded by *pdc* and alcohol dehydrogenase encoded by *adh*, to complete the conversion of pyruvate to ethanol. In 1999, *pdc* and *adh* were introduced into *Synechococcus elongatus* PCC 7942 for ethanol production (Deng and Coleman 1999). A promoter, *PrbcL*, from the cyanobacterial ribulose-1,5-bisphosphate carboxylase gene operon was used to control the transcription of *pdc* and *adh*. This engineered strain resulted in a titer of 0.23 g L^{-1} after 4 weeks cultivation in shaking flasks (Deng and Coleman 1999).

Ten years later, these two genes were transformed into the *psbA2* locus in *Synechocystis* PCC 6803 genome, under the control of the strong light enhanced promoter *PpsbA2* (Dexter and Fu 2009). An average production rate of 0.0766 g L^{-1} day^{-1} was obtained using a computerized photobioreactor over a period of 6 days cultivation. Since then, numerous alcohol dehydrogenases from heterogeneous organisms have been examined along with different promoters towards higher ethanol production in cyanobacteria. *Synechocystis* PCC 6803 is the most widely used model organism to produce ethanol because it has a higher ethanol tolerance compared to that of *Synechococcus elongatus* 7942 (Dexter and Fu 2009).

Several patent applications from e.g., Algenol (Inc 2013) indicate that ethanol production can be enhanced by displacing the *adh* from *Zymomonas mobilis* to endogenous *Synechocystis* alcohol dehydrogenase (*slr1192*). This may due to the reversible catalysis or the strong cofactor preference of alcohol dehydrogenase from *Zymomonas mobilis*. In 2012, a study showed a production rate of a 0.212 g L^{-1} day^{-1} in column photobioreactors coupled with condensation device when using an

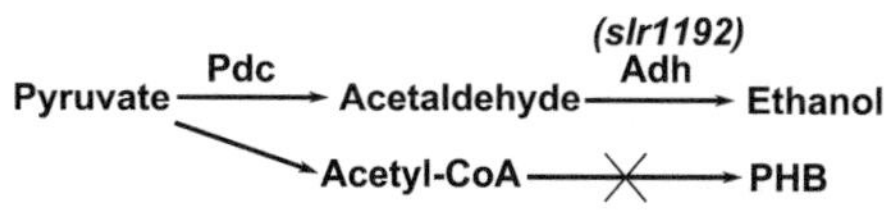

Fig. 2 The synthetic pathways for ethanol production both in *Synechococcus elongatus* PCC 7942 and *Synechocystis* PCC 6803. Pdc: pyruvate decarboxylase (*pdc*) from *Zymomonas mobilis*, Adh: alcohol dehydrogenase (*adh*) from *Zymomonas mobilis* and *Synechocystis* endogenous alcohol dehydrogenase (*slr1192*), and PHB: poly-β-hydroxybutyrate. The cross stands for deletion

engineered *Synechocystis* PCC 6803 strain, which had *pdc/1192*-pathway overexpressed twice on the chromosome, in the neutral site *slr0168* and in the suspected competing PHB synthesis pathway locus (*slr1993, slr1994)*, both under the control of promoter *PrbcL* (Gao et al. 2012). In the same study, an engineered strain with pdc/1192-pathway overexpression in neutral site was used for tap water cultivation test. There was no significant effect on both cell growth and ethanol production between the distilled water culture and tap water culture. Anoxic condition was also tested for the same strain and the results showed a prolonged (from 15 days to 30 days) productivity when the cells were cultivated in 5 % CO_2 and 95 % N_2 compared to when the cells were grown in 5 % CO_2 and 95 % air.

Moreover, along with the studies on ethanol production, ethanol tolerance of host chassis has been investigated for a long time. For example, more lately, an ATP-binding cassette transporter (*slr0982*) was identified to be involved in ethanol tolerance in *Synechocystis* PCC 6803 (Zhang et al. 2015). A *slr0982* deletion strain showed retared growth compared to wild type cells which was recovered when *slr0982* was complemented. However, most of these studies were performed with extracellular addition of ethanol. Recently, an interesting study was conducted to monitor the transcriptomic responses to endogenous ethanol production from an engineered *Synechocystis* strain (Dienst et al. 2014). The results demonstrated that endogenous ethanol production gave a very narrow range response at the transcriptional level as compared to the response from extracellular ethanol stress. The most significant response was the discordant expression of *cpc*BA operon which strongly reduced the light harvesting apparatus.

Metabolic Engineering for Isopropanol Production

Isopropanol is a widely used compound in the chemical, medical, and the automotive industry. It can be produced naturally using *Clostridium* species with highest titer of 1.8 g L^{-1} (Chen and Hiu 1986). One of the most well studied native isobutanol pathway is from *Clostridium beijerinckii*, which catalyzes acetyl-CoA to isopropanol using 4 enzymes: acetyl-CoA acetyltransferase (Thl), acetoacetyl-CoA transferase (CtfAB), acetoacetate decarboxylase (Adc), and primary–secondary alcohol dehydrogenase (Adh) (Wiesenborn et al. 1988). Several different synthetic pathways for isopropanol production have also been demonstrated in *E. coli*. A synthetic pathway overexpressing *Clostridium acetobutylicum thl*, *E. coli atoAD*, *Clostridium acetobutylicum adc*, and *Clostridium beijerinckii adh* produced isopropanol with a titer of 4.9 g L^{-1}, under anaerobic condition, which is higher than the titer reported from any native producer to date (Hanai et al. 2007). To date, the highest titer of isopropanol production in *E. coli* is 143 g L^{-1} in 240 h (Inokuma et al. 2010).

Because of the high isopropanol production in *E. coli*, it became worthwhile to try to produce it in cyanobacteria, which can convert fixed CO_2 to isopropanol directly using the energy from photosynthesis. Amongst the first trials, a synthetic pathway was introduced into the genome of *Synechococcus elongatus* PCC 7942.

Fig. 3 The synthetic pathway for isopropanol production in *Synechococcus elongatus* PCC 7942. ACoAAT: acetyl-CoA acetyltransferase (thl) from *Clostridium acetobutylicum* ATCC 824, ACoAT: acetoacetyl-CoA transferase (atoAD) from *E. coli* K-12 MG1655, ADC: acetoacetate decarboxylase (adc) from *Clostridium acetobutylicum* ATCC 824, and SADH: primary–secondary alcohol dehydrogenase (adh) from *Clostridium beijerinckii*

The pathway was designed to express acetyl-CoA acetyltransferase (ACoAAT) encoded by *thl* and acetoacetate decarboxylase (ADC) encoded by *adc* from *Clostridium acetobutylicum* ATCC 824, acetoacetyl-CoA transferase (ACoAT) encoded by *atoAD* from *E. coli* K-12 MG1655, and primary–secondary alcohol dehydrogenase (SADH) encoded by *adh* from *Clostridium beijerinckii* (Fig. 3). The entire pathway was constructed as two parts and introduced into two neutral sites in the *Synechococcus* genome (Kusakabe et al. 2013). This engineered strain failed to produce isopropanol under aerobic, prolonged light conditions. However, 21.7 mg L^{-1} isopropanol was produced after addition of 90 mM potassium acetate to BG11 HEPES buffered media, under fluorescent light 50 μmol photon m^2 s^{-1} in shaking flasks. In a further improvement, 26.5 mg L^{-1} isopropanol was produced, after 9 days, by cultivating the cells anaerobically in the dark in a specialized BG11 media deprived of nitrogen and phosphate (Kusakabe et al. 2013). In the following study, this engineered strain produced a titer of 146 mg L^{-1} with optimized process parameters (Hirokawa et al. 2015). Instead of inducing the cells for production during the log phase, the culture was induced in the early stationary phase. In addition, utilization of dark-light cycles and pH monitoring contributed to the enhanced yields.

Metabolic Engineering for 1-Butanol Production

1-butanol is an alcohol used in diverse industrial applications. It is also considered as a promising gasoline supplement or replacement. Compared to ethanol, 1-butanol has lower hygroscopicity, higher energy density, and higher compability with the existing infrastructure (Machado and Atsumi 2012). In nature, 1-butanol production exists in the ABE (acetone, butanol, and ethanol) pathway in *Clostridium* species, for instance, in *Clostridium acetobutylicum* (Jones and Woods 1986). The *Clostridium* 1-butanol producing pathway, also called acetyl-CoA-dependent pathway, contains six enzymes: acetyl-CoA acetyltransferase (Thl), β-hydroxybutyryl-CoA dehydrogenase (Hbd), 3-hydroxybutyryl-CoA dehydratase (Crt), butyryl-CoA dehydrogenase (Bcd), electron transferring protein A and B (EtfAB), and bifunctional butyraldehyde dehydrogenase (AdhE2).

In a pioneering work, the native clostridial pathway for 1-butanol production was heterologously expressed in *E. coli* and a titer of 13.9 mg L^{-1} was obtained after 40 h of anaerobic cultivation (Atsumi et al. 2008). In the same study, expression of

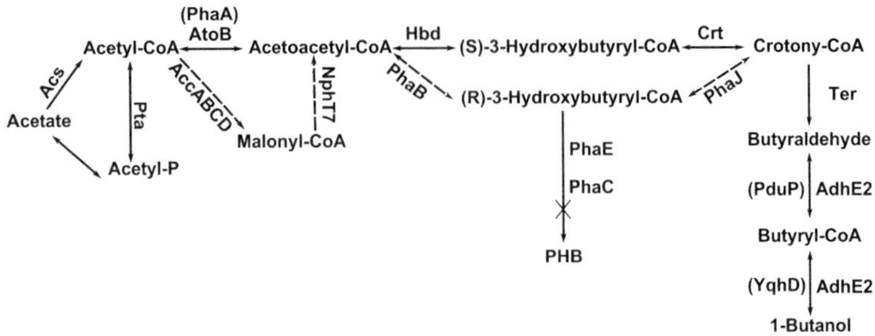

Fig. 4 The synthetic pathways for 1-butanol production in cyanobacteria. The enzymes in brackets are the alternative ones that have been tested to catalyze the same reactions as the enzymes without brackets. The reactions labeled with dotted lines are the alternative pathways that have been studied. The heterogenous enzymes are AtoB: acetyl-CoA acetyltransferase (*atoB*) from *E. coli,* Hbd: β-hydroxybutyryl-CoA dehydrogenase (*hbd*) from *Clostridium acetobutylicum*, Crt: 3-hydroxybutyryl-CoA dehydratase (*crt*) from *Clostridium acetobutylicum*, Ter: trans-enoyl-CoA-reductase (*ter*) from *Trepomena denticola*, AdhE2: bifunctional butyraldehyde dehydrogenase (*adhE*) from *Clostridium acetobutylicum,* PduP: CoA-acylating aldehyde dehydrogenase (*pduP*) from *Salmonella enterica,* Bldh: butyraldehyde dehydrogenase (*bldh*) from *Clostridium saccharoperbutylacetonicum* N1-4, and YqhD: alcohol dehydrogenase (*yqhD*) from *E. coli*

acetyl-CoA acetyltransferase (AtoB) from *E. coli* gave a more than threefold production enhancement. The titer was further increased to 373 mg L^{-1} by deleting of competing native pathways and further to 552 mg L^{-1} by using a nutrient rich media. More investigations were done in the following years to enhance the driving force towards 1-butanol production in *E. coli*. By using the previous engineered strain as background, a trans-enoyl-CoA-reductase (Ter) from *Trepomena denticola* was introduced to replace the oxygen sensitive Bcd-EtfAB complex (Shen et al. 2011). The conversion from crontonyl-CoA to butyryl-CoA catalyzed by Ter is irreversible. This improvement gave a titer of 1.8 g L^{-1} after 24 h cultivation. Furthermore, coupled with the deletion of phosphate acetyltransferase (Pta) and the overexpression of formate dehydrogenase (Fdh), the titer increased to 15 g L^{-1} after 3 days.

This modified pathway was introduced into the genome of *Synechococcus elongatus* PCC 7942 and generated a titer of 3.04 mg L^{-1} after 7 days (Fig. 4) (Lan and Liao 2011). Due to the considerable low enzyme activity of Ter, several Ter from different sources were investigated and a His-tagged Ter from *Trepomena denticola* showed the highest activity. This improvement was attributed to protein stability and folding validity. The study also investigated the effects of light and oxygen on titer values. The best production (14.5 mg L^{-1}) was obtained after 7 days from a culture incubated in the dark under anoxic conditions. In a later study, another driving force was introduced by utilizing endogenous AccABCD and NphT7 from *Streptomyces* CL190 to catalyze the first step that was initially catalyzed by AtoB. This was considered since the condensation of two acetyl-CoA molecules is not a favorable reaction in accordance with thermodynamics. This approach coupled with the change of NADH-dependent alcohol dehydrogenase (AdhE2) to NAD(P)H-dependent butyraldehyde dehydrogenase (Bldh) from *Clostridium saccharoperbutylacetonicum* N1-4 and alcohol dehydrogenase YqhD from *E. coli* increased 1-butanol production to 30 mg L^{-1} (Lan and Liao 2012). Further modification was targeted to replace the oxygen sensitive butyraldehyde dehydrogenase (Bldh) from *Clostridium saccharoperbutylacetonicum* N1-4 with the oxygen tolerant CoA-acylating aldehyde dehydrogenase (PduP) from different organisms. The engineered strain containing *pduP* from *Salmonella enterica* achieved the highest 1-butanol productivity, 51 mg L^{-1} day^{-1} (Lan et al. 2013).

The first 1-butanol producing *Synechocystis* PCC 6803 engineered strain was obtained by applying a modified acetyl-CoA dependent pathway which contains *Synechocystis* endogenous beta-ketothiolase (PhaA) and acetoacetyl-CoA reductase (PhaB), trans-enoyl-CoA reductase (Ter) from *Trepomena denticola*, and bifunctional butyraldehyde dehydrogenase (AdhE2) from *Clostridium acetobutylicum*, together with the phosphoketolase (Xfpk) from *Bifidobacterium breve* in a PHB pathway deficient strain (Anfelt et al. 2015). Nitrogen starvation was applied in this study to increase the productivity of each cell. After 8 days of cultivation under nitrogen subsistent condition, a titer of 37 mg L^{-1} was achieved. However, the highest specific productivity (mg DCW^{-1} day^{-1}) was achieved from cultures deprived of nitrogen.

Even though the production of 1-butanol in cyanobacteria has been improved step by step, it is still much lower than the productivity of the optimized natural producer *Clostidium*, which is over 30 g L^{-1} in batch fermentation culture. More studies are required to investigate the bottleneck of 1-butanol production in cyanobacteria. A recent quantitative target analysis and kinetic profiling of acyl-CoA indicate the requirement of a larger acetyl-CoA pool intracellularly to drive the conversion from pyruvate to acetyl-CoA (Noguchi et al. 2016).

Fatty Acid Derived Chemical Production in Cyanobacteria

Fatty acid derivatives have attracted great attention as potential biofuel molecules. As essential metabolites, they are ubiquitously found in the bacterial and eukaryotic domains of life. These compounds possess highly reduced aliphatic moieties that are used by the cells for energy storage and chemical production. In fact these reduced moieties are equivalent to the hydrocarbon components of petroleum-derived fuels. It has been suggested that an ideal biofuel should be chemically similar to petroleum and hence the vested interest in the fatty acid molecules as prospective biofuel candidates. However, the properties of the carboxylic moieties of fatty acids such as polarity and reactivity make them less attractive as an ideal fuel (Beller et al. 2015). Removal of the carboxylic moiety by derivatization restores model fuel properties. In the last decade, a number of metabolic pathways have been elucidated to facilitate the conversion of fatty acids to a range of industrially relevant compounds, such as fatty acid alkyl esters, fatty alcohols, alkanes, and alkenes

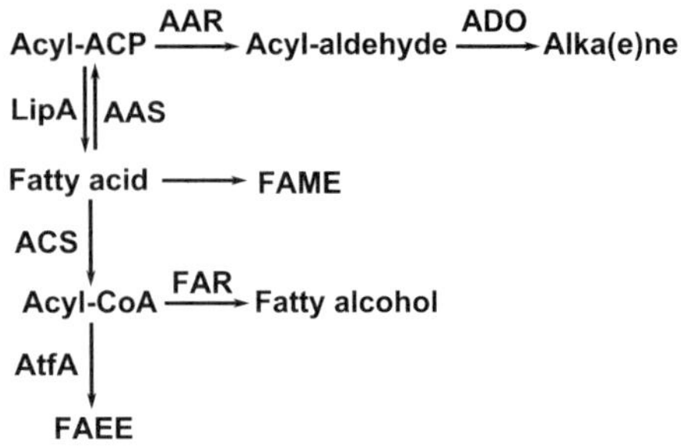

Fig. 5 The fatty acid derived biofuel production pathways in cyanobacteria. *AAR* acyl-ACP reductase (*aar*), *ADO* aldehyde-deformylating oxygenase (*ado*), *FAR* fatty aldehyde reductase (*far*), *AAS* acyl-ACP synthetase (*aas*), *AtfA* acyltransferase (*atfA*)

(Beller et al. 2015; Lennen and Pfleger 2013; Zhou et al. 2014). This section summarizes the recent engineering of cyanobacteria, to produce a range of compounds that may serve as biofuels (Fig. 5) (Table 4).

Although fatty acid synthesis (FAS) occurs both in eukaryotes and prokaryotes, they fundamentally differ at genetic level. Non-plant eukaryotes possess type I FAS while bacteria, plants and algae possess type II FAS (Beld et al. 2015). To date, type II FAS in cyanobacteria is not well characterized. Only three enzymes in FAS pathway (MCAT, ACP, and KASII) have been characterized in cyanobacteria and these enzymes show high homology to other bacterial fatty acid synthases. Therefore, in the following text, type II fatty acid biosynthesis is described as characterized in *E. coli* where it has been most extensively studied (Chan and Vogel 2010).

The first step of type II fatty acid biosynthesis is the carboxylation of acetyl-CoA to malonyl-CoA, catalyzed by acetyl-CoA carboxylase (ACC, or AccABCD*)*. In the second step, malonyl-CoA:ACP transacylase (FabD) converts malonyl-CoA to malonyl-ACP. The fatty acid chains are extended in a cyclic manner using ACP activated intermediates. The primary condensation step, which is catalyzed by β-ketoacyl-ACP synthase III (FabH), uses acetyl-CoA as a primer and malonyl-ACP as the acceptor. The cycle initiates from β-ketoacyl-ACP to β-hydroxyacyl-ACP, catalyzed by β-ketoacyl-ACP reductase (FabG). Then β-hydroxyacyl-ACP will be dehydrated by FabA or FabZ to form enoyl-ACP. An enoyl-ACP reductase (FabI) converts enoyl-ACP to acyl-ACP, which then begins the next elongation cycle by decarboxylative Claisen condensation with malonyl-ACP catalyzed by a β-ketoacyl-ACP synthase II (FabB or FabF) (Fig. 6).

Investigations in *E. coli* revealed that FAS is under feedback regulation of three key enzymes in its pathway, including acetyl-CoA carboxylase (ACC), 3-oxoacyl-ACP-synthase (FabH), and enoyl-ACP reductase (FabI) (Heath and Rock 1996a, b; Davis and Cronan 2001). For successful commercial application, it is imperative to overcome the factors triggering such feedback inhibition. In view of the same, thioesterase, which cleaves acyl-ACP to yield ACP and FFA, was overexpressed in *Synechocystis* PCC 6803 that resulted in a significant increase of FFA (Voelker and Davis 1994; Jiang and Cronan Jr 1994).

Table 4 Fatty acid derived biofuel production in cyanobacteria

Strain	Modification	Yield	Comment	Reference
Fatty alcohol production				
Synechocystis PCC 6803	Contains jojoba plant fatty acyl-CoA reductase (*far*) gene under the control of P_{rbc} promoter	9.73 ± 2.73 μg OD^{-1} L^{-1}	Strains were grown in shake flasks under standard lab culture conditions	(Tan et al. 2011)
Synechocystis PCC 6803	Contains the jojoba plant fatty acyl-CoA reductase (*far*) gene under the control of P_{rbc} promoter	200.44 ± 8.07 μg L^{-1}	Strains were grown in bubble column photobioreactors with a 5 % (v/v) CO_2-enriched airflow	(Tan et al. 2011)
Synechocystis PCC 6803	Expression of engineered FAR protein from *A. thaliana*	426 ± 49.5 μg g^{-1} DCW		(Qi et al. 2013)
Synechocystis PCC 6803	Contains two copies of far_jojoba and one copy of *far* gene from *A. thaliana* in a background of *agp* and *phaAB* gene deletion	761 ± 216 μg g^{-1}DCW	The strain integrates several genetic modifications including the expression of multiple copies of fatty acyl-CoA reductase (FAR) under the control of strong promoters, disruption of the competing pathways for poly-β-hydroxy butyrate and glycogen synthesis, and expression of protein engineered FAR.	(Qi et al. 2013)
Synechocystis PCC 6803	Expression of fatty acyl-Coenzyme A (acyl-CoA) reductase gene maqu_2220 from the marine bacterium *Marinobacter aquaeolei* VT8. The engineered strain included knock-out of *sll0208* and *sll0209*	2.87 mg g^{-1} DCW	Maqu_2220 catalyzes both reduction of fatty acyl-CoA or acyl-Acyl Carrier Protein (acyl-ACP) and reduction of fatty aldehyde to fatty alcohol. Knockout of *sll0208* blocked hydrocarbon accumulation while knockout of *sll0209* removed the activity of acyl-ACP reductase	(Yao et al. 2014)
Alka(e)ne production				
Synechocystis PCC 6803	Endogenous over expression of acyl-ACP reductase (AAR) and aldehyde-deformylating oxygenase (ADO) genes	5-fold increase in the intracellular levels of heptadecane, and a threefold increase in 9-heptadecene		(Hu et al. 2013)
Synechocystis PCC 6803	Overexpression of two copies of *sll0208* and *sll0209 in slr0168* and *slr1556* site	2.3 mg L^{-1} $OD_{730-}{}^{1}$	Overexpressing alkane biosynthetic genes in multiple gene loci can significantly improve the efficiency of alka(e)ne production in cyanobacteria	(Wang et al. 2013)
Synechococcus NKBG15041c	Expression of AAR/ADO pathway genes from *Synechococcus elongatus* PCC 7942	4.2 ± 1.2 μg g^{-1} DCW		(Yoshino et al. 2014)

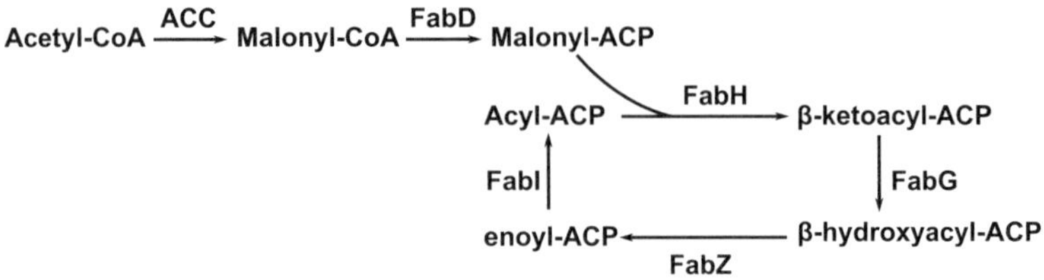

Fig. 6 The elongation cycle of fatty acid biosynthesis in *E. coli.* ACC: acetyl-CoA carboxylase (*accABCD*), FabD: malonyl-CoA:ACP transacylase (*fabD*), FabH: β-Ketoacyl-ACP synthase III (*fabH*), FabG: β-ketoacyl-ACP reductase (*fabG*), FabZ: β-hydroxyacyl-ACP dehydrogenase (*fabz*), and FabI: enoyl-ACP reductase (*fabI*)

Metabolic Engineering for Fatty Alcohol Production

Fatty alcohols may serve as good biofuel additives because of their ideal fuel properties. Myriad organisms including bacteria, plants, and mammals produce fatty alcohols that exist as free fatty alcohols or waxes (oxygen esters of primary fatty alcohols and fatty acids) (Metz et al. 2000; Cheng and Russell 2004; Teerawanichpan and Qiu 2010; Willis et al. 2011). Two independent pathways can generate fatty alcohols in the nature. One pathway involves the reduction of fatty acyl-CoA (or acyl-ACP) to the corresponding fatty alcohol while the other pathway employs two discrete enzymes, acyl-CoA reductase, which catalyzes the reduction of fatty acyl-CoA to a fatty aldehyde intermediate, and a fatty aldehyde reductase (FAR) which further catalyzes the reduction of fatty aldehyde to fatty alcohol (Metz et al. 2000).

Natively, cyanobacteria do not possess fatty alcohol synthesis pathways; however, recent attempts have been made to use cyanobacteria as chassis to produce fatty alcohols. In their pioneer studies Tan et al. (2011) reported non-native fatty alcohol synthesis in *Synechocystis* PCC 6803 by heterologous expression of FAR from mouse, jojoba (*Simmondsia chinensis*), and Arabidopsis (*Arabidopsis thaliana*) (Tan et al. 2011). Up to $9.73 \pm 2.73\ \mu g^{-1}\ L^{-1}\ OD^{-1}$ of fatty alcohol was obtained in the *Synechocystis* PCC 6803 strain expressing the jojoba plant *far* under the control of the constitutive P*rbcL* promoter. The yields were significantly improved to $200.44 \pm 8.07\ \mu g\ L^{-1}$ in the bubble column photobioreactors. However, compared to *E. coli*, the fatty alcohol yields in cyanobacteria are still lagging behind. Thus, further research on developing genetically engineered strains is required to improve the cyanobacterial fatty alcohol production. So far, emphasis has been laid on introduction of novel *far* genes into cyanobacterial hosts or the disruption of the competitive pathways.

Towards this, further improvements have been done at the genetic level using stronger promoters and co-expression of multi-copies of FAR from Arabidopsis at3g11980 and jojoba (Qi et al. 2013). A protein engineered FAR from Arabidopsis at3g11980 was tested along with the deletion of glycogen synthesis and poly-3-hydroxybutyrate synthesis by knocking out *agp* and *phaAB*. Integrating all these genetic modifications, the photosynthetic production of fatty alcohols in *Synechocystis* PCC 6803 was dramatically improved. As an independent modification, the best results, 426 ± 49.5 μg g^{-1} DCW production, were obtained using the truncated version of *A. thaliana* FAR (at3g11980-trunc) under the regulation of the strong promoter P*psbD13*. Considering, the cumulative effect of the modifications, a yield of 761 ± 216 μg g^{-1} DCW total fatty alcohols, the highest ever reported to date, was obtained in shake flask, from the knockout mutant of *agp* and *pha*AB genes which in addition expressed two copies of FAR from jojoba and one copy of FAR from *A. thaliana* at3g11980 as well.

In the most recent published work, fatty alcohols production in *Synechocystis* PCC 6803 was improved by heterologous overexpression of fatty acyl-Coenzyme A (acyl-CoA) reductase *maqu_2220* from the marine bacterium *Marinobacter aquaeolei* VT8 (Yao et al. 2014). The molecular advantage of this enzyme is its twin capacity to catalyze both the four-electron reduction of fatty acyl-CoA or acyl-Acyl Carrier Protein (acyl-ACP) and the two-electron reduction of fatty aldehyde to fatty alcohol. Towards further improvement in fatty alcohol titer values, the acyl-ACP reductase (*sll0209*) and aldehyde-deformylating oxygenase (*sll0208*) involved in the fatty alkane biosynthesis pathway were deleted in the *maqu_2220*-expressing *Synechocystis* PCC 6803 strain. These systematic approaches resulted in an increase of up to 2.87 mg g^{-1} DCW fatty alcohol production in cyanobacteria.

Thus over the years, systematic studies have significantly improved the titer of fatty alcohols production in cyanobacteria. To date, a maximum productivity of 598.6 mg L^{-1} of fatty alcohol has been achieved in *E. coli*, without detailed optimization of the production conditions (Zheng et al. 2012). Considering this, the photosynthetic yields are trailing far behind. Thus, considerable modifications in the metabolic network of fatty acid biosynthesis are still needed for any significant improvement of photosynthetic fatty alcohol production in cyanobacteria.

Metabolic Engineering for Alka(e)ne Production

Alka(e)nes are the major constituents of gasoline, diesel, and jet fuels. Long chain alkanes (C4-C23) are characterized with high energy density and hydrophobic property. They can be produced by a variety of organisms such as bacteria, yeasts, plants, and insects (Schirmer et al. 2010). Endogenous alka(e)ne production has been demonstrated in cyanobacteria since the 1960s (Winters et al. 1969). However, in some strains such as *Synechococcus* PCC 7002, no alka(e)nes were detectable, indicating that the responsible synthesis pathway is not present in these strains. The two-step alka(e)ne pathway comprises of an acyl-ACP reductase (AAR) that catalyzes the reaction from ACP to aldehyde. The aldehyde is further oxidized to alka(e)ne by an aldehyde-deformylating oxygenase (ADO) (Li et al. 2011, 2012). Alternatively, by-products of degraded lipid membrane can also be catalyzed to acyl-ACPs by acyl-ACP synthetase (AAS) (Kaczmarzyk and Fulda 2010). Indeed, deletion strain of *slr1609* encoding AAS gene in *Synechocystis* showed significantly reduced alka(e)ne production (Gao et al. 2012).

In a systematic study, *Synechocystis* was engineered in various approaches such as overexpression of desired enzymes in single or in multiple copies and deletion of competing pathways to improve photosynthetic production of alk(e)ne (Wang et al. 2013). Genes from the endogenous alkane biosynthetic pathway (*sll0208* and *sll0209*) were overexpressed in both *slr0168* and *slr1556* loci. This approach enhanced alkane production to 2.3 mg L^{-1} OD_{730}^{-1}, while no significant changes were detected in the engineered strains only expressing either *sll0208* or *sll0209*. In addition, redirecting the carbon flux from glyceraldehyde 3-phosphate (3-PGA) originating from CBB cycle to acyl-ACP precursor was investigated and the deletion of PHB synthesis pathway in order to boost acetyl-CoA and NAD(P)H was also tested. Similarly, another related study also showed enhanced hydrocarbon yield by overexpressing an endogenous acyl-ACP reductase (AAR) and aldehyde-deformylating oxygenase (ADO) in *Synechocystis* PCC 6803 (Hu et al. 2013). More recently, the production of heptadecane in *Synechococcus* NKBG15041c was achieved by expressing the AAR/ADO pathway from *Synechococcus elongatus* PCC 7942 (Yoshino et al. 2014).

Though these results show some improvements in yields there is still tremendous scope to increase the alka(e)ne production in cyanobacteria. However, it is imperative to holistically understand the physiological roles and regulatory mechanism(s) of native alka(e)nes in cyanobacterial cell in order to utilize cyanobacteria as cell factories for alka(e)ne synthesis. At present, the physiological roles of alka(e)nes in cyanobacteria still remain unclear. It is possible that alka(e)nes are required for proper membrane fluidity or function. Alternatively, they serve as carbon storage compounds under excess carbon and/or nutrient deficient conditions. After bridging this knowledge gap, we may be able to significantly improve cyanobacterial alka(e)ne titer values to meet commercial standards.

Metabolic Engineering for Fatty Acid Alkyl Ester Production

Fatty acid alkyl (methyl and ethyl) esters are popularly clubbed under the category of biodiesel. They can be used alone or blended with petro-diesel in standard diesel engines. It is typically made by transesterification of plant and animal-derived triacylglycerols with methanol under alkaline conditions. The popular biodiesels include fatty acid methyl esters (FAME) and fatty acid ethyl esters (FAEE). Currently, the high cost and the limited availability of plant oils have hindered large-scale commercial viability of biodiesel. To facilitate their production in a non-agro invasive manner, model organisms have now been engineered to produce FAME and FAEE with the hope of converting them into microbial biodiesel factories. The natural occurrence of FAEE has been reported in a range of organisms, including microalgae, insects, fungi, and protozoa (Laseter and Weete 1971). Efforts are ongoing to express the FAEE and FAME pathways in cyanobacteria. Recently, five cyanobacterial isolates from the genera *Synechococcus, Trichormus, Microcystis, Leptolyngbya*, and *Chlorogloea* were examined in terms of their quantity and quality of native lipid feedstock for biodiesel production (Da Rós et al. 2013). However to date, there is no published report on biodiesel production from any of them. Therefore, this section discusses recent developments of in vivo microbial production of FAEE and FAME in light of the studies conducted in *E. coli*. Although bacteria accumulate lower concentrations of lipids compared to microalgae, they offer secondary advantages such as easy cultivation and high specific growth rate.

Among the initial studies of FAEE production in *E. coli*, the growth media was exogenously supplied with fatty acids. Further, the strains were engineered to express genes related to alcohol production and esterification. The engineered strain produced ethanol by heterologous expression of pyruvate decarboxylase (Pdc) and alcohol dehydrogenase (AdhB) from *Zymomonas mobilis*. This was necessary primarily due to the insufficient production of endogenous ethanol in *E. coli*. Further, reliance on endogenous ethanol production would also restrict the production of the dependent FAEE to anaerobic conditions. Esterification was catalyzed by the acyltransferase (AtfA) from *Acinetobacter baylyi* strain ADP1 due to its native capacity to function with ethanol. The engineered strain was shown to produce up to 0.64 g^{-1} L^{-1} of FAEE (Kalscheuer et al. 2006). In a follow-up work, the issue of exogenous supply of fatty acids was overcome, but still, a similar production as 0.67 g^{-1} L^{-1} was obtained (Steen et al. 2010).

In a novel approach, FAEE production in *E. coli* was significantly enhanced using the dynamic sensor-regulator system (DSRS) that focuses on the acyl-CoA-responsive transcription factor, FadR. This approach involved overexpression of FadR and the design of hybrid promoters that included binding sites for both FadR and Lac*I*. The dual binding sites facilitated to detect the responses to the concentration change of both fatty acids and the exogenous inducer (IPTG). Such a unique genetic design enhanced the FAEE titer to 1.5 g L^{-1}, the maximum yield of FAEE in *E. coli* reported to date (Zhang et al. 2012).

Current biodiesel yields achieved in *E. coli* are moderate but must be improved for economic viability in the challenging biofuel market. However, with the rapid pace of development in the field of synthetic biology, significant improvements in production appear feasible. More so, it would be rather interesting to accomplish biodiesel production in photobiological cyanobacteria chassis where solar energy and CO_2 can be directly converted into fuel, thereby overcoming the need of using glucose as feedstock.

Conclusion

Synthetic biology offers the means to produce a wide range of high value energy dense fuels from cyanobacteria. Indeed, as described above, recent research efforts have demonstrated the ability to produce a wide variety of compounds from engineered cyanobacteria. However, considering the low titer values and yield of product, at the moment, it appears a daunting task to commercialize cyanobacterial biofuel production. Nonetheless, with the rapid pace of development in synthetic and systems biology including the advancements in omic technologies, significant improvement in production values may soon be feasible. Considering the immense benefits of using such a renewable system, it seems reasonable to invest both time and money into this research.

References

Abe K, Miyake K, Nakamura M, Kojima K, Ferri S, Ikebukuro K, Sode K (2013) Engineering of a green-light inducible gene expression system in *Synechocystis* sp. PCC 6803. Microb Biotechnol 7:177–183. doi:10.1111/1751-7915.12098

Ajlani G, Vernotte C (1998) Construction and characterization of a phycobiliprotein-less mutant of *Synechocystis* sp. PCC 6803. Plant Mol Biol 37:577–580. doi:10.1023/A:1005924730298

Anfelt J, Hallstrom B, Nielsen J, Uhlén M, Hudson EP (2013) Using transcriptomics to improve butanol tolerance of *Synechocystis* sp. Strain PCC 6803. Appl Environ Microbiol 79:7419–7427. doi:10.1128/AEM.02694-13

Anfelt J, Kaczmarzyk D, Shabestary K, Renberg B, Rockberg J, Nielsen J, Uhlén M, Hudson EP (2015) Genetic and nutrient modulation of acetyl-CoA levels in *Synechocystis* for n-butanol production. Microb Cell Fact 14:167. doi:10.1186/s12934-015-0355-9

Appel J, Phunpruch S, Steinmüller K, Schulz R (2000) The bidirectional hydrogenase of *Synechocystis* sp. PCC 6803 works as an electron valve during photosynthesis. Arch Microbiol 173:333–338. doi:10.1007/s002030000139

Atsumi S, Cann AF, Connor MR, Shen CR, Smith KM, Brynildsen MP, Chou KJ, Hanai T, Liao JC (2008) Metabolic engineering of *Escherichia coli* for 1-butanol production. Metab Eng 10:305–311. doi:10.1016/j.ymben.2007.08.003

Atsumi S, Higashide W, Liao JC (2009) Direct photosynthetic recycling of carbon dioxide to isobutyraldehyde. Nat Biotechnol 27:1177–1180. doi:10.1038/nbt.1586

Aubert-Jousset E, Cano M, Guedeney G, Richaud P, Cournac L (2011) Role of HoxE subunit in *Synechocystis* PCC6803 hydrogenase. FEBS J 278:4035–4043. doi:10.1111/j.1742-4658.2011.08308.x

Badger MR, Price GD (2003) CO_2 concentrating mechanisms in cyanobacteria: molecular components, their diversity and evolution. J Exp Bot 54:609–622. doi:10.1093/jxb/erg076

Bar-Even A, Noor E, Lewis NE, Milo R (2010) Design and analysis of synthetic carbon fixation pathways. Proc Natl Acad Sci USA 107:8889–8894. doi:10.1073/pnas.0907176107

Barney BM, Igarashi RY, Dos Santos PC, Dean DR, Seefeldt LC (2004) Substrate Interaction at an Iron-Sulfur Face of the FeMo-cofactor during Nitrogenase Catalysis. J Biol Chem 279:53621–53624. doi:10.1074/jbc.M410247200

Beiter P (2014) 2014 renewable energy data book. http://www.nrel.gov/docs/fy16osti/64720.pdf. Accessed 10 Feb 2016

Beld J, Lee DJ, Burkart MD (2015) Fatty acid biosynthesis revisited: structure elucidation and metabolic engineering. Mol BioSyst 11:38–59. doi:10.1039/c4mb00443d

Beller HR, Lee TS, Katz L (2015) Natural products as biofuels and bio-based chemicals: fatty acids and isoprenoids. Nat Prod Rep 32:1508–1526. doi:10.1039/c5np00068h

Benemann JR, Weare NM (1974) Hydrogen evolution by nitrogen-fixing *Anabaena cylindrica* cultures. Science 184:174–175. doi:10.1126/science.184.4133.174

Berla BM, Saha R, Immethun CM, Maranas CD, Moon TS, Pakrasi HB (2013) Synthetic biology of cyanobacteria: unique challenges and opportunities. Front Microbiol 4:246. doi:10.3389/fmicb.2013.00246

Bernát G, Waschewski N, Rögner M (2009) Towards efficient hydrogen production: the impact of antenna size and external factors on electron transport dynamics in *Synechocystis* PCC 6803. Photosynth Res 99:205–216. doi:10.1007/s11120-008-9398-7

Blankenship RE, Chen M (2013) Spectral expansion and antenna reduction can enhance photosynthesis for energy production. Curr Opin Chem Biol 17:457–461. doi:10.1016/j.cbpa.2013.03.031

Borirak O, de Koning L, van der Woude AD, Hoefsloot HC, Dekker HL, Roseboom W, de Koster CG, Hellingwerf KJ (2015) Quantitative proteomics analysis of an ethanol and a lactate-producing mutant strain of *Synechocystis* sp. PCC6803. Biotechnol Biofuels 8:111. doi:10.1186/s13068-015-0294-z

Bothe H, Schmitz O, Yates MG, Newton WE (2010) Nitrogen fixation and hydrogen metabolism in cyanobacteria. Microbiol Mol Biol Rev 74:529–551. doi:10.1128/MMBR.00033-10

Calvin M, Benson JA (1948) The Path of Carbon in Photosynthesis. Science 107:476–480. doi:10.1126/science.107.2784.476

Camsund D, Lindblad P (2014) Engineered transcriptional systems for cyanobacterial biotechnology. Front Bioeng Biotechnol 2:40. doi:10.3389/fbioe.2014.00040

Cano M, Volbeda A, Guedeney G, Aubert-Jousset E, Richaud P, Peltier G, Cournac L (2014) Improved oxygen tolerance of the *Synechocystis* sp. PCC 6803 bidirectional hydrogenase by site-directed mutagenesis of putative residues of the gas diffusion channel. Int J Hydrogen Energ 39:16872–16884. doi:10.1016/j.ijhydene.2014.08.030

Carrieri D, Wawrousek K, Eckert C, Yu J, Maness PC (2011) The role of the bidirectional hydrogenase in cyanobacteria. Bioresource Technol 102:8368–8377. doi:10.1016/j.biortech.2011.03.103

Carrieri D, Broadbent C, Carruth D, Paddock T, Ungerer J, Maness PC, Ghirardi M, Yu J (2015) Enhancing photo-catalytic production of organic acids in the cyanobacterium Synechocystis sp. PCC 6803 Δ glgC, a strain incapable of glycogen storage. Microb Biotechnol 8:275–280. doi:10.1111/1751-7915.12243

Chan DI, Vogel HJ (2010) Current understanding of fatty acid biosynthesis and the acyl carrier protein. Biochem J 430:1–19. doi:10.1042/BJ20100462

Chen JS, Hiu SF (1986) Acetone-butanol-isopropanol production by *Clostridium beijerinckii* (synonym, *Clostridium butylicum*). Biotechnol Lett 8:371–376. doi:10.1007/BF01040869

Chen LM, Omiya T, Hata S, Izui K (2002) Molecular characterization of a phosphoenolpyruvate carboxylase from a thermophilic cyanobacterium, *Synechococcus vulcanus* with unusual allosteric properties. Plant Cell Physiol 43:159–169. doi:10.1093/pcp/pcf019

Chen M, Hiller RG, Howe CJ, Larkum AW (2004) Unique origin and lateral transfer of prokaryotic chlorophyll-b and chlorophyll-d light-harvesting systems. Mol Biol Evol 22:21–28. doi:10.1093/molbev/msh250

Chen M, Schliep M, Willows RD, Cai ZL, Neilan BA, Scheer H (2010) A red-shifted chlorophyll. Science 329:1318–1319. doi:10.1126/science.1191127

Cheng JB, Russell DW (2004) Mammalian Wax Biosynthesis: I. Identification of two fatty acyl-coenzyme a reductases with different substrate specificities and tissue distributions. J Biol Chem 279:37789–37797. doi:10.1074/jbc.M406225200

Chollet R, Vidal J, O'Leary MH (1996) Phosphoenol Pyruvate Carboxylase: A ubiquitous, highly regulated enzyme in plants. Annu Rev Plant Physiol Plant Mol Biol 47:273–298. doi:10.1146/annurev.arplant.47.1.273

Da Rós PC, Silva CS, Silva-Stenico ME, Fiore MF, De Castro HF (2013) Assessment of chemical and physico-chemical properties of cyanobacterial lipids for biodiesel production. Mar Drugs 11:2365–2381. doi:10.3390/md11072365

Davis MS, Cronan JE Jr (2001) Inhibition of *Escherichia coli* Acetyl Coenzyme A Carboxylase by Acyl-Acyl Carrier Protein. J Bacteriol 183:1499–1503. doi:10.1128/JB.183.4.1499-1503.2001

De Rosa E, Checchetto V, Franchin C, Bergantino E, Berto P, Szabò I, Giacometti GM, Arrigoni G, Costantin P (2015) [NiFe]-hydrogenase is essential for cyanobacterium *Synechocystis* sp. PCC 6803 aerobic growth in the dark. Sci Rep 5:12424. doi:10.1038/srep12424

Dementin S, Leroux F, Cournac L, de Lacey AL, Volbeda A, Léger C, Burlat B, Martinez N, Champ S, Martin L, Sanganas O, Haumann M, Fernández VM, Guigliarelli B, Fontecilla-Camps JC, Rousset M (2009) Introduction of methionines in the gas channel makes [NiFe] hydrogenase aero-Tolerant. J Am Chem Soc 131:10156–10164. doi:10.1021/ja9018258

Deng M, Coleman JR (1999) Ethanol Synthesis by genetic engineering in cyanobacteria. Appl Environ Microbiol 65:523–528

Dexter J, Fu P (2009) Metabolic engineering of cyanobacteria for ethanol production. Energ Environ Sci 2:857–864. doi:10.1039/B811937F

Dienst D, Georg J, Abts T, Jakorew L, Kuchmina E, Börner T, Wilde A, Dühring U, Enke H, Hess WR (2014) Transcriptomic response to prolonged ethanol production in the cyanobacterium *Synechocystis* sp. PCC 6803. Biotechnol Biofuels 7:21. doi:10.1186/1754-6834-7-21

Dismukes GC, Carrieri D, Bennette N, Ananyev GM, Posewitz MC (2008) Aquatic phototrophs: efficient alternatives to land-based crops for biofuels. Curr Opin Biotechnol 19:235–240. doi:10.1016/j.copbio.2008.05.007

Ducat DC, Way JC, Silver PA (2011) Engineering cyanobacteria to generate high-value products. Trends Biotechnol 29:95–103. doi:10.1016/j.tibtech.2010.12.003
Durall CL, Lindblad P (2015) Mechanisms of carbon fixation and engineering for increased carbon fixation in cyanobacteria. Algal Res 11:263–270. doi:10.1016/j.algal.2015.07.002
Durall C, Rukminasari N, Lindblad P (2016) Enhanced growth at low light intensity in the cyanobacterium *Synechocystis* PCC 6803 by overexpressing phosphoenolpyruvate carboxylase. Algal Res 16:275–281. doi:10.1016/j.algal.2016.03.027
Eckert C, Boehm M, Carrieri D, Yu J, Dubini A, Nixon PJ, Maness PC (2012) Genetic analysis of the hox hydrogenase in the cyanobacterium *Synechocystis* sp. PCC 6803 reveals subunit roles in association, assembly, maturation, and function. J Biol Chem 287:43502–43515. doi:10.1074/jbc.M112.392407
English CM, Eckert C, Brown K, Seibert M, King PW (2009) Recombinant and in vitro expression systems for hydrogenases: new frontiers in basic and applied studies for biological and synthetic H_2 production. Dalton Trans 7:9970–9978. doi:10.1039/b913426n
Espie GS, Kimber MS (2011) Carboxysomes: cyanobacterial RubisCO comes in small packages. Photosynth Res 109:7–20. doi:10.1007/s11120-011-9656-y
Fairley P (2011) Introduction: Next generation biofuels. Nature 474:S2–S5. doi:10.1038/474S02a
Fukuzawa H, Suzuki E, Komukai Y, Miyachi S (1992) A gene homologous to chloroplast carbonic anhydrase (*icfA*) is essential to photosynthetic carbon dioxide fixation by *Synechococcus* PCC7942. Proc Natl Acad Sci USA 89:4437–4441. doi:10.1073/pnas.89.10.4437
Gao Z, Zhao H, Li Z, Tan X, Lu X (2012) Photosynthetic production of ethanol from carbon dioxide in genetically engineered cyanobacteria. Energy Environ Sci 5:9857–9865. doi:10.1039/C2EE22675H
Ghirardi ML (2015) Implementation of photobiological H_2 production: the O_2 sensitivity of hydrogenases. Photosynth Res 125:383–393. doi:10.1007/s11120-015-0158-1
Gierga G, Voss B, Hess WR (2009) The Yfr2 ncRNA family, a group of abundant RNA molecules widely conserved in cyanobacteria. RNA Biol 6:222–227. doi:10.4161/rna.6.3.8921
Guo M, Song W, Buhain J (2015) Bioenergy and biofuels: History, status, and perspective. Renew Sustainable Energy Rev 42:712–725. doi:10.1016/j.rser.2014.10.013
Gutekunst K, Chen X, Schreiber K, Kaspar U, Makam S, Appel J (2013) The bidirectional nifehydrogenase in *Synechocystis* sp. PCC 6803 is reduced by flavodoxin and ferredoxin and is essential under mixotrophic, nitrate-limiting conditions. J Biol Chem 289:1930–1937. doi:10.1074/jbc.M113.526376
Haake V, Zrenner R, Sonnewald U, Stitt M (1998) A moderate decrease of plastid aldolase activity inhibits photosynthesis, alters the levels of sugars and starch, and inhibits growth of potato plants. Plant J 14:147–157. doi:10.1046/j.1365-313X.1998.00089.x
Hanai T, Atsumi S, Liao JC (2007) Engineered Synthetic Pathway for Isopropanol Production in *Escherichia coli*. Appl Environ Microbiol 73:7814–7818. doi:10.1128/AEM.01140-07
Happe T, Schutz K, Bohme H (2000) Transcriptional and mutational analysis of the uptake hydrogenase of the filamentous cyanobacterium *Anabaena variabilis* ATCC 29413. J Bacteriol 182:1624–1631. doi:10.1128/JB.182.6.1624-1631.2000
Heath RJ, Rock CO (1996a) Inhibition of beta-Ketoacyl-Acyl Carrier Protein Synthase III (FabH) by Acyl-Acyl Carrier Protein in *Escherichia coli*. J Biol Chem 271:10996–11000. doi:10.1074/jbc.271.18.10996
Heath RJ, Rock CO (1996b) Regulation of Fatty Acid Elongation and Initiation by Acyl-Acyl Carrier Protein in *Escherichia coli*. J Biol Chem 271:1833–1836. doi:10.1074/jbc.271.4.1833
Heinhorst S, Cannon GC, Shively J (2014) Carboxysomes and carboxysome-like inclusions in nanomicrobiology, 1st edn. Springer, Berlin Heidelberg, Germany
Henkes S, Sonnewald U, Badur R, Flachmann R, Stitt M (2001) A small decrease of plastid transketolase activity in antisense tobacco transformants has dramatic effects on photosynthesis and phenylpropanoid metabolism. Plant Cell 13:535–551. doi:10.1105/tpc.13.3.535
Herrero A, Flores E (2008) The cyanobacteria: molecular biology, genomics, and evolution. Caister Academic Press, Norfolk, UK. ISBN 978-1-904455-15-8

Hirokawa Y, Suzuki I, Hanai T (2015) Optimization of isopropanol production by engineered cyanobacteria with a synthetic metabolic pathway. J Biosci Bioeng 119:585–590. doi:10.1016/j.jbiosc.2014.10.005

Hu P, Borglin S, Kamennaya N, Chen L, Park H, Mahoney L, Kijac A, Shan G, Chavarría KL, Zhang C, Quinn NWT, Wemmer D, Holman HY, Jansson C (2013) Metabolic phenotyping of the cyanobacterium *Synechocystis* 6803 engineered for production of alkanes and free fatty acids. Appl Energ 102:850–859. doi:10.1016/j.apenergy.2012.08.047

Huang HH, Lindblad P (2013) Wide-dynamic-range promoters engineered for cyanobacteria. J Biol Eng 7:10. doi:10.1186/1754-1611-7-10

Huang HH, Camsund D, Lindblad P, Heidorn T (2010) Design and characterization of molecular tools for a Synthetic Biology approach towards developing cyanobacterial biotechnology. Nucleic Acids Res 38:2577–2593. doi:10.1093/nar/gkq164

Ihara M, Nakamoto H, Kamachi T, Okura I, Maeda M (2006) Photoinduced hydrogen production by direct electron transfer from photosystem I cross-linked with cytochrome c3 to [NiFe]-hydrogenase. Photochem Photobiol 82:1677–1685. doi:10.1111/j.1751-1097.2006.tb09830.x

Iijima H, Nakaya Y, Kuwahara A, Hirai MY, Osanai T (2015) Seawater cultivation of freshwater cyanobacterium *Synechocystis* sp. PCC 6803 drastically alters amino acid composition and glycogen metabolism. Front Microbiol 6:326. doi:10.3389/fmicb.2015.00326

Inc AB (2013) Methods and compositions for ethanol producing cyanobacteria. US8372613 B2 (16 Jan 2007)

Inokuma K, Liao JC, Okamoto M, Hanai T (2010) Improvement of isopropanol production by metabolically engineered *Escherichia coli* using gas stripping. J Biosci Bioeng 110:696–701. doi:10.1016/j.jbiosc.2010.07.010

Iwaki T, Haranoh K, Inoue N, Kojima K, Satoh R, Nishino T, Wada S, Ihara H, Tsuyama S, Kobayashi H, Wadano A (2006) Expression of foreign type I ribulose-1,5-bisphosphate carboxylase/oxygenase (EC 4.1.1.39) stimulates photosynthesis in cyanobacterium *Synechococcus* PCC7942 cells. Photosynth Res 88:287–297. doi:10.1007/s11120-006-9048-x

Jia XH, Zhang PP, Shi DJ, Mi HL, Zhu JC, Huang XW, He PM (2014) Regulation of *pepc* gene expression in *Anabaena* sp. PCC 7120 and its effects on cyclic electron flow around photosystem I and tolerances to environmental stresses. J Integr Plant Biol 57:468–476. doi:10.1111/jipb.12241

Jiang P, Cronan JE Jr (1994) Inhibition of fatty acid synthesis in *Escherichia coli* in the absence of phospholipid synthesis and release of inhibition by thioesterase action. J Bacteriol 176:2814–2821

Jin H, Chen L, Wang J, Zhang W (2014) Engineering biofuel tolerance in non-native producing microorganisms. Biotechnol Adv 32:541–548. doi:10.1016/j.biotechadv.2014.02.001

Jones PR (2014) Genetic Instability in Cyanobacteria-An Elephant in the Room? Front Bioeng Biotechnol. doi:10.3389/fbioe.2014.00012

Jones DT, Woods DR (1986) Acetone-butanol fermentation revisited. Microbiol Rev 50:484–524

Kaczmarzyk D, Fulda M (2010) Fatty acid activation in cyanobacteria mediated by acyl-acyl carrier protein synthetase enables fatty acid recycling. Plant Physiol 152:1598–1560. doi:10.1104/pp.109.148007

Kalscheuer R, Stolting T, Steinbuchel A (2006) Microdiesel: *Escherichia coli* engineered for fuel production. Microbiology 152:2529–2536. doi:10.1099/mic.0.29028-0

Kamennaya N, Ahn S, Park H, Bartal R, Sasaki KA, Holman HY, Jansson C (2015) Installing extra bicarbonate transporters in the cyanobacterium *Synechocystis* sp. PCC 6803 enhances biomass production. Metab Eng 29:76–85. doi:10.1016/j.ymben.2015.03.002

Kaplan A, Hagemann M, Bauwe H, Bauwe H, Kahlon S, Ogawa T (2008) Carbon acquisition by cyanobacteria: mechanisms, comparative genomics and evolution. In: The cyanobacteria: molecular biology, genomics, and evolution. Caister Academic Press, Norfolk, UK

Kasahara M, Obmori M (1997) Carbamoyl Phosphate Synthetase of the cyanobacterium *Anabaena cylindrica*. Plant Cell Physiol 38:734–739. doi:10.1093/oxfordjournals.pcp.a029227

Khanna N, Lindblad P (2015) Cyanobacterial hydrogenases and hydrogen metabolism revisited: recent progress and future prospects. Int J Mol Sci 16:10537–10561. doi:10.3390/ijms160510537

Khanna N, Raleiras P, Lindblad P (2016) Fundamentals and recent advances in hydrogen production and nitrogen fixation in cyanobacteria. In: The Physiology of Microalgae. Springer International Publishing, Switzerland. ISBN 978-3-319-24943-8

Khetkorn W, Lindblad P, Incharoensakdi A (2012) Inactivation of uptake hydrogenase leads to enhanced and sustained hydrogen production with high nitrogenase activity under high light exposure in the cyanobacterium *Anabaena siamensis* TISTR 8012. J Biol Eng 6:19. doi:10.1186/1754-1611-6-19

Khetkorn W, Khanna N, Incharoensakdi A, Lindblad P (2013) Metabolic and genetic engineering of cyanobacteria for enhanced hydrogen production. Biofuels 4:535–561. doi:10.4155/bfs.13.41

Khozaei M, Fisk S, Lawson T, Gibon Y, Sulpice R, Stitt M, Lefebvre SC, Raines CA (2015) Overexpression of plastid transketolase in tobacco results in a thiamine auxotrophic phenotype. Plant Cell 27:432–447. doi:10.1105/tpc.114.131011

Kothe M, Purcarea C, Guy HI, Evans DR, Powers-Lee SG (2009) Direct demonstration of carbamoyl phosphate formation on the C-terminal domain of carbamoyl phosphate synthetase. Protein Sci 14:37–44. doi:10.1110/ps.041041305

Kusakabe T, Tatsuke T, Tsuruno K, Hirokawa Y, Atsumi S, Liao JC, Hanai T (2013) Engineering a synthetic pathway in cyanobacteria for isopropanol production directly from carbon dioxide and light. Metab Eng 20:101–108. doi:10.1016/j.ymben.2013.09.007

Lan EI, Liao JC (2011) Metabolic engineering of cyanobacteria for 1-butanol production from carbon dioxide. Metab Eng 13:353–363. doi:10.1016/j.ymben.2011.04.004

Lan EI, Liao JC (2012) ATP drives direct photosynthetic production of 1-butanol in cyanobacteria. Proc Natl Acad Sci USA 109:6018–6023. doi:10.1073/pnas.1200074109

Lan EI, Ro SY, Liao JC (2013) Oxygen-tolerant coenzyme A-acylating aldehyde dehydrogenase facilitates efficient photosynthetic n-butanol biosynthesis in cyanobacteria. Energ Environ Sci 6:2672–2681. doi:10.1039/C3EE41405A

Laseter JL, Weete JD (1971) Fatty Acid Ethyl Esters of *Rhizopus arrhizus*. Science 172:864–865. doi:10.1126/science.172.3985.864

Lea-Smith DJ, Bombelli P, Dennis JS, Scott SA, Smith AG, Howe CJ (2014) Phycobilisome-deficient strains of *Synechocystis* sp. PCC 6803 have reduced size and require carbon-limiting conditions to exhibit enhanced productivity. Plant Physiol 165:705–714. doi:10.1104/pp.114.237206

Lennen RM, Pfleger BF (2013) Microbial production of fatty acid-derived fuels and chemicals. Curr Opin Biotech 24:1044–1053. doi:10.1016/j.copbio.2013.02.028

Lepiniec L, Keryer E, Philippe H, Gadal P, Crétin C (1993) Sorghum phosphoenolpyruvate carboxylase gene family: structure, function and molecular evolution. Plant Mol Biol 21:487–502. doi:10.1007/BF00028806

Li N, Nørgaard H, Warui DM, Booker SJ, Krebs C, Bollinger JM Jr (2011) Conversion of fatty aldehydes to alka(e)nes and formate by a cyanobacterial aldehyde decarbonylase: cryptic redox by an unusual dimetal oxygenase. J Am Chem Soc 133:6158–6161. doi:10.1021/ja2013517

Li N, Chang WC, Warui DM, Booker SJ, Krebs C, Bollinger JM Jr (2012) Evidence for only oxygenative cleavage of aldehydes to alk(a/e)nes and formate by cyanobacterial aldehyde decarbonylases. Biochemistry 51:7908–7916. doi:10.1021/bi300912n

Li X, Shen C, Liao JC (2014) Isobutanol production as an alternative metabolic sink to rescue the growth deficiency of the glycogen mutant of *Synechococcus elongatus* PCC 7942. Photosynth Res 120:301–310. doi:10.1007/s11120-014-9987-6

Liang F, Lindblad P (2016) Effect of overexpressing photosynthetic carbon flux control enzymes in the cyanobacterium *Synechocystis* PCC 6803. Metab Eng 38:56–64. doi:10.1016/j.ymben.2016.06.005

Lindberg P, Schutz K, Happe T, Lindblad P (2002) A hydrogen-producing, hydrogenase-free mutant strain of *Nostoc punctiforme* ATCC 29133. Int J Hydrogen Energ 27:1291–1296. doi:10.1016/S0360-3199(02)00121-0

Lindberg P, Devine E, Stensjo K, Lindblad P (2011) HupW protease specifically required for processing of the catalytic subunit of the uptake hydrogenase in the cyanobacterium *Nostoc* sp. Strain PCC 7120. Appl Environ Microbiol 78:273–276. doi:10.1128/AEM.05957-11

Liu J, Chen L, Wang J, Qiao J, Zhang W (2012) Proteomic analysis reveals resistance mechanism against biofuel hexane in *Synechocystis* sp. PCC 6803. Biotechnol Biofuels 5:68. doi:10.1186/1754-6834-5-68

Long BM, Tucker L, Badger MR, Price GD (2010) Functional cyanobacterial carboxysomes have an absolute requirement for both long and short forms of the CcmM protein. Plant Physiol 153:285–293. doi:10.1104/pp.110.154948

Luinenburg I, Coleman JR (1990) A requirement for phosphoenolpyruvate carboxylase in the cyanobacterium *Synechococcus* PCC 7942. Arch Microbiol 154:471–474. doi:10.1007/BF00245230

Ma W, Shi D, Wang Q, Wei L, Chen H (2005) Exogenous expression of the wheat chloroplastic fructose-1,6-bisphosphatase gene enhances photosynthesis in the transgenic cyanobacterium, *Anabaena* PCC7120. J Appl Phycol 17:273–280. doi:10.1007/s10811-005-4850-y

Machado I, Atsumi S (2012) Cyanobacterial biofuel production. J Biotechnol 162:50–56. doi:10.1016/j.jbiotec.2012.03.005

Madgwick PJ, Parmar S, Parry MA (1998) Effect of mutations of residue 340 in the large subunit polypeptide of Rubisco from Anacystis nidulans. Eur J Biochem 253:476–479. doi:10.1046/j.1432-1327.1998.2530476.x

Maeda S, Badger MR, Price GD (2002) Novel gene products associated with NdhD3/D4-containing NDH-1 complexes are involved in photosynthetic CO_2 hydration in the cyanobacterium, *Synechococcus* sp. PCC7942. Mol Microbiol 43:425–435. doi:10.1046/j.1365-2958.2002.02753.x

Marcus Y, Altman-Gueta H, Finkler A, Gurevitz M (2003) Dual role of cysteine 172 in redox regulation of ribulose 1,5-bisphosphate carboxylase/oxygenase activity and degradation. J Bacteriol 185:1509–1517. doi:10.1128/JB.185.5.1509-1517.2003

Marcus Y, Altman-Gueta H, Finkler A, Gurevitz M (2005) Mutagenesis at two distinct phosphate-binding sites unravels their differential roles in regulation of rubisco activation and catalysis. J Bacteriol 187:4222–4228. doi:10.1128/JB.187.12.4222-4228.2005

Marcus Y, Altman-Gueta H, Wolff Y, Gurevitz M (2011) Rubisco mutagenesis provides new insight into limitations on photosynthesis and growth in *Synechocystis* PCC 6803. J Exp Bot 62:4173–4182. doi:10.1093/jxb/err116

Masukawa H, Mochimaru M. Sakurai H (2002) Disruption of the uptake hydrogenase gene, but not of the bidirectional hydrogenase gene, leads to enhanced photobiological hydrogen production by the nitrogen-fixing cyanobacterium *Anabaena* sp. PCC 7120. Appl Microbiol Biotechnol 58:618–624. doi:10.1007/s00253-002-0934-7

Masukawa H, Inoue K, Sakurai H, Wolk CP, Hausinger RP (2010) Site-directed mutagenesis of the *Anabaena* sp. Strain PCC 7120 nitrogenase active site to increase photobiological hydrogen production. Appl Environ Microbiol 76:6741–6750. doi:10.1128/AEM.01056-10

Masukawa H, Sakurai H, Hausinger RP, Inoue K (2014) Sustained photobiological hydrogen production in the presence of N_2 by nitrogenase mutants of the heterocyst-forming cyanobacterium *Anabaena*. Int J Hydrogen Energ 39:19444–19451. doi:10.1016/j.ijhydene.2014.09.090

Mclntosh CL, Germer F, Schulz R, Appel J, Jones AK (2011) The [NiFe]-hydrogenase of the cyanobacterium *Synechocystis* sp. PCC 6803 works bidirectionally with a bias to H_2 production. J Am Chem Soc 133:11308–11319. doi:10.1021/ja203376y

Metz JG, Pollard MR, Anderson L, Hayes TR, Lassner MW (2000) Purification of a jojoba embryo fatty acyl-coenzyme a reductase and expression of its cdna in high erucic acid rapeseed. Plant Physiol 122:635–644. doi:10.1104/pp.122.3.635

Mi H, Endo T, Ogawa T, Asada K (1995) Thylakoid membrane-bound, NADPH-specific pyridine nucleotide dehydrogenase complex mediates cyclic electron transport in the cyanobacterium *Synechocystis* sp. PCC 6803. Plant Cell Physiol 36:661–668

Mitschke J, Georg J, Scholz I, Sharma CM, Dienst D, Bantscheff J, Voss B, Steglich C, Wilde A, Vogel J, Hess WR (2011) An experimentally anchored map of transcriptional start sites in the

model cyanobacterium *Synechocystis* sp. PCC 6803. Proc Natl Acad Sci USA 108:2124–2129. doi:10.1073/pnas.1015154108
Miyagawa Y, Tamoi M, Shigeoka S (2001) Overexpression of a cyanobacterial fructose-1,6-/sedoheptulose-1,7-bisphosphatase in tobacco enhances photosynthesis and growth. Nat Biotechnol 19:965–969. doi:10.1038/nbt1001-965
Miyashita H, Adachi K, Kurano N, Ikemoto H, Chihara M, Miyach S (1997) Pigment composition of a novel oxygenic photosynthetic prokaryote containing chlorophyll d as the major chlorophyll. Plant Cell Physiol 38:274–281. doi:10.1093/oxfordjournals.pcp.a029163
Nakahara K, Yamamoto H, Miyake C, Yokota A (2003) Purification and characterization of class-I and class-II fructose-1,6-bisphosphate aldolases from the cyanobacterium *Synechocystis* sp. PCC 6803. Plant Cell Physiol 44:326–333. doi:10.1093/pcp/pcg044
Nakajima Y, Ueda R (1997) Improvement of photosynthesis in dense microalgal suspension by reduction of light harvesting pigments. J Appl Phycol 9:503–510. doi:10.1023/A:1007920025419
Niederholtmeyer H, Wolfstadter BT, Savage DF, Silver PA, Way JC (2010) Engineering cyanobacteria to synthesize and export hydrophilic products. Appl Environ Microb 76:6023–6023. doi:10.1128/AEM.00202-10
Nogales J, Gudmundsson S, Knight E, Palsson BO, Thiele I (2012) Detailing the optimality of photosynthesis in cyanobacteria through systems biology analysis. Proc Natl Acad Sci USA 109:2678–2683. doi:10.1073/pnas.1117907109
Noguchi S, Putri SP, Lan EI, Lavina WA, Dempo Y, Bamba T, Liao JC, Fukusaki E (2016) Quantitative target analysis and kinetic profiling of acyl-CoAs reveal the rate-limiting step in cyanobacterial 1-butanol production. Metabolomics 12:26. doi:10.1007/s11306-015-0940-2
Nyberg M, Heidorn T, Lindblad P (2015) Hydrogen production by the engineered cyanobacterial strain *Nostoc* PCC 7120 ΔhupW examined in a flat panel photobioreactor system. J Biotechnol 215:35–43. doi:10.1016/j.jbiotec.2015.08.028
Ogawa T, Tamoi M, Kimura A, Mine A, Sakuyama H, Yoshida E, Maruta T, Suzuki K, Ishikawa T, Shigeoka S (2015) Enhancement of photosynthetic capacity in *Euglena gracilis* by expression of cyanobacterial fructose-1,6-/sedoheptulose-1,7-bisphosphatase leads to increases in biomass and wax ester production. Biotechnol Biofuels 8:80. doi:10.1186/s13068-015-0264-5
Omata T, Price GD, Badger MR, Okamura M, Gohta S, Ogawa T (1999) Identification of an ATP-binding cassette transporter involved in bicarbonate uptake in the cyanobacterium *Synechococcus* sp. strain PCC 7942. Proc Natl Acad Sci USA 96:13571–13576. doi:10.1073/pnas.96.23.13571
Owttrim GW, Colman B (1986) Purification and characterization of phosphoenolpyruvate carboxylase from a cyanobacterium. J Bacteriol 168:207–212
Pinto F, van Elburg KA, Pacheco CC, Lopo M, Noirel J, Montagud A, Urchueguía JF, Wright PC, Tamagnini P (2011) Construction of a chassis for hydrogen production: physiological and molecular characterization of a *Synechocystis* sp. PCC 6803 mutant lacking a functional bidirectional hydrogenase. Microbiol 158:448–464. doi:10.1099/mic.0.052282-0
Price GD (2011) Inorganic carbon transporters of the cyanobacterial CO_2 concentrating mechanism. Photosynth Res 109:47–57. doi:10.1007/s11120-010-9608-y
Price GD, Badger MR (1989) Expression of human carbonic anhydrase in the cyanobacterium *Synechococcus* PCC7942 creates a high co_2-requiring phenotype : evidence for a central role for carboxysomes in the co_2 concentrating mechanism. Plant Physiol 91:505–513. doi:10.1104/pp.91.2.505
Price GD, Maeda S, Omata T, Badger MR (2002) Modes of active inorganic carbon uptake in the cyanobacterium, *Synechococcus* sp. PCC7942. Funct Plant Biol 29:131–149. doi:10.1071/PP01229
Price GD, Badger MR, Woodger FJ, Long BM (2007) Advances in understanding the cyanobacterial CO_2-concentrating-mechanism (CCM): functional components, Ci transporters, diversity, genetic regulation and prospects for engineering into plants. J Exp Bot 59:1441–1461. doi:10.1093/jxb/erm112

Qi F, Yao L, Tan X, Lu X (2013) Construction, characterization and application of molecular tools for metabolic engineering of *Synechocystis* sp. Biotechnol Lett 35:1655–1661. doi:10.1007/s10529-013-1252-0

Qiao J, Wang J, Chen L, Tian X, Huang S, Ren X, Zhang W (2012) Quantitative iTRAQ LC–MS/MS Proteomics Reveals Metabolic Responses to Biofuel Ethanol in Cyanobacterial *Synechocystis* sp. PCC 6803. J Proteome Res 11:5286–5300. doi:10.1021/pr300504w

Qiao J, Shao M, Chen L, Wang J, Wu G, Tian X, Liu J, Huang S, Zhang W (2013) Systematic characterization of hypothetical proteins in *Synechocystis* sp. PCC 6803 reveals proteins functionally relevant to stress responses. Gene 512:6–15. doi:10.1016/j.gene.2012.10.004

Quintana N, Van der Kooy F, Van de Rhee MD, Voshol GP, Verpoorte R (2011) Renewable energy from Cyanobacteria: energy production optimization by metabolic pathway engineering. Appl Microbiol Biotechnol 91:471–490. doi:10.1007/s00253-011-3394-0

Raleiras P, Khanna N, Miranda H, Mészáros LS, Krassen H, Ho F, Battchikova N, Aro E, Magnuson A, Lindblad P, Styring S (2016) Turning around the electron flow in an uptake hydrogenase. EPR spectroscopy and in vivo activity of a designed mutant in HupSL from *Nostoc punctiforme*. Energy Environ Sci 9:581–594. doi:10.1039/C5EE02694F

Rivoal J, Dunford R, Plaxton WC, Turpin DH (1996) Purification and properties of four phosphoenolpyruvate carboxylase isoforms from the green alga *Selenastrum minutum*: evidence that association of the 102-kda catalytic subunit with unrelated polypeptides may modify the physical and kinetic properties of the enzyme. Arch Biochem Biophys 332:47–57. doi:10.1006/abbi.1996.0315

Rivoal J, Plaxton WC, Turpin DH (1998) Purification and characterization of high and low-molecular-mass isoforms of phosphoenolpyruvate carboxylase from *Chlamydomonas reinhardtii*. Kinetic, structural and immunological evidence that the green algal enzyme is distinct from the prokaryotic and higher plant enzymes. Biochem J 331:201–209. doi:10.1042/bj3310201

Ruffing AM (2013) Borrowing genes from Chlamydomonas reinhardtii for free fatty acid production in engineered cyanobacteria. J Appl Phycol 25:1495–1507. doi:10.1007/s10811-013-9993-7

Ruffing A (2014) Improved free fatty acid production in cyanobacteria with *Synechococcus* sp. PCC 7002 as host. Front Bioeng Biotechnol 2:17. doi:10.3389/fbioe.2014.00017

Savakis P, Hellingwerf K (2015) Engineering cyanobacteria for direct biofuel production from CO_2. Curr Opin Biotechnol 33:8–14. doi:10.1016/j.copbio.2014.09.007

Schenk G, Duggleby RG, Nixon PF (1998) Properties and functions of the thiamin diphosphate dependent enzyme transketolase. Int J Biochem Cell Biol 30:1297–1318. doi:10.1016/S1357-2725(98)00095-8

Schirmer A, Rude MA, Li X, Popova E, del Cardayre SB (2010) Microbial biosynthesis of alkanes. Science 329:559–562. doi:10.1126/science.1187936

Shen G, Boussiba S, Vermaas WF (1993) *Synechocystis* sp PCC 6803 Strains lacking photosystem i and phycobilisome function. Plant Cell 5:1853–1863. doi:10.1105/tpc.5.12.1853

Shen CR, Lan EI, Dekishima Y, Baez A, Cho KM, Liao JC (2011) Driving Forces Enable High-Titer Anaerobic 1-Butanol Synthesis in *Escherichia coli*. Appl Environ Microbiol 77:2905–2915. doi:10.1128/AEM.03034-10

Shibata M, Ohkawa H, Katoh H, Shimoyama M, Ogawa T (2002) Two CO_2 uptake systems in cyanobacteria: four systems for inorganic carbon acquisition in *Synechocystis* sp. strain PCC6803. Funct Plant Biol 29:123–129. doi:10.1071/PP01188

Sigfridsson KG, Leidel N, Sanganas O, Chernev P, Lenz O, Yoon KS, Nishihara H, Parkin A, Armstrong FA, Dementin S, Rousset M, De Lacey AL, Haumann M (2015) Structural differences of oxidized iron–sulfur and nickel–iron cofactors in O_2-tolerant and O_2-sensitive hydrogenases studied by X-ray absorption spectroscopy. Biochim Biophys Acta 1847:162–170. doi:10.1016/j.bbabio.2014.06.011

Stal LJ, Moezelaar R (2006) Fermentation in cyanobacteria. FEMS Microbiol Rev 21:179–211. doi:10.1111/j.1574-6976.1997.tb00350.x

Steen EJ, Kang Y, Bokinsky G, Hu Z, Schirmer A, McClure A, Del Cardayre SB, Keasling JD (2010) Microbial production of fatty-acid-derived fuels and chemicals from plant biomass. Nature 463:559–562. doi:10.1038/nature08721

Suzuki E, Ohkawa H, Moriya K, Matsubara T, Nagaike Y, Iwasaki I, Fujiwara S, Tsuzuki M, Nakamura Y (2010) Carbohydrate Metabolism in Mutants of the Cyanobacterium *Synechococcus elongatus* PCC 7942 Defective in Glycogen Synthesis. Appl Environ Microb 76:3153–3159. doi:10.1128/AEM.00397-08

Takahama K, Matsuoka M, Nagahama K, Ogawa T (2003) Construction and analysis of a recombinant cyanobacterium expressing a chromosomally inserted gene for an ethylene-forming enzyme at the *psbAI* locus. J Biosci Bioeng 95:302–305. doi:10.1016/S1389-1723(03)80034-8

Tamagnini P, Axelsson R, Lindberg P, Oxelfelt F, Wünschiers R, Lindblad P (2002) Hydrogenases and hydrogen metabolism of cyanobacteria. Microbiol Mol Biol Rev 66:1–20. doi:10.1128/MMBR.66.1.1-20.2002

Tamagnini P, Leitão E, Oliveira P, Ferreira D, Pinto F, Harris DJ, Heidorn T, Lindblad P (2007) Cyanobacterial hydrogenases: diversity, regulation and applications. FEMS Microbiol Rev 31:692–720. doi:10.1111/j.1574-6976.2007.00085.x

Tamoi M, Ishikawa T, Takeda T, Shigeoka S (1996) Molecular characterization and resistance to hydrogen peroxide of two fructose-1,6-bisphosphatases from *Synechococcus* PCC 7942. Arch Biochem Biophys 334:27–36. doi:10.1006/abbi.1996.0425

Tan X, Yao L, Gao Q, Wang W, Qi F, Lu X (2011) Photosynthesis driven conversion of carbon dioxide to fatty alcohols and hydrocarbons in cyanobacteria. Metab Eng 13:169–176. doi:10.1016/j.ymben.2011.01.001

Teerawanichpan P, Qiu X (2010) Fatty Acyl-CoA reductase and wax synthase from *euglena gracilis* in the biosynthesis of medium-chain wax esters. Lipids 45:263–273. doi:10.1007/s11745-010-3395-2

Tian X, Chen L, Wang J, Qiao J, Zhang W (2013) Quantitative proteomics reveals dynamic responses of *Synechocystis* sp. PCC 6803 to next-generation biofuel butanol. J Proteomics 78:326–345. doi:10.1016/j.jprot.2012.10.002

Uematsu K, Suzuki N, Iwamae T, Inui M, Yukawa H (2012) Increased fructose 1,6-bisphosphate aldolase in plastids enhances growth and photosynthesis of tobacco plants. J Exp Bot 63:3001–3009. doi:10.1093/jxb/ers004

Vignais PM, Billoud B (2007) Occurrence, Classification, and Biological Function of Hydrogenases: An Overview. Chem Rev 107:4206–4272. doi:10.1002/chin.200750259

Vincent KA, Parkin A, Armstrong FA (2007) Investigating and exploiting the electrocatalytic properties of hydrogenases. Chem Rev 107:4366–4413. doi:10.1002/chin.200751261

Voelker TA, Davis HM (1994) Alteration of the specificity and regulation of fatty acid synthesis of *Escherichia coli* by expression of a plant medium- chain acyl-acyl carrier protein thioesterase. J Bacteriol 176:7320–7327

Wang W, Liu X, Lu X (2013) Engineering cyanobacteria to improve photosynthetic production of alka(e)nes. Biotechnol Biofuels 6:69. doi:10.1186/1754-6834-6-69

Weyman PD, Pratte B, Thiel T (2010) Hydrogen production in nitrogenase mutants in *Anabaena variabilis*. FEMS Microbiol Lett 304:55–61. doi:10.1111/j.1574-6968.2009.01883.x

Weyman PD, Vargas WA, Tong Y, Yu J, Maness PC, Smith HO, Xu Q (2011) Heterologous Expression of *Alteromonas macleodii* and *Thiocapsa roseopersicina* [NiFe] Hydrogenases in *Synechococcus elongatus*. PLoS ONE 6, e20126. doi:10.1371/journal.pone.0020126

Wiesenborn DP, Rudolph FB, Papoutsakis ET (1988) Thiolase from *Clostridium acetobutylicum* ATCC 824 and Its Role in the Synthesis of Acids and Solvents. Appl Environ Microbiol 54:2717–2722

Willis RM, Wahlen BD, Seefeldt L, Barney B (2011) Characterization of a fatty acyl-coa reductase from *Marinobacter aquaeolei* VT8: a bacterial enzyme catalyzing the reduction of fatty acyl-coa to fatty alcohol. Biochemistry 50:10550–10558. doi:10.1021/bi2008646

Winters K, Parker PL, Van Baalen C (1969) Hydrocarbons of blue-green algae: geochemical significance. Science 163:467–468. doi:10.1126/science.163.3866.467

Woodger FJ, Badger MR, Price GD (2003) Inorganic carbon limitation induces transcripts encoding components of the co_2-concentrating mechanism in *Synechococcus* sp. PCC7942 through a redox-independent pathway. Plant Physiol 133:2069–2080. doi:10.1104/pp.103.029728

Yao L, Qi F, Tan X, Lu X (2014) Improved production of fatty alcohols in cyanobacteria by metabolic engineering. Biotechnol Biofuels 7:94. doi:10.1186/1754-6834-7-94

Yoshino F, Ikeda H, Masukawa H, Sakurai H (2006) High Photobiological hydrogen production activity of a *Nostoc* sp. PCC 7422 uptake hydrogenase-deficient mutant with high nitrogenase activity. Mar Biotechnol 9:101–112. doi:10.1007/s10126-006-6035-3

Yoshino T, Liang Y, Arai D, Maeda Y, Honda T, Muto M, Kakunaka N, Tanaka T (2014) Alkane production by the marine cyanobacterium *Synechococcus* sp. NKBG15041c possessing the α-olefin biosynthesis pathway. Appl Microbiol Biotechnol 99:1521–1529. doi:10.1007/s00253-014-6286-2

Zhang F, Rodriguez S, Keasling J (2011) Metabolic engineering of microbial pathways for advanced biofuels production. Curr Opin Biotechnol 22:775–783. doi:10.1016/j.copbio.2011.04.024

Zhang F, Carothers JM, Keasling JD (2012) Design of a dynamic sensor-regulator system for production of chemicals and fuels derived from fatty acids. Nat Biotechnol 30:354–359. doi:10.1038/nbt.2149

Zhang Y, Niu X, Shi M, Pei G1, Zhang X, Chen L, Zhang W (2015) Identification of a transporter *Slr0982* involved in ethanol tolerance in cyanobacterium *Synechocystis* sp. PCC 6803. Front Microbiol 6:487. doi:10.3389/fmicb.2015.00487

Zheng YN, Li LL, Liu Q, Yang JM, Wang XW, Liu W, Xu X, Liu H, Zhao G, Xian M (2012) Optimization of fatty alcohol biosynthesis pathway for selectively enhanced production of C12/14 and C16/18 fatty alcohols in engineered *Escherichia coli*. Microb Cell Fact 11:65. doi:10.1186/1475-2859-11-65

Zhou YJ, Buijs NA, Siewers V, Nielsen J (2014) Fatty acid-derived biofuels and chemicals production in *Saccharomyces cerevisiae*. Front Bioeng Biotechnol 2:32. doi:10.3389/fbioe.2014.00032

Zhu XG, de Sturler E, Long SP (2007) Optimizing the distribution of resources between enzymes of carbon metabolism can dramatically increase photosynthetic rate: a numerical simulation using an evolutionary algorithm. Plant Physiol 145:513–526. doi:10.1104/pp.107.103713

Zhu H, Ren X, Wang J, Song Z, Shi M, Qiao J, Tian X, Liu J, Chen L, Zhang W (2013) Integrated OMICS guided engineering of biofuel butanol-tolerance in photosynthetic *Synechocystis* sp. PCC 6803. Biotechnol Biofuels 6:106. doi:10.1186/1754-6834-6-106

Engineering Photosynthetic α-Proteobacteria for the Production of Recombinant Proteins and Terpenoids

Achim Heck and Thomas Drepper

Abstract Phototrophic non-sulfur purple α-proteobacteria are able to harvest sunlight and to fix atmospheric carbon dioxide and dinitrogen. Consequently, these microbes are used as model organisms for the investigation of regulation and activity of the photosynthesis complexes, the ribulose-1,5-bisphosphate carboxylase/oxygenase (RuBisCO) as well as the nitrogenase and hydrogenase enzyme complexes. In addition, this group of prokaryotic phototrophs has emerged as microbial production chassis for the synthesis of recombinant proteins and natural products. To this end, a versatile set of different expression tools has been developed allowing the functional expression of single genes as well as the transfer of complete metabolic pathways. This review provides an overview of different strategies to engineer photosynthetic α-proteobacteria, especially the two most commonly employed representatives *Rhodobacter capsulatus* and *Rhodobacter sphaeroides*, for the production of difficult-to-express proteins and terpenoids. Unique physiological properties of these alternative production hosts are discussed in the context of respective production processes. Furthermore, synthetic biology tools applicable for heterologous gene expression and establishment of combinatorial biosynthetic pathways in phototrophic α-proteobacteria are described. Finally, the potential of phototrophic bacteria in future bioeconomic production routes is briefly discussed.

Keywords *Rhodobacter* • *Rhodospirillum* • *Rhodovulum* • α-Proteobacteria • ICM • Protein expression • Expression vector • Expression of clustered genes • Promoter • Synthetic biology • Membrane proteins • GPCR • Redox proteins • Membrane anchor • Combinatorial biosynthesis • Terpenoids • Carotenoids

A. Heck
Institute of Bio- and Geosciences, IBG-1: Biotechnology,
Forschungszentrum Jülich, 52425 Jülich, Germany
e-mail: a.heck@fz-juelich.de

T. Drepper (✉)
Institute of Molecular Enzyme Technology, Heinrich Heine University
Düsseldorf, Forschungszentrum Jülich, 52425 Jülich, Germany
e-mail: t.drepper@fz-juelich.de

© Springer International Publishing Switzerland 2017

P.C. Hallenbeck (ed.), *Modern Topics in the Phototrophic Prokaryotes*,
DOI 10.1007/978-3-319-46261-5_12

Introduction

The group of α-proteobacteria constitutes the largest and phenotypically most diverse subdivision of the domain bacteria with huge variations in genome size, habitat, and lifestyle (Kersters et al. 2006; Ettema and Andersson 2009). Among the members of the α-proteobacteria, non-sulfur anoxygenic phototrophs and especially the species of the genus *Rhodobacter* (*Rba.*) have drawn particular attention as platform organisms for biotechnological applications. *Rhodobacter* species are metabolically highly versatile as they are able to grow under phototrophic conditions (in the absence of oxygen) by performing anoxygenic photosynthesis and under chemotrophic conditions (in the presence of atmospheric oxygen down to microaerobic concentrations) (Madigan and Jung 2009). They are able to use a wide spectrum of carbon and electron sources (Jaschke et al. 2011) and can fix atmospheric dinitrogen and carbon dioxide, although organic compounds are preferred over CO_2 (Madigan 1995; Romagnoli and Tabita 2009). To maintain these different lifestyles, *Rhodobacter* owns a complex metabolic network with highly specialized enzyme complexes and regulatory mechanisms. Upon reduction of oxygen tension, *Rhodobacter* cells form an intracytoplasmic membrane (ICM) from invaginations of the cytoplasmic membrane which then houses the photosynthesis apparatus (Chory et al. 1984). The photosynthetic complexes, i.e., the reaction centers (RC), the core light harvesting complexes (LHI), and the peripheral light harvesting complexes (LHII), coordinate bacteriochlorophyll *a* and the carotenoids spheroidene and spheroidenone as photopigments. Thus, the pathways generating the respective pyrrole and isoprenoid scaffolds are interesting targets to utilize *Rhodobacter* for the production of vitamin B_{12}, terpenoids, and derivatives thereof. Furthermore, the phototrophic lifestyle offers new approaches to employ these microbes as alternative hosts for the production of otherwise difficult-to-express proteins including membrane proteins and metalloenzymes. Remarkably, due to their unique non-toxic and non-pyrogenic lipopolysaccharides (LPS), *Rba. sphaeroides* and *Rba. capsulatus* can further be used for the production of pharmaceutically relevant compounds as well as additives for nutrition and feed (Strittmatter et al. 1983; Kawata et al. 1999).

This review first describes developed expression tools which are key to engineered *Rhodobacter* as photosynthetic cell factories. Because of the broad spectrum of potential products that can be manufactured in these organisms, the production of membrane proteins and terpenoids is further outlined including the description of specific challenges as well as advantageous features of the alternative hosts.

Expression Tools

In the last decades, many different vectors have been created that can be applied for the *in trans* expression of genes and synthetic operons in phototrophic α-proteobacteria. Most of these recombinant plasmids employ vector backbones that harbor broad-host range origins of replication (ori), thereby allowing the cloning and amplification of the expression constructs in standard laboratory *Escherichia coli* strains. In contrast to *E. coli* and many other Gram-negative expression hosts, no transformation or electroporation protocol for the direct transfer of DNA into phototrophic bacteria such as *Rhodobacter* could be established so far. Therefore, all vector backbones carry a second ori (oriT, also called mob site) which allows the introduction of plasmid DNA via conjugational transfer using an appropriate *E. coli* donor strain (e.g., *E. coli* S17-1, Simon et al. 1983).

Many different intrinsic and recombinant promoters have been used in these bacteria for constitutive or inducible expression of target genes including the prominent P_{lac} (Ind et al. 2009) and P_{T7} promoters (Katzke et al. 2010; Loeschcke et al. 2013) that have been derived from well-known and broadly applied *E. coli* expression systems. In most cases, however, promoters from *Rhodobacter* photosynthesis genes *puc* and *puf* were used for the expression of homologous and heterologous genes. These promoters combine a strong expression with tight control by oxygen availability and can be induced upon oxygen limitation. As it is not possible to cover all plasmids and genetic elements that are available for expression in different genera of phototrophic α-proteobacteria in this chapter, a list of mostly used expression systems referenced throughout the following review is summarized in Table 1.

Established procedures for cloning of appropriate expression vectors and the production of recombinant membrane proteins in *Rba. sphaeroides* have been recently reviewed by Jaschke et al. (2011). Similarly, a detailed description for vector construction and expression of soluble and membrane proteins in *Rba. capsulatus* can be found in Katzke et al. (2012). In the following section we will thus only briefly highlight some of the recently developed synthetic biology tools and engineering approaches to further enhance the properties of *Rhodobacter* species as expression chassis.

The group of Schmidt-Dannert presented a set of optimized BioBrick™ compatible expression vectors suitable for high-level protein production in *Rba. sphaeroides* (Tikh et al. 2014). In this approach, the *puf* promoter region was stepwise optimized by consecutive truncations to reach strong expression levels that can be tuned by changing the oxygen concentrations. Moreover, to achieve high protein yields in *Rba. capsulatus*, a series of expression plasmids (pRho) has been designed (Katzke et al. 2010; Katzke et al. 2012). Among them, a T7 RNA polymerase-based expression system was constructed consisting of the respective expression vector and a *Rba. capsulatus* T7 expression strain (B10S-T7) carrying the corresponding polymerase gene on its chromosome. Here, strong overexpression of target genes can be induced via a fructose-inducible promoter (Duport et al. 1994), which coordinates the expression of the T7 RNA polymerase gene. The same phage polymerase was employed for a synthetic biology tool that facilitates the transfer and concerted expression of clustered genes (Loeschcke et al. 2013) which is explained in more detail within the sections "Production of Complex Redox Proteins and Expression of Clustered Genes" and "Engineering of Recombinant Carotenoid Pathways in Photosynthetic α-Proteobacteria" of this chapter. Another study has targeted the ribosomal binding site (RBS) which apparently also plays a critical role for efficient

Table 1 Properties of vectors suitable for gene expression in phototrophic α-proteobacteria

Promoter	Induction	Host range	Vector backbone	Vector name	Organism[a]	References
P_{puc}	Low O_2 tension	Broad	pRK404[b]	pRKPLHT7	*Rba. sp.*	Ditta et al. (1985), Laible et al. (2009), Erbakan et al. (2014)
			pRK415[b]	pRKCBC1	*Rba. sp.*	Keen et al. (1988), Jones et al. (1992), Fowler et al. (1995)
				pRKpucC	*Rba. sp., Rhv. sul.*	Keen et al. (1988), Wang et al. (2009)
			pLAFR3[c]	pLEK	*Rba. caps.*	Staskawicz et al. (1987), Barbieri et al. (2002)
				pLEK	*Rba. sp., Rhv. sul.*	Staskawicz et al. (1987), Katsiou et al. (1998)
				pLEP	*Rba. caps.*	
				pLΔDE	*Rba. caps.*	
			pBBR1MCS-2	pDEST-E[d]	*Rba. sp.*	Kovach et al. (1995), Bernaudat et al. (2011)
			pBBR1MCS-3	pPUC	*Rsp. rub.*	Kovach et al. (1995), Butzin et al. (2010)
P_{puf}	Low O_2 tension	Broad	pRK404[b]	pRKPLHT1	*Rba. sp.*	Ditta et al. (1985), Laible et al. (2009), Erbakan et al. (2015)
				pRKEH10D	*Rba. sp.*	Ditta et al. (1985), Roy et al. (2008)
			pRK415[b]	pRKY415	*Rhv. sul.*	Keen et al. (1988), Mukoyama et al. (2006)
			pBBRBB (BioBrick™)	pBBRBB-$P_{puf1\text{-}1,200}$	*Rba. sp.*	Vick et al. (2011), Tikh et al. (2014)
				pBBRBB-$P_{puf1\text{-}842}$	*Rba. sp.*	
				pBBRBB-$P_{puf843\text{-}1,200}$	*Rba. sp.*	
$P_{lac/puc}$	IPTG upon low O_2 tension	Broad	pRK415[b]	pRK*lac*I[q]*puc*P*puc*BHis$_{10}$AC	*Rba. sp.*	Keen et al. (1988), Hu et al. (2010), Nie et al. (2015)
P_{lac}	Constitutive	Broad	pBBR1MCS-2	pBBR1MCS-2	*Rba. caps.*	Kovach et al. (1995), Khan et al. (2015)
	IPTG	Broad	pBBR1MCS-2	pBK	*Rba. sp.*	Kovach et al. (1995), Lu et al. (2015)
P_{tac}	IPTG	Broad	pBBR1MCS-2	pMCS	*Rba. sp.*	

$P_{A1/04/03}$	IPTG	Narrow	pMG170	pIND4[e]	*Rba. sp.*, *Rba. caps.*	Inui et al. (2003), Ind et al. (2009), Cheng et al. (2015a), Lu et al. (2015)
P_{trc}	Constitutive, self-regulating together with *lacI*[q]	Broad	pBBR1MCS-2	pBsrid	*Rba. sp.*	Kovach et al. (1995), Lu et al. (2014)
P_{sqr}	Na_2S, O_2 tension	Broad	pRK415[b]	pRK415[b]	*Rba. caps.*	Keen et al. (1988), Han and Perner (2016)
P_{dor}	O_2 tension, DMSO, molybdate	Broad	pRK415[b]	pRK-sorex	*Rba. caps.*	Keen et al. (1988), Kappler and McEwan (2002)
P_{coxII}	O_2 tension	Broad	pRK415[b]	pMEG	*Rba. sp.*	Keen et al. (1988), Graichen et al. (1999)
P_{crtD}	Low O_2 tension	Broad	pMHE	pMHEcrt	*Rba. caps.*	Fodor et al. (2004)
P_{nif}	Nitrogen depletion	Broad	pMHE	pMHEnif	*Rba. caps.*	
P_{aphII}	Constitutive	Broad	pBBR1MCS	pRhokHi	*Rba. caps.*	Kovach et al. (1994), Katzke et al. (2010), Katzke et al. (2012)
				pRhokS	*Rba. caps.*	
P_{T7}	Fructose (with *Rba. caps.* B10S-T7)	Broad	pBBR1MCS	pRhotHi	*Rba. caps.*	
				pRhotS	*Rba. caps.*	
P_{fru}	Fructose	Broad	pBBR1MCS	pRhofHi-2	*Rba. caps.*	Kovach et al. (1994), Duport et al. (1994); Loeschcke and Drepper, unpublished

(continued)

Table 1 (continued)

P_{T7}	IPTG or fructose (host-dependent)	Broad	Integrative	TREX	*Rba. caps.*	Loeschcke et al. (2013)
				ΩSp-P_{T7}	*Rba. caps.*	Arvani et al. (2012)
P_{lac}	IPTG	Broad	pML5	pML5-P_{lac}T7[f]	*Rba. caps.* (with TREX or ΩSp-P_{T7})	Labes et al. (1990), Arvani et al. (2012), Loeschcke et al. (2013)
P_{fru}	Fructose	Broad	pML5	pML5-P_{fru}T7[f]		

[a]*Rba. sp.*: *Rhodobacter sphaeroides*; *Rba. caps.*: *Rhodobacter capsulatus*; *Rhv. sul.*: *Rhodovulum sulfidophilum*; *Rsp. rub.*: *Rhodospirillum rubrum*
[b]Derived from naturally occurring low copy plasmid RK2 (Ditta et al. 1980)
[c]Derived from recombinant plasmid pRK290 (Ditta et al. 1980)
[d]Cloning using the Gateway system seems to have a negative effect on protein formation due to the additionally fused amino acid sequence of the attB1 attachment site to the desired target gene product
[e]High segregational stability in *Rba. sp.* but not in *Rba. caps.*(Nybo et al. 2015)
[f]Plasmids for *in trans* expression of T7 RNA polymerase gene; designed for application together with TREX and ΩSp-P_{T7}

protein accumulation (Lu et al. 2014). Here, a set of five different RBS with various strengths were used as suitable biobricks to adjust translation of the repressor LacI for a self-regulating expression system. This system could successfully be applied to optimize coenzyme Q_{10} production in *Rba. sphaeroides* (section "Recent Advances in Producing Terpenoids with *Rhodobacter*"). Finally, it is well known that many heterologous genes are difficult to overexpress due to the different codon usage of the original and the production host (e.g., reviewed by Elena et al. 2014). Therefore, a plasmid-borne system was recently developed encompassing eight rare *Rba. sphaeroides* tRNA genes (Cheng et al. 2015a). The results of these studies clearly demonstrated that co-expression of those tRNA genes can significantly elevate the final yield of rare-codon containing target proteins.

In summary, the described tools and techniques for heterologous expression in phototrophic α-proteobacteria already have reached an outstanding level of development. Together with the increasing knowledge about their unique physiology and the decoded genome sequences, these bacteria have the potential to become future model organisms for synthetic biology and biotechnological approaches. In the following, the applications of phototrophic α-proteobacteria expression toolkits for the production of membrane proteins and terpenes are described in detail.

Target Proteins

Production of Membrane Proteins

Membrane proteins play an essential role in the physiology of all living cells. They are key players in signal transduction, response to external stimuli, cell–cell interaction, energy metabolism, and transport phenomena including uptake of nutrients and excretion of wastes (Elofsson and von Heijne 2007; Laible et al. 2011). Therefore, it is little surprising that a significant proportion of genes, approximately 30 % of known prokaryotic as well as eukaryotic genomes, encodes membrane proteins (McLuskey et al. 2010). Moreover, about 60–70 % of all currently marketed pharmaceuticals directly or indirectly target membrane proteins (Lundstrom 2007; Yin 2008). Among these membrane proteins, rhodopsin-like G protein-coupled receptors (GPCRs) constitute the largest class of drug targets, followed by nuclear receptors and ligand-gated ion channels (Overington et al. 2006). Consequently, the biochemical and structural characterization of membrane proteins is an important task in fundamental sciences as well as drug discovery research.

Despite their tremendous importance, the function of most membrane proteins has not been assigned yet. Likewise, the number of approximately 650 known unique membrane protein structures[1] is remarkably low compared to the number of nearly 115,000

---

[1]According to Stephen White, Department of Physiology and Biophysics, School of Medicine, University of California at Irvine: http://blanco.biomol.uci.edu/mpstruc/, October 10th, 2016.

deposited protein structures[2] in the Protein Data Bank of the Research Collaboratory for Structural Bioinformatics (RCSB; Berman et al. 2000). This discrepancy is attributed to the fact that membrane proteins only represent about 1 % of the total cellular protein content under physiological conditions (Bernaudat et al. 2011). Thus, it is challenging to isolate large amounts of membrane proteins from natural sources. Researchers have therefore put much effort in establishing heterologous production approaches to gain sufficient amounts of functional membrane proteins for biochemical characterization and structural elucidation. Therefore, many different systems enabling the heterologous production of large amounts of a desired membrane protein are available. Bacteria-based expression systems, including the most widely used *E. coli* systems (Sahdev et al. 2008), are easy to handle and operate at low costs. Additionally, cell-free expression systems based on *E. coli* extracts have emerged over the last years as alternative production platforms for determination of membrane protein structures (Mus-Veteau et al. 2014). However, difficulties are encountered as most of the in vivo expression systems suffer from insufficient membrane space to incorporate larger amounts of heterologously produced membrane proteins (Arechaga et al. 2000). Although certain *E. coli* strains are known to form intracellular membrane proliferations and incorporate some heterologously produced membrane proteins, limited membrane surface remains one of the major bottlenecks (Miroux and Walker 1996; Hattab et al. 2010). Moreover, these systems usually lack efficient protein translocation machineries to correctly fold and translocate heterologous membrane proteins (Essen 2002). As a result, overexpression often leads to cell death or formation of insoluble aggregates in which inactive membrane proteins accumulate in the so-called inclusion bodies. Refolding and subsequent solubilization is, at least in most cases, neither trivial nor straightforward (Laible et al. 2004). The in vitro systems, on the other hand, are usually difficult to handle and to maintain. Moreover, compared to the in vivo systems they are usually more expensive and thus are not used for biotechnological applications. Therefore, novel hosts are urgently needed to circumvent the above mentioned bottlenecks and thus allow for the high-level synthesis of functional membrane proteins.

Photosynthetic α-Proteobacteria as Novel Expression Hosts for the Production of Membrane Proteins

Photosynthetic α-proteobacteria are especially suited to cope with the challenges associated with the production of membrane proteins. They form an ICM system from invaginations of the cytoplasmic membrane in response to changes in O_2 availability (Fedotova and Zeilstra-Ryalls 2014). The ICM enlarges the surface of the cellular membrane by up to one order of magnitude (Adams and Hunter 2012; Erbakan et al. 2015), forming a cell compartment which naturally harbors the proteins of the

[2] According to Research Collaboratory for Structural Bioinformatics (RCSB) Protein Data Bank (PDB) statistics: http://www.rcsb.org/pdb/statistics/holdings.do, October 10th, 2016.

photosynthetic apparatus including the RC, LHI and LHII, and all associated redox and energy transferring cofactors (Kiley et al. 1988). Therefore, the group of photosynthetic α-proteobacteria is, in contrast to standard hosts such as *E. coli*, naturally equipped with a large storage capacity for membrane proteins. Moreover, the ability of this group of bacteria to grow under aerobic as well as phototrophic conditions offers another very interesting feature. During the adaptation process from the aerobic to the phototrophic growth mode, the developing membrane of the ICM contains significantly elevated levels of general membrane assembly factors (Woronowicz et al. 2013; Niederman 2016) which assist in incorporating the protein complexes of the photosynthetic apparatus. This machinery, however, can also contribute to efficiently incorporate foreign membrane proteins into the emerging membrane vesicles. These aspects render facultative photosynthetic bacteria promising host candidates for heterologous production of this class of difficult-to-express proteins. The ICM vesicles can even easily be harvested after cell disruption via density gradient ultracentrifugation and are available for downstream applications after recovery from the gradient. In the following, selected achievements in the production of homologous and heterologous membrane proteins in *Rhodobacter* species, as well as in the closely related *Rhodospirillum rubrum* are further discussed.

Production of Homologous Membrane Proteins in *Rba. sphaeroides* and *Rba. capsulatus*

Due to their capability to conduct photosynthesis, *Rba. sphaeroides* and *Rba. capsulatus* have been historically employed to produce homologous membrane proteins for structural studies. Hence, the crystal structures of membrane proteins such as porin (Weiss and Schulz 1992, 2POR) or cytochrome bc_1 (Berry et al. 2004, 1ZRT) of *Rba. capsulatus* as well as cytochrome bc_1 (Esser et al. 2006, 2FYN), cytochrome *c* oxidase (Svensson-Ek et al. 2002, 1M56; Qin et al. 2006, 2GSM) or the photosynthetic reaction center of *Rba. sphaeroides* (e.g., Allen et al. 1987, 4RCR; Chang et al. 1991, 2RCR; Chirino et al. 1994, 1PSS; Ermler et al. 1994, 1PCR; Stowell et al. 1997, 1AIJ; Katona et al. 2003, 1OGV; Koepke et al. 2007, 2J8C, or Saer et al. 2014, 4N7K) were solved after production and purification of the respective proteins from the homologous hosts.

More recently, both species have also been used for the high-level production of homologous membrane proteins. Erbakan and colleagues, for instance, reported the production of the homologous aquaporin (RsAqpZ) in *Rba. sphaeroides* (Erbakan et al. 2014). Approximately 20 mg/l culture of fully active, highly permeable aquaporin could be produced and characterized. In a second report, the same group describes the production of homologous cellulose synthase enzyme complex BcsAB in *Rba. sphaeroides* (Erbakan et al. 2015). This membrane-associated protein, which is composed of the BcsA and BcsB subunits, was purified to yield 1.5 mg active protein per liter culture.

Examples for the Production of Heterologous Membrane Proteins in *Rhodobacter*

Various studies applied heterologous membrane protein expression in *Rhodobacter*. In the following, examples are presented, highlighting applied expression and engineering strategies. In one of the first studies, Fowler and coworkers utilized *Rba. sphaeroides* for the heterologous production of the peripheral light harvesting complex LHII of *Rhodopseudomonas acidophila* and *Rubrivivax gelatinosus* (Fowler et al. 1995). To this end, the authors used different *Rba. sphaeroides* mutants carrying deletions within the photosynthesis gene clusters and restored their function by introducing the respective plasmid-encoded heterologous genes. These plasmids originated from plasmid RK2 described by Ditta et al. (1980) which has been modified manifold meanwhile (cf. Table 1). For example, plasmid-borne gene expression can nowadays be conducted using several photosynthetic promoters, such as *puc* or *puf* promoters. Here, the induction of heterologous gene expression together with ICM formation is mainly triggered by the absence of oxygen. In another study, Wang and colleagues utilized *Rba. sphaeroides* strain DD13 in which the majority of the photosynthetic genes had been deleted by transposon mutagenesis (Simon et al. 1983) for heterologous expression of photosynthesis proteins (Wang et al. 2009). The authors managed to assemble the LHII of *Rhodovulum sulfidophilum* in its active conformation in *Rba. sphaeroides*. Following a similar approach in *Rba. capsulatus*, Katsiou et al. (1998) and Barbieri et al. (2002) functionally assembled the LHII of *Rhv. sulfidophilum* in a *Rba. capsulatus* LHII deletion strain.

In addition, *Rba. capsulatus* was also used to analyze the activities of three sulfide-quinone reductases (SQRs) from *Sulfurimonas denitrificans* in analogous complementation studies (Han and Perner 2016). Here, an accordingly modified *Rba. capsulatus sqr*-deletion strain was employed. By expression of all three enzyme homologs, SQR function could successfully be restored. Moreover, Tikh and colleagues managed to utilize *Rhb. sphaeroides* and the BioBrick™ system to produce the membrane model protein proteorhodopsin (Tikh et al. 2014). However, the yield was less than 1 mg of cofactor containing assembled protein after solubilization and purification. Remarkably, when using a *Rba. sphaeroides* ΔRCLH deletion strain that lacks the ability to form LHI, LHII, and RC (Tehrani and Beatty 2004), approximately 3 mg proteorhodopsin per liter expression culture were produced, indicating that the removal of the photosynthetic machinery may provide unoccupied space within the host membrane that can be used to incorporated larger amounts of heterologous membrane protein.

Besides the above described in vivo complementation studies, different eukaryotic membrane proteins could successfully be produced and subsequently characterized in *Rhodobacter* cells, enriched membrane fractions, or as purified proteins, respectively. For example, Bernaudat and coworkers were able to express the quinone oxidoreductase ceQORH from *A. thaliana* (<0.1 mg/l culture), the human apoptosis regulator Bcl-xL (<0.1 mg/l culture), and the uncoupling protein UCP1 from *Rattus norvegicus* (0.1–0.5 mg/l culture) by using a modified pBBR1MCS broad-host range plasmid harboring the *puc* promoter for expression (Bernaudat et al. 2011). More importantly, *Rhodobacter* also proved suitable as alternative expression host for the synthesis of human membrane proteins. For example, Roy and coworkers employed *Rba. sphaeroides* to functionally produce human GPCRs (Roy et al. 2008). Therefore, target genes encoding the adenosine A2a receptor (A2aR), the angiotensin AT1a receptor (AT1aR), and the bradykinin B2 receptor (B2R) were cloned into a derivative of the aforementioned RK2 plasmid under the control of the photosynthetic *puf* promoter. In order to provide sufficient space for the accommodation of the foreign receptor proteins, the authors also utilized the *Rba. sphaeroides* photosynthesis mutant strain DD13 that has been described above. Surprisingly, the use of specific vector variants that enabled co-expression of target proteins with the two proteins PufL and PufM of the photosynthesis RC yielded higher amounts of GPCR proteins. The activities of the GPCRs were analyzed using radio-ligand binding assays allowing to estimate amounts of approximately 0.01 mg of functional B2R and 0.1 mg of AT1aR produced per liter culture, which is significantly more than could be retrieved from native tissues. Similar expression levels were observed for A2aR. In another study, *Rba. sphaeroides* was employed to produce aquaporin 9 (hAQP9) and the tight junction protein occludin (Occ) from *Homo sapiens* (Erbakan et al. 2015). The authors were able to gain titers of 7.5 mg/l culture for purified Occ and 0.5 mg/l for purified hAQP9. Moreover, the functionality of hAQP9 was successfully proven. Finally, the functional expression of microbial and human membrane proteins including membrane-bound enzymes (e.g., P450 monooxygenases), GPCRs, and GPCR-like proteins was also successfully conducted in *Rba. capsulatus* (Heck, Özgür, Malach, Jaeger and Drepper; unpublished data).

***Rhodospirillum rubrum*: An Emerging Host for the Production of Heterologous Membrane Proteins**

In addition to *Rba. sphaeroides* and *Rba. capsulatus*, *Rhodospirillum rubrum*, a third species of the α-proteobacteria group, was recently established as expression host for the production of heterologous membrane proteins. Like *Rhodobacter*, *Rsp. rubrum* is also a facultative photosynthetic bacterium that forms an ICM. Therefore, Butzin and colleagues created deletion mutants lacking the ability to form the photosystem (Butzin et al. 2010). By using pBBR1MCS-3 broad-host range plasmid derivatives, the mechanosensitive ion channel MscL (22–23 mg/l culture) and the heme peroxidase CycB (6–7 mg/l culture) of *Pseudomonas aeruginosa* as well as the potassium ion channel KcsA (13–14 mg/l) of *Streptomyces lividans* could successfully be synthesized and translocated into the ICM. CycB activity and its spectral characteristics indicated that the enzyme assembled with the heme cofactor and thus reached its native conformation. While the activity of MscL and KcsA was not analyzed, the apparent oligomerization of the proteins suggests that they folded into the correct conformation.

According to the described case studies, *Rba. sphaeroides, Rba. capsulatus*, and *Rsp. rubrum* can provide homologous as well as heterologous membrane proteins and, in particular, novel pharmaceutically relevant human drug targets for future applications. Moreover, the target proteins can directly be characterized in a natural biological membrane or can alternatively be solubilized and subsequently purified for in vitro studies.

Artificial Anchoring of Soluble Proteins to the ICM of Rhodobacter

In addition to the storage of native integral membrane proteins, the ICM of *Rhodobacter* can also be targeted for soluble proteins that are fused to an appropriate membrane anchor. In this context, two different short membrane-associated polypeptides of the photosynthesis complex, namely the β subunit of the LHII complex (Nie et al. 2015) as well as PufX that is known to form the photosynthetic core complex together with the reaction center and LHI (Sznee et al. 2014; Olsen et al. 2014) have been used in *Rba. sphaeroides* as appropriate membrane anchors.

Nie and coworkers succeeded to functionally produce the human neutrophil peptide 3 (HNP3) in *Rba. sphaeroides* by fusing it to the β subunit of the LHII complex (Nie et al. 2015). To this end, a recombinant *puc* promoter was used which additionally contains the *E. coli lac* operator. The novel hybrid promoter therefore allows tight, LacI-mediated repression which can be relieved by the addition of isopropyl-β-d-thiogalactopyranosid (IPTG). Using this strategy, the soluble HNP3 protein was produced as a fusion protein with the LHII-β-polypeptide and thereby was effectively linked to the ICM. Remarkably, the HNP3-LHII-β fusion protein, although integrated into the ICM, fully retained its activity. Furthermore, formation of LHII develops characteristic absorption maxima at approximately 800 nm and 850 nm. Therefore, the synthesis of recombinant fusion proteins and their subsequent incorporation into the cell membrane could rapidly be detected via the elevated absorption at these wavelengths (Zhao et al. 2011, Cheng et al. 2015b).

In contrast to proteins of the LHII, PufX is a component of the *Rhodobacter* RC-LHI complex. The PufX polypeptide consists of 78 amino acids in *Rba. capsulatus* and 82 amino acids in *Rba. sphaeroides*, respectively (Youvan et al. 1984; Lee et al. 1989). The N-terminus of the *Rba. sphaeroides* PufX polypeptide is exposed on the cytoplasmic side of the membrane, equivalent to that of the α- and β-polypeptides of the LHI and LHII which are reported to span the membrane once (Pugh et al. 1998; Holden-Dye et al. 2008). Therefore, PufX is also suited as fusion partner to immobilize soluble proteins in the *Rhodobacter* ICM. By using this membrane anchor, Lu and coworkers could show that the artificial sequestering of soluble enzymes to the membrane can assist in bringing the biocatalyst in close proximity to hydrophobic substrates as demonstrated in *R. sphaeroides* for quinone-modifying enzymes (Lu et al. 2015, see also section "Recent Advances in Producing Terpenoids with *Rhodobacter*" for more detail).

Production of Complex Redox Proteins and Expression of Clustered Genes

Many enzymes require complex cofactors and prosthetic groups to achieve active conformation and therefore to warrant their proper metabolic functions. As described for membrane proteins, heterologous protein production is the method of choice to enable detailed studies. However, the heterologous proteins have to be equipped with the appropriate cofactors and/or prosthetic groups to achieve their active conformation. By using standard expression hosts, high-level production of such cofactor-dependent enzymes can lead to the formation of target proteins with a large proportion accumulating as apoproteins without cofactor, in contrast to cofactor-loaded holoproteins. In this context, *Rhodobacter* is able to produce most types of metal-containing prosthetic groups including the iron-containing tetrapyrrole heme, numerous different iron–sulfur clusters, the molybdopterin, and the *bis*-molybdopterin guanine dinucleotide as well as more specific cofactors such as the FeMoco and FeFeco of the conventional and alternative nitrogenases, the nickel-containing [NiFe] cofactors of the uptake hydrogenase, and the cobalt-containing tetrapyrrole cobalamin (Leimkühler et al. 1998; Solomon et al. 1997; Zappa et al. 2010; Masepohl and Hallenbeck 2010; Vignais and Billoud 2007). This wealth of different metal-cofactors renders the alternative expression hosts useful for the functional expression of metalloproteins. Furthermore, the production of many of the named cofactors can even be triggered in *Rhodobacter* by applying appropriate cultivation condition. For instance, synthesis of FeS clusters, FeMoco and [NiFe] cofactor is significantly boosted or even exclusively induced under diazotrophic growth conditions (Vignais 2009, Masepohl et al. 2002).

In addition to the cofactor supply, a crucial aspect for functionality of many redox enzymes is their location in the periplasmic space. Therefore, translocation of theses enzymes is essential which can be conducted in *Rhodobacter* in case of readily assembled proteins via the twin-arginine-translocation pathway (Tat; Müller and Klösgen 2005; Lindenstrauß and Brüser 2006) as well as for unfolded proteins by the general secretory system (Sec). To demonstrate the functional assembly and secretion of complex heterologous redox proteins in *Rba. capsulatus*, Kappler and McEwan expressed the sulfite:cytochrome *c* oxidoreductase (SorAB) from *Starkeya novella* (Kappler and McEwan 2002). The active SorAB enzyme requires both a molybdopterin-containing subunit (SorA) that is translocated via the Tat pathway and a *c*-type cytochrome heme-containing subunit (SorB) which is translocated by the Sec system into the periplasmic space, where the enzyme complex is finally assembled. The plasmid pRK-sorex (again a derivative of the vector RK2) was used to express the genes *sorA* and *sorB* from the homologous dimethyl sulfoxide reductase promoter P_{dor} (Shaw et al. 1996). Accumulation of the active enzyme complex in the periplasm could be shown at a concentration of approximately 0.6 mg/l culture. The biochemical properties of the recombinant SorAB were nearly identical to those of the native enzyme. Therefore, the production of active SorAB thus included (a) the concerted translocation of different subunits of the enzyme complex through different secretion pathways, (b) the correct folding of the subunits including (c) the incorporation of respective cofactors as well as (d) the final assembly in *Rba. capsulatus*.

Other studies revealed that the flavocytochrome *c*-sulfide dehydrogenases (FCSD) from *Ectothiorhodospira vacuolata* and *Allochromatium vinosum* could be produced in *Rba. sphaeroides* and *Rba. capsulatus* by the expression of their respective genes from broad-host range plasmid pGV910-derivatives under the control of the *A. vinosum rbcA* RuBisCO-promoter, yielding amounts of 0.1–0.5 mg holoprotein per liter culture (de Smet et al. 2001). Similar to the before mentioned example, functional expression of FCSD required its translocation via the Sec pathway and a correct assembly in the periplasm. Thus, both reports emphasize the application of *Rhodobacter* for the production and translocation of enzymes with complex cofactors.

In other cases, however, complex redox enzymes are assembled in large multi-subunit proteins or require the expression of a number of accessory genes that encode, e.g., specific chaperones or further cofactors. The biosynthesis of methylamine dehydrogenase (MADH) from *Paracoccus denitrificans* represents an interesting example, where four accessory genes need to be expressed in addition to the structural genes *mauA* and *mauB* (Graichen et al. 1999). While one of the respective gene products (MauD) is probably required for the specific processing of disulfide bonds, MauE appears to be a membrane protein and MauG was predicted to be a heme-bearing peroxidase (Chistoserdov et al. 1994; van der Palen et al. 1997). However, all these gene products play an essential role in translocation and assembly of the MADH small subunit or are involved in the biosynthesis of the required cofactor tryptophan tryptophylquinone (TTQ) which appears to be one of the few cofactors that are not provided by *Rba. sphaeroides*. To achieve the co-expression of all *mau* genes in *Rba. sphaeroides*, a broad-host range expression plasmid was used allowing for the concerted gene expression from the cox_{II} promoter derived from the cytochrome oxidase subunit II-encoding gene (Graichen et al. 1999). After heterologous expression, the authors were able to localize MADH exclusively in the periplasm of the host organism. Spectroscopic, physical, kinetic, and redox properties were identical to the properties of the enzyme produced in the homologous host.

Other multi-domain structures as well as many biosynthetic pathways are frequently organized in even larger gene clusters that can easily reach a size of up to 100 kb (Wenzel and Müller 2005). The functional expression of such clustered genes is, however, challenging since it requires an expression system that facilitates the concerted transcription of numerous genes irrespective of their natural promoters and terminators. To accomplish this task, the uptake hydrogenase gene cluster of *Rba. capsulatus* was chosen as a proof of concept (Drepper et al. 2005). The uptake hydrogenase enables the bacterium to utilize H_2 as an electron donor and is perfectly suited as candidate for cluster expression studies. The enzyme complex is composed of the two subunits HupS, which carries three [Fe-S] clusters as cofactors, and HupL, which binds the hydrogenase-specific [NiFe] cluster (Frey 2002). All hydrogenase genes, i.e., the two structural genes *hupS* and *hupL* as well as four regulatory genes and 13 accessory genes (required for cofactor synthesis and enzyme assembly), are clustered in a 25-kb genomic region (Vignais et al. 2000) encompassing different transcriptional units. To overexpress the uptake hydrogenase gene cluster of *Rba. capsulatus,* the T7 RNA polymerase was used assuming that this processive phage enzyme ignores bacterial transcription termination sites and produces large transcripts encompassing all hydrogenase genes. To prove this assumption, an interposon cassette containing the omega fragment of plasmid pHP45Ω (Prentki and Krisch 1984) and the T7 promoter of the vector plasmid pET22b were inserted into the *hupV* gene. This gene codes for the large subunit of the regulatory sensor hydrogenase and is involved in H_2-dependent induction of uptake hydrogenase synthesis (Elsen et al. 1996). Therefore, the integration of the interposon cassette into *hupV* resulted in a *Rhodobacter* strain where the expression of all uptake hydrogenase genes is solely controlled by the T7 promoter. The T7 RNA polymerase was then expressed from a broad-host range plasmid using the fructose-inducible promoter P_{fru} (Daniels et al. 1988, Duport et al. 1994). Analysis of hydrogenase transcripts subsequently corroborated that the T7 RNA polymerase is in principle able to fully transcribe long gene regions consisting of multiple transcriptional units (Arvani et al. 2012). Meanwhile, the idea of an expression tool that allows the concerted expression of large gene clusters with the aid of the T7 RNA polymerase was realized by the creation of the pathway transfer and expression system (TREX; Loeschcke et al. 2013). The TREX system can not only be utilized to produce complex multi-subunit redox enzymes but also to express large pathways of secondary metabolites in *Rhodobacter*. Hence, the TREX system is discussed in more detail in section "Engineering of Recombinant Carotenoid Pathways in Photosynthetic α-Proteobacteria."

Terpenoids

Phototrophic Bacteria as Chassis for Terpenoid Production

Terpenoids (also referred to as terpenes or isoprenoids) are the largest and most diverse class of natural products with more than 40,000 structurally different compounds (Breitmaier 2006; Marienhagen and Bott 2013; Tholl 2015; Schrader and Bohlmann 2015). Members of this class of metabolites are mainly produced by plants, where they are either involved in basic cellular processes (e.g., the regulation of membrane fluidity or photosynthetic metabolism) or exhibit specialized functions essential for the interaction of plants with their respective environment such as the defense against pathogens and herbivores (Vranová et al. 2012; Tholl 2015). Many terpenoids have important applications such as pharmaceuticals, nutraceuticals, agrochemicals, flavors, fragrances, and colorants and thus possess a tremendous value for the pharmaceutical and biotechnological industry. The importance of terpenoids for biomedical applications became particularly obvious in the year 2015, when Prof. Youyou Tu received the Nobel Prize in Physiology or Medicine for her discovery of the sesquiterpene artemisinin, a natural antimalarial drug isolated from *Artemisia annua* (Kong and Tan 2015). In times of steadily emerging numbers of sequenced genes encoding new terpene biosynthesis routes, constant effort is put into the refinement and development of techniques to transfer, express, and engineer entire pathways with the help of fast growing, easy-to-handle, and metabolically suited microbes. In this part of the review, we thus will describe the current status and new developments in using phototrophic α-proteobacteria as alternative terpene production hosts.

Biosynthesis of Terpenoids

Terpenoids can be classified based on the number of carbon atoms into monoterpenes (C_{10}), sesquiterpenes (C_{15}), triterpenes (C_{30}), and tetraterpenes (i.e., carotenoids, C_{40}). In contrast to the structural diversity of terpenoid end products, the biosynthetic routes leading to plant terpenoid scaffolds are organized rather simple, since they are all derived from the same isoprene C_5 unit isopentenyl pyrophosphate (IPP) and its isomer dimethylallyl pyrophosphate (DMAPP) (Fig. 1). Most bacteria, including phototrophic α-proteobacteria and cyanobacteria, use the 2-methyl-D-erythritol 4-phosphate (MEP) pathway for IPP and DMAPP synthesis (Vranová et al. 2013). This pathway is fed by pyruvate and glyceraldehyde-3-phosphate (GAP) that are, in phototrophic bacteria, both directly derived from CO_2 fixation (Calvin–Benson cycle). Beside the MEP pathway, plants and a minor number of bacteria additionally harbor a second, the so-called mevalonate (MVA) pathway starting with acetyl-CoA to build the respective isoprenoid precursors (Lombard and Moreira 2011). In the following catalytic steps, IPP and DMAPP are condensed to geranyl pyrophosphate (GPP, C_{10}) which is sequentially further converted with IPP into farnesyl pyrophosphate (FPP, C_{15}) and geranyl geranyl pyrophosphate (GGPP, C_{20}). For mono-, sesqui- and diterpenes, the respective linear C_{10}, C_{15}, and C_{20} prenyl pyrophosphate precursors are subsequently rearranged by terpene synthases (TPSs) leading to the formation of terpene backbones with (multiple) ring structures. Tri- and tetraterpenes, however, are not directly synthesized via cyclisation of the respective pyrophosphate precursors. For the biosynthesis of triterpenes, a further two-step catalytic reaction is required prior to cyclisation. First, two molecules of FPP are condensed to squalene by the squalene synthase (SQS). The consecutive second step is the epoxidation of squalene catalyzed by the squalene epoxidase (SQE). Subsequently, the resulting shared key triterpene precursor 2,3-oxidosqualene is further converted by different oxidosqualene cyclases (OSCs) resulting in the formation of numerous sterol and triterpene backbones (Thimmappa et al. 2014; Sawai and Saito 2011). Analogously, the synthesis of tetraterpenes, which is catalyzed by enzymes of the carotenoid pathway, requires the synthesis of the common precursor phytoene, which is formed by a phytoene synthase (CrtB)-catalyzed condensation reaction of two GGPP molecules. Finally, all terpene scaffolds can be further modified, for example, by P450 monooxygenases, resulting in the above mentioned vast structural diversity within this class of natural compounds.

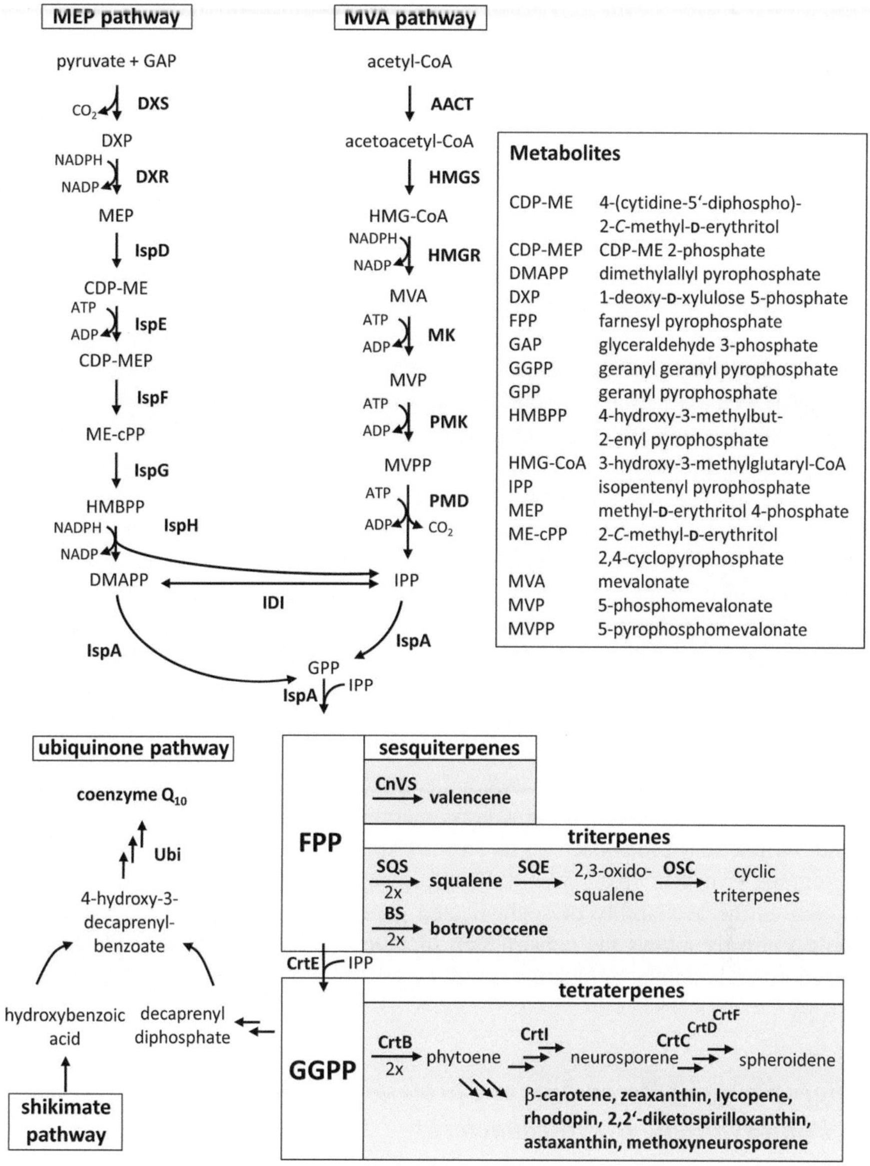

Fig. 1 Schematic overview of terpenoid pathways established and engineered in photosynthetic α-proteobacteria

Unique Properties of Phototrophic Bacteria for Combinatorial Terpenoid Biosynthesis

The isolation from the original or genetically modified plants is usually hampered by slow plant growth and low abundance of target compounds in combination with a complex composition of naturally occurring secondary metabolites. In addition, the availability, quality, and prizes of required feedstocks are subjected to environmental, seasonal, and regional variations. Therefore, production of terpenoids in phototrophic bacteria is a promising alternative to plant-based production routes. These microbial hosts exhibit an intrinsic capability of producing C_{40} carotenoids in high amounts which provides an optimal metabolic background for engineered terpene production. Photosynthetic α-proteobacteria can furthermore optionally use solar energy, carbon dioxide as well as renewable organic carbon sources and even atmospheric dinitrogen in order to gain fast growth, short production times, and high production yields in variable cultivation scales (Davies et al. 2015). In this context, it is worth to note that carotenoids play an essential role in protecting the cell against photooxidative stress that occurs exclusively in the presence of O_2 during photosynthesis by the formation of reactive oxygen species (ROS) (Fraser et al. 2001; Glaeser and Klug 2005). Since the photoprotective function of carotenoids is not essential in anoxygenic phototrophic bacteria in the absence of molecular oxygen, carotenoid pathways can much easier be utilized for the combinatorial biosynthesis of high-value terpenes in these organisms as compared to cyanobacteria, for instance (Frigaard 2016). Furthermore, one limitation in the efficient terpene biosynthesis is posed by typically limited membrane space available for storage of both, membrane-associated biosynthetic enzymes as well as hydrophobic intermediates and end products. *Rhodobacter* species offer a strong advantage here providing a large membrane space as discussed in section "Photosynthetic α-Proteobacteria as Novel Expression Hosts for the Production of Membrane Proteins." In addition, the genetic accessibility of these microorganisms as well as the availability of sophisticated genetic tools (see section "Expression Tools") already allows the construction of tailor-made phototrophic cell factories according to individual production needs.

Engineering of Recombinant Carotenoid Pathways in Photosynthetic α-Proteobacteria

Phototrophic α-proteobacteria have drawn considerable interest as alternative production hosts for various biotechnological applications. However, although this group of organisms offers a tremendous potential for the bio-production of high-value terpenoids, until today, only a small number of species have already been used for this purpose. Combinatorial biosynthesis of foreign isoprenoids in phototrophic α-proteobacteria was first successfully applied for different carotenoids. This class of natural compounds is broadly used in industry as feed and food additives. Moreover, their health-related beneficial properties further raise the demand for structurally diverse carotenoids for nutraceutical and pharmaceutical applications (Sandmann 2015).

First attempts to produce industrially relevant C_{40} carotenoids have been conducted in species of the genus *Rhodobacter, Rhodovulum*, and *Rhodospirillum* which naturally accumulate carotenoids in their cytoplasmic membranes in high amounts during O_2 limitation under non-phototrophic and phototrophic conditions (Bélanger and Gingras 1988; Gregor and Klug 1999; Elsen et al. 2005). In the 1990s of the last century, Hunter and colleagues were among the first who engineered *Rba. sphaeroides* in order to synthesize β-carotene and zeaxanthin (Fig. 1). The authors could demonstrate that different mutations in the intrinsic spheroidene/spheroidenone biosynthesis genes in combination with the heterologous expression of respective carotenoid (*crt*) genes isolated from the non-phototrophic bacterium *Pantoea agglomerans* (formerly *Erwinia herbicola*) could implement effective redirection of terpene biosynthesis. Interestingly, the analysis of pigment composition showed that *Rba. sphaeroides* was able to form LH complexes that incorporated the heterologously produced photopigments (Hunter et al. 1994). A similar approach was used to redirect the natural carotenoid biosynthesis route of *Rba. sphaeroides* to the spirilloxanthin pathway. Here, the host desaturase CrtI, which catalyzes the 3-step desaturation of phytoene to neurosporene, was replaced by the CrtI homologue from *P. agglomerans*. Expression of this enzyme in different *Rba. sphaeroides crt* mutants, in turn, led to the accumulation of lycopene, rhodopin, and 2,2′-diketospirilloxanthin, respectively (Garcia-Asua et al. 2002; Chi et al. 2015). In addition, the marine photosynthetic α-proteobacterium *Rhv. sulfidophilum* was used for the production of astaxanthin (Mukoyama et al. 2006). Similar to the previous studies with *Rba. sphaeroides*, the interruption of host-specific *crt* genes resulting in the accumulation of phytoene was key to achieve elevated levels of the foreign carotenoid. In the mentioned study, the corresponding *crt* genes from *Erythrobacter longus* were used for combinatorial biosynthesis of the new carotenoid.

The successful reconstitution of a recombinant terpene pathway, however, not only depends on an appropriate host bacterium with advantageous metabolic properties, but also requires the stable maintenance and functional expression of all functionally coupled genes within the chosen host. To this end, the above mentioned synthetic biology expression tool TREX (Loeschcke et al. 2013; section "Production of Complex Redox Proteins and Expression of Clustered Genes") was developed to facilitate the cloning, transfer, and concerted expression of clustered genes. The applicability of the TREX system in this context was demonstrated, among others, by transplanting the *Pantoea ananatis crt* gene cluster into *Rba capsulatus*. After transposon-assisted integration of the β-carotene biosynthesis genes into the chromosome of a *Rba. capsulatus* mutant lacking the neurosporene 1,2-hydratase-encoding gene *crtC*, heterologous genes could be fully expressed yielding up to 4.0 mg/g dry cell weight (DCW) of β-carotene.

Besides transferring foreign biosynthesis genes, combinatorial biosynthesis of novel carotenoids can also be achieved by engineering homologous pathway genes. In this case, high-throughput screening of enzyme variants can help to identify amino acid residues, which are crucial to change the catalytic properties (e.g., elevated substrate conversion rates, change of substrate specificity, or formation of new products). In a recent study, for instance, the *crtD* gene of *Rsp. rubrum,* encoding the rhodopin desaturase of the native spirilloxanthin pathway, was randomly mutated (Autenrieth and Ghosh 2015) and expressed in a *Rsp. rubrum crtC-crtD* deletion mutant which accumulates lycopene in high amounts (Wang et al. 2012). Clones exhibiting unusual color phenotypes after the expression of certain CrtD derivatives were investigated in detail, revealing the formation of the non-natural asymmetric carotenoids 3,4,3′,4′-tetradehydrorhodopin and 3′,4′-didehydroanhydrorhodovibrin. In another study, the neurosporene 1,2-hydratase-encoding gene *crtC* of *Rba. sphaeroides* was altered by random chemical mutagenesis (Wu and Liu 2011). High-resolution NMR analysis of photopigment-containing methanol extracts of clones showing unusual coloration demonstrated that, besides the original spheroidenone, the novel carotenoid methoxyneurosporene accumulated in high amounts (Wang et al. 2016). Remarkably, this extract, called Lycogen™, offers promising biotechnological and pharmaceutical properties including antioxidation, anti-inflammation, inhibition of melanogenesis, and prevention of cisplatin-induced renal injury (Wu and Liu 2011; Yang et al. 2014; Wang et al. 2014; Wang et al. 2016).

An additional interesting aspect in the context of carotenoid pathway engineering is the utilization of *crt* genes as reporters for colorimetric whole-cell biosensors. By using either a *crtA*⁻ mutant strain of *Rhv. sulfidophilum* or a *crtI*-deficient strain of *Rhodopseudomonas palustris* containing a plasmid that encompasses the corresponding *crt* (reporter) gene *in trans* under the control of an arsenite- or dimethyl sulfide-responsive promoter, environmental pollutants could easily be monitored by the change of color phenotypes (Fujimoto et al. 2006; Yoshida et al. 2007; Yoshida et al. 2008; Maeda et al. 2006). While the cultures appeared yellow or blue-green due to the incomplete carotenoid pathway in the absence of the toxic compound, cells regained their original deep red coloration under increasing concentrations of the respective analyte due to the induction of *crt* reporter gene expression.

Recent Advances in Producing Terpenoids with Rhodobacter

In the last few years, the spectrum of industrially relevant terpenoids that can successfully be produced in *Rhodobacter* was further extended. Examples include the bicyclic oxidized sesquiterpene valencene and the triterpene hydrocarbons botryococcene and squalene (Fig. 1). As an aroma component of citrus fruits, valencene is industrially used in flavorings and fragrances. Isolation from citrus oil usually leads to low valencene yields and prizes of raw material depend on seasonal influences. Therefore, new biotechnological production routes are needed. Consequently, a codon-optimized gene encoding the highly active valencene synthase CnVS derived from cypress *Callitropsis nootkatensis* was heterologously expressed in *Rba. sphaeroides* (Beekwilder et al. 2014). First, CnVS expression in the wild-type strain only resulted in moderate accumulation of valencene. Remarkably, the co-expression of CnVS together with all genes from *Paracoccus zeaxanthinifaciens* which are needed to reconstitute the MVA pathway (Fig. 1; Hümbelin et al. 2002) in *R. sphaeroides* resulted in a significant increase of valencene production that reached a final titer of 352 mg/l.

Another sophisticated approach was described by the group of Curtis, where *Rba. capsulatus* metabolic fluxes were modelled and engineered to produce two C_{30} isoprenoids (Khan et al. 2014, Khan et al. 2015), namely the food supplement and vaccine adjuvant squalene (Naziri and Tsimidou 2013; Tagliabue and Rappuoli 2008), and the methylated triterpene botryococcene (C_{30+}) which constitutes a promising alternative biofuel, because of its high energy density and good processability in petroleum industry (Hillen et al. 1982). To catalytically convert the C_{15} precursor FPP into the respective triterpenes, the genes encoding the SQS and the recombinant botryococcene synthase (BS)—both originating from the colonial algae *Botryococcus braunii* race B (Niehaus et al. 2011)—were constitutively expressed in the facultative phototrophic bacterium. As observed for efficient sesquiterpene synthesis in *R. sphaeroides*, also triterpene production benefited from an improved precursor supply. However, in this study the goal was achieved by co-expressing rate limiting enzymes of the MEP pathway encoded by *dxs* and *idi* (Fig. 1). In addition, the *ispA* gene was cloned together with MEP genes and the corresponding codon-optimized synthase gene into one expression vector. To investigate, if the induction of the carotenoid pathway upon phototrophic growth supports efficient production of foreign triterpenes, squalene and botryococcene formation was analyzed in different growth modes (i.e., aerobic heterotrophic, aerobic chemoautotrophic, and anaerobic photoheterotrophic growth). Surprisingly, observed triterpene levels were similar under all tested conditions which might reflect that the metabolite pool of the central metabolism predominantly limits the metabolic flux through the combinatorial terpene pathway in this organism.

The coenzyme Q_{10} (CoQ_{10} or ubiquinone) biosynthetic route is another example, where efficient supply of the isoprene C_5 building blocks IPP and DMAPP plays a crucial role (Fig. 1). Besides its natural function as an electron carrier, CoQ_{10} attained considerable biotechnological importance as an effective antioxidant that can be used as food and cosmetic additive as well as for the treatment of mitochondrial diseases, arterial hypertension, diabetes as well as Parkinson's and Alzheimer's disease (Jeya et al. 2010; Yang et al. 2010). A microbial CoQ_{10} production strain was constructed by stepwise engineering the respective metabolic pathway of *Rba. sphaeroides*, which naturally accumulates this electron carrier in high amounts. In the course of their study, the group of Yu identified (a) UbiG, UbiE, and UbiH, which are part of the ubiquinone pathway ending in the formation of CoQ_{10}, as well as (b) DXS, DXR, IDI, and IspD, four enzymes responsible for IPP and DMAPP synthesis within the MEP pathway, as rate limiting enzymes (Fig. 1; Lu et al. 2013, Lu et al. 2014, Lu et al. 2015). It is worth mentioning that the expression levels of the latter enzymes had to be carefully fine-tuned for improved fluxes of the above mentioned toxic isoprene intermediates through the pathway, which was achieved by applying a LacI repressor-based self-regulated expression system (Lu et al. 2014, see section "Expression Tools"). The combination of all optimization steps finally resulted in a product titer of 138 mg CoQ_{10} per liter culture, which was three times as high as the corresponding control strain (Lu et al. 2015).

In summary, the unique characteristics of α-proteobacteria, paired with sophisticated engineering approaches and tuning of isoprene precursor supply, have led to very successful high-value terpenoid biosynthesis. The suitability of *Rhodobacter* species as future industrial scale terpenoid producers is yet difficult to evaluate based on a so far limited number of studies. However, the already achieved yields undisputedly suggest a promising potential.

Future Applications and Outlook

Anoxygenic phototrophic α-proteobacteria are a metabolic versatile group of microorganisms which had long been used for different biotechnological applications such as biohydrogen formation, production of bioplastic, and bioremediation (for example, described by Kim and Kim 2011; Kobayashi et al. 2011; Trchounian 2015; Keskin et al. 2011; Jin and Nikolau 2014; Frigaard 2016; some of these aspects are also covered by different chapters of this book). In addition, as described in this review, members of this group of bacteria are promising candidates for the production of difficult-to-express membrane proteins by harnessing the protein synthesis machinery in combination with the ICM system. Those two systems are, due to the facultative photosynthetic lifestyle of these alternative hosts, already naturally evolved for the efficient production, translocation, assembly, and storage of large amounts of membrane proteins especially during the differentiation process that occurs when cells are shifted from aerobic to phototrophic growth mode. In this context, *Rhodobacter* has already been targeted as protein production platform in a structural genomics initiative allowing high-throughput expression, labeling with selenomethionine and purification of membrane proteins (Laible et al. 2004, Laible et al. 2009) as well as in the industrial expression platform ExpressO (http://www.clib2021.de/en/technologyplatforms/expression; Drepper, Heck, Jaeger, unpublished). Furthermore, the phototrophic lifestyle can also assist to facilitate sequential conversion of hydrophobic substrates, such as plant-derived terpenoids, through natural and engineered biosynthetic pathways. In this context, *Rhodobacter* with its ability to form ICM vesicles has the potential to play a pivotal role for producing new natural compounds that might serve as drugs or high-value additives in food and health care products in near future. The ICM-derived vesicles could further be exploited as nano-bioreactors that contain immobilized enzymes of entire biosynthetic pathways, which may naturally be membrane-associated proteins, integral membrane proteins, or soluble proteins. Moreover, the capacity to efficiently process hydrophobic metabolites and macromolecules under anoxygenic conditions can be combined in a plug-and-play-like fashion together with the use of redox enzymes that are either sensitive towards oxygen or require metal-containing cofactors (e.g., cytochrome P450 monooxygenases) to expand the diversity of chemical scaffolds. To this end, new advances in synthetic biology (as described, for example, in Keasling 2012, Way et al. 2014) have recently been adapted and transferred to the world of photosynthetic bacteria for the first time, now allowing scientists to start creating tailor-made microbial production chassis that can harness sunlight as biocatalytic driving force. In addition, the newly established expression tools, although still in their early stages of development for phototrophic microbes, will now be available for exploiting the so far still almost untapped wealth of environmental genomes through function-based metagenome screening approaches (Liebl et al. 2014, Katzke et al. 2016). Also here, the unique metabolic capacity of α-proteobacteria provides exciting opportunities and allows for the identification of novel enzymes and compounds upon screening of metagenomics libraries.

The use of phototrophic bacteria for capturing solar energy to sustainably drive the biocatalytic conversion of biocommodities into valuable biomaterials is inherently appealing. However, inefficient precursor supply, toxicity of metabolites or metabolic imbalances as well as difficulties in upscaling of culture volumes of phototrophic microbes to industrially relevant levels will be future challenges for bringing commercial biosynthesis of high-value compounds such as terpenoids within reach (Lee and Schmidt-Dannert 2002, Davies et al. 2015, Arendt et al. 2016). Furthermore, the development of next-generation bioeconomy approaches additionally could include the design of complex multi-species microbial communities, which will ultimately expand the potential of multi-step bioprocesses. To this end, degradation of renewable raw material or the utilization of waste streams and industrial flue gases as alternative carbon sources, for instance, could be coupled with photosynthetic processes in order to set up sophisticated light-driven, community-based production platforms in the future.

Acknowledgements Research in the authors' laboratories is supported by the Ministry of Innovation, Science and Research of the German State of North Rhine-Westphalia MIWF (Research and Technology Platform ExpressO, NRW Strategieprojekt BioSC), the German Federal Ministry of Education and Research BMBF (ExpresSys, HT-ENZ), and the German Research Foundation DFG (Cluster of Excellence on Plant Sciences CEPLAS). Further, the authors would like to thank Anita Loeschcke, Stephan Thies, and Karl-Erich Jaeger for the critical revision of the manuscript.

References

Adams PG, Hunter CN (2012) Adaptation of intracytoplasmic membranes to altered light intensity in *Rhodobacter sphaeroides*. Biochim Biophys Acta 1817:1616–1627

Allen JP, Feher G, Yeates TO, Komiya H, Rees DC (1987) Structure of the reaction center from *Rhodobacter sphaeroides* R-26: the protein subunits. Proc Natl Acad Sci U S A 84:6162–6166

Arechaga I, Miroux B, Karrasch S, Huijbregts R, de Kruijff B, Runswick MJ, Walker JE (2000) Characterisation of new intracellular membranes in *Escherichia coli* accompanying large scale over-production of the b subunit of F_1F_0 ATP synthase. FEBS Lett 482:215–219

Arendt P, Pollier J, Callewaert N, Goossens A (2016) Synthetic biology for production of natural and new-to-nature terpenoids in photosynthetic organisms., Plant J 87:16–37

Arvani S, Markert A, Loeschcke A, Jaeger K-E, Drepper T (2012) A T7 RNA polymerase-based toolkit for the concerted expression of clustered genes. J Biotechnol 159:162–171

Autenrieth C, Ghosh R (2015) Random mutagenesis and overexpression of rhodopin-3,4-desaturase allows the production of highly conjugated carotenoids in *Rhodospirillum rubrum*. Arch Biochem Biophys 572:134–141

Barbieri MR, Kerber NL, Pucheu NL, Tadros MH, García AF (2002) Effect of light and oxygen and adaptation to changing light conditions in a photosynthetic mutant in which the LHII complex of *Rhv. sulfidophilum* was heterologously expressed in a strain of *Rb. capsulatus* whose *puc* operon was deleted. Curr Microbiol 45:209–216

Beekwilder J, van Houwelingen A, Cankar K, van Dijk AD, de Jong RM, Stoopen G, Bouwmeester H, Achkar J, Sonke T, Bosch D (2014) Valencene synthase from the heartwood of Nootka cypress (*Callitropsis nootkatensis*) for biotechnological production of valencene. Plant Biotechnol J 12:174–182

Bélanger G, Gingras G (1988) Structure and expression of the *puf* operon messenger RNA in *Rhodospirillum rubrum*. J Biol Chem 263:7639–7645

Berman HM, Westbrook J, Feng Z, Gilliland G, Bhat TN, Weissig H, Shindyalov IN, Bourne PE (2000) The Protein Data Bank. Nucleic Acids Res 28:235–242

Bernaudat F, Frelet-Barrand A, Pochon N, Dementin S, Hivin P, Boutigny S, Rioux JB, Salvi D, Seigneurin-Berny D, Richaud P, Joyard J, Pignol D, Sabaty M, Desnos T, Pebay-Peyroula E, Darrouzet E, Vernet T, Rolland N (2011) Heterologous expression of membrane proteins: choosing the appropriate host. PLoS One 6, e29191

Berry EA, Huang L-S, Saechao LK, Pon NG, Valkova-Valchanova M, Daldal F (2004) X-ray structure of *Rhodobacter capsulatus* cytochrome bc_1: comparison with its mitochondrial and chloroplast counterparts. Photosynth Res 81:251–275

Breitmaier E (2006) Terpenes: flavors, fragrances, pharmaca, pheromones. Wiley-VCH Verlag GmbH & Co, KGaA

Butzin NC, Owen HA, Collins MLP (2010) A new system for heterologous expression of membrane proteins: *Rhodospirillum rubrum*. Protein Expr Purif 70:88–94

Chang C-H, el-Kabbani O, Tiede D, Norris J, Schiffer M (1991) Structure of the membrane-bound protein photosynthetic reaction center from *Rhodobacter sphaeroides*. Biochemistry 30:5352–5360

Cheng D, Wang R, Prather KJ, Chow KL, Hsing I-M (2015a) Tackling codon usage bias for heterologous expression in *Rhodobacter sphaeroides* by supplementation of rare tRNAs. Enzyme Microb Technol 72:25–34

Cheng L, Chen G, Ding G, Zhao Z, Dong T, Hu Z (2015b) Overexpression of *pucC* improves the heterologous protein expression level in a *Rhodobacter sphaeroides* expression system. Genet Mol Res 14:4058–4067

Chi SC, Mothersole DJ, Dilbeck P, Niedzwiedzki DM, Zhang H, Qian P, Vasilev C, Grayson KJ, Jackson PJ, Martin EC, Li Y, Holten D, Hunter CN (2015) Assembly of functional photosystem complexes in *Rhodobacter sphaeroides* incorporating carotenoids from the spirilloxanthin pathway. Biochim Biophys Acta 1847:189–201

Chirino AJ, Lous EJ, Huber M, Allen JP, Schenck CC, Paddock ML, Feher G, Rees DC (1994) Crystallographic analyses of site-directed mutants of the photosynthetic reaction center from *Rhodobacter sphaeroides*. Biochemistry 33:4584–4593

Chistoserdov AY, Chistoserdova LV, McIntire WS, Lidstrom ME (1994) Genetic organization of the mau gene cluster in *Methylobacterium extorquens* AM1: complete nucleotide sequence and generation and characteristics of *mau* mutants. J Bacteriol 176:4052–4065

Chory J, Donohue TJ, Varga AR, Staehelin LA, Kaplan S (1984) Induction of the photosynthetic membranes of *Rhodopseudomonas sphaeroides*: biochemical and morphological studies. J Bacteriol 159:540–554

Daniels GA, Drews G, Saier MH Jr (1988) Properties of a Tn5 insertion mutant defective in the structural gene (*fruA*) of the fructose-specific phosphotransferase system of *Rhodobacter capsulatus* and cloning of the *fru* regulon. J Bacteriol 170:1698–1703

Davies FK, Jinkerson RE, Posewitz MC (2015) Toward a photosynthetic microbial platform for terpenoid engineering. Photosynth Res 123:265–284

de Smet L, Kostanjevecki V, Guisez Y, van Beeumen J (2001) A novel system for heterologous expression of flavocytochrome *c* in phototrophic bacteria using the *Allochromatium vinosum rbcA* promoter. Arch Microbiol 176:19–28

Ditta G, Stanfield S, Corbin D, Helinski DR (1980) Broad host range DNA cloning system for Gram-negative bacteria: construction of a gene bank of *Rhizobium meliloti*. Proc Natl Acad Sci U S A 77:7347–7351

Ditta G, Schmidhauser T, Yakobson E, Lu P, Liang X-W, Finlay DR, Guiney D, Helinski DR (1985) Plasmids related to the broad host range vector, pRK290, useful for gene cloning and for monitoring gene expression. Plasmid 13:149–153

Drepper T, Arvani S, Rosenau F, Wilhelm S, Jaeger K-E (2005) High-level transcription of large gene regions: a novel T_7 RNA-polymerase-based system for expression of functional hydrogenases in the phototrophic bacterium *Rhodobacter capsulatus*. Biochem Soc Trans 33:56–58

Duport C, Meyer C, Naud I, Jouanneau Y (1994) A new gene expression system based on a fructose-dependent promoter from *Rhodobacter capsulatus*. Gene 145:103–108

Elena C, Ravasi P, Castelli ME, Peiru S, Menzella HG (2014) Expression of codon optimized genes in microbial systems: current industrial applications and perspectives. Front Microbiol 5:21

Elofsson A, von Heijne G (2007) Membrane protein structure: prediction versus reality. Annu Rev Biochem 76:125–140

Elsen S, Colbeau A, Chabert J, Vignais PM (1996) The *hupTUV* operon is involved in negative control of hydrogenase synthesis in *Rhodobacter capsulatus*. J Bacteriol 178:5174–5181

Elsen S, Jaubert M, Pignol D, Giraud E (2005) PpsR: a multifaceted regulator of photosynthesis gene expression in purple bacteria. Mol Microbiol 57:17–26

Erbakan M, Shen Y-X, Grzelakowski M, Butler PJ, Kumar M, Curtis WR (2014) Molecular cloning, overexpression and characterization of a novel water channel protein from *Rhodobacter sphaeroides*. PLoS One 9, e86830

Erbakan M, Curtis BS, Nixon BT, Kumar M, Curtis WR (2015) Advancing *Rhodobacter sphaeroides* as a platform for expression of functional membrane proteins. Protein Expr Purif 115:109–117

Ermler U, Fritzsch G, Buchanan SK, Michel H (1994) Structure of the photosynthetic reaction centre from *Rhodobacter sphaeroides* at 2.65 Å resolution: cofactors and protein-cofactor interactions. Structure 2:925–936

Essen L-O (2002) Structural genomics of "non-standard" proteins: a chance for membrane proteins? Gene Funct Dis 3:39–48

Esser L, Gong X, Yang S, Yu L, Yu C-A, Xia D (2006) Surface-modulated motion switch: capture and release of iron–sulfur protein in the cytochrome bc_1 complex. Proc Natl Acad Sci U S A 103:13045–13050

Ettema TJG, Andersson SGE (2009) The α-proteobacteria: the Darwin finches of the bacterial world. Biol Lett 5:429–432

Fedotova Y, Zeilstra-Ryalls J (2014) Analysis of the role of PrrA, PpsR, and FnrL in intracytoplasmic membrane differentiation of *Rhodobacter sphaeroides* 2.4.1 using transmission electron microscopy. Photosynth Res 119:283–290

Fodor BD, Kovács ÁT, Csáki R, Hunyadi-Gulyás É, Klement É, Maróti G, Mészáros LS, Medzihradszky KF, Rákhely G, Kovács KL (2004) Modular broad-host-range expression vectors for single-protein and protein complex purification. Appl Environ Microbiol 70:712–721

Fowler GJ, Gardiner AT, Mackenzie RC, Barratt SJ, Simmons AE, Westerhuis WH, Cogdell RJ, Hunter CN (1995) Heterologous expression of genes encoding bacterial light-harvesting complexes in *Rhodobacter sphaeroides*. J Biol Chem 270:23875–23882

Fraser NJ, Hashimoto H, Cogdell RJ (2001) Carotenoids and bacterial photosynthesis: the story so far… Photosynth Res 70:249–256

Frey M (2002) Hydrogenases: hydrogen-activating enzymes. Chembiochem 3:153–160

Frigaard NU (2016) Biotechnology of anoxygenic phototrophic bacteria. Adv Biochem Eng Biotechnol

Fujimoto H, Wakabayashi M, Yamashiro H, Maeda I, Isoda K, Kondoh M, Kawase M, Miyasaka H, Yagi K (2006) Whole-cell arsenite biosensor using photosynthetic bacterium *Rhodovulum sulfidophilum—Rhodovulum sulfidophilum* as an arsenite biosensor. Appl Microbiol Biotechnol 73:332–338

Garcia-Asua G, Cogdell RJ, Hunter CN (2002) Functional assembly of the foreign carotenoid lycopene into the photosynthetic apparatus of *Rhodobacter sphaeroides*, achieved by replacement of the native 3-step phytoene desaturase with its 4-step counterpart from *Erwinia herbicola*. Mol Microbiol 44:233–244

Glaeser J, Klug G (2005) Photo-oxidative stress in *Rhodobacter sphaeroides*: protective role of carotenoids and expression of selected genes. Microbiology 151:1927–1938

Graichen ME, Jones LH, Sharma BV, van Spanning RJ, Hosler JP, Davidson VL (1999) Heterologous expression of correctly assembled methylamine dehydrogenase in *Rhodobacter sphaeroides*. J Bacteriol 181:4216–4222
Gregor J, Klug G (1999) Regulation of bacterial photosynthesis genes by oxygen and light. FEMS Microbiol Lett 179:1–9
Han Y, Perner M (2016) Sulfide consumption in *Sulfurimonas denitrificans* and heterologous expression of its three sulfide-quinone reductase homologs. J Bacteriol 198:1260–1267
Hattab G, Suisse AYT, Ilioaia O, Casiraghi M, Dezi M, Warnet XL, E. WD, Moncoq, K., Zoonens M, Miroux B (2010) Membrane protein production in *Escherichia coli*: overview and protocols. In: Mus-Veteau I (ed) Membrane proteins production for structural analysis. Springer Science+Business Media, New York, pp 87–106
Hillen LW, Pollard G, Wake LV, White N (1982) Hydrocracking of the oils of *Botryococcus braunii* to transport fuels. Biotechnol Bioeng 24:193–205
Holden-Dye K, Crouch LI, Jones MR (2008) Structure, function and interactions of the PufX protein. Biochim Biophys Acta 1777:613–630
Hu Z, Zhao Z, Pan Y, Tu Y, Chen G (2010) A powerful hybrid *puc* operon promoter tightly regulated by both IPTG and low oxygen level. Biochemistry (Mosc) 75:519–525
Hümbelin M, Thomas A, Lin J, Li J, Jore J, Berry A (2002) Genetics of isoprenoid biosynthesis in *Paracoccus zeaxanthinifaciens*. Gene 297:129–139
Hunter CN, Hundle BS, Hearst JE, Lang HP, Gardiner AT, Takaichi S, Cogdell RJ (1994) Introduction of new carotenoids into the bacterial photosynthetic apparatus by combining the carotenoid biosynthetic pathways of *Erwinia herbicola* and *Rhodobacter sphaeroides*. J Bacteriol 176:3692–3697
Ind AC, Porter SL, Brown MT, Byles ED, de Beyer JA, Godfrey SA, Armitage JP (2009) Inducible-expression plasmid for *Rhodobacter sphaeroides* and *Paracoccus denitrificans*. Appl Environ Microbiol 75:6613–6615
Inui M, Nakata K, Roh JH, Vertes AA, Yukawa H (2003) Isolation and molecular characterization of pMG160, a mobilizable cryptic plasmid from *Rhodobacter blasticus*. Appl Environ Microbiol 69:725–733
Jaschke PR, Saer RG, Noll S, Beatty JT (2011) Modification of the genome of *Rhodobacter sphaeroides* and construction of synthetic operons. Methods Enzymol 497:519–538
Jeya M, Moon H-J, Lee J-L, Kim I-W, Lee J-K (2010) Current state of coenzyme Q_{10} production and its applications. Appl Microbiol Biotechnol 85:1653–1663
Jin H, Nikolau BJ (2014) Evaluating PHA productivity of bioengineered *Rhodosprillum rubrum*. PLoS One 9, e96621
Jones MR, Visschers RW, van Grondelle R, Hunter CN (1992) Construction and characterization of a mutant of *Rhodobacter sphaeroides* with the reaction center as the sole pigment-protein complex. Biochemistry 31:4458–4465
Kappler U, McEwan AG (2002) A system for the heterologous expression of complex redox proteins in *Rhodobacter capsulatus*: characterisation of recombinant sulphite:cytochrome *c* oxidoreductase from *Starkeya novella*. FEBS Letters 529:208–214
Katona G, Andreasson U, Landau EM, Andreasson LE, Neutze R (2003) Lipidic cubic phase crystal structure of the photosynthetic reaction centre from *Rhodobacter sphaeroides* at 2.35 Å resolution. J Mol Biol 331:681–692
Katsiou E, Sturgis JN, Robert B, Tadros MH (1998) Heterologous expression of genes encoding bacterial light-harvesting complex II in *Rhodobacter capsulatus* and *Rhodovulum sulfidophilum*. Microbiol Res 153:189–204
Katzke N, Arvani S, Bergmann R, Circolone F, Markert A, Svensson V, Jaeger K-E, Heck A, Drepper T (2010) A novel T7 RNA polymerase dependent expression system for high-level protein production in the phototrophic bacterium *Rhodobacter capsulatus*. Protein Expr Purif 69:137–146
Katzke N, Bergmann R, Jaeger K-E, Drepper T (2012) Heterologous high-level gene expression in the photosynthetic bacterium *Rhodobacter capsulatus*. Methods Mol Biol 824:251–269
Katzke N, Knapp A, Loeschcke A, Drepper T, Jaeger K-E (2016) Novel tools for the functional expression of metagenomic DNA. Meth Mol Biol

Kawata T, Bristol JR, Rossignol DP, Rose JR, Kobayashi S, Yokohama H, Ishibashi A, Christ WJ, Katayama K, Yamatsu I, Kishi Y (1999) E5531, a synthetic non-toxic lipid A derivative blocks the immunobiological activities of lipopolysaccharide. Br J Pharmacol 127:853–862

Keasling JD (2012) Synthetic biology and the development of tools for metabolic engineering. Metab Eng 14:189–195

Keen NT, Tamaki S, Kobayashi D, Trollinger D (1988) Improved broad-host-range plasmids for DNA cloning in Gram-negative bacteria. Gene 70:191–197

Kersters K, de Vos P, Gillis M, Swings J, Vandamme P, Stackebrandt E (2006) Introduction to the proteobacteria. Prokaryotes 5:3–37

Keskin T, Abo-Hashesh M, Hallenbeck PC (2011) Photofermentative hydrogen production from wastes. Bioresour Technol 102:8557–8568

Khan NE, Myers JA, Tuerk AL, Curtis WR (2014) A process economic assessment of hydrocarbon biofuels production using chemoautotrophic organisms. Bioresour Technol 172:201–211

Khan NE, Nybo SE, Chappell J, Curtis WR (2015) Triterpene hydrocarbon production engineered into a metabolically versatile host—*Rhodobacter capsulatus*. Biotechnol Bioeng 112:1523–1532

Kiley PJ, Varga A, Kaplan S (1988) Physiological and structural analysis of light-harvesting mutants of *Rhodobacter sphaeroides*. J Bacteriol 170:1103–1115

Kim D-H, Kim M-S (2011) Hydrogenases for biological hydrogen production. Bioresour Technol 102:8423–8431

Kobayashi J, Yoshimune K, Komoriya T, Kohno H (2011) Efficient hydrogen production from acetate through isolated *Rhodobacter sphaeroides*. J Biosci Bioeng 112:602–605

Koepke J, Krammer E-M, Klingen AR, Sebban P, Ullmann GM, Fritzsch G (2007) pH modulates the quinone position in the photosynthetic reaction center from *Rhodobacter sphaeroides* in the neutral and charge separated states. J Mol Biol 371:396–409

Kong LY, Tan RX (2015) Artemisinin, a miracle of traditional Chinese medicine. Nat Prod Rep 32:1617–1621

Kovach ME, Phillips RW, Elzer PH, Roop RM II, Peterson KM (1994) pBBR1MCS: a broad-host-range cloning vector. Biotechniques 16:800–802

Kovach ME, Elzer PH, Hill DS, Robertson GT, Farris MA, Roop RM II, Peterson KM (1995) Four new derivatives of the broad-host-range cloning vector pBBR1MCS, carrying different antibiotic-resistance cassettes. Gene 166:175–176

Labes M, Pühler A, Simon R (1990) A new family of RSF1010-derived expression and *lac*-fusion broad-host-range vectors for Gram-negative bacteria. Gene 89:37–46

Laible PD, Scott HN, Henry L, Hanson DK (2004) Towards higher-throughput membrane protein production for structural genomics initiatives. J Struct Funct Genomics 5:167–172

Laible PD, Mielke DL, Hanson DK (2009) Foreign gene expression in photosynthetic bacteria. In: Hunter CN, Daldal F, Thurnauer MC, Beatty JT (eds) The purple phototrophic bacteria. Springer Science + Business Media B.V., Dordrecht, pp 839–860

Laible PD, Mielke DL, Hanson DK (2011) Membrane protein production using photosynthetic bacteria: a practical guide. In: Robinson AS (ed) Production of Membrane Proteins: Strategies for Expression and Isolation, 1st edn. Wiley-VCH Verlags GmbH & Co, KGaA, pp 167–198

Lee PC, Schmidt-Dannert C (2002) Metabolic engineering towards biotechnological production of carotenoids in microorganisms. Appl Microbiol Biotechnol 60:1–11

Lee JK, DeHoff BS, Donohue TJ, Gumport RI, Kaplan S (1989) Transcriptional analysis of *puf* operon expression in *Rhodobacter sphaeroides* 2.4.1 and an intercistronic transcription terminator mutant. J Biol Chem 264:19354–19365

Leimkühler S, Kern M, Solomon PS, McEwan AG, Schwarz G, Mendel RR, Klipp W (1998) Xanthine dehydrogenase from the phototrophic purple bacterium *Rhodobacter capsulatus* is more similar to its eukaryotic counterparts than to prokaryotic molybdenum enzymes. Mol Microbiol 27:853–869

Liebl W, Angelov A, Juergensen J, Chow J, Loeschcke A, Drepper T, Classen T, Pietruzska J, Ehrenreich A, Streit WR, Jaeger K-E (2014) Alternative hosts for functional (meta)genome analysis. Appl Microbiol Biotechnol 98:8099–8109

Lindenstrauß U, Brüser T (2006) Conservation and variation between *Rhodobacter capsulatus* and *Escherichia coli* Tat systems. J Bacteriol 188:7807–7814

Loeschcke A, Markert A, Wilhelm S, Wirtz A, Rosenau F, Jaeger K-E, Drepper T (2013) TREX: a universal tool for the transfer and expression of biosynthetic pathways in bacteria. ACS Synth Biol 2:22–33

Lombard J, Moreira D (2011) Origins and early evolution of the mevalonate pathway of isoprenoid biosynthesis in the three domains of life. Mol Biol Evol 28:87–99

Lu W, Shi Y, He S, Fei Y, Yu K, Yu H (2013) Enhanced production of CoQ_{10} by constitutive overexpression of 3-demethyl ubiquinone-9 3-methyltransferase under *tac* promoter in *Rhodobacter sphaeroides*. Biochem Eng J 72:42–47

Lu W, Ye L, Xu H, Xie W, Gu J, Yu H (2014) Enhanced production of coenzyme Q_{10} by self-regulating the engineered MEP pathway in *Rhodobacter sphaeroides*. Biotechnol Bioeng 111:761–769

Lu W, Ye L, Lv X, Xie W, Gu J, Chen Z, Zhu Y, Li A, Yu H (2015) Identification and elimination of metabolic bottlenecks in the quinone modification pathway for enhanced coenzyme Q_{10} production in *Rhodobacter sphaeroides*. Metab Eng 29:208–216

Lundstrom K (2007) Structural genomics and drug discovery. J Cell Mol Med 11:224–238

Madigan MT (1995) Microbiology of nitrogen fixation by anoxygenic photosynthetic bacteria. In: Blankenship RE, Madigan MT, Bauer CE (eds) Anoxygenic photosynthetic bacteria. Springer Netherlands, Dordrecht, pp 915–928

Madigan MT, Jung D (2009) An overview of purple bacteria: systematics, physiology, and habitats. In: Hunter CN, Daldal F, Thurnauer MC, Beatty JT (eds) The purple phototrophic bacteria. Springer Science + Business Media B.V., pp 1–15

Maeda I, Yamashiro H, Yoshioka D, Onodera M, Ueda S, Kawase M, Miyasaka H, Yagi K (2006) Colorimetric dimethyl sulfide sensor using *Rhodovulum sulfidophilum* cells based on intrinsic pigment conversion by CrtA. Appl Microbiol Biotechnol 70:397–402

Marienhagen J, Bott M (2013) Metabolic engineering of microorganisms for the synthesis of plant natural products. J Biotechnol 163:166–178

Masepohl B, Hallenbeck PC (2010) Nitrogen and molybdenum control of nitrogen fixation in the phototrophic bacterium *Rhodobacter capsulatus*. Adv Exp Med Biol 675:49–70

Masepohl B, Drepper T, Paschen A, Gross S, Pawlowski A, Raabe K, Riedel K-U, Klipp W (2002) Regulation of nitrogen fixation in the phototrophic purple bacterium *Rhodobacter capsulatus*. J Mol Microbiol Biotechnol 4:243–248

McLuskey K, Roszak AW, Zhu Y, Isaacs NW (2010) Crystal structures of all-alpha type membrane proteins. Eur Biophys J 39:723–755

Miroux B, Walker JE (1996) Over-production of proteins in *Escherichia coli*: mutant hosts that allow synthesis of some membrane proteins and globular proteins at high levels. J Mol Biol 260:289–298

Mukoyama D, Takeyama H, Kondo Y, Matsunaga T (2006) Astaxanthin formation in the marine photosynthetic bacterium *Rhodovulum sulfidophilum* expressing *crtI*, *crtY*, *crtW* and *crtZ*. FEMS Microbiol Lett 265:69–75

Müller M, Klösgen RB (2005) The Tat pathway in bacteria and chloroplasts. Mol Membr Biol 22:113–121

Mus-Veteau I, Demange P, Zito F (2014) Membrane protein production for structural analysis. In: Mus-Veteau I (ed) Membrane proteins production for structural analysis. Springer Science + Business Media, New York, pp 1–44

Naziri E, Tsimidou MZ (2013) Formulated squalene for food related applications. Recent Pat Food Nutr Agric 5:83–104

Nie X, Zhang L, Hu Z, Liu Y, Zhao Z (2015) Human neutrophil peptide 3 could be functionally expressed in *Rhodobacter sphaeroides*. Acta Biochim Pol 62:259–263

Niederman RA (2016) Development and dynamics of the photosynthetic apparatus in purple phototrophic bacteria. Biochim Biophys Acta 1857:232–246

Niehaus TD, Okada S, Devarenne TP, Watt DS, Sviripa V, Chappell J (2011) Identification of unique mechanisms for triterpene biosynthesis in *Botryococcus braunii*. Proc Natl Acad Sci U S A 108:12260–12265

Nybo SE, Khan NE, Woolston BM, Curtis WR (2015) Metabolic engineering in chemolithoautotrophic hosts for the production of fuels and chemicals. Metab Eng 30:105–120

Olsen JD, Adams PG, Jackson PJ, Dickman MJ, Qian P, Hunter CN (2014) Aberrant assembly complexes of the reaction center light-harvesting 1 PufX (RC-LH1-PufX) core complex of *Rhodobacter sphaeroides* imaged by atomic force microscopy. J Biol Chem 289:29927–29936

Overington JP, Al-Lazikani B, Hopkins AL (2006) How many drug targets are there? Nat Rev Drug Discov 5:993–996

Prentki P, Krisch HM (1984) *In vitro* insertional mutagenesis with a selectable DNA fragment. Gene 29:303–313

Pugh RJ, McGlynn P, Jones MR, Hunter CN (1998) The LH1-RC core complex of *Rhodobacter sphaeroides*: interaction between components, time-dependent assembly, and topology of the PufX protein. Biochim Biophys Acta 1366:301–316

Qin L, Hiser C, Mulichak A, Garavito RM, Ferguson-Miller S (2006) Identification of conserved lipid/detergent-binding sites in a high-resolution structure of the membrane protein cytochrome *c* oxidase. Proc Natl Acad Sci U S A 103:16117–16122

Romagnoli S, Tabita FR (2009) Carbon dioxide metabolism and its regulation in nonsulfur purple photosynthetic bacteria. In: Daldal F, Thurnauer MC, Beatty JT (eds) Hunter CN. The Purple Phototrophic Bacteria, Springer Science + Business Media B.V., pp 563–576

Roy A, Shukla AK, Haase W, Michel H (2008) Employing *Rhodobacter sphaeroides* to functionally express and purify human G protein-coupled receptors. Biol Chem 389:69–78

Saer RG, Pan J, Hardjasa A, Lin S, Rosell F, Mauk AG, Woodbury NW, Murphy ME, Beatty JT (2014) Structural and kinetic properties of *Rhodobacter sphaeroides* photosynthetic reaction centers containing exclusively Zn-coordinated bacteriochlorophyll as bacteriochlorin cofactors. Biochim Biophys Acta 1837:366–374

Sahdev S, Khattar SK, Saini KS (2008) Production of active eukaryotic proteins through bacterial expression systems: a review of the existing biotechnology strategies. Mol Cell Biochem 307:249–264

Sandmann G (2015) Carotenoids of biotechnological importance. Adv Biochem Eng Biotechnol 148:449–467

Sawai S, Saito K (2011) Triterpenoid biosynthesis and engineering in plants. Front Plant Sci 2:25

Schrader J, Bohlmann J (2015) Biotechnology of isoprenoids. Advances in biochemical engineering/biotechnology, vol 148. Springer International Publishing

Shaw AL, Hanson GR, McEwan AG (1996) Cloning and sequence analysis of the dimethylsulfoxide reductase structural gene from *Rhodobacter capsulatus*. Biochim Biophys Acta 1276:176–180

Simon R, Priefer U, Pühler A (1983) A broad host range mobilization system for *in vivo* genetic engineering: transposon mutagenesis in Gram negative bacteria. Nat Biotechnol 1:784–791

Solomon PS, Lane I, Hanson GR, McEwan AG (1997) Characterisation of the pterin molybdenum cofactor in dimethylsulfoxide reductase of *Rhodobacter capsulatus*. Eur J Biochem 246:200–203

Staskawicz B, Dahlbeck D, Keen N, Napoli C (1987) Molecular characterization of cloned avirulence genes from race 0 and race 1 of *Pseudomonas syringae* pv. *glycinea*. J Bacteriol 169:5789–5794

Stowell MH, McPhillips TM, Rees DC, Soltis SM, Abresch E, Feher G (1997) Light-induced structural changes in photosynthetic reaction center: implications for mechanism of electron-proton transfer. Science 276:812–816

Strittmatter W, Weckesser J, Salimath PV, Galanos C (1983) Nontoxic lipopolysaccharide from *Rhodopseudomonas sphaeroides* ATCC 17023. J Bacteriol 155:153–158

Svensson-Ek M, Abramson J, Larsson G, Törnroth S, Brzezinski P, Iwata S (2002) The X-ray crystal structures of wild-type and EQ(I-286) mutant cytochrome *c* oxidases from *Rhodobacter sphaeroides*. J Mol Biol 321:329–339

Sznee K, Crouch LI, Jones MR, Dekker JP, Frese RN (2014) Variation in supramolecular organisation of the photosynthetic membrane of *Rhodobacter sphaeroides* induced by alteration of PufX. Photosynth Res 119:243–256

Tagliabue A, Rappuoli R (2008) Vaccine adjuvants: the dream becomes real. Hum Vaccin 4:347–349

Tehrani A, Beatty JT (2004) Effects of precise deletions in *Rhodobacter sphaeroides* reaction center genes on steady-state levels of reaction center proteins: a revised model for reaction center assembly. Photosynth Res 79:101–108

Thimmappa R, Geisler K, Louveau T, O'Maille P, Osbourn A (2014) Triterpene biosynthesis in plants. Annu Rev Plant Biol 65:225–257

Tholl D (2015) Biosynthesis and biological functions of terpenoids in plants. Adv Biochem Eng Biotechnol 148:63–106

Tikh IB, Held M, Schmidt-Dannert C (2014) BioBrick™ compatible vector system for protein expression in *Rhodobacter sphaeroides*. Appl Microbiol Biotechnol 98:3111–3119

Trchounian A (2015) Mechanisms for hydrogen production by different bacteria during mixed-acid and photo-fermentation and perspectives of hydrogen production biotechnology. Crit Rev Biotechnol 35:103–113

van der Palen CJNM, Reijnders WNM, de Vries S, Duine JA, van Spanning RJM (1997) MauE and MauD proteins are essential in methylamine metabolism of *Paracoccus denitrificans*. Antonie Van Leeuwenhoek 72:219–228

Vick JE, Johnson ET, Choudhary S, Bloch SE, Lopez-Gallego F, Srivastava P, Tikh IB, Wawrzyn GT, Schmidt-Dannert C (2011) Optimized compatible set of BioBrick™ vectors for metabolic pathway engineering. Appl Microbiol Biotechnol 92:1275–1286

Vignais PM (2009) Regulation of hydrogenase gene expression. In: Daldal F, Thurnauer MC, Beatty JT (eds) Hunter CN. The Purple Phototrophic Bacteria, Springer Science+Business Media B.V., pp 743–757

Vignais PM, Billoud B (2007) Occurrence, classification, and biological function of hydrogenases: an overview. Chem Rev 107:4206–4272

Vignais PM, Dimon B, Zorin NA, Tomiyama M, Colbeau A (2000) Characterization of the hydrogen-deuterium exchange activities of the energy-transducing HupSL hydrogenase and H_2-signaling HupUV hydrogenase in *Rhodobacter capsulatus*. J Bacteriol 182:5997–6004

Vranová E, Coman D, Gruissem W (2012) Structure and dynamics of the isoprenoid pathway network. Mol Plant 5:318–333

Vranová E, Coman D, Gruissem W (2013) Network analysis of the MVA and MEP pathways for isoprenoid synthesis. Annu Rev Plant Biol 64:665–700

Wang W, Hu Z, Chen X, Zhao Z, Li J, Chen G (2009) Heterologous synthesis and assembly of functional LHII antenna complexes from *Rhodovulum sulfidophilum* in *Rhodobacter sphaeroides* mutant. Mol Biol Rep 36:1695–1702

Wang GS, Grammel H, Abou-Aisha K, Sägesser R, Ghosh R (2012) High-level production of the industrial product lycopene by the photosynthetic bacterium *Rhodospirillum rubrum*. Appl Environ Microbiol 78:7205–7215

Wang C-C, Liu W-S, Chang F-H, Tsai P-Y, Tsai M-K, Shiang J-C, Wen Z-H (2014) *Rhodobacter sphaeroides* extract improves glucose homeostasis in streptozotocin-induced diabetic mice. J Microb Biochem Technol 6:38–42

Wang C-C, Ding S, Chiu K-H, Liu W-S, Lin T-J, Wen Z-H (2016) Extract from a mutant *Rhodobacter sphaeroides* as an enriched carotenoid source. Food Nutr Res 60:29580

Way JC, Collins JJ, Keasling JD, Silver PA (2014) Integrating biological redesign: where synthetic biology came from and where it needs to go. Cell 157:151–161

Weiss MS, Schulz GE (1992) Structure of porin refined at 1.8 Å resolution. J Mol Biol 227:493–509

Wenzel SC, Müller R (2005) Recent developments towards the heterologous expression of complex bacterial natural product biosynthetic pathways. Curr Opin Biotechnol 16:594–606

Woronowicz K, Harrold JW, Kay JM, Niederman RA (2013) Structural and functional proteomics of intracytoplasmic membrane assembly in *Rhodobacter sphaeroides*. J Mol Microbiol Biotechnol 23:48–62

Wu W-T, Liu W-S (2011) Anti-inflammatory property of biomaterial carotenoids production by *Rhodobacter sphaeroides* WL-APD911. Adv Mat Res 287–290:2028–2031

Yang X, Dai G, Li G, Yang ES (2010) Coenzyme Q10 reduces β-amyloid plaque in an APP/PS1 transgenic mouse model of Alzheimer's disease. J Mol Neurosci 41:110–113

Yang T-H, Lai Y-H, Lin T-P, Liu W-S, Kuan L-C, Liu C-C (2014) Chronic exposure to *Rhodobacter sphaeroides* extract Lycogen™ prevents UVA-induced malondialdehyde accumulation and procollagen I down-regulation in human dermal fibroblasts. Int J Mol Sci 15:1686–1699

Yin H (2008) Exogenous agents that target transmembrane domains of proteins. Angew Chem Int Ed 47:2744–2752

Yoshida K, Yoshioka D, Inoue K, Takaichi S, Maeda I (2007) Evaluation of colors in green mutants isolated from purple bacteria as a host for colorimetric whole-cell biosensors. Appl Microbiol Biotechnol 76:1043–1050

Yoshida K, Inoue K, Takahashi Y, Ueda S, Isoda K, Yagi K, Maeda I (2008) Novel carotenoid-based biosensor for simple visual detection of arsenite: characterization and preliminary evaluation for environmental application. Appl Environ Microbiol 74:6730–6738

Youvan DC, Bylina EJ, Alberti M, Begusch H, Hearst JE (1984) Nucleotide and deduced polypeptide sequences of the photosynthetic reaction-center, B870 antenna, and flanking polypeptides from *R. capsulata*. Cell 37:949–957

Zappa S, Li K, Bauer CE (2010) The tetrapyrrole biosynthetic pathway and its regulation in *Rhodobacter capsulatus*. Adv Exp Med Biol 675:229–250

Zhao Z, Hu Z, Nie X, Cheng L, Ding G, Luo M, Pan Y, Liang Y, Chen G (2011) A novel *Rhodobacter sphaeroides* expression system for real-time evaluation of heterologous protein expression levels. Protein Pept Lett 18:568–572

Anoxygenic Phototrophic Bacteria from Extreme Environments

Johannes F. Imhoff

Abstract Anoxygenic phototrophic bacteria have conquered various types of habitats of extreme conditions of temperature, pH and salinity. A short overview is given on the species found, on the properties of their extreme habitats and their phylogenetic relations. Hot spring habitats include springs from the USA, Russia, Japan, Tibet, Greenland with the Yellowstone hot springs being the most prominent and best studied. *Chloroflexus aurantiacus* occurring worldwide in hot springs is tolerant to the highest temperatures up to 70 °C. Cold adapted phototrophic bacteria forms have been studied in Antarctic fjords and lakes and in sea ice with an example from the Baltic Sea. *Chlorobaculum tepidum* and *Rhodopila globiformis* are prominent examples of the small number of phototrophic bacteria adapted to acidic pH. Though anoxygenic phototrophic bacteria are prominent inhabitants of alkaline soda lakes within the whole range of salt concentrations, only a few groups have successfully conquered this type of habitats. Most widely distributed and diverse are the representatives of phototrophic as well as chemotrophic Ectothiorhodospiraceae. These form often massive colored developments in shallow waters and sediments. Different types of phototrophic purple bacteria inhabit hypersaline coastal lagoons, salterns and salt lakes. Purple sulfur bacteria with prominent representatives of *Halochromatium* and *Thiohalocapsa* species and purple nonsulfur bacteria with *Rhodovibrio*, *Rhodothalassium* species are regularly found in microbial mats occurring in these habitats. A short review is also included on mechanisms of adaptation to high temperatures and salt concentrations.

Keywords Anoxygenic phototrophic bacteria • Compatible solute • Extreme environments • Hot springs • Sea ice • Cold waters • Acidic habitats • Soda lakes • Salt lakes

---

J.F. Imhoff (✉)
GEOMAR Helmholtz Center for Ocean Research Kiel, Marine Microbiology, 24105 Kiel, Germany
e-mail: jimhoff@geomar.de

© Springer International Publishing Switzerland 2017

P.C. Hallenbeck (ed.), *Modern Topics in the Phototrophic Prokaryotes*,
DOI 10.1007/978-3-319-46261-5_13

Introduction

Life in extreme environments always has fascinated scientists and major motivations to investigate microorganisms from such habitats are the desire to learn about the borderlines of life and the mechanisms to adapt to and to live at such extreme conditions. Extreme habitats present harsh conditions for life and a general tendency is a decrease in diversity with increasing harsh conditions, because only few microorganisms are well adapted to live at the most extreme conditions. Under increasingly harsh and extreme conditions competition is considered to be of less importance because the number of microorganisms able to develop is dramatically reduced. Instead, the ability to cope with the harsh environmental conditions and the extreme parameters are of primary relevance to live and survive. In order to thrive under extreme conditions it is not necessarily required that the microorganisms find optimal growth conditions, but they might instead tolerate these extreme conditions and cope better with them than others. Some non-halophilic or slightly halophilic bacteria, e.g., are extremely salt tolerant and can thrive at extremely high salt concentrations thereby being able to compete with moderately halophilic species as was shown for some marine phototrophic bacteria (Imhoff 2001; see below).

Consequently, two types of organisms are able to thrive under extreme conditions, those that are well adapted to these conditions, the extremophiles (e.g., extreme thermophiles and extreme halophiles), and those that can tolerate the extreme conditions, the extremotolerant microorganisms (e.g., the thermotolerant and halotolerant microorganisms). The extremophilic property is defined by the growth optimum of the organisms, the tolerance by the upper (or in the case of pH and temperature also the lower) limit at which growth occurs.

The ecological niches of a particular bacterium and its ability to successfully compete are defined by its physiological properties, on one hand, and by the complete set of physicochemical properties of the habitat, on the other hand, which include temperature, pH, presence of oxygen and/or sulfide, light conditions, and mineral salts composition and concentration. Thus different types of microorganisms are expected to thrive at acidic and alkaline pH, at low and high temperatures, at different concentration ranges of salts and also at different concentrations of individual ions.

Representatives of anoxygenic phototrophic bacteria are the masters of life at extremely high concentrations of salts and often are blooming at extremely high salinity in salt and soda lakes, where they regularly form visible mass developments (see Figs. 3, 6, 8 and 9). A few phototrophic bacteria also can grow at high temperatures in hot springs and form visible colored microbial mats (see Figs. 1 and 2). However, in general anoxygenic phototrophic bacteria have not conquered the most extreme conditions at which life is possible. In the following, we will consider the limits for growth of different anoxygenic phototrophic bacteria under a variety of physicochemical conditions such as temperature, pH, and salinity. Phototrophic bacteria adapted to these habitats and the diversity of their communities in representative examples of such habitats will be discussed.

Due to the application and rapid development of molecular methods to approach diversity and composition of environmental communities over the past decades, we gained new options to study communities of anoxygenic phototrophic bacteria. The classical approach, using microscopic studies, the establishment of pure cultures and their morphological and physiological characterization continues to be the central approach to study bacterial properties in relation to their environment. With the establishment of molecular analyses in microbial ecology for the first time whole bacterial (microbial) communities could be systematically studied. Sequence information of the 16S rRNA gene soon not only was established as an essential tool for the identification of pure cultures, but also became important as a link between pure culture studies and environmental community analyses, in which the rRNA gene was used to characterize the community diversity.

For studies on communities of phototrophic bacteria, both of these approaches have different advantages and disadvantages. The classical approach is useful for the characterization of individual and possibly important members of the community but hardly can depict the diverse community structure. The molecular approach using ribosomal RNA gene sequences can depict the overall composition of a bacterial community, with a certain threshold level (depending on the sequencing depth of the metagenome analyses) but without specificity to the phototrophic (or any other functional) property of the community members. Consequently, in all cases where phototrophic and non-phototrophic bacteria are closely related and have high similarity in ribosomal gene sequences, a clear assignment of environmental sequences to a phototrophic bacterium is problematic or not possible. The tremendous advances in sequencing technologies and the enormous sequence depth in ribosomal gene analyses of metagenomes today have not changed this basic problem. This was reason to introduce functional genes related to bacterial photosynthesis first of all into systematic studies with available pure cultures (Alexander et al. 2002; Tank et al. 2009) and then also to the specific analysis of communities of phototrophic bacteria. The *fmoA* gene was used as a specific tool for the green sulfur bacteria (Alexander and Imhoff 2006) and the almost complete *pufLM* gene sequences for the study of phototrophic purple bacteria (Thiel et al. 2010; Tank et al. 2011). In a number of studies a short part of the *pufM* gene was used to characterize purple sulfur bacterial communities (Achenbach et al. 2001; Karr et al. 2003; Asao et al. 2011). With a comprehensive and growing backbone of complete *pufLM* sequences from identified pure cultures in databases, genes specifically approaching the desired functional group of microorganisms qualify as tools to establish inventories of anoxygenic phototrophic bacteria in environments and to study their dynamic response to changes of the environmental conditions (Imhoff 2016).

Hot Springs and Thermophilic Phototrophic Bacteria

Thermophilic phototrophic bacteria are considered as being among the most ancient organisms on earth (Cavalier-Smith 2010). In the Precambrian Ocean (3.5–2 billion years) the average water temperature was probably about 55–85 °C (Knauth 2005; Robert and Chaussidon 2006) and thereby provided worldwide suitable conditions for the ancestors of thermophilic filamentous phototrophic bacteria as known today, including *Chloroflexus*, *Heliothrix*, and *Roseiflexus* (Grouzdev et al. 2015). Most likely the first microbial mats in the Precambrian Ocean were formed by ancestors of *Roseiflexus* which are the most deeply branching within this phylum (Tice and Lowe 2004). They lack chlorosomes, which might have been acquired later, by ancestors of *Chloroflexus*.

Thermophilic anoxygenic phototrophic bacteria have not conquered the most extreme ranges of high temperatures and extremely thermophilic or hyperthermophilic forms are not known to date. The most thermotolerant representative is *Chloroflexus aurantiacus* which is growing best at 55 °C and tolerates temperatures up to 70 °C (Figs. 1 and 2). The temperature relations and habitats of thermophilic anoxygenic phototrophic bacteria are shown in Table 1. Often microorganisms that have growth optima between 65 and 80 °C are considered as thermophiles and are distinguished from the hyperthermophiles growing best above 80 °C. Thermophilic phototrophic bacteria were classified according to their upper limits of growth, i.e., their temperature tolerance, as highly thermophilic if growing up to 63–73 °C, as moderately thermophilic if growing up to 53–63 °C, and as slightly thermophilic if growing up to 43–53 °C (Castenholz and Pierson 1995). As such definitions are helpful as guides for orientation we follow largely to use these temperature ranges to define the thermophily and thermotolerance of the bacteria. However, we rather would use the growth optima to define the thermophilic character and consider those bacteria with temperature optima higher than 48–50 °C as moderately thermophilic and those with slightly increased temperature optima between 38 and 48 °C, relative to most mesophilic phototrophic bacteria (optima at 20–35 °C), as slightly thermophilic or tepidophilic bacteria.

Hot springs are found all over the world alongside the spreading axes of the tectonic plates on land as well as in the oceans. An important difference exists between hot springs containing sufficient sulfide to feed sulfur-oxidizing phototrophic bacteria in the effluent and springs that lack sufficient amounts of sulfide. This is in particular relevant because with a few exceptions cyanobacteria are inhibited by sulfide and *Chloroflexus aurantiacus* may grow either photoheterotrophically in the absence or photoautotrophically in the presence of sulfide (Madigan and Brock 1975, 1977; Madigan et al. 1974). The presence of sulfide also has significant influence on the development of green and purple sulfur bacteria in hot springs. In addition to the temperature, the presence of sulfide and the pH are major factors determining the composition of phototrophic microbial mats and communities in thermal springs.

Nonsulfidic, alkaline (pH 8.0–8.5) hot springs are represented, e.g., by Octopus Hot spring (lower geyser basin in Yellowstone National Park) and Hunter's Hot Springs (southern Oregon). In many nonsulfidic hot springs *Chloroflexus* develops as an orange mat underneath a top layer of cyanobacteria (Castenholz and Pierson 1995). Because of the lack of sulfide and reduced sulfur compounds as potential electron donors for photoautotrophic growth, it was concluded that *Chloroflexus aurantiacus* grows in these hot springs either photo- or chemoheterotrophically on the extent of organic molecules excreted by the cyanobacteria, which include among others succinate, acetate, fructose, adonitol, and beta-hydroxybutyrate (Castenholz and Pierson 1995). In Hunter's Hot Springs, sulfide is primarily absent but the presence of 2.7 mM sulfate enables sulfate reduction which is initiated below 58 °C and supports dense populations of *Thermochromatium tepidum* to develop in the lower temperature range under a cover of *Oscillatoria terebriformis* and *Chloroflexus aurantiacus* (Castenholz and Pierson 1995).

Sulfidic hot springs are widespread and in these springs sulfide-dependent photoautotrophic growth of *Chloroflexus aurantiacus* is possible in mats that develop without cyanobacteria. The filaments are green in the absence of oxygen but of orange color in its presence. In upper terraces of Mammoth Hot Spring at pH 6.2–6.8, dull green mats of *Chloroflexus aurantiacus* were found in the temperature range of 51–66 °C (Castenholz and Pierson 1995).

Prominent Hot Springs in Different Countries

In the USA, Yellowstone National Park is a hot spot in research on hot springs and thermophilic, phototrophic bacteria. Initiated in the 1960s with ecological studies on the hot springs (Brock 1978), this research continued to deliver exiting results on phototrophic bacteria over almost 50 years, in particular by the work of T. Brock and M. Madigan, of B. Pierson and D. Castenholz, and more recently also by the group of D. Bryant (Bryant et al. 2007; Tank and Bryant 2015a). Indeed, most of the thermophilic phototrophic bacteria known to date (see below) have been isolated from Yellowstone hot springs (Castenholz and Pierson 1995; Madigan 2003).

In Russia, a prominent alkaline hot spring is Bol'sherechenskii Hot Spring in the Bol'shoi River Valley near Baikal (Namsaraev et al. 2003), which has a number of remarkable properties. This spring aroused special interest because of the high temperature of the effluent water of 72–74°, the high pH from 9.2 to 9.8, and a remarkable high sulfide content of >12 mg/l (Kompantseva and Gorlenko 1988; Namsaraev et al. 2003). In the spring effluents microbial mats extend from 62 °C down to <30 °C with cyanobacteria (*Phormidium valderiae* and *Synechococcus elongatus*) and *Chloroflexus aurantiacus* as dominant bacteria. Despite of the high pH and sulfide content, cyanobacteria together with *Chloroflexus aurantiacus* are major components of the mats at 62–51 °C. In thin microbial mats at lower temperatures from 39 to 51 °C again *Chloroflexus aurantiacus* together with cyanobacteria was dominant, but in addition purple nonsulfur bacteria, heliobacteria, and aerobic phototrophic bacteria (*Roseococcus thiosulfatophilus*) were found. *Rhodopseudomonas palustris* and *Rhodomicrobium vannielii* were most abundant and widely occurring species and in addition *Rubrivivax gelatinosus*, *Blastochloris viridis*, and *Blastochloris sulfoviridis* were identified (Namsaraev et al. 2003). Despite their presence at temperatures around 50 °C and pH >9 in the springs, in laboratory cultures photosynthetic rates were highest at neutral pH and the growth range was between 20 and 45 °C. Two new heliobacteria, *Heliobacterium sulfidophilum* and *Heliobacterium undosum* (optimum growth at 30–35 °C and pH 7–8) were first isolated from these springs (Bryantseva et al. 2000b). Chromatiaceae were found only in the lower temperature range around 30 °C and green sulfur bacteria were absent.

In Japan, numerous hot springs are distributed throughout the country and many are well taken by the Japanese people. Well studied hot springs in Japan are Nakabusa and Naganoyu hot springs, in which mats of *Chloroflexus* develop together with cyanobacteria at 50–65 °C (Hisada et al. 2007; Kubo et al. 2011) and from which *Roseiflexus castenholzii* was isolated for the first time (Hanada et al. 2002). From a larger number of these Japanese hot springs *Chloroflexus aurantiacus* has been isolated and detailed studies on the isolated cultures revealed two strains of a new species, *Chloroflexus aggregans* (Hanada et al. 1995b).

In central Tibet, the diversity of several microbial mats of thermal habitats was studied using 16S rRNA gene analysis with samples recovered from three hydrothermal locations, the Daggyai Tso Geyser field, a geyser field near Yibbug Caka, and an unnamed location (Lau et al. 2009). All samples originated from waters of 60–65 °C and pH 7.0–7.4 and sulfide could not be detected at any of the sites. Most interesting, all samples showed a different composition of the communities, including the potentially phototrophic phyla (Cyanobacteria, Chlorobi, Chloroflexi, and Proteobacteria) and also the highly diverse chemotrophic bacteria. The relevance and importance of the phototrophic organism in the mats/springs remained unclear, in particular due to the application of the ribosomal rRNA gene, which in general is problematic in analyses of functional groups (Tank et al. 2009, 2011). Only a few of the sequences are associated with the phylum Chloroflexi. Some are related to those of *Roseiflexus castenholzii* but appear to be different from the type strain and from those found in Yellowstone National Park (Hanada et al. 2002; Boomer et al. 2002). Others show some association with *Heliothrix oregonensis*. None of the sequences presented of Chlorobi is matching the established lines represented by the green sulfur bacteria of *Chlorobium*, *Chlorobaculum*, and *Prosthecochloris* species (Lau et al. 2009; Alexander et al. 2002; Imhoff 2003). Therefore the presence and diversity of phototrophic bacteria in these hot springs remains unclear.

In bacterial diversity studies of hot springs in Greenland unfortunately also primers for 16S rRNA gene analysis were used. Although they were regarded as specific for different groups of phototrophic prokaryotes, in most instances the sequences were distantly related to those of known phototrophic bacteria and therefore, the phototrophic nature of the corresponding bacteria remains obscure (Roeselers et al. 2007).

The Genus **Chloroflexus** *and* **Chloroflexus aurantiacus** *(55 °C)*

The description of *Chloroflexus aurantiacus* isolated from a Yellowstone hot spring (Pierson and Castenholz 1974a, b) and the work at these hot springs over the 1970s by the group of T. Brock (see Brock 1978) set a hallmark for the thermophilic phototrophic bacteria and up to date *Chloroflexus aurantiacus* is the most thermotolerant phototrophic bacterium known, with a temperature maximum of 70 °C (Figs. 1 and 2).

Table 1 Thermophilic anoxygenic phototrophic bacteria

Species	T-optimum	T-range	Habitat	References
		Upper limit		
Chloroflexi				
Chloroflexus aurantiacus	55	70	Yellowstone hot springs	Pierson and Castenholz (1974a, b)
Chloroflexus aggregans	55	60	Metobuchi and Yufuin hot springs, Japan	Hanada et al. (1995b)
Roseiflexus castenholzii	45–55	60	Nakabusa hot spring, Japan	Hanada et al. (2002)
Heliothrix oregonensis		56–60	Octopus hot spring, Yellowstone; Hunter's hot spring, Oregon	Pierson et al. (1985)
Heliobacteria				
Heliobacterium modesticaldum	52	56	Yellowstone hot springs	Kimble et al. (1995)
Acidobacteria				
Chloroacidobacterium thermophilum	51	44–58	Octopus hot spring, Yellowstone	Tank and Bryant (2015a, b)
Chlorobi				
Chlorobaculum tepidum	48	55	Roturoa Travelodge spring, New Zealand	Wahlund et al. (1991)
Gammaproteobacteria				
Thermochromatium tepidum	48–50	34–56	Mammoth hot springs, Yellowstone	Madigan (1986)
Alphaproteobacteria				
Rhodopseudomonas cryptolactis	40	46	Thermopolis hot spring, Wyoming	Stadtwald-Demchick et al. (1990)
Rhodocista centenaria	40–42	45	Thermopolis hot spring, Wyoming	Favinger et al. (1989)
Rhodopseudomonas pentothenatexigens	30	25–50	Paddy soil, Andhra Pradesh, India	Kumar et al. (2013)
Rhodopseudomonas thermotolerans	30–35	25–60	Paddy soil, Andhra Pradesh, India	Kumar et al. (2013)

After first observations on colored microbial mats in hot springs of the Yellowstone National Park by T. Brock (see Brock 1978), soon first cultures of *Chloroflexus aurantiacus* were obtained (Pierson and Castenholz 1974a, b) and a number of important key properties of these bacteria were established. M. Madigan in his Ph.D. thesis (in the lab of T. Brock) performed first intensive studies on the physiology of *Chloroflexus aurantiacus* including their ability to use sulfide, and fix

Fig. 1 *Chloroflexus aurantiacus* thrives at the borderline of high temperatures in hot spring effluents. This hot spring effluent in Yellowstone National Park nicely shows the development of phototrophic microbial mats with *Chloroflexus aurantiacus* reaching up to the maximum tolerated temperature of approx. 70 °C, first developing at the rim of the effluent where the water cools down faster (photo: JF Imhoff)

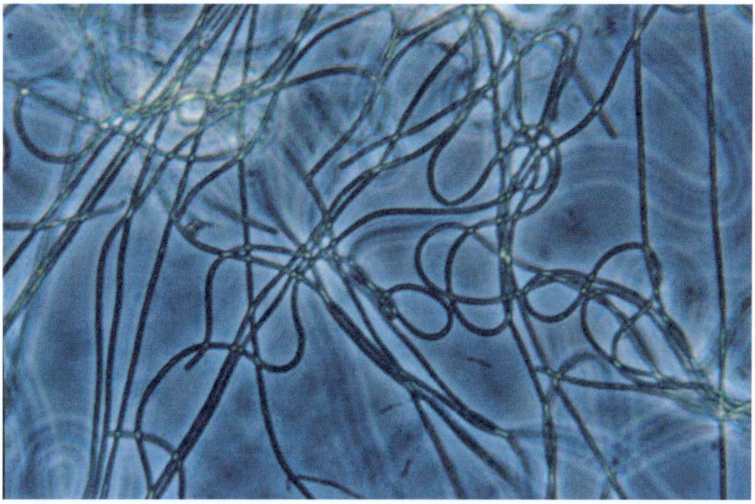

Fig. 2 Microscopic picture of a culture of the filamentous gliding *Chloroflexus aurantiacus* from Aix-les-Thermes hot spring in southern France (photo: JF Imhoff)

CO_2 (Madigan and Brock 1975, 1977; Madigan et al. 1974). *Chloroflexus aurantiacus* can grow either photoorganotrophically under anoxic conditions in the light or chemoheterotrophically in the presence of oxygen. Also slow growth under photoautotrophic conditions is possible with sulfide as electron donor. This metabolic flexibility together with the adaptation to high temperatures is an important prerequisite for the development of *Chloroflexus* under different conditions in hot spring effluents. Similar to the green sulfur bacteria, *Chloroflexus aurantiacus* has chlorosomes, but its photosystem and the photosynthetic energy transfer are different from those of the Chlorobi. Unlike the green sulfur bacteria, it has a type-2 photosynthetic reaction center and lacks the bacteriochlorophyll-a binding protein (Fenna–Matthews–Olson protein) involved in energy transfer from chlorosomes to the reaction center in green sulfur bacteria (Fenna et al. 1974; Blankenship et al. 1995). *Chloroflexus* was the first representative of a new family and new phylum (Oyaizu et al. 1987) now known as the Chloroflexaceae and Chloroflexi.

Representatives of *Chloroflexus aurantiacus* and related phototrophic species are found worldwide in alkaline hot springs with pH of 7.5–8.5 (>6.2). From a larger number of hot springs throughout Japan, *Chloroflexus aurantiacus* and relatives were isolated from approx. 80 % of the springs and they were found at temperatures of 50–70 °C and pH 6.4–8.2 (Hanada et al. 1995a). A second thermophilic species of this genus, *Chloroflexus aggregans*, is growing well between 50 and 60 °C and was isolated from hot springs in Japan (Hanada et al. 1995b).

Heliothrix oregonensis

Another thermophilic filamentous gliding bacterium phylogenetically related to *Chloroflexus aurantiacus* is *Heliothrix oregonensis*. It is an aerotolerant, highly motile photoheterotrophic bacterium with upper temperature limits at 56–60 °C (Pierson et al. 1985). It has no chlorosomes and contains bchl a, but no bchl c or d (Pierson et al. 1985). It is found in Hunter's Hot Springs and Octopus Spring (Yellowstone) below 56 °C (Castenholz and Pierson 1995). So far, no pure cultures are available.

Roseiflexus castenholzii *(45–55 °C)*

Distinct red-colored mats consisting of filamentous phototrophic bacteria first have been observed in neutral and alkaline hot springs in Yellowstone National Park (Boomer et al. 2000; Castenholz and Pierson 1995) and in Japan (Hanada et al. 2002), but over many years could not be grown in pure culture.

Roseiflexus castenholzii was originally isolated from a microbial mat of Nakabusa hot spring in Japan (Hanada et al. 2002). The cells lack chlorosomes (and bchl c), contain *bchl a* and gamma-carotene as photosynthetic pigments. *Roseiflexus castenholzii* grows not only photoheterotrophically under anaerobic conditions in the light, but also chemoheterotrophically under aerobic-dark conditions (Hanada et al. 2002). Good growth was observed between 45 and 55 °C, not above 60 °C and below 40 °C. The pH range for growth was from 6.0 to 9.0, the optimum at pH 7.5–8.0 (Hanada et al. 2002). Additional strains related to but distinct from *Roseiflexus castenholzii*, *Roseiflexus* sp. RS-1 and RS-2, were isolated from Yellowstone Octopus hot spring with temperature optima at 55–60 °C (van der Meer et al. 2010).

Red-colored mats observed in several alkaline hot springs of Yellowstone National Park showed a remarkable microheterogeneity of 16S rRNA gene sequences related to *Roseiflexus castenholzii* (Boomer et al. 2002; Van der Meer et al. 2010). All 44 sequences obtained from five different hot springs formed a large group clearly separated from other filamentous gliding phototrophic bacteria including representatives of *Chloroflexus*, *Heliothrix*, and *Oscillochloris*. *Roseiflexus castenholzii* has only 83.8 % similarity of its 16S rRNA gene sequence to *Chloroflexus aurantiacus* (Hanada et al. 2002). The *Roseiflexus* group itself is separated into at least two major clusters (YRL-A and YRL-B) with the tendency of subclusters being formed of sequences from the same spring. One of the clusters contained, e.g., in separated groupings all ten sequences from Spray Geyser and eight sequences from Shoshone spring (Boomer et al. 2002). The presence of phylotypes related to *Roseiflexus castenholzii* in Alla hot springs in the Baikal rift zone (Russia) was established using both 16S rRNA and *pufLM* gene sequences (Gaisin et al. 2015). Altogether, bacteria of the *Roseiflexus* type appear widespread in hot springs all over the world.

Though many hot springs occur on all of our continents, all of these are isolated "islands" surrounded by ambient temperature environments. As these hot places are several 1000 km apart from each other, a central question always has been the spreading of the thermophilic bacteria inhabiting these hot springs and the phylogenetic relationship between populations at different sites. An important contribution to this aspect is the comparison of the phylogeny of geographical separated mats of *Roseiflexus* around the world and the correlation found between geographic distance and genetic divergence (Gaisin et al. 2016). 16S rRNA gene sequences with similarities higher than 95 % to *Roseiflexus castenholzii* and *Roseiflexus* RS-1 were used for phylogeographic considerations and a comparison was made of 184 sequences from 26 thermal springs (12 areas of hydrothermal activity) with 11 sequences from hot springs in Yellowstone National Park (Gaisin et al. 2016). In this study, phylotype sequences of *Roseiflexus* were compared from hot springs of Japan (Nakabusa, Jinsham), USA (Yellowstone National Park), Kamchatka (Thermophilny hot spring), Thailand, Tibet, Russia (Alla hot spring), Mongolia (Tensher hot spring), Bulgaria (Rupin Basin), and Chile (El Tatio, Copahue area) (Gaisin et al. 2016).

It is amazing to see that sequences from the geographic separated sites form individual clusters and their relationships fit well the geographic distribution. Only from Yellowstone hot springs two different clusters were identified which grouped also in two geographic subgroups. A clear distinction can be made between the American cluster (representatives from Chile and Yellowstone type 1) and the Eurasian cluster with one subgroup containing representatives from Kamchatka, Japan and the Yellowstone type 2 sequences, a second subgroup containing representatives from hot springs in Thailand, Tibet, Russia, and Mongolia, and a third subgroup with *Roseiflexus* from hot springs in Bulgaria and Eastern China (Gaisin et al. 2016).

The authors conclude that the *Roseiflexus* group forms a single ecotype of obligate thermophilic bacteria with ecophysiological dependence on cyanobacterial mats and different geotypes dependent on the geographic region of the hydrothermal vents (Gaisin et al. 2016). As *Roseiflexus* is restricted to highly thermophilic habitats, and the branching pattern of *Roseiflexus* phylotypes correlates with the geological record of plate formation, it was proposed that *Roseiflexus* geotypes were maintained in their biosphere over geological time scales (Gaisin et al. 2016).

Chloroacidobacterium thermophilum *(51 °C)*

Microbial mats from Octopus and Mushroom springs in the Yellowstone National Park were again the source for another new thermophilic phototrophic bacterium being the first phototrophic representative of the Acidobacteria (Bryant et al. 2007). *Chloroacidobacterium thermophilum* contains bacteriochlorophyll *a* and *c* and has chlorosomes and the bacteriochlorophyll-a binding protein FmoA (Fenna–Metthew–Olson protein) together with type-1 reaction centers (Tank and Bryant 2015a). These properties are shared with the strictly anaerobic green sulfur bacteria, the phototrophic Chlorobi. However, *Chloroacidobacterium* is a photoheterotrophic, microaerophilic bacterium and synthesis of bacteriochlorophyll requires the presence of oxygen. It is strictly dependent on light and oxygen. It is a moderately thermophilic bacterium growing well between 46 and 55 °C (optimum at 51 °C) and pH 5.5–9.0 (optimum at 7.0) (Tank and Bryant 2015b).

Chlorobaculum tepidum *(48 °C)*

Chlorobaculum tepidum (formerly *Chlorobium tepidum*, Wahlund et al. 1991; Imhoff 2003) is the only green sulfur bacterium adapted to high temperatures of hot springs, though genes of photosynthetic reaction centers in metagenome data from Yellowstone hot springs indicate the presence of another green sulfur bacterium somehow related to *Chloroherpeton thalassium* (Bryant et al. 2007), which has been isolated from a marine coastal habitat (Gibson et al. 1984).

Chlorobaculum tepidum is found in hot springs at high sulfide concentrations (0.3–1.8 mM), at moderate temperatures (40–55 °C), and at pH 4.3–6.2 (optimum is 6.8–7.0). Such conditions were found in Roturoa Travelodge Spring, New Zealand and exclude cyanobacteria due to the sulfide content and *Chloroflexus* due to the low pH and allow *Chlorobaculum tepidum* to develop as sole phototrophic bacterium (Madigan 2003).

Chlorobaculum tepidum grows best at 48 °C and up to temperatures of 52–55 °C with sulfide (1–4 mM) as photosynthetic electron donor and also uses thiosulfate (which avoids the tedious feeding of cultures with sulfide). The pH range is 5.9–7.2 (optimum 6.8–7.0). It has been intensively studied (Wahlund et al. 1991; Madigan 2003).

Heliobacteria and Heliobacterium modesticaldum *(52 °C)*

Heliobacteria are Firmicutes and represent unique phototrophic bacteria which form endospores and are highly oxygen-sensitive. The presence of bchl *g* and chl *a* distinguishes them from all other phototrophic bacteria (Gest and Favinger 1983). A single thermophilic species has been described of this group of phototrophic bacteria, which is *Heliobacterium modesticaldum* (Kimble et al. 1995). It has been isolated from Yellowstone hot springs and hot springs on Iceland (Stevenson et al. 1997). The temperature optimum is at 52 °C and the temperature maximum at 56 °C (Kimble et al. 1995).

In enrichments from Hunter's Hot Springs (Lakeview, Oregon) regularly bacteria resembling another heliobacterium, *Heliobacillus mobilis* were enriched at temperatures from 40 to 53 °C with an upper temperature limit at 53 °C (Castenholz and Pierson 1995).

Thermochromatium tepidum *(50 °C)*

Thermochromatium tepidum (formerly *Chromatium tepidum* Madigan 1986; Imhoff et al. 1998b) was the first thermophilic purple sulfur bacterium with a growth optimum of approx. 50 °C brought into pure culture. It was isolated from Mammoth hot springs in Yellowstone National Park and can grow photoautotrophically with sulfide as electron donor. Acetate and pyruvate are the only organic compounds that are photoassimilated. The optimum temperature for growth is at 48–50 °C and the temperature range is from 34 to 57 °C. The pH optimum is at pH 7.0 and it is a typical freshwater bacterium without requirement for NaCl, which is growth inhibitory at >1 % (Madigan 1986). At sulfide concentrations >25 μM, a pH >6.2, and temperatures from 50 to 55 °C *Thermochromatium tepidum* may be the sole photoautotroph in hot springs (Madigan 2003). Although such conditions are not common, they are found in a number of North American hot springs, including Thermopolis Hot Springs and Mammoth Hot Springs, where a mat with *Chloroflexus* is present (Castenholz and Pierson 1995). *Thermochromatium tepidum* may be more widely distributed in suitable hot springs worldwide.

Marichromatium gracile *(< 44 °C)*

A single strain of a purple sulfur bacterium identified as a *Marichromatium* species with high tolerance of sulfide (up to 16 mM) and growth up to 44 °C was isolated from sediments of the German Waddensea (Serrano et al. 2009). It had the highest sequence similarity (>99 %) of the 16S rRNA gene to *Marichromatium gracile* but on basis of its different phenotypic characteristics in particular the tolerance of elevated temperatures was assigned to a biotype of this species (*Marichromatium gracile* biotype *thermosulfidiphilum*) (Serrano et al. 2009). The slight thermophilic properties of marine phylotypes of *Marichromatium gracile* were substantiated by molecular analysis of temperature dependent enrichment cultures using *pufLM* gene sequences, which demonstrated that *Marichromatium gracile* became the dominant part of the community at 41 and 44 °C, but was not detected in the sediment source sample from a Baltic Sea coastal lagoon and in enrichments below 40 °C (Tank et al. 2009). Based on the sequence identity with the *Marichromatium gracile* type, it was concluded that the thermophilic property was inherent to this species or at least to some strains of this species (Tank et al. 2009).

Among the thermophilic phototrophic bacteria mentioned so far, *Marichromatium gracile* is the only one not obtained from a hot spring habitat. Although it is only a borderline slightly halophilic bacterium, its general presence in shallow marine coastal pools points out the importance of temperature adaptation in these habitats that are heated by the sun to temperatures often exceeding 40 °C.

Purple Nonsulfur Bacteria (38–42 °C)

Purple nonsulfur bacteria have been regularly observed as companions in microbial mats of hot springs, though their temperature optima and tolerances were much lower and generally growth was not obtained >45 °C. Enrichments for purple nonsulfur bacteria from hot springs of western North America invariably resulted in success at 40 °C, sometimes at 45 °C, but never at 50 °C (Castenholz and Pierson 1995). From microbial mats developing at 50–65 °C in Nakabusa and Nakanoyu hot springs in Japan several purple nonsulfur bacteria were isolated at 37 °C (Hisada et al. 2007). They showed optimal growth at around but not exceeding 40 °C and had upper temperature limits at 41–45 °C. The isolates were phylogenetically identified as *Blastochloris sulfoviridis*, *Phaeospirillum molischianum*, *Rhodopseudomonas palustris*, *Rhodomicrobium vannielii*, and *Rubrivivax gelatinosus*. Other isolates were most closely related to *Rhodoplanes roseus* and to *Rhodoplanens cryptolactis* (Hisada et al. 2007). Interestingly, all of these species were found also in other hot springs around the world. In a hot spring in southern France at Aix-Les-Thermes a *Chloroflexus* mat developed and several purple nonsulfur bacteria were isolated from the mat tentatively identified as *Blastochloris viridis*, *Rhodopseudomonas palustris*, *Rhodobacter sphaeroides*, and *Rubrivivax gelatinosus* (Imhoff, unpublished). All of these isolates were not thermophilic. A budding purple nonsulfur bacterium containing bacteriochlorophyll *b* and growing optimally at 42 °C and up to 47 °C was isolated from a hot spring microbial mat in New Mexico (Resnick and Madigan 1989). The properties of this strain GI resemble those of *Blastochloris sulfoviridis*. An isolate of *Rhodomicrobium vannielii* from the sulfidic Gadek hot spring in Malaysia (50–58 °C, pH 6.8–8.0) showed good growth at 38–40 °C and pH 7.0 (Ainon et al. 2006).

Also two new species of purple nonsulfur bacteria were isolated from hot springs. *Rhodoplanes cryptolactis,* formerly *Rhodopseudomonas cryptolactis* was isolated from Thermopolis hot spring (Wyoming) with a temperature optimum at 40 °C and a maximum at 46 °C (Stadtwald-Demchick et al. 1990).

Rhodocista centenaria (formerly *Rhodospirillum centenum*, Favinger et al. 1989; Kawasaki et al. 1992) also was isolated from the edge of the source pool of Thermopolis hot spring (Wyoming) and later also from a number of hot springs in Yellowstone National Park (Nickens et al. 1996). Its outstanding properties include formation of heat-resistant cysts, formation of the so-called R-bodies inside the cells, and a special swarming behavior on agar plates related to phototactic responses (Favinger et al. 1989; Ragatz et al. 1995). These properties and in addition its phylogenetic position clearly separate this bacterium from all other *Rhodospirillum* species and therefore it was transferred into a new genus and is recognized now as *Rhodocista centenaria* (Kawasaki et al. 1992), which has uncertain affiliation among the Alphaproteobacteria. [It should be noted that properties of this species are absolutely not conform to those specified for the genus *Rhodospirillum*. The transfer of this species into a new genus with the name

Rhodocista was formally correctly proposed and published by Kawasaki and coauthors. Confusion is caused in the literature as a number of scientists working with this bacterium do not accept these facts.] *Rhodocista centenaria* grows photoheterotrophically as well as chemoheterotrophically and has a temperature optimum at 40–42 °C and a maximum at 45 °C, though the temperature of the sample source was around 55 °C (Favinger et al. 1989).

Apparently, purple nonsulfur bacteria whose upper temperature limits do not exceed 45 °C in laboratory cultures are regularly found at temperatures up to 65 °C in microbial mats of hot springs. Only a few borderline slightly thermophilic/tepidophilic species are known. The general presence of purple nonsulfur bacteria at temperatures that exceed their upper temperature limit measured in laboratory cultures provokes questions concerning the accurate measurements of the in situ temperature at the microscale level as well as towards the adaptation of these bacteria to the in situ conditions. In this context it is interesting that two recently described mesophilic new *Rhodopseudomonas* species have growth optima at 30–35 °C but upper limits of temperature tolerance at 50 °C (*Rhodopseudomonas pentothenatexigens*) and 60 °C (*Rhodopseudomonas thermotolerans*) (Kumar et al. 2013).

Adaptation to High Temperatures

In general, it is apparent that there is a discrepancy between the temperature range of growth in laboratory cultures, in particular of purple nonsulfur bacteria from hot springs and their occurrence in the environmental settings. Therefore, a few remarks should be made to possibly explain this discrepancy.

First of all, the hot effluent is cooling down horizontally and vertically. It is in contact with the underlying soil or ground (as well as to the air) of ambient temperature which is much below the temperature of the hot waters and gradients from 70 °C to 20 °C (or lower) may occur from the top of the mat to the ground. Therefore, steep temperature gradients exist that possibly may allow slightly or nonthermophilic bacteria to grow at the lower and cooler part of the microbial mat. This aspect may well apply for purple nonsulfur bacteria as long as they can cope with the limited light available at the bottom of the mat, but it can not fully explain the large discrepancies between the indicated in situ temperature and the measured temperature limits in the laboratory.

Secondly, it should be realized that two mechanisms of adaptation to life at high temperature exist, an intrinsic part associated with structural adaptation of the cell components and an extrinsic part associated with accumulation of compatible solutes to protect cellular structures. The role of the extrinsic part is well recognized (Chin et al. 2010), but has been completely neglected in discussions on adaptation of less thermophilic microorganisms to habitats with higher temperatures.

Structural Adaptation of Malate Dehydrogenase

Work on temperature response of microorganisms has focused on the intrinsic part, in particular on the modified structure of proteins in microorganisms adapted to live in extreme environments. Much emphasis has been given to the structural adaptation of proteins to extremely high temperatures of extremely thermophilic and hyperthermophilic microorganisms (Szilágyi and Závodszky 2000; Argos et al. 1979). Interestingly, no new amino acids, covalent modifications, or structural motifs were found in proteins of extremophilic microorganisms to explain the ability of their enzymes to function under the extreme conditions (Fields 2001). Instead, appropriate redistribution of the same intramolecular interactions active in protein stabilization at moderate temperatures are sufficient to maintain structural integrity at hot or cold extremes. Most important, the key for the proper function of the proteins is a balance between molecular stability and structural flexibility to ensure ligand binding as well as catalysis at appropriate rates (Fields 2001).

An interesting case study has compared the thermal stability of a single enzyme, the malate dehydrogenase of three phototrophic green sulfur and nonsulfur bacteria with different temperature responses, of *Chloroflexus aurantiacus*, *Chlorobaculum tepidum*, and "*Chlorobium vibrioforme*" (Dalhus et al. 2002). Despite the interesting aspects regarding the protein structure, a critical comment must be added in respect to the lack of identification of the investigated strains. Because no strain designations were given in the publication, the identity of the organisms and their phenotypic properties cannot clearly be delineated. Especially the identity of "*Chlorobium vibrioforme*" remains unclear, because of misclassification of a number of strains designated to this species but recognized after phylogenetic reclassification as strains of the different species *Chlorobium luteolum*, *Chlorobium phaeovibrioides*, *Prosthecochloris vibrioformis*, and *Chlorobaculum parvum*, depending on the strain (Imhoff 2003).

The authors indicate optimum temperatures of 32 °C for the "*Chlorobium vibrioforme*," 47 °C for *Chlorobaculum tepidum*, and 55 °C for *Chloroflexus aurantiacus*. The enzyme of "*Chl. vibrioforme*" is rapidly inactivated at 55 °C, that of *Cba. tepidum* has a half-lifetime >30 min at this temperature, but is inactivated at temperatures of 60 and 65 °C rapidly, at which the enzyme from *Chloroflexus aurantiacus* still has a half-lifetime of 25 min. This result is well in accord with the temperature responses of the three bacteria. The thermal stability of the MDH oligomers arises from the combined action of increased numbers of salt bridges, hydrogen bonds, and aromatic interactions (Dalhus et al. 2002). The increased thermal stability of the MDH of *Chlorobaculum tepidum* as compared to that of "*Chlorobium vibrioforme*" was considered to relate to the presence of specific polar residues in *Chlorobaculum tepidum* that form additional hydrogen bonds and stabilize the monomeric subunit. The MDA of *Chloroflexus aurantiacus* contains more proline and alanine residues, which are known to stabilize the protein backbone. This MDH also has a larger number of ionic-pair and hydrogen bond interactions that stabilize the quaternary structure of the protein (Dalhus et al. 2002).

Accumulation of Compatible Solutes

In addition to the structural adaptation, a more flexible mechanism involves the accumulation of compatible solutes that are accumulated in response to heat stress. Again, focus of research has been given to the solutes accumulated by extremely thermophilic and hyperthermophilic archaea and bacteria (Santos and daCosta 2002), but little is known about temperature dependent accumulation of compatible solutes in slightly or moderately thermophilic bacteria. Detailed studies on these interactions and on the accumulation of compatible solutes in response to increased temperatures in phototrophic bacteria are lacking. It is well known that in halophilic microorganisms compatible solutes are accumulated as osmotica in order to maintain not only cell turgor but also stability and flexibility of cellular structures and functions (Brown 1976; Imhoff and Rodriguez-Valera 1984; Imhoff 1986, 1993, 2001). It was noted that in halophilic phototrophic bacteria, the adaptation to high salt concentrations, which triggers the accumulation of high concentrations of compatible solutes, comes along with an increased tolerance to higher temperatures (Imhoff 1986, 1993, 2001). Solutes accumulated to one stress factor (salt) could well give resistance also to other factors (high temperature). In fact, it could be demonstrated that accumulation of compatible solutes triggered by increased salinities can also extend the temperature range of activity and growth of microorganisms (Chin et al. 2010). Therefore, it is well conceivable that the accumulation of compatible solutes may be triggered by elevated temperatures and can extend the upper temperature limits for growth. Trehalose has been found to accumulate as a response to different low to moderate stress conditions, such as elevated temperature and osmolarity/salinity and to prevent proteins from denaturation thereby contributing to protection against several stress conditions (Singer and Linquist 1998).

Generally, low temperatures promote noncovalent interactions and rigidity of cellular macromolecules, while high temperatures reduce these interactions and promote structural disorders of the macromolecules. Accordingly, structural adaptations should allow less interaction at lower temperatures but stronger interaction at high temperatures to compensate these effects (Chin et al. 2010). In consequence, temperature effects can be counterbalanced by solutes that promote molecular disorder (chaotropic solutes) at low temperatures and solutes that promote molecular interactions at high temperatures (kosmotropic solutes) (Chin et al. 2010).

Such solutes might be accumulated by biosynthesis or by uptake from the external environment. In microbial mats they may be produced by part of the community and leaked out of the cells into the mat. Here, they are available also for non-producing bacteria and may be accumulated by uptake mechanisms and thereby give increased tolerance to high temperatures to those bacteria unable to perform their biosynthesis. In the absence of such solutes in pure laboratory cultures, accumulation is not possible and accordingly the upper limit of growth will be lower. Such scenario at least in part could explain the findings of non-thermophilic or slightly thermophilic purple nonsulfur bacteria in waters exceeding their upper temperature limit determined in laboratory cultures. The same holds for other slightly and moderately thermophilic phototrophic bacteria.

Sea Ice and Cold Waters

Most prominent cold habitats on earth are represented by the ice-covered polar region and the deep ocean waters. While the deep ocean is an entirely dark environment, if we neglect the black body radiation emitted by hot vents, which is insufficient to support phototrophic life (see van Dover et al. 1994) and not suited for life of phototrophic organisms, the polar ice or sea ice in general might well be inhabited by phototrophic microorganisms. During the freezing of seawater, salt is excluded from the ice and increased salt concentrations are found within the small channels that establish in the frozen ice (Eicken et al. 2000). Though colored layers have been observed in polar ice (E. Helmke, AWI, pers. comm.) and the dense microbial communities that develop in distinct layers of the sea ice may produce anoxic conditions and can be expected to receive sufficient light to promote phototrophic life, only few studies dealing with anoxygenic phototrophic bacteria in this environment are known to date. Based on the analysis of clone libraries, aerobic anoxygenic phototrophic bacteria related to the *Roseobacter* clade have been identified in Antarctic seawater and sea ice using 16S rRNA gene sequences (Maas et al. 2012) and *pufM* sequences (Koh et al. 2011). Bacterial communities of sea ice that occasionally is formed in the southern Baltic Sea during the late winter season have been studied by molecular genetic approaches and found to develop a distinct layer containing phototrophic purple sulfur bacteria within weeks (Petri and Imhoff 2001).

Baltic Sea Ice

It is well established that high concentrations of dissolved organic matter accumulate in the internal brines of sea ice and that active microbial communities are living within these brines that may reach twice the salinity of the freezing seawater (Mock and Thomas 2005). Denitrification has been demonstrated in certain zones of sea ice from the Baltic Sea (Kaartokallio 2001) and from the Arctic (Thomas and Dieckmann 2002) and it is tempting to assume that the degradation of the organic matter may provide oxygen-deficient or even anoxic zones within layers of the sea ice. Little attention has been paid to these processes and the presence of anoxygenic phototrophic bacteria.

In winter 1995/1996 from the end of January on, an ice cover was formed in the southern Baltic Sea of the Kiel Fjord (the last such event since two decades by now) that lasted approx. 2 month and reached a thickness of approx. 70 cm (Petri and Imhoff 2001). Molecular genetic analysis using 16S rRNA gene sequences as the target revealed a characteristic distribution for each of the analyzed horizons and also significant differences to the bacterial community in the underlying water (Petri and Imhoff 2001). In the 50–60 cm horizons of the ice core, three bands in highly resolved DGGE gels, including the dominant one, were related to purple sulfur bacteria. An almost complete 16S rRNA sequence obtained from one of these bacteria was similar to those of *Rhabdochromatium marinum* (94.4 %) and *Thiorhodovibrio winogradskyi* (92.2 %) (Petri and Imhoff 2001). The relations of these 16S rRNA gene sequences to *Thiorhodovibrio winogradskyi* resemble quite well phylogenetic relations of *pufLM* sequences to *Thiorhodovibrio* (phylotypes 15–18) found in a coastal lagoon of the Kiel bight (approx. 2 km distance to the ice core sampling) in a study a few years later (Tank et al. 2011). It is assumed that these bacteria are around in the oxic waters of Kiel Bight in low numbers to seed a growing ice cover. The clear stratification of the sea ice horizons and the distribution of bacteria within the different horizons indicated the presence of anoxic parts within the sea ice. Although both purple sulfur bacteria were not considered to be specifically adapted to the cold sea ice environment, they apparently were the fittest to adapt to and grow under the conditions provided within the sea ice. This view was supported by temperature- and salt-dependent experiments with samples from the Baltic Sea coastal lagoon, in which these phylotypes had a specific advantage at low temperature (below 15 °C) and elevated salinity (approx. 5 %) (Tank et al. 2011). Both bacteria represent true marine phototrophic bacteria with salt optima at 1.5–5 % (*Rhabdochromatium marinum*) and 2–3 % NaCl (*Thiorhodovibrio winogradskyi*) and are phylogenetically related species (Imhoff et al. 1998b). While *Thiorhodovibrio winogradskyi* (first isolated from Mahoney Lake and also found in North Sea and Baltic Sea) can grow under microoxic conditions and exhibits high specific respiration rates (Overmann et al. 1992), *Rhabdochromatium marinum* is an obligate phototrophic and strictly anaerobic bacterium (Dilling et al. 1995). Although the sequences from the sea ice samples are clearly associated with these two known species, the phylogenetic distances indicate that conclusions in regard to their physiological properties have to be taken with care. However, a purple sulfur bacterium with preference to grow at lower temperatures and capacities to use and remove oxygen by respiratory activity, such as *Thiorhodospira winogradskyi*, would be a perfect candidate to settle in the sea ice.

Antarctic Lakes and Fjords

In Antarctica several lakes are found that exclusively have microbial life and some are almost permanently frozen with an ice cover of 4–6 m depth (Priscu et al. 1987). Anoxygenic phototrophic bacteria have been studied in the waters of some of these lakes.

Lake Fryxell in Taylor Valley (Antarctica) is a stratified freshwater lake and has slightly saline waters at the bottom (Madigan 2003). Sulfide is produced in the sediments (19 m depth) and maybe present up to 9 m depth. The temperature remains low all over the year and is constant at 2 °C in 10 m depth (Karr et al. 2003). Although the conditions in this lake would be expected to favor their development, no evidence was found for the presence of phototrophic green and purple sulfur bacteria (Achenbach et al. 2001). However, high diversity of purple nonsulfur bacteria was found by analyzing the water column of the lake using *pufM* primers and sequence analysis of the bands separated by DGGE (Karr et al. 2003). Though the sequence information (only 228 nt in length) was not reliable for a clear species assignment by the sequences, it appears that a diverse array of sequences was related to purple nonsulfur bacteria and aerobic phototrophic bacteria such as *Roseateles*, *Rubrivivax*, *Rhodoferax, Acidiphilium*, and also *Bradyrhizobium* (Karr et al. 2003). The dominance of Betaproteobacteria in this lake is in accord with a general preference of Betaproteobacteria for freshwater habitats, though a number of unique and interesting aspects are associated with these communities. Two isolates of this lake are the first representatives of the purple nonsulfur bacteria forming gas vesicles and thus appear well adapted to the life in the water column of this lake. One of these strains is characterized as *Rhodoferax antarcticus* growing best at 15–18 °C and, though slowly, down to 0 °C, but not above 25 °C (Madigan et al. 2000). It grows photoheterotrophically and is remarkably tolerant to sulfide up to 4 mM (Jung et al. 2004). These properties demonstrate that this bacterium is well adapted to the conditions in Lake Fryxell.

In the Vestfold hills in East Antarctica several permanently frozen hypersaline lakes exist with high concentrations of sulfide in the hypolimnion (Burke and Burton 1988a, b). These lakes and several fjords in the area were formed approx. 5700 years ago (Gibson 1999). Sulfate reduction is active in a number of these lakes and a chemocline develops (Burke and Burton 1988a, b). In Ellis Fjord, which is 10 km long, up to 100 m deep, and stratified due to the limited opening to the open ocean, the oxygen/sulfide interface is at approx. 40 m depth. In Burton Lake and Ellis fjord, green sulfur bacteria develop at the oxygen/sulfide interface and were the dominant phototrophic bacteria. They were accompanied by lower numbers of purple bacteria tentatively identified as *Thiocapsa roseopersicina* and *Rhodopseudomonas palustris* (Burke and Burton 1988a, b). Phototrophic sulfur bacteria with properties similar to *Chlorobium* and *Thiocapsa* were isolated, respectively, enriched from these lakes by M. Madigan (Madigan 1998). They had temperature optima around 20 °C and therefore cannot be regarded as psychrophilic. As many bacteria growing in cold environments and according to Madigan (2003) they may be tolerant to the low temperatures and adapted to grow at temperatures well below their growth optimum.

Microbial mats surrounding the almost permanently frozen lakes in McMurdo Dry Valleys (Antarctica) are habitats of phototrophic bacteria (Madigan et al. 2000; Achenbach et al. 2001). When the ice melts during the summer, microbial mats develop at the edges of the lakes and in these mats filamentous cyanobacteria, *Chloroflexus*-like bacteria, and also purple bacteria were found (Madigan 2003).

In the meromictic Lake Vanda phototrophic purple sulfur bacteria of the Chromatiaceae were found (Kriss et al. 1976) and purple nonsulfur bacteria were grown after enrichment and identified as *Rhodomicrobium vannielii*, *Rhodopseudomonas palustris*, and some spirilloid forms of purple nonsulfur bacteria (Takii et al. 1986).

Green sulfur bacteria also represent the dominant phototrophic organisms at the chemocline of Ace Lake which is covered by ice for more than 10 month of the year. With a "highly enriched sample" from this lake metagenomic and metaproteomic analyses were performed and properties of the green sulfur bacteria were inferred from these data (Ng et al. 2010). Several traits of importance for the adaptation to the cold have been identified in this metagenome, including synthesis of monounsaturated fatty acids, biosynthesis of specific polysaccharides, and of proteins active in folding/refolding of proteins (Ng et al. 2010). Because work on isolated green sulfur bacteria from such a lake revealed that they were not psychrophilic but tolerant to the cold temperatures (Madigan 2003), conclusions from the environmental metagenome analysis in regard to the cold adaptation of the green sulfur bacteria have to be taken with care (Table 1).

Acidophilic Phototrophic Bacteria

There are not many acidic habitats that provide proper conditions for anoxygenic phototrophic bacteria, as many acidic environments either are in the dark or represent quite oxic places that are unsuitable for life of anaerobic anoxygenic phototrophic bacteria. Accordingly, there are only few examples of phototrophic bacteria isolated from acidic habitats, such as acidic hot springs, acidic lakes, bogs, and marshes.

With the exception of *Chlorobaculum tepidum* (see paragraph on hot springs) massive developments of phototrophic bacteria at acidic habitats have not been reported. *Chlorobaculum tepidum* has a preference for slightly acidic conditions (as other green sulfur bacteria) and has been isolated from green mats developing in the outflow of acidic hot springs in Iceland, New Zealand, and Yellowstone National Park which have pH values from 4.5 to 6.0, temperatures from 45 to 55 °C, and sulfide concentrations from 0.2 to 2.0 mM (Castenholz 1988; Castenholz et al. 1990; Madigan 2003).

A few purple nonsulfur bacteria are known to prefer slightly acidic pH for growth. *Rhodoblastus acidophilus* (formerly *Rhodopseudomonas acidophila*) and *Rhodomicrobium vannielii* preferably grow at slightly acidic pH in succinate-mineral salts medium (Pfennig 1969). *Rhodoblastus acidophilus* was isolated from acidic lakes, bogs, and marshes and has a growth range from pH 4.8–7.2 with an optimum at pH 5.8 (Pfennig 1969). The purple nonsulfur bacterium *Rhodopila globiformis* (formerly *Rhodopseudomonas globiformis,* Pfennig 1974; Imhoff et al.

1984) was isolated from an acidic warm spring in Yellowstone National Park (aside the Gibbon River) containing sulfide and with a pH from 3.5 to 4 (Pfennig 1974) and from a warm acidic lake (Nymph Lake) also in Yellowstone National Park (Madigan 2003). Growth of *Rhodopila globiformis* with gluconate and mannitol as substrates occurred only at acidic pH 4.9 (Pfennig 1974).

A remarkable group of bacteriochlorophyll-containing aerobic Alphaproteobacteria of the genera *Acidiphilium* and *Acidisphaera* can grow in the pH-range of 3.5–6.0 and has optima around pH 4.5–5.0. Species of the genus *Acidiphilium* represent acidophilic chemotrophic bacteria growing aerobically in a pH-range from 2.5 to 5.9 (some as low as 1.5) and producing bacteriochlorophyll only in the presence of oxygen (Hiraishi and Shimada 2001). They are living in strongly acidic environments and are aerobic phototrophic bacteria that have a special form of bacteriochlorophyll chelating zinc instead of magnesium (Wakao et al. 1993). The Zn-bchl complex is much more stable under acidic conditions as the Mg-bchl-complex and therefore the chelation of Zn is considered as a special adaptation of these bacteria to the acidic conditions (Hiraishi and Shimada 2001). One of these species, *Acidiphilium acidophilum* (formerly known as *Thiobacillus acidophilus*) often accompanies acidophilic sulfur-oxidizing *Acidithiobacillus ferrooxidans* in mine drainage and metal leaching habitats (Hiraishi et al. 1998). Another acidophilic aerobic bacterium containing Zn-bchl-complexes was isolated from acidic hot springs and mine drainage, growing well at pH 4.5–5.0 (Hiraishi et al. 2000). Quite interesting, this new species and genus, *Acidisphaera rubrifa*ciens, is phylogenetically a close relative of the anaerobic acidophilic phototrophic *Rhodopila globiformis* and together with the *Acidiphilium* species these species form a major cluster of acidophilic Alphaproteobacteria, all having pH optima around 4.5–5.0 (Hiraishi and Shimada 2001).

Alkaline Soda Lakes and Alkaliphilic Phototrophic Bacteria

Much attention has been paid to alkaline soda lakes. They are characterized by high concentrations of carbonates, are devoid of magnesium and calcium ions, and have pH values of more than pH 8, often higher than pH 10 and sometimes even pH 11. Soda lakes with salinities ranging from fresh-water to highly saturated brines are widely distributed over the globe and many such lakes have been studied. Highly saline and alkaline soda lakes belong to the most extreme environments on the planet. Prominent examples of such soda lakes are lakes Bogoria, Magadi, and Nakuru in the east African Rift Valley (Kenya), the soda lakes of the Wadi El Natrun in the Libyan Desert (Egypt) and Mono Lake, Soap Lake, and Big Soda Lake in the USA. Many other soda lakes have lower or even negligible content of salts. In the semiarid landscapes of Siberia and Mongolia numerous soda lakes of low salinity exist in a number of well recognized locations with examples in Mongolia, in the

Transbaikal region and the Kulunda steppe (Sorokin et al. 2004; Kompantseva et al. 2007, 2010a).

Alkaline soda lakes are prominent habitats of phototrophic bacteria and often harbor massive developments of phototrophic microorganisms, oxygenic as well as anoxygenic ones, though only few species are specifically adapted to the life under alkaline and saline conditions. The first extremely halophilic purple sulfur bacterium described and the most halophilic anoxygenic phototrophic bacterium known is *Halorhodospira halophila* (formerly *Ectothiorhodospira halophila* Raymond and Sistrom 1969; Imhoff and Süling 1996) first isolated from the highly alkaline and saline Summer Lake (Oregon, Raymond and Sistrom 1967). It has later been isolated from soda lakes of the Wadi El Natrun with high salt content up to saturation and alkalinity with pH values exceeding pH 11 (Imhoff et al. 1978, 1979) and meanwhile has been shown to be widely distributed globally in highly saline soda lakes.

It is quite remarkable that those phototrophic sulfur bacteria common to many alkaline and saline soda lakes belong to the Ectothiorhodospiraceae (*Ectothiorhodospira, Halorhodospira*, *Thiorhodospira*, and *Ectothiorhodosinus* species) and are phylogenetically related to prominent alkaliphilic chemotrophic sulfur-oxidizing bacteria such as species of *Thioalkalivibrio* and *Thioalkalispira*, which are members of the Ectothiorhodospiraceae as well and apparently represent a prominent group of sulfur-oxidizing chemotrophic bacteria in many alkaline soda lakes (Tourova et al. 2013; Sorokin et al. 2004).

In addition to Ectothiorhodospiraceae, a few prominent alkaliphilic phototrophic Gammaproteobacteria are represented by the Chromatiaceae species *Thiorhodococcus modestalkalophilus, Thioalkalicoccus limnaeus*, and *Thiocapsa imhoffii*.

A larger group of alkaliphilic phototrophic Alphaproteobacteria has been isolated from soda lakes worldwide (e.g., from Siberia, Antarctica, and Mono Lake). This group is associated with the genera *Rhodobaca/Rhodovulum* and includes anaerobic photosynthetic as well as aerobic bacteriochlorophyll-a containing species. Prominent members are the anaerobic phototrophic species *Rhodobaca bogorensis* (Milford et al. 2000), *Rhodobaca barguzinensis* (Boldareva et al. 2008), *Rhodovulum steppense* (Kompantseva et al. 2010b), *Rhodovulum tesquicolum* (Kompantseva et al. 2012), *Rhodobaculum claviforme* (Bryantseva et al. 2015), and *Rubribacterium polymorphum* (Boldareva and Gorlenko 2014), together with the aerobic phototrophic bacteria *Roseinatronobacter thiooxidans* (Sorokin et al. 2004), *Roseinatronobacter monicus* (Boldareva et al. 2007), and *Roseibacula alcaliphila* (Boldareva and Gorlenko 2014). In the heliobacteria, three species of the genus *Heliorestis* (*Hrs. daurensis*, *Hrs. convoluta*, and *Hrs. baculata*) occur in alkaline and slightly saline habitats. Most of these species were isolated from soda lakes of low salinity and show good growth at pH 8.5–10, at temperatures from 20 to 35 °C and salt concentrations from 1 to 6 % NaCl, reflecting salinities of brackish water and seawater salinities. Salt and pH relations of alkaliphilic anoxygenic phototrophic bacteria and the habitats they have been isolated from are summarized in Table 2.

Table 2 Alkaliphilic anoxygenic phototrophic bacteria

Species	pH-range	pH optimum	Salt range	Salt optimum	Habitat	Reference
Rhodobaca bogoriensis	7.5–10	9	1–3 %	1 %	Lake Bogoria, Crater Lake, Kenya	Milford et al. (2000)
Rhodobaca barguzinensis	7.5–9	8.2	1–8 %	2–3 %	Soda lake in Barguzin Valley, Russia	Boldareva et al. (2008)
Rhodovulum tesquicolum	7.5–10	8.5–9	0.3–10	1–3	Sul'fatnoe soda lake, Siberia	Kompantseva et al. (2012)
Rhodovulum steppense		8.5	0.3–10 %	1–5 %	Lake Khilganta, Siberia	Kompantseva et al. (2010b)
Rhodobaculum claviforme	7.5–9.7	8–8.8	0.5–4	2	Soda lake Doroninskoe, Siberia	Bryantseva et al. (2015)
Rubribacterium polymorphum	7–10	8.5–9.5	0.5–4	1	Soda lake, Barguzin Valley, Siberia	Boldareva et al. (2009)
Roseibaca ekhonensis		7–9.5	0–4	2.5	Ekho Lake Antarctica/Vestfold Hills	Labrenz et al. (2009)
Roseibacula alcaliphila	8–10	9.8	0.5–5	1	Soda lake Doroninskoe, Siberia	Boldareva and Gorlenko (2014)
Roseinatronobacter thiooxidans	8.5–10.5	10	1–10	3	Lake Gorbunka, Siberia	Sorokin et al. (2000)
Roseinatronobacter monicus	8–10	8.5–9.5	0–10	4–6	Mono Lake, CA	Boldareva et al. (2007)
Thiorhodococcus modestalkaliphilus	6.5–9.5	8.5	0.5–4	1.5	Chilika Lagoon, Orissa, India	Sucharita et al. (2010)
Thiocapsa imhoffii	7.5–9.5	8.5	0–4 %	0 %	Soap Lake, Washington	Asao et al. (2007)
Thioalkalicoccus limnaeus	8–10	8.8–9.5	0.5–6 %	0.5–5 %	Lakes Dabasa-Nur, Gorbunka, Verkhneye Beloe, Tsaidam, Siberia	Bryantseva et al. (2000a)
Heliorestis convoluta	7.5–10	8.5	Up to 3 %	0–1 %	El Hamra, Wadi El Natrun, Egypt	Asao et al. (2006)
Heliorestis daurensis	7.5–10	9	nd	nd	Lake Barun Torey, Daur Steppe, Siberia	Bryantseva et al. (1999a)
Heliorestis baculata	8–10.5	8.5–9.5	Up to 3 %	0.5–1	Lake Ostozhe, Siberia	Bryantseva et al. (2000b)

Mongolian Soda Lakes

A larger number of small, shallow soda lakes is found in north-eastern Mongolia, which is a flat, dry steppe in a semiarid climate (200 mm rain per year) with temperatures around the year between −40 and +40 °C and pH values from 9.0 to 10.6 (Sorokin et al. 2004). These authors identified the salinity as a key factor for the development and composition of microbial communities in these lakes, which were studied by culture-dependent approaches. The extremely halophilic *Halorhodospira halophila* and two morphological distinct types assigned to the genus *Ectothiorhodospira* were found in the soda lakes with high salinities (>15 % salts). Different types of *Ectothiorhodospira* and *Thiorhodospira* species as well as representatives of the *Rhodobaca*/*Rhodobacter*/*Rhodovulum* group were present at salinities up to 5.5–6.0 % salts. From one of these lakes the new borderline moderately halophilic *Ectothiorhodosinus mongolicus* (salt optimum 1–7 %) was isolated for the first time (Gorlenko et al. 2004). Some Chromatiaceae and also *Oscillochloris* type filaments were found only at lower salinities (<3.5 % salts, Sorokin et al. 2004). *Chloroflexus*-like filaments were observed occasionally. Green-colored, bchl b-containing *Halorhodospira* species and green sulfur bacteria were not found.

Soda Lakes of the Kulunda Steppe (Altai Krai)

Phototrophic bacteria were found in a larger number of saline and alkaline soda lakes in the Kulunda Steppe, most of which had a pH around 10 and salinities ranging from 1.5 to 28 % total salts. In those lakes with higher rates of sulfate reduction also visible mass developments of phototrophic bacteria were observed (Kompantseva et al. 2010a). In all of these soda lakes representatives of Ectothiorhodospiraceae were the dominant representatives of the anoxygenic phototrophic bacteria. In lakes with low and moderate salt concentrations members of *Ectothiorhodospira* and in those with high salt concentrations of more than 15 % *Halorhodospira* species predominated. In these culture-dependent studies, a larger diversity of *Ectothiorhodospira* species was anticipated and several morphological types were distinguished (Kompantseva et al. 2010a). There was a clear shift of the different morphological types of Ectothiorhodospiraceae in correlation with the lake's salinity with the dominance of one *Ectothiorhodospira* type at low (type 2 at 1.5–4 % salts) and another type at moderate to high salinities (type 1 at 6–12 % salts, abundant also in some lakes with 20 % salts). In addition, purple nonsulfur bacteria similar to *Rhodobaca* or *Rhodovulum* species were abundant in all lakes with salinities up to 16 %. Purple sulfur bacteria resembling *Halochromatium* and *Thiocapsa* were observed in some of them and were abundant in lakes of intermediate salinity (6–16 % salts).

Soda Lakes in the Transbaikal Region

A large number of soda lakes in the southeastern Transbaikal dry steppe is characterized by alkaline pH between 8.5 and 10.5 and by low salinity, in most of them in the range of freshwater and brackish-water habitats (below 1 % salts), in six between 1 and 4 % total salts (Kompantseva et al. 2007). Due to unstable water regimes there were considerable fluctuations in the salinity. The phototrophic communities of 24 of these soda lakes were studied by culture-dependent methods (Kompantseva et al. 2007). In particular, in the shallow coastal part of many of these lakes, colored mats developed at the surface of the sediments and a pink-red layer of phototrophic purple bacteria occurred below a cyanobacterial top layer. In the saline lakes Kholvo-Torom and Khilganta (3.5 and 4 % salts), the mats covered the whole bottom. An estimation of the number of phototrophic bacteria was made on the basis of direct counts of the samples directly inoculated into agar dilution series, a method also used in the author's lab and described with its specific advantages by Imhoff and Trüper (1992).

Representatives of Chromatiaceae, Ectothiorhodospiraceae, and purple nonsulfur bacteria (*Rhodobacter*/*Rhodobaca*/*Rhodovulum* group) were found in almost all lakes, regardless of the salt content. The bacteriochlorophyll *b* containing *Thioalkalicoccus limnaeus* was found only in four brackish-water lakes (0.5–2 % salts). A clear tendency pointed to a high abundance of Chromatiaceae at low salinities up to 1 % and their dominance below 0.5 % salts, but a high abundance of Ectothiorhodospiraceae already at salinities as low as 0.5 % and their dominance at more than 0.5 % salts (Kompantseva et al. 2007). This distribution correlates well with the alkaliphilic character of *Ectothiorhodospira*, *Thiorhodospira*, and *Halorhodospira* species with growth optima from pH 8.5–10.5. Several new alkaliphilic species for the first time were isolated from these lakes. These include *Thiorhodospira sibirica* from Lake Malyi Kasytui and several other lakes (Bryantseva et al. 1999b), *Thioalkalicoccus limnaeus* from lakes Dabasa-Nur, Gorbunka, Verkhneye Beloe, and Tsaidam (Bryantseva et al. 2000a), *Rhodovulum steppense* from Lake Khilganta (Kompantseva et al. 2010b), the alkaliphilic aerobic phototrophic *Roseinatronobacter thiooxidans* from Lake Gorbunka (Sorokin et al. 2000) as well as the heliobacteria *Heliorestis daurensis* from Lake Barun Torey, and *Heliorestis baculata* from Lake Ostozhe (Bryantseva et al. 1999a, 2000b) (see Table 2). No green sulfur bacteria were found in these soda lakes.

Soda Lakes of the Wadi El Natrun

Among the first intensively studied soda lakes were those in the Wadi El Natrun in Egypt (Abd-el-Malek and Rizk 1963; Jannasch 1957; Imhoff et al. 1978, 1979). Intense red coloration within the water and the top layers of the sediments indicated mass development of phototrophic purple bacteria in those parts that are anoxic and sulfidic (Figs. 3–5), whereas in the oxic, highly saline parts alkaliphilic halophilic archaea produced pink coloration (Jannasch 1957; Imhoff et al. 1979). First observations of mass accumulations of purple sulfur bacteria by Holger Jannasch (1957) led us to organise an expedition in 1976 (Fig. 6) and start investigations on the purple sulfur bacteria from these lakes years later (Imhoff et al. 1978, 1979). The appearance of massive developments of halophilic *Halorhodospira* species is found in red coloration of some shallow parts of the soda lakes (Fig. 3), in a separation of the green-colored species and the red-colored ones in differnt small puddles (Fig. 4) and in separated colored layers within the sediments (Fig. 5).

The Wadi El Natrun is located alongside the desert road from Alexandria to Kairo in Egypt and forms a depression approx. 38 m below the Nile level (Imhoff

Fig. 3 Colored mass development of *Halorhodospira halophila* in shallow waters of the strongly alkaline soda lake Hamra in the Wadi El Natrun, Egypt (photo: JF Imhoff)

Fig. 4 Red and green puddles were found in the almost dried Lake Muluk showing the separate development of green-colored (*Halorhodospira halochloris* and *Halorhodospira abdelmalekii*) and red-colored (*Halorhodospira halophila*) species of these highly halophilic phototrophic bacteria (photo: JF Imhoff)

Fig. 5 Unusual microbial mats in sandy sediments of Lake Muluk in the Wadi El Natrun. Microscopic examination, absorption spectra of the distinct layers and isolation of pure cultures revealed that separate green-colored and red-colored *Halorhodospira* species form distinct layers. This demonstrates their occupation of different ecological niches (Imhoff et al. 1979) (photo: JF Imhoff)

Fig. 6 The expedition team of an excursion to the soda lakes of the Wadi El Natrun in 1976 with local people from Bir Hooker, which were excellent guides being familiar with all properties and local aspects of the soda lakes and three PhD students from the University of Bonn (Germany): our native Egyptian Gaber Soliman (3rd from left), Hans Georg Sahl (2nd from right) and Johannes F. Imhoff (2nd from left) (photo: JF Imhoff)

et al. 1979). Water enters the lakes by infiltration of Nile water through the ground. A chain of small, shallow soda lakes extends along the deepest part of the Wadi El Natrun. The hydrochemistry of the lakes is governed by high evaporation rates and active sulfate reduction in the sediments that are generally anoxic and rich in organic matter (Abd-el-Malek and Rizk 1963; Imhoff et al. 1979). The lakes are strongly alkaline with pH values around 10.8–11.2 and salinity ranges from 16 % to saturation. An exceptional low salinity of approx. 9 % is found in Lake Gabara, where the massive development of *Chromatium* type cells was observed underneath a top layer of Cyanobacteria (Imhoff et al. 1979). In the other lakes the salt concentrations range from 16 % (Lake Muluk) and 23.8 % (Lake Hamra) to almost 40 % (Lakes Zugm, Gaar, and Rizunia).

All attempts of isolation of anoxygenic phototrophic bacteria yielded various representatives of the Ectothiorhodospiraceae, for which a special medium was designed on the basis of the mineral salts composition of the lakes (Imhoff and Trüper 1977), being strongly buffered at pH 9 with carbonates and containing different salt concentrations (Imhoff 1992a, b). The majority of more than 20 isolates of purple sulfur bacteria from these lakes were strictly adapted to the alkaline conditions and mineral salts composition as provided by the lake's water and were identified as *Halorhodospira halophila*. All of these isolates had pH optima at 8.5–9.0, while pH optima of pH 7.4–7.9 were reported for the type strain SL1, which originates from the highly alkaline and saline Summer Lake in Oregon (Raymond and Sistrom 1969). Two more haloalkaliphilic species of this genus with bacteriochlorophyll *b* as major bacteriochlorophyll: *Halorhodospira halochloris* (Imhoff and Trüper 1977; Imhoff and Süling 1996) and *Halorhodospira abdelmalekii* (Imhoff and Trüper 1981; Imhoff and Süling 1996) originated from these lakes. *Halorhodospira halochloris* is an alkaliphilic and extremely halophilic bacterium (isolated from lakes Gabara, Gaar, and Hamra) with optimum growth at pH 8–9.5, salt concentrations of 14–27 %, and temperatures above 40 °C up to 47–50 °C (Imhoff and Trüper 1977). *Halorhodospira abdelmalekii* is another haloalkaliphilic species with bacteriochlorophyll *b* which was first isolated from lakes Muluk and Gabara in the Wadi El Natrun (Imhoff and Trüper 1981).

In addition, a few strains of alkaliphilic *Ectothiorhodospira* species with much lower salt requirement (optimum at 5 %) were isolated from soda lakes of the Wadi El Natrun and assigned to the new species *Ectothiorhodospira haloalkaliphila* (Imhoff et al. 1978; Imhoff and Süling 1996). More recently, another new *Ectothiorhodospira* species, *Ectothiorhodospira variabilis*, which is most closely related to *Ectothiorhodospira haloalkaliphila*, was isolated from soda lakes of the Wadi El Natrun, from Siberia and Mongolia (Gorlenko et al. 2009). The bacterium showed optimum growth at 5–8 % salts and pH 9.0–9.5 and was extremely salt tolerant with a salt range up to 20 % (Gorlenko et al. 2009). Also another alkaliphilic Heliobacterium, *Heliorestis convoluta*, was isolated from Lake Hamra in the Wadi El Natrun, though astonishingly this bacterium has no distinct salt requirement with an upper salt tolerance at 3 % NaCl and growth optima at 30–35 °C and pH 8.5 (Asao et al. 2006).

The analysis of 16S rRNA gene clone libraries from Lake Fazda, Lake Hamra, and Lake UmRisha of the Wadi El Natrun almost three decades later demonstrated that a diverse array of sequences related to *Halorhodospira* species including *Halorhodospira halochloris* and *Ectothiorhodospira haloalkaliphila* were present. Also a larger number of phylotypes related to *Rhodobaca bogoriensis* were identified (Mesbah et al. 2007). The phototrophic nature of some of the bacteria retrieved by the sequences, in particular some with distant relationship to known Chromatiaceae, remains unclear as 16S rRNA gene sequences include nonphototrophic relatives. A list of species isolated from alkaline soda lakes is shown in Table 2, those of the Ectothiorhodospiraceae are included into Table 3.

Soap Lake

Remarkable are the studies on Soap Lake (Washington), which is a small soda lake with a distinct chemocline (at 20 m depth), a maximum depth of 24 m, a pH of 9.7, and high concentrations of sulfide reaching up to 175 mM near the sediment (Asao et al. 2011). The hypolimnion is hypersaline (14 % salts, 6–8 °C) and the epilimnion is more dilute (1.4 % salts). Both culture-based and culture-independent approaches were used to study the community of phototrophic bacteria in this lake.

A number of strains have been isolated from this lake that all grow optimally at alkaline pH (Asao et al. 2011) but have different salt responses. Their phylogenetic relationship was established on the basis of 16S rRNA gene sequences. *Thiocapsa imhoffii* is an alkaliphilic purple sulfur bacterium first isolated from this lake, which showed optimal growth at pH 8.5 and at 32 °C and grows well from pH 7.5–9.5 (Asao et al. 2007). Optimum growth is in the absence of salt but as other *Thiocapsa* strains and species, it is salt tolerant and grows well up to 4 % NaCl (Imhoff 2001). Another non-motile coccoid purple sulfur bacterium resembling *Thiocapsa marina* was also enriched and revealed good growth at 25 °C, pH 8, and 1.5 % NaCl (Asao et al. 2011). Two isolates of *Ectothiorhodospira* were closely related to *Ectothiorhodospira haloalkaliphila,* to *Ectothiorhodospira marina*, and to other isolates from Lake Bogoria (Kenya), Big Soda Lake (Nevada), and Mono Lake (California) (Asao et al. 2011). Another bacterium isolated from Soap Lake was related to *Thiorhodospira sibirica* originally described from Siberian soda lakes (Bryantseva et al. 1999b). Optimum growth of this new isolate was at pH 8–9 and 1.5–3 % NaCl (tolerance up to 7 %).

Also a new isolate of *Rhodobaca bogoriensis* was obtained. But similar to the soda lakes of the African Rift Valley, no other purple nonsulfur bacteria could be detected (Asao et al. 2011; Milford et al. 2000). As in other soda lakes, green sulfur bacteria were not detected in Soap Lake even by specific amplification from environmental DNA (Asao et al. 2011).

The diversity of phototrophic purple bacteria of Soap Lake was approached by partial sequences of the *pufM* gene from environmental DNA and from isolated bacteria (Asao et al. 2011). Though two out of a total of 12 environmental *pufM* sequences (136 nt long) were identical to the dominant *Thiocapsa imhoffii* at the chemocline of Soap Lake, the others were not in accord with the strains isolated from the lake (Asao et al. 2011) and in spite of the short sequences their assignment to related species may not be clear. The sequences do, however, depict the diversity of phototrophic bacteria present in Soap Lake and indicate that only a small part of the diversity of phototrophic purple bacteria of this extreme habitat was brought into culture.

Big Soda Lake

Big Soda Lake (Nevada) is a small meromictic crater lake with a surface salinity of less than 2 % salts, a pH of 9.7 and a constant temperature in the monimolimnion of 12 °C, which is permanently anoxic and has high concentrations of sulfide (Cloern et al. 1983a, b). A dense plate of phototrophic purple sulfur bacteria developed at the chemocline and one of these was tentatively identified as *Ectothiorhodospira vacuolata* (Cloern et al. 1983a). Due to its preference to low salinity conditions and the production of gas vesicles it appears well adapted to grow under the conditions prevailing in Big Soda Lake and it might well be the dominant phototrophic bacterium in this lake.

Mono Lake

Mono Lake (California) is a soda lake with 7–10 % salinity and a pH around 10. Though results of a microbial diversity study of Mono Lake (California), in which clone libraries of 16S rRNA gene sequences were analyzed, allow limited conclusions on the identity of phototrophic bacteria, with some confidence a larger group of sequences from this lake can be related to the alkaliphilic purple nonsulfur bacterium *Rhodobaca bogoriensis* (Humayoun et al. 2003). A close relative to *Rhodobaca bogoriensis*, the aerobic bacteriochlorophyll-a containing *Roseinatronobacter monicus* was isolated from Mono Lake (Boldareva et al. 2007). In addition, sequences shown to be related to *Thioalkalivibrio jannaschii* might also relate to phototrophic *Ectothiorhodospira* species. In fact, later studies demonstrated the oxidation of monothioarsenate, arsenite, and thioarsenate mixtures by *Halorhodospira halophila*, *Ectothiorhodospira vacuolata* as well as a number of *Ectothiorhodospira* strains including isolates from Soap Lake and Mono Lake (Edwardson et al. 2014). Reevaluation of phylogenetic relationships revealed closest relations of a large number of clone sequences from Mono Lake to *Ectothiorhodospira variabilis* and of those from Soap Lake to *Ectothiorhodospira vacuolata* and *Ectothiorhodospira magna* (Edwardson et al. 2014).

Böddi-szék Soda Lake in Ungary

In the very shallow soda lakes Böddi-szék in Ungary (1.2 % salinity, 33 °C, and less than 10 cm water depth at the sampling time) a number of alkaliphilic phototrophic bacteria were identified, that were known from other soda lakes. *Rhodobaca bogoriensis* was identified after cultivation and 16S rRNA gene-based clone sequences were obtained with relationship to the alkaliphilic *Rhodobaca barguzinensis*, *Ectothiorhodospira shaposhnikovii*, *Thiorhodospira sibirica*, *Heliorestis daurensis,* and in addition to *Rhodobacter capsulatus* and *Rhodovulum strictum* (Borsodi et al. 2013).

Marine Salterns and Salt Lakes

Marine salterns are man-made habitats in which seawater is concentrated in a succession of separated shallow ponds to a final stage where NaCl is crystallized. In this way a series of evaporation ponds is created with increasing concentration of seawater. Thereby, a stepwise gradient of salinity is achieved starting with seawater and ending up with saturated salt brines with a succession of microbial communities adapted to the different ranges of salinity. Such salterns exist around the world and

Table 3 Moderately and extremely halophilic phototrophic bacteria and haloalkaliphilic Ectothiorhodospiraceae

Species	Salt-range [%]	Salt optimum [%]	pH optimum	T-optimum [°C]	Habitat	References
Rhodovibrio salinarum	2–20	12–18	7.5–8	42	Portuguese saltern	Nissen and Dundas (1984)
Rhodothalassium salexigens	5–20	7	6.6–7.4	35–40	Coastal pond, Oregon	Drews (1981)
Rhodovibrio sodomensis	6–21	12	7	35–40	Dead Sea near Ein Gedi	Mack et al. (1993)
Thiohalocapsa halophila	3–20	4–10	7	20–30	Salin-de-Giraud, France	Caumette et al. (1991)
Halochromatium glycolicum	2–20	3–10	7.2–7.4	30–35	Solar Lake, Sinai	Caumette et al. (1997)
Halochromatium salexigens	4–20	8–11	7.4–7.6	20–30	Salin-de-Giraud, France	Caumette et al. (1988)
Lamprobacter modestohalophilus	1–9	2–4	7.4–7.6	25–34	Lake Sivash, Crimea/Lake Shunet, Khakassia	Gorlenko et al. (2014)
Ectothiorhodospira haloalkaliphila	2.5–15	5	8.5–10	26–40	Wadi El Natrun, Abu Gabra	Imhoff and Süling (1996)
Ectothiorhodospira variabilis	2–20	5–8	9–9.5	30–35	Wadi El Natrun, Egypt	Gorlenko et al. (2009)
Ectothiorhodospira salini	0.5–12	5–6	7.5	25–35	Saltern, Kanyakumari, India	Ramana et al. (2010)
Ectothiorhodospira marina	0.5–10	2–6	7.5–8.5	30–40	Marine coastal waters and sediments	Imhoff and Süling (1996)
Ectothiorhodospira marismortui	1–20	3–8	7–8	35–45	Dead Sea	Oren et al. (1989)
Ectothiorhodospira mobilis	1–5	2–3	7.6–8	25–30	Salt marshes, coastal waters	Pelsh (1936)
Ectothiorhodospira shaposhnikovii	0–7	3	8–8.5	30–35	Saline waters and sediments	Cherni et al. (1969)
Ectothiorhodospira vacuolata	0.5–10	1–6	7.5–9.5	30–40	Salt and soda lakes	Imhoff et al. (1981)
Ectothiorhodospira magna	0–8	0.5–1.5	9–10	30–35	Lake Doroninskoe, Siberia	Bryantseva et al. (2010)
Ectothiorhodosinus mongolicus	0.5–9 %	1–7 %	8.3–9.1	30–35	Lake Dzun Uldziit Nur, Mongolia	Gorlenko et al. (2004)
Thiorhodospira sibirica	0–5 % NaCl	0–5 %	8.5–9.5	25–30	Lake Malyi Kasytui, Siberia	Bryantseva et al. (1999b)
Halorhodospira halophila	3–30	11–32	8.5–9	30–47	Summer Lake, Oregon	Raymond and Sistrom (1967)
Halorhodospira halochloris	10–34	14–27 %	8.5–9	45–47	Wadi El Natrun, Gaar, Hamra	Imhoff and Trüper (1977)
Halorhodospira abdelmalekii	5–30	14–16	8–9	30–40	Wadi El Natrun, Muluk, Gabara	Imhoff and Trüper (1981)
Halorhodospira neutriphila	7–25	9–12	6.8–7	30–35	Salin-de-Giraud, France	Hirschler-Réa et al. (2003)

are characterized by a uniform thalassohaline salts composition. The sediments of salterns are regularly anoxic and, depending on the content of organic matter, sulfate reduction is active and produces sulfide that serves sulfur-oxidizing chemotrophic and phototrophic bacteria as an energy source. In the shallow ponds light generally penetrates down into the sediment and in consequence photosynthetic microbial mats including anoxygenic phototrophic bacteria are expected to develop in or on top of the sediments.

The situation in the salterns very much resembles coastal lagoons in which, due to evaporation, salts can be concentrated quite effectively. In consequence, such coastal lagoons and marine salterns are expected to share common representatives of marine, halophilic, and halotolerant bacteria. Here we will focus on the moderate to high concentrated ponds of salterns in which halophilic phototrophic purple and green bacteria are found. Though typical marine, slightly halophilic bacteria (salt optima 2–7 % NaCl) were found in the first ponds, moderately halophilic purple sulfur bacteria with optimal growth at 7–15 % salts were observed and isolated from hypersaline ponds of marine salterns. Similar to the thermophilic phototrophic bacteria, halophilic bacteria are classified according to their growth optima (as defined earlier by Imhoff 1986, 2001) as slightly halophilic, marine bacteria (salt optima 2–7 % salts), as moderately halophilic bacteria (salt optima between 7 and 15 % salts), and extremely halophilic bacteria (salt optima higher than 15 % salts). The same salt ranges are used to characterize the salt tolerance. Thus, *Thiohalocapsa halophila* growing optimal at 7 % salts and in a range from 3 to 20 % salts is a borderline moderately halophilic and extremely halotolerant bacterium.

The most commonly found anoxygenic phototrophic purple bacteria in evaporation ponds of marine salterns are *Halochromatium salexigens* (Caumette et al. 1988), *Thiohalocapsa halophila* (Caumette et al. 1991), *Rhodovibrio salinarum* first isolated from a Portuguese saltern (Nissen and Dundas 1984), and *Rhodothalassium salexigens* first isolated from concentrated coastal pools in Oregon (Drews 1981) and later from Spanish salterns (Rodriguez-Valera et al. 1985). Other moderately halophilic phototrophic purple bacteria are *Rhodovibrio sodomensis* first isolated from the Dead Sea (Mack et al. 1993) and *Halochromatium glycolicum* isolated from Solar Lake (Caumette et al. 1997). A list of halophilic anoxygenic phototrophic bacteria including haloalkaliphilic Ectothiorhodospiraceae is given in Table 3. In neutral saline lakes and marine salterns, the diversity of phototrophic bacteria is generally higher compared to alkaline soda lakes and different types of phototrophic bacteria occur in both types of environments. These findings were highlighted by investigations on a number of saline lakes in the Crimean region (Kompantseva et al. 2010a).

Salin-de-Giraud Salterns (France)

In the Salin-de-Giraud salterns in southern France red layers of phototrophic purple bacteria have been found below a cyanobacterial mat underneath a gypsum crust at salinities of 13–20 % (Caumette et al. 1994). This is a common picture often found in evaporation ponds of marine salterns with elevated salinity (Fig. 7). From the purple layer two moderately halophilic purple sulfur bacteria were isolated, *Halochromatium salexigens* (former *Chromatium salexigens*, Caumette et al. 1988; Imhoff et al. 1998b) and *Thiohalocapsa halophila* (former *Thiocapsa halophila*, Caumette et al. 1991; Imhoff et al. 1998b), growing between 4–20 % (optimum at 10 %) and 3–20 % NaCl (optimum 4–10 %), respectively. Quite interesting, both of these bacteria not only synthesize, but also very effectively take up glycine betaine (Caumette 1993), which is known as a widely distributed and most effective compatible solute in halophilic and halotolerant bacteria (Imhoff and Rodriguez-Valera 1984; Imhoff 1986, 1993). Later also the new moderately halophilic *Halorhodospira neutriphila* was isolated from these salterns with growth optima at neutral pH 6.8–7.0 and temperatures of 30–35 °C (Hirschler-Réa et al. 2003). The optimum salt concentration of this bacterium is 9–12 % NaCl, though it is extremely halotolerant with growth ranges from 6 % to almost 30 % NaCl. It appears that this is the first *Halorhodospira* species with a clear preference for growth at neutral pH. Also halophilic purple nonsulfur bacteria closely related to *Rhodovibrio sodomensis* were isolated from these salterns (Hirschler-Réa et al. 2003). Two sites of this saltern, covered by cyanobacterial mats (salinities of 15–20 % and 25–32 %), apparently not only differed in salinity, but also had a different composition of the bacterial communities (Caumette et al. 1994; Hirschler-Réa et al. 2003). *Halochromatium* and *Thiohalocapsa* species were major constituents of the purple layer underneath the cyanobacteria at the lower salinity (15–20 %) and the *Halorhodospira neutriphila* and *Rhodovibrio sodomensis* were the dominant forms at the higher salinity. Molecular studies that were performed on both sites using 16S rRNA gene sequences, however, did not show representatives of the phototrophic Gammaproteobacteria, but instead revealed a higher diversity of the halophilic spirilloid Alphaproteobacteria related to the known *Rhodovibrio* species and *Rhodothalassium salexigens* (Mouné et al. 2003).

Salinas at Santa Pola (Spain)

The salterns of Santa Pola near Alicante (Spain) have been studied intensively over the past decades, including culture-based studies as well as metagenomics analyses. They represent a multi-pond system, in which the salinity in the sequence of ponds is maintained at a specific increasing level from one pond to the next by the workers of the salinas. In the range from 15 to 30 % salts, a high bacterial diversity was observed and eventually anoxygenic phototrophic bacteria developed, while in the ponds with the highest salinity clearly archaeal halobacteria dominated the communities (Rodriguez-Valera et al. 1985). Cells similar to small-celled rod-shaped Chromatiaceae with elemental sulfur stored inside the cells were observed and might be related to *Halochromatium* species (Rodriguez-Valera et al. 1985; Imhoff 2001).

In the range from 10 to 23 % salts unicellular cyanobacteria regularly formed extended mats on top of the sediment or within the precipitated gypsum crusts. These mats were associated with heterotrophic bacteria and over large areas a separate pink-red-colored layer of spiral-shaped phototrophic purple nonsulfur bacteria developed underneath the cyanobacteria. From this rose-red layer, phototrophic purple nonsulfur bacteria were isolated, all of which were identified as strains of *Rhodothalassium salexigens* (Rodriguez-Valera et al. 1985; Imhoff et al. 1998a; Imhoff 2001), an extremely halotolerant species that is well adapted to life at high salt concentrations (Drews 1981). This is one of the rare cases, where the mass development of purple nonsulfur bacteria is well documented. Similar observations made in other highly saline habitats and the extremely high salt tolerance of these spirilloid species are tempting to assume that they also form massive developments at other similar habitats (Fig. 7).

In a metagenomic study on these salterns using 16S rRNA gene sequences, a high bacterial diversity was found at salt concentrations around 10 % and up to 20 % total salts. At higher salt concentrations the bacterial diversity in the pond water decreased significantly and in the crystallizer ponds bacteria were almost completely replaced by a variety of halobacteria (Ventosa et al. 2004; Fernández et al. 2014). As in other similar studies (e.g., Dillon et al. 2013) and due to the phylogenetic distance to phototrophic reference sequences, the metagenomic 16S rRNA gene sequences could not be clearly correlated to phototrophic representatives. Certainly, more specific investigations are needed to unravel there abundance and diversity of anoxygenic phototrophic bacteria alongside the salinity gradient in these salterns.

Salterns at Guerrero Negro, Baja California

Obviously, Chloroflexus-like bacteria are not restricted to hot spring habitats but much wider distributed, much more diverse, and their ecological importance is much greater than recognized so far. *Chloroflexus*-like bacteria have for long been observed in microbial mats of freshwater, marine, and also hypersaline habitats (Gorlenko 1975, 1988; Pierson et al. 1994). Photoautotrophic and strictly anaerobic *Chloroflexus* type bacteria were highly enriched from microbial mats of Lake Sivash and a White Sea estuary with salinities up to 10 %, but not obtained in pure culture (Gorlenko 1988). *Chloroflexus*-like bacteria were also abundant in hypersaline mats in Laguna Figueroa, in Solar Lake (Sinai, Egypt) and in lagoons of the Sea of Azov (see Pierson et al. 1994). The discovery of *Chloroflexus*-like organisms in marine and hypersaline environments was given much importance because many Precambrian stromatolites are of marine or lacustrine origin and *Chloroflexus*-like bacteria are considered being most likely the major components of these stromatolites (Pierson et al. 1994).

Most detailed studies have been performed on hypersaline mats occurring in the evaporation ponds of the salterns of Guerrero Negro (Baja California Sur, Mexico) in which Chloroflexus-like bacteria play a dominant role (Pierson et al. 1994). Three morphological types of *Chloroflexus*-like bacteria were differentiated, grown, and maintained in enrichment cultures in the lab (Pierson et al. 1994). The mats were typically several centimeters thick and constituted a multilayered system. A high diversity of bacteria related to *Chloroflexus* in these mats was found and a new group of sequences related to *Chloroflexus* and *Oscillochloris* was identified (Nübel et al. 2001). A highly enriched culture was used to characterize a bacterium belonging to this group and to describe the Candidatus *Chlorothrix halophila* (Klappenbach and Pierson 2004). It is a mesophilic (optimum 35–38 °C, pH 7.5), moderately halophilic (salinity range 5–12 %, salt optimum 10 % salts), filamentous anoxygenic phototrophic bacterium, performing sulfide oxidation and carbon dioxide fixation. It has chlorosomes and contains bchl c and also bchl a (Klappenbach and Pierson 2004). Phylogenetically it forms a separate branch together with *Chloroflexus* and *Oscillochloris* sequences, but distinct from the *Roseiflexus* group (Klappenbach and Pierson 2004).

The extensive analysis of clone libraries of 16S rRNA gene sequences from the community DNA of these mats revealed an unexpected high bacterial diversity including 15 novel candidate phyla (Ley et al. 2006). Chloroflexi sequences dominated the clone libraries from different layers of these mats and approximately a quarter of these sequences were closely related to phototrophic relatives such as *Chloroflexus* and *Chlorothrix*, the presence of which in these mats had been established earlier (Nübel et al. 2001; Klappenbach and Pierson 2004). From the phylogenetic relationships of the sequences these authors concluded that another so far unrecognized group of potentially phototrophic *Chloroflexus*-like bacteria is present in these mats (Ley et al. 2006). This supports the high abundance and relative proportion of Chloroflexus-like bacteria in the mats of Guerrero Negro as well as a hitherto not realized diversity of this group of bacteria. Apparently, Chloroflexus-like bacteria cope well with the elevated salt concentrations in evaporation ponds of salterns and may be more widely distributed at such habitats than realized so far.

Also two types of moderately halophilic and extremely halotolerant types of phototrophic purple bacteria were almost regularly found in salterns, if it was looked for. The purple nonsulfur bacteria *Rhodovibrio salinarum* (Nissen and Dundas 1984), *Rhodovibrio sodomensis*, and *Rhodothalassium salexigens*, on the one hand, and the purple sulfur bacteria *Halochromatium salexigens*, *Halochromatium glycolicum*, and *Thiohalocapsa halophila*, on the other hand (Caumette et al. 1988, 1991; Caumette 1993) are prominent examples of these two types of phototrophic bacteria. The presence of sulfide may be the determinative factor for the development of members of either the one or the other group.

Lake Chiprana

Lake Chiprana is a permanently hypersaline lake in northeastern Spain with a maximum depth of 5 m and a total salinity of 7.8 %. It is particularly rich in magnesium (0.35 M) and sulfate (0.5 M) (Jonkers et al. 2003). In the anoxic hypolimnion green sulfur bacteria develop and a thin surface layer of the benthic mats is composed mainly out of *Chloroflexus*-like bacteria, as demonstrated by microscopic evidence (Jonkers et al. 2003). The phylogenetic analysis of this mat with primers specific for the 16S rRNA gene of Chloroflexi revealed a high diversity of Chloroflexi in this mat, presumably phototrophic as well as non-phototrophic representatives (Bachar et al. 2007). From 16 Chloroflexi phylotypes identified (sequences >96 % sequence similarity) four are related to *Chloroflexus*, *Chlorothrix*, and to an uncultured bacterium from a hypersaline microbial mat related to *Heliothrix* (Bachar et al. 2007) and can be expected to represent phototrophic relatives.

Great Salt Lake and Dead Sea

Though both of these salt lakes are highly prominent and were well studied, communities of anoxygenic phototrophic bacteria are not obvious and have not been specifically analyzed.

In Great Salt Lake (Utah) 16S rRNA clone libraries revealed the presence of phylotypes with some relationship to the halophilic Chromatiaceae including *Halochromatium salexigens* and also to halophilic purple nonsulfur bacteria including *Rhodothalassium salexigens* as well as *Rhodovibrio salinarum*. However, the phototrophic property of the corresponding bacteria remains an open question and some of the sequences may represent chemotrophic bacteria related to *Silicibacter* (Tazi et al. 2014).

The Dead Sea has not been considered a favorable place for phototrophic bacteria, though two new species have been isolated. The moderately halophilic *Rhodovibrio sodomensis* has been isolated from sediments of the Dead Sea (Mack et al. 1993) and *Ectothiorhodospira marismortui* was isolated from a spring given support to the Dead Sea (Oren et al. 1989) (therefore its properties may not be representative for a bacterium living in the Dead Sea itself).

Salt Lakes of the Salar de Atacama (Chile)

The salt lakes of Chilean salares represent an extraordinary and extreme habitat with special conditions regarding salt concentration and composition, irradiation, and drastic diurnal changes (Dorador et al. 2008, 2009). The salts composition of the lakes is greatly different depending on the mineral salts contained in the source waters in the Chilean highlands and is generally of the athalassohaline type.

Like in other hypersaline environments, extended purple-red-colored microbial layers develop in and on the sediment surface of the shallow lakes such as Laguna Chaxa and Laguna Tebenquiche of the Salar de Atacama. These mats may be found below a top layer of cyanobacteria and algae in shallow waters or below precipitated gypsum crusts covering wide areas of the dried sediments (Figs. 7 and 8). They also may cover the top of sediments (Fig. 9). Anoxygenic phototrophic purple bacteria, including diverse groups purple sulfur and purple nonsulfur bacteria, but also aerobic phototrophic bacteria of the Roseobacter/Rhodobacter clade appear to be common to these lakes (Dorador et al. 2013; Tank et al. 2009; Thiel et al. 2010). A comprehensive study of communities of anoxygenic phototrophic purple bacteria of these salt lakes using almost the complete sequence of the *pufLM* genes (Tank et al. 2009) demonstrated the presence of a highly diverse community of purple sulfur bacteria (Thiel et al. 2010). The great majority of purple sulfur bacterial phylotypes of this habitat

Fig. 7 Such microbial mats underlying salt or gypsum crusts in highly saline lakes with cyanobacteria developing on top and phototrophic purple bacteria below are widely distributed and typically occurring in salt lakes and salterns. In Laguna Chaxa of the Atacama Desert in Chile they extend over large areas underneath the salt dry crust (photo: V Thiel)

could be related to known purple sulfur bacteria, but was supposed to be new at the genus level or even at higher taxonomic rank (Thiel et al. 2010).

The communities of purple sulfur bacteria from both salt lakes were characterized by the presence of representatives related to the type strains of the moderately halophilic *Chromatiaceae* such as *Halochromatium salexigens*, *Halochromatium glycolicum*, *Thiohalocapsa halophila,* and Ectothiorhodospiraceae such as *Ectothiorhodospira mobilis*, *Ectothiorhodospira variabilis*, and *Halorhodospira halophila* as closest relatives (Caumette et al. 1988, 1991, 1997; Gorlenko et al.

Fig. 8 Different massive microbial mats were found within small pools of Laguna Chaxa, Atacama Desert in Chile. They are covered by shallow water (photo: JF Imhoff)

Fig. 9 Red-colored microbial mats of phototrophic purple bacteria extend over a large area on the sediment surface in the shallow waters of Laguna Tebenquiche, Atacama Desert in Chile (photo: JF Imhoff)

2009). Evidence was also obtained for the presence of several phylotypes of BChl *b*-containing anoxygenic phototrophic bacteria distantly related to (<80 % sequence similarity) the genera *Thiococcus*, *Thioflavicoccus*, and *Thioalkalicoccus*, that form a separate phylogenetic branch among the purple sulfur bacteria (Imhoff et al. 1998b; Bryantseva et al. 2000a; Tank et al. 2009). These bacteria are known as inhabitants of marine sediments and have a particular advantage in sandy sediment habitats due to their special pigment content with *bchl-b* (Imhoff and Pfennig 2001; Nicholson et al. 1987; Pfennig et al. 1997). Therefore, their presence in these salt lake sediments is not surprising, though they have been rarely isolated from salt and soda lakes (Bryantseva et al. 2000a).

Two out of 24 phylotypes identified as phototrophic Gammaproteobacteria could be clearly identified on the species level (*Ectothiorhodospira mobilis* and *Thiohalocapsa halophila*), whereas the great majority of phylotypes had a *pufLM* sequence of low similarity (<80 %) to known purple sulfur bacteria and quite likely might represent new species or genera. Most remarkable was the dominance and diversity (11 phylotypes) of a novel, so far unknown lineage of *pufLM* containing Gammaproteobacteria, which was highly diverse and prevalent in different lakes of the Salar de Atacama (Thiel et al. 2010).

Phototrophic Betaproteobacteria are rare in the salares of the Chilean Altiplano, but clones of Betaproteobacteria distantly related to *Rhodoferax fermentans* were identified (Thiel et al. 2010; Dorador et al. 2013).

In conclusion, most of the purple sulfur bacteria recognized by *pufLM* sequences in these salt lakes are related to the known halophilic groups of purple sulfur bacteria but represent new bacteria at the genus level. This depicts quite well the extraordinary situation of the habitats in the Chilean highlands with extreme climatic and environmental conditions as well as the great geographic distance to all so far investigated habitats of phototrophic bacteria and points out the uniqueness of their bacterial communities.

Though in the two salt lakes situated in close distance (Laguna Chaxa, Laguna Tebenquiche) species and phylotypes were identified that can be expected to be specifically adapted to the high salt concentrations in the habitat (relatives to *Halochromatium*, *Thiohalocapsa*, *Ectothiorhodospira*, and *Halorhodospira*), it is remarkable to see that different structures of microbial mats develop in locations close to each other (Figs. 7–9) and at a small spatial scale different samples even from the same lake may demonstrate largely different composition of the communities of purple sulfur bacteria (Thiel et al. 2010).

Osmotic Adaptation

As osmotic adaptation and accumulation of compatible solutes are essential aspects for life at elevated salt concentrations, we will shortly summarize some of these aspects here. The term compatible solute was first introduced by Brown and Simpson (1972), to describe solutes that are accumulated in response to osmotic or salt stress and are compatible with cellular structures and functions even at high concentrations. The compatibility includes protection against inactivation, inhibition, and denaturation of macromolecules. With increasing osmotic stress also increasing concentrations of compatible solutes have to be accumulated inside the cells and the number of solutes that fulfil this function also at high concentrations is low. In particular glycine betaine, ectoine, glutamate, and proline are widely distributed as compatible solutes in halophilic bacteria (Imhoff and Rodriguez-Valera 1984; Imhoff 1986, 1993). Glycine betaine and ectoine are prominent solutes of extremely halophilic and halotolerant phototrophic bacteria (Galinski and Trüper 1982; Galinski et al. 1985; Imhoff and Rodriguez-Valera 1984; Imhoff 1986, 1988, 1993; Trüper and Galinski 1990). The first bacterium shown to synthesize ectoine was *Halorhodospira halochloris* which accumulates glycine betaine to molar concentrations and in addition to considerable amounts of the new amino acid ectoine and also trehalose (Galinski and Trüper 1982; Galinski et al. 1985). Several species which are adapted to grow at hypersaline salt concentrations accumulate a combination of glycine betaine, ectoine, *N*-acetyl-glutaminylglutamine amide (or other glutamine amide derivatives) with the one or the other sugar (e.g., trehalose or sucrose) in addition. Glycine betaine and Na-carbamoyl glutamine amide are compatible solutes in *Ectothiorhodospira marismortui* (Oren et al. 1991). Trehalose and sucrose apparently are compatible solutes found in the non-halophilic phototrophic purple and green sulfur bacteria, while the halophilic species *Thiohalocapsa halophila* and *Halochromatium salexigens* accumulate glycine betaine and N-acetyl-glutaminylglutamine amide (Welsh and Herbert 1993).

Two possibilities exist for the accumulation of compatible solutes, biosynthesis, or transport. Transport is the energetically favored process and many bacteria have transport systems for compatible solutes such as proline, glycine betaine, trehalose, and ectoine (e.g., Jebbar et al. 1997). All tested non-halophilic and halophilic purple and green sulfur bacteria did accumulate exogenous supplied glycine betaine in response to the salt stress (Welsh and Herbert 1993). The transport of compatible solutes is of strategic importance for bacteria living in microbial mats and of particular relevance for those bacteria unable to synthesize these compounds. They can achieve osmotic adaptation by uptake of such solutes from the environment and thereby take advantage from the fact that those bacteria accumulating the substances by biosynthesis are leaking part of the solutes to the environment. Even non-halophilic freshwater bacteria such as *Rhodobacter capsulatus* gain salt tolerance in the presence of glycine betaine and growth inhibition at 0.2 M NaCl could be relieved by uptake and accumulation of glycine betaine in the cells (Igeno et al. 1995). A correlation was found between the type of compatible solute accumulated and the degree of salt tolerance in a number of cyanobacteria with different salt responses (Mackay et al. 1984; Reed et al. 1986). These findings suggested that a correlation exists between the upper limit of tolerated salt concentrations and the solutes accumulated inside the cells, i.e., the more compatible the accumulated solute the higher the salt or osmotic tolerance (Imhoff 1986). Therefore, it was concluded that the accumulation of compatible solutes in the cells, though it is a necessary prerequisite to cope with increased osmotic pressures in the saline environment and to adapt to elevated salt concentrations, defines the upper limit and the tolerance of salts rather than the requirement (Imhoff 1986, 1993).

Conclusions

The picture emerging from studies on phototrophic bacteria from extreme environments shows that only a few specialists or groups of specialist species have conquered habitats with extremes of physicochemical parameters and different types of specialists exist in various types of extreme habitats. On one hand, several phylogenetically defined groups of phototrophic bacteria are well adapted to hot vents, to alkaline and saline soda lakes, to moderately saline and almost neutral salt lakes, or to acidic environments. On the other hand, a number of single species of different genera are known that have gained properties to cope with the extreme conditions.

Thermophiles

Thermophilic representatives occur in all phylogenetic phyla of phototrophic bacteria (see Table 1). A single major phylogenetic group of moderately thermophilic bacteria is associated with hot spring habitats and found at the upper temperature limits at which phototrophic bacteria grow. This group is represented by the phototrophic Chloroflexi with *Chloroflexus*, *Roseiflexus*, *Chlorothrix*, and *Heliothrix* and their relatives, most of which have temperature optima between 50 and 55 °C and upper temperature limits of 60 °C and exceptionally for *Chloroflexus aurantiacus* at approx. 70 °C.

Single representative species of moderately thermophilic phototrophic bacteria with growth optima between 48 and 52 °C are known in the Acidobacteria with *Chloroacidobacterium thermophilum* (optimum temperature 51 °C), in the Heliobacteria with *Heliobacterium modesticaldum* (optimum temperature 52 °C), in the Chlorobi with *Chlorobaculum tepidum* (optimum temperature 48 °C), and in the Proteobacteria with *Thermochromatium tepidum* (optimum temperature 50 °C). In addition, a number of purple nonsulfur bacteria have temperature optima significantly higher compared to most mesophilic phototrophic bacteria (growing well between 20 and 35 °C) and in laboratory cultures show growth optima at 35–45 °C. However, in hot springs these and also several non-thermophilic species are regularly found at temperatures exceeding 55 °C and even up to 65 °C. Therefore, they may be considered as borderline slightly thermophilic and moderately thermotolerant bacteria.

Acidophiles

Among all anaerobic anoxygenic phototrophic bacteria only *Rhodopila globiformis* is able to cope well with pH values below pH 5. If aerobic bacteriochlorophyll-containing representatives are included into the consideration, a whole phylogenetic group of Alphaproteobacteria exists that has adapted to acidic environments. This group is formed by *Rhodopila globiformis* and representatives of the genera *Acidiphilium* and *Acidisphaera,* all of which have optimal growth conditions in the pH-range from 4.5 to 5.0.

Alkaliphiles

Phototrophic bacterial communities of soda lakes are unique and a small number of phototrophic bacteria are specifically adapted to the alkaline conditions (see Tables 2 and 3). The pH is a clearly discriminating factor in particular in saline and hypersaline habitats. It appears that under alkaline conditions the increase of salinity further reduces the diversity of adapted forms to a small number of species and with a clear advantage of *Halorhodospira* species in the alkaline, highly saline environments. Those phototrophic bacteria that cope best with alkaline conditions are representatives of the Ectothiorhodospiraceae, which are adapted to different ranges of salinity. They form a major phylogenetic group of alkaliphilic Gammaproteobacteria with phototrophic representatives (*Ectothiorhodospira*, *Halorhodospira*, *Thiorhodospira*, and *Ectothiorhodosinus*) as well as chemotrophic representatives (*Thioalkalivibrio* and *Thioalkalispira*). The results of a comparative study of a larger number of salt and soda lakes may be representative for the significant advantage of Ectothiorhodospiraceae under alkaline conditions (Kompantseva et al. 2010a). Though most representatives of Ectothiorhodospiraceae are adapted to alkaline waters at pH >8.5, a few species grow quite well near neutrality at pH 7–7.5 (*Halorhodospira neutriphila*, *Ectothiorhodospira salini*). Different species of this group have growth optima in different ranges of salinity and dominate over Chromatiaceae in neutral lakes at high salt concentrations (20–25 % salts). In soda lakes they predominate already at low salt concentrations (0.5 % salts) which indicates their advantage in particular at alkaline pH (Kompantseva et al. 2010a). These observations are in line with competition experiments with cultures of *Allochromatium vinosum* and *Ectothiorhodospira mobilis*, in which the *Ectothiorhodospira* rapidly outcompeted the *Allochromatium* at pH higher than 8.3 (Trüper and Imhoff 1981).

Also, a larger group of purple nonsulfur bacteria of the *Rhodobaca/Rhodovulum* group with *Rhodobaca bogoriensis*, *Rhodobaca barguzinensis*, *Rhodovulum tesquicolum*, and *Rhodovulum steppense* as representatives is coping quite well with both alkalinity and low levels of salinity. This group also includes a number of alkaliphilic or alkalitolerant aerobic bacteriochlorophyll-containing bacteria such as *Roseinatronobacter thiooxidans*. Notable a group of three *Heliorestis* species is adapted to alkaline environments, as are a few representatives of the Chromatiaceae. Green sulfur bacteria have not been found and apparently are absent from alkaline lakes, which correlates well with their preference for slightly acidic conditions (below pH 7). The limited potential for adaptation to elevated salt concentrations is an additional reason for the lack of green sulfur bacteria in saline soda lakes.

Halophiles

Although the whole spectrum of salt concentrations is inhabited by anoxygenic phototrophic bacteria, individual species are adapted to a selected range of salt concentrations. It is evident that different species and phylotypes of phototrophic bacteria are associated with different types of lakes. In particular the clear distinction between species associated with soda lakes and neutral salt lakes is quite obvious. In addition, the diversity of phototrophic bacteria is significantly higher in neutral saline lakes compared to saline and alkaline soda lakes (Kompantseva et al. 2010a).

With the exception of *Ectothiorhodospira* and *Halorhodospira* species that are present in both alkaline and neutral salt lakes, different groups of phototrophic bacteria, in particular phototrophic purple bacteria, are thriving in neutral salt lakes of moderate salinities and in soda lakes. In moderately saline neutral salt lakes and salterns worldwide a group of purple sulfur bacteria phylogenetically related to *Halochromatium salexigens* and *Thiohalocapsa halophila* are important players. In addition, a group of purple nonsulfur bacteria related to *Rhodothalassium salexigens*, *Rhodovibrio sodomensis*, and *Rhodovibrio salinarum* are important inhabitants and may form massive visible developments in neutral salt lakes and salterns. They represent slightly or moderately halophilic species and are extremely halotolerant (Table 3).

Though green sulfur bacteria are common in marine coastal sediments and *Prosthecochloris aestuarii* is a widely distributed marine and moderately halotolerant bacterium (Imhoff 2001), they rarely have been recognized in salterns and salt lakes and may be present in low numbers. *Prosthecochloris aestuarii* is a typical representative of marine and slightly hypersaline habitats and apparently the most salt tolerant species of this group of bacteria. *Prosthecochloris* species do form a major component in the stratified hypersaline Solar Lake (Sinai) in which clearly separate layers of green sulfur and purple sulfur bacteria develop (Cohen et al. 1977). The increased salt tolerance of *Prosthecochloris aestuarii* is considered responsible for their regular occurrence in moderately hypersaline waters (Puchkova 1984).

References

Abd-el-Malek Y, Rizk SG (1963) Bacterial sulphate reduction and the development of alkalinity. III. Experiments under natural conditions. J Appl Bacteriol 26:20–26

Achenbach LA, Carey JR, Madigan MT (2001) Photosynthesis and phylogenetic primers for the detection of anoxygenic phototrophs in natural environments. Appl Environ Microbiol 67:2922–2926

Ainon H, Tan CJ, Vikineswary S (2006) Biological characterization of *Rhodomicrobium vannielii* isolated from a hot spring at Gadek, Malacca, Malaysia. Malaysian J Microbiol 2:15–21

Alexander B, Imhoff JF (2006) Communities of green sulfur bacteria in different marine and saline habitats analyzed by gene sequences of 16S rRNA and of the Fenna-Matthews-Olson protein. Int Microbiol 9:259–266

Alexander B, Andersen JH, Cox RP, Imhoff JF (2002) Phylogeny of green sulfur bacteria on the basis of gene sequences of 16S rRNA and of the Fenna-Matthews-Olson protein. Arch Microbiol 178:131–140

Argos P, Rossmann MG, Grau UM, Zuber H, Frank G, Tratschin JD (1979) Thermal stability and protein structure. Biochemistry 18:5698–5703

Asao M, Jung DO, Achenbach LA, Madigan MT (2006) *Heliorestis convoluta* sp. nov., a coiled, alkaliphilic heliobacterium from the Wadi El Natroun, Egypt. Extremophiles 10:403–410

Asao M, Takaichi S, Madigan MT (2007) *Thiocapsa imhoffii*, sp. nov., an alkaliphilic purple sulfur bacterium of the family Chromatiaceae from Soap Lake, Washington (USA). Arch Microbiol 188:665–675

Asao M, Pinkart HC, Madigan MT (2011) Diversity of extremophilic purple phototrophic bacteria in Soap Lake, a central Washington (USA) soda lake. Environ Microbiol 13:2146–2157

Bachar A, Omoregie E, deWit R, Jonkers HM (2007) Diversity and function of *Chloroflexus*-like bacteria in a hypersaline microbial mat: phylogenetic characterization and impact on aerobic respiration. Appl Environ Microbiol 73:3975–3983

Blankenship RE, Olson JM, Miller M (1995) Antenna complexes from green photosynthetic bacteria. In: Blankenship RE, Madigan MT, Bauer CE (eds) Anoxygenic photosynthetic bacteria. Kluwer, Dordrecht, pp 399–435

Boldareva EN, Gorlenko VM (2014) *Roseibacula alcaliphilum* gen. nov. sp. nov., a new alkaliphilic aerobic anoxygenic phototrophic bacterium from a meromictic soda lake Doroninskoe (East Siberia, Russia). Microbiology 83:381–390

Boldareva EN, Bryantseva IA, Tsapin A, Nelson K, Sorokin D, Turova TP, Boichenko VA, Stadnichuk IN, Gorlenko VM (2007) The new alkaliphilic bacteriochlorophyll a-containing bacterium *Roseinatronobacter monicus* sp. nov. from the hypersaline soda Mono Lake (California, USA). Microbiology 76:82–92

Boldareva EN, Akimov VN, Boychenko VA, Stadnichuk IN, Moskalenko AA, Makhneva ZK, Gorlenko VM (2008) *Rhodobaca barguzinensis* sp. nov., a new alkaliphilic purple nonsulfur bacterium isolated from a soda lake of the Barguzin Valley (Buryat Republic, Eastern Siberia). Mikrobiologiia 77:206–218

Boldareva EN, Moskalenko AA, Makhneva ZK, Tourova TP, Kolganova TV, Gorlenko VM (2009) *Rubribacterium polymorphum* gen. nov., sp. nov., a novel alkaliphilic nonsulfur purple bacterium from an eastern Siberian soda lake. Microbiology 78:732–740

Boomer SM, Pierson BK, Austinhirst R, Castenholz RW (2000) Characterization of novel bacteriochlorophyll-a—containing red filaments from alkaline hot springs in Yellowstone National Park. Arch Microbiol 174:152–161

Boomer SM, Lodge DP, Dutton BE, Pierson BK (2002) Molecular characterization of novel red green nonsulfur bacteria from five distinct hot spring communities in Yellowstone National Park. Appl Environ Microbiol 68:346–355

Borsodi AK, Knab M, Czeibert K, Marialegeti K, Vörös L, Somogyi B (2013) Planktonic bacterial community of an extremely shallow soda pond during a phytoplankton bloom revealed by cultivation and molecular cloning. Extremophiles 17:575–584

Brock TD (1978) Thermophilic microorganisms and life at high temperatures. Springer, New York

Brown AD (1976) Microbial water stress. Bacteriol Rev 40:803–846

Brown AD, Simpson JR (1972) Water relations of sugar-tolerant yeasts: the role of intracellular polyols. J Gen Microbiol 72:589–591

Bryant DA, Costas AMG, Maresca JA, Chew AGM, Klatt CG, Bateson MM, Tallon LJ, Hostetler J, Nelson WC, Heidelberg JF, Ward DM (2007) *Candidatus Chloroacidobacterium* thermophilum: an aerobic phototrophic acidobacterium. Science 317:523–526

Bryantseva IA, Gorlenko VM, Kompantseva EI, Achenbach LA, Madigan MT (1999a) *Heliorestis daurensis* gen. nov. sp. nov., an alkaliphilic rod-to-coiled-shaped phototrophic heliobacterium from a Siberian soda lake. Arch Microbiol 172:167–174

Bryantseva IA, Gorlenko VM, Kompantseva EI, Imhoff JF, Süling J, Mityushina L (1999b) *Thiorhodospira sibirica* gen. nov., sp. nov., a new alkaliphilic purple sulfur bacterium from a Siberian soda lake. Int J Syst Bacteriol 49:697–703

Bryantseva IA, Gorlenko VM, Kompantseva EI, Imhoff JF (2000a) *Thioalkalicoccus limnaeus* gen. nov., sp. nov., a new alkaliphilic purple sulfur bacterium with bacteriochlorophyll *b*. Int J Syst Bacteriol 50:2157–2163

Bryantseva IA, Gorlenko VM, Tourova TP, Kuznetsov BB, Lysenko AM, Bykova SA, Gal'chenko VF, Mityushina LL, Osipov GA (2000b) *Heliobacterium sulfidophilum* sp. nov. and *Heliobacterium undosum* sp. nov.: sulfide-oxidizing heliobacteria from thermal sulfide springs. Mikrobiologyia 69:396–406

Bryantseva IA, Gaisin VA, Gorlenko VM (2015) *Rhodobaculum claviforme* gen. nov., sp. nov., a new alkaliphilic nonsulfur purple bacterium. Microbiology 84:247–255

Bryantseva IA, Tourova TP, Kovaleva OL, Kostrikina NA, Gorlenko VM (2010) Ectothiorhodospira magna sp. nov., a new large alkaliphilic purple sulfur bacterium. Microbiology 79:782–792

Burke CM, Burton HR (1988a) The ecology of photosynthetic bacteria in Burton Lake, Vestfold Hills, Antarctica. Hydrobiologia 165:1–11

Burke CM, Burton HR (1988b) Photosynthetic bacteria in meromictic lakes and stratified fjords of the Vestfold Hills, Antarctica. Hydrobiologia 165:13–23

Castenholz RW (1988) The green sulfur and nonsulfur bacteria from hot springs. In: Olson JM, Ormerod JG, Amesz J, Stackebrandt EE, Trüper HG (eds) Green photosynthetic bacteria. Plenum Press, New York, pp 243–255

Castenholz RW, Pierson BK (1995) Ecology of thermophilic anoxygenic phototrophs. In: Blankenship RE, Madfigan MT, Bauer CE (eds) Anoxygenic photosynthetic bacteria. Kluwer Publ, Dordrecht, pp 87–103

Castenholz RW, Bauld J, Jørgensen BB (1990) Anoxygenic microbial mats of hot springs: thermophilic *Chlorobium* sp. FEMS Microbiol Ecol 74:325–336

Caumette P (1993) Ecology and physiology of phototrophic bacteria and sulfate-reducing bacteria in marine salterns. Experientia 49:473–481

Caumette P, Baulaigue R, Matheron R (1988) Characterization of *Chromatium salexigens* sp. nov., a halophilic Chromatiaceae isolated from Mediterranean salinas. Syst Appl Microbiol 10:284–292

Caumette P, Baulaigue R, Matheron R (1991) *Thiocapsa halophila* sp. nov., a new halophilic phototrophic purple sulfur bacterium. Arch Microbiol 155:170–176

Caumette P, Matheron R, Raymond N, Relexans J-C (1994) Microbial mats in the hypersaline ponds of Mediterranean saltern (Salins-de-Giraud, France). FEMS Microbiol Ecol 13:273–286

Caumette P, Imhoff JF, Süling J, Matheron R (1997) *Chromatium glycolicum* sp. nov, a moderately halophilic purple sulfur bacterium that uses glycolate as substrate. Arch Microbiol 167:11–18

Cavalier-Smith T (2010) Deep phylogeny, ancestral groups and the four ages of life. Philos Trans R Soc London B Biol Sci 365:111–132

Cherni NE, Solovjeva JU, Fedorov VD, Kondratieva EN (1969) Ultrastructure of two species of purple sulfur bacteria. Microbiologiya 38:479–484

Chin JP, Megaw J, Magill CL, Nowotarski K, Williams JP, Bhagann P, Linton M, Patterson MF, Underwood GJC, Mswaka AY, Hallsworth JE (2010) Solutes determine the temperature window for microbial survival and growth. Proc Nat Acad Sci U S A 107:7835–7840

Cloern JE, Cole BE, Oremland RS (1983a) Autotrophic processes in meromictic Big Soda Lake, Nevada. Limnol Oceanogr 28:1049–1061

Cloern JE, Cole BE, Oremland RS (1983b) Seasonal changes in the chemistry and biology of a meromictic lake (Big Soda Lake, Nevada, U.S.A.). Hydrobiologia 105:195–206

Cohen Y, Krumbein WE, Shilo M (1977) Solar lake (Sinai). 2. Distribution of photosynthetic microorganisms and primary production. Limnol Oceanogr 22:609–620

Dalhus B, Saarinen M, Sauer UH, Eklund P, Johansson K, Karlsson A, Ramaswamy S, Bjørk A, Synstad B, Naterstad K, Sirevåg R, Eklund H (2002) Structural basis for thermophilic protein stability: structures of thermophilic and mesophilic malate dehydrogenases. J Mol Biol 318:707–721

Dilling W, Liesack W, Pfennig N (1995) *Rhabdochromatium marinum* gen. nom. rev., sp. nov., a purple sulfur bacterium from a salt marsh microbial mat. Arch Microbiol 164:125–131

Dillon JG, Carlin M, Gutierrez A, Nguyen V, McLain N (2013) Patterns of microbial diversity along a salinity gradient in the Guerrero Negro solar saltern, Baja CA Sur, Mexico. Front Microbiol 4:399. doi:10.3389/fmicb.2013.00399

Dorador C, Vila I, Witzel K-P, Imhoff JF (2008) Cyanobacterial diversity in Salar de Huasco, a high altitude saline wetland in Northern Chile, are highly similar to Antarctic cyanobacteria. FEMS Microbiol Ecol 64:419–432

Dorador C, Meneses D, Urtuvia V, Demergasso C, Vila I, Witzel K-P, Imhoff JF (2009) Diversity of *Bacteroidetes* in high altitude saline evaporitic basins in northern Chile. J Geophys Res Biogeosci 114:35–44

Dorador C, Vila I, Witzel K-P, Imhoff JF (2013) Bacterial and archaeal diversity in high altitude wetlands of the Chilean Altiplano. Fundam Appl Limnol 182:135–159

Drews G (1981) *Rhodospirillum salexigens,* spec. nov., an obligatory halophilic phototrophic bacterium. Arch Microbiol 130:325–327

Edwardson CF, Planer-Friedrich B, Hollibaugh JT (2014) Transformation of monothioarsenate by haloalkaliphilic, anoxygenic phototrophic purple sulfur bacteria. FEMS Microbiol Ecol 90:858–868

Eicken H, Bock C, Wittig R, Miller H, Poertner HO (2000) Magnetic resonance imaging of sea-ice pore fluids: methods and thermal evolution of pore microstructure. Cold Reg Sci Technol 31:207–225

Favinger J, Stadtwald R, Gest H (1989) *Rhodospirillum centenum* sp. nNov., a thermotolerant cyst-forming photosynthetic bacterium. Ant Leeuwenhoek 55:291–296

Fenna RE, Matthews BW, Olson JM, Shaw EK (1974) Structure of a bacteriochlorophyll-protein from the green photosynthetic bacterium *Chlorobium limicola*: crystallographic evidence for a trimer. J Mol Biol 84:231–240

Fernández AB, Ghai R, Martin-Cuadrado AB, Sánchez-Porro C, Rodriguez-Valera F, Ventosa A (2014) Prokaryotic taxonomic and metabolic diversity of an intermediate salinity hypersaline habitat assessed by metagenomics. FEMS Microbiol Ecol 88:623–635

Fields PA (2001) Protein function at thermal extremes: balancing stability and flexibility. Comp Biochem Physiol Part A 129:417–431

Gaisin VA, Kalashnikov AM, Sukhacheva MV, Namsaraev ZB, Barhutova DD, Gorlenko VM, Kuznetsov BB (2015) Filamentous anoxygenic phototrophic bacteria from cyanobacterial mats of Alla hot springs (Barguzin Valley, Russia). Extremophiles 19:1067–1076

Gaisin V, Grouzdev D, Namsaraev Z, Sukhacheva M, Gorlenko V, Kusnetsov B (2016) Biogeography of thermophilic phototrophic bacteria belonging to *Roseiflexus* genus. FEMS Microbiol Ecol 92(3): online first. doi:10.1093/femsec/fiw012

Galinski E, Trüper HG (1982) Betaine, a compatible solute in the extremely halophilic phototrophic bacterium *Ectothiorhodospira halochloris*. FEMS Lett 13:357–360

Galinski EA, Pfeiffer HP, Trüper HG (1985) 1,4,5,6-tetrahydro-2-methyl-4-pyrimidinecarboxylic acid. A novel cyclic amino acid from halophilic phototrophic bacteria of the genus *Ectothiorhodospira*. Eur J Biochem 149:135–139

Gest H, Favinger JL (1983) *Heliobacterium chlorum*, an anoxygenic brownish-green photosynthetic bacterium containing a new form of bacteriochlorophyll. Arch Microbiol 136:11–16

Gibson JAE (1999) The meromictic lakes and stratified marine basins of the Vestfold Hills, East Antarctica. Antarct Sci 11:175–192

Gibson J, Pfennig N, Waterbury JB (1984) *Chloroherpeton thalassium* gen. nov. et spec. nov., a non-filamentous, flexing and gliding green sulfur bacterium. Arch Microbiol 138:96–101

Gorlenko VM (1975) Characteristics of filamentous phototrophic bacteria from freshwater lakes. Microbiologiya 44:682–684

Gorlenko VM (1988) Ecological niches of green sulfur and gliding bacteria. In: Olson JM, Ormerod JG, Amesz J, Stackebrandt EE, Trüper HG (eds) Green photosynthetic bacteria. Plenum Press, New York, pp 257–267

Gorlenko VM, Bryantseva IA, Lunina ON, Tourova TP (2014) Phylogenetic position of the purple sulfur bacterium Lamprobacter modestohalophilus determined based on the data on new strains of the species. Microbiology 83:829–837

Gorlenko VM, Bryantseva IA, Panteleeva EE, Tourova TP, Kolganova TV, Makhneva ZK, Moskalenko AA (2004) *Ectothiorhodosinus mongolicum* gen. nov., sp. nov., a new purple bacterium from a soda lake in Mongolia. Microbiology 73:80–88

Gorlenko VM, Bryantseva IA, Rabold S, Tourova TP, Rubtsova D, Smirnova E, Thiel V, Imhoff JF (2009) *Ectothiorhodospira variabilis* sp. nov., an alkaliphilic and halophilic purple sulfur bacterium from soda lakes. Int J Syst Evol Microbiol 59:658–664

Grouzdev DS, Kuznetsov BB, Keppen OI, Krasil'nikova EN, Lebedeva NV, Ivanovsky RN (2015) Reconstruction of bacteriochlorophyll biosynthesis pathways in the filamentous anoxygenic phototrophic bacterium *Oscillochloris trichoides* DG-6 and evolution of anoxygenic phototrophs of the order Chloroflexales. Microbiology 161:120–130

Hanada S, Hiraishi A, Shimada K, Matsuura K (1995a) Isolation of *Chloroflexus aurantiacus* and related thermophilic phototrophic bacteria from Japanese hot springs using an improved isolation procedure. J Gen Appl Microbiol 41:119–130

Hanada S, Hiraishi A, Shimada K, Matsuura K (1995b) *Chloroflexus aggregans* sp. nov., a filamentous phototrophic bacterium which forms dense cell aggregates by active gliding movement. Int J Syst Evol Microbiol 45:676–681

Hanada S, Takaichi S, Matsuura K, Nakamura K (2002) *Roseiflexus castenholzii* gen. nov., sp. nov., a thermophilic, filamentous, photosynthetic bacterium that lacks chlorosomes. Int J Syst Evol Microbiol 52:187–193

Hiraishi A, Shimada K (2001) Aerobic anoxygenic photosynthetic bacteria with zinc-bacteriochlorophyll. J Gen Appl Microbiol 47:161–180

Hiraishi A, Nagashima KVP, Matsuura K, Shimada K, Takaichi S, Wakao N, Katayama Y (1998) Phylogeny and photosynthetic features of *Thiobacillus acidophilus* and related acidophilic bacteria: its transfer to the genus *Acidiphilium* as *Acidiphilium acidophilum* comb. nov. Int J Syst Bacteriol 48:21389–21398

Hiraishi A, Matsuzawa Y, Kanbe T, Wakao N (2000) *Acidisphaera rubrifaciens* gen. nov., sp. nov., an aerobic bacteriochlorophyll-containing bacterium isolated from acidic environments. Int J Syst Evol Microbiol 50:1539–1546

Hirschler-Réa A, Matheron R, Riffaud C, Mouné S, Eatock C, Herbert RA, Willison JC, Caumette P (2003) Isolation and characterization of spirilloid purple phototrophic bacteria forming red layers in microbial mats of Mediterranean salterns; description of *Halorhodospira neutriphila* sp. nov. and emendation of the genus *Halorhodospira*. Int J Syst Evol Microbiol 53:153–163

Hisada T, Okamura K, Hiraishi A (2007) Isolation and characterization of phototrophic purple nonsulfur bacteria from *Chloroflexus* and cyanobacterial mats in hot springs. Microbes Environ 22:405–411

Humayoun SB, Bano N, Hollibaugh JT (2003) Depth distribution of microbial diversity in Mono Lake, a meromictic soda lake in California. Appl Environ Microbiol 69:1030–1042

Igeno MI, DelMoral GC, Castillio F, Caballero FJ (1995) Halotolerance of the phototrophic bacterium *Rhodobacter capsulatus* E1F1 is dependent on the nitrogen source. Appl Environ Microbiol 61:2970–2975

Imhoff JF (1986) Osmoregulation and compatible solutes in eubacteria. FEMS Microbiol Rev 39:57–66

Imhoff JF (1988) Halophilic phototrophic bacteria. In: Rodriguez-Valera F (ed) Halophilic bacteria, vol 1. CRC Press, Boca Raton, pp 85–108

Imhoff JF (1992a) Taxonomy, phylogeny and general ecology of anoxygenic phototrophic bacteria. In: Carr NG, Mann NH (eds) Biotechnology handbook photosynthetic prokaryotes. Plenum Press, London, New York, pp 53–92

Imhoff JF (1992b) The family ectothiorhodospiraceae. In: Balows A, Trüper HG, Dworkin M, Harder W, Schleifer KH (eds) The Prokaryotes. A handbook on the biology of bacteria. Ecophysiology, isolation, identification, applications, 2nd edn. Springer Verlag, New York, pp 3222–3229

Imhoff JF (1993) Osmotic adaptation in halophilic and halotolerant microorganisms. In: Vreeland RH, Hochstein LJ (eds) The biology of halophilic bacteria. The CRC Press, Boca Raton, pp 211–253

Imhoff JF (2001) Minireview—true marine and halophilic anoxygenic phototrophic bacteria. Arch Microbiol 176:243–254
Imhoff JF (2003) A phylogenetic taxonomy of the family Chlorobiaceae on the basis of 16S rRNA and fmo (Fenna-Matthews-Olson protein) gene sequences. Int J Syst Evol Microbiol 53:941–951
Imhoff JF (2016) New dimensions in microbial ecology—functional genes in studies to unravel the biodiversity and role of functional microbial groups in the environment. Microorganisms (in press)
Imhoff JF, Pfennig N (2001) *Thioflavicoccus mobilis* gen. nov., sp. nov., a novel purple sulfur bacterium with bacteriochlorophyll *b*. Int J Syst Evol Microbiol 51:105–110
Imhoff JF, Rodriguez-Valera F (1984) Betaine is the main compatible solute of halophilic eubacteria. J Bacteriol 160:478–479
Imhoff JF, Süling J (1996) The phylogenetic relationship among Ectothiorhodospiraceae. A reevaluation of their taxonomy on the basis of 16S rDNA analysis. Arch Microbial 165:106–113
Imhoff JF, Trüper HG (1977) *Ectothiorhodospira halochloris* sp. nov., a new extremely halophilic phototrophic bacterium containing bacteriochlorophyll b. Arch Microbiol 114:115–121
Imhoff JF, Trüper HG (1981) *Ectothiorhodospira abdelmalekii* sp. nov., a new halophilic and alkaliphilic phototrophic bacterium. Zbl Bakt I Abt Orig C2:228–234
Imhoff JF, Trüper HG (1992) The genus *Rhodospirillum* and related genera. In: Balows A, Trüper HG, Dworkin M, Harder W, Schleifer KH (eds) The Prokaryotes. A handbook on the biology of bacteria. Ecophysiology, isolation, identification, applications, 2nd edn. Springer Verlag, New York, pp 2141–2155
Imhoff JF, Hashwa F, Trüper HG (1978) Isolation of extremely halophilic phototrophic bacteria from the alkaline Wadi Natrun, Egypt. Arch Hydrobiol 84:381–388
Imhoff JF, Sahl HG, Soliman GSH, Trüper HG (1979) The Wadi Natrun: chemical composition and microbial mass developments in alkaline brines of eutrophic desert lakes. Geophys J R Astron Soc 1:219–234
Imhoff JF, Tindall BJ, Grant WD, Trüper HG (1981) *Ectothiorhodospira vacuolata* sp. nov., a new phototrophic bacterium from soda lakes. Arch Microbiol 130:238–242
Imhoff JF, Trüper HG, Pfennig N (1984) Rearrangement of the species and genera of the phototrophic "Purple Nonsulfur Bacteria". Int J Syst Bacteriol 34:340–343
Imhoff JF, Petri R, Süling J (1998a) Reclassification of species of the spiral-shaped phototrophic purple nonsulfur bacteria of the alpha-proteobacteria: description of the new genera *Phaeospirillum* gen. nov., *Rhodovibrio* gen. nov., *Rhodothalassium* gen. nov. and *Roseospira* gen. nov. as well as transfer of *Rhodospirillum fulvum* to *Phaeospirillum fulvum* comb. nov., of *Rhodospirillum molischianum* to *Phaeospirillum molischianum* comb. nov., of *Rhodospirillum salinarum* to *Rhodovibrio salinarum* comb. nov., of *Rhodospirillum sodomense* to *Rhodovibrio sodomensis* comb. nov., of *Rhodospirillum salexigens* to *Rhodothalassium salexigens* comb. nov., and of *Rhodospirillum mediosalinum* to *Roseospira mediosalina* comb. nov. Int J Syst Bacteriol 48:793–798
Imhoff JF, Süling J, Petri R (1998b) Phylogenetic relationships among the *Chromatiaceae*, their taxonomic reclassification and description of the new genera *Allochromatium, Halochromatium, Isochromatium, Marichromatium, Thiococcus, Thiohalocapsa* and *Thermochromatium*. Int J Syst Bacteriol 48:1129–1143
Jannasch HW (1957) Die bakterielle Rotfärbung der Salzseen des Wadi Natrun. Arch Hydrobiol 53:425–433
Jebbar M, Blohn C, Bremer E (1997) Ectoine functions as an osmoprotectant in *Bacillus subtilis* and is accumulated via the ABC-transport system OpuC. FEMS Microbiol Lett 154:325–330
Jonkers HM, Ludwig R, deWit R, Pringault O, Muyzer G, Niemann H, Finke N, deBeer D (2003) Structural and functional analysis of a microbial mat ecosystem from a unique permanent hypersaline inland lake: La Salada de Chiprana (NE Spain). FEMS Microbiol Ecol 44:175–189
Jung DO, Achenbach LA, Karr EA, Takaichi S, Madigan MT (2004) A gas vesiculate planktonic strain of the purple nonsulfur bacterium *Rhodoferax antarcticus* isolated from Lake Fryxell, Dry Valleys, Antarctica. Arch Microbiol 182:236–243

Kaartokallio H (2001) Evidence for active microbial nitrogen transformations in sea ice (Gulf of Bothnia, Baltic Sea) in mid winter. Polar Biol 24:21–28

Karr EA, Sattley WM, Jung DO, Madigan MT, Achenbach LA (2003) Remarkable diversity of phototrophic purple bacteria in a permanently frozen Antarctic lake. Appl Environ Microbiol 69:4910–4914

Kawasaki H, Hoshino Y, Kuraishi H, Yamasato K (1992) *Rhodocista centenaria* gen. nov., sp. nov., a cyst-forming anoxygenic photosynthetic bacterium and its phylogenetic position in the Proteobacteria alpha group. J Gen Appl Microbiol 38:541–551

Kimble LK, Mandelco L, Woese CR, Madigan MT (1995) *Heliobacterium modesticaldum*, sp. nov., a thermophilic heliobacterium of hot springs and volcanic soils. Arch Microbiol 163:259–267

Klappenbach JA, Pierson BK (2004) Phylogenetic and physiological characterization of a filamentous anoxygenic photoautotrophic bacterium '*Candidatus* Chlorothrix halophila' gen. nov., sp. nov., recovered from hypersaline microbial mats. Arch Microbiol 181:17–25

Knauth LP (2005) Temperature and salinity history of the Precambrian ocean: implications for the course of microbial evolution. Palaeogeogr Palaeoclimatol Palaeoecol 219:53–69

Koh EY, Phua W, Ryan KG (2011) Aerobic anoxygenic phototrophic bacteria in Antarctic sea ice and seawater. Environ Microbiol Rep 3:710–716

Kompantseva EI, Gorlenko VM (1988) Phototrophic communities of some thermal springs of Lake Baikal. Mikrobiologiya 57:841–846

Kompantseva EI, Bryantseva IA, Komova AV, Namsaraev BB (2007) The structure of phototrophic communities of soda lakes of the southeastern Transbaikal region. Microbiology 76:211–219

Kompantseva EI, Komova AV, Kostrikina NA (2010a) *Rhodovulum steppense* sp. nov., an obligatory haloalkaliphilic purple nonsulfur bacterium widespread in saline soda lakes of Central Asia. Int J Syst Evol Microbiol 60:1210–1214

Kompantseva EI, Komova AV, Sorokin DY (2010b) Communities of anoxygenic phototrophic bacteria in the saline soda lakes of the Kunlunda Steppe (Altai Krai). Microbiology 79:89–95

Kompantseva EI, Komova AV, Novikov AA, Kostrikina NA (2012) *Rhodovulum tesquicola* sp. nov., a haloalkaliphilic purple non-sulfur bacterium from brackish steppe soda lakes. Int J Syst Evol Microbiol 62:2962–2966

Kriss AE, Mitskevich IN, Rozanova EP, Osnitskaya L (1976) Microbiological investigations of Lake Vanda (Antarctica). Microbiology 45:917–922

Kubo K, Knittel K, Amann R, Fukui M, Matsuura K (2011) Sulfur-metabolizing bacterial populations in microbial mats of the Nakabusa hot spring, Japan. Syst Appl Microbiol 34:293–302

Kumar BV, Ramprasad EVV, Sasikala C, Ramana CV (2013) *Rhodopseudomonas pentothenatexigens* sp. nov. and *Rhodopseudomonas thermotolerans* sp. nov., isolated from paddy soils. Int J Syst Evol Microbiol 63:200–207

Labrenz M, Lawson PA, Tindall BJ, Hirsch P (2009) *Roseibaca ekhonensis* gen. nov., sp. nov., an alkalitolerant and aerobic bacteriochlorophyll a-producing alphaproteobacterium from hypersaline Ekho Lake. Int J Syst Evol Microbiol 59:1935–1940

Lau MCY, Aitchison JC, Pointing SB (2009) Bacterial community composition in thermophilic microbial mats from five hot springs in central Tibet. Extremophiles 13:139–149

Ley RE, Harris JK, Wilcox J, Spear JR, Miller SR, Bebout BM, Maresca JA, Bryant DA, Sogin ML, Pace NR (2006) Unexpected diversity and complexity of the Gurrero Negro hypersaline microbial mat. Appl Environ Microbiol 72:3685–3695

Maas EW, Simpson AM, Martin A, Thompson S, Koh WY, Davy SK, Ryan KG, O'Toole RF (2012) Phylogenetic analyses of bacteria in sea ice at Cape Hallet, Antarctica. N Z J Mar Freshwater Res 46:3–12

Mack EE, Mandelco L, Woese CR, Madigan MT (1993) *Rhodospirillum sodomense,* sp. nov., a Dead Sea *Rhodospirillum* species. Arch Microbiol 160:363–371

Mackay MA, Norton RS, Borowitzka LJ (1984) Organic osmoregulatory solutes in cyanobacteria. J Gen Microbiol 130:2177–2191

Madigan MT (1986) *Chromatium tepidum* sp. nov. a thermophilic photosynthetic bacterium of the family Chromatiaceae. IJSEM 36:222–227

Madigan MT (1998) Isolation and characterization of psychrophilic bacteria from Antarctica. In: Pescheck GA, Löffelhardt W, Schmetterer G (eds) The phototrophic prokaryotes. Plenum Press, New York, pp 699–706

Madigan MT (2003) Anoxygenic phototrophic bacteria from extreme environments. Photosyn Res 76:157–171

Madigan MT, Brock TD (1975) Photosynthetic sulfide oxidation by *Chloroflexus aurantiacus*, a filamentous, photosynthetic, gliding bacterium. J Bacteriol 122:782–784

Madigan MT, Brock TD (1977) CO_2 fixation in photosynthetically grown *Chloroflexus aurantiacus*. FEMS Microbiol Lett 1:301–304

Madigan MT, Petersen SR, Brock TD (1974) Nutritional studies on *Chloroflexus*, a filamentous photosynthetic, gliding bacterium. Arch Microbiol 100:97–103

Madigan MT, Jung DO, Woese CR, Achenbach LA (2000) *Rhodoferax antarcticus* sp. nov., a moderately psychrophilic purple nonsulfur bacterium isolated from an Antarctic microbial mat. Arch Microbiol 173:269–277

Mesbah NM, Abou-El-Ela SH, Wiegel J (2007) Novel and unexpected prokaryotic diversity in water and sediments of the alkaline, hypersaline lakes of the Wadi Natrun, Egypt. Microbial Ecol 54:598–617

Milford AD, Achenbach LA, Jung DO, Madigan MT (2000) *Rhodobaca bogoriensis* gen. nov. and sp. nov., an alkaliphilic purple nonsulfur bacterium from African Rift Valley soda lakes. Arch Microbiol 174:18–27

Mock T, Thomas DN (2005) Recent advances in sea-ice microbiology. Environ Microbiol 7:605–619

Moune S, Caumette P, Matheron R, Willison JC (2003) Molecular sequence analysis of prokaryotic diversity in the anoxic sediments underlying cyanobacterial mats of two hypersaline ponds in Mediterranean salterns. FEMS Microbiol Ecol 44:117–130

Namsaraev ZB, Gorlenko VM, Namsaraev BB, Buryukhaev SP, Yurkow VV (2003) The structure and biogeochemical activity of the phototrophic communities from the Bol'sherechenskii alkaline hot spring. Mikrobiologiia 72:228–238

Ng C, Demaere MZ, Williams TJ, Lauro FM, Raftery M, Gibson JAE, Andres-Pfannkoch C, Lewis M, Hoffman JM, Thomas T, Cavicchioli R (2010) Metaproteogenomic analysis of a dominant green sulfur bacterium from Ace Lake, Antarctica. ISME J 4:1002–1019

Nicholson JAM, Stolz JF, Pierson BK (1987) Structure of a microbial mat at Great Sippewissett Marsh, Cape Cod, Massachusetts. FEMS Microbiol Ecol 45:343–364

Nickens D, Fry CJ, Ragatz L, Bauer CE, Gest H (1996) Biotype of the purple nonsulfur bacterium, *Rhodospirillum centenum*. Arch Microbiol 165:91–96

Nissen H, Dundas ID (1984) *Rhodospirillum salinarum* sp. nov., a halophilic photosynthetic bacterium isolated from a Portuguese saltern. Arch Microbiol 138:251–256

Nübel U, Bateson MM, Madigan MT, Kühl M, Ward DM (2001) Diversity and distribution in hypersaline microbial mats of bacteria related to *Chloroflexus* spp. Appl Environ Microbiol 67:4365–4371

Oren A, Kessel M, Stackebrandt E (1989) *Ectothiorhodospira marismortui* sp. nov., an obligately anaerobic, moderately halophilic purple sulfur bacterium from a hypersaline sulfur spring at the shore of the Dead Sea. Arch Microbiol 151:524–529

Oren A, Simon G, Galinski EA (1991) Intracellular salt and solute concentrations in *Ectothiorhodospira marismortui*: glycine betaine and Na-carbamoyl glutamineamide as osmotic solutes. Arch Microbiol 156:350–355

Overmann J, Fischer U, Pfennig N (1992) A new purple sulfur bacterium from saline littoral sediments, *Thiorhodovibrio winogradskyi* gen. nov. and sp. nov. Arch Microbiol 157:329–335

Oyaizu H, Debrunner-Vossbrinck B, Mandelco L, Studier JA, Woese CR (1987) The green nonsulfur bacteria: a deep branching in the eubacterial line of descent. Syst Appl Microbiol 9:47–53

Pelsh AD (1936) Hydrobiology of Karabugaz Bay of the Caspian Sea. Tr. Solyanoi Lab. Vses. Inst. Metallurgii NIS NKT. Izdatelstvo Akad. Nauk SSSR (Moscow-Leningrad). 5:49–126

Petri R, Imhoff JF (2001) Genetic analysis of sea-ice bacterial communities of the Western Baltic Sea using an improved double gradient method. Polar Biol 24:252–257

Pfennig N (1969) *Rhodopseudomonas acidophila*, sp. n., a new species of the budding purple nonsulfur bacteria. J Bacteriol 99:597–602

Pfennig N (1974) *Rhodopseudomonas globiformis*, sp. n., a new species of the Rhodospirillaceae. Arch Microbiol 100:197–206

Pfennig N, Lünsdorf H, Süling J, Imhoff JF (1997) *Rhodospira trueperi*, gen. nov. and spec. nov., a new phototrophic Proteobacterium of the alpha group. Arch Microbiol 168:39–45

Pierson BK, Castenholz RW (1974a) A phototrophic gliding filamentous bacterium from hot springs, *Chloroflexus aurantiacus* gen. and sp. nov. Arch Microbiol 100:5–24

Pierson BK, Castenholz RW (1974b) Studies on pigments and growth in *Chloroflexus aurantiacus*, a phototrophic filamentous bacterium. Arch Microbiol 100:283–305

Pierson BK, Giovannoni SJ, Stahl DA, Castenholz RW (1985) *Heliothrix oregonensis*, gen. nov. spec. nov., a phototrophic filamentous gliding bacterium containing bacteriochlorophyll a. Arch Microbiol 142:164–167

Pierson BK, Valdez D, Larsen M, Morgan E, Mack EE (1994) *Chloroflexus*-like organisms from marine and hypersaline environments: distribution and diversity. Photosynth Res 41:35–52

Priscu JC, Priscu LR, Vincent WF, Howard-Williams C (1987) Photosynthate distribution by microplankton in permanently ice-covered Antarctic desert lakes. Limnol Oceanogr 32:260–270

Puchkova NN (1984) Green sulfur bacteria as a component of the sulfureta of shallow saline waters of the Crimea and northern Caucasus. Mikrobiologyia 53:324–328

Ragatz L, Jiang ZY, Bauer CE, Gest H (1995) Macroscopic phototactic behavior of the purple nonsulfur bacterium *Rhodospirillum centenum*. Arch Microbiol 163:1–6

Ramana VV, Sasikala C, Ramaprasad EVV, Ramana CV (2010) Description of *Ectothiorhodospira salini* sp. nov. J Gen Appl Microbiol 56:313–319

Raymond JC, Sistrom WR (1967) The isolation and preliminary characterization of a halophilic photosynthetic bacterium. Arch Mikrobiol 59:255–268

Raymond JC, Sistrom WR (1969) *Ectothiorhodospira halophila*: a new species of the genus Ectothiorhodospira. Arch Mikrobiol 69:121–126

Reed RH, Borowitzka LJ, Mackay MA, Chudek JA, Foster R, Warr SC, Moort DJ, Stewart WDP (1986) Organic solute accumulation in osmotically stressed cyanobacteria. FEMS Microbiol Rev 39:51–56

Resnick SM, Madigan MT (1989) Isolation and characterization of a mildly thermophilic nonsulfur purple bacterium containing bacteriochlorophyll b. FEMS Microbiol Lett 65:165–170

Robert F, Chaussidon MA (2006) A paleotemperature curve for the Precambrian oceans based on silicon isotopes in cherts. Nature 443:969–972

Rodriguez-Valera F, Ventosa A, Juez G, Imhoff JF (1985) Variation of environmental features and microbial populations with salt concentrations in a multi-pond saltern. Microbiol Ecol 11:107–115

Roeselers G, Norris TB, Castenholz RW, Rysgaard S, Glud RN, Kühl M, Muyzer G (2007) Diversity of phototrophic bacteria in microbial mats from Arctic hot springs (Greenland). Environ Microbiol 9:26–38

Santos H, daCosta MS (2002) Compatible solutes of organisms that live in hot saline environments. Environ Microbiol 4:501–509

Serrano W, Amann R, Fischer U (2009) A new moderately thermophilic and high sulfide tolerant biotype of *Marichromatium gracile*, isolated from tidal sediments of the German Wadden Sea: *Marichromatium gracile* biotype thermosulfidiphilum. Syst Appl Microbiol 32:1–7

Singer MA, Linquist S (1998) Multiple effects of trehalose on protein folding in vitro and in vivo. Mol Cell 1:639–648

Sorokin DY, Tourova TP, Kuznetsov BB, Bryantseva IA, Gorlenko VM (2000) *Roseinatronobacter thiooxidans* gen. nov., sp. nov., a new alkaliphilic aerobic bacteriochlorophyll a-containing bacterium isolated from a soda lake. Microbiology 69:75–82

Sorokin DY, Gorlenko VM, Namsaraev BB, Namsaraev ZB, Lysenko AM, Eshinimaev BT, Khmelenina VN, Trotsenko YA, Kuenen R (2004) Prokaryotic communities of the north-eastern Mongolian soda lakes. Hydrobiologia 522:235–248

Stadtwald-Demchick R, Turner FR, Gest H (1990) *Rhodopseudomonas cryptolactis*, sp. nov., a new thermotolerant species of budding phototrophic purple bacteria. FEMS Microbiol Lett 71:117–121

Stevenson AK, Kimble LK, Woese CR, Madigan MT (1997) Characterization of new heliobacteria and their habitats. Photosynth Res 53:1–12

Sucharita K, Sasikala Ch, Ramana ChV (2010) Thiorhodococcus modestohalophilus sp. nov. a phototrophic gammaproteobacterium from Chilika salt water lagoon, India. J Gen Appl Microbiol 56:93–99

Szilágyi A, Závodszky P (2000) Structural differences between mesophilic, moderately thermophilic and extremely thermophilic protein subunits: results of a comprehensive survey. Structure 8:493–504

Takii S, Konda T, Hiraishi A, Matsumoto GI, Kawano T, Torii T (1986) Vertical distribution in and isolation of bacteria from Lake Vanda: an Antarctic Lake. Hydrobiologia 135:15–21

Tank M, Bryant DA (2015a) *Chloroacidobacterium thermophilum* gen nov., sp. nov.: an anoxygenic microaerophilic chlorophotoheterotrophic acidobacterium. Int J Syst Evol Microbiol 65:1426–14230

Tank M, Bryant DA (2015b) Nutrient requirements and growth physiology of the photoheterotrophic acidobacterium, *Chloroacidobacterium thermophilum*. Front Microbiol 6, 226. doi:10.3389/fmicb.2015.00226

Tank M, Thiel V, Imhoff JF (2009) Phylogenetic relationship of phototrophic purple sulfur bacteria according to *pufL* and *pufM* genes. Int Microbiol 12:175–185

Tank M, Blümel M, Imhoff JF (2011) Communities of purple sulfur bacteria in a Baltic Sea coastal lagoon analyzed by *pufLM* gene libraries and the impact of temperature and NaCl concentration in experimental enrichment cultures. FEMS Microbiol Ecol 78:428–438

Tazi L, Breakwell DP, Harker AR, Crandall KA (2014) Life in extreme environments: microbial diversity in Great Salt Lake, Utah. Extremophiles 18:525–535

Thiel V, Tank M, Neulinger SC, Gehrmann L, Dorador C, Imhoff JF (2010) Unique communities of anoxygenic phototrophic bacteria in saline lakes of Salar de Atacama (Chile). Evidence for a new phylogenetic lineage of phototrophic Gammaproteobacteria from *pufLM* gene analyses. FEMS Microbiol Ecol 74:510–522

Thomas DN, Dieckmann GS (2002) Antarctic sea ice—a habitat for extremophiles. Science 295:641–644

Tice MM, Lowe DR (2004) Photosynthetic microbial mats in the 4,416, Myr old ocean. Nature 431:549–552

Tourova TP, Slobodova NV, Bumazhkin BK, Kolganova TV, Muyzer G, Sorokin DY (2013) Analysis of community composition of sulfur-oxidizing bacteria in hypersaline and soda lakes using soxB as a functional marker. FEMS Microbiol Ecol 84:280–289

Trüper HG, Galinski EA (1990) Biosynthesis and fate of compatible solutes in extremely halophilic phototrophic eubacteria. FEMS Microbiol Rev 75:247–254

Trüper HG, Imhoff JF (1981) The genus *Ectothiorhodospira*. In: Stolp H, Trüper HG, Balows A, Schlegel HG (eds) The prokaryotes. A handbook on habitats, isolation and identification of bacteria. Starr MP Starr. Springer, New York, pp 274–278

Van der Meer MTJ, Klatt CG, Wood J, Bryant DA, Bateson MM, Lammerts L, Schouten S, Damste JSS, Madigan MT, Ward DM (2010) Cultivation and genomic, nutritional, and lipid biomarker characterization of *Roseiflexus* strains closely related to prominent in situ populations inhabiting Yellowstone hot spring microbial mats. J Bacteriol 192:3033–3042

Van Dover CL, Cann JR, Cavanaugh C, Chamberlain S, Delaney JR, Janecky D, Imhoff JF, Tyson A (1994) Light at deep sea hydrothermal vents. Eos Trans AGU 75:44–45

Ventosa A, Fernández AB, León MJ, Sánchez-Porro C, Rodriguez-Valera F (2004) The Santa Pola saltern as a model for studying the microbiota of hypersaline environments. Extremophiles 18:811–824

Wahlund TM, Castenholz RW, Woese CR, Madigan MT (1991) A thermophilic green sulfur bacterium from New Zealand hot springs. *Chlorobium tepidum* sp. nov. Arch Microbiol 156:81–91

Wakao N, Shiba T, Hiraishi A, Ito M, Sakurai Y (1993) Distribution of bacteriochlorophyll a in species of the genus *Acidiphilium*. Curr Microbiol 27:277–279

Welsh DT, Herbert RA (1993) Identification of organic solutes accumulated by purple and green sulfur bacteria during osmotic stress using natural abundance ^{13}C nuclear magnetic resonance spectroscopy. FEMS Microbiol Ecol 13:145–150

Index

© Springer International Publishing Switzerland 2017

P.C. Hallenbeck (ed.), *Modern Topics in the Phototrophic Prokaryotes*,
DOI 10.1007/978-3-319-46261-5

B

C